Advances in Experimental Medicine and Biology

Volume 916

More information about this series at http://www.springer.com/series/5584

Advances in Experimental Medicine and Biology

[illegible]

[illegible]

David M. Langenau
Editor

Cancer and Zebrafish

Mechanisms, Techniques, and Models

Springer

Editor
David M. Langenau
Molecular Pathology and Cancer Center
Massachusetts General Hospital
Charlestown, MA, USA

ISSN 0065-2598 ISSN 2214-8019 (electronic)
Advances in Experimental Medicine and Biology
ISBN 978-3-319-30652-0 ISBN 978-3-319-30654-4 (eBook)
DOI 10.1007/978-3-319-30654-4

Library of Congress Control Number: 2016934705

Printed on acid-free paper

This Springer imprint is published by Springer Nature
The registered company is Springer International Publishing AG Switzerland

Prelude: Cancer and Zebrafish

Cancer and Its History

Cancer is the second leading cause of death in the United States after heart disease and claimed the lives of approximately 585,000 Americans in 2014. In total, nearly 1.7 million people will be diagnosed with cancer in the United States this year and approximately 14.1 million worldwide. Although cancer rates increase with advanced age, many cancers arise in children, adolescents, and young adults. Moreover, distinct cancer subtypes are uniquely found within specific age classes, ethnicity, gender, and regions of the world—avowing that cancer can touch the life of anyone.

Cancer arises when cells within the body acquire uncontrolled growth. Cancer is not a single disease, but rather a heterogeneous compilation of diseases that are defined by the anatomical location within the body, the cells within a specific organ system that are transformed, and, more recently, by the expression of oncogenic driver mutations. These abnormal, malignant cells eventually seed distant sites, leading to metastatic spread and destruction of organs, tissues, and impairment of normal body function.

The Egyptians were the first to document the existence of cancer around 1600 BC, with our modern notions of cancer, including its name, being derived from Hippocrates around 400 BC. Termed "carcinos," a Greek work for crab, Hippocrates studies were limited to external evaluation of tumors and largely focused on detailing the rapid growth of masses that were commonly associated with the establishment of vascular networks that predominate in solid tumors. To Hippocrates, these vessels looked like the sprawling and long legs of a crab, hence the coining of "carcinos" for these diseases. Celsus later translated "carcinos" into Latin around the time of Christ, defining these growths as "cancer" and cancer swelling as "carcinoma."

Rudimentary treatments were initially used to treat cancer before the 1900s, which largely focused on excising the cancer mass and ridding the body of black humors through bloodletting. Not surprisingly, cancer was nearly universally lethal prior to 1900. However, over the past century, we have waged a significant battle in the war on cancer. Using chemotherapy, radiation, and surgery, oncologists had successfully defeated many pediatric cancers by the middle of this century that for the first time led to long-term and durable cures. These initial successes were largely confined to pediatric cancer patients and eventually paved the way for novel therapies in adults. To date, we have seen some successes in the treatment of subsets of adult cancers. Yet, despite these initial wins, the battle to suppress and kill cancer continues. Industrialization, obesity, changes in diet, and increased life expectancy have invariably led to increased cancer rates, erasing many perceived gains in our war on cancer. Moreover, as more patients survive cancer, the long-term effects of therapy have become an enormous clinical challenge—fueling elevated rates of secondary cancer, mental cognition effects, and sterility. Despite many advances made over the last century, it is clear that cancer will continue to be a public health problem and its study will predominant the scientific arena for years to come.

The Modern View of Cancer

Cancer is increasingly viewed as a disease driven by genetic mutation and which ultimately leads to the acquisition of targeted changes within affected cells. These genetic lesions impart new phenotypic changes including elevated growth, ability to recruit vasculature, suppression of cell death, growth factor independence, therapy resistance, and metastasis. Initial studies have identified dominant oncogenic drivers, classically defined as tumor suppressor genes and oncogenes. Tumor suppressors are defined by inactivation of a specific gene through deletion, inactivating point mutations, early gene termination, or gene silencing due to methylation or epigenetic regulation. Oncogenes are amplified, overexpressed, or acquire genetic mutations to enhance and/or provide new function to a specific gene. Oncogenes are ideal drug targets since they are highly expressed in tumors and many are required for continued tumor growth. For example, Herceptin is an antibody with potent antitumor activity in HER2-amplified breast cancer, and antiestrogen therapies have led to durable responses in ER+ disease. More recently, novel therapies have now been developed to target specific activating mutations including those in EGFR-mutant lung cancer, BCR-ABL fusions in acute myelogenous leukemia, and BRAF in melanoma. Yet other classic oncogenes including MYC and RAS have been difficult to target directly.

To date, targeted therapies have become a powerful tool for treating a subset of cancers. Yet for others including lung adenocarcinoma and melanoma, cancer is now viewed as a chronic disease that requires constant surveillance to identify new resistance mutations. These evolved tumors are then treated with second- and third-generation targeted therapies. It is clear that the next few years will continue to witness the emergence of a wide range of drugs that target the most prevalent, recurrent mutations in cancer. Yet for rare and underrepresented oncogenic driver mutations, it will be difficult to match clinically available drugs with actionable targets.

Finally, the promise of targeted therapy will likely rely on combinatorial approaches that also utilize chemotherapy, radiation, and immunotherapy.

Building on these initial successes in targeted therapy and improvements in sequencing technologies, the 2010s have been defined by identifying predisposing oncogenes and tumor suppressors in a wide range of cancer subtypes, attempting to define patient-specific mutations and couple them with novel, targeted therapies. These studies have uncovered that many cancers and subtypes of disease often mutate the same dominant oncogenes and tumor suppressors including tyrosine kinases, RAS, MYC, and P53. Moreover, it has become increasingly recognized that not all cancers are alike and will require personalized therapy approaches to kill cancer. Such approaches require detailed understanding of each person's cancer and uncovering the genetic vulnerabilities in the molecular pathways that drive sustained tumor growth. This new appreciation of cancer identifies increasingly smaller subsets of "cancers," with treatment no longer exclusively defined by standard assignment based on cell of origin or anatomical location of the primary tumor.

Cancer subtype-specific mutations are common and are often confined to only a small fraction of human tumors. These infrequent mutations often converge on common underlying molecular pathways to drive transformation, with many of these same pathways being critically important for normal development. In an era of whole genome sequencing, a plethora of genes have now been identified that are mutated in cancer and yet are found in only a small fraction of patients. Our encyclopedia of possible oncogenic driver mutations now numbers thousands of genes, yet defining the role of these genes in cancer and if they are in fact oncogenic drivers rather than passenger mutations is an ever-increasing problem for cancer biologists. Moreover, defining how best to exploit these rare mutations for targeted therapy remains a challenge, especially for the pharmaceutical industry where only 1 % of drugs entering the clinical evaluation pipeline go on to receive FDA approval, with an estimated cost of over one billion dollars to bring a new drug to market.

In parallel to genetic studies that have identified a litany of mutations found in human cancer, cancer researchers have increasingly focused on reeducating the host immune system to attack and kill cancer cells. Immunotherapy is a burgeoning area of research that has exploded on the scene over the last decade. From developing novel antibodies to CTLA4, PD-1, and PD-L1 to stimulate T-cell killing of tumor to engineering CAR T cells from a patient's own blood cells that recognize tumor-specific antigens and effectively mount an immune attack, it is clear that immunotherapy will continue to revolutionize patient care and treatment in the coming years. Yet, sadly, cancer is highly mutable and ever-changing. Thus, it is inevitable that a subset of tumors will acquire resistance mechanisms to immunotherapies, requiring continued treatment with second-line targeted therapies and combination drug treatments.

Intratumoral heterogeneity and clonal evolution in cancer are also now widely recognized as a major clinical challenge. Again whole genome sequencing and large-scale analysis of individual cancer cells have uncovered a wide diversity of genetic mutations found within clones. This intratumoral heterogeneity provides a larger pool of cells on which Darwinian evolution can occur, providing a rich diversity of ways that cells evade therapy and drive progression. Discovering how individual cells and pathways drive cancer evolution will remain a challenge to our field for years to come.

Zebrafish Models of Cancer

Animal models of cancer, especially mouse models, have been invaluable to our understanding of human cancer. Aided by targeted gene inactivation, gene replacement technologies, and conditional transgenic approaches, mouse models of cancer have become increasingly sophisticated and accurately mimic many of the underpinning of human disease. Most notably, mice have been integral in defining the underlying genetic mutations that drive cancer initiation, progression, and therapy responses. Beyond the use of genetically engineered mouse models of cancer, immunocompromised mice have fast become a reliable and useful tool to study human cancer growth in vivo. For example, mice can efficiently engraft a wide range of human cancers and engrafted animals studied for responses to chemotherapy. Such approaches have transformed our understanding of human cancer and provided systematic approaches to identify drugs and combination therapies for the treatment of cancer.

Yet, mouse models are not the only animal models of cancer. For example, elegant models of cancer exist in *Drosophila, C. elegans*, and fish. For example, flies have been studied for defects that lead to growth defects in the eye, imaginal discs, and hemolymph. Moreover, transgenic approaches can be used to drive oncogenes in developing tissue, most notably the use of PAX7-FKHR in fly models of premalignant alveolar rhabdomyosarcoma. *C. elegans* have also been used for assessing cancer processes, most notably the outgrowth of uveal tissues. These invertebrate models offer many advantages including genetic screening tools, short generation time, and facile manipulation of animals; yet tumors do not, by and large, recapitulate disease found in humans, and many organ systems are not found in these animals. By contrast, fish have emerged as a powerful model of cancer, mimicking many of the classic cancer phenotypes and molecular ontogeny as found in their human counterpart cancers. The first models of cancer in fish included chemical- and carcinogen-induced tumors, while the first genetic fish model of cancer was melanoma in the platyfish, *Xiphophorous*. Building on the unique attributes afforded to fish models, many laboratories have now developed zebrafish, medaka, and other piscine cancer models that transgenically overexpress oncogenes and/or harbor inactivating mutations in tumor suppressors. Moreover, many laboratories are working toward the goal of implanting human cancers into immunodeficient zebrafish—prior to establishment of a fully formed immune system by 21 days of life.

Zebrafish have dominated in the arena of using fish to study cancer, in large part to the powerful genetic and developmental tools available and the long history of using zebrafish for genetic screens. Zebrafish have many attributes that make them ideal models for the study of cancer. Most importantly, zebrafish cancers are

molecularly and histopathologically similar to human disease, a prerequisite for modeling cancer processes in any animal model. Zebrafish are also small, allowing rearing of tens of thousands of animals in a relatively small space and permitting economy of scale at reduced costs when compared with mouse models. Zebrafish also exhibit high fecundity with each female mating weekly and capable of producing 100–200 eggs per week. Zebrafish are optical clear during early larval development, and genetic strains have been produced that allow direct visualization of organs in nonpigmented, *Casper*-strain fish. Capitalizing on the optical translucency of zebrafish, facile transgenic approaches have been used to track cancer cells over time and to dynamically visualize cancer processes in vivo. Such approaches are severely limited in mouse models of cancer. Finally, zebrafish are a robust model for drug discovery and large-scale in vivo drug screening.

This book will help define the unique attributes of using the zebrafish and other piscine cancer models for discovering important cancer pathways and drugs for the treatment of human disease. This volume will cover the use of zebrafish for oncogene and tumor suppressor discovery, chemical genetic approaches, genomics, epigenetics, cancer imaging, and cell transplantation. Sections of this book will outline how the zebrafish will be useful for characterizing the long list of possible oncogenic drivers identified by whole genome sequencing of human cancer and define the importance heterogeneity has in cancer evolution. These contributed chapters represent a cross-section of our community and define the contribution of prominent laboratories working in this area. Latter chapters will provide detailed analysis of the state of the field concerning specific zebrafish cancer models including T-cell leukemia, rhabdomyosarcoma, liver and pancreatic cancer, melanoma, neuroblastoma, germ cell tumors, and malignant peripheral sheath tumors. A chapter is dedicated to defining the development and use of alternative piscine models of cancer. This book is the most comprehensive work published to date detailing the use of fish to study cancer and provides a much-needed resource to define the powerful attributes of this model system.

I thank our chapter contributors for their unique insights and time dedicated to making this project a success. I am particularly indebted to my colleagues at Dana-Farber Cancer Institute, Children's Hospital Boston, Massachusetts General Hospital Boston, and Harvard Medical School for the rich environment I have grown up in as a cancer biologist over the past 15 years. I also thank those bold scientists in my laboratory who continue to wonder, search, and strive to find new understanding of human cancer including my past and present fellows, Drs. Myron Ignatius, Eleanor Chen, Madeline Hayes, Jessica Blackburn, Finola Moore, John Moore, and Riadh Lobbardi; my PhD and master's students, Ines Tenente, Elaine Garcia, Qin Tang, and Manon de Waard; technical staff, Karin McCarthy, Ashwin Ramakrishnan, Aleksey Molodtsov, Ryan Clagg, Sarah Martinez, Sali Liu, and Aubrey Raimondi; and undergraduate researchers and interns too numerous to list. I thank my wife Brenna and two children Daniel and Derek for their constant support and my mother, Diana Langenau, and the late Clarence Yount and Doris Langenau who define to me what it means to live a full and happy life as a cancer survivor. I thank the patients, their loved ones, and advocates for their continued support of cancer research, especially in the pediatric cancer arena including but not limited to Alex's Lemonade Stand Foundation for Childhood Cancer, St. Baldrick's Foundation, Live Like Bella Foundation, Cure4Cam, and QuadW Foundation. Finally, I thank Fiona Sarne and Joy Bramble from Springer who have provided vision and guidance in compiling and editing this book.

David Langenau

Contents

Part I Powers of the Zebrafish Model

Uncharted Waters: Zebrafish Cancer Models Navigate a Course for Oncogene Discovery 3
Craig J. Ceol and Yariv Houvras

The Toolbox for Conditional Zebrafish Cancer Models 21
Marie Mayrhofer and Marina Mione

Approaches to Inactivate Genes in Zebrafish 61
John M. Parant and Jing-Ruey Joanna Yeh

Tumor Suppressors in Zebrafish: From TP53 to PTEN and Beyond 87
Jeroen den Hertog

Identifying Novel Cancer Therapies Using Chemical Genetics and Zebrafish 103
Michelle Dang, Rachel Fogley, and Leonard I. Zon

Genomic Approaches to Zebrafish Cancer 125
Richard M. White

Transcriptomic Analyses in Zebrafish Cancer Models for Global Gene Expression and Pathway Discovery 147
Xiaoqian Huang, Ira Agrawal, Zhen Li, Weiling Zheng, Qingsong Lin, and Zhiyuan Gong

Zebrafish Discoveries in Cancer Epigenetics 169
Yelena Chernyavskaya, Brandon Kent, and Kirsten C. Sadler

Lymphatics, Cancer and Zebrafish 199
Jonathan W. Astin and Philip S. Crosier

In Vivo Imaging of Cancer in Zebrafish 219
Myron S. Ignatius, Madeline Hayes, and David M. Langenau

Imaging Cancer Angiogenesis and Metastasis in a Zebrafish Embryo Model 239
C. Tulotta, S. He, W. van der Ent, L. Chen, A. Groenewoud, H.P. Spaink, and B.E. Snaar-Jagalska

Allograft Cancer Cell Transplantation in Zebrafish 265
John C. Moore and David M. Langenau

The Zebrafish Xenograft Platform: Evolution of a Novel Cancer Model and Preclinical Screening Tool 289
Jaime Wertman, Chansey J. Veinotte, Graham Dellaire, and Jason N. Berman

Automation of Technology for Cancer Research 315
Wietske van der Ent, Wouter J. Veneman, Arwin Groenewoud, Lanpeng Chen, Claudia Tulotta, Pancras C.W. Hogendoorn, Herman. P. Spaink, and B. Ewa Snaar-Jagalska

Part II Cancer Models in Fish

Zebrafish Models of Human Leukemia: Technological Advances and Mechanistic Insights 335
Nicholas R. Harrison, Fabrice J.F. Laroche, Alejandro Gutierrez, and Hui Feng

Zebrafish Rhabdomyosarcoma 371
Michael Phelps and Eleanor Chen

Baiting for Cancer: Using the Zebrafish as a Model in Liver and Pancreatic Cancer 391
Katie L. Hwang and Wolfram Goessling

Focusing the Spotlight on the Zebrafish Intestine to Illuminate Mechanisms of Colorectal Cancer 411
Viola H. Lobert, Dmitri Mouradov, and Joan K. Heath

Zebrafish Melanoma 439
Charles K. Kaufman

Neuroblastoma and Its Zebrafish Model 451
Shizhen Zhu and A. Thomas Look

Zebrafish Germ Cell Tumors 479
Angelica Sanchez and James F. Amatruda

Malignant Peripheral Nerve Sheath Tumors 495
Adam D. Durbin, Dong Hyuk Ki, Shuning He, and A. Thomas Look

Xiphophorus and Medaka Cancer Models 531
Manfred Schartl and Ronald B. Walter

Part I
Powers of the Zebrafish Model

Part I
Powers of the Zebrafish Model

Uncharted Waters: Zebrafish Cancer Models Navigate a Course for Oncogene Discovery

Craig J. Ceol and Yariv Houvras

Abstract Over a decade has elapsed since the first genetically-engineered zebrafish cancer model was described. During this time remarkable progress has been made. Sophisticated genetic tools have been built to generate oncogene expressing cancers and characterize multiple models of solid and blood tumors. These models have led to unique insights into mechanisms of tumor initiation and progression. New drug targets have been identified, particularly through the functional analysis of cancer genomes. Now in the second decade, zebrafish cancer models are poised for even faster growth as they are used in high-throughput genetic analyses to elucidate key mechanisms underlying critical cancer phenotypes.

Keywords Zebrafish • Cancer • Transgenic • Ras • BRAF • p53

Introduction

It has been less than 40 years since pivotal work established that defects in our own genes are a major cause of cancer (reviewed in [1]). Thanks to The Cancer Genome Atlas and other cancer sequencing studies we now have comprehensive catalogs of genetic changes across major cancer types. These databases have revealed new oncogenes and tumor suppressor genes as well as polygenic signatures that are sometimes shared across different types of cancers. Yet, these catalogs are in many ways a beginning. Functional studies of recurrently mutated genes are critical in determining the mechanisms by which any new cancer genes are involved in tumorigenesis. In many cases, genes are mutated at a frequency that borders on the edge of

C.J. Ceol (✉)
Program in Molecular Medicine and Department of Molecular, Cell and Cancer Biology, University of Massachusetts Medical School, 368 Plantation Street, Worcester, MA 01605, USA
e-mail: Craig.Ceol@umassmed.edu

Y. Houvras (✉)
Departments of Surgery and Medicine, Weill Cornell Medical College, 1300 York Avenue, New York, NY 10065, USA
e-mail: yah9014@med.cornell.edu

D.M. Langenau (ed.), *Cancer and Zebrafish*, Advances in Experimental Medicine and Biology 910, DOI 10.1007/978-3-319-30654-4_1

significance, and for these genes it will be important to study the relevance to tumor progression. Other types of genetic changes—recurrent copy number variation and chromosomal aberrations—do not always nominate single genes, and follow-up studies are necessary to pinpoint the genes that are critical for promoting tumorigenesis. Lastly, while genome sequences are undoubtedly a boon to cancer research, a great deal of tumor progression is not addressed by sequence, and complementary approaches are required to identify and understand causative mechanisms.

Animal models have played a key role in deconstructing tumor progression. Over the past decade zebrafish have emerged as an innovative model that is particularly well suited for functional genetic studies in cancer, especially as it pertains to oncogene discovery. As a model system zebrafish possess multiple unique attributes may be harnessed for cancer research. Using zebrafish, large-scale genetic screens are possible. In addition, the zebrafish is optically transparent, enabling real-time and longitudinal studies of cancer cells in an intact microenvironment. Drug treatments are straightforward, and libraries of drugs can be screened for compounds that affect tumor-relevant phenotypes. This review highlights transgenic zebrafish models of oncogene-induced solid tumors, with an emphasis on how the unique attributes of zebrafish as an experimental system have been and can be applied to cancer research.

Zebrafish Cancer Models

A major reason why zebrafish was originally adopted and continues to grow as a model system is the conservation of tissues and organs with higher vertebrates. Zebrafish vasculature, heart, bone, blood, liver, kidney, pancreas, intestine, thyroid, skin, thymus, and gonads share anatomic organization, cellular makeup, and developmental origins with their mammalian counterparts. This conservation of tissues and organs led Langenau, Look, and colleagues, to test whether expression of a well characterized oncogene, *c-myc*, could lead to neoplastic phenotypes in zebrafish. Transgenic expression of *c-myc* under control of a lymphocyte specific promoter led to T-cell lymphoblastic leukemia in zebrafish, in a landmark paper published in 2003 [2]. Subsequent studies demonstrated that bcl-2 overexpression confers protection from radiation induced apoptosis in this model [3], and examined clonal evolution after transplantation of single T-ALL cells into recipient zebrafish [4]. These studies demonstrate that transgenic cancer models in zebrafish may recapitulate human disease, and lead to genetic and phenotypic insights. Since the initial model of leukemia was created in zebrafish, multiple diverse solid tumor models have been described (Table 1). These include models of rhabdomyosarcoma, liver cancer, pancreatic cancers, neuroblastoma, melanoma and others (reviewed in [4, 53]). These transgenic models have been created with a range of tissue specific promoters, disease relevant oncogenes, and tumor suppressors. Tumor phenotypes arising in zebrafish cancer models often bear striking similarity to human cancer histopathology, and reveal conserved changes in gene expression and critical

Table 1 Development and applications of cancer models in zebrafish using transgenic tools

Tumor	Cell type	Transgene	Key features	References
T-cell leukemia/lymphoma (T-ALL)	Lymphocytes	*rag2:c-myc*	Highly malignant T-ALL develops after induction of *Myc*	Langenau et al. [2]
			Inducible oncogene expression using Cre-lox technology in T-ALL model	Langenau et al. [5] and Feng et al. [6]
			Expression of *bcl-2* confers resistance to radiation induced apoptosis	Langenau et al. [3]
			bcl-2 expression inhibits progression to T-ALL and blocks vascular invasion	Feng et al. [7]
			Single cell transplantation to examine clonal evolution	Blackburn et al. [4]
			Chemical screen identifies drugs that synergize with Notch inhibitors	Gutierrez et al. [8]
		rag2:NOTCH1	Expression of *NOTCH1* leads to T-ALL and cooperates with *bcl-2* loss	Chen et al. [9] and Blackburn et al. [10]
B-cell leukemia	Lymphocytes	*ef1alpha:TEL-AML1*	Expression of a *TEL-AML1* fusion gene leads to B cell leukemia	Sabaawy et al. [11]
Acute myeloid leukemia (AML)	Myeloid cells	*spi1:MYST3-NCOA2*	Expression of a *MYST3-NCOA2* fusion leads to acute myeloid leukemia	Zhuravleva et al. [12]
Myeloproliferative disorder	Myeloid cells	*spi1:NUP98-HOXA9*	Expression of *NUP98-HOXA9* leads to a myeloproliferative disorder	Forrester et al. [13] and Deveau et al. [14]
Myelodysplastic syndrome (MDS)	Myeloid cells	*CRISPR/Cas9 editing of tet2*	First CRISPR/Cas9 disease model in fish demonstrates *tet2* loss leads to MDS	Gjini et al. [15]
Systemic mastocytosis	Mast cells	*actb2:KIT(D816V)*	Expression of oncogenic *KIT* leads to aggressive mastocytosis	Balci et al. [16]
Melanoma	Melanocytes	*mitfa:BRAF(V600E)*	*BRAF(V600E)* expression leads to melanoma on a *tp53* mutant background	Patton et al. [17] and Yen et al. [18]
			Zebrafish melanoma model used for functional genetic screens to identify new oncogenes	Ceol et al. [19] and Iyengar et al. [20]

(continued)

Table 1 (continued)

Tumor	Cell type	Transgene	Key features	References
			Genetic modulation of mitf leads to melanoma regression	Lister et al. [21]
			Zebrafish melanoma model used to screen for novel anti-melanoma drugs	White et al. [22]
		mitfa:NRAS(Q61K)	Transgenic expression of *NRAS(Q61K)* leads to melanomas	Dovey et al. [23]
		mitfa:HRAS(G12V)	*HRAS* acts through PI3K signaling to induce melanomas	Michailidou et al. [24]
		kita:Gal4 x UAS:HRAS(G12V)	*HRAS* expression in *kit*-expressing cells led to highly penetrant and invasive phenotypes	Anelli et al. [25] and Santoriello et al. [26]
Hepatocellular carcinoma	Hepatocyte	*fabp10:MYC*	Liver tumors with a conserved gene signature of *MYC* overexpression	Li et al. [27]
		fabp10:KRAS(G12V)	Development of a *KRAS* liver tumor model and chemical screening platform	Nguyen et al. [28, 29]
		lfabp:UHRF1	Expression of *UHRF1* leads to hypomethylation and development of liver tumors	Mudbhary et al. [30]
		fabp10:beta-catenin	Liver cancer models reveals JNK pathway activation and novel chemical inhibitors	Evason et al. [31]
		lfabp:HBx	Expression of hepatitis B virus *HBx* gene promotes hepatocellular carcinoma	Lu et al. [32]
		lfabp:src	*src* expression combined with loss of *tp53* causes hepatocellular carcinoma	Lu et al. [32]
Cholangiocarcinoma	Biliary epithelia	*lfabp:tTA x TRE-CMV:HBx x TRE-CMV:HCP*	Viral genes *HBx* and HCP cooperate to form cholangiocarcinoma	Liu et al. [33]
Pancreatic carcinoma	Exocrine pancreas	*ptf1:KRAS(G12D)*	*KRAS* induces pancreatic carcinoma and a block in differentiation	Park et al. [34] and Liu et al. [35]
			Identification of a ribosomal protein gene (*rpl36*) as a tumor suppressor in *KRAS(G12V)* pancreatic cancer	Provost et al. [36]

			Visualization of pathway specific reporters in a model of pancreatic cancer	Schiavone et al. [37]
	Endocrine pancreas	*myod:MYCN*	Neuroendocrine pancreatic tumors arise after expression of *MYC*	Yang et al. [38]
Testicular germ cell tumor	Spermatagonia	*alk6b* mutation	Heritable model of germ cell tumors arising in fish harboring a BMP receptor mutation	Neumann et al. [39, 40]
		F.r. flck:SV40 large T antigen	Expression of SV40 Large T antigen leads to a germ cell tumor in zebrafish	Gill et al. [41]
Glioma	Neurons	*krt5:smoa1*	Hedgehog signaling promotes gliomas in the optic pathway	Ju et al. [42]
		krt5:RAS(G12D), *gfap:RAS(G12D)*	Brain tumors arise after tissue-specific expression of *RAS(G12D)*	Ju et al. [43]
Malignant peripheral nerve sheath tumor (MPNST)	Schwann cell	*p53(M214K)*	Point mutation in *tp53* leads to MPNSTs and reveal conserved *tp53* function, cooperation with *nf1*	Berghmans et al. [44] and Shin et al. [45]
Rhabdomyosarcoma	Satellite cell	*rag2:KRAS(G12D)*	Conserved gene signature with embryonal rhabdomyosarcomas	Langenau et al. [46]
Ewing's sarcoma	Osteoblast	*hsp70:EWS-FLI1*	Development of tumors resembling human Ewing's sarcoma after expression of *EWS-FLI1* oncogene	Leacock et al. [47]
Chordoma	Notochord cell	*mu4465:Gal4 x UAS:HRAS(G12V)*	Chordomas, notochord-derived tumors, arise from oncogenic *HRAS* expression	Burger et al. [48]
Lipoma	Adipocyte	*krt4:myrAKT1*	Lipoma formation after expression of activated *AKT*	Chu et al. [49]
Neuroblastoma	Neurons	*dbh:MYCN*	*MYCN* and *ALK* cooperate in neuroblastoma development	Zhu et al. [50]
Pituitary adenomas	Corticotrophs	*pomc:pttg*	Expression of the human pituitary oncogene leads to hyperplasia and adenoma formation	Liu et al. [35]
Diverse tumors	Multiple	*KRAS(G12D)*	Heat shock induction of Cre recombinase used to express *KRAS(G12D)*	Le et al. [51]
		myr-AKT; *smoa1*	Cooperation of *AKT* and Hedgehog pathways lead to diverse tumor types	Ju et al. [52]

signaling and developmental pathways. These models have been used to perform chemical-genetic screens [8, 22], examine migration and metastasis in vivo [7, 54, 55], and identify cooperating genetic lesions that contribute to tumor initiation and progression [19, 24, 50].

Some basic knowledge about transgenic zebrafish cancer models emerges from the body of work produced over the past decade. Critical to the success of zebrafish models has been the identification of tissue specific promoters. The developmental biology community has built a deep resource of whole mount in situ patterns in early development [56, 57], which can be used to identify genes with tissue restricted expression. With the maturation of the zebrafish genome sequence, candidate promoters can be rapidly cloned and tested using fluorescent reporters. Selection of appropriate oncogenes may be guided by knowledge of the specific human disease. Activating mutations in canonical oncogenes, such as RAS-G12V/D, overexpression of wild-type oncogenes, such as c-myc, and expression of fusion proteins arising after translocation, such as TEL-AML1 [11], have all been successfully utilized to model cancer in zebrafish. Several tumor suppressor mutants have been characterized in zebrafish, including pten and nf1 [45, 58–61]. The most experience is with a zebrafish p53 point mutant (M214K) that was generated using chemical mutagenesis [44]. This is a loss-of-function mutation and zebrafish harboring this mutation spontaneously develop malignant peripheral nerve sheath tumors (MPNSTs). Multiple cancer models have utilized this p53(lf) mutant to demonstrate cooperativity with oncogenes. In the zebrafish melanoma model, for example, expression of BRAF(V600E) in melanocytes only leads to melanoma on a p53 mutant background [17]. Transgenic tools in zebrafish are aided by the relative simplicity of microinjection at the 1 cell stage and the embryo's rapid developmental program.

Building on this basic toolkit of transgenic tools, some investigators have added even more sophisticated approaches. Expression of Cre and other recombinases has been utilized to achieve temporal control over oncogene induction [5, 51, 62]. A zebrafish heat-shock promoter has been employed to allow external warming to activate specific genes, including oncogenes or Cre [5]. Transplantation has been increasingly used in zebrafish to address questions related to cell autonomy and tumor initiating cell populations. Initial tumor models in zebrafish utilized transplantation to demonstrate that tumors possessed tumor-initiating potential in sublethally irradiated recipients [2, 63]. These approaches have advanced to a point where engraftment after single cell transplantation is possible [4] and where immunocompromised *rag* mutant hosts are available [64]. Pathway specific reporters have been created to monitor the activity of TGF-B, Notch, Bmp, and Shh during tumor development [37]. Transgenic zebrafish cancer models may be employed in chemical screens to identify small molecule inhibitors of particular pathways or lineages [8, 22]. Because zebrafish are optically clear as embryos, and various pigment mutants have been developed for adult studies [65], imaging tumor development over time offers significant advantages in understanding tumor initiation and progression.

Transgenic approaches to model cancer in zebrafish are progressing rapidly thanks to recent breakthroughs in CRISPR/Cas9 technology. Prior to CRISPR/Cas9 and other genome editing technologies the only reliable way to generate loss-of-function mutations in desired genes was to perform ENU mutagenesis and screen large pools of animals to identify a random mutation in a gene of interest. While this approach was successful and generated several important mutants, CRISPR/Cas9 is significantly less resource intensive. Although genome editing in zebrafish using CRISPR/Cas9 is still being optimized, soon it will likely be routinely possible to edit and recombine into the fish genome to create specific changes in protein coding genes, examine the importance of regulatory regions, and introduce protein tags for immunoaffinity purification or immunofluorescence. Tissue specific expression of Cas9 has already been utilized to restrict genome editing to specific lineages and organs [66]. These approaches will revolutionize zebrafish cancer models as we gain the ability to alter any nucleotide sequence in the genome in any cell type.

Tumors arising in zebrafish cancer models share many histopathologic features with human cancers. Tumors are often highly vascular, and fluorescent lines that label the vascular endothelium have been used to study the process of tumor angiogenesis. Many tumors, such as the melanomas and pancreatic adenocarcinomas are highly invasive [26, 34]. Models of metastasis are being actively developed, aided by the development of a nearly transparent adult zebrafish, *casper*. Tumors display cytopathologic features consistent with high grade neoplasms, such as a high nuclear to cytoplasmic ratio, nuclear pleomorphism, and frequent mitotic figures. In characterizing the melanoma model we have worked closely with expert human pathologists, which has been essential in understanding the similarities between the human disease and the zebrafish model. Further efforts to expand the use of antibodies for immunostaining may identify tumor and pathway specific markers that are shared across species.

Zebrafish cancer models have shown conservation of the mechanisms involved in tumor initiation and progression. Oncogenes and tumors suppressors that drive cancer formation in humans act similarly in zebrafish models. *BRAF(V600E)* is a major oncogene in human melanoma and can drive melanoma formation in zebrafish in cooperation with a p53 loss of function mutation [17]. Expression of BRAF(V600E) alone in melanocytes leads to fish nevi, a benign lesion characterized by oncogene induced senescence. This is similar to human nevi that also harbor a high frequency of BRAF(V600E) mutations. NF1 mutations predispose patients to peripheral and central nervous system tumors, and additional abnormalities involving neural crest, neural tube and mesoderm-derived tissues [67]. Loss-of-function mutations in the two zebrafish NF1 orthologs synergize to cause a similar spectrum of defects, in particular peripheral nerve sheath tumors and gliomas [45], to those prevalent in NF1 patients.

Gene expression studies from zebrafish cancer models have revealed activation of conserved pathways between fish and human cancers. Langenau, Zon and colleagues created a rhabdomyosarcoma model by expressing activated Ras^{G12D} in muscle tissue [46]. Expression profiles of these tumors were most similar to human embryonal rhabdomyosarcomas, which are commonly driven by *RAS* mutant oncogenes, and less similar to alveolar rhabdomyosarcomas, which arise from forkhead transcription factor translocations but not *RAS* mutations. In broader comparisons, the zebrafish rhabdomyosarcoma profiles were used to define a core gene signature that is shared by human pancreatic and other cancers that have *RAS* mutations. Using a similar approach, Gong and colleagues created a liver tumor model by inducibly expressing the *MYC* oncogene under the control of the hepatocyte-specific *fabp10* promoter [27]. Gene expression profiles of liver tumors were obtained, and they most closely matched profiles of human hepatocellular carcinomas, which are often driven by high levels of *MYC*. Together these studies have established that cancer models created in zebrafish recapitulate key signaling pathways and gene expression patterns found in human tumors.

We acknowledge that there are some limitations to modeling cancer in zebrafish. For example, zebrafish do not possess lung or mammary tissues. In these cases it may be possible to study lung and breast tumors using xenotransplantation. Alternatively, it is possible that some related tissues may be appropriate surrogates. For example, we are unaware of a zebrafish prostate organ, yet highly specific prostate epithelial genes are conserved in zebrafish, leading us to speculate that there may be urogenital cells with conserved function amenable to disease modeling. In some cases key tissue specific tumor suppressors are conserved in zebrafish, for example, the breast-ovarian cancer susceptibility genes, BRCA1 and BRCA2 have zebrafish orthologs. Genetic perturbation of these genes may reveal essential insights into their function in ovarian cells, for example. Some zebrafish organs exhibit differences in cellular organization or tissue architecture that may influence the range of possible tumor phenotypes. It is vital for scientists modeling cancer in zebrafish to collaborate with human pathologists expert in the specific disease in order to define specific histologic features that are conserved across species.

Beyond Model Building: What Have We Learned?

In recent years, a transition has occurred from building zebrafish cancer models to using these models to understand aspects of disease. These studies focus on a range of genetic, molecular and cellular defects in cancers. A theme of many studies has been their application of unique features of the zebrafish toward cancer biology. Below we highlight several of these studies and, with each, describe the topic of study and how attributes of the zebrafish were leveraged to enable novel discoveries.

Identifying Relevant Cancer Genes: Somatic Cell Genetics

With the appropriate design, genetic analyses can be performed in the somatic cells of zebrafish. A major advantage of somatic cell genetics is that it forgoes the time and space-consuming breeding schemes involved in manipulating germline-transmitted mutations or transgenes. Somatic cell genetic approaches often create distinct genetic modifications (e.g. integrations, shRNA knockdowns, indels) in many different cells of a single animal [68–70]. Therefore, in a single animal many different genetic perturbations can be sampled. Because of the prolific nature and ease of transgenesis of zebrafish, many different individuals, along with different cells in an individual, can be used to probe distinct genetic modifications. Furthermore, transgenic methods in zebrafish are relatively uncomplicated, especially as compared with higher vertebrates, such as the mouse. These approaches may utilize CRISPR/Cas9 to perform loss-of-function studies [66], or they may harness tissue specific promoters to express transgenes in specific cell lineages or organs [20]. Because of the prolific nature, ease of transgenesis, and ability to manipulate somatic cells of zebrafish, a major strength of these approaches is that each animal may possess multiple different genetic perturbations, and thus a range of phenotypes can be efficiently characterized.

Somatic cell genetics has been applied in the zebrafish to deconstruct the effects of copy number changes in human melanoma. Melanomas and other solid tumors have rampant gene copy number variation. Certain regions of the genome are recurrently copy number varied, and recurrently amplified regions often contain oncogenes whereas recurrently deleted regions contain tumor suppressors. However, determining which genes in a recurrently varied interval are involved in tumorigenesis can be challenging, especially when the interval is large and contains many candidate genes. To test candidate genes in a recurrently varied interval, Ceol, Houvras and Zon developed a transgenic approach in somatic cells in which melanocytes were generated in animals without the need for germline transgenesis [19]. In this system, a wild-type *mitfa* minigene was introduced into melanocyte-deficient *mitfa* mutants. The *mitfa* minigene acts cell autonomously, so every rescued melanocyte contained the minigene. An analysis of copy number alterations in human melanoma was used to select candidate melanoma oncogenes residing in genomic regions that are recurrently amplified [71]. Candidate melanoma oncogenes were coupled to the *mitfa* minigene so they too were expressed in rescued melanocytes. Rescued melanocytes were generated on a *mitfa(lf)* background that also contained the *BRAF*^{V600E} oncogene and a *p53* loss-of-function mutation, and the effect of a candidate gene on melanoma onset was measured. One gene in the recurrently amplified interval, *SETDB1*, accelerated tumor onset and increased melanoma cell invasion. *SETDB1* encodes a histone methyltransferase, and its amplification likely acts to bypass oncogene-induced senescence and alter expression of Hox and other genes. In support of the work done in zebrafish, genome-wide associated studies of melanoma patients have also implicated *SETDB1* as important in melanoma susceptibility [72, 73], and as an essential gene and potential drug target in a murine model of leukemia [74].

Somatic cell genetics has also been used to define putative cancer genes through mutagenesis. McGrail, Essner and colleagues adapted a system, previously developed in the mouse [75], to mobilize mutagenic transposons in zebrafish somatic cells [76]. Fish with tumors in different tissues were isolated, and transposon integration sites identified through ligation-mediated PCR. Many of the genes with transposon integrations were also sites of integration in mouse cancer studies, prioritizing this group of genes for further investigation. In such studies, common sites of integration are critical in defining new cancer genes, and the large number of tumor-bearing fish that can be isolated makes this a powerful approach.

Interactions Between Tumor Genes: Combinatorial Genetics

The complexity of solid tumor genomes leads to a challenging question: how do the different genes altered in a given tumor interact to drive tumor progression? One way to answer this question is through combinatorial genetics, that is, testing the different alterations alone and in combination with each other. Zhu and Look adopted this approach to determine the role of *ALK* mutations in neuroblastoma pathogenesis [50]. A subset of neuroblastomas are driven by *MYCN* amplifications, and activation of *ALK*, either by gain-of-function mutation or amplification, is frequently associated with *MYCN* amplification [77]. Overexpression of *MYCN* in the zebrafish peripheral sympathetic nervous system (PSNS) led to neuroblastomas, although these tumors arose with partial penetrance and late onset. Most of the cells that experienced *MYCN* overexpression were eliminated via apoptosis, with only a fraction surviving and giving rise to tumors. When an activated *ALK* variant was co-expressed with *MYCN*, more PSNS cells survived, neuroblastoma penetrance was increased and onset was accelerated. Thus, activated *ALK* appears to synergize with *MYCN* by preventing the *MYCN*-induced cellular death in peripheral sympathetic neurons, enabling neuroblastoma initiation.

Combinatorial genetics have also been used to probe the role of the MITF transcription factor in melanoma. MITF is the key regulator of melanocyte cell fate, and its activity is altered in human melanomas. Lister, Patton and coworkers used a temperature-sensitive allele of *mitfa*, the zebrafish MITF ortholog, to determine its role in melanoma initiation and progression [21]. This temperature-sensitive allele was bred with a transgenic strain expressing the $BRAF^{V600E}$ melanoma oncogene in melanocytes. $BRAF^{V600E}$ transgenic animals never developed melanomas. However, partial *mitfa* activity, created by holding *mitfa* temperature-sensitive mutants at an intermediate temperature, synergized with $BRAF^{V600E}$ to cause melanoma formation. The reason for this synergism is unclear, but it was suggested that partial levels of *mitfa* hold cells in a progenitor-like state where they are more able to serve as cells of origin for tumors. Remarkably, the authors also utilized the mitfa temperature-sensitive allele to demonstrate that established melanomas regress when mitf activity is abrogated. Together these studies highlight how rapid breeding and ease of transgenesis can be leveraged for multigenic cancer studies in zebrafish.

Deconstructing Tumor Genomes: Evolutionary Distance

Completion of the zebrafish genome has facilitated broad comparative genomic studies between zebrafish and man [78]. Cancer genomes exhibit significant aberrations manifested as copy number variation, focal amplifications and deletions, translocations, and aneuploidy. These alterations have also been identified in zebrafish cancer models. Freeman, Lee, and colleagues described the first bacterial artificial chromosome (BAC)-based platform for studying copy number alterations in zebrafish [79]. This approach led to the identification of copy number changes in zebrafish rhabdomyosarcomas, T-ALLs, and melanomas. Subsequent studies using

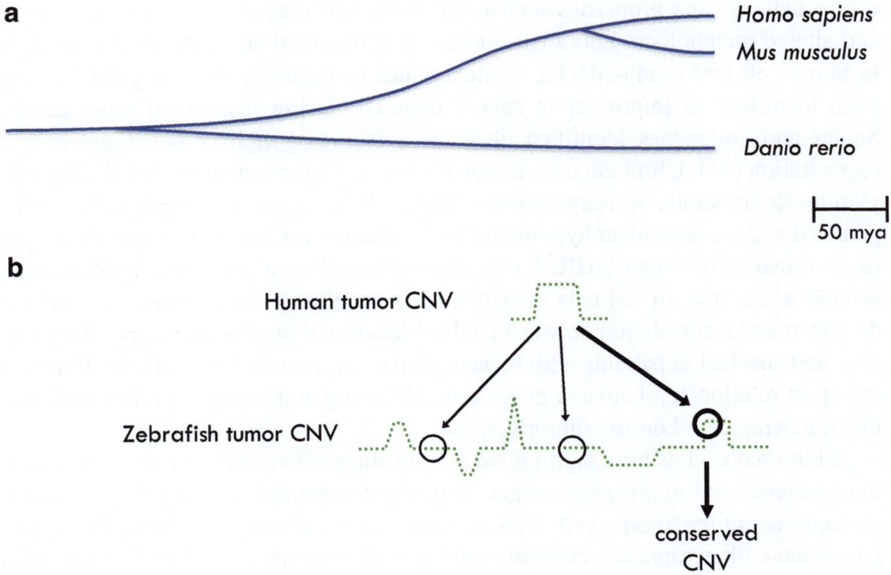

Fig. 1 Conservation and comparative genomics. (**a**) Evolutionary relationship between zebrafish, mouse and humans. Mice and humans share a last common ancestor 90 mya, whereas zebrafish and humans diverged 420 mya. (**b**) Comparative genomics of copy number changes in cancer. Syntenic dispersal between humans and zebrafish can be used to winnow candidate oncogenic drivers present in regions of copy number variation (CNV)

microarray-based approaches confirmed that zebrafish MPNSTs also exhibit a high degree of aneuploidy [80]. Rudner and colleagues examined copy number alterations across several T-ALL models in zebrafish and identified a set of genes that was amplified in both zebrafish and human ALLs [81]. The authors found significant overlap in genes between aggressive zebrafish T-ALLs and T-ALL samples from patients with poor clinical outcomes. Yen, White, and colleagues performed DNA sequencing from zebrafish melanomas and identified recurrent amplification of a gene encoding the catalytic subunit of protein kinase A, suggesting cooperativity with $BRAF^{V600E}$ and p53 loss of function [18]. These studies demonstrate that genomic aberrations are also common in zebrafish cancers and point to pools of genes that are altered in tumors across species as being enriched for cancer drivers (Fig. 1). Indeed genes subject to conserved copy number changes have been implicated as tumor drivers in both MPNST and rhabdomyosarcoma models [82, 83].

Mechanistic Study of Cancer Genes: Functional Study in Early Development

Forward genetic screens have played a key role in establishing zebrafish as a preeminent model organism for studying vertebrate development [84]. Many pathways involved in development are critically deregulated during tumorigenesis as cancer cells acquire properties such as unlimited self-renewal, invasion, metastasis, and altered metabolism. Zebrafish studies have identified developmental regulators in many cell and organ-specific contexts, and increasingly these regulators have been identified as important in cancer contexts. Using insertional mutagenesis, Sadler and colleagues identified uhrf1 as a regulator of liver development and regeneration [85]. Uhrf1 encodes a protein essential for maintenance of DNA methylation. Remarkably, overexpression of human UHRF1 using a liver-specific transgene led to dose dependent hypomethylation, senescence, and reduction in liver size in zebrafish [30]. When UHRF1 was overexpressed on a p53 heterozygous background a majority of animals developed hepatocellular carcinomas. The authors demonstrated a correlation between UHRF1 levels and disease recurrence after surgery and survival in patients with hepatocellular carcinoma. These studies illustrate the close relationship between genes critical for organ development and pathways that are deregulated during tumorigenesis.

Additional studies have highlighted relationships between developmental signaling pathways and tumor progression. Topczewska, Hendrix, and colleagues transplanted human melanoma cells into zebrafish embryos and identified a phenotype consistent with ectopic secretion of nodal, a TGF-β family ligand [86]. Subsequent experiments found that melanoma cells produced abundant nodal activity, which promoted melanoma cell invasion via SMAD2/3 signaling. Schiavone, Argenton, and colleagues focused on the activities of multiple developmental signaling pathways in a model on pancreatic adenocarcinoma [37]. They used live imaging to monitor TGF-β, Notch, Bmp, and Shh pathways during tumor progression in vivo. This study and others illustrate how investigating developmental regulators can illuminate important pathways in cancer biology.

Further Leveraging of Zebrafish for Cancer Research: What Does the Future Hold?

The first decade of transgenic approaches to model cancer in zebrafish was characterized by remarkable innovation and scientific progress. Now, in the second decade challenges at hand are to address the many questions raised by cancer genomics studies and, beyond questions raised by sequence, solve fundamental issues in cancer biology that are best studied with intact tumors in vivo. Zebrafish cancer models hold significant promise in identifying key pathways, novel drug targets, and elucidating mechanisms of tumorigenesis. Clearly, the advent of CRISPR/Cas9 genome editing offers significant technical capability to perform reverse genetics more efficiently, and in a tissue-restricted manner. Inducible Cas9 and the ability to precisely recombine in the fish genome may herald a golden age of zebrafish transgenics, in which it is possible to alter any base in the genome. These approaches will allow researchers to examine the consequences of an edited oncogene in its native genomic context, perhaps even in a cell type restricted manner. The pace of innovation and discovery using zebrafish transgenic tools to study fundamental problems in cancer biology is likely to experience continued significant growth.

Acknowledgments We sincerely apologize to those colleagues whose work was not included in this review. Funding for this effort was supported in part by the NIH (R01AR063850 to C.C.), the Department of Defense (CA120099 to C.C.), the Kimmel Scholar Award to C.C., and the American Cancer Society Research Scholar Award to C.C.

References

1. Varmus HE (1984) The molecular genetics of cellular oncogenes. Annu Rev Genet 18:553–612
2. Langenau DM, Traver D, Ferrando AA, Kutok JL, Aster JC, Kanki JP, Lin S, Prochownik E, Trede NS, Zon LI et al (2003) Myc-induced T cell leukemia in transgenic zebrafish. Science 299:887–890
3. Langenau DM, Jette C, Berghmans S, Palomero T, Kanki JP, Kutok JL, Look AT (2005) Suppression of apoptosis by bcl-2 overexpression in lymphoid cells of transgenic zebrafish. Blood 105:3278–3285
4. Blackburn JS, Liu S, Wilder JL, Dobrinski KP, Lobbardi R, Moore FE, Martinez SA, Chen EY, Lee C, Langenau DM (2014) Clonal evolution enhances leukemia-propagating cell frequency in T cell acute lymphoblastic leukemia through Akt/mTORC1 pathway activation. Cancer Cell 25:366–378
5. Langenau DM, Feng H, Berghmans S, Kanki JP, Kutok JL, Look AT (2005) Cre/lox-regulated transgenic zebrafish model with conditional myc-induced T cell acute lymphoblastic leukemia. Proc Natl Acad Sci U S A 102:6068–6073
6. Feng H, Langenau DM, Madge JA, Quinkertz A, Gutierrez A, Neuberg DS, Kanki JP, Look AT (2007) Heat-shock induction of T-cell lymphoma/leukaemia in conditional Cre/lox-regulated transgenic zebrafish. Br J Haematol 138:169–175
7. Feng H, Stachura DL, White RM, Gutierrez A, Zhang L, Sanda T, Jette CA, Testa JR, Neuberg DS, Langenau DM et al (2010) T-lymphoblastic lymphoma cells express high levels of BCL2, S1P1, and ICAM1, leading to a blockade of tumor cell intravasation. Cancer Cell 18:353–366
8. Gutierrez A, Pan L, Groen RW, Baleydier F, Kentsis A, Marineau J, Grebliunaite R, Kozakewich E, Reed C, Pflumio F et al (2014) Phenothiazines induce PP2A-mediated apoptosis in T cell acute lymphoblastic leukemia. J Clin Invest 124:644–655
9. Chen J, Jette C, Kanki JP, Aster JC, Look AT, Griffin JD (2007) NOTCH1-induced T-cell leukemia in transgenic zebrafish. Leukemia 21:462–471
10. Blackburn JS, Liu S, Raiser DM, Martinez SA, Feng H, Meeker ND, Gentry J, Neuberg D, Look AT, Ramaswamy S et al (2012) Notch signaling expands a pre-malignant pool of T-cell acute lymphoblastic leukemia clones without affecting leukemia-propagating cell frequency. Leukemia 26:2069–2078
11. Sabaawy HE, Azuma M, Embree LJ, Tsai HJ, Starost MF, Hickstein DD (2006) TEL-AML1 transgenic zebrafish model of precursor B cell acute lymphoblastic leukemia. Proc Natl Acad Sci U S A 103:15166–15171
12. Zhuravleva J, Paggetti J, Martin L, Hammann A, Solary E, Bastie JN, Delva L (2008) MOZ/TIF2-induced acute myeloid leukaemia in transgenic fish. Br J Haematol 143:378–382
13. Forrester AM, Grabher C, McBride ER, Boyd ER, Vigerstad MH, Edgar A, Kai FB, Da'as SI, Payne E, Look AT et al (2011) NUP98-HOXA9-transgenic zebrafish develop a myeloproliferative neoplasm and provide new insight into mechanisms of myeloid leukaemogenesis. Br J Haematol 155:167–181
14. Deveau AP, Forrester AM, Coombs AJ, Wagner GS, Grabher C, Chute IC, Leger D, Mingay M, Alexe G, Rajan V et al (2015) Epigenetic therapy restores normal hematopoiesis in a zebrafish model of NUP98-HOXA9-induced myeloid disease. Leukemia 29:2086–2097
15. Gjini E, Mansour MR, Sander JD, Moritz N, Nguyen AT, Kesarsing M, Gans E, He S, Chen S, Ko M et al (2015) A zebrafish model of myelodysplastic syndrome produced through tet2 genomic editing. Mol Cell Biol 35:789–804

16. Balci TB, Prykhozhij SV, Teh EM, Da'as SI, McBride E, Liwski R, Chute IC, Leger D, Lewis SM, Berman JN (2014) A transgenic zebrafish model expressing KIT-D816V recapitulates features of aggressive systemic mastocytosis. Br J Haematol 167:48–61
17. Patton EE, Widlund HR, Kutok JL, Kopani KR, Amatruda JF, Murphey RD, Berghmans S, Mayhall EA, Traver D, Fletcher CD et al (2005) BRAF mutations are sufficient to promote nevi formation and cooperate with p53 in the genesis of melanoma. Curr Biol 15:249–254
18. Yen J, White RM, Wedge DC, Van Loo P, de Ridder J, Capper A, Richardson J, Jones D, Raine K, Watson IR et al (2013) The genetic heterogeneity and mutational burden of engineered melanomas in zebrafish models. Genome Biol 14:R113
19. Ceol CJ, Houvras Y, Jane-Valbuena J, Bilodeau S, Orlando DA, Battisti V, Fritsch L, Lin WM, Hollmann TJ, Ferre F et al (2011) The histone methyltransferase SETDB1 is recurrently amplified in melanoma and accelerates its onset. Nature 471:513–517
20. Iyengar S, Houvras Y, Ceol CJ (2012) Screening for melanoma modifiers using a zebrafish autochthonous tumor model. J Vis Exp. e50086
21. Lister JA, Capper A, Zeng Z, Mathers ME, Richardson J, Paranthaman K, Jackson IJ, Patton EE (2014) A conditional zebrafish MITF mutation reveals MITF levels are critical for melanoma promotion vs. regression in vivo. J Invest Dermatol 134:133–140
22. White RM, Cech J, Ratanasirintrawoot S, Lin CY, Rahl PB, Burke CJ, Langdon E, Tomlinson ML, Mosher J, Kaufman C et al (2011) DHODH modulates transcriptional elongation in the neural crest and melanoma. Nature 471:518–522
23. Dovey M, White RM, Zon LI (2009) Oncogenic NRAS cooperates with p53 loss to generate melanoma in zebrafish. Zebrafish 6:397–404
24. Michailidou C, Jones M, Walker P, Kamarashev J, Kelly A, Hurlstone AF (2009) Dissecting the roles of Raf- and PI3K-signalling pathways in melanoma formation and progression in a zebrafish model. Dis Model Mech 2:399–411
25. Anelli V, Santoriello C, Distel M, Koster RW, Ciccarelli FD, Mione M (2009) Global repression of cancer gene expression in a zebrafish model of melanoma is linked to epigenetic regulation. Zebrafish 6:417–424
26. Santoriello C, Gennaro E, Anelli V, Distel M, Kelly A, Koster RW, Hurlstone A, Mione M (2010) Kita driven expression of oncogenic HRAS leads to early onset and highly penetrant melanoma in zebrafish. PLoS One 5, e15170
27. Li Z, Zheng W, Wang Z, Zeng Z, Zhan H, Li C, Zhou L, Yan C, Spitsbergen JM, Gong Z (2013) A transgenic zebrafish liver tumor model with inducible Myc expression reveals conserved Myc signatures with mammalian liver tumors. Dis Model Mech 6:414–423
28. Nguyen AT, Emelyanov A, Koh CH, Spitsbergen JM, Lam SH, Mathavan S, Parinov S, Gong Z (2011) A high level of liver-specific expression of oncogenic Kras(V12) drives robust liver tumorigenesis in transgenic zebrafish. Dis Model Mech 4:801–813
29. Nguyen AT, Emelyanov A, Koh CH, Spitsbergen JM, Parinov S, Gong Z (2012) An inducible kras(V12) transgenic zebrafish model for liver tumorigenesis and chemical drug screening. Dis Model Mech 5:63–72
30. Mudbhary R, Hoshida Y, Chernyavskaya Y, Jacob V, Villanueva A, Fiel MI, Chen X, Kojima K, Thung S, Bronson RT et al (2014) UHRF1 overexpression drives DNA hypomethylation and hepatocellular carcinoma. Cancer Cell 25:196–209
31. Evason KJ, Francisco MT, Juric V, Balakrishnan S, Lopez Pazmino Mdel P, Gordan JD, Kakar S, Spitsbergen J, Goga A, Stainier DY (2015) Identification of chemical inhibitors of beta-catenin-driven liver tumorigenesis in zebrafish. PLoS Genet 11, e1005305
32. Lu JW, Yang WY, Tsai SM, Lin YM, Chang PH, Chen JR, Wang HD, Wu JL, Jin SL, Yuh CH (2013) Liver-specific expressions of HBx and src in the p53 mutant trigger hepatocarcinogenesis in zebrafish. PLoS One 8, e76951
33. Liu W, Chen JR, Hsu CH, Li YH, Chen YM, Lin CY, Huang SJ, Chang ZK, Chen YC, Lin CH et al (2012) A zebrafish model of intrahepatic cholangiocarcinoma by dual expression of hepatitis B virus X and hepatitis C virus core protein in liver. Hepatology 56:2268–2276
34. Park SW, Davison JM, Rhee J, Hruban RH, Maitra A, Leach SD (2008) Oncogenic KRAS induces progenitor cell expansion and malignant transformation in zebrafish exocrine pancreas. Gastroenterology 134:2080–2090

35. Liu NA, Jiang H, Ben-Shlomo A, Wawrowsky K, Fan XM, Lin S, Melmed S (2011) Targeting zebrafish and murine pituitary corticotroph tumors with a cyclin-dependent kinase (CDK) inhibitor. Proc Natl Acad Sci U S A 108:8414–8419
36. Provost E, Bailey JM, Aldrugh S, Liu S, Iacobuzio-Donahue C, Leach SD (2014) The tumor suppressor rpl36 restrains KRAS(G12V)-induced pancreatic cancer. Zebrafish 11:551–559
37. Schiavone M, Rampazzo E, Casari A, Battilana G, Persano L, Moro E, Liu S, Leach SD, Tiso N, Argenton F (2014) Zebrafish reporter lines reveal in vivo signaling pathway activities involved in pancreatic cancer. Dis Model Mech 7:883–894
38. Yang HW, Kutok JL, Lee NH, Piao HY, Fletcher CD, Kanki JP, Look AT (2004) Targeted expression of human MYCN selectively causes pancreatic neuroendocrine tumors in transgenic zebrafish. Cancer Res 64:7256–7262
39. Neumann JC, Chandler GL, Damoulis VA, Fustino NJ, Lillard K, Looijenga L, Margraf L, Rakheja D, Amatruda JF (2011) Mutation in the type IB bone morphogenetic protein receptor Alk6b impairs germ-cell differentiation and causes germ-cell tumors in zebrafish. Proc Natl Acad Sci U S A 108:13153–13158
40. Neumann JC, Dovey JS, Chandler GL, Carbajal L, Amatruda JF (2009) Identification of a heritable model of testicular germ cell tumor in the zebrafish. Zebrafish 6:319–327
41. Gill JA, Lowe L, Nguyen J, Liu PP, Blake T, Venkatesh B, Aplan PD (2010) Enforced expression of Simian virus 40 large T-antigen leads to testicular germ cell tumors in zebrafish. Zebrafish 7:333–341
42. Ju B, Chen W, Spitsbergen JM, Lu J, Vogel P, Peters JL, Wang YD, Orr BA, Wu J, Henson HE et al (2014) Activation of Sonic hedgehog signaling in neural progenitor cells promotes glioma development in the zebrafish optic pathway. Oncogenesis 3, e96
43. Ju B, Chen W, Orr BA, Spitsbergen JM, Jia S, Eden CJ, Henson HE, Taylor MR (2015) Oncogenic KRAS promotes malignant brain tumors in zebrafish. Mol Cancer 14:18
44. Berghmans S, Murphey RD, Wienholds E, Neuberg D, Kutok JL, Fletcher CD, Morris JP, Liu TX, Schulte-Merker S, Kanki JP et al (2005) tp53 mutant zebrafish develop malignant peripheral nerve sheath tumors. Proc Natl Acad Sci U S A 102:407–412
45. Shin J, Padmanabhan A, de Groh ED, Lee JS, Haidar S, Dahlberg S, Guo F, He S, Wolman MA, Granato M et al (2012) Zebrafish neurofibromatosis type 1 genes have redundant functions in tumorigenesis and embryonic development. Dis Model Mech 5:881–894
46. Langenau DM, Keefe MD, Storer NY, Guyon JR, Kutok JL, Le X, Goessling W, Neuberg DS, Kunkel LM, Zon LI (2007) Effects of RAS on the genesis of embryonal rhabdomyosarcoma. Genes Dev 21:1382–1395
47. Leacock SW, Basse AN, Chandler GL, Kirk AM, Rakheja D, Amatruda JF (2012) A zebrafish transgenic model of Ewing's sarcoma reveals conserved mediators of EWS-FLI1 tumorigenesis. Dis Model Mech 5:95–106
48. Burger A, Vasilyev A, Tomar R, Selig MK, Nielsen GP, Peterson RT, Drummond IA, Haber DA (2014) A zebrafish model of chordoma initiated by notochord-driven expression of HRASV12. Dis Model Mech 7:907–913
49. Chu CY, Chen CF, Rajendran RS, Shen CN, Chen TH, Yen CC, Chuang CK, Lin DS, Hsiao CD (2012) Overexpression of Akt1 enhances adipogenesis and leads to lipoma formation in zebrafish. PLoS One 7, e36474
50. Zhu S, Lee JS, Guo F, Shin J, Perez-Atayde AR, Kutok JL, Rodig SJ, Neuberg DS, Helman D, Feng H et al (2012) Activated ALK collaborates with MYCN in neuroblastoma pathogenesis. Cancer Cell 21:362–373
51. Le X, Langenau DM, Keefe MD, Kutok JL, Neuberg DS, Zon LI (2007) Heat shock-inducible Cre/Lox approaches to induce diverse types of tumors and hyperplasia in transgenic zebrafish. Proc Natl Acad Sci U S A 104:9410–9415
52. Ju B, Spitsbergen J, Eden CJ, Taylor MR, Chen W (2009) Co-activation of hedgehog and AKT pathways promote tumorigenesis in zebrafish. Mol Cancer 8:40
53. White R, Rose K, Zon L (2013) Zebrafish cancer: the state of the art and the path forward. Nat Rev Cancer 13:624–636
54. Chapman A, Fernandez del Ama L, Ferguson J, Kamarashev J, Wellbrock C, Hurlstone A (2014) Heterogeneous tumor subpopulations cooperate to drive invasion. Cell Rep 8:688–695

55. Ignatius MS, Chen E, Elpek NM, Fuller AZ, Tenente IM, Clagg R, Liu S, Blackburn JS, Linardic CM, Rosenberg AE et al (2012) In vivo imaging of tumor-propagating cells, regional tumor heterogeneity, and dynamic cell movements in embryonal rhabdomyosarcoma. Cancer Cell 21:680–693
56. Ruzicka L, Bradford YM, Frazer K, Howe DG, Paddock H, Ramachandran S, Singer A, Toro S, Van Slyke CE, Eagle AE et al (2015) ZFIN, the zebrafish model organism database: updates and new directions. Genesis 53:498–509
57. Thisse B, Thisse C (2014) In situ hybridization on whole-mount zebrafish embryos and young larvae. Methods Mol Biol 1211:53–67
58. Choorapoikayil S, Kuiper RV, de Bruin A, den Hertog J (2012) Haploinsufficiency of the genes encoding the tumor suppressor Pten predisposes zebrafish to hemangiosarcoma. Dis Model Mech 5:241–247
59. Croushore JA, Blasiole B, Riddle RC, Thisse C, Thisse B, Canfield VA, Robertson GP, Cheng KC, Levenson R (2005) Ptena and ptenb genes play distinct roles in zebrafish embryogenesis. Dev Dyn 234:911–921
60. Faucherre A, Taylor GS, Overvoorde J, Dixon JE, Hertog J (2008) Zebrafish pten genes have overlapping and non-redundant functions in tumorigenesis and embryonic development. Oncogene 27:1079–1086
61. Padmanabhan A, Lee JS, Ismat FA, Lu MM, Lawson ND, Kanki JP, Look AT, Epstein JA (2009) Cardiac and vascular functions of the zebrafish orthologues of the type I neurofibromatosis gene NFI. Proc Natl Acad Sci U S A 106:22305–22310
62. Park JT, Leach SD (2013) TAILOR: transgene activation and inactivation using lox and rox in zebrafish. PLoS One 8, e85218
63. Taylor AM, Zon LI (2009) Zebrafish tumor assays: the state of transplantation. Zebrafish 6:339–346
64. Tang Q, Abdelfattah NS, Blackburn JS, Moore JC, Martinez SA, Moore FE, Lobbardi R, Tenente IM, Ignatius MS, Berman JN et al (2014) Optimized cell transplantation using adult rag2 mutant zebrafish. Nat Methods 11:821–824
65. White RM, Sessa A, Burke C, Bowman T, LeBlanc J, Ceol C, Bourque C, Dovey M, Goessling W, Burns CE et al (2008) Transparent adult zebrafish as a tool for in vivo transplantation analysis. Cell Stem Cell 2:183–189
66. Ablain J, Durand EM, Yang S, Zhou Y, Zon LI (2015) A CRISPR/Cas9 vector system for tissue-specific gene disruption in zebrafish. Dev Cell 32:756–764
67. Cichowski K, Jacks T (2001) NF1 tumor suppressor gene function: narrowing the GAP. Cell 104:593–604
68. Livshits G, Lowe SW (2013) Accelerating cancer modeling with RNAi and nongermline genetically engineered mouse models. Cold Spring Harb Protoc 2013
69. van Lohuizen M, Verbeek S, Scheijen B, Wientjens E, van der Gulden H, Berns A (1991) Identification of cooperating oncogenes in E mu-myc transgenic mice by provirus tagging. Cell 65:737–752
70. Zender L, Xue W, Zuber J, Semighini CP, Krasnitz A, Ma B, Zender P, Kubicka S, Luk JM, Schirmacher P et al (2008) An oncogenomics-based in vivo RNAi screen identifies tumor suppressors in liver cancer. Cell 135:852–864
71. Lin WM, Baker AC, Beroukhim R, Winckler W, Feng W, Marmion JM, Laine E, Greulich H, Tseng H, Gates C et al (2008) Modeling genomic diversity and tumor dependency in malignant melanoma. Cancer Res 68:664–673
72. Amos CI, Wang LE, Lee JE, Gershenwald JE, Chen WV, Fang S, Kosoy R, Zhang M, Qureshi AA, Vattathil S et al (2011) Genome-wide association study identifies novel loci predisposing to cutaneous melanoma. Hum Mol Genet 20:5012–5023
73. Macgregor S, Montgomery GW, Liu JZ, Zhao ZZ, Henders AK, Stark M, Schmid H, Holland EA, Duffy DL, Zhang M et al (2011) Genome-wide association study identifies a new melanoma susceptibility locus at 1q21.3. Nat Genet 43:1114–1118
74. Shi J, Wang E, Milazzo JP, Wang Z, Kinney JB, Vakoc CR (2015) Discovery of cancer drug targets by CRISPR-Cas9 screening of protein domains. Nat Biotechnol 33:661–667

75. Dupuy AJ, Akagi K, Largaespada DA, Copeland NG, Jenkins NA (2005) Mammalian mutagenesis using a highly mobile somatic Sleeping Beauty transposon system. Nature 436:221–226
76. McGrail M, Hatler JM, Kuang X, Liao HK, Nannapaneni K, Watt KE, Uhl JD, Largaespada DA, Vollbrecht E, Scheetz TE et al (2011) Somatic mutagenesis with a Sleeping Beauty transposon system leads to solid tumor formation in zebrafish. PLoS One 6, e18826
77. Cheung NK, Dyer MA (2013) Neuroblastoma: developmental biology, cancer genomics and immunotherapy. Nat Rev Cancer 13:397–411
78. Howe K, Clark MD, Torroja CF, Torrance J, Berthelot C, Muffato M, Collins JE, Humphray S, McLaren K, Matthews L et al (2013) The zebrafish reference genome sequence and its relationship to the human genome. Nature 496:498–503
79. Freeman JL, Ceol C, Feng H, Langenau DM, Belair C, Stern HM, Song A, Paw BH, Look AT, Zhou Y et al (2009) Construction and application of a zebrafish array comparative genomic hybridization platform. Genes Chromosomes Cancer 48:155–170
80. Zhang G, Hoersch S, Amsterdam A, Whittaker CA, Lees JA, Hopkins N (2010) Highly aneuploid zebrafish malignant peripheral nerve sheath tumors have genetic alterations similar to human cancers. Proc Natl Acad Sci U S A 107:16940–16945
81. Rudner LA, Brown KH, Dobrinski KP, Bradley DF, Garcia MI, Smith AC, Downie JM, Meeker ND, Look AT, Downing JR et al (2011) Shared acquired genomic changes in zebrafish and human T-ALL. Oncogene 30:4289–4296
82. Chen EY, Dobrinski KP, Brown KH, Clagg R, Edelman E, Ignatius MS, Chen JY, Brockmann J, Nielsen GP, Ramaswamy S et al (2013) Cross-species array comparative genomic hybridization identifies novel oncogenic events in zebrafish and human embryonal rhabdomyosarcoma. PLoS Genet 9, e1003727
83. Zhang G, Hoersch S, Amsterdam A, Whittaker CA, Beert E, Catchen JM, Farrington S, Postlethwait JH, Legius E, Hopkins N et al (2013) Comparative oncogenomic analysis of copy number alterations in human and zebrafish tumors enables cancer driver discovery. PLoS Genet 9, e1003734
84. Haffter P, Granato M, Brand M, Mullins MC, Hammerschmidt M, Kane DA, Odenthal J, van Eeden FJ, Jiang YJ, Heisenberg CP et al (1996) The identification of genes with unique and essential functions in the development of the zebrafish, Danio rerio. Development 123:1–36
85. Sadler KC, Krahn KN, Gaur NA, Ukomadu C (2007) Liver growth in the embryo and during liver regeneration in zebrafish requires the cell cycle regulator, uhrf1. Proc Natl Acad Sci U S A 104:1570–1575
86. Topczewska JM, Postovit LM, Margaryan NV, Sam A, Hess AR, Wheaton WW, Nickoloff BJ, Topczewski J, Hendrix MJ (2006) Embryonic and tumorigenic pathways converge via Nodal signaling: role in melanoma aggressiveness. Nat Med 12:925–932
87. Liu S, Leach SD (2011) Screening pancreatic oncogenes in zebrafish using the Gal4/UAS system. Methods Cell Biol 105:367–381

The Toolbox for Conditional Zebrafish Cancer Models

Marie Mayrhofer and Marina Mione

Abstract Here we describe the conditional zebrafish cancer toolbox, which allows for fine control of the expression of oncogenes or downregulation of tumor suppressors at the spatial and temporal level. Methods such as the Gal4/UAS or the Cre/lox systems paved the way to the development of elegant tumor models, which are now being used to study cancer cell biology, clonal evolution, identification of cancer stem cells and anti-cancer drug screening. Combination of these tools, as well as novel developments such as the promising genome editing system through CRISPR/Cas9 and clever application of light reactive proteins will enable the development of even more sophisticated zebrafish cancer models. Here, we introduce this growing toolbox of conditional transgenic approaches, discuss its current application in zebrafish cancer models and provide an outlook on future perspectives.

Keywords Cancer models • Cre/lox • Gal4/UAS • Heat shock • Inducible systems • ER/tamoxifen • Tet-On • Transgenic methods • Zebrafish

Abbreviations

4-OHT	4-Hydroxy-Tamoxifen
AD	Activation Domain
CRISPR/Cas9	Clustered Regularly Interspaced, Short Palindromic Repeats/ CRISPR Associated System

While this manuscript was under review, a novel tool for genome editing in somatic cells was described in: A CRISPR/Cas9 Vector System for Tissue-Specific Gene Disruption in Zebrafish, by Ablain J, Durand EM, Yang S, Zhou Y, and Zon LI, Developmental Cell 32, 756–764, March 23, 2015 (doi: 10.1016/j.devcel.2015.01.032) and a description of in-frame knock-in without integration of plasmid backbone was reported in: A precise in-frame integration of exogenous DNA mediated by CRISPR/Cas9 system in zebrafish, by Hisano Y, Sakuma T, Nakade S, Ohga R, Ota S, Okamoto H, Yamamoto T and Kawahara A, Sci Rep. 5, 8841. March 5, 2015 (doi: 10.1038/srep08841).

M. Mayrhofer • M. Mione (✉)
Institute of Toxicology and Genetics, Karlsruhe Institute of Technology, Eggenstein-Leopoldshafen, Germany
e-mail: marie.spitzner@kit.edu; maria.mione@kit.edu

D.M. Langenau (ed.), *Cancer and Zebrafish*, Advances in Experimental Medicine and Biology 910, DOI 10.1007/978-3-319-30654-4_2

CRY2-CIB1	Cryptochrome 2—Cryptochrome-Interacting Basic helix-loop-helix
DBD	DNA-Binding Domain
Dox	Doxycycline
EcR	Ecdysone Receptor
ER	Estrogen Receptor
GE	Genome Editing
GFP	Green Fluorescent Protein
GR	Glucocorticoid Receptor
gRNA	Guide RNA
HCC	Hepatocellular Carcinoma
HSE	Heat Shock Element
HSF	Heat Shock Factor
LOV	Light, Oxygen and Voltage Domain
miR-shRNA	Fusion construct between micro RNA and short hairpin RNA
MPD	Myeloproliferative Disorder
MPNST	Malignant Peripheral Nerve Sheath Tumor
NHEJ	Non-Homologous End Joining
PAM	Protospacer Adjacent Motif
PhyB-PIF	Phytochrome B—Phytochrome Interacting Factor
PR	Progesterone Receptor
RFP	Red Fluorescent Protein
RMCE	Recombinase Mediated Cassette Exchange
rtTA	Reverse Tetracycline Transactivator
shRNA	Short Hairpin RNA
SSR	Site Specific Recombinase
TALEN	Transcription Activator-Like Effector Nucleases
T-ALL	T-cell Acute Lymphatic Leukemia
tetO	Tetracycline Operator
tetR	Tetracycline Repressor
TrpR	Tryptophan Repressor
tTA	Tetracycline Transactivator
UAS	Upstream Activating Sequence
UVR8	UV-resistance locus 8
ZFN	Zink Finger Nucleases

Introduction

As a vertebrate, the zebrafish spontaneously develops tumors and together with its fast, transparent and ex utero embryonic development attracted the attention of researchers as a disease model organism [1]. The first cancer studies used chemicals and induced mainly liver and intestinal tumors in zebrafish [2–4]. Shortly after, the first stable transgenic zebrafish line was generated in 1988 [5]. However, due to difficulties with stable germline transmission and silencing of the integrated transgenes it took several years to establish reliable transgenic protocols [6].

Thereby, application of endonucleases such as the meganuclease I-SceI and transposases, including Sleeping Beauty or the tol2 system, improved the transgenic efficiency and stability. Especially, the *medaka* derived tol2 system, developed by Kawakami, greatly improved transgenesis as it integrates mostly single copies through a cut-and-paste mechanism without rearrangement or modification at the target site [7]. Its unbiased integration into nearly any genomic site and reliable expression of integrated transgenes made transgenesis in zebrafish easy, fast, efficient and stable [7].

A straight forward method to model a specific cancer type in zebrafish is to generate transgenes with oncogenes placed directly under the control of a tissue specific promoter. However, strong oncogenes will induce deleterious effects that may lead to lethality before the fish reach sexual maturity thus impeding the generation of stable lines. To overcome this drawback and to improve spatial and temporal control over oncogene expression, a toolbox of conditional transgenic approaches has been established in the zebrafish that provides the potential to (1) perform isolated gene analysis in specific tissues or cell type subsets, (2) determine the time point of gene expression onset and (3) offset and (4) control the strength of oncogene expression. The bipartite Gal4/*UAS* system facilitated the development of a wide range of aggressive cancer models [8–13] as it allows for independent inheritance of tissue specificity through Gal4 in one parental line and silent oncogenes in the second parental line. Other conditional approaches include the Cre/*lox* system, that generates stable recombination events but that in zebrafish still suffers of inefficient activity leading to a mosaic pattern [14, 15], and temporal control systems such as the *heat shock promoter*, the tetracycline dependent Tet-On and Tet-Off system or chemically inducible hormone receptors that can be activated or inactivated upon external stimuli.

Spatial Control

Genetic methods for restricting the expression of oncogenes to a specific tissue or cell type are typically based on the use of tissue specific promoters. Several systems have been developed that increase flexibility, including the possibility of turning oncogene expression on and off. Importantly, conditional transgenic approaches allow maintenance of lines, which are embryonic-lethal because of embryonic activity of the promoter. These include the Gal4/*UAS* system and its derivatives, the site specific recombinases (CRE/*lox* system & Co) and more recently, genome editing through CRISPR/Cas9.

The Gal4/UAS System

The Gal4/*UAS* system is one of the first conditional transgenic methods applied to zebrafish [16]. This binary system, derived from the yeast *Saccharomyces cerevisia,* consists of the transcriptional activator Gal4 that controls gene expression through its DNA-binding motif, *UAS* (Upstream Activating Sequence) (Fig. 1a) [17].

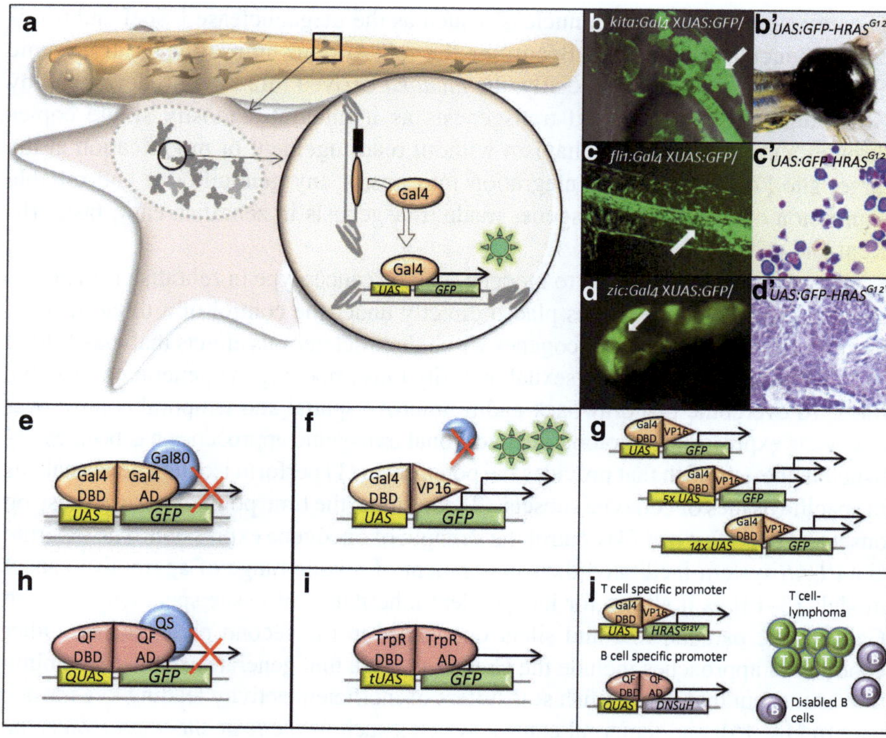

Fig. 1 The Gal4/*UAS* system and its variants enable cell type specific oncogene expression. (**A**) Schematic of the Gal4/*UAS* system applied to a 5 dpf larva melanocyte (*box*): tissue specific gene expression is enabled through promoter dependent Gal4 expression (*orange oval*) which, upon binding to *UAS* in the same cell, activates expression of a *UAS* driven transgene such as green fluorescent protein (GFP) leading to fluorescent protein production (*green stars*). (**B–D**) Expression of Gal4 under tissue specific promoters allows for *UAS* controlled GFP expression in different tissues as indicated by *white arrows*: melanocytes (*kita:Gal4*, **B**), endothelial cells (*fli1:Gal4*, **C**) and brain (*zic:Gal4*, **D**) while *UAS* controlled expression of the oncogene $HRAS^{G12V}$ in these lines induces formation of melanoma (**B'**), leukemia (**C'**) and glioma (**D'**). (**E**) Inhibition of Gal4 activity can be mediated by Gal80 which binds to the activation domain (AD) of Gal4. (**F**) In the fusion protein Gal4-VP16, the AD of Gal4 is exchanged with a VP16 variant (*pink triangle*) disabling Gal80 mediated Gal4 inhibition while increasing the transcriptional activity of Gal4. (**G**) Increased transgene expression is achieved by application of up to 5 *UAS* repeats, while a much higher copy number reduces transcription efficiency. Systems similar to Gal4/*UAS* include the Q-system (**H**) and the TrpR/*tUAS* system (**I**) composed of the transactivator (*pink oval*) QF/TrpR and the upstream activation sequence (*yellow rectangle*) *QUAS*/*tUAS*, respectively. The QF can be inhibited by QS (*blue half circle*). (**J**) As these systems do not cross-react they can be combined in one model to study cell non-autonomous effects for example by expressing Gal4 and QF under different cell type specific promoters to elucidate the effect of disabled Notch-signaling (through expression of dominant-negative suppressor of hairless, *grey rectangle*) in B-cells, and assess the effect on oncogenic behavior of T-cells. *Abbreviations*: *AD* activation domain, *DBD* DNA-binding domain, *DNSuH* dominant negative suppressor of hairless, *GFP* green fluorescent protein, *UAS* upstream activating sequence

To achieve tissue specific gene expression, the gene of interest is placed downstream of the *UAS* motif and the transcriptional activator Gal4 is expressed under the control of a tissue specific promoter. As the *UAS* motif is exclusively active in presence of Gal4, the gene of interest will be expressed exclusively in the tissue where the chosen promoter is active. Separate integration of the two components in an activator line (Gal4) and an effector line (*UAS*-gene x) allows for silent inheritance of the gene of interest and its conditional activation only in double transgenic offsprings from crosses between activator and effector lines. A third component of this system that has been adopted for further control is the Gal4 inhibitor Gal80. Gal80 binds and inhibits the activating domain of Gal40 and thus inhibits expression of the *UAS* controlled gene (Fig. 1e) [18, 19].

The Gal4/*UAS* system has been used in a variety of organisms including plants [20], drosophila [21], xenopus [22] and mice [23]. Scheer and Campos-Ortega were the first to apply it to zebrafish [16]. They developed activator lines that express the full-length Gal4 gene under control of one of the two ubiquitous promoters SV40/thymidine kinase (*svtk*) or carp β-actin and effector lines that express an active form of the Notch1 receptor under *UAS* control. Double transgenic offspring from crosses between the stable transgenic activator and effector lines expressed the *myc-notch1a:intra* in the same regions where Gal4 was expressed and thus confirmed that Gal4 can regulate the expression of a *UAS* controlled gene. Still the system suffered from low gene induction, low transgenic efficiency and variegated/mosaic expression. Several alterations of the constructs and discoveries of new techniques have improved the system and made it widely used in zebrafish. Specifically, the discovery of the transposase Tol2 from *medaka* revolutionized the genomic integration of transgenes and reduced the mosaic and variegated expression pattern of the Gal4 system and facilitated its rapid development.

Further optimization of the system aimed to increase Gal4 activity through fusion of the Gal4 DNA-binding domain (DBD) with a strong transcriptional activation domain (AD). Intensive efforts have been undertaken to optimize the transcriptional AD from the VP16 protein of the herpes simplex virus. In a first attempt the Gal4-DBD was fused to the full-length VP16 [24, 25] which improved the strength of the *UAS* induction and maintained the promoter activity on Gal4 expression but also revealed massive toxic effects reflected in developmental defects and high embryonic lethality [26, 27]. These effects were thought to appear due to a phenomenon called "squelching" [28, 29]. VP16, as one of the most potent activation domains known, interacts with a number of transcriptional components and thus through a titration effect can lead to non-specific inhibition of transcription [28]. Therefore, to reduce toxicity several modifications were developed to reduce toxicity while keeping Gal4 activity as strong as when it is combined with full-length VP16. Such attempts include the fusion with the truncated form $VP16_{413–470}$ that lacks the last 20 amino acids important for its transcriptional activity. Albeit it is less active than full length VP16, $VP16_{413–470}$ still induces developmental defects. Asakawa et al. used an even smaller fragment of VP16 containing the 12 amino acids 436–447 and an N-terminal prolin [30, 31]. Two copies fused to Gal4 (Gal4FF) allowed sufficient and safe Gal4 expression while a further increase in copy numbers would increase Gal4 efficiency but also leads to side effects [32]. Using GFP (green fluorescent protein) and the newly developed Gal4FF, Kawakami et al. [33] generated a large library of diverse gene trap and enhancer trap lines which are published and available to the zebrafish community under the name zTrap (see Box 1 for web sites). Distel et al. applied a third version of the VP16 AD, called TA4 [24]. It contains three copies of $VP16_{436–447}$ but two of them with weakening mutations in the position 442, a position crucial for its function [34, 35]. A Kozac sequence before the Gal4 and a β-globin intron positioned 3′ of TA4 improves the translation of the mRNA and gives the construct the new name KalTA4GI. This construct generates robust non-deleterious and non-variegated transactivation in a variety of lines [34].

Unfortunately, the use of the Gal4 inhibitor Gal80 to modulate the activity of Gal4 is limited by the modifications imposed to Gal4 to increase its efficiency. As Gal80 targets the C-terminal AD of Gal4 [19], which is missing in the Gal4-VP16 fusion variants, it is unable to bind and inhibit Gal4 activity in the vast majority of existing activator lines (Fig. 1f) [36, 37]. In alternative to VP16, the carboxyl terminal AD of the protein NFkB p65 was shown to significantly improve Gal4 activity with only minor toxic effects, but this combination has not been used by the zebrafish community [32].

To further optimize the binary Gal4/*UAS* system, the UAS promoter has been tested for its efficiency according to the number of *UAS* repeats. Thereby, a nearly linear dose effect was observed when increasing the *UAS* number from 1 to 5 (19.5 times stronger) repeats while further increase to 10 or even 14 repeats again reduces effects to a level comparable to 4 repeats (Fig. 1g) [34]. One drawback of the system is the gene silencing occurring over time through methylation probably due to the high *GC* content of the *UAS* sequence and the use of high copy numbers. Indeed, comparison of 4x*UAS* with 14x*UAS* showed that low copy number *UAS* constructs are less vulnerable to methylation induced silencing [38].

Since its development the Gal4/*UAS* system has been used for a variety of diverse zebrafish studies ranging from studies on early developmental processes to investigation of organization and function of individual brain structures [39–42].

The high flexibility of the Gal4/*UAS* system has promoted a number of cancer studies in zebrafish. In our lab, different cancer models have been established that utilize the same *UAS* controlled oncogene $HRAS^{G12V}$ fused to *GFP*, but depend on different Gal4 lines. Gal4 expression under control of the *kita* or *fli1* promoter for example induces development of melanoma [9] or leukemia [43], respectively. Remarkably, while the melanoma model develops aggressive tumors already in 20 % of 1 month old double transgenic fish, most of them survive to fecundity, which allows to keep the line in a double transgenic state [9]. The developing melanomas resemble the human phenotype not only histologically through infiltrative behavior and polyploid nuclei but also immunologically through expression of human melanoma markers such as Tyrosinase, Melan-a, s100 and HMB45. The double transgenic larvae are also characterized by a hyperpigmentation phenotype at 3 dpf, which make them ideal for chemical screens. The leukemia model instead is highly lethal in double transgenic larvae and requires tol2 mediated mosaic integration of the oncogene construct in somatic cells to allow development of leukemia in juveniles [43]. In larvae the disease is characterized by prominent expansion of the caudal hematopoietic tissue and disruption of the vascular system and in adults appears through an increased number of immature cells and arrest of myeloid maturation in kidney marrow. Applying the same *GFP* fused $HRAS^{G12V}$ oncogene, two further studies established novel cancer models. Burger and colleagues describe the first chordoma model through Gal4/*UAS* controlled $HRAS^{G12V}$ expression under the notochord prominent promoter *tiggywinkle hedgehog* (*twhh*) which shows histological features comparable to the human disease [13]. We also describe the first tissue specific glioma model in zebrafish based on $HRAS^{G12V}$ expression under control of the *zic4* enhancer [11]. Expressing the oncogene exclusively in the dorsal central nervous system of double transgenic larvae induces an increased number of dedifferentiated neural cells and increased brain size due to hyper proliferation, which leads to early lethality. Juvenile tumor carriers, generated through somatic oncogene expression, reveal massive telencephalic tumors resembling human malignant glioma. In another glioma study, Ju et al describe the induction of malignant brain tumors in zebrafish by expression of mCherry fused $KRAS^{G12V}$ under two different neural promoters, *krt5* and *gfap* [44]. While both models develop brain tumors in about 50 % of fish at 1 year of age, tumor type and histology are different with the *krt5* promoter inducing mostly MPNST (malignant peripheral nerve sheath tumor)—tumors and *gfap* promoter mainly tumors of the brain parenchyma. These studies using the same oncogenic line reveal the diverse impacts the same oncogene can have on different cell types (Fig. 1b–d). The availability of many diverse Gal4 lines and tested oncogenic constructs therefore allows to screen the existing libraries for new driver lines suitable for cancer modeling.

Additionally, the same GAl4 line can be used to study the effects of different oncogenes on cancer type/progression. In a pancreas cancer model the oncogenic Ras variant $KRAS^{G12V}$ was placed under *UAS* control and induced by Gal4 under control of the *ptf1a* promoter. While this promoter activates gene expression in the cerebellum, hindbrain and pancreas, double transgenic $KRAS^{G12V}$ expressing fish developed only pancreatic cancer [10]. Overexpression of another oncogene (dominant active Akt) under the *ptf1a* promoter instead leads to glioma development in nearly $^1/_3$ of double transgenic 6 months old fish [12]. The *ptf1a:Gal4* line was further used to evaluate different *KRAS* mutants for their oncogenic potential through injection of the constructs into one-cell stage *pft1a:GAL4* carriers [45]. From the 12 mutant *KRAS* forms used, all the 7 variants found in human pancreatic cancer also induced pancreatic cancer in zebrafish, while only 1 of the 5 mutant *KRAS* forms found in human non-pancreatic cancers was able to induce pancreatic cancer in zebrafish [45]. While confirming the similarity between zebrafish and human oncogenic processes, this study also revealed that the Gal4/*UAS* system is suitable to screen the oncogenic potential of different mutant proteins in a tissue specific manner. The possibility to apply different transgenes to the Gal4 line directly through injection into one-cell stage Gal4 carriers obviates the generation of large numbers of stable transgenic lines, although the expression of the transgenes will be mosaic and therefore differences are difficult to interpret.

The Gal4/*UAS* system can also be utilized for gene interaction studies. In a crossing intensive approach the pancreas cancer model has been tested in mutant background lines for the ribosomal proteins $rpl36^{-/-}$, $rpl23a^{-/-}$ or $rpl36^{-/-}/rpl23a^{-/-}$ which showed that rpl36 but not rpl23a functions as haploinsufficient tumor suppressor [46]. By far easier is the co-expression of different transgenes. In an activator line driving Gal4 expression under the cytokeratin promoter *krt4*, Ju et al. investigated the oncogenic effect of *UAS* controlled mutant *sonic hedgehog* (SHH) and *mutant AKT1* (myrhAkt1) [8]. While neither of them alone induce tumors, injection of double transgenic carriers *krt4:Gal4, UAS:SHH* with the construct *UAS:myrhAKT1* at one-cell stage led to development of diverse tumor types in the trunk, the eye and the head region that were identified as spindle cell sarcoma, rhabdomyoma, ocular melanoma, myoma, astrocytoma and glioblastoma. This latter study illustrates the strength of the Gal4/*UAS* system for easy and fast combined directed gene expression studies without the need of long and costly transgenic approaches, besides the promiscuity of the *krt4* promoter.

Site directed gene silencing instead still remains an obstacle in zebrafish. Help may come from a technique tested by Dong et al. [47]. Through combination of inhibitory short hairpin RNA (shRNA) with the backbone of the micro RNA *miR-30* they generated a stable transgenic line (miR-shRNA), which is able to inhibit translation through mRNA degradation. Similar approaches have been described using artificial microRNAs [48], but have not yet been exploited in zebrafish. Controlled by a *UAS* promoter, these effects on mRNAs can be directed to specific cells or tissues [47]. While these methods have not been tested in zebrafish cancer studies yet, they have the potential to revolutionize model development as it not only accomplishes gene silencing, but also allows for spatial and temporal control of gene silencing. A schematic representation of the GAL4/*UAS* system (and its derivatives) applied to zebrafish cancer models is depicted in Fig. 1.

Gal4/UAS Alternative Systems

Two alternative systems to Gal4/*UAS* have been tested in zebrafish that promise to undergo less silencing and will increase transgenic potential as they are not interacting with the Gal4/*UAS* system (Fig. 1h, j). The TrpR/*tUAS* system [49] and the Q-system [50] are derived from *Escherichia coli* and the fungus *Neurospora crassa*, respectively. Due to a lower *CG* content in their UAS sequences both systems are less vulnerable to methylation induced silencing than the Gal4/*UAS* system. In a pioneering zebrafish study the wildtype tryptophan repressor (TrpR) and two mutants with 5x (T81M) and 11x (T81A) reduced transcriptional activity have been tested [49]. While no silencing occurred over four generations, TrpR appeared toxic to zebrafish larvae when expressed under strong promoters, thus requiring further optimization steps. In addition to the transcription factor (QF) and its binding site (*QUAS*) the Q-system includes a repressor (QS) which itself can be inhibited in some organisms by quinic acid [51, 52]. However, while QF and *QUAS* are functional in zebrafish, efficient inhibition of the system could only be achieved through injection of *QS mRNA* in embryos [50]. Stable effective QS expression instead could not be achieved, probably due to titration effects. Thus, the functionality of quinic acid has not been tested in zebrafish. As both systems have been generated through the Gateway cloning system they can easily be adapted to different needs and thus provide a novel platform for transgenic zebrafish models.

Site Specific Recombinases

Site specific recombinases (SSR) allow for site specific genome manipulation. Currently, three main systems are used for site specific recombination: the Cre/*loxP*, Flp/*frt* and phiC31 system, plus the newly discovered Dre/*rox*-system. While the Cre, Flp and Dre are similar and recombine between a pair of identical sites, the phiC31 system recombines between two different sites. However, all these systems allow for three different reactions: excision, insertion and inversion.

The Cre/lox System

The Cre/lox system relies on the 343 amino acid long Cre recombinase of the bacteriophage P1 that catalyzes site specific recombination between two 34 bp long *loxP* (locus of cross-over in P1) or simply *lox* sites. The *lox* sites consist of an asymmetric 8 bp spacer that determines site orientation and is flanked by two 13 bp inverted repeats, to which Cre binds [53]. By manipulating the orientation of two *lox* sites, Cre functionality can be controlled in the way that head-to-tail orientation induces almost irreversible insert excision, while head-to-head orientation leads to insert inversion. As after inversion the two *lox* sites still reside in *cis*, this reaction is reversible and thus inefficient (Fig. 2a, b, d).

Different variants of the wildtype *loxP* sites exist, which increase the efficiency and flexibility of the system and allow for more sophisticated transgenic approaches such as orientation coordinated "recombinase-mediated cassette exchange" (RMCE) or stable insert inversion through the "flip-excision" (FLEx) approach or the inverted-repeat method that depends on two partly mutated *lox* sites (reviewed in [54, 55]). This set of techniques made the Cre/*lox* system the first widely used transgenic approach in fly [56], mice [57] and mammalian cell cultures [58].

In zebrafish the Cre/*lox* system was first tested for its excision ability. In 2005 Pan et al. applied *Cre mRNA* to one-cell stage eggs of a transgenic zebrafish line expressing *lox* flanked GFP, which leads to strong mosaic GFP reduction through excision of the GFP sequence from the zebrafish genome [59]. Few years later Liu et al. proved that the Cre/*lox* system can also be applied for gene inversion in zebrafish, although with very low efficiency [60]. Using the LE (left excision)/RE (right excision) mutant sites *lox71* and *lox66*, respectively, they showed that a *lox66* flanked *RFP* (red fluorescent protein) coding sequence can be integrated into a stable single-site *lox71* line through co-injection of *Cre mRNA* and the vector *pZk-lox66RFP*. Shortly after, it was shown that—just as in mouse—Cre can be stably

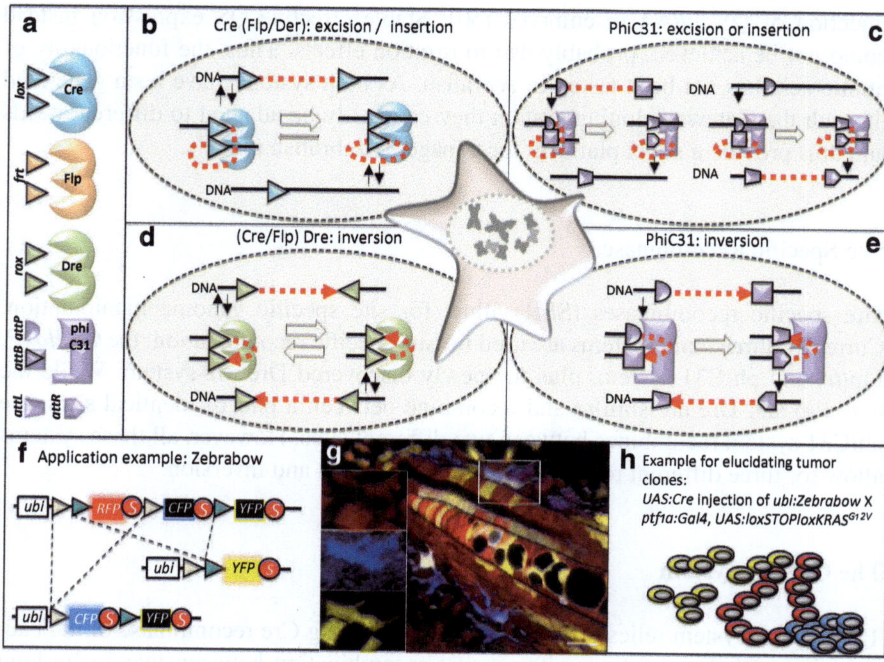

Fig. 2 Site specific recombinases (SSRs) enable cell type specific stable transgene integration. (**A**) The group of available SSR systems comprises the proteins Cre (*blue shape*), Flp (*pink shape*) and Dre (*green shape*) that target pairs of the respective sites *lox*, *frt* and *dre* (*small shapes with corresponding colors*), and the protein phiC31 (*violet shape*) that recombines between *attP* and *attB* sites transforming them into nonreactive *attL* and *attR* sites. (**B–E**) To recombine, all SSRs require a pair of target loci in close proximity and catalyze different reactions depending on loci orientation. (**B, C**) Excision and insertion of fragments between target sites in head-to-tail orientation can be catalyzed reversibly by Cre/Flp/Dre (**B**) or irreversibly by phiC31 (**C**). (**D, E**) Inversion of fragments between target sites in head-to-head orientation can be catalyzed reversibly by Cre/Flp/Dre (**D**) or irreversibly by phiC31 (**E**). (**F, H**) Smart combination of different mutants of loci of the same SSR system allow for enhanced reaction control. (**F**) A promising application example for zebrafish cancer studies is the *ubi:zebrabow* line that encodes the three fluorescent proteins RFP (*red*), CFP (*blue*) and YFP (*yellow*) flanked by pairs of the mutant *lox2272* sites (*white triangles*) and the wildtype *loxP* sites (*blue triangles*) in head-to-tail orientation. Cre can either be inactive, labeling cells in *red*, or can be active and induce recombination between pairs of homologous *lox* sites, labeling cells and their respective clones either in *blue* (*RFP* excision, *blue rectangles with blue halo*) or *yellow* (*RFP* and *CFP* excision, *yellow rectangles with yellow halo*). (**G**) Tail region of a 2 dpf *ubi:zebrabow* larva injected with the ubiquitous Cre expressing plasmid *CMV:Cre*. *Insets* show single channel expression (RFP, CFP and YFP) of the boxed melanocyte. (**H**) An example for the application of the zebrabow tool to study tumor origin by injecting *UAS* controlled Cre into crosses between the *ubi:zebrabow* and a *ptf1a:Gal4*, *UAS:loxSTOPloxKRAS*^{G12V} line (Gal4 expression in the pancreas and *UAS* controlled oncogenic *KRAS* which must be activated by Cre dependent excision of the *STOP* codon) that would allow to distinguish different tumor clones in a somatic pancreas cancer model. Calibration bar: 25 μm. *Abbreviations*: *CFP* cyan fluorescent protein, *dpf* days post fertilization, *RFP* red fluorescent protein, *SSR* site specific recombinase, *UAS* upstream activating sequence, *YFP* yellow fluorescent protein, *ubi* ubiquitin promoter

expressed under specific promoters in zebrafish without significant toxic effects [61, 62] which allows for promoter dependent tissue and time specific onset of Cre expression in this organism.

Despite these promising results until today only few groups developed cancer models using the Cre/*lox* system. The first group to apply it for cancer modeling was Langenau et al. in order to develop their early lethal, *c-Myc* dependent T-ALL (T-cell acute lymphatic leukemia) model into a stable transgenic line [15]. Briefly, this model depends on the construct *rag2-EGFP-mMyc* that drives expression of *mouse c-Myc* under the *rag2* promoter and leads to GFP labeled, *m-Myc* expressing T-lymphocytes. These fish develop clonal lymphoblastic leukemia that expresses the human T-ALL oncogenes *tal1/scl* and *lmo2* and can be transplanted into irradiated recipients. Due to its aggressiveness the fish die before reaching sexual maturity, and studies on the model require injections of the transgene, which leads to variability in integration. To overcome this hurdle, Langenau et al. generated a stable line expressing a modified construct containing the *loxP* flanked red fluorescent protein *dsRED* between the *rag2* promoter and *GFP*. Without Cre this line expresses *rag2*-controlled *dsRed*, while *Cre mRNA* injection mediates *dsRed* excision and allows *GFP* and *m-Myc* expression as well as development of leukemia even though with much reduced frequency compared to the Cre/*lox* independent version of this model. In another study Seok et al. used the Cre/*lox* system to investigate the ability of oncogenic $KRAS^{G12V}$ under control of the *nestin* promoter to cause brain tumors, but instead found that activation of $KRAS^{G12V}$ expression through injection of *Cre mRNA* only induces neural cell differentiation, apoptosis or brain edema which leads to early lethality [63].

Besides conditional tumor induction, the Cre/*lox* system also proved its value for detailed lineage tracing [64, 65], which is especially interesting for studies on tumor clones. The so called zebrabow line is derived from the mouse Brainbow model [66, 67] and is based on a construct that contains a promoter, the three red, blue and yellow fluorescent proteins *RFP*, *CFP* and *YFP*, respectively and pairs of *lox2272* and *loxP* sites spanning the first or the first two fluorescent proteins, respectively [68]. In this way Cre/*lox* activity leads to different excision events and allows only one fluorescent protein to be expressed per recombination event, hence, enabling differentiation between cells belonging to different clones. Further variability is added through copy numbers as each copy can add a different color to the cells increasing color variability in the tissue. Therefore, hetero- or homozygosity of the fish severely influences cell labeling (Fig. 2f, g). Pan et al. generated two zebrabow lines. The first line known as *ubi:zebrabow* uses the *ubi* promoter, established by Mosimann et al. [69], to label cells during all stages of development. The second line known as *UAS:zebrabow* uses the Gal4 specific promoter sequence *UAS*, hence allowing tissue specific cell labeling. This latter line may be a valuable tool in zebrafish cancer studies to examine tumor heterogeneity, clone development and progression and decision fate in metastatic models [70]. Examples of the flexibility of the CRE/*lox* system are shown in Fig. 2.

The Flp/frt System

The Flp/*frt* system derived from *Saccharomyces cerevisiae* is highly comparable to the Cre/*lox* system in terms of recognition sites and functionality (Fig. 2a, b, d). The recombinase Flp (flipase) recognizes 34 bp sequences, called *frt* (*f*lp *r*ecognition *t*arget). Even though they have a similar design to the *lox* sites, these *frt* sites are inert to Cre, which allows simultaneous application of both systems in a line, thus enhancing transgenic flexibility. As the best activity of Flp is achieved at around 30 °C, an enhanced version (FLPe) was developed which retains its activity at higher temperatures, useful for applications in mammals [71]. The Flp/*frt* system has been intensively used over a long time in drosophila, where it is an established tool (reviewed in [72, 73]) and in the mouse both alone [74, 75] and in combination with the Cre/*lox* system [76, 77]. Only in 2011 it was tested in zebrafish through an excision experiment to proof its functionality in this model organism [78]. Since then Cre/Flp combination tools have been developed allowing highly sophisticated gene expression control. The FlipTrap was developed by Trinh et al. as a novel tool for fluorophore labeling of endogenous proteins for analysis without the bias of overexpression approaches [79]. For their study, they generated a library of about 170 FlipTrap lines (see Box 1 for web site). The plasmid to label endogenous proteins contains the yellow fluorophore *Citrine* flanked by splice acceptor and donor sequence instead of initiation and stop codon and an adjacent inverted *mCherry* sequence. Both sequences are embedded in two heterologous pairs of *lox* sites that allow simultaneous Cre induced excision of citrine and inversion and hence activation of *mCherry*, which results in truncated *mCherry* labeled endogenous proteins. *Frt* sites flanking the construct allow for exchange of the construct at the very same locus. The FT1 vector developed by the Chen lab allows conditional sequential activation and inhibition of gene expression [80, 81]. The Tol2 based vector combines both Flp and Cre sensitivity by strategically flanking a gene and a five repeat stop sequence with pairs of heterotypical *lox* and *frt* sites so that Cre and Flp activity in either order can invert the gene and so activate or inactivate its expression. Dependent on the expression pattern of Cre and Flp this allows for (1) tissue specific or general knock-out experiments followed by tissue specific rescue or, if the construct is initially integrated in active orientation, it allows for (2) tissue specific or general rescue followed by tissue specific knock-out. Even though none of these systems has been used so far in cancer studies, they illustrate the gain in flexibility obtained through combination of SSR systems and may serve as inspiration for the development of more sophisticated zebrafish cancer models.

The Dre/rox System

A third recombination system is the Dre/*rox* system, which has recently been discovered in the phage D6, related to bacteriophage P1 (Fig. 2a, b, d). It recognizes sequences called *rox*, which resemble *lox* sites in size, sequence and structure but are not compatible with Cre. Equally, also Dre does not function on *lox* sites [82].

Until now this system has been tested in embryonic stem cells and mouse [83, 84], and also in zebrafish, where it proved its flexibility and compatibility with the Cre system [85]. Thus, it provides an additional tool for SSR dependent conditional transgenesis allowing for more sophisticated lineage labeling and gene expression control, for example through intersection of partially overlapping promoter activities.

The phiC31 System

The phiC31 integrase system offers highly specific spatial control over any transgene integration. It is derived from the PhiC31 bacteriophage, which uses it to integrate its genome into the *Streptomyces* chromosome. The recombination is highly specifically performed between the *attP* site of the phage and the *attB* site of the *Streptomyces* and requires no cofactors [86]. Successful recombination results in hybrid *attL* and *attR* sites. As these are incompatible with phiC31 the reaction is irreversible and provides a unique characteristic over the reversible reactions catalyzed by Cre and Flipase (Fig. 2a, c, e). This makes the phiC31 integrase preferable for transgene insertion strategies. Only recently, a recombination directionality factor has been found which allows reversion of the reaction when combined to the phiC31 integrase [87]. Of note, despite the nomenclature is the same, the phiC31 system sites are different from the *attP* and *attB* sites of the lambda phage-derived Gateway system and thus, are not interacting with those.

Since its discovery, the phiC31 system has been applied to various eukaryotic systems including human cells [88], mouse [89], drosophila [90] and frog embryos [91]. In these, it was established as an advantageous system for site directed insertional transgenesis, while in mice and humans pseudo-*attP* sites in the genome complicate the application of phiC31 in these models [92]. Indeed, gene function analysis in transgenics suffers from unpredictable integration sites of diverse transgenes making quantitative and qualitative analyses difficult and labor intensive. With the phiC31 system, comparable transgenic lines can easily be generated. Well-characterized transgenic lines with a single *attP* site integration can be used to integrate different transgenes into a predefined region to compare them under similar genomic conditions. This facilitates for example comparison between native and mutant proteins. Further, as it naturally recombines sequences of 100 kb, the phiC31 integrase sets basically no limit to insert size allowing a wide range of applications.

Despite its success in other organisms, its functionality in zebrafish has first been shown in 2010 [93, 94]. As in other model organisms, the phiC31 integrase also proved its advantage for comparative transgene analyses. After Lu et al. [93], also Lister showed that phiC31 is generally active in zebrafish and able to excise inserts flanked by *attP* and *attB* sites [93, 94]. Hu et al. proposed a method for RMCE in zebrafish through phiC31 [95]. Using a tol2 construct they expressed *attP* flanked *GFP* under a tissue specific promoter and could exchange it with an *attB* flanked *mCherry*. A drawback however is the relatively low efficiency of phiC31 integrase in genomic approaches. Another approach described by Mosimann et al. aims to establish single-insertion zebrafish lines that harbor only one *attP* site [96].

They developed three functional lines with single *attP* integration sites of which they characterized two with homogenous gene expression. Using the *phiC31 mRNA* inserts as large as 8 kb flanked by *attB* sites could be inherited. They showed that for at least four generations the signal is stable appearing with the same intensity among all larvae of a clutch. Similarly, Roberts et al. confirmed the site specificity of phiC31 mediated transgene integration as well as a low rate of unspecific integration. This leads to the conclusion that the fidelity of the recombination is high [97]. Further development of this system has been applied in a medaka study by Kirchmaier et al. who constructed an elegant example for the combination of the phiC31 system with the Cre/*lox* system to integrate different genes and promoters at specific sites and described a fluorescence based control mechanism for efficient recombination [98]. This system allows for directed gene integration into a predefined locus through the phiC31 integrase, while fluorescence confirms successful integration. In a second step, the lox sites allow for locus cleanup removing fluorescent proteins and promoters and leaving only the insert behind [98].

To summarize, during the last 5 years, the phiC31 system has been developed as a powerful tool for highly controlled spatial transgenesis, which may improve comparison between gene effects independently of integration site but also allows the study of enhancer and promoter characteristics. Its ability to integrate constructs of yet unlimited size adds additional value to this system. However, in order to gain full use of the phiC31 integrase system, more *attP* site single-insertion zebrafish need to be developed and characterized. Here, the novel discovery of the CRISPR/Cas9 system could be used to develop designer lines. Also concentration dependent toxicity [95] could be reduced and activity improved through modifications of the phiC31 integrase as it has been already done for the mouse.

Conditional Cancer Gene Manipulation Through Genome Editing

Genome editing includes a range of methods for targeted genome manipulation. Combination of an engineered nuclease containing a sequence specific DBD with a generic DNA cleavage module enables highly locus specific knock-out and knock-in of individual genes. This provides a great advantage over timely limited knock-down methods such as morpholinos or mRNA injections and unspecific mutagenesis through chemicals or radiation. The three main systems applied are known under the acronyms ZFN, TALEN and CRISPR/Cas9.

Zink-finger nucleases (ZFNs) were the first genome editing tool applied to zebrafish in 2008 [99]. Only 3 years later the discovery of the transcription activator-like effector nucleases (TALENs) provided a new and more efficient tool for targeted zebrafish mutagenesis [100, 101]. Both ZFNs and TALENs rely on protein-DNA interaction with relatively restrictive combinations between grouped amino acids and nucleotides. For detailed information we refer the readers to the many excellent reviews published on these methods [102–105]. Both ZFNs and TALENs effectively allow genome manipulation through mutation, but also deletions and insertions and are even applied for therapeutic approaches to combat HIV/

AIDS currently in clinical trials [106]. However, reports of deletions or insertions of cancer-related genes using ZFNs or TALENs in zebrafish are lacking, thus testifying the complexity in tool design, and restricting their use to experienced labs.

The clustered, regularly interspaced, short palindromic repeats (CRISPR)/CRISPR associated (Cas) system (CRISPR/Cas) provides a great alternative to ZFNs and TALENs as it relies on Watson-Crick base pairing instead of the potentially less specific protein-DNA interaction [106]. The CRISPR/Cas system can be found in many bacteria and archaea as an adaptive immune system against viral infection where it integrates viral DNA into the CRISPR loci as so called protospacer. The protospacers are recognized through their PAM sites (*p*rotospacer *a*djacent *m*otif), which flank them at their 5′ or 3′ end as *GGN* or *NGG*, where *N* stands for any nucleotide. At the second infection, protospacers are transcribed into *CRISPR RNA* (crRNA), which anneals to *transactivating crRNA* (tracrRNA) that in combination with the endonuclease Cas specifically digests foreign DNA sequences. The CRISPR/Cas system II is the system of choice as it is the only of the three known CRISPR/Cas systems that requires only a single Cas endonuclease (Cas9) [107].

Since its first application in mammalian cells in 2013 [108, 109], it has already been applied to a variety of species including plants [110], mice [111, 112], rats [113], rabbits [114], goats [115], pigs [116] and zebrafish [117, 118], to name some. The main advantage of the CRIPSR/Cas9 system lies in its simple design. It relies on two components (1) a ~20 bp long guide RNA (*gRNA*) sequence, encoding the sequence of the target and (2) the Cas9 which can be applied either as protein or as mRNA. The Cas9 endonuclease contains two enzymatic activities that cleave the complementary and non-complementary strands, allowing Cas9 to induce double strand breaks in any target sequence [107] (Fig. 3a). Thus, for transgenesis only the *gRNA* needs to be designed. In zebrafish the CRISPR/Cas9 system is applied by injection of a *gRNA* and the *Cas9 mRNA* from *Streptococcus pyogenes* into one-cell stage embryos and reveals high efficiency for knock-out of single genes (70–90 % [117–119]), as well as for simultaneous knock-out of multiple genes which allows analysis of both single and combinatorial knock-out effects already in the F0 generation. Through injection of two different *gRNA*s and *Cas9 mRNA* even huge chromosomal deletions are possible [120]. To reduce mutation events the knock-out efficiency can also be titrated through varying RNA concentrations [118]. As the mutations are inherited to the next generation the CRISPR/Cas9 system allows to generate stable knock-out lines with a fast and easy protocol [121]. While off-target effects have been found to appear in only few cases, still deeper investigation is required [119]. However, the system displays variable efficiencies for different genomic targets with some genes yielding effects in <1 % of injected embryos [117, 118, 122]. To optimize the CRISPR/Cas9 system, a zebrafish codon-optimized version of Cas9 (zCas9) was constructed and flanked with nuclear localization signals that showed a superior performance compared to wildtype and human optimized Cas9 [118, 123]. Application of different Cas9 variants from other prokaryotes or generated through mutation may lead to further improvements in efficiency and specificity. The *gRNA* sequence is especially important for specific and directed Cas9 activity and can be optimized by various means [122]. For design and specificity

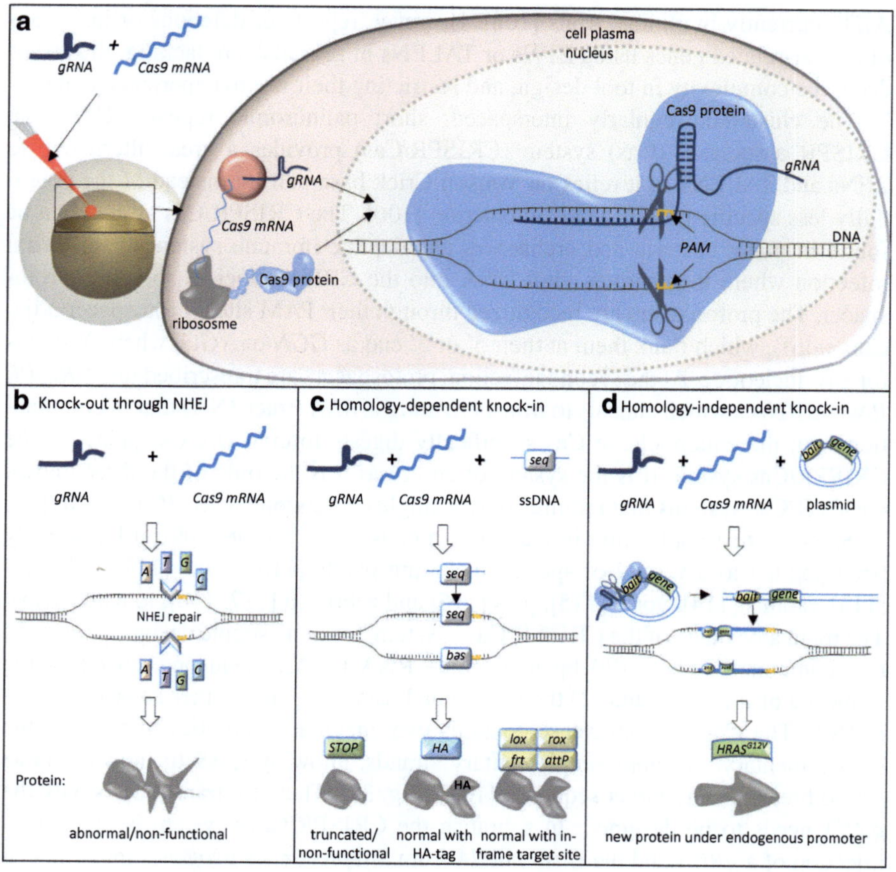

Fig. 3 CRISPR/Cas9 facilitates genome manipulation on demand. (**A**) Schematic of CRISPR/Cas9 technology applied to zebrafish. For Cas9 dependent genome editing the *gRNA* and *Cas9 mRNA* are injected into one-cell stage embryos, where the *Cas9 mRNA* is translated into Cas9 protein and together with the *gRNA* transported into the cell nucleus. In the nucleus the *gRNA* binds the corresponding DNA sequence which allows Cas9 to digest the DNA at the adjacent *PAM* sequences and induce a double strand break through its two enzymatic activities (scissors). The cell can repair this damage through different mechanisms that allow various applications (**B**–**D**). (**B**) Through Non-Homologous End Joining (NHEJ) the double strand break is repaired by random integration of nucleotides (boxes with *A*, *T*, *G* or *C*) which often induces a frame shift and generates abnormal and mostly non-functional proteins. (**C**) Coinjection of gRNA and *Cas9 mRNA* with small single stranded DNA (ssDNA) allows strand repair through homology-dependent recombination which leads to the integration of a desired small sequence such as a *stop codon* (*green box*) to generate a truncated protein, tags of human influenza hemagglutinin (*HA*) (*blue box*) to allow easy protein detection, or target sites for SSRs such as *lox*, *frt*, *rox* or *attP* (*yellow boxes*) to allow to use this line for site specific integration of constructs into a defined locus. (**D**) Co-injection of *gRNA* and *Cas9 mRNA* with a plasmid encoding a bait sequence (*yellow oval*) and a gene (*green rectangle*) allows strand repair through homology-independent recombination and insertion. *gRNA* dependent linearization of the plasmid at the bait sequence allows its integration into the genomic DNA at the site of *gRNA* induced double strand break resulting in expression of a transgene under an endogenous promoter. *Abbreviations*: *A* adenine, *C* cytosine, *G* guanine, *gRNA* guide RNA, *HA* human influenza hemagglutinin, *NHEJ* non-homologous end joining, *PAM* protospacer adjacent motif, *ssDNA* single strand DNA, *SSR* site specific recombinase, *T* thymidine

testing several freely available software tools have been developed (see Box 1 for web sites) to enable even inexperienced labs to implement the CRISPR/Cas9 system in their daily routine.

Besides successful knock-out, also knock-in strategies have been developed with the CRISPR/Cas9 system (Fig. 3b–d). Homology directed recombination through injection of single strand DNA together with *gRNA* and *Cas9 mRNA* allows insertion of short fragments of ~50 bp such as *loxP* or *HA* tags into the zebrafish genome but frequently leads to inverted or partial insertion [119, 123]. Homology-independent knock-in methods instead, make use of an insert flanked by one or two *gRNA* target sites (baits) as double strand DNA or as plasmid [124–126]. As plasmid DNA is less toxic than linearized DNA, it can be used at higher concentrations for more efficient integration [125]. In several studies, integration efficiencies of 20–30 % have been reached in the F0 generation with inheritance to the next generation by 10 % of carriers [124–126]. A step-by-step protocol for homologous-independent knock-in is available from Auer and colleagues [127].

While none of the genome editing approaches has been used to generate zebrafish cancer models, yet, they have so far found their application in gene function analyses [128, 129] and disease modeling [130–132]. Hence, we propose that genome editing systems provide a valuable tool for detailed analysis of tumor suppressor genes. With the CRISPR/Cas9 system, the zebrafish research community now gained a flexible, cheap and easy tool to generate mosaic knock-out or lines on demand which shall improve the analysis of otherwise difficult to investigate tumor suppressor genes.

Temporal Control

Timing of oncogene expression or inactivation of tumor suppressors may be critical in cancer modeling, as it may trap specific cell types which are more capable of cancer initiation. In addition, duration of oncogenic stimuli or withdrawal from oncogene expression may reveal the strength of oncogene addiction at different stages of tumor progression. Therefore, systems for temporal control of expression of oncogenes or tumor suppressors have been developed in zebrafish. Some of these systems will also allow spatial control, or may be used in combination with spatial control methods. The most popular systems for temporal control of gene expression are listed below.

Heat Shock Promoters

Control of gene expression through temperature regulation has been achieved in zebrafish thanks to promoters of heat shock proteins [133]. Understandable, this approach is difficult in mammals due to their tight regulation of body temperature [134]. In poikilotherm organisms instead, such as the zebrafish, temperature controllable systems have become widely implemented. The zebrafish *heat shock promoter 70l* is a 1.5 kb sequence that contains tandem arrays of 5 bp DNA consensus-sequences named heat shock elements (HSE). The transcription factor to activate the HSE is called heat shock factor (HSF). At physiological temperatures HSF is bound in a protein complex containing the protein hsp90. Increase in temperature forces the proteins in the complex to change their conformation and release HSF to bind and activate the HSE [135, 136]. While its application in zebrafish is easy and straight forward, it also holds a significant drawback. Besides its activation through a lot of stressors such as toxins or starvation, it is also active during early development [137]. However, this did not impede its wide use in zebrafish models. The *heat shock promoter* has been used instead of ubiquitous promoters to drive general gene expression as well as in combination with laser irradiation for tight local expression control to study gene function, search for transcription enhancers and for lineage labeling (reviewed in [138]).

Also cancer models have been generated in zebrafish based on the *heat shock promoter*. Due to its easy timely controllable induction, the *heat shock promoter* allows for development of stable lines that encode otherwise lethal oncogenes. The induction of $HRAS^{G12V}$ expression during larval stages induces hyperproliferation and cellular senescence [139], similar to the expression of oncogenic *HRAS* in human fibroblasts [140]. Additionally, the *heat shock promoter* has been used to construct inducible leukemia models. Yeh and colleagues expressed the human oncogenic fusion protein AML1-ETO under control of the *heat shock promoter 70l* [141]. Activation of this oncogene in larvae resulted in a lack of circulating blood cells despite an intact cardiovascular system, accumulation of immature hematopoietic blast cells, and gene expression comparable to the myeloid leukemia in human with upregulation of *lmo1*, *hoxa9* and *hoxa10* and downregulation of *cmyb*, *gata* and *scl*. Remarkably, injection of *scl mRNA* and treatment with the drug Trichostatin A could partially rescue the *AML1-ETO* induced phenotype which makes the line suitable for drug screens. Shen et al. developed a line that expresses the oncogene *Mycn* and the fluorescence protein *GFP* under control of a sequence of bicistronic *HSE* [142]. Already one single heat shock at larval stage induces the development of leukemia like lesions that are characterized by leukemia typical changes in blood cell populations to more immature types and upregulation of *scl* and *lmo2*, genes that have also been found upregulated in human leukemia. Additionally, heat shocked fish have a significantly reduced life span (9.5 months compared to 2.5 years in adults) and even at 8 months show a similar histopathology as the 2 month old juveniles.

A model for Ewing sarcoma, a malignant tumor characterized as small-round-blue cell tumor, depends on *heat shock promoter* controlled expression of the human EWS-FLI1, a fusion oncoprotein of the 5′ transactivation domain of EWS and the 3′ Ets domain of FLI-1 [143]. Induction of this construct in a p53-/- background through a heat shock at 50 % epiboly induces development of solid tumors, leukemia like tumors and malignant peripheral nerve sheath tumors with the typical histology of small, round, blue cells similar to human Ewing's sarcoma and mouse models, which could also be transplanted.

Besides *heat shock promoters* also heat sensitive splicing has been described. An example is the $mitf^{vc7}$ mutant which allows inactivation of *mitf* at temperatures above 28 °C. This variant has been used in a melanoma model to elucidate the function of *mitf* in melanoma development [144].

Hormones

Hormone dependent activation of transgenes depends on so called nuclear receptors. A large group of steroid and thyroid receptors, which are located in the cytoplasm, upon ligand binding translocate to the nucleus where they bind specific genomic regions through their DBD and function as transcription factors. The homodimeric functional steroid receptors estrogen receptor (ER), progesterone receptor (PR) and glucocorticoid receptor (GR) are the most frequently used ones for conditional transgenic approaches. To utilize their hormone specificity, the ligand-binding domain is fused to a transcription factor of interest thus rendering the activity of the transcription factor hormone dependent [145]. To activate the ligand-binding domains of ER, PR and GR the most frequently applied hormone analogs are tamoxifen, mifeprestone (also known as RU486) and dexamethasone, respectively. While mifeprestone and dexamethasone are directly effective, tamoxifen is a prodrug, which is metabolized to 4-hydroxy-tamoxifen (4-OHT) through the liver enzyme CYP 450 2D6.

In the zebrafish, 4-OHT can be applied directly to the water to avoid irregularities due to enzymatic interactions and reduce variations in dose response curves. In different studies it was shown to reliably control gene expression in a dose dependent manner and without measurable leakiness [146, 147]. In a further evolution of the *rag2* controlled leukemia model developed in the lab of T. Look [15] the c*Myc* oncogene is fused to the ligand binding domain of ER [147]. This allows for conditional activation of the oncogenic effect of *cMyc* through application of 4-OHT which leads to development of fully established T-ALL already after 5 weeks of treatment, while withdrawal of 4-OHT blocks *cMyc* activity and leads to apoptosis of cancer cells. The untreated siblings instead do not develop blood abnormalities. Also in an in vitro study of a zebrafish derived liver cell line transfected with ER bound Raf1 only cells treated with 4-OHT showed hyperactivation of the MEK-ERK cascade and increased proliferation in cell culture as well as after transplantation while untreated cells showed no alterations [146]. The other two systems, namely PR/mifeprestone and GR/dexamethason, are also functional in zebrafish [85, 148]. As all three systems are not cross reacting [85], they allow for simultaneous or sequential activation of different target genes and thus add enormous flexibility in transgenic zebrafish cancer models.

Tet-On/Tet-Off

The Tet-system is derived from *Escherichia coli*, where it controls expression of genes mediating tetracycline resistance. Its activity is controlled through the tet-Repressor (tetR), which is reactive to tetracycline or its more stable derivative doxycycline (Dox). Through fusion of tetR to the activation domain of VP16 the Dox controlled transactivator (tTA) is generated. In absence of Dox tTA binds specifically to tetracycline operator (tetO) sequences and activates transcription from minimal promoters adjacent to tetO. As tTA is inhibited through Dox binding, this system is called Tet-Off [149]. The Tet-On system instead relies on a mutant transactivator (rtTA) with four substituted amino acids and is physiologically inactive. It can bind and activate the tetO element only upon Dox binding [150]. This system is mostly preferred for gene silencing and controlled activation of gene expression [151] as it circumvents problems that may arise with the Tet-Off system such as slow withdrawal of Dox from the organism and thus delays in induction of gene expression. The Tet-On system has been applied to many different organism including plants [152], drosophila [153], mice [154, 155] and zebrafish. However, in the first studies the system appeared leaky, which hampered its further use [156]. Through viral mutagenesis, variants have been designed with increased stability and reduced leakiness [157]. One of them is the Tet-On Advanced (rtTA2^{S}-M2), which was achieved through random mutation of the tetracycline repressor gene and screening for alleles with reverse binding properties. Additionally, the VP16 activation domain is replaced with a minimal activation domain (12 amino acids long) and a coding sequence optimized for expression in human cells [158]. These changes make rtTA2^{S}-M2 10 times more sensitive to Dox than rtTA, while it has severely reduced background activity and is more stable in eukaryotic cells. However, as in zebrafish some background activity is measureable, a fusion system has been developed. Binding of rtTA to the ligand binding domain of GR or the ecdysone receptor (EcR') from the silkworm (*Bombix mori*) allows for additional control through dexamethasone or tebufenozide, respectively [148]. Both hormone systems reduce leakiness to undetectable levels. While dexamethasone at higher concentrations may deplete T-cells [159], side effects of tebufenozide are not known. Additionally, the EcR' bound rtTA system exhibits higher sensitivity which makes it preferable.

In zebrafish cancer studies, the basic rtTA system has been sufficient. The initially Gal4/*UAS* based glioma model from Ju and colleagues [44] was also adapted for conditional oncogene expression through transversion to the Tet-On system. Expression of rtTA under either of the promoters *krt5* and *gfap* and control of the mCherry fused *KRAS*^{G12V} oncogene through a tetO allowed for analysis of early effects and thus for chemical treatment approaches [44]. Additionally, different Tet-On dependent liver cancer models have been generated with oncogenes expressed under the control of the liver specific promoter *fabp10*. The most aggressive model depends on conditional expression of *Xmrk*, the *Xiphophorus* derived mutant form of the epithelial growth factor receptor (EGFR) [160] and leads to hepatocellular carcinomas (HCC) already after 3 weeks of treatment accompanied by diminished growth and increased lethality, while withdrawal of Dox leads to a complete rescue after only 4 weeks. A less severe model relies on overexpression of *Myc* [161], a gene, often amplified and associated with bad prognosis in human HCC. While it can lead to induction of liver tumors with gene expression profiles comparable with human HCC and mouse models, it shows less malignancy than models depending on expression of X*mrk* or *KRAS*^{G12V}. However, this can be an advantage, as *Myc* expression can generate a sensitized background that allows to discovery and evaluate new, yet undetermined oncogenes in HCC.

In a different approach the Tet-On system was used to investigate the complicated relationship between oncogenic *kras* and *rhoA* [162]. Through Dox dependent co-expression of both $kras^{G12V}$ and wt, dominant active (DA) or dominant negative (DN) *rhoA*, Chew et al. [162] found that *rhoA* has an inhibitory effect on the oncogenic effects of $kras^{G12V}$. This underlines the flexibility of the Tet-On system for gene interaction studies.

Optogenetics

Optogenetic tools only emerged during the last 10 years. Nonetheless, optogenetics gained tremendous interest and has been established in a wide range of organisms. In a broad definition optogenetics is the science of genetic encoding of light-gated proteins for targeted control of cellular behavior, which has been shown to provide a superior tool for precise spatial and temporal control of stimulation or inhibition of cellular activities. However, optogenetics is often restricted to the application of light-sensitive microbial membrane proteins which can be classified into the two groups of "optogenetic activators" (light sensitive channels and pumps) and "optogenetic sensors" (fluorescent proteins that sense Ca^{2+} concentrations or membrane potentials). Mainly applied in neuroscience and cardiac studies, optogenetic activators and sensors have intensively furthered our understanding in these fields. Especially the zebrafish, as a transparently developing organism, provides easy application of optogenetic activators and sensors [163–165]. However, as this subset of optogenetic tools provides only limited benefit for cancer research it will not be further discussed in this chapter. The interested reader is referred to recent reviews [166–170].

Another branch of optogenetics makes use of a range of light-sensitive non-channel proteins to spatiotemporally control cell signaling pathways and gene transcription (reviewed in [171]). Among the most frequently applied protein systems are (1) the LOV (light, oxygen and voltage) domain, a small domain that reacts to blue light (450 nm) and requires the cofactor flavin mononucleotide (FMN), ubiquitously expressed in mammalian cells and zebrafish, (2) the PhyB (phytochrome B)—PIF (phytochrome interacting factor) system, a binary system that heterodimerizes through red light (650 nm) and dissociates through infrared light (750 nm), (3) the CRY2 (cryptochrome 2)-CIB1 (cryptochrome-interacting basic helix-loop-helix) system that heterodimerizes under blue light exposure and (4) the UVR8 (UV-resistance locus 8), a small molecule that forms homodimers which dissociate under UV-light. These tools allow a broad range of applications. The LOV domain can be used to mask protein domains and release them upon blue light illumination, allowing for example light-induced apoptosis in a caspase7-LOV human cell model [172] or light-induced protein degradation when fused to a transgenic protein capped with degrons, small sequences that induce protein degradation [173]. In another mammalian study UVR8 fusion proteins have been used to study secretion processes. As UVR8 forms dimers, it leads to protein aggregation in the endoplasmatic reticulum, which prevents further secretion. Only upon UV-light exposure UVR8 dissociates allowing further trafficking of the protein [174]. Protein localization can also be affected through both, the PhyB-PIF and the CRY2-CIB1 system, which allows close and direct control over cell signaling pathways. By anchoring PhyB or CIB1 to the cell membrane and fusing PIF or Cry2 to a membrane active protein, activation or inhibition of for example the mitogen-activated protein kinase (MAPK) signaling pathway [175] or the phosphatidylinositol 3-kinase (PI3K) signaling pathway [176, 177] is possible. To control gene transcription, the AD and the DBD of the transcription activator Gal4 can be fused to either component of the PhyB-PIF [178] or CRY2-CIB1 [179] system. Light-induced fusion of the systems allows dimerization of the AD and DBD and activates Gal4 dependent gene expression. In drosophila, a model of CRY2-CIB1 controlled Cre activation allows for highly specific neural cell labeling [180].

Besides their application in a wide range of organisms, optogenetic non-channel proteins have also been used in zebrafish. Currently evaluated systems include the CYP2-CIB1 controlled Gal4 activation [179] and a single-protein transcription regulator that uses the fast cycling LOV containing protein EL222 which additionally contains a helix-turn-helix (HTH) DBD that is covered by the LOV domain in the dark to prevent DNA binding [181]. In both systems gene expression is activated through exposure of target cells to blue light. Thereby, the strength of gene expression can be regulated through intensity and length of laser exposure, which allows fine tuning of the target gene. Another interesting light sensitive protein tested in zebrafish is the KRed (killerRed), an optogenetically activatable inducer of reactive oxygen species [182]. Several lines of tissue specific KRed expression are provided on the line database ZETRAP2.0 [183] that may be helpful in elucidating the role of oxidative stress in cancer development or cure.

In summary, optogenetics provides a tool with high spatial, temporal and dose control with broad applications in zebrafish due to the transparent nature of its early developmental stages. Several studies in mammalian systems manipulating potential cancerous pathways (MAPK, PI3K) through optogenetic techniques promise to be valuable also for cancer models in the zebrafish. Additionally, bipartite systems such as the CYP2-CIB1 system, which has already been tested in zebrafish, offer the chance to target distinct subpopulations of cells through expression of both components under promoters with intersectional activity. This would for example allow to elucidate effects of different subpopulation of hematopoietic cells on leukemia development at different ages. However, as expression dependent systems such as Gal4 activation would rely on continues light exposure, more stable tools such as the Cre system should be implement for zebrafish studies to achieve stable oncogenic activity while implementing the red light dependent PhyB-PIF system would allow for more versatile combination studies such as studies on tumor clone competition or consecutive oncogene expression.

Box 1: Webtools
Databases for conditional transgenic lines

Gal4/UAS
zTrap (http://kawakami.lab.nig.ac.jp/ztrap/)
Distel & Köster (http://www.helmholtz-muenchen.de/en/idg/groups/neuroimaging/lines_distel/)

SSR
CreZoo (https://crezoo.crt-dresden.de/crezoo)
FlipTrap (http://www.fliptrap.org/static/index_new.html)

Optogenetics
ZETRAP2.0 (http://plover.imcb.a-star.edu.sg/)
Webtools for genome editing through CRISPR/Cas9

Software for gRNA design
E-CRISP (http://www.e-crisp.org/E-CRISP/designcrispr.html)
CRISPR Design (http://crispr.mit.edu/)
Cas9 Design (http://cas9.cbi.pku.edu.cn/)
CHOPCHOP (https://chopchop.rc.fas.harvard.edu)
CRISPRdirect (http://crispr.dbcls.jp/)
ZiFiT Targeter (http://zifit.partners.org/ZiFiT/)

Software to search for gRNA target sites
Cas-OFFinder (http://www.rgenome.net/cas-offinder/)
GGGenome (http://GGGenome.dbcls.jp/)
CasOT (http://eendb.zfgenetics.org/casot/)

Combination Examples from Zebrafish

The diversity of available conditional transgenic systems already provides high flexibility in stage, tissue and cell type specific gene expression control. Smart combination of spatial regulation systems under control of different gene regulatory elements enables precise regulation of endogenous and exogenous gene expression in primarily spatial but also temporal manner. Temporal regulation systems instead allow for flexible gene expression independent of native gene expression timing. However, for many studies, such as evaluation of oncogene effects at different stages, high flexibility is required for both, spatial and temporal control of gene expression. In such cases, limitation to either of these control techniques would require intensive cloning and crossing workload. Thus, the combination of spatial control with temporal control systems is an important step in order to unleash the full potential of the large set of conditional transgenic approaches already established in zebrafish.

Temporal Control of SSRs

Since the first application of SSRs in zebrafish transgenesis, Cre & co have intensively been developed and combined in order to increase flexibility in gene expression manipulation. Thereby, the early combination with temporal control mechanisms boosted their improvement and increased their application field. One of the first attempts was to set the expression of Cre-GFP fusion proteins under control of a *heat shock* promoter [184]. This study not only confirmed the functionality of Cre in zebrafish, but additionally showed that it is not toxic when expressed systemically and can be induced through a rise in temperature. However, as the *heat shock promoter* is not exclusively heat sensitive but also active during early zebrafish development, leakiness was observed. This did not prevent its use in inducible cancer models.

Taking advantage of the heat shock controllable Cre activity Feng et al. further improved their zebrafish model of acute T-cell leukemia [14]. This model relies on a line encoding the construct *rag2-LoxP-dsRED2-LoxP-EGFP-mMyc* and was previously activated through injection of *Cre mRNA*. Crossing the line with a *hsp:Cre* line [14] allows for controlled induction of oncogene expression at different larval stages which leads to development of T-ALL in 81 % of heat shocked larvae in contrast to the injection approach (13 %) thereby showing that Cre expression from chromosomal integration is far more effective than from mRNA injection. Despite the positive outcomes, the leakiness of the *heat shock promoter* also leads to few problems. Thus, it also causes leukemia in 13 % of non-heat shocked fish and the attempt to produce a line encoding both, the Cre and the *lox* constructs, reveals reduced aggressiveness of the developing leukemia already in the third generation. This occurs most likely due to Cre induced loss of copy numbers during the breeding through the leaky *heat shock promoter* and thus requires continued control of the double transgenic line. Instead keeping both constructs in separate lines helps to overcome this problem. The leukemia model of Forrester and colleagues uses the same heat shock inducible system to conditionally activate expression of the human leukemia associated gene *NUP98-HOXA9* under the promoter *spi1* [185]. Despite its expression in myeloid cells, central nervous system and musculature, they found exclusively myeloid proliferative like malignancies in 23 % of 19–23 months old fish. In another zebrafish cancer model the oncogene *KRAS*^{G12D} is expressed ubiquitously under the *β-actin* promoter using the construct *β-actin-LoxP-EGFP-pA--LoxP-KRAS*^{G12D} [186]. Through crossing with a *hsp:Cre* line [14], oncogene induction can be timely controlled which leads to increased lethality and development of hyperplasia and multiple tumor types including skeletal muscle tumors, myeloproliferative disorder (MPD), intestinal epithelial hyperplasia and rarely malignant peripheral nerve sheath tumors. Also in this line, background expression of Cre can be observed which leads to tumor development in non-heat shocked fish but at later time points and with lower frequency than in their heat shocked siblings. The MPD developed in this model is not only transplantable, forming the same disease in irradiated recipients but can also be induced through resection of hematopoietic kidney cells from non-heat shocked double transgenic adults and transplantation after heat shock into irradiated recipients. Hence, both models proved that despite the leakiness, control of Cre expression through a *heat shock promoter* improves survival of oncogene carriers and allows for time controlled induction of oncogene expression. Additionally, it enables easy transplantation studies with freshly induced tumor cells.

As the leakiness of the heat shock-Cre system provides severe limitations, more reliable methods have been developed. In contrast to the *heat shock promoter*, fusion of Cre with the estrogen receptor ER^{T2} leads to permanent Cre expression, but strictly tamoxifen dependent Cre activity in the nucleus [187]. Thereby, Cre activity can also be titrated in a concentration dependent way but must be tested for each *CreER*^{T2} line separately. Until now the tamoxifen inducible Cre system has been successfully applied for developmental studies, mainly for imaging approaches [188, 189]. Additionally, a large number of existing *CreER*^{T2} lines offers application of this system also for cancer studies. Since its establishment several *CreER*^{T2} lines have been developed with different global or tissue specific *CreER*^{T2} expression pattern. Hans et al. developed two different *pax2a* dependent lines that express *CreER*^{T2} (1) at the prospective diencephalon of the developing forebrain and (2) at the future rhombomere 3 and 5 of the hindbrain anlage, respectively. The *ubi:CreER*^{T2} line [69] instead allows general induction of Cre activity at any stage which makes it preferable over most general promoters used such as *XlEef1a1* (strongest expression during gastrulation and midsomitogenesis) [187], *β-actin* (low expression in erythrocytes, fins) [96, 190]. These and many more *CreER*^{T2} expressing lines have been compiled in the database CreZoo (see Box 1 for web site), a large library that contains over 30 different tissue and cell specific *CreER*^{T2} expressing lines together with well organized molecular and expression data [191]. This database provides a great platform for cancer screen studies when combined with a corresponding set of lox lines for Cre- mediated recombination of tumor suppressors or oncogenes.

Another hormone dependent control mechanism relies on the progesterone analog mifepristone. As it does not cross interact with the tamoxifen receptor ER^{T2} both systems can be used in combination enhancing complexity of transgenic approaches. In zebrafish this system has been tested by Park et al. who combined the tamoxifen inducible *CreER*^{T2} with the mifepristone inducible *DrePR* to the system TAILOR (*t*ransgene *a*ctivation and *i*nactivation through *lox* and *rox*) [85] which not only proved the functionality of both systems in zebrafish but also their compatibility. Hence, it can be used for time controlled activation and inactivation of gene expression as well as for sequential transgene activation through *lox* and *rox*. Thus, it provides additional flexibility for SSR dependent conditional transgenesis which will not only improve cell lineage tracing, temporal and spatial gene expression control through intersection of partially overlapping promoter activities, but may also facilitate mimicking of consecutive events in cancer development through precise and independent activation or inhibition of oncogenes or tumor suppressor genes, respectively.

Temporal Control of the Gal4/UAS System

The Gal4/*UAS* system has proved its value in a variety of cancer studies [8–10, 43, 45, 192]. Although no cancer model in zebrafish applied temporally controllable Gal4/*UAS*-systems yet, several other studies revealed the advantageous characteristics of temporally controllable Gal4/*UAS* as they allowed studying gene transcription effects at later time points of development. Coupling Gal4 to the *heat shock promoter* allows to activate Gal4 expression in the whole organism fast and easy through an incubation at 38 °C for 30 min or spatially restricted through laser exposure [193, 194].

Several heat shock controlled Gal4 lines that have been developed by Asakawa and Kawakami can be found in the zebrafish line database zTrap [195]. However, *heat shock promoters* driving Gal4 expression exclude the use of tissue specific promoters, possess leakiness and permit only limited control of expression kinetics. Hormone bound Gal4 expression instead retains the possibility of tissue specific Gal4 expression and provides dose dependent kinetic control. While glucocorticoid receptor regulated Gal4 has already been published in 1998 [196], further applications were only described in 2013 using ER^{T2} fused Gal4 [197, 198]. In these lines, tamoxifen could induce Gal4 expression in larvae (already 3–4.5 h post treatment) and in adults. Even though slight leakiness was observed, tissue specific Gal4 expressing lines could be generated and allowed to study for example tissue specific effects of elevated Notch expression on notochord development.

The Gal4/*UAS* system has also been combined with optogenetic tools. Placing optogenetic activators and sensors under the *UAS* promoter allowed Umeda et al. to take advantage of existing Gal4 lines and to control channel protein expression in specific neural cell types [199]. Non-channel protein systems such as the CRY2-CIB1 instead have been used to control Gal4 activity through fusion of the Gal4 components AD and DBD to CRY2 and CIB1, respectively which allows for light controlled Gal4 activation through light induced fusion of the two components [179].

Besides combination with temporal control tools, the Gal4/*UAS* system has also been connected with the Cre/*lox* system such as in the promising zebrabow model developed by Pan and colleagues. In this model *lox* flanked fluorescent proteins are under *UAS* control which allows Gal4 dependent tissue specific expression in contrast to the ubiquitously expressing *ubi:zebrabow* line [68].

Another combination of several tools is referred to as MAZe (*M*osaic *A*nalysis in *Ze*brafish) [200, 201]. This system combines a heat shock controlled Cre and a promoter controlled Gal4 which is inhibited by a *loxP* flanked stop codon. As the *heat shock promoter* activates Cre only inefficiently, this combination results in mosaic expression of Gal4 which allows studies on sparse somatic cells or in a highly restricted spatial control of target gene activation by laser induced heat shock. Together with the previously descried Gal4 combination tools and the wide range of *UAS* encoding effector lines and constructs available, this system could provide new possibilities for detailed studies on somatic expression of oncogenes.

Mifepristone/LexPR

The mifepristone inducible LexPR system allows both temporal and spatial control of gene expression. Similar to the Gal4/*UAS* system, the LexPR system consists of a transactivator, which can be set under tissue specific promoter control but additionally functions only upon binding of mifepristone, and an operator that controls expression of a subsequent gene. The transcription factor LexPR is a fusion of the DBD of the bacterial *LexA* protein, a truncated ligand binding domain of the human progesterone receptor and the AD of the human p65 protein. The operator promoter LexOP consists of a synthetic *LexA* operator fused to the minimal 35S promoter from Cauliflower Mosaic Virus [202].

Emelyanov and Parinov were the first to adopt and test this system in zebrafish [202]. For a proof of principle, they developed a line containing both, the LexPR transactivator and the gene of interest under control of the LexOP. In this line they showed that the LexPR system is not toxic to zebrafish, allows tight control of gene expression without basal expression, can reliably and concentration dependent be turned on by mifepristone and turned off again through withdrawal of mifepristone by easily moving fish into fresh water. Mifepristone itself was tested to be not toxic in larvae at concentrations sufficient for target gene expression and to be applicable to adult females to gain maternal expression in embryos. Additionally, mifepristone is a relatively cheap drug and safe to use due to the low concentrations required for system activation [202]. Using a two line system composed of a driver line that express the transactivator LexPR and an effector line encoding the LexOP controlled gene of interest, Emelyanov and Parinov also demonstrated the possibility to control gene expression in trans, to control expression of two independent genes in the same line and to efficiently express oncogenic *KRAS* under the cytokeratin promoter *krt8* which caused strong developmental abnormalities. One drawback of the system, as observed also in other binary gene expression system, is the diversity in expression strength not only between larvae of the same clutch but also between cells of the same larva, thus caution is required with quantitative analyses. Nguyen et al. took advantage of the system to timely control oncogene expression in a zebrafish liver cancer model using an activator line expressing LexPR under control of the promoter *fabp10* and an effector line expressing *kras*^{G12V} under LexOP control [203]. This approach allowed not only to recapitulate the phenotype of the previous not-inducible model [204] but also to delay the oncogene expression to 1 month of age at which 100 % of fish treated with 2 μM mifepristone developed tumors after 1 month of treatment, while lower concentrations induced tumors later and less frequently. Additionally, the reversibility of the system allowed investigators to turn off *kras*^{G12V} expression and demonstrated oncogene addiction of tumors. As the LexPR and the Gal4 system do not cross react, it is possible to combine them in the same line to develop more sophisticated models [202].

Table 1 shows an updated list of cancer models developed through conditional methods in zebrafish.

Table 1 Application of the conditional transgenic toolbox in zebrafish cancer models

Spatial control	Temporal control	Cancer	Promoter	Oncogene	References
Gal4/UAS	–	Melanoma	*kita*	*GFP-HRAS*^{G12V}	[9]
Gal4/UAS	–	Leukemia	*fli.1*	*GFP-HRAS*^{G12V}	[43]
Gal4/UAS	–	Chordoma	*twhh*	*GFP-HRAS*^{G12V}	[13]
Gal4/UAS	–	Glioma	*zic1/4*	*GFP-HRAS*^{G12V}	[11]
Gal4/UAS	–	Glioma	*ptf1a*	*marhAKT1* *GFP-RAC*^{G12V}	[12]
Gal4/UAS	–	Glioma	*krt5* *gfap*	*mCherry-KRAS*^{G12V}	[44]
Gal4/UAS	–	Pancreas	*ptf1a*	*GFP-KRAS*^{mut}	[10, 45, 192]
Gal4/UAS	–	Multiple	*krt4*	*smoa1-GFP* *myrhAKT1*	[8]
Cre	– or Heat Shock	Leukemia (T-ALL)	*rag2*	*GFP-mMyc*	[14, 15]
Cre	Heat Shock	MPN	*spi*	*NUP98-HOXA9*	[185]
Cre	Heat Shock	Multiple	*b-actin*	*KRAS*^{G12D}	[186]
–	Heat Shock	Leukemia	*hsp70*	*AML1-ETO*	[141]
–	Heat Shock	Leukemia like lesions	*hse*	*Mycn*	[142]
–	Heat Shock	Ewing's sarcoma	*hsp70*	*EWS-FLI1*	[143]
–	Tet-On (Dox)	Glioma	*krt5* *gfap*	*mCherryKRAS*^{G12V}	[44]
–	Tet-On (Dox)	Liver	*fabp10*	*Xmrk*	[160]
–	Tet-On (Dox)	Liver	*fabp10*	*Myc*	[161]
–	Tet-On (Dox)	Liver	*fabp10*	*GFP-kras*^{G12V} *mCherry-rhoA* (WT) *mCherry-rhoA*^{G14V} (DA) *mCherry-rhoA*^{T19N} (DN)	[162]
–	Tamoxifen	Leukemia (T-ALL)	*rag2*	*mMyc-ER*	[147]
–	Tamoxifen	Liver (zebrafish liver cell line)		*deltaRaf1-ER*	[146]
LexPR	Mifeprestone	Liver	*fabp10*	*GFP-kras*^{G12V}	[203]

Collection of available cancer models in zebrafish that are based on conditional transgenic methods stating the conditional methods applied, the type of cancer induced and the promoter and oncogene used

Abbreviations: *DA* dominant active, *DN* dominant negative, *Dox* doxycycline, *MPN* myeloproliferative neoplasia, *T-ALL* T-cell acute lymphatic leukemia, *WT* wildtype

Future Perspectives

The development of conditional transgenic approaches in zebrafish is already offering a broad variety of tools for diverse expression controls. Moreover, new tools are continuously developed adding even more flexibility to the already large tool box, including the Gal4/*UAS* analog systems TrpR/*tUAS* and QF/*QUAS*, the novel gene silencing *shRNA-miRNA*, the dre/*rox* system, CRISPR/Cas9 knock-in system and optogenetic approaches. Additionally, the possibility of a synergistic application of the large variety of tools will expand exponentially the repertoire of conditional approaches in zebrafish cancer models. For example, the CRISPR/Cas9 system is a promising easy tool to generate *attP* landing sites at desired loci either in non-coding regions with limited effect on adjacent genes or directly adjacent to enhancers or in coding regions in order to generate a library of lines that allows locus independent comparison of gene activities, enhancer studies or endogenous gene labeling/disruption, respectively. Additionally, these *attP* sites can be used to integrate sequences encoding for (1) stable or reversible, (2) light, hormone or heat tunable (3) single or multicomponent systems. Current studies in zebrafish using mi-siRNA also promise to achieve targeted silencing of genes at will [47]. Thus, the field of zebrafish cancer studies will soon make full use of an immense set of tools for controlling gene expression specifically in organs, tissues, cells or time, or all at once.

Acknowledgements We thank World Wide Cancer Research (grant no: 11_0624) and the "Studienstiftung des deutschen Volkes" for their financial support.

We thank Addgene for the plasmid CMV:cre (31308) [205] and Martin Distel, Reinhard W. Köster, Markus Affolter and Alexander Schier for the transgenic lines shown in Figs. 1 and 2.

References

1. Matthews JL (2004) Common diseases of laboratory zebrafish. Methods Cell Biol 77:617–643
2. Stanton MF (1965) Diethylnitrosamine-induced hepatic degeneration and neoplasia in the aquarium fish, Brachydanio rerio. J Natl Cancer Inst 34:117–130
3. Pliss GB, Khudoley VV (1975) Tumor induction by carcinogenic agents in aquarium fish. J Natl Cancer Inst 55(1):129–136
4. Khudoley VV (1984) Use of aquarium fish, Danio rerio and Poecilia reticulata, as test species for evaluation of nitrosamine carcinogenicity. Natl Cancer Inst Monogr 65:65–70
5. Stuart GW, McMurray JV, Westerfield M (1988) Replication, integration and stable germline transmission of foreign sequences injected into early zebrafish embryos. Development 103(2):403–412
6. Udvadia AJ, Linney E (2003) Windows into development: historic, current, and future perspectives on transgenic zebrafish. Dev Biol 256(1):1–17
7. Kawakami K (2007) Tol2: a versatile gene transfer vector in vertebrates. Genome Biol 8(Suppl 1):S7. doi:10.1186/gb-2007-8-s1-s7
8. Ju B, Spitsbergen J, Eden CJ, Taylor MR, Chen W (2009) Co-activation of hedgehog and AKT pathways promote tumorigenesis in zebrafish. Mol Cancer 8:40. doi:10.1186/1476-4598-8-40

9. Santoriello C, Gennaro E, Anelli V, Distel M, Kelly A, Koster RW, Hurlstone A, Mione M (2010) Kita driven expression of oncogenic HRAS leads to early onset and highly penetrant melanoma in zebrafish. PLoS One 5(12), e15170. doi:10.1371/journal.pone.0015170
10. Liu S, Leach SD (2011) Screening pancreatic oncogenes in zebrafish using the Gal4/UAS system. Methods Cell Biol 105:367–381. doi:10.1016/B978-0-12-381320-6.00015-1
11. Spitzner M, Reischl M, Mione M (2014) A novel zebrafish model of glioma reveals cell fate alteration through expression of oncogenic RAS. Eur J Cancer 50(5):S159–S160. doi:10.1016/S0959-8049(14)50586-2
12. Jung IH, Leem GL, Jung DE, Kim MH, Kim EY, Kim SH, Park HC, Park SW (2013) Glioma is formed by active Akt1 alone and promoted by active Rac1 in transgenic zebrafish. Neuro Oncol 15(3):290–304. doi:10.1093/neuonc/nos387
13. Burger A, Vasilyev A, Tomar R, Selig MK, Nielsen GP, Peterson RT, Drummond IA, Haber DA (2014) A zebrafish model of chordoma initiated by notochord-driven expression of HRASV12. Dis Model Mech 7(7):907–913. doi:10.1242/dmm.013128
14. Feng H, Langenau DM, Madge JA, Quinkertz A, Gutierrez A, Neuberg DS, Kanki JP, Look AT (2007) Heat-shock induction of T-cell lymphoma/leukaemia in conditional Cre/lox-regulated transgenic zebrafish. Br J Haematol 138(2):169–175. doi:10.1111/j.1365-2141.2007.06625.x
15. Langenau DM, Feng H, Berghmans S, Kanki JP, Kutok JL, Look AT (2005) Cre/lox-regulated transgenic zebrafish model with conditional myc-induced T cell acute lymphoblastic leukemia. Proc Natl Acad Sci U S A 102(17):6068–6073. doi:10.1073/pnas.0408708102
16. Scheer N, Campos-Ortega JA (1999) Use of the Gal4-UAS technique for targeted gene expression in the zebrafish. Mech Dev 80(2):153–158
17. Brand AH, Perrimon N (1993) Targeted gene expression as a means of altering cell fates and generating dominant phenotypes. Development 118(2):401–415
18. Johnston SA, Salmeron JM Jr, Dincher SS (1987) Interaction of positive and negative regulatory proteins in the galactose regulon of yeast. Cell 50(1):143–146
19. Ma J, Ptashne M (1987) The carboxy-terminal 30 amino acids of GAL4 are recognized by GAL80. Cell 50(1):137–142
20. Johnson AA, Hibberd JM, Gay C, Essah PA, Haseloff J, Tester M, Guiderdoni E (2005) Spatial control of transgene expression in rice (Oryza sativa L.) using the GAL4 enhancer trapping system. Plant J 41(5):779–789. doi:10.1111/j.1365-313X.2005.02339.x
21. Fischer JA, Giniger E, Maniatis T, Ptashne M (1988) GAL4 activates transcription in Drosophila. Nature 332(6167):853–856. doi:10.1038/332853a0
22. Hartley KO, Nutt SL, Amaya E (2002) Targeted gene expression in transgenic Xenopus using the binary Gal4-UAS system. Proc Natl Acad Sci U S A 99(3):1377–1382. doi:10.1073/pnas.022646899
23. Ornitz DM, Moreadith RW, Leder P (1991) Binary system for regulating transgene expression in mice: targeting int-2 gene expression with yeast GAL4/UAS control elements. Proc Natl Acad Sci U S A 88(3):698–702
24. Koster RW, Fraser SE (2001) Tracing transgene expression in living zebrafish embryos. Dev Biol 233(2):329–346. doi:10.1006/dbio.2001.0242
25. Sadowski I, Ma J, Triezenberg S, Ptashne M (1988) GAL4-VP16 is an unusually potent transcriptional activator. Nature 335(6190):563–564. doi:10.1038/335563a0
26. Grabher C, Wittbrodt J (2004) Efficient activation of gene expression using a heat-shock inducible Gal4/Vp16-UAS system in medaka. BMC Biotechnol 4:26. doi:10.1186/1472-6750-4-26
27. Scott EK, Mason L, Arrenberg AB, Ziv L, Gosse NJ, Xiao T, Chi NC, Asakawa K, Kawakami K, Baier H (2007) Targeting neural circuitry in zebrafish using GAL4 enhancer trapping. Nat Methods 4(4):323–326. doi:10.1038/nmeth1033
28. Triezenberg SJ, Kingsbury RC, McKnight SL (1988) Functional dissection of VP16, the trans-activator of herpes simplex virus immediate early gene expression. Genes Dev 2(6):718–729
29. Gill G, Ptashne M (1988) Negative effect of the transcriptional activator GAL4. Nature 334(6184):721–724. doi:10.1038/334721a0

30. Asakawa K, Kawakami K (2008) Targeted gene expression by the Gal4-UAS system in zebrafish. Dev Growth Differ 50(6):391–399. doi:10.1111/j.1440-169X.2008.01044.x
31. Asakawa K, Kawakami K (2009) The Tol2-mediated Gal4-UAS method for gene and enhancer trapping in zebrafish. Methods 49(3):275–281. doi:10.1016/j.ymeth.2009.01.004
32. Ogura E, Okuda Y, Kondoh H, Kamachi Y (2009) Adaptation of GAL4 activators for GAL4 enhancer trapping in zebrafish. Dev Dyn 238(3):641–655. doi:10.1002/dvdy.21863
33. Kawakami K, Abe G, Asada T, Asakawa K, Fukuda R, Ito A, Lal P, Mouri N, Muto A, Suster ML, Takakubo H, Urasaki A, Wada H, Yoshida M (2010) zTrap: zebrafish gene trap and enhancer trap database. BMC Dev Biol 10:105. doi:10.1186/1471-213X-10-105
34. Distel M, Wullimann MF, Koster RW (2009) Optimized Gal4 genetics for permanent gene expression mapping in zebrafish. Proc Natl Acad Sci U S A 106(32):13365–13370. doi:10.1073/pnas.0903060106
35. Baron U, Gossen M, Bujard H (1997) Tetracycline-controlled transcription in eukaryotes: novel transactivators with graded transactivation potential. Nucleic Acids Res 25(14):2723–2729
36. Fujimoto E, Gaynes B, Brimley CJ, Chien CB, Bonkowsky JL (2011) Gal80 intersectional regulation of cell-type specific expression in vertebrates. Dev Dyn 240(10):2324–2334. doi:10.1002/dvdy.22734
37. Faucherre A, Lopez-Schier H (2011) Delaying Gal4-driven gene expression in the zebrafish with morpholinos and Gal80. PLoS One 6(1), e16587. doi:10.1371/journal.pone.0016587
38. Akitake CM, Macurak M, Halpern ME, Goll MG (2011) Transgenerational analysis of transcriptional silencing in zebrafish. Dev Biol 352(2):191–201. doi:10.1016/j.ydbio.2011.01.002
39. Scott EK (2009) The Gal4/UAS toolbox in zebrafish: new approaches for defining behavioral circuits. J Neurochem 110(2):441–456. doi:10.1111/j.1471-4159.2009.06161.x
40. Seebald JL, Szeto DP (2011) Zebrafish eve1 regulates the lateral and ventral fates of mesodermal progenitor cells at the onset of gastrulation. Dev Biol 349(1):78–89. doi:10.1016/j.ydbio.2010.10.005
41. Nakayama S, Ikenaga T, Kawakami K, Ono F, Hatta K (2012) Transgenic line with gal4 insertion useful to study morphogenesis of craniofacial perichondrium, vascular endothelium-associated cells, floor plate, and dorsal midline radial glia during zebrafish development. Dev Growth Differ 54(2):202–215. doi:10.1111/j.1440-169X.2011.01322.x
42. Satou C, Kimura Y, Hirata H, Suster ML, Kawakami K, Higashijima S (2013) Transgenic tools to characterize neuronal properties of discrete populations of zebrafish neurons. Development 140(18):3927–3931. doi:10.1242/dev.099531
43. Alghisi E, Distel M, Malagola M, Anelli V, Santoriello C, Herwig L, Krudewig A, Henkel CV, Russo D, Mione MC (2013) Targeting oncogene expression to endothelial cells induces proliferation of the myelo-erythroid lineage by repressing the Notch pathway. Leukemia 27(11):2229–2241. doi:10.1038/leu.2013.132
44. Ju B, Chen W, Orr BA, Spitsbergen JM, Jia S, Eden CJ, Henson HE, Taylor MR (2015) Oncogenic KRAS promotes malignant brain tumors in zebrafish. Mol Cancer 14(1):18. doi:10.1186/s12943-015-0288-2
45. Park JT, Johnson N, Liu S, Levesque M, Wang YJ, Ho H, Huso D, Maitra A, Parsons MJ, Prescott JD, Leach SD (2014) Differential in vivo tumorigenicity of diverse KRAS mutations in vertebrate pancreas: a comprehensive survey. Oncogene. doi:10.1038/onc.2014.223
46. Provost E, Bailey JM, Aldrugh S, Liu S, Iacobuzio-Donahue C, Leach SD (2014) The tumor suppressor rpl36 restrains KRAS(G12V)-induced pancreatic cancer. Zebrafish 11(6):551–559. doi:10.1089/zeb.2014.1024
47. Dong Z, Peng J, Guo S (2013) Stable gene silencing in zebrafish with spatiotemporally targetable RNA interference. Genetics 193(4):1065–1071. doi:10.1534/genetics.112.147892
48. Lagana A, Shasha D, Croce CM (2014) Synthetic RNAs for gene regulation: design principles and computational tools. Front Bioeng Biotechnol 2:65. doi:10.3389/fbioe.2014.00065
49. Suli A, Guler AD, Raible DW, Kimelman D (2014) A targeted gene expression system using the tryptophan repressor in zebrafish shows no silencing in subsequent generations. Development 141(5):1167–1174. doi:10.1242/dev.100057

50. Subedi A, Macurak M, Gee ST, Monge E, Goll MG, Potter CJ, Parsons MJ, Halpern ME (2014) Adoption of the Q transcriptional regulatory system for zebrafish transgenesis. Methods 66(3):433–440. doi:10.1016/j.ymeth.2013.06.012
51. Potter CJ, Tasic B, Russler EV, Liang L, Luo L (2010) The Q system: a repressible binary system for transgene expression, lineage tracing, and mosaic analysis. Cell 141(3):536–548. doi:10.1016/j.cell.2010.02.025
52. Wei X, Potter CJ, Luo L, Shen K (2012) Controlling gene expression with the Q repressible binary expression system in Caenorhabditis elegans. Nat Methods 9(4):391–395. doi:10.1038/nmeth.1929
53. Abremski K, Hoess R, Sternberg N (1983) Studies on the properties of P1 site-specific recombination: evidence for topologically unlinked products following recombination. Cell 32(4):1301–1311
54. Bode J, Schlake T, Iber M, Schubeler D, Seibler J, Snezhkov E, Nikolaev L (2000) The transgeneticist's toolbox: novel methods for the targeted modification of eukaryotic genomes. Biol Chem 381(9-10):801–813. doi:10.1515/BC.2000.103
55. Schnutgen F, Doerflinger N, Calleja C, Wendling O, Chambon P, Ghyselinck NB (2003) A directional strategy for monitoring Cre-mediated recombination at the cellular level in the mouse. Nat Biotechnol 21(5):562–565. doi:10.1038/nbt811
56. Siegal ML, Hartl DL (1996) Transgene Coplacement and high efficiency site-specific recombination with the Cre/loxP system in Drosophila. Genetics 144(2):715–726
57. Branda CS, Dymecki SM (2004) Talking about a revolution: the impact of site-specific recombinases on genetic analyses in mice. Dev Cell 6(1):7–28
58. Sauer B, Henderson N (1988) Site-specific DNA recombination in mammalian cells by the Cre recombinase of bacteriophage P1. Proc Natl Acad Sci U S A 85(14):5166–5170
59. Pan X, Wan H, Chia W, Tong Y, Gong Z (2005) Demonstration of site-directed recombination in transgenic zebrafish using the Cre/loxP system. Transgenic Res 14(2):217–223
60. Liu WY, Wang Y, Qin Y, Wang YP, Zhu ZY (2007) Site-directed gene integration in transgenic zebrafish mediated by cre recombinase using a combination of mutant lox sites. Mar Biotechnol (NY) 9(4):420–428. doi:10.1007/s10126-007-9000-x
61. Wang L, Zhang Y, Zhou T, Fu YF, Du TT, Jin Y, Chen Y, Ren CG, Peng XL, Deng M, Liu TX (2008) Functional characterization of Lmo2-Cre transgenic zebrafish. Dev Dyn 237(8):2139–2146. doi:10.1002/dvdy.21630
62. Liu X, Li Z, Emelyanov A, Parinov S, Gong Z (2008) Generation of oocyte-specifically expressed cre transgenic zebrafish for female germline excision of loxP-flanked transgene. Dev Dyn 237(10):2955–2962. doi:10.1002/dvdy.21701
63. Seok SH, Na YR, Han JH, Kim TH, Jung H, Lee BH, Emelyanov A, Parinov S, Park JH (2010) Cre/loxP-regulated transgenic zebrafish model for neural progenitor-specific oncogenic Kras expression. Cancer Sci 101(1):149–154. doi:10.1111/j.1349-7006.2009.01393.x
64. Yoshikawa S, Kawakami K, Zhao XC (2008) G2R Cre reporter transgenic zebrafish. Dev Dyn 237(9):2460–2465. doi:10.1002/dvdy.21673
65. Zhou Y, Cashman TJ, Nevis KR, Obregon P, Carney SA, Liu Y, Gu A, Mosimann C, Sondalle S, Peterson RE, Heideman W, Burns CE, Burns CG (2011) Latent TGF-beta binding protein 3 identifies a second heart field in zebrafish. Nature 474(7353):645–648. doi:10.1038/nature10094
66. Lichtman JW, Livet J, Sanes JR (2008) A technicolour approach to the connectome. Nat Rev Neurosci 9(6):417–422. doi:10.1038/nrn2391
67. Livet J, Weissman TA, Kang H, Draft RW, Lu J, Bennis RA, Sanes JR, Lichtman JW (2007) Transgenic strategies for combinatorial expression of fluorescent proteins in the nervous system. Nature 450(7166):56–62. doi:10.1038/nature06293
68. Pan YA, Freundlich T, Weissman TA, Schoppik D, Wang XC, Zimmerman S, Ciruna B, Sanes JR, Lichtman JW, Schier AF (2013) Zebrabow: multispectral cell labeling for cell tracing and lineage analysis in zebrafish. Development 140(13):2835–2846. doi:10.1242/dev.094631
69. Mosimann C, Kaufman CK, Li P, Pugach EK, Tamplin OJ, Zon LI (2011) Ubiquitous transgene expression and Cre-based recombination driven by the ubiquitin promoter in zebrafish. Development 138(1):169–177. doi:10.1242/dev.059345

70. Blackburn JS, Langenau DM (2014) Zebrafish as a model to assess cancer heterogeneity, progression and relapse. Dis Model Mech 7(7):755–762. doi:10.1242/dmm.015842
71. Buchholz F, Angrand PO, Stewart AF (1998) Improved properties of FLP recombinase evolved by cycling mutagenesis. Nat Biotechnol 16(7):657–662. doi:10.1038/nbt0798-657
72. Theodosiou NA, Xu T (1998) Use of FLP/FRT system to study Drosophila development. Methods 14(4):355–365. doi:10.1006/meth.1998.0591
73. Bischof J, Basler K (2008) Recombinases and their use in gene activation, gene inactivation, and transgenesis. Methods Mol Biol 420:175–195. doi:10.1007/978-1-59745-583-1_10
74. Ludwig DL, Stringer JR, Wight DC, Doetschman HC, Duffy JJ (1996) FLP-mediated site-specific recombination in microinjected murine zygotes. Transgenic Res 5(6):385–395
75. Kanki H, Suzuki H, Itohara S (2006) High-efficiency CAG-FLPe deleter mice in C57BL/6J background. Exp Anim 55(2):137–141
76. Meyers EN, Lewandoski M, Martin GR (1998) An Fgf8 mutant allelic series generated by Cre- and Flp-mediated recombination. Nat Genet 18(2):136–141. doi:10.1038/ng0298-136
77. Yamamoto M, Shook NA, Kanisicak O, Yamamoto S, Wosczyna MN, Camp JR, Goldhamer DJ (2009) A multifunctional reporter mouse line for Cre- and FLP-dependent lineage analysis. Genesis 47(2):107–114. doi:10.1002/dvg.20474
78. Wong AC, Draper BW, Van Eenennaam AL (2011) FLPe functions in zebrafish embryos. Transgenic Res 20(2):409–415. doi:10.1007/s11248-010-9410-9
79. le Trinh A, Hochgreb T, Graham M, Wu D, Ruf-Zamojski F, Jayasena CS, Saxena A, Hawk R, Gonzalez-Serricchio A, Dixson A, Chow E, Gonzales C, Leung HY, Solomon I, Bronner-Fraser M, Megason SG, Fraser SE (2011) A versatile gene trap to visualize and interrogate the function of the vertebrate proteome. Genes Dev 25(21):2306–2320. doi:10.1101/gad.174037.111
80. Ni TT, Lu J, Zhu M, Maddison LA, Boyd KL, Huskey L, Ju B, Hesselson D, Zhong TP, Page-McCaw PS, Stainier DY, Chen W (2012) Conditional control of gene function by an invertible gene trap in zebrafish. Proc Natl Acad Sci U S A 109(38):15389–15394. doi:10.1073/pnas.1206131109
81. Maddison LA, Li M, Chen W (2014) Conditional gene-trap mutagenesis in zebrafish. Methods Mol Biol 1101:393–411. doi:10.1007/978-1-62703-721-1_19
82. Sauer B, McDermott J (2004) DNA recombination with a heterospecific Cre homolog identified from comparison of the pac-c1 regions of P1-related phages. Nucleic Acids Res 32(20):6086–6095. doi:10.1093/nar/gkh941
83. Anastassiadis K, Fu J, Patsch C, Hu S, Weidlich S, Duerschke K, Buchholz F, Edenhofer F, Stewart AF (2009) Dre recombinase, like Cre, is a highly efficient site-specific recombinase in E. coli, mammalian cells and mice. Dis Model Mech 2(9-10):508–515. doi:10.1242/dmm.003087
84. Birling MC, Gofflot F, Warot X (2009) Site-specific recombinases for manipulation of the mouse genome. Methods Mol Biol 561:245–263. doi:10.1007/978-1-60327-019-9_16
85. Park JT, Leach SD (2013) TAILOR: transgene activation and inactivation using lox and rox in zebrafish. PLoS One 8(12), e85218. doi:10.1371/journal.pone.0085218
86. Khan MS, Khalid AM, Malik KA (2005) Phage phiC31 integrase: a new tool in plastid genome engineering. Trends Plant Sci 10(1):1–3. doi:10.1016/j.tplants.2004.11.001
87. Farruggio AP, Chavez CL, Mikell CL, Calos MP (2012) Efficient reversal of phiC31 integrase recombination in mammalian cells. Biotechnol J 7(11):1332–1336. doi:10.1002/biot.201200283
88. Groth AC, Olivares EC, Thyagarajan B, Calos MP (2000) A phage integrase directs efficient site-specific integration in human cells. Proc Natl Acad Sci U S A 97(11):5995–6000. doi:10.1073/pnas.090527097
89. Belteki G, Gertsenstein M, Ow DW, Nagy A (2003) Site-specific cassette exchange and germline transmission with mouse ES cells expressing phiC31 integrase. Nat Biotechnol 21(3):321–324. doi:10.1038/nbt787
90. Groth AC, Fish M, Nusse R, Calos MP (2004) Construction of transgenic Drosophila by using the site-specific integrase from phage phiC31. Genetics 166(4):1775–1782
91. Allen BG, Weeks DL (2005) Transgenic Xenopus laevis embryos can be generated using phiC31 integrase. Nat Methods 2(12):975–979. doi:10.1038/nmeth814

92. Smith MC, Brown WR, McEwan AR, Rowley PA (2010) Site-specific recombination by phiC31 integrase and other large serine recombinases. Biochem Soc Trans 38(2):388–394. doi:10.1042/BST0380388
93. Lister JA (2010) Transgene excision in zebrafish using the phiC31 integrase. Genesis 48(2):137–143. doi:10.1002/dvg.20595
94. Lu J, Maddison LA, Chen W (2011) PhiC31 integrase induces efficient site-specific excision in zebrafish. Transgenic Res 20(1):183–189. doi:10.1007/s11248-010-9394-5
95. Hu G, Goll MG, Fisher S (2011) PhiC31 integrase mediates efficient cassette exchange in the zebrafish germline. Dev Dyn 240(9):2101–2107. doi:10.1002/dvdy.22699
96. Mosimann C, Puller AC, Lawson KL, Tschopp P, Amsterdam A, Zon LI (2013) Site-directed zebrafish transgenesis into single landing sites with the phiC31 integrase system. Dev Dyn 242(8):949–963. doi:10.1002/dvdy.23989
97. Roberts JA, Miguel-Escalada I, Slovik KJ, Walsh KT, Hadzhiev Y, Sanges R, Stupka E, Marsh EK, Balciuniene J, Balciunas D, Muller F (2014) Targeted transgene integration overcomes variability of position effects in zebrafish. Development 141(3):715–724. doi:10.1242/dev.100347
98. Kirchmaier S, Hockendorf B, Moller EK, Bornhorst D, Spitz F, Wittbrodt J (2013) Efficient site-specific transgenesis and enhancer activity tests in medaka using PhiC31 integrase. Development 140(20):4287–4295. doi:10.1242/dev.096081
99. Doyon Y, McCammon JM, Miller JC, Faraji F, Ngo C, Katibah GE, Amora R, Hocking TD, Zhang L, Rebar EJ, Gregory PD, Urnov FD, Amacher SL (2008) Heritable targeted gene disruption in zebrafish using designed zinc-finger nucleases. Nat Biotechnol 26(6):702–708. doi:10.1038/nbt1409
100. Chen S, Oikonomou G, Chiu CN, Niles BJ, Liu J, Lee DA, Antoshechkin I, Prober DA (2013) A large-scale in vivo analysis reveals that TALENs are significantly more mutagenic than ZFNs generated using context-dependent assembly. Nucleic Acids Res 41(4):2769–2778. doi:10.1093/nar/gks1356
101. Moore FE, Reyon D, Sander JD, Martinez SA, Blackburn JS, Khayter C, Ramirez CL, Joung JK, Langenau DM (2012) Improved somatic mutagenesis in zebrafish using transcription activator-like effector nucleases (TALENs). PLoS One 7(5), e37877. doi:10.1371/journal.pone.0037877
102. Handel EM, Cathomen T (2011) Zinc-finger nuclease based genome surgery: it's all about specificity. Curr Gene Ther 11(1):28–37
103. Ekker SC (2008) Zinc finger-based knockout punches for zebrafish genes. Zebrafish 5(2):121–123. doi:10.1089/zeb.2008.9988
104. Wright DA, Li T, Yang B, Spalding MH (2014) TALEN-mediated genome editing: prospects and perspectives. Biochem J 462(1):15–24. doi:10.1042/BJ20140295
105. Clark KJ, Voytas DF, Ekker SC (2011) A TALE of two nucleases: gene targeting for the masses? Zebrafish 8(3):147–149. doi:10.1089/zeb.2011.9993
106. Gaj T, Gersbach CA, Barbas CF 3rd (2013) ZFN, TALEN, and CRISPR/Cas-based methods for genome engineering. Trends Biotechnol 31(7):397–405. doi:10.1016/j.tibtech.2013.04.004
107. Wiedenheft B, Sternberg SH, Doudna JA (2012) RNA-guided genetic silencing systems in bacteria and archaea. Nature 482(7385):331–338. doi:10.1038/nature10886
108. Cong L, Ran FA, Cox D, Lin S, Barretto R, Habib N, Hsu PD, Wu X, Jiang W, Marraffini LA, Zhang F (2013) Multiplex genome engineering using CRISPR/Cas systems. Science 339(6121):819–823. doi:10.1126/science.1231143
109. Mali P, Yang L, Esvelt KM, Aach J, Guell M, DiCarlo JE, Norville JE, Church GM (2013) RNA-guided human genome engineering via Cas9. Science 339(6121):823–826. doi:10.1126/science.1232033
110. Jiang W, Zhou H, Bi H, Fromm M, Yang B, Weeks DP (2013) Demonstration of CRISPR/Cas9/sgRNA-mediated targeted gene modification in Arabidopsis, tobacco, sorghum and rice. Nucleic Acids Res 41(20), e188. doi:10.1093/nar/gkt780
111. Shen B, Zhang J, Wu H, Wang J, Ma K, Li Z, Zhang X, Zhang P, Huang X (2013) Generation of gene-modified mice via Cas9/RNA-mediated gene targeting. Cell Res 23(5):720–723. doi:10.1038/cr.2013.46

112. Wang H, Yang H, Shivalila CS, Dawlaty MM, Cheng AW, Zhang F, Jaenisch R (2013) One-step generation of mice carrying mutations in multiple genes by CRISPR/Cas-mediated genome engineering. Cell 153(4):910–918. doi:10.1016/j.cell.2013.04.025
113. Hu X, Chang N, Wang X, Zhou F, Zhou X, Zhu X, Xiong JW (2013) Heritable gene-targeting with gRNA/Cas9 in rats. Cell Res 23(11):1322–1325. doi:10.1038/cr.2013.141
114. Yan Q, Zhang Q, Yang H, Zou Q, Tang C, Fan N, Lai L (2014) Generation of multi-gene knockout rabbits using the Cas9/gRNA system. Cell Regen (Lond) 3(1):12. doi:10.1186/2045-9769-3-12
115. Ni W, Qiao J, Hu S, Zhao X, Regouski M, Yang M, Polejaeva IA, Chen C (2014) Efficient gene knockout in goats using CRISPR/Cas9 system. PLoS One 9(9), e106718. doi:10.1371/journal.pone.0106718
116. Whitworth KM, Lee K, Benne JA, Beaton BP, Spate LD, Murphy SL, Samuel MS, Mao J, O'Gorman C, Walters EM, Murphy CN, Driver J, Mileham A, McLaren D, Wells KD, Prather RS (2014) Use of the CRISPR/Cas9 system to produce genetically engineered pigs from in vitro-derived oocytes and embryos. Biol Reprod 91(3):78. doi:10.1095/biolreprod.114.121723
117. Hwang WY, Fu Y, Reyon D, Maeder ML, Tsai SQ, Sander JD, Peterson RT, Yeh JR, Joung JK (2013) Efficient genome editing in zebrafish using a CRISPR-Cas system. Nat Biotechnol 31(3):227–229. doi:10.1038/nbt.2501
118. Jao LE, Wente SR, Chen W (2013) Efficient multiplex biallelic zebrafish genome editing using a CRISPR nuclease system. Proc Natl Acad Sci U S A 110(34):13904–13909. doi:10.1073/pnas.1308335110
119. Hruscha A, Krawitz P, Rechenberg A, Heinrich V, Hecht J, Haass C, Schmid B (2013) Efficient CRISPR/Cas9 genome editing with low off-target effects in zebrafish. Development 140(24):4982–4987. doi:10.1242/dev.099085
120. Xiao A, Wang Z, Hu Y, Wu Y, Luo Z, Yang Z, Zu Y, Li W, Huang P, Tong X, Zhu Z, Lin S, Zhang B (2013) Chromosomal deletions and inversions mediated by TALENs and CRISPR/Cas in zebrafish. Nucleic Acids Res 41(14), e141. doi:10.1093/nar/gkt464
121. Talbot JC, Amacher SL (2014) A streamlined CRISPR pipeline to reliably generate zebrafish frameshifting alleles. Zebrafish 11(6):583–585. doi:10.1089/zeb.2014.1047
122. Gagnon JA, Valen E, Thyme SB, Huang P, Ahkmetova L, Pauli A, Montague TG, Zimmerman S, Richter C, Schier AF (2014) Efficient mutagenesis by Cas9 protein-mediated oligonucleotide insertion and large-scale assessment of single-guide RNAs. PLoS One 9(5), e98186. doi:10.1371/journal.pone.0098186
123. Chang N, Sun C, Gao L, Zhu D, Xu X, Zhu X, Xiong JW, Xi JJ (2013) Genome editing with RNA-guided Cas9 nuclease in zebrafish embryos. Cell Res 23(4):465–472. doi:10.1038/cr.2013.45
124. Auer TO, Duroure K, De Cian A, Concordet JP, Del Bene F (2014) Highly efficient CRISPR/Cas9-mediated knock-in in zebrafish by homology-independent DNA repair. Genome Res 24(1):142–153. doi:10.1101/gr.161638.113
125. Irion U, Krauss J, Nusslein-Volhard C (2014) Precise and efficient genome editing in zebrafish using the CRISPR/Cas9 system. Development 141(24):4827–4830. doi:10.1242/dev.115584
126. Kimura Y, Hisano Y, Kawahara A, Higashijima S (2014) Efficient generation of knock-in transgenic zebrafish carrying reporter/driver genes by CRISPR/Cas9-mediated genome engineering. Sci Rep 4:6545. doi:10.1038/srep06545
127. Auer TO, Duroure K, Concordet JP, Del Bene F (2014) CRISPR/Cas9-mediated conversion of eGFP- into Gal4-transgenic lines in zebrafish. Nat Protoc 9(12):2823–2840. doi:10.1038/nprot.2014.187
128. Zhang D, Golubkov VS, Han W, Correa RG, Zhou Y, Lee S, Strongin AY, Dong PD (2014) Identification of Annexin A4 as a hepatopancreas factor involved in liver cell survival. Dev Biol 395(1):96–110. doi:10.1016/j.ydbio.2014.08.025
129. Tatsumi Y, Takeda M, Matsuda M, Suzuki T, Yokoi H (2014) TALEN-mediated mutagenesis in zebrafish reveals a role for r-spondin 2 in fin ray and vertebral development. FEBS Lett 588(24):4543–4550. doi:10.1016/j.febslet.2014.10.015

130. Zhang Y, Ear J, Yang Z, Morimoto K, Zhang B, Lin S (2014) Defects of protein production in erythroid cells revealed in a zebrafish Diamond-Blackfan anemia model for mutation in RPS19. Cell Death Dis 5, e1352. doi:10.1038/cddis.2014.318
131. Zada D, Tovin A, Lerer-Goldshtein T, Vatine GD, Appelbaum L (2014) Altered behavioral performance and live imaging of circuit-specific neural deficiencies in a zebrafish model for psychomotor retardation. PLoS Genet 10(9), e1004615. doi:10.1371/journal.pgen.1004615
132. Borck G, Hog F, Dentici ML, Tan PL, Sowada N, Medeira A, Gueneau L, Thiele H, Kousi M, Lepri F, Wenzeck L, Blumenthal I, Radicioni A, Schwarzenberg TL, Mandriani B, Fischetto R, Morris-Rosendahl DJ, Altmuller J, Reymond A, Nurnberg P, Merla G, Dallapiccola B, Katsanis N, Cramer P, Kubisch C (2015) BRF1 mutations alter RNA polymerase III-dependent transcription and cause neurodevelopmental anomalies. Genome Res 25(2):155–166. doi:10.1101/gr.176925.114
133. Halloran MC, Sato-Maeda M, Warren JT, Su F, Lele Z, Krone PH, Kuwada JY, Shoji W (2000) Laser-induced gene expression in specific cells of transgenic zebrafish. Development 127(9):1953–1960
134. Christians ES, Benjamin IJ (2005) The stress or heat shock (HS) response: insights from transgenic mouse models. Methods 35(2):170–175. doi:10.1016/j.ymeth.2004.08.008
135. Sorger PK (1991) Heat shock factor and the heat shock response. Cell 65(3):363–366
136. Zou J, Guo Y, Guettouche T, Smith DF, Voellmy R (1998) Repression of heat shock transcription factor HSF1 activation by HSP90 (HSP90 complex) that forms a stress-sensitive complex with HSF1. Cell 94(4):471–480
137. Yeh FL, Hsu T (2000) Detection of a spontaneous high expression of heat shock protein 70 in developing zebrafish (Danio rerio). Biosci Biotechnol Biochem 64(3):592–595. doi:10.1271/bbb.64.592
138. Shoji W, Sato-Maeda M (2008) Application of heat shock promoter in transgenic zebrafish. Dev Growth Differ 50(6):401–406. doi:10.1111/j.1440-169X.2008.01038.x
139. Santoriello C, Deflorian G, Pezzimenti F, Kawakami K, Lanfrancone L, d'Adda di Fagagna F, Mione M (2009) Expression of H-RASV12 in a zebrafish model of Costello syndrome causes cellular senescence in adult proliferating cells. Dis Model Mech 2(1-2):56–67. doi:10.1242/dmm.001016
140. Di Micco R, Fumagalli M, Cicalese A, Piccinin S, Gasparini P, Luise C, Schurra C, Garre M, Nuciforo PG, Bensimon A, Maestro R, Pelicci PG, d'Adda di Fagagna F (2006) Oncogene-induced senescence is a DNA damage response triggered by DNA hyper-replication. Nature 444(7119):638–642. doi:10.1038/nature05327
141. Yeh JR, Munson KM, Chao YL, Peterson QP, Macrae CA, Peterson RT (2008) AML1-ETO reprograms hematopoietic cell fate by downregulating scl expression. Development 135(2):401–410. doi:10.1242/dev.008904
142. Shen LJ, Chen FY, Zhang Y, Cao LF, Kuang Y, Zhong M, Wang T, Zhong H (2013) MYCN transgenic zebrafish model with the characterization of acute myeloid leukemia and altered hematopoiesis. PLoS One 8(3), e59070. doi:10.1371/journal.pone.0059070
143. Leacock SW, Basse AN, Chandler GL, Kirk AM, Rakheja D, Amatruda JF (2012) A zebrafish transgenic model of Ewing's sarcoma reveals conserved mediators of EWS-FLI1 tumorigenesis. Dis Model Mech 5(1):95–106. doi:10.1242/dmm.007401
144. Lister JA, Capper A, Zeng Z, Mathers ME, Richardson J, Paranthaman K, Jackson IJ, Patton EE (2014) A conditional zebrafish MITF mutation reveals MITF levels are critical for melanoma promotion vs. regression in vivo. J Invest Dermatol 134(1):133–140. doi:10.1038/jid.2013.293
145. Webster NJ, Green S, Jin JR, Chambon P (1988) The hormone-binding domains of the estrogen and glucocorticoid receptors contain an inducible transcription activation function. Cell 54(2):199–207
146. He S, Krens SG, Zhan H, Gong Z, Hogendoorn PC, Spaink HP, Snaar-Jagalska BE (2011) A DeltaRaf1-ER-inducible oncogenic zebrafish liver cell model identifies hepatocellular carcinoma signatures. J Pathol 225(1):19–28. doi:10.1002/path.2936

147. Gutierrez A, Grebliunaite R, Feng H, Kozakewich E, Zhu S, Guo F, Payne E, Mansour M, Dahlberg SE, Neuberg DS, den Hertog J, Prochownik EV, Testa JR, Harris M, Kanki JP, Look AT (2011) Pten mediates Myc oncogene dependence in a conditional zebrafish model of T cell acute lymphoblastic leukemia. J Exp Med 208(8):1595–1603. doi:10.1084/jem.20101691
148. Knopf F, Schnabel K, Haase C, Pfeifer K, Anastassiadis K, Weidinger G (2010) Dually inducible TetON systems for tissue-specific conditional gene expression in zebrafish. Proc Natl Acad Sci U S A 107(46):19933–19938. doi:10.1073/pnas.1007799107
149. Gossen M, Bujard H (1992) Tight control of gene expression in mammalian cells by tetracycline-responsive promoters. Proc Natl Acad Sci U S A 89(12):5547–5551
150. Gossen M, Freundlieb S, Bender G, Muller G, Hillen W, Bujard H (1995) Transcriptional activation by tetracyclines in mammalian cells. Science 268(5218):1766–1769
151. Jost M, Kari C, Rodeck U (1997) An episomal vector for stable tetracycline-regulated gene expression. Nucleic Acids Res 25(15):3131–3134
152. Weinmann P, Gossen M, Hillen W, Bujard H, Gatz C (1994) A chimeric transactivator allows tetracycline-responsive gene expression in whole plants. Plant J 5(4):559–569
153. Mortlock A, Low W, Crisanti A (2003) Suppression of gene expression by a cell-permeable Tet repressor. Nucleic Acids Res 31(23), e152
154. Furth PA, St Onge L, Boger H, Gruss P, Gossen M, Kistner A, Bujard H, Hennighausen L (1994) Temporal control of gene expression in transgenic mice by a tetracycline-responsive promoter. Proc Natl Acad Sci U S A 91(20):9302–9306
155. Yu Z, Redfern CS, Fishman GI (1996) Conditional transgene expression in the heart. Circ Res 79(4):691–697
156. Huang CJ, Jou TS, Ho YL, Lee WH, Jeng YT, Hsieh FJ, Tsai HJ (2005) Conditional expression of a myocardium-specific transgene in zebrafish transgenic lines. Dev Dyn 233(4):1294–1303. doi:10.1002/dvdy.20485
157. Zhou X, Vink M, Klaver B, Berkhout B, Das AT (2006) Optimization of the Tet-On system for regulated gene expression through viral evolution. Gene Ther 13(19):1382–1390. doi:10.1038/sj.gt.3302780
158. Koponen JK, Kankkonen H, Kannasto J, Wirth T, Hillen W, Bujard H, Yla-Herttuala S (2003) Doxycycline-regulated lentiviral vector system with a novel reverse transactivator rtTA2S-M2 shows a tight control of gene expression in vitro and in vivo. Gene Ther 10(6):459–466. doi:10.1038/sj.gt.3301889
159. Langenau DM, Ferrando AA, Traver D, Kutok JL, Hezel JP, Kanki JP, Zon LI, Look AT, Trede NS (2004) In vivo tracking of T cell development, ablation, and engraftment in transgenic zebrafish. Proc Natl Acad Sci U S A 101(19):7369–7374. doi:10.1073/pnas.0402248101
160. Li Z, Huang X, Zhan H, Zeng Z, Li C, Spitsbergen JM, Meierjohann S, Schartl M, Gong Z (2012) Inducible and repressable oncogene-addicted hepatocellular carcinoma in Tet-on xmrk transgenic zebrafish. J Hepatol 56(2):419–425. doi:10.1016/j.jhep.2011.07.025
161. Li Z, Zheng W, Wang Z, Zeng Z, Zhan H, Li C, Zhou L, Yan C, Spitsbergen JM, Gong Z (2013) A transgenic zebrafish liver tumor model with inducible Myc expression reveals conserved Myc signatures with mammalian liver tumors. Dis Model Mech 6(2):414–423. doi:10.1242/dmm.010462
162. Chew TW, Liu XJ, Liu L, Spitsbergen JM, Gong Z, Low BC (2014) Crosstalk of Ras and Rho: activation of RhoA abates Kras-induced liver tumorigenesis in transgenic zebrafish models. Oncogene 33(21):2717–2727. doi:10.1038/onc.2013.240
163. Wyart C, Del Bene F (2011) Let there be light: zebrafish neurobiology and the optogenetic revolution. Rev Neurosci 22(1):121–130. doi:10.1515/RNS.2011.013
164. Arrenberg AB, Stainier DY, Baier H, Huisken J (2010) Optogenetic control of cardiac function. Science 330(6006):971–974. doi:10.1126/science.1195929
165. Mickoleit M, Schmid B, Weber M, Fahrbach FO, Hombach S, Reischauer S, Huisken J (2014) High-resolution reconstruction of the beating zebrafish heart. Nat Methods 11(9):919–922. doi:10.1038/nmeth.3037

166. Portugues R, Severi KE, Wyart C, Ahrens MB (2013) Optogenetics in a transparent animal: circuit function in the larval zebrafish. Curr Opin Neurobiol 23(1):119–126. doi:10.1016/j.conb.2012.11.001
167. Del Bene F, Wyart C (2012) Optogenetics: a new enlightenment age for zebrafish neurobiology. Dev Neurobiol 72(3):404–414. doi:10.1002/dneu.20914
168. Simmich J, Staykov E, Scott E (2012) Zebrafish as an appealing model for optogenetic studies. Prog Brain Res 196:145–162. doi:10.1016/B978-0-444-59426-6.00008-2
169. Rein ML, Deussing JM (2012) The optogenetic (r)evolution. Mol Genet Genomics 287(2):95–109. doi:10.1007/s00438-011-0663-7
170. Hausser M (2014) Optogenetics: the age of light. Nat Methods 11(10):1012–1014. doi:10.1038/nmeth.3111
171. Zhang K, Cui B (2015) Optogenetic control of intracellular signaling pathways. Trends Biotechnol 33(2):92–100. doi:10.1016/j.tibtech.2014.11.007
172. Mills E, Chen X, Pham E, Wong S, Truong K (2012) Engineering a photoactivated caspase-7 for rapid induction of apoptosis. ACS Synth Biol 1(3):75–82. doi:10.1021/sb200008j
173. Bonger KM, Rakhit R, Payumo AY, Chen JK, Wandless TJ (2014) General method for regulating protein stability with light. ACS Chem Biol 9(1):111–115. doi:10.1021/cb400755b
174. Chen D, Gibson ES, Kennedy MJ (2013) A light-triggered protein secretion system. J Cell Biol 201(4):631–640. doi:10.1083/jcb.201210119
175. Strickland D, Lin Y, Wagner E, Hope CM, Zayner J, Antoniou C, Sosnick TR, Weiss EL, Glotzer M (2012) TULIPs: tunable, light-controlled interacting protein tags for cell biology. Nat Methods 9(4):379–384. doi:10.1038/nmeth.1904
176. Toettcher JE, Gong D, Lim WA, Weiner OD (2011) Light-based feedback for controlling intracellular signaling dynamics. Nat Methods 8(10):837–839. doi:10.1038/nmeth.1700
177. Idevall-Hagren O, Dickson EJ, Hille B, Toomre DK, De Camilli P (2012) Optogenetic control of phosphoinositide metabolism. Proc Natl Acad Sci U S A 109(35):E2316–E2323. doi:10.1073/pnas.1211305109
178. Hughes RM, Bolger S, Tapadia H, Tucker CL (2012) Light-mediated control of DNA transcription in yeast. Methods 58(4):385–391. doi:10.1016/j.ymeth.2012.08.004
179. Liu H, Gomez G, Lin S, Lin C (2012) Optogenetic control of transcription in zebrafish. PLoS One 7(11), e50738. doi:10.1371/journal.pone.0050738
180. Boulina M, Samarajeewa H, Baker JD, Kim MD, Chiba A (2013) Live imaging of multicolor-labeled cells in Drosophila. Development 140(7):1605–1613. doi:10.1242/dev.088930
181. Motta-Mena LB, Reade A, Mallory MJ, Glantz S, Weiner OD, Lynch KW, Gardner KH (2014) An optogenetic gene expression system with rapid activation and deactivation kinetics. Nat Chem Biol 10(3):196–202. doi:10.1038/nchembio.1430
182. Teh C, Chudakov DM, Poon KL, Mamedov IZ, Sek JY, Shidlovsky K, Lukyanov S, Korzh V (2010) Optogenetic in vivo cell manipulation in KillerRed-expressing zebrafish transgenics. BMC Dev Biol 10:110. doi:10.1186/1471-213X-10-110
183. Kondrychyn I, Teh C, Garcia-Lecea M, Guan Y, Kang A, Korzh V (2011) Zebrafish Enhancer TRAP transgenic line database ZETRAP 2.0. Zebrafish 8(4):181–182. doi:10.1089/zeb.2011.0718
184. Thummel R, Burket CT, Brewer JL, Sarras MP Jr, Li L, Perry M, McDermott JP, Sauer B, Hyde DR, Godwin AR (2005) Cre-mediated site-specific recombination in zebrafish embryos. Dev Dyn 233(4):1366–1377. doi:10.1002/dvdy.20475
185. Forrester AM, Grabher C, McBride ER, Boyd ER, Vigerstad MH, Edgar A, Kai FB, Da'as SI, Payne E, Look AT, Berman JN (2011) NUP98-HOXA9-transgenic zebrafish develop a myeloproliferative neoplasm and provide new insight into mechanisms of myeloid leukaemogenesis. Br J Haematol 155(2):167–181. doi:10.1111/j.1365-2141.2011.08810.x
186. Le X, Langenau DM, Keefe MD, Kutok JL, Neuberg DS, Zon LI (2007) Heat shock-inducible Cre/Lox approaches to induce diverse types of tumors and hyperplasia in transgenic zebrafish. Proc Natl Acad Sci U S A 104(22):9410–9415. doi:10.1073/pnas.0611302104
187. Hans S, Kaslin J, Freudenreich D, Brand M (2009) Temporally-controlled site-specific recombination in zebrafish. PLoS One 4(2), e4640. doi:10.1371/journal.pone.0004640

188. Jopling C, Sleep E, Raya M, Marti M, Raya A, Izpisua Belmonte JC (2010) Zebrafish heart regeneration occurs by cardiomyocyte dedifferentiation and proliferation. Nature 464(7288):606–609. doi:10.1038/nature08899
189. Matsuda H, Parsons MJ, Leach SD (2013) Aldh1-expressing endocrine progenitor cells regulate secondary islet formation in larval zebrafish pancreas. PLoS One 8(9), e74350. doi:10.1371/journal.pone.0074350
190. Burket CT, Montgomery JE, Thummel R, Kassen SC, LaFave MC, Langenau DM, Zon LI, Hyde DR (2008) Generation and characterization of transgenic zebrafish lines using different ubiquitous promoters. Transgenic Res 17(2):265–279. doi:10.1007/s11248-007-9152-5
191. Jungke P, Hans S, Brand M (2013) The zebrafish CreZoo: an easy-to-handle database for novel CreER(T2)-driver lines. Zebrafish 10(3):259–263. doi:10.1089/zeb.2012.0834
192. Schiavone M, Rampazzo E, Casari A, Battilana G, Persano L, Moro E, Liu S, Leach SD, Tiso N, Argenton F (2014) Zebrafish reporter lines reveal in vivo signaling pathway activities involved in pancreatic cancer. Dis Model Mech 7(7):883–894. doi:10.1242/dmm.014969
193. Scheer N, Riedl I, Warren JT, Kuwada JY, Campos-Ortega JA (2002) A quantitative analysis of the kinetics of Gal4 activator and effector gene expression in the zebrafish. Mech Dev 112(1-2):9–14
194. Zhan H, Gong Z (2010) Delayed and restricted expression of UAS-regulated GFP gene in early transgenic zebrafish embryos by using the GAL4/UAS system. Mar Biotechnol (NY) 12(1):1–7. doi:10.1007/s10126-009-9217-y
195. Asakawa K, Suster ML, Mizusawa K, Nagayoshi S, Kotani T, Urasaki A, Kishimoto Y, Hibi M, Kawakami K (2008) Genetic dissection of neural circuits by Tol2 transposon-mediated Gal4 gene and enhancer trapping in zebrafish. Proc Natl Acad Sci U S A 105(4):1255–1260. doi:10.1073/pnas.0704963105
196. de Graaf M, Zivkovic D, Joore J (1998) Hormone-inducible expression of secreted factors in zebrafish embryos. Dev Growth Differ 40(6):577–582
197. Gerety SS, Breau MA, Sasai N, Xu Q, Briscoe J, Wilkinson DG (2013) An inducible transgene expression system for zebrafish and chick. Development 140(10):2235–2243. doi:10.1242/dev.091520
198. Akerberg AA, Stewart S, Stankunas K (2014) Spatial and temporal control of transgene expression in zebrafish. PLoS One 9(3), e92217. doi:10.1371/journal.pone.0092217
199. Umeda K, Shoji W, Sakai S, Muto A, Kawakami K, Ishizuka T, Yawo H (2013) Targeted expression of a chimeric channelrhodopsin in zebrafish under regulation of Gal4-UAS system. Neurosci Res 75(1):69–75. doi:10.1016/j.neures.2012.08.010
200. Blackburn JS, Langenau DM (2010) aMAZe-ing tools for mosaic analysis in zebrafish. Nat Methods 7(3):188–190. doi:10.1038/nmeth0310-188
201. Collins RT, Linker C, Lewis J (2010) MAZe: a tool for mosaic analysis of gene function in zebrafish. Nat Methods 7(3):219–223. doi:10.1038/nmeth.1423
202. Emelyanov A, Parinov S (2008) Mifepristone-inducible LexPR system to drive and control gene expression in transgenic zebrafish. Dev Biol 320(1):113–121. doi:10.1016/j.ydbio.2008.04.042
203. Nguyen AT, Emelyanov A, Koh CH, Spitsbergen JM, Parinov S, Gong Z (2012) An inducible kras(V12) transgenic zebrafish model for liver tumorigenesis and chemical drug screening. Dis Model Mech 5(1):63–72. doi:10.1242/dmm.008367
204. Nguyen AT, Emelyanov A, Koh CH, Spitsbergen JM, Lam SH, Mathavan S, Parinov S, Gong Z (2011) A high level of liver-specific expression of oncogenic Kras(V12) drives robust liver tumorigenesis in transgenic zebrafish. Dis Model Mech 4(6):801–813. doi:10.1242/dmm.007831
205. Ryffel GU, Werdien D, Turan G, Gerhards A, Goosses S, Senkel S (2003) Tagging muscle cell lineages in development and tail regeneration using Cre recombinase in transgenic Xenopus. Nucleic Acids Res 31(8), e44. doi:10.1093/nar/gng044

Approaches to Inactivate Genes in Zebrafish

John M. Parant and Jing-Ruey Joanna Yeh

Abstract Animal models of tumor initiation and tumor progression are essential components toward understanding cancer and designing/validating future therapies. Zebrafish is a powerful model for studying tumorigenesis and has been successfully exploited in drug discovery. According to the zebrafish reference genome, 82 % of disease-associated genes in the Online Mendelian Inheritance in Man (OMIM) database have clear zebrafish orthologues. Using a variety of large-scale random mutagenesis methods developed to date, zebrafish can provide a unique opportunity to identify gene mutations that may be associated with cancer predisposition. On the other hand, newer technologies enabling targeted mutagenesis can facilitate reverse cancer genetic studies and open the door for complex genetic analysis of tumorigenesis. In this chapter, we will describe the various technologies for conducting genome editing in zebrafish with special emphasis on the approaches to inactivate genes.

Keywords CRISPR • Cas9 • Zinc finger nuclease • Transcription activator-like effector nuclease • Mutagenesis • Zebrafish • Genome engineering • Cancer • Gene targeting • Homologous recombination

Non-targeted Methods

Chemical Mutagenesis by N-*Ethyl*-N-*Nitrosourea (ENU)*

Forward genetic screen has been a fruitful approach for discovering new biological pathways. While primarily utilized by non-vertebrate organisms like yeast, *C. elegans* or *Drosophila*, zebrafish provides a unique vertebrate model in which near

J.M. Parant (✉)
Department of Pharmacology and Toxicology, UAB Comprehensive Cancer Center, University of Alabama at Birmingham School of Medicine, Birmingham, AL 35294, USA
e-mail: jparant@uab.edu

J.-R. J. Yeh (✉)
Cardiovascular Research Center, Massachusetts General Hospital, Charlestown, MA 02129, USA

Department of Medicine, Harvard Medical School, Boston, MA 02115, USA
e-mail: jyeh1@mgh.harvard.edu

D.M. Langenau (ed.), *Cancer and Zebrafish*, Advances in Experimental Medicine and Biology 910, DOI 10.1007/978-3-319-30654-4_3

saturation, forward genetic screens can be achieved [1]. ENU (*N*-ethyl-*N*-nitrosourea) is an alkylating agent commonly used to mutate zebrafish and causes A to T transversions and AT to GC transitions, as well as other less frequent nucleotide alterations. By soaking an adult male zebrafish (referred to as a G0 fish) in a non-toxic dose of ENU, mutations are randomly generated in the germ stem cells, which can be passed down to the offspring (often referred to as F1). Importantly, while the G0 fish carry mosaic germline mutations, the F1 offspring carry heterozygous mutations, and each F1 fish is unique for the mutations it contains. A number of approaches have been developed to determine the recessive phenotypes of these mutations including traditional F3 genetic screen, F2 haploid screen and F2 early pressure gynogenetic diploid screen (Fig. 1).

As shown in Fig. 1, with the traditional F3 genetic screen, F2 families are generated from the unique F1 animals, such that 50 % of the F2 population will be heterozygous for a mutation contained in the F1. By breeding a number of F2 pairs (~8 pairs) to generate F3 animals, all possible homozygous phenotypes derived from mutations within this family can be observed in different F3 clutches. In principle, by screening a large number of F2 families, all potential genes involved in a particular phenotype can be identified. While this approach is schematically straightforward and does not require any special technique, it does require a significant fish space which can be inhibitory to small labs. Thus, two innovative screening methods have been developed as described below.

The haploid screen takes advantage of the fact that UV-treated sperms can fertilize eggs from a F1 female without providing genomic equivalents to the egg, thereby generating haploid F2 animals that contain only 1N of DNA from the maternal genome (Fig. 1). Importantly, haploid zebrafish embryos can survive well up to 3 days post-fertilization (dpf) allowing for phenotypic analysis prior to this time. The key advantage to this method is that phenotypic analysis can be achieved from the progeny of the F1 population, hence requiring considerably less fish space than a F3 screen. For example, 20 F1 fish (equivalent to one housing tank) screened by haploid is comparable to 20 F2 families (equivalent to 40–60 housing tanks) screened through a traditional F3 screen. Nonetheless, the caveats to this approach are that the process is mildly laborious, the female must survive the egg squeezing procedure, and there are a number of underlying cellular defects not well understood occurring in these haploid embryos [2, 3].

The Early Pressure (EP) gynogenetic diploid screen is an offshoot of the haploid screen, however following fertilization with UV-treated sperm, the egg is placed under pressure, not allowing the second polar body to extrude, resulting in a 2N embryo which in concept is homozygous for one of the two sets of maternal chromosomes (Fig. 1). In reality, meiotic recombination can occur during this process, creating heterozygosity near the telomere but homozygosity near the centromere; therefore this screen primarily identifies recessive phenotypes from centromeric mutations [4]. Importantly, a large number of these 2N gynogenetic embryos are viable and can be raised to adulthood. Compared to the other two screening methods, clear advantages of this type of screen are not only that the screen can be performed directly from the F1 animals, but also that it is not limited to phenotypes in <3 dpf window (e.g. adult screen can be performed). However, this procedure can

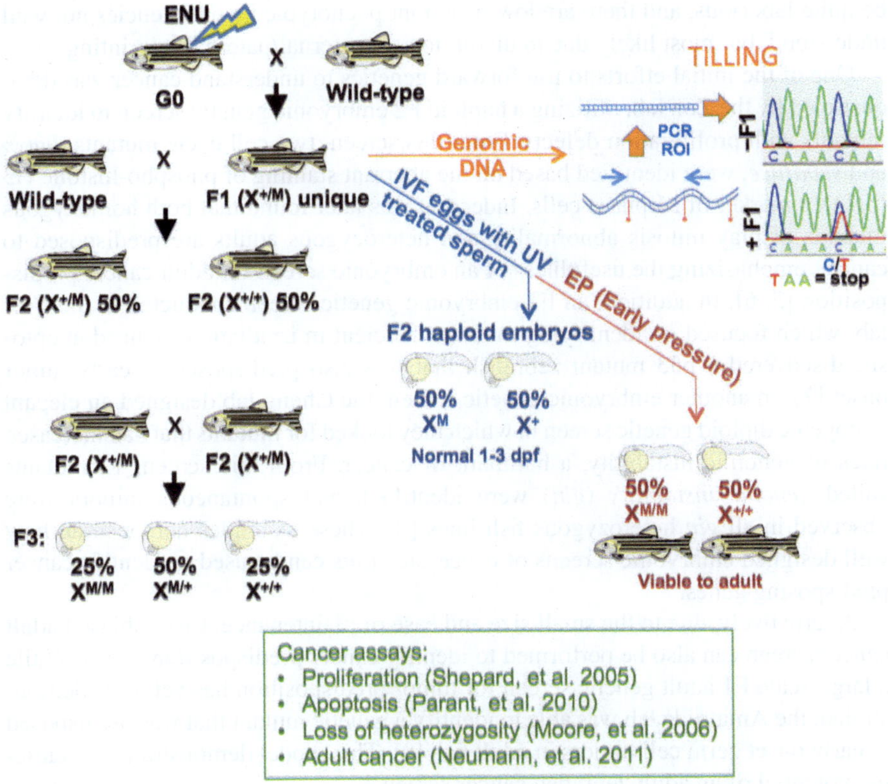

Fig. 1 Depiction of the different genetic screening methods as well as a targeted screening method called Targeting Induced Local Lesions IN Genomes (TILLING). Amongst all of these genetic screens, adult male zebrafish (G0) are exposed to ENU mutagen. Mutations induced in the germ cells can be propagated into the next generation through breeding to a wild-type female, producing an F1 offspring. Each F1 offspring contains unique mutations only found in that F1. In the traditional F3 recessive genetic screen (scheme marked in *black letters*), in order to generate additional animals with the F1 mutations, these animals are bred to a wild-type animal thus producing offspring in which half will be heterozygous for any individual mutation. By crossing these F2 animals, 25 % of the offspring (F3) will be homozygous for a mutant allele. Alternatively, haploid embryos can be generated by in vitro fertilization (IVF) of eggs from a female F1 with UV-treated sperms, which can fertilize but not contribute DNA content, thereby generating 50 % of embryos that are recessive for a F1 derived mutation (scheme marked in *blue letters*). Screening these embryos for cancer phenotypes alleviates the need to generate and maintain F2 families. However, these embryos develop normally until 3 dpf, therefore limited to early embryonic screens. An extension of the haploid screen is the early pressure-derived gynogenic diploid screen in which the fertilized eggs from the haploid procedure are exposed to high pressure (scheme marked in *magenta letters*). This procedure prevents the second polar body from extruding, hence a gynogenic diploid embryo is generated which will be homozygous for a maternal allele. Importantly these diploid embryos can be raised to adults. Finally, examples of assays for cancer-related phenotypes that have been conducted are shown at the bottom in the *green text box*. For TILLING (scheme marked in *orange letters*), genomic DNA is isolated from large cohorts of F1 adult fish, each harboring a unique collection of mutations. In this approach, first a PCR amplicon across a genomic region of interest (ROI) is generated from each of these F1 genomic DNA samples. Utilizing a variety of mutation detection methods described in this chapter, F1 animals containing novel mutations are identified and ultimately bred to generate a cohort of heterozygous animals for further analysis

be quite laborious, and there are low penetrant phenotypic inconsistencies not well understood, but most likely due to disruption of maternal/paternal imprinting.

One of the initial efforts to use forward genetics to understand cancer was demonstrated by the Zon lab, utilizing a haploid F2 embryonic genetic screen to identify mutants with proliferation defects. From this screen, two cell cycle mutants, *bmyb* and *separase*, were identified based on the aberrant staining of phospho-histone H3 (pH3), a marker of M-phase cells. Indeed, it was later found that both homozygous mutants display mitosis abnormality and heterozygous adults are predisposed to cancer, emphasizing the usefulness of an embryonic screen for adult cancer predisposition [5, 6]. In addition, an F3 embryonic genetic screen conducted in the Yost lab, which focused on identifying mutants deficient in irradiation-induced apoptosis, discovered a p53 mutant zebrafish that was also predisposed to early tumor onset [7]. In another embryonic genetic screen, the Cheng lab designed an elegant gynogenic diploid genetic screen in which they looked for mutants that had increased rates of genomic instability, a hallmark of cancer. From this screen, 12 mutants called *genome instability* (*gin*) were identified, and spontaneous tumors were observed in all *gin* heterozygous fish lines [8]. These examples demonstrate how well designed embryonic screens of cancerous traits can be used to identify cancer predisposing genes.

Alternatively, due to the small size and ease of maintenance, true unbiased adult cancer screen can also be performed to identify cancer predisposition genes. While a large-scale F1 adult genetic screen for tumor predisposition has yet to be demonstrated, the Amatruda lab was able to identify a genetic mutant that was predisposed to early onset germ cell tumors in adult fish [9]. This model demonstrates the enticing potential of an adult zebrafish cancer screen.

Insertional Mutagenesis by Retroviruses

In the early days ENU mutagenesis was the most effective method for mutating zebrafish genes, however mapping the ENU-derived mutation was burdensome. In 1996, pioneering work from the Hopkins lab established that high-titer mouse retroviruses could be used as a germline mutagen in zebrafish (Fig. 2a, b) [10]. The Hopkins group utilized this approach to generate a collection of 315 retroviral insertion mutant lines in which homozygotes were embryonic lethal prior to 5 dpf. These embryonic lethal mutants were identified through a breeding scheme similar to an ENU F3 screen. However, unique to this approach because of the retroviral insertion tag, the exact insertion site within the genome could be identified through inverse PCR methods. Therefore, the causative genes for all 315 embryonic lethal lines were identified and represented the largest mutant collection in which all the mutated genes were known [11]. Interestingly, the retrovirus has an unclear preference for the 5′ ends of genes encompassing promoters and early exons or introns, which allows for a useful tool to mutate genes (Fig. 2c). Follow-up from the Burgess lab, assessing a large unbiased cohort of insertions established that a majority of

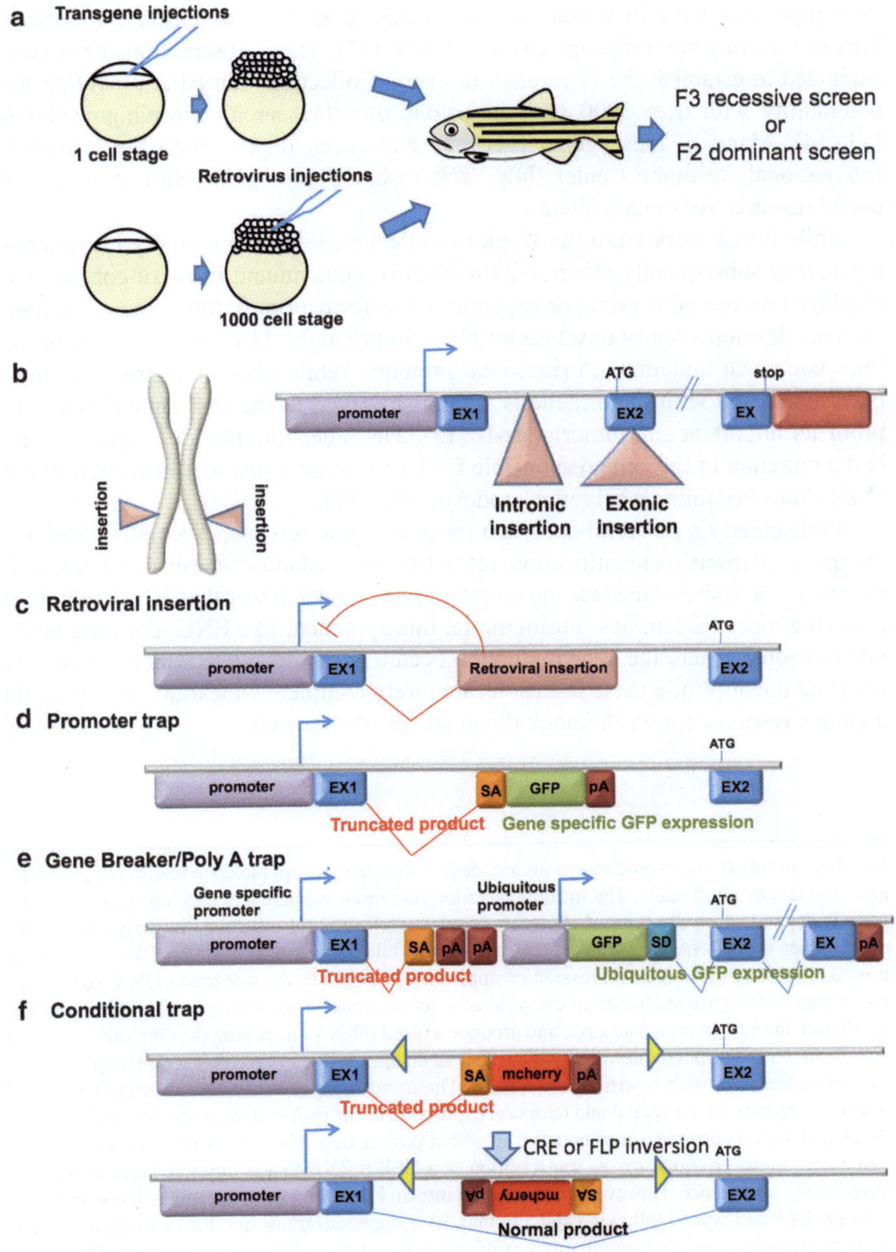

Fig. 2 Insertional mutagenesis. (**a**) Transgenic insertions are generated by injecting transposase mRNA plus transposable plasmid DNA into the 1 cell egg, while retroviral insertions are generated by injecting high-titer retroviruses into the 1000-cell stage embryos. Ultimately an adult F0 chimeric animal is generated, in which breeding scheme similar to ENU (Fig. 1) can be used to identify F2 dominant phenotypes or F3 recessive phenotypes. (**b**) While many insertions are in extragenic regions of the genome, some disruptive insertions are located in exons or introns of genes.

viral insertions were in intron 1 of genes, and that 80 % of intron 1 insertions knocked down gene transcript levels by 80 % [12]. These observations have been expanded to establish the "Zebrafish Insertion Collection" aimed at providing the community with over 3000 gene insertions (http://research.nhgri.nih.gov/ZInC/) [13, 14]. Many of these gene insertions have been deposited to the Zebrafish International Resource Center (http://zebrafish.org/home/guide.php), providing a useful resource for mutant alleles.

While initial work from the Hopkins collection focused on embryonic phenotypes, they subsequently monitored the heterozygous mutant lines for cohorts that displayed decreased survival or apparent tumor formation. In this same collection, 12 lines developed early onset tumor [11]. Surprisingly, 11 of the 12 tumor-prone lines contained mutations in ribosomal proteins. While ribosomal stress has long been associated with tumorigenesis, zebrafish served as the first animal model to proof its importance in tumorigenesis [15]. The other line that developed tumors had a mutation in the gene responsible for human cancer disorder neurofibromin 2 (NF2), thus becoming a zebrafish model of NF2 [11].

While clearly a powerful approach for phenotypic screens, especially considering the rapid insertion identification, retrovirus-induced mutagenesis is not practical for the average sized lab since the mutation rate is much lower than ENU, therefore requiring more F2 families. Furthermore, this approach, like ENU, does not allow one to choose where the gene disruption occurs or the degree to which the gene is knocked down. While these insertions are rarely complete knockouts, they provide a unique resource for weak/knock-down alleles for research.

Fig. 2 (continued) The exonic insertions are ideal in that they disrupt the open reading frame of the gene and create a null allele. The intronic insertions are more variable and their ability to disrupt a gene is dependent on the type of vector employed. (**c**) Retroviral insertions in early introns can often reduce the mRNA levels for the inserted gene. While not completely clear of the mechanism, it is likely through promoter repression or improper splicing into the integrated DNA. (**d**) A promoter trap is designed such that an exogenous splice acceptor sequence inserted within an intron will hijack the endogenous transcript and produce a fused mRNA containing the GFP cDNA, which may be identified through microscopy. The clear advantage of this system is that GFP expression in strongly correlative with insertion into a gene. The disadvantage is that positive identification of insertion depends on the spatial and temporal expression of the endogenous promoter. (**e**) The Gene Breaker design incorporates a promoter trap with a ploy A trap. The poly A trap is designed such that a ubiquitous promoter drives transcription of a GFP mRNA that is unstable since it does not have a poly A sequence. However when inserted into an intron of a gene the splice donor sequence after the GFP can hijack, following splicing into the endogenous transcript, the endogenous poly A sequence thereby providing stability to the GFP mRNA and ultimately GFP expression. The advantage of this system is that identifying insertion events by GFP expression does not depend on the expression pattern of the endogenous promoter. (**f**) The conditional trap can use the promoter trap or the Gene Breaker design, but importantly these cassettes are flanked by either Lox or FRT sites in the opposite orientation such that when Cre or Flp is expressed the insert may be inverted. When inverted the splice acceptor no longer disrupts the endogenous transcript. Importantly expression of the Cre or Flp again will invert the construct making it disruptive again. By using spatial or temporal control of the recombinase conditional rescue or knockout can be achieved

Insertional Mutagenesis by Transgenes

Transgene insertion is an attractive alternative to retrovirus-based insertional mutagenesis due to the increased cargo size, flexible design, and lack of need for cell culture to produce viral titers. Traditionally naked DNA injection approaches in zebrafish were very inefficient and not feasible for genetic screens. However, the application of transposon-based insertions, through Sleeping Beauty (SB) or tol2 transposon systems opened up the door for transgene insertion-mediated mutagenesis [16, 17]. These transposon systems allowed for high numbers of single copy insertions, resulting in high mutagenic rates. Since this chapter focuses on gene disruption, we will not discuss enhancer trap or other less deleterious transgenic constructs.

The first gene trap approach utilized the tol2 system to insert a splice acceptor-GFP-polyA construct randomly into the zebrafish genome [18]. This approach, also known as a promoter trap, exploited the trapped gene promoter and splicing machinery to generate GFP expression (Fig. 2d). Gene trap efficiencies were very high in that approximately 50 % of F0 fish, derived from tol2 transposition, produced GFP-positive F1 progeny. Meanwhile, as expected, a majority of the mapped GFP-positive insertions were within intragenic regions. While establishing a proof of principle for transgenic insertion-based mutagenesis, early studies showed that gene knockdown was often incomplete and did not consistently result in loss-of-function phenotypes [18, 19]. In addition, only insertions into genes that are expressed during early embryogenesis can be identified through this approach. Subsequently, an optimized "gene-breaking" vector was generated using a more proficient splice accepter and polyA termination sequence, to more effectively truncate the endogenous gene product (Fig. 2e). In addition, this vector contains a polyA trap for ubiquitous GFP expression, which allows for identification of all gene insertions irrespective of if the gene is active during embryogenesis. From this system, 10 % of injected embryos produced a germline polyA trap event (denoted by ubiquitous GFP expression) and of these 14 % produced a homozygous phenotype [20, 21]. Both SB and tol2 delivery systems seem to produce similar efficiencies [19]. Further improvements by the Chen and Ekker labs have made these insertions conditional, because they are gene-disrupting only when inserted in one of the two orientations. By design, the orientation of the integrated DNA may be reversed using CRE or FLP recombinase (Fig. 2f) [21, 22]. Only a few examples of genetic screens have been derived from these transgenic insertions [23], however they are good resource for knockout alleles and the only resources for conditional genetic mutants in zebrafish to date (http://kawakami.lab.nig.ac.jp/ztrap/; http://zfishbook.org/).

The key attraction to this approach is the technical amenability to almost all zebrafish labs and the relative easiness in mapping the insertion loci, although the latter has become less significant due to the advances in Next-Generation sequencing (NGS). Due to the abilities to customize the insertion vectors and make the mutations reversible, insertional mutagenesis by transgenes certainly stands out to be an advantageous platform for innovatively designed experiments.

Targeted Methods

Targeting Induced Local Lesions IN Genomes (TILLING)

TILLING was first described for identification of mutations in *Arabidopsis thaliana* in 2000, however quickly became adopted in many organisms, including zebrafish as the ideal mutation generation and identification approach [24–26]. The concept is to screen through individual genomic DNA samples from a large cohort of ENU/EMS (ethyl methanesulfonate)-derived F1 animals to identify an individual that contains a mutation that either alters a key amino acid or generates a premature stop codon in the desired gene (Fig. 1). The detection methods for these mutations have evolved from denaturing high performance liquid chromatography (DHPLC) analysis [24], then CEL1 heteroduplex endonuclease followed by electrophoresed using LI-COR IR^2 gel analyzer [27], to high-throughput NGS of the whole exome of F1 animals [28]. After examining 1673 exomes using NGS, it was found that the frequency of non-sense and disruptive splice site mutations was approximately 10 per exome.

While this approach is advantageous for identifying gene specific knockouts, the TILLING technology was not very amenable to most individual labs. Therefore two main TILLING centers evolved—the FHCRC Zebrafish TILLING project (https://webapps.fhcrc.org/science/tilling/) and the Sanger Zebrafish Mutation Project (http://www.sanger.ac.uk/Projects/D_rerio/zmp/). Eventually, the Wellcome Trust Sanger Institute took it upon itself the mission to identify at least one gene-disrupting mutation in every zebrafish gene. Presently the project is at 45 % null-mutation coverage (11,892 of 26,206 protein-coding zebrafish genes; http://www.sanger.ac.uk/Projects/D_rerio/zmp/) [28]. TILLING mutants are available from various zebrafish stock centers (http://zebrafish.org/home/guide.php; http://www.ezrc.kit.edu/). Looking forward, while initially the rate of identifying a gene-disrupting mutation in a gene, where such a mutation had not been previously identified, was high (10 per F1), this rate has declined dramatically as more exomes were sequenced. This decline may be caused by a combination of several factors including: (1) saturation of the discovery limit, in that by random chance there will be an increasing proportion of previously identified null alleles; (2) small-sized genes are less likely to be mutated by random chance; (3) mutagenic hotspot/coldspot effects of chemical mutagenesis; and (4) some sequences have a lower mutagenic potential.

The Plasterk group was the first to apply this TILLING technology to identify novel zebrafish mutant alleles [26, 29–31]. Among tumor suppressor genes, in 2005, a p53 missense mutant zebrafish derived from TILLING was the first zebrafish tumor suppressor gene mutant, which developed early onset tumors and initiated zebrafish as a useful model for cancer studies [29]. Subsequently in 2006, a TILLING-derived mutant heterozygous for the *adenomatous polyposis coli* (*apc*) gene was demonstrated to develop adenomas in the intestine, which now serves as a zebrafish model of familial adenomas polyposis (FAP) [30]. The teleost linage underwent a duplication event such that there are two *phosphatase and tensin homolog* (*PTEN*) orthologs—*ptena* and *ptenb*, in zebrafish. Mutations in both of these genes have been identified by TILLING. In mice, loss of *PTEN* leads to early embryonic lethality [32, 33]. Double homozygous pten-null zebrafish are also embryonic lethal. Interestingly, homozygous deletion of *ptenb* alone is viable, however adults have early onset tumor, suggesting through this gene duplication the embryonic and tumor suppressor roles of PTEN can be separated [34]. Human hereditary non-polyposis colon cancer (HNPCC) is caused by germline mutations in various key mismatch repair genes [35–40]. Truncated mutations in 3 of the HNPCC genes—MLH1, MSH2, and MSH6, have also been found by TILLING. Homozygous mutant zebrafish in either one of these three genes developed tumor at an early age, establishing a zebrafish model of HNPCC [41]. In sum, TILLING was the first technology that established zebrafish as an animal model for targeted gene inactivation and identification of specific tumor suppressor genes. Despite these successes, when compared to the newer technologies such as engineered endonucleases, one of the key disadvantages of TILLING is that a researcher cannot choose the type or location of the mutation.

si/shRNA

RNA interference (RNAi) using small interfering RNA (siRNA) or short hairpin RNA (shRNA) was a technological revolution of the 2000s. In concept a 20–22 nt double stranded RNA molecule, either supplied directly or generated by the Dicer-mediated cleavage of double stranded RNA or microRNA, can specifically target homologous sequences in endogenous mRNA and prevent post-transcriptional protein production [42]. The beauty of this system in studying tumorigenesis is that one can create various degrees of knockdown dependent on the amount of RNAi present, as well as the potential to have temporal or spatially restricted expression of the RNAi. While this RNAi technology was found very effective in multiple animal systems, it was less effective in zebrafish. However when incorporated into an endogenous zebrafish microRNA structure, like miR30, a number of labs were able to achieve as high as 75 % knockdown with certain sequences [42–44]. While enticing, a majority of target sequences derived less than 50 % knockdown which is not generally desirable. To our knowledge this technology has not been applied to cancer predisposition studies. However, partial knockdown of genes that are homozygous null lethal or in theory conditional knockdown would be an attractive application of this technology.

Engineered Endonuclease (EEN) Approaches

Undoubtedly, the most versatile and efficient approaches to modify the sequence of any predetermined target gene are the methods utilizing engineered endonucleases (EENs). EENs are designed to induce double-strand DNA breaks (DSBs) only at their user-specified genomic target sites. Thereafter, several types of genetic modifications may be generated simply by the work of the cellular DNA repair machineries. For example, the non-homologous end joining (NHEJ) mechanism can often result in a small random insertion or deletion (indel) at the cut site [45–53]. Since this is the predominant DSB repair mechanism during early embryonic development in zebrafish, EENs are mostly used to introduce frameshift or knockout mutations [54]. Nonetheless, EEN-induced DSBs can also facilitate the generation of many other types of genetic modifications, such as precise point mutations, targeted knock-in or even long-range chromosomal alterations. The use of EENs in those applications will be discussed later in this chapter.

To date, all three major types of EENs that have been applied to the mammalian cells have also been successfully implemented in the zebrafish system. These are zinc-finger nucleases (ZFNs), transcription activator-like effector nucleases (TALENs) and clustered regularly interspaced short palindromic repeats (CRISPR) RNA-guided Cas9 nucleases (CRISPR/Cas) [55–58]. In the following sections, we will discuss what they are, their efficiencies, how to make them, and how to choose an approach based on the user's need. Essentially, all EENs are introduced into 1-cell stage zebrafish embryos by microinjection of in vitro-transcribed, EEN-coding mRNAs. It is very common for a skilled person to inject hundreds of embryos in one morning. Thus, this is a fairly high-throughput process.

EEN: Zinc-Finger Nucleases (ZFNs)

ZFNs are the first class of EENs that can confer customizable DNA recognition specificities (Fig. 3a) [59]. These artificial restriction enzymes contain a tandem array of Cys2-His2 zinc finger motifs, one of the most common types of DNA recognition motifs utilized by the eukaryotes [60, 61]. While each finger motif binds to approximately 3 bps of DNA, an extended array of 3–6 finger motifs gives a ZFN the ability to recognize 9–18 bps of DNA [48, 51, 62]. In addition, ZFNs contain a nuclease domain of the *Fok*I enzyme. The *Fok*I nuclease domain cleaves DNA outside of the enzyme recognition sequence, so itself is sequence-independent and can

Fig. 3 (continued) The N-terminus of the TAL repeat domain (represented by *blue hexagons*) can recognize a "T" nucleotide. Both homodimeric and heterodimeric FokI nuclease domains have been used in TALENs. (**c**) CRISPR/Cas9 is a programmable RNA-guided nuclease system. The 5′ end of the guide RNA (represented by a *curvy blue line*) forms base-complementary interactions with its target DNA, while the 3′ end of the guide RNA interacts with the Cas9 nuclease (represented by a *blue bi-lobular shape*). Cas9 has two nuclease activities and can create a DSB at its target site. (**d**) Cas9 nickase is the Cas9 D10A mutant that can cut only one strand of DNA. Thus, paired nickases can be used to generate DSBs, which can be >1000-fold more specific than wild-type Cas9. (**e**) FokI-dCas9 is a dimeric RNA-guided *Fok*I nuclease. It is the fusion product of a *Fok*I nuclease domain and dCas9, a Cas9 D10A and H840A double mutant devoid of any DNA cleavage activity. Thus, FokI-dCas9 can bind to a DNA target via a guide RNA, but it cuts DNA using its *Fok*I nuclease domain. It has been shown that FokI-dCas9 has lower off-target mutation frequencies than Cas9 nickase

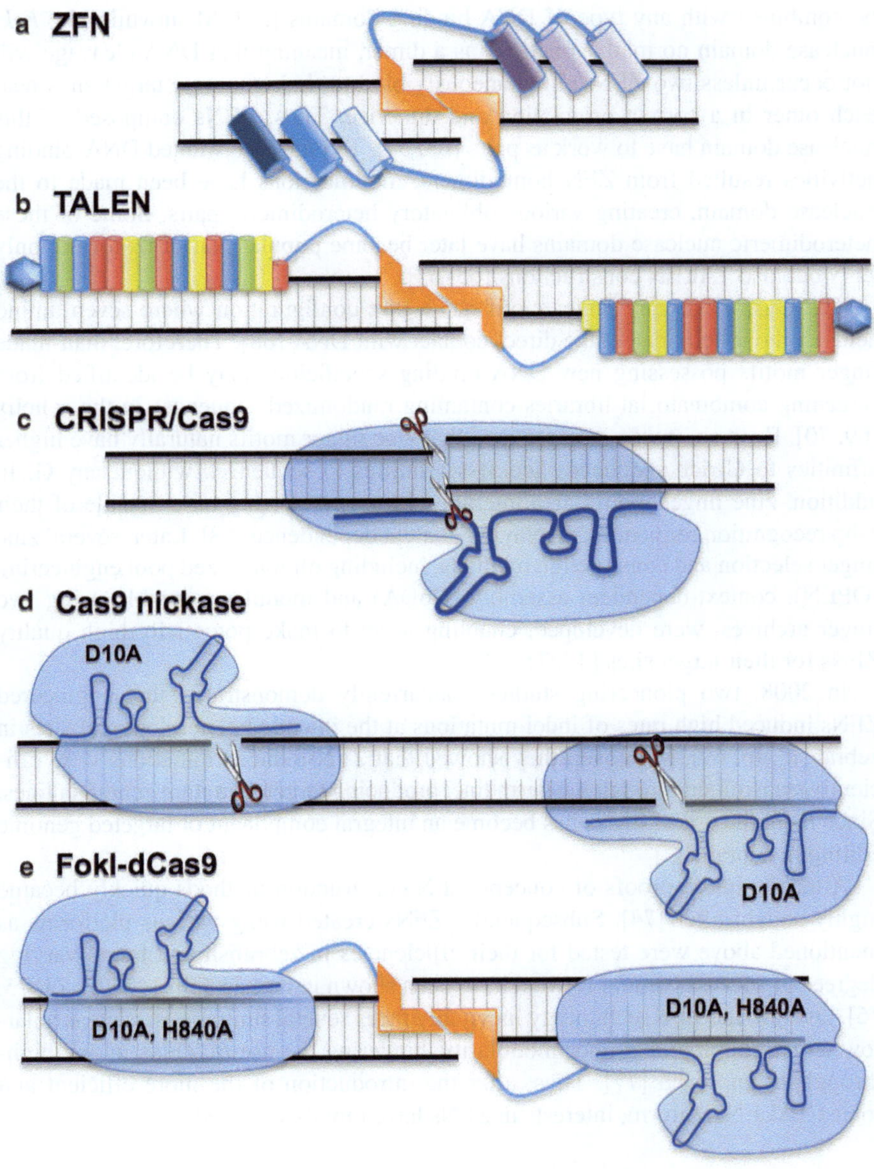

Fig. 3 Types of customizable engineered endonucleases (EENs). (**a**) Zinc finger nucleases (ZFNs) are EENs consisting of a zinc finger DNA-binding domain and a *Fok*I nuclease domain. The DNA-binding domain generally contains 3–6 zinc finger motifs (represented by *purple* and *blue cylinders*). The nuclease domains of a pair of ZFNs (represented by *orange right-angled shapes*) must dimerize before creating a double-strand DNA break (DSB). Although wild-type *Fok*I contains a homodimeric nuclease domain, obligatory heterodimeric *Fok*I nuclease pairs have also been developed. (**b**) Transcription activator-like effector nucleases (TALENs) are EENs consisting of a TAL repeat DNA-binding domain and a *Fok*I nuclease domain. Each TAL repeat (represented by a *red*, *yellow*, *green* or *blue cylinder*) binds to a specific DNA nucleotide, namely "A", "T", "C" or "G". A TALEN generally contains 15–17 TAL repeats where the last repeat is shorter than all the other ones.

be combined with any type of DNA-binding domains [59]. Meanwhile, the *Fok*I nuclease domain normally functions as a dimer, meaning that DNA cleavage will not occur unless two ZFNs simultaneously bind to their genomic target sites near each other in a correct orientation and dimerize. Thus, EENs composed of this nuclease domain have to work as pairs [63, 64]. To reduce unwanted DNA-binding activities resulted from ZFN homodimers, modifications have been made to the nuclease domain, creating various obligatory heterodimeric pairs. Some of these heterodimeric nuclease domains have later become popular platforms for not only ZFN but also TALEN constructions [65–67].

The Cys2-His2 zinc finger motif has a $\beta\beta\alpha$ configuration where seven amino acids in the α helix can make direct contact with DNA [68]. Therefore, man-made finger motifs possessing new DNA binding specificities may be identified from screening combinatorial libraries containing randomized sequences in the α helix [69, 70]. Even so, it was soon realized that zinc finger motifs naturally have higher affinities to G-rich sequences and less affinities to sequences without any G. In addition, zinc finger motifs also interact with a neighboring base outside of their 3-bp recognition sequences, displaying context dependence [68]. Later, several zinc finger selection and construction methods, including oligomerized pool engineering (OPEN), context-dependent assembly (CoDA) and modular assembly using two finger archives, were developed, enabling users to make potentially high quality ZFNs for their target sites [47, 71–73].

In 2008, two pioneering studies concurrently demonstrated that engineered ZFNs induced high rates of indel mutations at the intended genomic target sites in zebrafish [46, 49]. Moreover, they showed that ZFN-induced mutations were efficiently transmitted through the germ line, producing targeted mutant zebrafish lines. Since then, the use of EENs has become an integral component of targeted genome editing in zebrafish.

After the initial proofs of concept, ZFN construction methods quickly became highly sought-after [74]. Subsequently, ZFNs created using various platforms as mentioned above were tested for their efficiencies in zebrafish and found varying degrees of success, similar to what have been shown in human cells [47, 72, 73, 75, 76]. Overall, the ZFN technology inevitably faces several limitations, such as a narrow targeting range and poor modularity, hindering the use of ZFNs in any high-throughput approach [77]. Thus, after the introduction of the more efficient and robust TALEN platform, interests in ZFNs have rapidly declined.

EEN: Transcription Activator-Like Effector Nucleases (TALENs)

Transcription activator-like (TAL) effectors are the virulence proteins used by *Xanthomonas*, a genus of pathogenic bacteria in plants, to control host gene transcription. It was discovered in 2009 that these proteins had evolved a simple, modular DNA recognition system where each DNA-binding module, or TAL repeat, recognizes one DNA nucleotide. All TAL repeats consist of 32–35 almost identical amino acid sequences. However, two variable amino acids at position 12 and 13, termed repeat-variable diresidue (RVD), determine the DNA binding specificity [78, 79]. For example, TAL repeats that have RVD sequence NI, NG, HD and NN will bind to A, T, C and G, respectively. To date, customized TAL effector nucleases (TALENs) composed of a TAL repeat domain and a *Fok*I nuclease domain have been shown to induce high rates of mutations in their intended target sites in a wide variety of model systems including zebrafish (Fig. 3b) [80–84].

In 2011, Miller et al. identified an efficient TALEN framework based on TALE13 of *Xanthomonas axonopodis* [85]. Specifically, they found that shortening the TAL repeat domain to NΔ152/C63 significantly increased the nuclease activity. In 2011, using this basic framework, two studies independently demonstrated that modularly assembled TALENs could efficiently target endogenous zebrafish genes [83, 86]. The TALEN construction platform used by Sander et al. can be obtained through the nonprofit organization Addgene (https://www.addgene.org/talengineering/) [83]. Its associated protocols can be found on the TALengineering website (http://www.talengineering.org). Overall, this TALEN framework has shown effectiveness in *C. elegans*, zebrafish, rats, human somatic cells and human pluripotent stem cells [80, 82–85, 87, 88]. Other TALEN frameworks have also been shown to work well in plants, zebrafish, *Xenopus* and pigs, such as the ones originally developed by the Voytas group based on PthXo1 of *Xanthomonas oryzae* and later modified by other groups using a truncated TAL repeat domain (NΔ152/C63) [89–93]. TALEN construction kits for these platforms are also available through Addgene (https://www.addgene.org/talen/).

In addition, several studies have provided important insights into various components in a TALEN framework that may enhance its efficiencies. For example, it has been shown that TALENs with 15–17 TAL repeats can produce high success rates and efficiencies in human cells and in zebrafish, whereas TALENs with smaller numbers of TAL repeats are more likely to elicit cytotoxicity [80, 82, 83]. In addition, differences in the *Fok*I nuclease domain used in the TALEN framework may also affect its efficiency. For example, Cade et al. showed that replacing a homodimeric *Fok*I nuclease domain with a heterodimeric *Fok*I nuclease domain could in some cases increase TALEN activities [80]. On the other hand, it was found that the Sharkey version of *Fok*I nuclease domain severely reduced TALEN activities in zebrafish [94]. Moreover, in the original TALEN architecture described by Miller et al., which was made publicly available by Joung and colleagues through Addgene, non-RVD variations were incorporated in the TAL repeats to reduce DNA sequence repetition and increase DNA stability [85]. Interestingly, Sakuma et al. later found that non-RVD variations in the TAL repeats could indeed enhance TALEN activities [95].

The TALEN technology offers high efficiencies, success rates and can be applied on a large scale. TALEN construction may be completed within a few days to a couple of weeks manually or by automation. In addition, due to their robust activities, TALENs have facilitated the generation of the first targeted, precise sequence modifications in zebrafish via homology-mediated DNA repair mechanisms [89, 96]. These strategies and applications will be discussed later. Nonetheless, it is still costly and takes some efforts to engineer customized TALENs for each target site of interest. In this regard, the CRISPR/Cas technology is even cheaper, easier to implement, and may provide even more versatile utilities.

EEN: RNA-Guided Cas Nucleases

The identification of a customizable RNA-guided site-specific nuclease system is considered by many one of the greatest discoveries in the twenty-first century [97]. Its potential applications and impacts may extend beyond basic science, modern medicine, biotechnology and agriculture. A clustered regularly interspaced short palindromic repeats (CRISPR) RNA-Cas, which stands for CRISPR-associated protein, (CRISPR-Cas) system, can be found in approximately 40 % and 90 % of all sequenced bacterial and archaeal genomes, respectively. It is an adaptive defense mechanism for degrading alien DNAs. The CRISPR RNA (crRNA), composed of a variable "spacer" region and a short constant region, forms a complex with a trans-acting RNA (tracrRNA) and the Cas protein. Once the spacer of a crRNA binds to a foreign DNA due to sequence complementation, Cas is activated, resulting in a double-strand DNA break and eventually the destruction of the foreign DNA [98–100].

Among all CRISPR/Cas systems characterized to date, the DNA recognition system of Cas9 of *Streptococcus pyogenes* (SpCas9) seemed to be both simple and adaptable [100]. Thus, in 2012, Jinek et al. showed that artificial single guide RNA (sgRNA), which is a chimera of crRNA and tracrRNA, could be used to direct the SpCas9 nuclease activities toward user-specified target sequences in vitro [101]. After this first demonstration, five follow-up studies quickly established the successful use of engineered SpCas9 and sgRNA for targeted mutagenesis in human cells, bacteria and zebrafish [102–106]. Since then, customized sgRNA/Cas9 have also successfully targeted genes in mice, rats, *C. elegans*, *Drosophila*, pigs, goats, rice, *Arabdopsis* and yeast, bringing highly efficient and sometimes unprecedented genome editing capabilities to a wide variety of experimental model organisms, crops and livestock (Fig. 3c) [107–120].

Customizable sgRNA/Cas9 construction kits successfully used in various zebrafish studies can be obtained through Addene (https://www.addgene.org/CRISPR/cut/) under a separate section for "Zebrafish". It should be noted that human codon-optimized Cas9 (plasmid JDS246) has been found to be more effective than SpCas9 (plasmid MLM3616) in zebrafish [104, 121]. In zebrafish, engineered sgRNA/Cas9 can sometimes induce mutation rates reaching nearly 100 % [104, 122–124]. Meanwhile, the overall success rate is roughly 75–85 % based on two recently published large-scale evaluations, although about a quarter of them may have relatively low efficiencies (<5 %) [124, 125]. Interestingly, these two studies also found a similar trend suggesting that target sites with higher GC contents were more likely to have high efficiencies compared to target sites with lower GC contents, although a high GC content by itself did not guarantee a success in inducing mutations.

To generate high rates of mutations in a developing embryo or to create a knockout zebrafish line, both sgRNAs and the Cas9-coding mRNAs are often transcribed in vitro and co-injected. However, in vitro transcribed sgRNAs can also be mixed in with purified Cas9 protein and then co-injected [124]. Both of these methods will ensure ubiquitous expression of the sgRNA/Cas9 complex within an embryo. However, in some cases, when investigating the roles of tumor suppressor genes, local or tissue-specific knockout strategies may be particularly useful if a candidate gene happens to be essential during early embryonic development. Several strategies for creating conditional knockouts using sgRNA/Cas9 have been employed in murine models. For example, a sgRNA and Cas9 expression vector can be introduced locally and directly into somatic tissues via injection or viral particles [126, 127]. Alternatively, Cas9 can be integrated in an animal as a transgene and controlled by a conditional promoter [128]. Conceivably, similar strategies may also be applicable to zebrafish. For example, DNA or RNA may be introduced to zebrafish adults or larvae by in vivo electroporation or local injection [129–132]. Alternatively, zebrafish cells may be isolated, electroporated *ex vivo*, and then transplanted back into adult zebrafish [133, 134]. It should also be a fairly straightforward process to generate transgenic Cas9 zebrafish controlled by a tissue-specific or an inducible promoter. Although in zebrafish such an example has not been reported at the time this manuscript is prepared, we expect that various types of conditional approaches will soon become available to the zebrafish research community.

Choosing a EEN Platform Based on User's Need

Engineered TALENs and sgRNA/Cas9 can both be highly efficient, hence which method to choose between these two can be based on user's preference and other considerations, such as the target site frequency, the nuclease specificity and the need of targeting multiple genes at once.

First, TALENs are superior to sgRNA/Cas9 in terms of target site selection. A full target site for a TALEN pair should have a 5′-T and a 3′-A, but there can be a range of distances between them. It has been suggested that the theoretical targeting frequency based on random DNA sequences is three TALENs for every bp of DNA [56, 122]. On the other hand, the theoretical targeting frequency for sgRNA/Cas9 is one in every eight bps [122]. This is because a sgRNA target site has to be immediately followed by a sequence called protospacer adjacent motif (PAM). This sequence is specific to each Cas protein. For SpCas9, the PAM sequence is NGG, although it may also have some activities when PAM is NNGG or NNAG.

Second, regarding the nuclease specificities, some sgRNAs have reported high off-target activities, more so than TALENs have [135, 136]. However, several groups have come up with different strategies to mitigate this potential problem. For example, Cas9 nickases are Cas9 proteins that can cut only one but not both strands of DNA. Thus, only paired Cas9 nickases that successfully target sequences near each other and cleave two different strands of DNA can induce NHEJ and indel mutations (Fig. 3d) [137, 138]. In addition, leveraging the necessity of dimerization for *Fok*I nuclease activities, dCas9 (catalytically inactive Cas9)-FokI fusion proteins have also been created to reduce potential off-target effects from monomeric Cas9 (Fig. 3e) [139, 140]. Interestingly, it has also been shown that off-target activities can sometimes be reduced by 5000-fold or more simply by shortening the spacer region of a sgRNA from 20 to 17 bps [141]. Certainly, for germline mutations in zebrafish, a few outcrosses should quickly segregate intended mutations from any unlinked off-target mutations.

Third, sgRNA/Cas9 is far superior to TALENs in regards to multiplexing. Similar to what has been shown in human cells and mouse models, multiple sgRNAs targeting different genes can be co-injected into a zebrafish embryo, inducing mutations in all target genes effectively. Thus, combining with the impressive imaging power and the adaptability to high-throughput approaches of various zebrafish cancer models, a facile method for simultaneously targeting two or more candidate genes such as sgRNA/Cas9 promises to enable rapid assessment of cooperating mutations in cancer.

Applications of EENs in Addition to Inactivating Genes

Genome Engineering Using Homology-Dependent DNA Repair Mechanisms

Importantly, EENs can facilitate the generation of various types of genome modifications in addition to indel mutations (Fig. 4a). In mice, genome engineering using embryonic stem cells and homologous recombination (HR) has been the standard method for creating targeted mutations in the last two decades. However, in early zebrafish embryos, the basal level of HR seems to be too low to be useful for this purpose. Nonetheless, it has been shown that the rate of HR can be enhanced by EENs in human cells and several model organisms (Fig. 4b) [142–145]. The first successful example of gene targeting by HR in zebrafish generated a GFP knockin allele of *tyrosine hydroxylase* (*th*) using a TALEN pair targeting the *th* gene and a targeting vector containing two <1-kb DNA fragments homologous to the sequences flanking the genomic cut site [96]. Later, via a systematic evaluation of several different targeting vector configurations, Shin et al. proposed the guidelines for an efficient HR strategy in zebrafish, such as the optimal lengths of homologous arms [146]. Interestingly, the authors also found that linearization of a targeting vector, especially when it was linearized at the shorter homologous arm, could increase HR efficiencies. On the other hand, in another study, Irion et al. used a circular donor DNA containing two sgRNA target sites so that it can be linearized in vivo [147]. In both studies, around 10 % of the injected animals transmitted the modified alleles containing either a

Fig. 4 (continued) enhance the efficiencies of this approach. (**c**) A DSB can also promote small but predetermined sequence alterations in the presence of single stranded oligodeoxynucleotides (ssODN) containing ~20–30 base pairs of homologous arms to a target locus. In zebrafish, the end results of this approach can be precise but are sometimes a mix of predetermined sequence alterations and random indels near the DNA break. (**d**) Targeted integration in a homology-independent manner may be achieved at high rates at EEN target sites. Linearization of the donor construct in vivo by EENs (represented by a scissor cutting the plasmid DNA) can significantly enhance the efficiencies of this approach. (**e**) Long-range DNA deletions, inversions or even chromosomal translocations may be induced by EENs at medium to low efficiencies

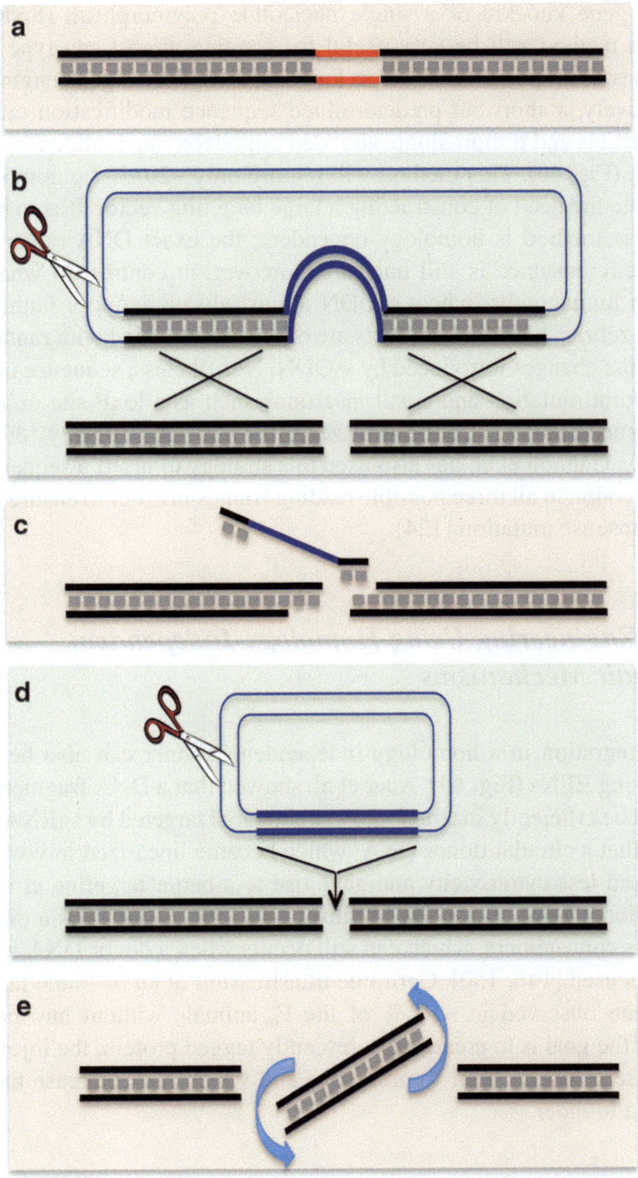

Fig. 4 Genome editing approaches enabled by engineered endonucleases (EENs). (**a–e**) *Red lines* indicate random sequence alterations. *Thick black lines with gray dashes* represent genomic DNA sequences. *Thick blue lines* represent pre-designed insertions, while *thin blue lines* represent vector DNA. (**a**) Random insertions or deletions (indels) may be introduced at an EEN-induced double-strand DNA break (DSB) via an error-prone DNA repair mechanism called non-homology end joining (NHEJ). (**b**) A DSB can promote targeted sequence exchange with a DNA donor construct via homologous recombination. This method can generate long insertions/deletions and other sequence alterations in a predetermined manner. Some data suggest that linearization of the donor construct in vivo by EENs (represented by a scissor cutting the plasmid DNA) can further

fluorescent gene knockin or a single nucleotide polymorphism (SNP) exchange. Indeed, this method will be very useful for creating almost any type of sequence modifications, from point mutations to long and complicated gene engineering.

Alternatively, a short but predetermined sequence modification may be generated using EENs and a short single stranded oligodeoxynucleotides (ssODN) as the donor DNA (Fig. 4c). This method will require only ~20–30 bp homologous arms and eliminate the need of constructing a large targeting vector. It is to be noted that although this method is homology-dependent, the exact DNA repair mechanism under this circumstance is still unclear. Moreover, in contrast to what have been observed in human cells, where ssODN almost always induces faithful sequence changes, in zebrafish the end products are often accompanied with random indels in addition to the changes introduced by ssODN. Nonetheless, sequence modifications such as a point mutation and small insertions such as a loxP site or a protein tag have been successfully introduced to zebrafish with this strategy [89, 122, 148]. Interestingly, Gagnon et al. has also used this strategy to insert a sequence that contains a stop codon in all three possible reading frames in order to ensure the outcome will be a nonsense mutation [124].

Genome Engineering Using Homology-Independent DNA Repair Mechanisms

Targeted integration in a homology-independent manner can also be achieved in zebrafish using EENs (Fig. 4d). Auer et al. showed that a DNA fragment as large as >5 kb could be efficiently inserted into genomic loci targeted by sgRNA/Cas9. They also found that a circular donor DNA, which became linearized in vivo by sgRNA/Cas9, induced less cytotoxicity and gave rise to a better targeting efficiency compared to a donor DNA linearized in vitro [149, 150]. Sometimes a donor DNA is integrated as concatemers, which can still occurs when a donor DNA with homologous arms is used [146, 150]. Germline transmission of an in-frame targeted insertion has been observed in ~10 % of the F_0 animals without any pre-selection. However, if the goal is to create a fluorescently tagged protein, the injected animals can be pre-screened by their fluorescence and which may increase the chance of identifying a founder.

Genome Engineering of Long-Range DNA Deletion or Inversion

Long-range DNA deletions or inversions in zebrafish may be generated using two pairs of ZFNs, TALENs or multiplexing sgRNA/Cas9 (Fig. 4e) [151, 152]. Deletions less than 70 kb can be generated efficiently, for which germline transmission has been observed in 1–15 % of the F_0 animals. Inversions and larger deletions up to several megabase pairs are less frequent [151]. Using EENs to induce a long-range DNA deletion will be very useful when investigating the functions of tandemly duplicated genes and non-coding RNAs, as well as the generation of balancers alleles for stock maintenance and genetic screens.

Summary

Zebrafish has proven to be an advantageous model organism for cancer studies for various reasons. To name a few, zebrafish require less space, providing more statistical power to cohort tumor analysis, and are more economical compared to mice. In addition, zebrafish cancer models may enable high-throughput experimentation, chemical screening and facile real-time live imaging analyses that cannot be done as easily if using mouse models. While oncogenic transformation can be modeled through transgenesis, developing cancer models of tumor suppressor genes requires genomic alterations of the endogenous genes. Historically reverse genetic has been technically challenging in zebrafish. Nevertheless, forward genetic screens have identified various zebrafish cancer models of known human tumor suppressor genes and have shed light on the roles of cancer-causing mutations that were not known previously. Currently, TILLING in zebrafish has generated an invaluable, publicly available collection of zebrafish mutants in nearly half of the protein-coding genes. Using the approaches discussed in this chapter, conditional knockouts and targeted mutations in any gene-of-interest may be generated in zebrafish. Due to the high efficiencies of the CRISPR/Cas system, and ease in generating complex genetic lines, collaborating mutations in cancer may be dissected rapidly in zebrafish. In the future, dCas9 tethered to other functional protein domains may help understand the roles of epigenetic regulation in cancer. In closing, the wealth of genetic and cancer focused tools, advanced technologies, and applied chemical biology has pushed cancer research using the zebrafish model system to the forefront, and will continue to facilitate the fight against cancer.

Acknowledgments This work is supported by the National Institutes of Health (J.-R. J. Yeh, GM088040 and CA140188; J. M. Parant, NS092105) and the Hassenfeld Scholar's Award (J.-R. J. Yeh).

References

1. Mullins MC, Hammerschmidt M, Haffter P, Nusslein-Volhard C (1994) Large-scale mutagenesis in the zebrafish: in search of genes controlling development in a vertebrate. Curr Biol 4:189–202
2. Kroeger PT Jr, Poureetezadi SJ, McKee R, Jou J, Miceli R, Wingert RA (2014) Production of haploid zebrafish embryos by in vitro fertilization. J Vis Exp
3. Walker C (1999) Haploid screens and gamma-ray mutagenesis. Methods Cell Biol 60:43–70

4. Streisinger G, Singer F, Walker C, Knauber D, Dower N (1986) Segregation analyses and gene-centromere distances in zebrafish. Genetics 112:311–319
5. Shepard JL, Amatruda JF, Stern HM, Subramanian A, Finkelstein D, Ziai J, Finley KR, Pfaff KL, Hersey C, Zhou Y et al (2005) A zebrafish bmyb mutation causes genome instability and increased cancer susceptibility. Proc Natl Acad Sci U S A 102:13194–13199
6. Shepard JL, Amatruda JF, Finkelstein D, Ziai J, Finley KR, Stern HM, Chiang K, Hersey C, Barut B, Freeman JL et al (2007) A mutation in separase causes genome instability and increased susceptibility to epithelial cancer. Genes Dev 21:55–59
7. Parant JM, George SA, Holden JA, Yost HJ (2010) Genetic modeling of Li-Fraumeni syndrome in zebrafish. Dis Model Mech 3:45–56
8. Moore JL, Rush LM, Breneman C, Mohideen MA, Cheng KC (2006) Zebrafish genomic instability mutants and cancer susceptibility. Genetics 174:585–600
9. Neumann JC, Chandler GL, Damoulis VA, Fustino NJ, Lillard K, Looijenga L, Margraf L, Rakheja D, Amatruda JF (2011) Mutation in the type IB bone morphogenetic protein receptor Alk6b impairs germ-cell differentiation and causes germ-cell tumors in zebrafish. Proc Natl Acad Sci U S A 108:13153–13158
10. Gaiano N, Amsterdam A, Kawakami K, Allende M, Becker T, Hopkins N (1996) Insertional mutagenesis and rapid cloning of essential genes in zebrafish. Nature 383:829–832
11. Amsterdam A, Nissen RM, Sun Z, Swindell EC, Farrington S, Hopkins N (2004) Identification of 315 genes essential for early zebrafish development. Proc Natl Acad Sci U S A 101:12792–12797
12. Wang D, Jao LE, Zheng N, Dolan K, Ivey J, Zonies S, Wu X, Wu K, Yang H, Meng Q et al (2007) Efficient genome-wide mutagenesis of zebrafish genes by retroviral insertions. Proc Natl Acad Sci U S A 104:12428–12433
13. Varshney GK, Lu J, Gildea DE, Huang H, Pei W, Yang Z, Huang SC, Schoenfeld D, Pho NH, Casero D et al (2013) A large-scale zebrafish gene knockout resource for the genome-wide study of gene function. Genome Res 23:727–735
14. Varshney GK, Huang H, Zhang S, Lu J, Gildea DE, Yang Z, Wolfsberg TG, Lin S, Burgess SM (2013) The Zebrafish Insertion Collection (ZInC): a web based, searchable collection of zebrafish mutations generated by DNA insertion. Nucleic Acids Res 41:D861–D864
15. Teng T, Thomas G, Mercer CA (2013) Growth control and ribosomopathies. Curr Opin Genet Dev 23:63–71
16. Kawakami K, Shima A, Kawakami N (2000) Identification of a functional transposase of the Tol2 element, an Ac-like element from the Japanese medaka fish, and its transposition in the zebrafish germ lineage. Proc Natl Acad Sci U S A 97:11403–11408
17. Davidson AE, Balciunas D, Mohn D, Shaffer J, Hermanson S, Sivasubbu S, Cliff MP, Hackett PB, Ekker SC (2003) Efficient gene delivery and gene expression in zebrafish using the Sleeping Beauty transposon. Dev Biol 263:191–202
18. Kawakami K, Takeda H, Kawakami N, Kobayashi M, Matsuda N, Mishina M (2004) A transposon-mediated gene trap approach identifies developmentally regulated genes in zebrafish. Dev Cell 7:133–144
19. Balciunas D, Ekker SC (2005) Trapping fish genes with transposons. Zebrafish 1:335–341
20. Sivasubbu S, Balciunas D, Davidson AE, Pickart MA, Hermanson SB, Wangensteen KJ, Wolbrink DC, Ekker SC (2006) Gene-breaking transposon mutagenesis reveals an essential role for histone H2afza in zebrafish larval development. Mech Dev 123:513–529
21. Clark KJ, Balciunas D, Pogoda HM, Ding Y, Westcot SE, Bedell VM, Greenwood TM, Urban MD, Skuster KJ, Petzold AM et al (2011) In vivo protein trapping produces a functional expression codex of the vertebrate proteome. Nat Methods 8:506–515
22. Ni TT, Lu J, Zhu M, Maddison LA, Boyd KL, Huskey L, Ju B, Hesselson D, Zhong TP, Page-McCaw PS et al (2012) Conditional control of gene function by an invertible gene trap in zebrafish. Proc Natl Acad Sci U S A 109:15389–15394
23. Petzold AM, Balciunas D, Sivasubbu S, Clark KJ, Bedell VM, Westcot SE, Myers SR, Moulder GL, Thomas MJ, Ekker SC (2009) Nicotine response genetics in the zebrafish. Proc Natl Acad Sci U S A 106:18662–18667

24. McCallum CM, Comai L, Greene EA, Henikoff S (2000) Targeting induced local lesions IN genomes (TILLING) for plant functional genomics. Plant Physiol 123:439–442
25. Wienholds E, van Eeden F, Kosters M, Mudde J, Plasterk RH, Cuppen E (2003) Efficient target-selected mutagenesis in zebrafish. Genome Res 13:2700–2707
26. Wienholds E, Schulte-Merker S, Walderich B, Plasterk RH (2002) Target-selected inactivation of the zebrafish rag1 gene. Science 297:99–102
27. Colbert T, Till BJ, Tompa R, Reynolds S, Steine MN, Yeung AT, McCallum CM, Comai L, Henikoff S (2001) High-throughput screening for induced point mutations. Plant Physiol 126:480–484
28. Kettleborough RN, Busch-Nentwich EM, Harvey SA, Dooley CM, de Bruijn E, van Eeden F, Sealy I, White RJ, Herd C, Nijman IJ et al (2013) A systematic genome-wide analysis of zebrafish protein-coding gene function. Nature 496:494–497
29. Berghmans S, Murphey RD, Wienholds E, Neuberg D, Kutok JL, Fletcher CD, Morris JP, Liu TX, Schulte-Merker S, Kanki JP et al (2005) tp53 mutant zebrafish develop malignant peripheral nerve sheath tumors. Proc Natl Acad Sci U S A 102:407–412
30. Haramis AP, Hurlstone A, van der Velden Y, Begthel H, van den Born M, Offerhaus GJ, Clevers HC (2006) Adenomatous polyposis coli-deficient zebrafish are susceptible to digestive tract neoplasia. EMBO Rep 7:444–449
31. Wienholds E, Koudijs MJ, van Eeden FJ, Cuppen E, Plasterk RH (2003) The microRNA-producing enzyme Dicer1 is essential for zebrafish development. Nat Genet 35:217–218
32. Stambolic V, Suzuki A, de la Pompa JL, Brothers GM, Mirtsos C, Sasaki T, Ruland J, Penninger JM, Siderovski DP, Mak TW (1998) Negative regulation of PKB/Akt-dependent cell survival by the tumor suppressor PTEN. Cell 95:29–39
33. Di Cristofano A, Pesce B, Cordon-Cardo C, Pandolfi PP (1998) Pten is essential for embryonic development and tumour suppression. Nat Genet 19:348–355
34. Faucherre A, Taylor GS, Overvoorde J, Dixon JE, Hertog J (2008) Zebrafish pten genes have overlapping and non-redundant functions in tumorigenesis and embryonic development. Oncogene 27:1079–1086
35. Berends MJ, Wu Y, Sijmons RH, Mensink RG, van der Sluis T, Hordijk-Hos JM, de Vries EG, Hollema H, Karrenbeld A, Buys CH et al (2002) Molecular and clinical characteristics of MSH6 variants: an analysis of 25 index carriers of a germline variant. Am J Hum Genet 70:26–37
36. Akiyama Y, Sato H, Yamada T, Nagasaki H, Tsuchiya A, Abe R, Yuasa Y (1997) Germ-line mutation of the hMSH6/GTBP gene in an atypical hereditary nonpolyposis colorectal cancer kindred. Cancer Res 57:3920–3923
37. Papadopoulos N, Nicolaides NC, Wei YF, Ruben SM, Carter KC, Rosen CA, Haseltine WA, Fleischmann RD, Fraser CM, Adams MD et al (1994) Mutation of a mutL homolog in hereditary colon cancer. Science 263:1625–1629
38. Bronner CE, Baker SM, Morrison PT, Warren G, Smith LG, Lescoe MK, Kane M, Earabino C, Lipford J, Lindblom A et al (1994) Mutation in the DNA mismatch repair gene homologue hMLH1 is associated with hereditary non-polyposis colon cancer. Nature 368:258–261
39. Leach FS, Nicolaides NC, Papadopoulos N, Liu B, Jen J, Parsons R, Peltomaki P, Sistonen P, Aaltonen LA, Nystrom-Lahti M et al (1993) Mutations of a mutS homolog in hereditary nonpolyposis colorectal cancer. Cell 75:1215–1225
40. Fishel R, Lescoe MK, Rao MR, Copeland NG, Jenkins NA, Garber J, Kane M, Kolodner R (1993) The human mutator gene homolog MSH2 and its association with hereditary nonpolyposis colon cancer. Cell 75:1027–1038
41. Feitsma H, Kuiper RV, Korving J, Nijman IJ, Cuppen E (2008) Zebrafish with mutations in mismatch repair genes develop neurofibromas and other tumors. Cancer Res 68:5059–5066
42. Kelly A, Hurlstone AF (2011) The use of RNAi technologies for gene knockdown in zebrafish. Brief Funct Genomics 10:189–196
43. Dong M, Fu YF, Du TT, Jing CB, Fu CT, Chen Y, Jin Y, Deng M, Liu TX (2009) Heritable and lineage-specific gene knockdown in zebrafish embryo. PLoS One 4, e6125
44. De Rienzo G, Gutzman JH, Sive H (2012) Efficient shRNA-mediated inhibition of gene expression in zebrafish. Zebrafish 9:97–107

45. Beumer KJ, Trautman JK, Bozas A, Liu JL, Rutter J, Gall JG, Carroll D (2008) Efficient gene targeting in Drosophila by direct embryo injection with zinc-finger nucleases. Proc Natl Acad Sci U S A 105:19821–19826
46. Doyon Y, McCammon JM, Miller JC, Faraji F, Ngo C, Katibah GE, Amora R, Hocking TD, Zhang L, Rebar EJ et al (2008) Heritable targeted gene disruption in zebrafish using designed zinc-finger nucleases. Nat Biotechnol 26:702–708
47. Foley JE, Yeh JR, Maeder ML, Reyon D, Sander JD, Peterson RT, Joung JK (2009) Rapid mutation of endogenous zebrafish genes using zinc finger nucleases made by Oligomerized Pool ENgineering (OPEN). PLoS One 4, e4348
48. Maeder ML, Thibodeau-Beganny S, Osiak A, Wright DA, Anthony RM, Eichtinger M, Jiang T, Foley JE, Winfrey RJ, Townsend JA et al (2008) Rapid "open-source" engineering of customized zinc-finger nucleases for highly efficient gene modification. Mol Cell 31:294–301
49. Meng X, Noyes MB, Zhu LJ, Lawson ND, Wolfe SA (2008) Targeted gene inactivation in zebrafish using engineered zinc-finger nucleases. Nat Biotechnol 26:695–701
50. Perez EE, Wang J, Miller JC, Jouvenot Y, Kim KA, Liu O, Wang N, Lee G, Bartsevich VV, Lee YL et al (2008) Establishment of HIV-1 resistance in CD4+ T cells by genome editing using zinc-finger nucleases. Nat Biotechnol 26:808–816
51. Santiago Y, Chan E, Liu PQ, Orlando S, Zhang L, Urnov FD, Holmes MC, Guschin D, Waite A, Miller JC et al (2008) Targeted gene knockout in mammalian cells by using engineered zinc-finger nucleases. Proc Natl Acad Sci U S A 105:5809–5814
52. Moehle EA, Rock JM, Lee YL, Jouvenot Y, DeKelver RC, Gregory PD, Urnov FD, Holmes MC (2007) Targeted gene addition into a specified location in the human genome using designed zinc finger nucleases. Proc Natl Acad Sci U S A 104:3055–3060
53. Lombardo A, Genovese P, Beausejour CM, Colleoni S, Lee YL, Kim KA, Ando D, Urnov FD, Galli C, Gregory PD et al (2007) Gene editing in human stem cells using zinc finger nucleases and integrase-defective lentiviral vector delivery. Nat Biotechnol 25:1298–1306
54. Hagmann M, Bruggmann R, Xue L, Georgiev O, Schaffner W, Rungger D, Spaniol P, Gerster T (1998) Homologous recombination and DNA-end joining reactions in zygotes and early embryos of zebrafish (Danio rerio) and Drosophila melanogaster. Biol Chem 379: 673–681
55. Gaj T, Gersbach CA, Barbas CF 3rd (2013) ZFN, TALEN, and CRISPR/Cas-based methods for genome engineering. Trends Biotechnol 31:397–405
56. Joung JK, Sander JD (2013) TALENs: a widely applicable technology for targeted genome editing. Nat Rev Mol Cell Biol 14:49–55
57. Mali P, Esvelt KM, Church GM (2013) Cas9 as a versatile tool for engineering biology. Nat Methods 10:957–963
58. Urnov FD, Rebar EJ, Holmes MC, Zhang HS, Gregory PD (2010) Genome editing with engineered zinc finger nucleases. Nat Rev Genet 11:636–646
59. Kim YG, Cha J, Chandrasegaran S (1996) Hybrid restriction enzymes: zinc finger fusions to Fok I cleavage domain. Proc Natl Acad Sci U S A 93:1156–1160
60. Miller J, McLachlan AD, Klug A (1985) Repetitive zinc-binding domains in the protein transcription factor IIIA from Xenopus oocytes. EMBO J 4:1609–1614
61. Pavletich NP, Pabo CO (1991) Zinc finger-DNA recognition: crystal structure of a Zif268-DNA complex at 2.1 A. Science 252:809–817
62. Bibikova M, Golic M, Golic KG, Carroll D (2002) Targeted chromosomal cleavage and mutagenesis in Drosophila using zinc-finger nucleases. Genetics 161:1169–1175
63. Smith J, Bibikova M, Whitby FG, Reddy AR, Chandrasegaran S, Carroll D (2000) Requirements for double-strand cleavage by chimeric restriction enzymes with zinc finger DNA-recognition domains. Nucleic Acids Res 28:3361–3369
64. Mani M, Smith J, Kandavelou K, Berg JM, Chandrasegaran S (2005) Binding of two zinc finger nuclease monomers to two specific sites is required for effective double-strand DNA cleavage. Biochem Biophys Res Commun 334:1191–1197
65. Doyon Y, Vo TD, Mendel MC, Greenberg SG, Wang J, Xia DF, Miller JC, Urnov FD, Gregory PD, Holmes MC (2011) Enhancing zinc-finger-nuclease activity with improved obligate heterodimeric architectures. Nat Methods 8:74–79

66. Szczepek M, Brondani V, Buchel J, Serrano L, Segal DJ, Cathomen T (2007) Structure-based redesign of the dimerization interface reduces the toxicity of zinc-finger nucleases. Nat Biotechnol 25:786–793
67. Guo J, Gaj T, Barbas CF 3rd (2010) Directed evolution of an enhanced and highly efficient FokI cleavage domain for zinc finger nucleases. J Mol Biol 400:96–107
68. Dreier B, Segal DJ, Barbas CF 3rd (2000) Insights into the molecular recognition of the 5′-GNN-3′ family of DNA sequences by zinc finger domains. J Mol Biol 303:489–502
69. Rebar EJ, Pabo CO (1994) Zinc finger phage: affinity selection of fingers with new DNA-binding specificities. Science 263:671–673
70. Jamieson AC, Kim SH, Wells JA (1994) In vitro selection of zinc fingers with altered DNA-binding specificity. Biochemistry 33:5689–5695
71. Maeder ML, Thibodeau-Beganny S, Sander JD, Voytas DF, Joung JK (2009) Oligomerized pool engineering (OPEN): an 'open-source' protocol for making customized zinc-finger arrays. Nat Protoc 4:1471–1501
72. Sander JD, Dahlborg EJ, Goodwin MJ, Cade L, Zhang F, Cifuentes D, Curtin SJ, Blackburn JS, Thibodeau-Beganny S, Qi Y et al (2011) Selection-free zinc-finger-nuclease engineering by context-dependent assembly (CoDA). Nat Methods 8:67–69
73. Gupta A, Christensen RG, Rayla AL, Lakshmanan A, Stormo GD, Wolfe SA (2012) An optimized two-finger archive for ZFN-mediated gene targeting. Nat Methods 9:588–590
74. Sander JD, Yeh JR, Peterson RT, Joung JK (2011) Engineering zinc finger nucleases for targeted mutagenesis of zebrafish. Methods Cell Biol 104:51–58
75. Chen S, Oikonomou G, Chiu CN, Niles BJ, Liu J, Lee DA, Antoshechkin I, Prober DA (2013) A large-scale in vivo analysis reveals that TALENs are significantly more mutagenic than ZFNs generated using context-dependent assembly. Nucleic Acids Res 41:2769–2778
76. Gupta A, Meng X, Zhu LJ, Lawson ND, Wolfe SA (2011) Zinc finger protein-dependent and -independent contributions to the in vivo off-target activity of zinc finger nucleases. Nucleic Acids Res 39:381–392
77. Ramirez CL, Foley JE, Wright DA, Muller-Lerch F, Rahman SH, Cornu TI, Winfrey RJ, Sander JD, Fu F, Townsend JA et al (2008) Unexpected failure rates for modular assembly of engineered zinc fingers. Nat Methods 5:374–375
78. Boch J, Scholze H, Schornack S, Landgraf A, Hahn S, Kay S, Lahaye T, Nickstadt A, Bonas U (2009) Breaking the code of DNA binding specificity of TAL-type III effectors. Science 326:1509–1512
79. Moscou MJ, Bogdanove AJ (2009) A simple cipher governs DNA recognition by TAL effectors. Science 326:1501
80. Cade L, Reyon D, Hwang WY, Tsai SQ, Patel S, Khayter C, Joung JK, Sander JD, Peterson RT, Yeh JR (2012) Highly efficient generation of heritable zebrafish gene mutations using homo- and heterodimeric TALENs. Nucleic Acids Res 40:8001–8010
81. Kandavelou K, Chandrasegaran S (2009) Custom-designed molecular scissors for site-specific manipulation of the plant and mammalian genomes. Methods Mol Biol 544:617–636
82. Reyon D, Tsai SQ, Khayter C, Foden JA, Sander JD, Joung JK (2012) FLASH assembly of TALENs for high-throughput genome editing. Nat Biotechnol 30:460–465
83. Sander JD, Cade L, Khayter C, Reyon D, Peterson RT, Joung JK, Yeh JR (2011) Targeted gene disruption in somatic zebrafish cells using engineered TALENs. Nat Biotechnol 29:697–698
84. Tesson L, Usal C, Menoret S, Leung E, Niles BJ, Remy S, Santiago Y, Vincent AI, Meng X, Zhang L et al (2011) Knockout rats generated by embryo microinjection of TALENs. Nat Biotechnol 29:695–696
85. Miller JC, Tan S, Qiao G, Barlow KA, Wang J, Xia DF, Meng X, Paschon DE, Leung E, Hinkley SJ et al (2011) A TALE nuclease architecture for efficient genome editing. Nat Biotechnol 29:143–148
86. Huang P, Xiao A, Zhou M, Zhu Z, Lin S, Zhang B (2011) Heritable gene targeting in zebrafish using customized TALENs. Nat Biotechnol 29:699–700
87. Hockemeyer D, Wang H, Kiani S, Lai CS, Gao Q, Cassady JP, Cost GJ, Zhang L, Santiago Y, Miller JC et al (2011) Genetic engineering of human pluripotent cells using TALE nucleases. Nat Biotechnol 29:731–734

88. Wood AJ, Lo TW, Zeitler B, Pickle CS, Ralston EJ, Lee AH, Amora R, Miller JC, Leung E, Meng X et al (2011) Targeted genome editing across species using ZFNs and TALENs. Science 333:307
89. Bedell VM, Wang Y, Campbell JM, Poshusta TL, Starker CG, Krug RG 2nd, Tan W, Penheiter SG, Ma AC, Leung AY et al (2012) In vivo genome editing using a high-efficiency TALEN system. Nature 491:114–118
90. Cermak T, Doyle EL, Christian M, Wang L, Zhang Y, Schmidt C, Baller JA, Somia NV, Bogdanove AJ, Voytas DF (2011) Efficient design and assembly of custom TALEN and other TAL effector-based constructs for DNA targeting. Nucleic Acids Res 39, e82
91. Dahlem TJ, Hoshijima K, Jurynec MJ, Gunther D, Starker CG, Locke AS, Weis AM, Voytas DF, Grunwald DJ (2012) Simple methods for generating and detecting locus-specific mutations induced with TALENs in the zebrafish genome. PLoS Genet 8, e1002861
92. Carlson DF, Tan W, Lillico SG, Stverakova D, Proudfoot C, Christian M, Voytas DF, Long CR, Whitelaw CB, Fahrenkrug SC (2012) Efficient TALEN-mediated gene knockout in livestock. Proc Natl Acad Sci U S A 109:17382–17387
93. Lei Y, Guo X, Liu Y, Cao Y, Deng Y, Chen X, Cheng CH, Dawid IB, Chen Y, Zhao H (2012) Efficient targeted gene disruption in Xenopus embryos using engineered transcription activator-like effector nucleases (TALENs). Proc Natl Acad Sci U S A 109:17484–17489
94. Pillay LM, Selland LG, Fleisch VC, Leighton PL, Cheng CS, Famulski JK, Ritzel RG, March LD, Wang H, Allison WT et al (2013) Evaluating the mutagenic activity of targeted endonucleases containing a Sharkey FokI cleavage domain variant in zebrafish. Zebrafish 10:353–364
95. Sakuma T, Ochiai H, Kaneko T, Mashimo T, Tokumasu D, Sakane Y, Suzuki K, Miyamoto T, Sakamoto N, Matsuura S et al (2013) Repeating pattern of non-RVD variations in DNA-binding modules enhances TALEN activity. Sci Rep 3:3379
96. Zu Y, Tong X, Wang Z, Liu D, Pan R, Li Z, Hu Y, Luo Z, Huang P, Wu Q et al (2013) TALEN-mediated precise genome modification by homologous recombination in zebrafish. Nat Method
97. Pennisi E (2013) The CRISPR craze. Science 341:833–836
98. Horvath P, Barrangou R (2010) CRISPR/Cas, the immune system of bacteria and archaea. Science 327:167–170
99. Terns MP, Terns RM (2011) CRISPR-based adaptive immune systems. Curr Opin Microbiol 14:321–327
100. Wiedenheft B, Sternberg SH, Doudna JA (2012) RNA-guided genetic silencing systems in bacteria and archaea. Nature 482:331–338
101. Jinek M, Chylinski K, Fonfara I, Hauer M, Doudna JA, Charpentier E (2012) A programmable dual-RNA-guided DNA endonuclease in adaptive bacterial immunity. Science 337:816–821
102. Cho SW, Kim S, Kim JM, Kim JS (2013) Targeted genome engineering in human cells with the Cas9 RNA-guided endonuclease. Nat Biotechnol 31:230–232
103. Cong L, Ran FA, Cox D, Lin S, Barretto R, Habib N, Hsu PD, Wu X, Jiang W, Marraffini LA et al (2013) Multiplex genome engineering using CRISPR/Cas systems. Science 339:819–823
104. Hwang WY, Fu Y, Reyon D, Maeder ML, Tsai SQ, Sander JD, Peterson RT, Yeh JR, Joung JK (2013) Efficient genome editing in zebrafish using a CRISPR-Cas system. Nat Biotechnol 31:227–229
105. Jiang W, Bikard D, Cox D, Zhang F, Marraffini LA (2013) RNA-guided editing of bacterial genomes using CRISPR-Cas systems. Nat Biotechnol 31:233–239
106. Mali P, Yang L, Esvelt KM, Aach J, Guell M, DiCarlo JE, Norville JE, Church GM (2013) RNA-guided human genome engineering via Cas9. Science 339:823–826
107. Dicarlo JE, Norville JE, Mali P, Rios X, Aach J, Church GM (2013) Genome engineering in Saccharomyces cerevisiae using CRISPR-Cas systems. Nucleic Acids Res 41:4336–4343
108. Shan Q, Wang Y, Li J, Zhang Y, Chen K, Liang Z, Zhang K, Liu J, Xi JJ, Qiu JL et al (2013) Targeted genome modification of crop plants using a CRISPR-Cas system. Nat Biotechnol 31:686–688
109. Jiang W, Zhou H, Bi H, Fromm M, Yang B, Weeks DP (2013) Demonstration of CRISPR/Cas9/sgRNA-mediated targeted gene modification in Arabidopsis, tobacco, sorghum and rice. Nucleic Acids Res 41, e188

110. Wang H, Yang H, Shivalila CS, Dawlaty MM, Cheng AW, Zhang F, Jaenisch R (2013) One-step generation of mice carrying mutations in multiple genes by CRISPR/Cas-mediated genome engineering. Cell 153:910–918
111. Shen B, Zhang J, Wu H, Wang J, Ma K, Li Z, Zhang X, Zhang P, Huang X (2013) Generation of gene-modified mice via Cas9/RNA-mediated gene targeting. Cell Res 23:720–723
112. Li D, Qiu Z, Shao Y, Chen Y, Guan Y, Liu M, Li Y, Gao N, Wang L, Lu X et al (2013) Heritable gene targeting in the mouse and rat using a CRISPR-Cas system. Nat Biotechnol 31:681–683
113. Nakayama T, Fish MB, Fisher M, Oomen-Hajagos J, Thomsen GH, Grainger RM (2013) Simple and efficient CRISPR/Cas9-mediated targeted mutagenesis in Xenopus tropicalis. Genesis 51:835–843
114. Yu Z, Ren M, Wang Z, Zhang B, Rong YS, Jiao R, Gao G (2013) Highly efficient genome modifications mediated by CRISPR/Cas9 in Drosophila. Genetics 195:289–291
115. Bassett AR, Tibbit C, Ponting CP, Liu JL (2013) Highly efficient targeted mutagenesis of Drosophila with the CRISPR/Cas9 system. Cell Rep 4:220–228
116. Li W, Teng F, Li T, Zhou Q (2013) Simultaneous generation and germline transmission of multiple gene mutations in rat using CRISPR-Cas systems. Nat Biotechnol 31:684–686
117. Tan W, Carlson DF, Lancto CA, Garbe JR, Webster DA, Hackett PB, Fahrenkrug SC (2013) Efficient nonmeiotic allele introgression in livestock using custom endonucleases. Proc Natl Acad Sci U S A 110:16526–16531
118. Ni W, Qiao J, Hu S, Zhao X, Regouski M, Yang M, Polejaeva IA, Chen C (2014) Efficient gene knockout in goats using CRISPR/Cas9 system. PLoS One 9, e106718
119. Hai T, Teng F, Guo R, Li W, Zhou Q (2014) One-step generation of knockout pigs by zygote injection of CRISPR/Cas system. Cell Res 24:372–375
120. Jacobs JZ, Ciccaglione KM, Tournier V, Zaratiegui M (2014) Implementation of the CRISPR-Cas9 system in fission yeast. Nat Commun 5:5344
121. Gonzales AP, Yeh JR (2014) Cas9-based genome editing in zebrafish. Methods Enzymol 546:377–413
122. Hwang WY, Fu Y, Reyon D, Maeder ML, Kaini P, Sander JD, Joung JK, Peterson RT, Yeh JR (2013) Heritable and precise zebrafish genome editing using a CRISPR-Cas system. PLoS One 8, e68708
123. Jao LE, Wente SR, Chen W (2013) Efficient multiplex biallelic zebrafish genome editing using a CRISPR nuclease system. Proc Natl Acad Sci U S A 110:13904–13909
124. Gagnon JA, Valen E, Thyme SB, Huang P, Ahkmetova L, Pauli A, Montague TG, Zimmerman S, Richter C, Schier AF (2014) Efficient mutagenesis by Cas9 protein-mediated oligonucleotide insertion and large-scale assessment of single-guide RNAs. PLoS One 9, e98186
125. Thomas HR, Percival SM, Yoder BK, Parant JM (2014) High-throughput genome editing and phenotyping facilitated by high resolution melting curve analysis. PLoS One 9, e114632
126. Xue W, Chen S, Yin H, Tammela T, Papagiannakopoulos T, Joshi NS, Cai W, Yang G, Bronson R, Crowley DG et al (2014) CRISPR-mediated direct mutation of cancer genes in the mouse liver. Nature 514:380–384
127. Sanchez-Rivera FJ, Papagiannakopoulos T, Romero R, Tammela T, Bauer MR, Bhutkar A, Joshi NS, Subbaraj L, Bronson RT, Xue W et al (2014) Rapid modelling of cooperating genetic events in cancer through somatic genome editing. Nature 516:428–431
128. Platt RJ, Chen S, Zhou Y, Yim MJ, Swiech L, Kempton HR, Dahlman JE, Parnas O, Eisenhaure TM, Jovanovic M et al (2014) CRISPR-Cas9 knockin mice for genome editing and cancer modeling. Cell 159:440–455
129. Tawk M, Bianco IH, Clarke JD (2009) Focal electroporation in zebrafish embryos and larvae. Methods Mol Biol 546:145–151
130. Rao NM, Rambabu KM, Rao SH (2008) Electroporation of adult zebrafish. Methods Mol Biol 423:289–298
131. Kizil C, Brand M (2011) Cerebroventricular microinjection (CVMI) into adult zebrafish brain is an efficient misexpression method for forebrain ventricular cells. PLoS One 6, e27395

132. Holmes KE, Wyatt MJ, Shen YC, Thompson DA, Barald KF (2011) Direct delivery of MIF morpholinos into the zebrafish otocyst by injection and electroporation affects inner ear development. J Vis Exp
133. Tang Q, Abdelfattah NS, Blackburn JS, Moore JC, Martinez SA, Moore FE, Lobbardi R, Tenente IM, Ignatius MS, Berman JN et al (2014) Optimized cell transplantation using adult rag2 mutant zebrafish. Nat Methods 11:821–824
134. Blackburn JS, Langenau DM (2014) Zebrafish as a model to assess cancer heterogeneity, progression and relapse. Dis Model Mech 7:755–762
135. Fu Y, Foden JA, Khayter C, Maeder ML, Reyon D, Joung JK, Sander JD (2013) High-frequency off-target mutagenesis induced by CRISPR-Cas nucleases in human cells. Nat Biotechnol 31:822–826
136. Hsu PD, Scott DA, Weinstein JA, Ran FA, Konermann S, Agarwala V, Li Y, Fine EJ, Wu X, Shalem O et al (2013) DNA targeting specificity of RNA-guided Cas9 nucleases. Nat Biotechnol 31:827–832
137. Ran FA, Hsu PD, Lin CY, Gootenberg JS, Konermann S, Trevino AE, Scott DA, Inoue A, Matoba S, Zhang Y et al (2013) Double nicking by RNA-guided CRISPR Cas9 for enhanced genome editing specificity. Cell 154:1380–1389
138. Mali P, Aach J, Stranges PB, Esvelt KM, Moosburner M, Kosuri S, Yang L, Church GM (2013) CAS9 transcriptional activators for target specificity screening and paired nickases for cooperative genome engineering. Nat Biotechnol 31:833–838
139. Tsai SQ, Wyvekens N, Khayter C, Foden JA, Thapar V, Reyon D, Goodwin MJ, Aryee MJ, Joung JK (2014) Dimeric CRISPR RNA-guided FokI nucleases for highly specific genome editing. Nat Biotechnol 32:569–576
140. Guilinger JP, Thompson DB, Liu DR (2014) Fusion of catalytically inactive Cas9 to FokI nuclease improves the specificity of genome modification. Nat Biotechnol 32:577–582
141. Fu Y, Sander JD, Reyon D, Cascio VM, Joung JK (2014) Improving CRISPR-Cas nuclease specificity using truncated guide RNAs. Nat Biotechnol 32:279–284
142. Cui X, Ji D, Fisher DA, Wu Y, Briner DM, Weinstein EJ (2011) Targeted integration in rat and mouse embryos with zinc-finger nucleases. Nat Biotechnol 29:64–67
143. Shukla VK, Doyon Y, Miller JC, DeKelver RC, Moehle EA, Worden SE, Mitchell JC, Arnold NL, Gopalan S, Meng X et al (2009) Precise genome modification in the crop species Zea mays using zinc-finger nucleases. Nature 459:437–441
144. Urnov FD, Miller JC, Lee YL, Beausejour CM, Rock JM, Augustus S, Jamieson AC, Porteus MH, Gregory PD, Holmes MC (2005) Highly efficient endogenous human gene correction using designed zinc-finger nucleases. Nature 435:646–651
145. Bibikova M, Beumer K, Trautman JK, Carroll D (2003) Enhancing gene targeting with designed zinc finger nucleases. Science 300:764
146. Shin J, Chen J, Solnica-Krezel L (2014) Efficient homologous recombination-mediated genome engineering in zebrafish using TALE nucleases. Development 141:3807–3818
147. Irion U, Krauss J, Nusslein-Volhard C (2014) Precise and efficient genome editing in zebrafish using the CRISPR/Cas9 system. Development 141:4827–4830
148. Hruscha A, Krawitz P, Rechenberg A, Heinrich V, Hecht J, Haass C, Schmid B (2013) Efficient CRISPR/Cas9 genome editing with low off-target effects in zebrafish. Development 140:4982–4987
149. Auer TO, Duroure K, Concordet JP, Del Bene F (2014) CRISPR/Cas9-mediated conversion of eGFP- into Gal4-transgenic lines in zebrafish. Nat Protoc 9:2823–2840
150. Auer TO, Duroure K, De Cian A, Concordet JP, Del Bene F (2014) Highly efficient CRISPR/Cas9-mediated knock-in in zebrafish by homology-independent DNA repair. Genome Res 24:142–153
151. Gupta A, Hall VL, Kok FO, Shin M, McNulty JC, Lawson ND, Wolfe SA (2013) Targeted chromosomal deletions and inversions in zebrafish. Genome Res 23:1008–1017
152. Xiao A, Wang Z, Hu Y, Wu Y, Luo Z, Yang Z, Zu Y, Li W, Huang P, Tong X et al (2013) Chromosomal deletions and inversions mediated by TALENs and CRISPR/Cas in zebrafish. Nucleic Acids Res 41, e141

Tumor Suppressors in Zebrafish: From TP53 to PTEN and Beyond

Jeroen den Hertog

Abstract Zebrafish are increasingly being used to study cancer. Almost all tumor types have been found in zebrafish. However, tumor incidence is relatively low and tumors develop late in life. Functional inactivation of tumor suppressors is a crucial step in cancer progression and more and more tumor suppressor genes are being studied in zebrafish. Most often tumor suppressors have been inactivated by reverse genetics approaches using targeted disruption. However, some tumor suppressor mutants were identified by forward genetic screens for mutants with a particular phenotype. Some of the latter genes had not been recognized as tumor suppressors yet. Similarly, a screen for genes that suppress tumor formation in zebrafish in vivo led to the identification of a novel tumor suppressor gene. In this review, I will provide an overview of what the zebrafish has taught us about tumor suppressors.

Keywords Tumor suppressor • Zebrafish • tp53 • pten • apc • vhl

Introduction

> A tumor suppressor gene, or antioncogene, is a gene that protects a cell from one step on the path to cancer. When this gene mutates to cause a loss or reduction in its function, the cell can progress to cancer, usually in combination with other genetic changes (wikipedia).

Oncogenic activation of proto-oncogenes, for instance by mutation or overexpression of cancer driver genes promotes tumorigenesis. However, in the formation of human cancers, the loss of a tumor suppressor gene may be more important than the gain of an oncogene [1, 2]. Over the years, a large number of tumor suppressor genes have been identified in human cancer and often little is known about the cellular, biological or biochemical function of these genes. The zebrafish is an excellent experimental model to assess gene function in vivo and zebrafish is increasingly

J. den Hertog (✉)
Hubrecht Institute-KNAW and University Medical Center Utrecht,
Uppsalalaan 8, 3584 CT Utrecht, The Netherlands

Institute of Biology, Leiden University, 2300 RA Leiden, The Netherlands
e-mail: j.denhertog@hubrecht.eu

D.M. Langenau (ed.), *Cancer and Zebrafish*, Advances in Experimental Medicine and Biology 910, DOI 10.1007/978-3-319-30654-4_4

being used to model human disease [3], including cancer [4]. Particularly the advent of reverse genetics in zebrafish has facilitated the study of tumor suppressors and many tumor suppressors have since been inactivated in zebrafish, often resulting in developmental defects and/or enhanced tumor susceptibility.

TP53

tp53 is the most frequently mutated tumor suppressor in human cancers. More than 50 % of all solid tumors harbor mutations in *tp53* that inactivate TP53 protein function. TP53 has been described as the "guardian of the genome", because it induces cell cycle arrest or apoptosis upon various types of cellular stresses, thus maintaining the integrity of the genome [5–7]. DNA damage and other stressors induce post-translational TP53 protein stabilization, resulting in cell cycle arrest due to TP53-mediated transactivation of *p21* cyclin-dependent kinase inhibitor or to apoptosis in a transcription-dependent or -independent manner [8]. Loss of functional TP53 therefore can result in defective DNA repair and a reduced apoptotic response and hence to accumulation of more mutations in oncogenesis.

The TP53 protein is conserved in zebrafish (48 % at the amino acid level over the full length protein compared to human TP53) [9] and target selected gene inactivation led to the identification of several *tp53* mutants, two of which have missense mutations in orthologous residues as found in human cancer cells (N168K and M214K) [10]. Both mutants fail to transactivate the *p21* response element in SAOS-2 cells, demonstrating that these missense mutations functionally inactivate TP53. Incrosses of heterozygous *tp53*^{N168K} or *tp53*^{M214K} mutant fish result in normal sized clutches of embryos with Mendelian distribution of wild type, heterozygous and homozygous embryos. Development of the heterozygous and homozygous *tp53* mutant embryos is normal and also viability and fertility of these fish is indistinguishable from wild type fish. Yet, homozygous *tp53*^{M214K} mutant zebrafish embryos are resistant to γ-irradiation-induced apoptosis, consistent with TP53 loss of function. The TP53^{N168K} mutant protein turns out to be temperature-sensitive, like the human mutant protein in the orthologous residue. At the permissive temperature (28 °C), the *tp53*^{N168K} allele behaves like wild type, whereas at 37 °C, the non-permissive temperature, *tp53*^{N168K} becomes resistant to γ-irradiation-induced apoptosis. Cell cycle analysis demonstrated that the G1 checkpoint in the *tp53*^{M214K} mutant embryos is defective upon γ-irradiation, which is associated with a lack of upregulation of downstream targets, including *p21* and *mdm2* [10]. All in all, the *tp53*^{M214K} mutant embryos display the hallmarks of cancer cells lacking functional TP53.

Mouse mutants lacking functional TP53 protein are highly susceptible to tumor formation [11, 12]. In zebrafish, 28 % of the homozygous *tp53*^{M214K} mutant fish spontaneously develop tumors between 8 and 16.5 months of age. The majority of these tumors are diagnosed as malignant peripheral nerve sheath tumors (MPNSTs) [10]. This is surprising because MPNSTs are rare in humans and are not observed in mouse models lacking functional TP53 [12]. The difference in spectrum of spontaneous tumor formation between mammals and zebrafish remains to be determined [9].

Whereas the spontaneous tumors in *tp53* mutant zebrafish are predominantly MPNSTs, a broad spectrum of tumors has been identified in *tp53* mutant zebrafish upon expression of a range of oncogenes or cancer drivers. The first driver that was reported is $BRAF^{V600E}$ [13]. This mutation is commonly detected in human nevi and melanoma. Expression of $BRAF^{V600E}$ in wild type zebrafish melanocytes using the *mitfa* promoter induces nevi in ~10 % of the adult fish, but no melanoma. Expression of $BRAF^{V600E}$ in melanocytes of *tp53* mutant fish induces nevi in 13.6 % of the animals, of which almost half progressed to malignant melanoma. The difference in fates of $BRAF^{V600E}$-expressing melanocytes between wild type and *tp53* mutant zebrafish illustrates the tumor suppressive role of TP53 in wild type zebrafish. Histologically, zebrafish melanoma is highly similar to human melanoma, thus establishing the zebrafish as a melanoma model [13].

Similar to expression of $BRAF^{V600E}$, expression of oncogenic $NRAS^{Q61K}$ under the control of the *mitfa* promoter in wild type zebrafish also results in spots of hyperproliferative melanocytes from 3 to 4 weeks of age onwards. Malignant melanoma is not observed in these fish. However, expression of $NRAS^{Q61K}$ in homozygous *tp53* mutant zebrafish results in malignant melanoma in 100 % of the cases. Interestingly, melanoma is also enhanced in *tp53* heterozygous mutants and genotyping is indicative of *tp53* loss-of-heterozygosity in the tumors. Interestingly, comparison of the gene expression patterns from human melanomas to melanomas from *tp53* mutant zebrafish expressing $NRAS^{Q61K}$ indicates a striking similarity, confirming that zebrafish is a good model for human melanoma [14].

NRAS acts upstream of BRAF in the RAS/MAPK signaling pathway and activating mutations in either NRAS or BRAF are identified in 95 % of human melanomas [15, 16]. Therefore, it is perhaps not surprising that expression of mutant $BRAF^{V600E}$ or $NRAS^{Q61K}$ in melanocytes leads to melanoma in zebrafish. However, mutations in *tp53* are rare in human melanoma and the observation that melanoma occurs in a *tp53* mutant, but not wild type zebrafish, suggests that TP53 has a strong protective role. Moreover, the relatively late onset of melanoma—8 months in $BRAF^{V600E}$ expressing fish [13] and 10 weeks in $NRAS^{Q61K}$ expressing fish [14]—suggests that additional mutations are required in the *tp53* mutant background to proceed to malignant melanoma.

Whereas the role of TP53 in protection against malignant melanoma is clear in the $BRAF^{V600E}$ and $NRAS^{Q61K}$ expressing zebrafish described above, expression of oncogenic $HRAS^{G12V}$ in melanocytes under the control of the *kita* promoter induces malignant melanoma in the presence of functional *tp53* [17]. The difference may be due to higher expression of $HRAS^{G12V}$ from the *kita* promoter than $NRAS^{Q61K}$ from the *mitfa* promoter, to differences in the cell specificity of the *mitfa* and *kita* promoters, or to differences in signaling of the oncogenic RAS paralogs, $HRAS^{G12V}$ and $NRAS^{Q61K}$.

In this respect it is noteworthy that expression of oncogenic $KRAS^{G12D}$ in wild type zebrafish also induces tumors in the presence of functional TP53. $KRAS^{G12D}$ expression under the control of the *rag2* promoter, driving expression in satellite cells and myoblasts within skeletal muscle, induces tumors that resemble embryonal rhabdomyosarcoma [18]. *Tp53* mutations are associated with a subset of embryonal rhabdomyosarcomas in humans [19] and expression of $KRAS^{G12D}$ in homozygous *tp53* mutant embryos induces a significant increase in the incidence of embryonal rhabdomyosarcoma [18], demonstrating that inactivation of *tp53* cooperates with oncogenic activation of KRAS in the formation of embryonal rhabdomyosarcoma.

Whereas tissue-specific expression of $KRAS^{G12D}$ cooperates with loss of TP53 function in embryonal rhabdomyosarcoma formation, several other factors induce hepatocellular carcinoma in *tp53*-negative zebrafish. Liver-specific expression of the Hepatitis B virus X antigen (HBx) using the *l-fabp* promoter in wild type zebrafish causes steatosis, fibrosis and glycogen accumulation, but not tumor formation. Only in the *tp53*-negative background hepatocellular carcinoma is observed, which is accompanied by upregulation of *src* expression and activation of the Src tyrosine kinase signaling pathway. Interestingly, expression of exogenous Src in *tp53*-negative zebrafish also induces hepatocellular carcinoma, suggesting that upregulation of *src* expression is the mechanism underlying HBx-induced hepatocellular carcinoma in mutant zebrafish lacking functional TP53 [20].

Unlike HBx, which induces hepatocellular carcinoma in *tp53* mutant, but not wild type zebrafish, liver-specific, *fabp10* promoter-driven expression of UHRF1 (Ubiquitin-like with PHD and RING finger domains 1), an essential regulator of DNA methylation that is highly expressed in many cancers, induces hepatocellular carcinoma by itself in wild type zebrafish. However, the tumor incidence is enhanced in heterozygous *tp53* mutant zebrafish from 50 % to 87 % and in one case, the tumor type was distinct, suggesting that *tp53* does suppress tumor formation and may have a role in the spectrum of tumors caused by UHFR1 overexpression [21].

Well-differentiated liposarcoma is yet another tumor type in which TP53 is involved in humans. The majority (91 %) of tumors that are observed in response to expression of myristoylated Akt2, a constitutively active form of Akt2, in mesenchymal cells (*rag2* promoter) display hallmarks of well-differentiated liposarcoma. TP53 suppresses tumor formation, because expression of myristoylated Akt2 induced tumors in 8 % of wild type zebrafish, 6 % of *tp53* heterozygous fish and 29 % of *tp53* homozygous mutants. Akt2 is a kinase that acts downstream of phosphatidylinositol-3kinase in a signaling pathway that promotes cell survival and proliferation. The link of Akt signaling to well-differentiated liposarcoma is novel and may provide new leads to combat these tumors [22].

Not only expression of oncogenic proteins in the *tp53* mutant background induces tumors, but also the additional loss of another tumor suppressor in the *tp53* mutant background leads to tumor formation in zebrafish. *Early mitotic inhibitor 1*, *emi1*, is essential for genomic stability and the human homolog of *emi1* may have a function in leukemia. Zebrafish mutants with a proviral insertion in *emi1* that impairs its function display widespread developmental defects from 20 hpf onwards [23]. Whereas *tp53* mutant zebrafish develop MPNST tumors at a relatively low incidence, compound *tp53* mutants that also harbor an inactivating proviral insertion in *emi1* display a significant increase in MPNST formation. These results provide evidence that *emi1* is a tumor suppressor in *tp53* mutant zebrafish [24].

TSC1 and *TSC2* are two other tumor suppressor genes and loss of function mutations in *TSC1* and *TSC2* in humans are associated with tuberous sclerosis complex, a genetic disorder that is characterized by multi-organ hamartomas, i.e. benign tumors. mTORC1 signaling is elevated in patients that lack functional TSC1 or TSC2. Mutant zebrafish harboring a nonsense mutation in the *tsc2* gene that truncates the Tsc2 protein and impairs its function display increased cell size in the brain and liver and these mutants are not viable beyond the larval stage [25]. Whereas *tsc2* mutants do not develop cancer, compound heterozygous *tsc2*/homozygous *tp53* mutants display enhanced incidence of malignant tumors, evidenced by large sarcomatous tumors in multiple organs, compared to *tp53* homozygous mutants [26]. Therefore, loss of functional *tp53* is a genetic change that may be required for tuberous sclerosis complex to progress to malignancy on the one hand and in humans, loss of a single allele of *TSC1* or *TSC2* in a *TP53*-negative background may lead to severe forms of cancer too. Interestingly, treatment with the mTORC1 inhibitor rapamycin led to a rapid reduction in tumor size [26], suggesting that *TP53* negative tumors that are heterozygous for mutations in *TSC1* or *TSC2* may be treatable with rapamycin.

Human cancers that are associated with mutations in the tumor suppressor *BRCA2* (breast cancer 2) often also have mutations in *TP53*. BRCA2 is a component of the DNA repair machinery and mutations in *BRCA2* perturb double-strand DNA break repair, resulting in chromosomal aberrations. Target selected gene inactivation led to the identification of a nonsense mutation in the zebrafish gene encoding *brca2* that results in a Glutamine 658 to Stop mutation and hence to a severely truncated Brca2 protein that is not functional. Incrosses of heterozygous *brca2*^{Q658X} mutants result in Mendelian ratios of wild type, heterozygous and homozygous *brca2* mutants. Homozygous *brca2*^{Q658X} mutants are viable and at sexual maturity, all homozygous mutants are phenotypically male. At 10–16 months of age, 31 % of the homozygous *brca2*^{Q658X} mutants develop testicular neoplasia, consistent with the notion that *brca2* is a tumor suppressor gene. Compound *brca2*/*tp53* mutants display an accelerated onset of tumor formation, compared to homozygous *tp53* mutants. Interestingly, tumor onset in *tp53*-negative heterozygous and homozygous *brca2* mutants is similar. The majority of tumors are MPNSTs, but also some ovarian tumors are observed [27]. Interestingly, resequencing indicates loss-of-heterozygosity of both tumor suppressor genes in malignant tumors from zebrafish with heterozygous mutations in both *brca2* and *tp53* [28].

In humans, loss of TP53 is also known to cooperate with neurofibromin 1 (NF1). Human neurofibromatosis type 1 patients develop benign peripheral nerve sheath tumors and hyperpigmented skin lesions, caused by dominantly inherited mutations in NF1. These patients are predisposed to develop malignancies, including juvenile myelomonocytic leukemia, glioma and MPNSTs. Zebrafish have two *nf1* genes, *nf1a* and *nf1b*, and genetic models have been generated by target selected gene inactivation as well as using Zn-finger nuclease technology. Zebrafish lacking functional Nf1 display overproliferation of oligodendrocyte progenitor cells and these larvae do not live beyond 10 days post fertilization. Zebrafish that retain a single wild type *nf1* allele are viable and fertile and these fish do not develop tumors until 18 months of age. However, compound zebrafish mutants lacking functional Tp53 and Nf1b that retain a single wild type *nf1a* allele show a marked acceleration in tumor onset and increase in tumor penetrance, compared to zebrafish mutants lacking TP53 that are wild type for *nf1a* and *nf1b*. The observed brain tumors display hallmarks of diffuse high-grade gliomas, whereas the other tumors are consistent with MPNSTs [29].

Although *TP53* is the most frequently mutated tumor suppressor in human cancer, spontaneous tumor formation in *tp53* mutant zebrafish occurs relatively late with relatively low penetrance and the resulting tumors are predominantly a relatively rare tumor type, MPNST. Several tumor drivers have been expressed in *tp53* mutant zebrafish and compound mutants have been generated lacking functional Tp53 and other tumor suppressors, resulting in accelerated tumor onset, enhanced tumor penetrance and broadening of the tumor spectrum. Future analysis of additional tumor drivers and tumor suppressors that cooperate with loss of functional Tp53 in tumor formation in zebrafish will enhance our understanding of the role of Tp53 in tumorigenesis and may provide new leads how to combat cancer in humans.

PTEN

PTEN is the second most frequently mutated tumor suppressor gene in human cancer [30, 31]. Many different types of cancers have missense and/or nonsense mutations in PTEN [32–34]. Moreover, missense mutations have been identified in human syndromes that render carrier patients more prone to cancer [35–37]. Mouse models lacking functional PTEN are not viable beyond embryonic day 8.5 and heterozygous mice are tumor-prone [38–40]. In fact, a reduction of PTEN protein levels in mouse models by as little as 20 % already enhances tumor susceptibility [41].

Many functions have been ascribed to PTEN at the cellular and biochemical level [42]. However, PTEN is best known for its lipid phosphatase activity. PTEN dephosphorylates phosphatidylinositol(3,4,5)trisphosphates and is selective for the 3-position [43, 44]. Therefore, PTEN antagonizes phosphatidylinositol-3kinase and loss of PTEN results in hyperactivation of Akt (also known as PKB) [45, 46]. Further downstream signaling leads to enhanced cell survival and proliferation and hence, PTEN acts as a tumor suppressor gene.

The zebrafish genome harbors two *pten* genes, *ptena* and *ptenb* [47, 48]. Both *pten* genes encode functional Pten protein with lipid phosphatase activity, selective for the 3-position of phosphatidylinositol-polyphosphates. The two zebrafish *pten* genes are broadly expressed during embryonic development. Using target-selected gene inactivation, nonsense mutations were identified in each of the two *pten* genes, well upstream of the catalytic site. No functional Ptena or Ptenb protein is expressed in homozygous *ptena* or *ptenb* mutants, respectively. Homozygous *ptena* or *ptenb* single mutants do not display developmental defects and are viable and fertile, indicating functional redundancy between Ptena and Ptenb [48].

Whereas developmental defects are not observed in *ptena* or *ptenb* single mutants, homozygous *ptenb* mutant adults spontaneously developed eye tumors from 7 to 18 months of age at a relatively high incidence (33 %).

Immunohistochemistry demonstrated that the highly proliferative regions in these tumors displayed elevated levels of phospho-Akt, indicative of Akt activation [48]. Ocular tumors are extremely rare in wild type zebrafish and also in homozygous *ptena* mutant adult zebrafish, these eye tumors are never observed. Following several rounds of outcrossing, the ocular tumors in homozygous *ptenb* mutant zebrafish were not observed anymore. This may be due to the loss of an enhancer of the eye tumor phenotype in the *ptenb* mutant background, or to the gain of a suppressor of the phenotype. The identity of the supposed enhancer or suppressor remains elusive currently.

Ocular tumors are not observed in homozygous *ptenb* single mutants anymore. However, tumors are observed in compound *ptena/ptenb* mutants that retain a single wild type allele, i.e. *ptena+/-ptenb-/-* and *ptena-/-ptenb+/-* adult zebrafish. In a cohort of 294 *ptena+/-ptenb-/-* adults 10.2 % develop tumors in the first year of their life. Often the tumors are located close to the eye and pathological analysis indicates that the tumors consist of endothelial cells that form blood vessels that are actually filled with blood. These tumors are diagnosed as hemangiosarcomas [49]. Immunohistochemistry and immunoblotting demonstrates that these tumors display elevated Akt signaling. Why these hemangiosarcomas predominantly develop close to the eye remains to be determined. Under normal conditions, a meshwork of blood vessels, *rete mirabile*, is located close to the eye. This may result in production of high levels of proliferative signals for endothelial cells locally. In addition, due to lack of three of the four *pten* alleles, proliferative signaling in response to external signaling is already enhanced in *ptena+/-ptenb-/-* mutants. Together, high levels of signals and elevated intracellular signaling may lead to unusually enhanced endothelial cell proliferation and thus to tumor formation close to the eye [49]. It is noteworthy that *pten* mutant zebrafish are much less susceptible to tumor formation than mice. Double heterozygous mutant zebrafish (*ptena+/-ptenb+/-*), lacking 50 % pten expression, do not show an increase in spontaneous tumor formation [48, 49], whereas a 20 % reduction in PTEN expression in mouse mutants already enhances tumor susceptibility [41].

Despite the lack in enhanced spontaneous tumor formation in *pten* double heterozygous fish [48, 49], partial loss of pten does affect tumorigenesis in a model for T cell acute lymphoblastic leukemia (T-ALL) [50]. Guttierrez et al. expressed a fusion protein of the MYC oncogenic transcription factor and the estrogen receptor (MYC-ER) in the thymus under the control of the rag2 promoter. Treatment of these fish with ligand, 4-hydroxytamoxifen (4HT), from 5 days post fertilization onwards leads to fully penetrant T-ALL starting at 5 weeks post fertilization. Subsequent removal from 4HT leads to MYC downregulation after 4 days and to tumor regression in wild type zebrafish due to apoptosis. Ligand-induced MYC-ER activation does not accelerate the onset of T-ALL in double heterozygous *ptena+/-ptenb+/-* zebrafish. However, tumor regression following removal from 4HT is reduced from 69 % in wild type zebrafish to 10 % in *ptena+/-ptenb+/-* zebrafish [50]. Therefore, mutation of *pten* promotes loss of MYC oncogene dependence in T-ALL.

Pten double heterozygous fish and even mutant zebrafish that retain a single wild type *pten* allele are viable and fertile. However, complete loss of functional Pten is lethal around 6 days post fertilization. Double homozygous embryos display pleiotropic defects that are consistent with enhanced proliferation and survival. These developmental defects are largely rescued by microinjection of synthetic mRNA encoding Ptena or Ptenb at the one-cell stage, or by treatment of the embryos with the phosphatidylinositol-3kinase inhibitor, LY294002 [48]. Double mutant embryos display hyperbranching of blood vessels from 3 days post fertilization onwards, suggesting that endothelial cells are particularly sensitive to loss of functional Pten. This is consistent with the observation that *pten* mutant adult zebrafish develop hemangiosarcomas. Interestingly, double homozygous zebrafish embryos express elevated levels of *vegfaa*, encoding a ligand for VEGFRs that have crucial roles in endothelial cell proliferation [51]. Therefore, hyperbranching in *pten* mutant embryos may be caused by elevated VEGFR signaling because of enhanced levels of signal on the one hand and an increase in signaling due to the lack of Pten on the other.

In mammals, PTEN reportedly has an important role in many biological processes, including hematopoiesis. In homozygous double mutant zebrafish embryos lacking functional Pten, the number of hematopoietic stem and progenitor cells (HSPCs) is enhanced at 4 days post fertilization, consistent with enhanced proliferation in these cells. These HSPCs engage in all blood cell lineages, demonstrating that the commitment to differentiate is not impaired. Yet, fully mature blood cells are not observed, indicating that differentiation is blocked in embryos lacking functional Pten [52]. Similarly, increased numbers of myeloid cells are detected in *pten* mutant embryos, which are however immune-deficient [53]. Strikingly, inhibition of phosphatidylinositol-3kinase signaling at relatively late stages restores terminal differentiation of hematopoietic cells [52]. Taken together, loss of functional Pten results in enhanced proliferation of HSPCs and an arrest in terminal differentiation of progenitor cells.

Pten mutant zebrafish that retain a single wild type allele spontaneously develop tumors, hemangiosarcomas. Whether other tumor suppressors or cancer driver genes cooperate with the loss of Pten in tumor formation has not been addressed yet. Insight into which genes cooperate with *pten* in zebrafish tumor formation is highly interesting and will lead to fundamental insights into the function of the tumor suppressor, *pten*, which may provide the first step to combat cancer that is associated with loss of PTEN in human cancer.

APC

Adenomatous polyposis coli (APC) mutations are common in colorectal cancer. *APC* was found to be mutated in familial adenomatous polyposis syndrome in humans and in these patients, somatic mutation of the second allele leads to colorectal cancer [54]. APC has a central role in inhibition of Wnt/β-Catenin signaling. Truncation of APC by mutation leads to accumulation of nuclear β-Catenin and to constitutive further downstream signaling [55]. Mouse models lacking functional APC fail to complete gastrulation and heterozygous *Apc* mutant mice progressively develop intestinal tumors [56].

In zebrafish, a nonsense mutation was identified in *apc* by target selected gene inactivation. This mutation is located in the equivalent of the mutation cluster region in human APC. Heterozygous *apc* mutant zebrafish develop normally and they are viable and fertile. Unlike mice lacking functional APC, homozygous *apc* mutant zebrafish complete gastrulation. Yet, *apc* mutant zebrafish are embryonic lethal at 3–4 days post fertilization and these embryos display multiple developmental defects, of which the cardiac defects, cardiac edema, smaller eyes, and body curvature are most prominent. That these developmental defects are due to loss of functional Apc was confirmed by another mutant with a nonsense mutation in *apc* upstream of the mutation cluster region that was identified in an independent forward genetic screen. The hearts in homozygous *apc* mutants fail to loop properly and form excessive endocardial cushions [57]. Moreover, detailed analysis of the brain in developing embryos indicated that loss of Apc function impaired maintenance of local brain organizers and established brain subdivisions [58]. Whereas heterozygous *apc* mutant fish are viable and fertile, aged fish (>15 months of age) spontaneously develop neoplasias in the liver (18 %) and intestine (12 %). This is rarely (3 %) observed in siblings (liver neoplasia was observed in a single age-matched sibling) [59]. Whereas homozygous *apc* mutant zebrafish are embryonic lethal, they nevertheless provide an attractive model to test genetic cofactors and inhibitors. Transient expression of active $KRAS^{G12D}$ in homozygous *apc* mutant zebrafish embryos enhanced intestinal cell proliferation [60], and inhibition of mTORC1 partially rescued the early developmental defects in zebrafish embryos that lack functional Apc [61]. Further analysis of the *apc* model will provide leads for therapeutic intervention in cancer that is associated with loss of this prominent tumor suppressor.

VHL

Complete loss of the von Hippel-Lindau (VHL) tumor suppressor in heterozygous *VHL* patients by inactivation of the remaining allele in somatic cells predisposes these patients to the development of highly vascularized tumors and cysts in many organ systems [62]. VHL protein has a role in the adaptive response of cells to hypoxia. VHL is an E3 ubiquitin ligase and regulates hypoxia inducible factor (HIF) protein levels. Mouse *Vhl* knockouts die in utero around embryonic day 11.5–12.5 due to hemorrhagic lesions in the placenta [63]. Two zebrafish lines with nonsense mutations in *vhl* were identified by target selected gene inactivation. Transheterozygotes lack detectable Vhl protein levels and are larval-lethal at 8–11 days post fertilization. *Vhl* mutant embryos display a systemic hypoxic response and a marked increase in the number of red blood cells (polycythemia), reminiscent of the Chuvash form of Polycythemia (CP) that is observed in human patients with a R200W missense mutation in the extreme C-terminus of the VHL protein. It is noteworthy that CP patients are not predisposed to cancer and likewise, heterozygous *vhl* mutant zebrafish do not spontaneously develop tumors [64]. Whereas the *vhl* mutant zebrafish model does not develop tumors, it has been used successfully to model hypoxia, a central modulator of cellular physiology in cancer [65, 66].

Ribosomal Proteins

Heterozygous mutant zebrafish lines with viral insertions were generated in the course of a large-scale insertional mutagenesis screen and were screened for early mortality and tumor development later in live. Twelve lines were identified with enhanced tumor incidence and most tumors that were identified were diagnosed as MPNSTs. One line had an insertion in *nf2*, an established tumor suppressor gene, and surprisingly, all other lines had viral insertions in genes encoding ribosomal proteins (*rp*s), suggesting that these *rp* genes are haploinsufficient tumor suppressor genes [67]. MPNSTs from heterozygous zebrafish with viral insertions in *rp*s encode wild type TP53, but express very low to undetectable levels of TP53 protein. This, together with the notion that *tp53* mutant zebrafish spontaneously develop MPNSTs suggests that lack of TP53 protein may be the underlying mechanism for MPNST formation in the *rp* mutant lines [68].

Rpl36 and *rpl23a* are two of the genes encoding ribosomal proteins that were identified in the viral insertion screen above. Recently, *rpl36*, but not *rpl23a*, was confirmed to act as a tumor suppressor in pancreatic cancer. Expression of $KRAS^{G12V}$ in early pancreatic progenitor cells (*ptf1a* promoter) of heterozygous *rpl36* mutant fish increases pancreatic epithelial cell proliferation, accelerates tumor progression and decreases survival compared to wild type sibling controls, demonstrating that *rpl36* acts as a haploinsufficient tumor suppressor [69].

A mutation in *rps29* was identified in an independent forward genetic screen for hematopoietic mutants. Zebrafish embryos lacking functional Rps29 have less hematopoietic stem cells and display craniofacial defects, which is reminiscent of Diamond-Blackfan anemia in human patients that is associated with mutations in ribosomal proteins. Mutation of *tp53* almost completely rescues the developmental defects observed in *rps29* mutants [70]. Apparently, loss of Tp53 protein mediates MPNST formation in heterozygous rp mutants on the one hand and *tp53* mediates developmental defects caused by loss of Rps29 on the other.

Other Tumor Suppressors

Next to the tumor suppressor genes described above, several more have been knocked out in zebrafish by reverse genetics approaches, some tumor suppressor mutants were identified in forward genetic screens and a candidate tumor suppressor was identified in a screen for genes that suppressed tumor formation.

LKB1 is a tumor suppressor that encodes a serine/threonine kinase. Inactivating mutations in *LKB1* cause Peutz-Jeghers syndrome, an autosomal disorder characterized by multiple hamartomas and abnormal pigmentation of mucus membranes and predisposition to gastrointestinal cancers. Zebrafish mutants that lack functional Lkb1 develop normally up to 5 days post fertilization, but die at 7–8 days post fertilization. The *lkb1* mutant larvae exhibit a high metabolic rate and display a starvation response, but no increase in tumor susceptibility is reported in heterozygous *lkb1* mutant zebrafish [71].

Somatic loss-of-function mutations of the ten-eleven translocation 2 gene (*TET2*) are frequently observed in human myelodysplastic syndrome. *TET2* encodes a methylcytosine oxidase that has an important role in demethylation of DNA in CpG islands. Zn-finger nuclease technology was used to inactivate *tet2*. Homozygous *tet2* mutant zebrafish embryos do not display hematopoietic defects during embryonic development. Adult fish lacking Tet2 are viable and fertile, but develop progressive myelodysplasia, resulting in myelodysplastic syndrome at 2 years of age, which is highly reminiscent of myelodysplastic syndrome in humans due to loss of functional TET2 [72].

The *lgl* gene was originally identified as a tumor suppressor in *Drosophila*. A mutation in its zebrafish ortholog, *pen/lgl2*, was identified in a forward genetic screen. Homozygous *pen/lgl2* mutant embryos show overgrowth of endothelial cells, which is dependent on erbB signaling, and die at 4–5 days post fertilization and hence tumor formation cannot be assessed in homozygous *pen/lgl2* mutant zebrafish [73].

Space cadet is a mutant that was identified in a forward genetic screen for mutants with defective locomotor behavior and is caused by a nonsense mutation in the retinoblastoma susceptibility gene, *rb1*, a well-known tumor suppressor gene. Biallelic mutations in *RB1* cause intraocular childhood retinoblastoma. Zebrafish *rb1* mutants display defects in retinotectal tract development and visual function, resulting from a delay in cell cycle exit of retinotectal ganglion cells and initial failure of their axons to exit the retina, thus leading to optic nerve hypoplasia [74].

A forward genetic screen for mutants with defects in cilia function led to the identification of *hu255H*, a mutant that lacks motile cilia in the nose and neural tube. Positional cloning led to the identification of Leucine-rich repeat containing protein 50 (*lrrc50*) as the affected gene. Homozygous *lrrc50* mutants are larval-lethal [75]. Heterozygous *lrrc50* mutants are viable and fertile and are highly susceptible to the formation of seminomas. Loss-of-heterozygosity was observed in 44 % of tumor samples, which correlated with tumor progression. It turns out that *LRRC50* mutations are associated with seminomas in humans as well. This finding, together with the zebrafish data justifies the recognition of *LRRC50* as a tumor suppressor [76].

In search of novel tumor suppressors, 19 chromatin factors were selected for further analysis. As described above, expression of $KRAS^{G12D}$ under the control of the *rag2* promoter induces embryonal rhabdomyosarcoma formation [18]. To test for tumor suppressing activity, each of the 19 selected chromatin factors was co-expressed with $KRAS^{G12D}$ and tumor formation was assessed. One of the chromatin factors, SUV39H1, significantly suppressed tumor formation. Interestingly, SUV39H1 has a role in senescence and data in Oncomine is consistent with SUV39H1 being a tumor suppressor in humans [77]. Hence, this approach successfully led to the identification of a novel tumor suppressor gene, SUV39H1.

Conclusion

Zebrafish is maturing as a cancer model. Mutants lacking prominent tumor suppressor genes have been derived and some of these spontaneously develop tumors while others require additional mutations. It is noteworthy that the onset of spontaneous tumor formation is relatively late in some of the models, suggesting that additional mutations are required. Complete loss of activity of many tumor suppressors is not compatible with life and leads to embryonic or larval lethality. This provides insight into the developmental and cell biological function of these tumor suppressors. In addition, this provides an opportunity to screen for genes or chemical compounds that rescue these developmental defects and/or embryonic lethality. Despite all the work that has been done to date, there is room for discovery of tumor suppressor genes and cooperation between tumor suppressors as well as between tumor suppressors and cancer driver genes and—perhaps more importantly—for development of means of therapeutic intervention, for which the zebrafish is an ideal model system.

References

1. Hanahan D, Weinberg RA (2011) Hallmarks of cancer: the next generation. Cell 144:646–674
2. Sherr CJ (2004) Principles of tumor suppression. Cell 116:235–246
3. Lieschke GJ, Currie PD (2007) Animal models of human disease: zebrafish swim into view. Nat Rev Genet 8:353–367
4. Feitsma H, Cuppen E (2008) Zebrafish as a cancer model. Mol Cancer Res 6:685–694
5. Levine AJ (1997) p53, the cellular gatekeeper for growth and division. Cell 88:323–331
6. Vousden KH (2000) p53: death star. Cell 103:691–694
7. Vogelstein B, Lane D, Levine AJ (2000) Surfing the p53 network. Nature 408:307–310
8. Vousden KH, Lu X (2002) Live or let die: the cell's response to p53. Nat Rev Cancer 2:594–604
9. Storer NY, Zon LI (2010) Zebrafish models of p53 functions. Cold Spring Harb Perspect Biol 2:a001123
10. Berghmans S, Murphey RD, Wienholds E, Neuberg D, Kutok JL et al (2005) tp53 mutant zebrafish develop malignant peripheral nerve sheath tumors. Proc Natl Acad Sci U S A 102:407–412
11. Donehower LA, Harvey M, Slagle BL, McArthur MJ, Montgomery CA Jr et al (1992) Mice deficient for p53 are developmentally normal but susceptible to spontaneous tumours. Nature 356:215–221
12. Jacks T, Remington L, Williams BO, Schmitt EM, Halachmi S et al (1994) Tumor spectrum analysis in p53-mutant mice. Curr Biol 4:1–7
13. Patton EE, Widlund HR, Kutok JL, Kopani KR, Amatruda JF et al (2005) BRAF mutations are sufficient to promote nevi formation and cooperate with p53 in the genesis of melanoma. Curr Biol 15:249–254
14. Dovey M, White RM, Zon LI (2009) Oncogenic NRAS cooperates with p53 loss to generate melanoma in zebrafish. Zebrafish 6:397–404
15. Jiveskog S, Ragnarsson-Olding B, Platz A, Ringborg U (1998) N-ras mutations are common in melanomas from sun-exposed skin of humans but rare in mucosal membranes or unexposed skin. J Invest Dermatol 111:757–761
16. Omholt K, Platz A, Kanter L, Ringborg U, Hansson J (2003) NRAS and BRAF mutations arise early during melanoma pathogenesis and are preserved throughout tumor progression. Clin Cancer Res 9:6483–6488
17. Santoriello C, Gennaro E, Anelli V, Distel M, Kelly A et al (2010) Kita driven expression of oncogenic HRAS leads to early onset and highly penetrant melanoma in zebrafish. PLoS One 5, e15170
18. Langenau DM, Keefe MD, Storer NY, Guyon JR, Kutok JL et al (2007) Effects of RAS on the genesis of embryonal rhabdomyosarcoma. Genes Dev 21:1382–1395

19. Xia SJ, Pressey JG, Barr FG (2002) Molecular pathogenesis of rhabdomyosarcoma. Cancer Biol Ther 1:97–104
20. Lu JW, Yang WY, Tsai SM, Lin YM, Chang PH et al (2013) Liver-specific expressions of HBx and src in the p53 mutant trigger hepatocarcinogenesis in zebrafish. PLoS One 8, e76951
21. Mudbhary R, Hoshida Y, Chernyavskaya Y, Jacob V, Villanueva A et al (2014) UHRF1 overexpression drives DNA hypomethylation and hepatocellular carcinoma. Cancer Cell 25:196–209
22. Gutierrez A, Snyder EL, Marino-Enriquez A, Zhang YX, Sioletic S et al (2011) Aberrant AKT activation drives well-differentiated liposarcoma. Proc Natl Acad Sci U S A 108:16386–16391
23. Amsterdam A, Nissen RM, Sun Z, Swindell EC, Farrington S et al (2004) Identification of 315 genes essential for early zebrafish development. Proc Natl Acad Sci U S A 101:12792–12797
24. Rhodes J, Amsterdam A, Sanda T, Moreau LA, McKenna K et al (2009) Emi1 maintains genomic integrity during zebrafish embryogenesis and cooperates with p53 in tumor suppression. Mol Cell Biol 29:5911–5922
25. Kim SH, Speirs CK, Solnica-Krezel L, Ess KC (2011) Zebrafish model of tuberous sclerosis complex reveals cell-autonomous and non-cell-autonomous functions of mutant tuberin. Dis Model Mech 4:255–267
26. Kim SH, Kowalski ML, Carson RP, Bridges LR, Ess KC (2013) Heterozygous inactivation of tsc2 enhances tumorigenesis in p53 mutant zebrafish. Dis Model Mech 6:925–933
27. Shive HR, West RR, Embree LJ, Azuma M, Sood R et al (2010) brca2 in zebrafish ovarian development, spermatogenesis, and tumorigenesis. Proc Natl Acad Sci U S A 107: 19350–19355
28. Shive HR, West RR, Embree LJ, Golden CD, Hickstein DD (2014) BRCA2 and TP53 collaborate in tumorigenesis in zebrafish. PLoS One 9, e87177
29. Shin J, Padmanabhan A, de Groh ED, Lee JS, Haidar S et al (2012) Zebrafish neurofibromatosis type 1 genes have redundant functions in tumorigenesis and embryonic development. Dis Model Mech 5:881–894
30. Stokoe D (2001) Pten. Curr Biol 11:R502
31. Hollander MC, Blumenthal GM, Dennis PA (2011) PTEN loss in the continuum of common cancers, rare syndromes and mouse models. Nat Rev Cancer 11:289–301
32. Li DM, Sun H (1997) TEP1, encoded by a candidate tumor suppressor locus, is a novel protein tyrosine phosphatase regulated by transforming growth factor beta. Cancer Res 57:2124–2129
33. Li J, Yen C, Liaw D, Podsypanina K, Bose S et al (1997) PTEN, a putative protein tyrosine phosphatase gene mutated in human brain, breast, and prostate cancer. Science 275:1943–1947
34. Ali IU, Schriml LM, Dean M (1999) Mutational spectra of PTEN/MMAC1 gene: a tumor suppressor with lipid phosphatase activity. J Natl Cancer Inst 91:1922–1932
35. Marsh DJ, Dahia PL, Zheng Z, Liaw D, Parsons R et al (1997) Germline mutations in PTEN are present in Bannayan-Zonana syndrome. Nat Genet 16:333–334
36. Liaw D, Marsh DJ, Li J, Dahia PL, Wang SI et al (1997) Germline mutations of the PTEN gene in Cowden disease, an inherited breast and thyroid cancer syndrome. Nat Genet 16:64–67
37. Zhou XP, Waite KA, Pilarski R, Hampel H, Fernandez MJ et al (2003) Germline PTEN promoter mutations and deletions in Cowden/Bannayan-Riley-Ruvalcaba syndrome result in aberrant PTEN protein and dysregulation of the phosphoinositol-3-kinase/Akt pathway. Am J Hum Genet 73:404–411
38. Di Cristofano A, Pesce B, Cordon-Cardo C, Pandolfi PP (1998) Pten is essential for embryonic development and tumour suppression. Nat Genet 19:348–355
39. Podsypanina K, Ellenson LH, Nemes A, Gu J, Tamura M et al (1999) Mutation of Pten/Mmac1 in mice causes neoplasia in multiple organ systems. Proc Natl Acad Sci U S A 96:1563–1568
40. Suzuki A, de la Pompa JL, Stambolic V, Elia AJ, Sasaki T et al (1998) High cancer susceptibility and embryonic lethality associated with mutation of the PTEN tumor suppressor gene in mice. Curr Biol 8:1169–1178
41. Alimonti A, Carracedo A, Clohessy JG, Trotman LC, Nardella C et al (2010) Subtle variations in Pten dose determine cancer susceptibility. Nat Genet 42:454–458

42. Zhang S, Yu D (2010) PI(3)king apart PTEN's role in cancer. Clin Cancer Res 16:4325–4330
43. Maehama T, Dixon JE (1998) The tumor suppressor, PTEN/MMAC1, dephosphorylates the lipid second messenger, phosphatidylinositol 3,4,5-trisphosphate. J Biol Chem 273:13375–13378
44. Myers MP, Pass I, Batty IH, Van der Kaay J, Stolarov JP et al (1998) The lipid phosphatase activity of PTEN is critical for its tumor supressor function. Proc Natl Acad Sci U S A 95:13513–13518
45. Stambolic V, Suzuki A, de la Pompa JL, Brothers GM, Mirtsos C et al (1998) Negative regulation of PKB/Akt-dependent cell survival by the tumor suppressor PTEN. Cell 95:29–39
46. Song MS, Salmena L, Pandolfi PP (2012) The functions and regulation of the PTEN tumour suppressor. Nat Rev Mol Cell Biol 13:283–296
47. Croushore JA, Blasiole B, Riddle RC, Thisse C, Thisse B et al (2005) Ptena and ptenb genes play distinct roles in zebrafish embryogenesis. Dev Dyn 234:911–921
48. Faucherre A, Taylor GS, Overvoorde J, Dixon JE, Hertog J (2008) Zebrafish pten genes have overlapping and non-redundant functions in tumorigenesis and embryonic development. Oncogene 27:1079–1086
49. Choorapoikayil S, Kuiper RV, de Bruin A, den Hertog J (2012) Haploinsufficiency of the genes encoding the tumor suppressor Pten predisposes zebrafish to hemangiosarcoma. Dis Model Mech 5:241–247
50. Gutierrez A, Grebliunaite R, Feng H, Kozakewich E, Zhu S et al (2011) Pten mediates Myc oncogene dependence in a conditional zebrafish model of T cell acute lymphoblastic leukemia. J Exp Med 208:1595–1603
51. Choorapoikayil S, Weijts B, Kers R, de Bruin A, den Hertog J (2013) Loss of Pten promotes angiogenesis and enhanced vegfaa expression in zebrafish. Dis Model Mech 6:1159–1166
52. Choorapoikayil S, Kers R, Herbomel P, Kissa K, den Hertog J (2014) Pivotal role of Pten in the balance between proliferation and differentiation of hematopoietic stem cells in zebrafish. Blood 123:184–190
53. Dong ZW, Ren CG, Xia Y, Su D, Du TT et al (2014) Pten regulates homeostasis and inflammation-induced migration of myelocytes in zebrafish. J Hematol Oncol 7:17
54. Kinzler KW, Vogelstein B (1996) Lessons from hereditary colorectal cancer. Cell 87:159–170
55. Clevers H (2006) Wnt/beta-catenin signaling in development and disease. Cell 127:469–480
56. Fodde R, Edelmann W, Yang K, van Leeuwen C, Carlson C et al (1994) A targeted chain-termination mutation in the mouse Apc gene results in multiple intestinal tumors. Proc Natl Acad Sci U S A 91:8969–8973
57. Hurlstone AF, Haramis AP, Wienholds E, Begthel H, Korving J et al (2003) The Wnt/beta-catenin pathway regulates cardiac valve formation. Nature 425:633–637
58. Paridaen JT, Danesin C, Elas AT, van de Water S, Houart C et al (2009) Apc1 is required for maintenance of local brain organizers and dorsal midbrain survival. Dev Biol 331:101–112
59. Haramis AP, Hurlstone A, van der Velden Y, Begthel H, van den Born M et al (2006) Adenomatous polyposis coli-deficient zebrafish are susceptible to digestive tract neoplasia. EMBO Rep 7:444–449
60. Phelps RA, Chidester S, Dehghanizadeh S, Phelps J, Sandoval IT et al (2009) A two-step model for colon adenoma initiation and progression caused by APC loss. Cell 137:623–634
61. Valvezan AJ, Huang J, Lengner CJ, Pack M, Klein PS (2014) Oncogenic mutations in adenomatous polyposis coli (Apc) activate mechanistic target of rapamycin complex 1 (mTORC1) in mice and zebrafish. Dis Model Mech 7:63–71
62. Kaelin WG (2005) The von Hippel-Lindau tumor suppressor protein: roles in cancer and oxygen sensing. Cold Spring Harb Symp Quant Biol 70:159–166
63. Haase VH (2005) The VHL tumor suppressor in development and disease: functional studies in mice by conditional gene targeting. Semin Cell Dev Biol 16:564–574
64. van Rooijen E, Voest EE, Logister I, Korving J, Schwerte T et al (2009) Zebrafish mutants in the von Hippel-Lindau tumor suppressor display a hypoxic response and recapitulate key aspects of Chuvash polycythemia. Blood 113:6449–6460

65. van Rooijen E, Voest EE, Logister I, Bussmann J, Korving J et al (2010) von Hippel-Lindau tumor suppressor mutants faithfully model pathological hypoxia-driven angiogenesis and vascular retinopathies in zebrafish. Dis Model Mech 3:343–353
66. Santhakumar K, Judson EC, Elks PM, McKee S, Elworthy S et al (2012) A zebrafish model to study and therapeutically manipulate hypoxia signaling in tumorigenesis. Cancer Res 72:4017–4027
67. Amsterdam A, Sadler KC, Lai K, Farrington S, Bronson RT et al (2004) Many ribosomal protein genes are cancer genes in zebrafish. PLoS Biol 2, E139
68. MacInnes AW, Amsterdam A, Whittaker CA, Hopkins N, Lees JA (2008) Loss of p53 synthesis in zebrafish tumors with ribosomal protein gene mutations. Proc Natl Acad Sci U S A 105:10408–10413
69. Provost E, Bailey JM, Aldrugh S, Liu S, Iacobuzio-Donahue C et al (2014) The tumor suppressor rpl36 restrains KRAS(G12V)-induced pancreatic cancer. Zebrafish 11:551–559
70. Taylor AM, Humphries JM, White RM, Murphey RD, Burns CE et al (2012) Hematopoietic defects in rps29 mutant zebrafish depend upon p53 activation. Exp Hematol 40:228–237.e5
71. van der Velden YU, Wang L, Zevenhoven J, van Rooijen E, van Lohuizen M et al (2011) The serine-threonine kinase LKB1 is essential for survival under energetic stress in zebrafish. Proc Natl Acad Sci U S A 108:4358–4363
72. Gjini E, Mansour MR, Sander JD, Moritz N, Nguyen AT et al (2015) A zebrafish model of myelodysplastic syndrome produced through tet2 genomic editing. Mol Cell Biol 35:789–804
73. Reischauer S, Levesque MP, Nusslein-Volhard C, Sonawane M (2009) Lgl2 executes its function as a tumor suppressor by regulating ErbB signaling in the zebrafish epidermis. PLoS Genet 5, e1000720
74. Gyda M, Wolman M, Lorent K, Granato M (2012) The tumor suppressor gene retinoblastoma-1 is required for retinotectal development and visual function in zebrafish. PLoS Genet 8, e1003106
75. van Rooijen E, Giles RH, Voest EE, van Rooijen C, Schulte-Merker S et al (2008) LRRC50, a conserved ciliary protein implicated in polycystic kidney disease. J Am Soc Nephrol 19:1128–1138
76. Basten SG, Davis EE, Gillis AJ, van Rooijen E, Stoop H et al (2013) Mutations in LRRC50 predispose zebrafish and humans to seminomas. PLoS Genet 9, e1003384
77. Albacker CE, Storer NY, Langdon EM, Dibiase A, Zhou Y et al (2013) The histone methyltransferase SUV39H1 suppresses embryonal rhabdomyosarcoma formation in zebrafish. PLoS One 8, e64969

Identifying Novel Cancer Therapies Using Chemical Genetics and Zebrafish

Michelle Dang, Rachel Fogley, and Leonard I. Zon

Abstract Chemical genetics is the use of small molecules to perturb biological pathways. This technique is a powerful tool for implicating genes and pathways in developmental programs and disease, and simultaneously provides a platform for the discovery of novel therapeutics. The zebrafish is an advantageous model for in vivo high-throughput small molecule screening due to translational appeal, high fecundity, and a unique set of developmental characteristics that support genetic manipulation, chemical treatment, and phenotype detection. Chemical genetic screens in zebrafish can identify hit compounds that target oncogenic processes—including cancer initiation and maintenance, metastasis, and angiogenesis—and may serve as cancer therapies. Notably, by combining drug discovery and animal testing, in vivo screening of small molecules in zebrafish has enabled rapid translation of hit anti-cancer compounds to the clinic, especially through the repurposing of FDA-approved drugs. Future technological advancements in automation and high-powered imaging, as well as the development and characterization of new mutant and transgenic lines, will expand the scope of chemical genetics in zebrafish.

Keywords Chemical screen • Chemical genetics • Zebrafish • Cancer • Drug discovery • Therapeutics

Introduction

Chemical genetics is the use of small molecules to perturb biological pathways. Target-based and phenotype-driven chemical screens can elucidate mechanisms of developmental programs and signal transduction pathways, and simultaneously identify novel therapeutic drugs. Chemical screens may be designed with a

M. Dang • R. Fogley • L.I. Zon (✉)
Stem Cell Program and Division of Hematology/Oncology, Boston Children's Hospital, Boston, MA 02115, USA

Dana Farber Cancer Institute, Howard Hughes Medical Institute, Harvard Stem Cell Institute, Harvard Medical School, 1 Blackfan Circle, Boston, MA 02115, USA
e-mail: mdang@fas.harvard.edu; rfogley@enders.tch.harvard.edu; zon@enders.tch.harvard.edu

D.M. Langenau (ed.), *Cancer and Zebrafish*, Advances in Experimental Medicine and Biology 910, DOI 10.1007/978-3-319-30654-4_5

target-based approach, in which libraries of compounds are assayed in vitro for activation or inhibition of a known molecular target. Chemical hits identified to act on a specific target can then be evaluated in vivo. In a phenotype-driven chemical screen, small molecules are used to disrupt signal transduction pathways in vitro or in vivo, causing aberrant phenotypes. Targets of chemical hits can then be identified using biochemical and chemoinformatic techniques. Forward phenotype-driven genetic screening of large small molecule libraries is a powerful tool for implicating previously unidentified genes and pathways in disease.

While cell-based and biochemical small molecule screens have proven successful in identifying novel therapeutics, these methods often fail in vivo validation. For example, compounds can induce a quantifiable response in vitro, but are highly toxic or possess poor pharmaco-properties in vivo. Similarly, compounds may fail to elicit a detectable response in vitro, but cause a significant, robust response in vivo due to the presence of niche microenvironments or complex physiological signaling networks within the organism. For these reasons, the use of in vivo animal models enables more precise detection of physiologically relevant events and, in parallel, evaluations of drug efficacy and toxicity.

The zebrafish, *Danio rerio*, has been established as an indispensable in vivo model for chemical genetics and offers many benefits over traditional in vitro screening approaches. Genetic conservation of biological pathways between zebrafish and humans promotes the identification of clinically relevant drugs. High fecundity makes the zebrafish amenable to large-scale chemical screening, with females producing up to 300 embryos per week. Embryos develop *ex utero* on a scale of days, allowing for rapid screening. The aqueous environment of development is well-suited for chemical solubilization for high-throughput drug delivery and offers temporal control of pathway perturbation. Finally, embryo transparency enables visualization of all developmental stages and facilitates the detection of aberrant phenotypes. Taken together, the zebrafish has many advantages that can be scaled and optimized for chemical genetic screening.

The first high-throughput forward chemical genetic screen assayed 1100 synthetic small molecules to investigate vertebrate development and identify novel compounds that regulate embryogenesis [1]. Most importantly, the use of whole zebrafish embryos enabled researchers to simultaneously assess the effect of each small molecule on all developmental pathways. The screen identified several compounds that modulate development of the central nervous system, the cardiovascular system, the neural crest, and the ear [1]. Since this pioneering study, in vivo chemical screens using zebrafish have been utilized to study vertebrate development [2–5], behavior [6], and disease [7–10]. Notably, in vivo chemical screening using zebrafish has been instrumental in identifying novel cancer therapeutics [11–16].

This chapter outlines the general principles of chemical screening, with a focus on screen design, advantages of the zebrafish system, and future technologies to overcome current limitations. In addition, we highlight the important contribution of chemical genetics to cancer research, profiling three chemical screens performed in zebrafish to identify genes involved in oncogenic pathways and to discover novel melanoma and T-cell acute lymphoblastic leukemia (T-ALL) therapies.

Chemical Genetic Screens

Experimental Workflow

The workflow of in vivo small molecule screens using zebrafish embryos (Fig. 1) is straightforward and highly customizable to the scientific investigation. A small-scale trial screen is first performed with a limited number of embryos and compounds to

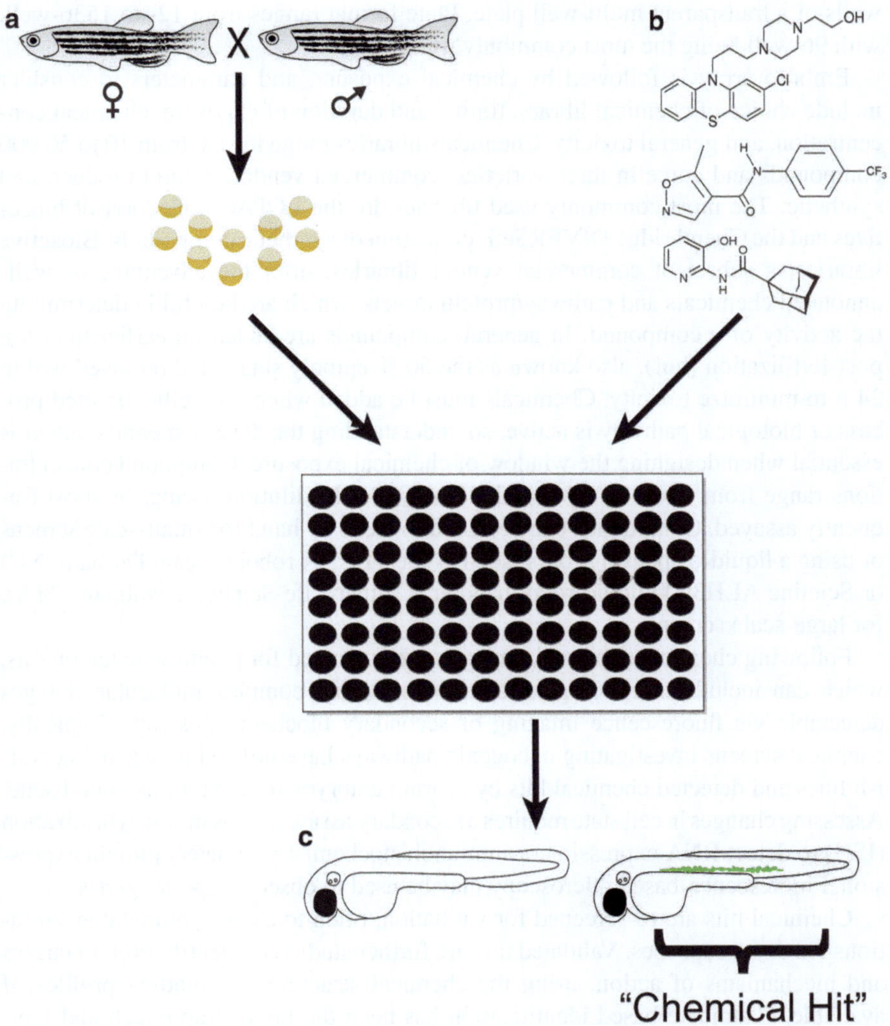

Fig. 1 Chemical screening workflow. (**a**) Developmentally synchronous embryos generated from adult zebrafish are collected and healthy embryos are arrayed into a multi-well format. (**b**) Chemicals are added to the multi-well plate at a specified dilution, and removed after the appropriate incubation period. (**c**) The embryos from individual wells are scored for the specified phenotypic endpoint

optimize experimental parameters. The number of embryos, choice of chemical library, timing and duration of chemical exposure, chemical concentration, and embryonic stage are important variables that require optimization [17–20].

In a full-scale screen, a large number of viable, developmentally synchronous embryos are first collected. This can be achieved through the setup of numerous individual breeding cages, typically containing 5 fish: 3 females per 2 males. Alternately, up to 200 adult zebrafish fish can be bred *en masse* using a specialized breeding vessel, producing up to 10,000 developmentally synchronous embryos within minutes [21]. Viable embryos are selected and re-distributed into individual wells of a transparent multi-well plate. Plate format ranges from 12- to 1536-well, with 96-well being the most commonly used.

Embryo array is followed by chemical exposure, and parameters to consider include choice of chemical library, timing and duration of exposure, chemical concentration, and general toxicity. Chemicals libraries range in size from 10 to 50,000 compounds and come in three varieties: commercial vendor, natural product, and synthetic. The most commonly used libraries are the LOPAC collection of bioactives and the Chembridge DIVERSetE collection of synthetic compounds. Bioactive libraries, a subset of commercial vendor libraries, offer the advantage of well-annotated chemicals and pathway/protein targets, which are helpful in determining the activity of a compound. In general, compounds are added no earlier than 6 h post-fertilization (hpf), also known as the 50 % epiboly stage, and removed within 24 h to minimize toxicity. Chemicals must be added when a specific targeted process or biological pathway is active, so understanding the developmental context is essential when designing the window of chemical exposure. Compound concentrations range from 1 to 100 mM, with 10 and 20 mM dilutions being the most frequently assayed. Compounds can be added to wells by hand for small-scale screens or using a liquid-dispensing robot, such as the TECAN robot (Tecan, Durham, NC) or Scicline ALH3000 liquid robot handler (Caliper Life Sciences, Waltham, MA), for large-scale screens.

Following chemical exposure, embryos are evaluated for phenotypic alterations, which can include morphological abnormalities and complex molecular changes detectable via fluorescence imaging or secondary biochemical assays. Typically, chemical screens investigating oncogenic pathways have utilized transgenic zebrafish lines and detected chemical hits by scoring embryos for alterations in cell state. Assessing changes in cell state requires a secondary assay, such as in situ hybridization (ISH) to detect RNA expression or immunohistochemistry to detect protein expression. Fluorescence-based microscopy may be used to observe reporter genes.

Chemical hits are re-screened for validation, often to assess optimal concentrations via dose responses. Validated hits are further studied to identify protein targets and mechanisms of action, using the chemical structure and binding profiles, if available. Candidate-based identification has been the favored approach and typically involves the use of annotated bioactive libraries, chemoinformatic analysis, and/or genetic studies. Annotated bioactive libraries are frequently used for screening because they contain information about the pathway and protein targets of small molecules. If the target of a small molecule is unknown, chemoinformatic analysis can be employed to identify structurally similar compounds with known activities. Alternately, the chemical of interest can be tagged with a magnetic bead, such that its endogenous bioactivity is unaltered, and isolated for the identification of binding partners via biochemical assays and computer simulations. Finally, genetic studies can be used to elucidate the activity of a compound. For example, gene expression can be measured by microarray analysis, real-time PCR (RT-PCR), or ISH, while candidate genes can be knocked down in an attempt to phenocopy the effect of the small molecule.

Advantages of In Vivo Small Molecule Screening Using a Zebrafish Model

The zebrafish offers a host of advantages for high-throughput small molecule screening, including translational appeal, high fecundity, and a unique set of developmental characteristics that support chemical treatment, genetic manipulation, and phenotype detection. High-quality genome sequencing and a complete annotation of zebrafish protein-coding genes have revealed that 71.4 % of human genes have at least one zebrafish ortholog, and, reciprocally, that 69 % of zebrafish genes have at least one human ortholog [22]. Notably, the conservation of functional domains, a category that includes drug-binding targets, nears 100 % [23]. In a study that assayed 23 compounds known to cause QT prolongation in humans, 22 compounds tested positive in zebrafish, underscoring a high conservation of drug responses [24]. Similarly, toxic reactions in zebrafish, including teratogenic effects [25–27], are comparable to those in mammalian models [28]. Conservation of carcinogenic effects has also been supported [29, 30]. Taken together, genetic similarities between zebrafish and humans and comparable responses to chemical treatment between zebrafish and mammalian models provide a promising translation of newly identified therapeutic compounds into the clinic.

The simultaneous screening of thousands of small molecules requires a large number of viable, developmentally synchronized embryos. Females are typically capable of producing 100–300 embryos per week. Multiple matings through the setup of numerous breeding cages, therefore, can amass a sufficient number of embryos for large-scale screens. However, independent mating of breeding pairs can result in a final pool of mildly asynchronous embryos and requires a large amount of space and input of time. The invention of a specialized breeding vessel, the iSpawn (Fig. 2), has effectively eliminated these obstacles [21]. While maintaining a small fingerprint of 2.92 ft^2, the vessel provides a platform for the simultaneous mating of nearly 200 fish within a short window of time, resulting in spawning rates of up to 860 embryos per minute and a collection count of nearly 10,000 highly developmentally synchronous embryos [21]. Embryo collection and equipment setup and breakdown are more efficient compared to conventional breeding methods, allowing large-scale screens to be completed in a period of weeks as opposed to months [21].

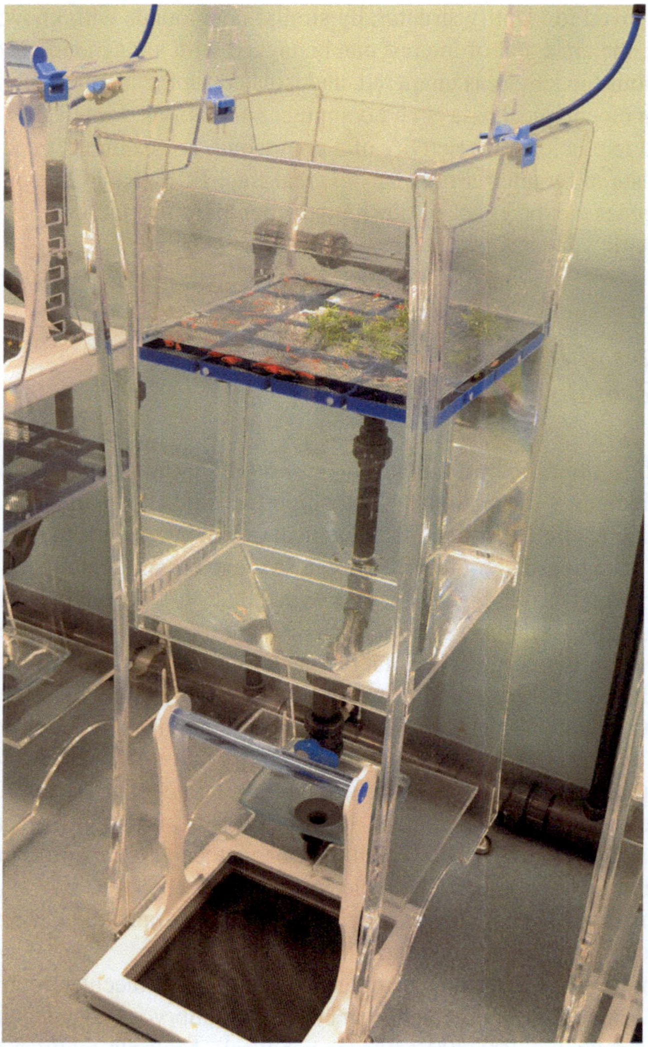

Fig. 2 The iSpawn is a specialized breeding tank designed to collect thousands of developmentally synchronous embryos. Approximately 200 adult zebrafish can breed in the shallow environment created by the mesh divider. Upwards of 10,000 embryos can be collected at the base of the vessel in minutes

External development of embryos in an aquatic environment provides the critical advantage of screening small molecules in vivo and affords a variable of temporal control not previously possible with classical genetic mutagenesis studies. Small molecules are easily solubilized in water and absorbed through the skin and gills where, once internal, they are subject to metabolic modifications. Indeed, exposure of some compounds to an active metabolic pathway has proven vital for efficacy. Murphey and colleagues screened 16,320 small molecules in zebrafish for mitotic inhibitors and found 14 novel compounds with cell cycle activity that, despite numerous screens in mammalian cell lines using the same chemical library, were previously unidentified [31].

In addition to enabling metabolic activation of small molecules, in vivo screening using zebrafish affords other benefits over cell-based screening, without compromising the magnitude of assayed libraries. Use of the entire organism enables exposure of small molecules to all biological pathways, rather than to a subset of pathways specific to a single cell line. This advantage has been especially important in forward phenotype-based screens, where the pathways and genes targeted by a small molecule are initially unknown. Similarly, use of an intact organism to screen compounds allows for the expression of phenotypes resulting from the interplay of multiple pathways, cell–cell interactions, and cell-niche interactions.

The *ex utero* development of transparent embryos facilitate visual assessment of broad phenotypic changes during chemical screening. Early chemical screens utilized wild type (WT) embryos and visually scored abnormalities ranging from developmental delays to deformities in the central nervous and cardiovascular systems [1]. Subsequently, the system was expanded to study cardiac function [24, 32], regeneration [4, 33], and animal behavior [6, 34]. For example, the photomotor response (PMR) is a behavior-based phenotypic endpoint that measures animal activity in response to light stimulation. Assessing the PMR in 36 hpf zebrafish embryos exposed to a large library of chemical compounds revealed complex behavioral patterns and identified novel psychotropic drugs [6, 34]. Another chemical screen performed in zebrafish identified bradycardia-inducing drugs, which later were shown to cause QT prolongation in humans [24]. Finally, a chemical screen of 2000 bioactive compounds revealed 17 chemicals that inhibit fin regeneration in zebrafish embryos and elucidated key pathways involved in vertebrate tissue regeneration [4].

In recent years, chemical genetic screening has grown in complexity due to the characterization of mutant zebrafish lines, making it possible to identify chemical inhibitors of mutant phenotypes through chemical suppressor screens. Stern and colleagues—full study featured below—utilized *crash&burn* (*crb*), a cell cycle arrest mutant, to identify oncogenic genes [35]. Chemical screening in a polycystic kidney disease mutant (*pkd2*) background identified inhibitory functions for histone deacetylase inhibitors trichostatin-A and valproic acid in kidney cyst formation [36]. Finally, chemical screening in *breakdance* (*kcnh2*) mutants revealed two compounds, flurandrenolide and 2-methoxy-*N*-(4-methylphenyl) benzamide (2-MMB), that rescue long QT cardiac phenotypes [9].

The development of transgenic zebrafish lines has similarly expanded the scope of chemical genetic screens by incorporating readouts detectable by advanced microscopy. Transgenic manipulation can specifically label physiological structures, cellular lineages, and mature cell types. For example, the first chemical screen to identify anti-angiogenic compounds utilized the Tg(*VEGFR2*:*GRCFP*) fluorescent transgenic line [37]. Additionally, Tg(*fli*:*EGFP*) and Tg(*flk*:*EGFP*) are common transgenic lines that fluorescently label the complete vasculature, and have been used in chemical screens to identify modulators of cardiovascular development [38, 39] and anti-angiogenic compounds [11]. Another transgenic fluorescent reporter, Tg(*lyve1:egfp*)^{nz150}, labels lymphatic vessels through the expression of lymphatic vessel endothelial hyaluronan receptor 1, and has been used to identify FDA-approved drugs with anti-lymphatic activity [16]. Finally, transgenic lines may mark specific cell lineages or types. Tg(*lck:EGFP*), which labels T-cells, enabled the discovery of Lenaldekar as a novel and selective anti-leukemic compound [13]. Future developments in zebrafish transgenesis, including the expansion of tissue specific lines and use of temporally controllable promoters, will contribute to the growing field of in vivo phenotype-based chemical genetics.

In vivo screening of molecules also provides important information on pharmacodynamics, pharmacokinetics, and toxicity, effectively combining drug discovery and animal testing. Bridging the gap between discovery and therapy has proven particularly fruitful through the screening and repurposing of FDA-approved drugs. For example, Gutierrez et al.—full study featured below—performed a screen in zebrafish to identify compounds amongst a library of FDA-approved drugs with activity against T-ALL. Perphenazine, an antipsychotic, was found to induce apoptosis in T-ALL cells through dephosphorylation of protein phosphatase 2A (PP2A) [15]. Similarly, White and colleagues—also featured below—identified leflunomide, an FDA-approved anti-arthritis drug [40], as a novel melanoma therapeutic [12].

An important advantage of the zebrafish model system that is manifested in many elements of high-throughput chemical screening is economy: of funds, time, and space. Zebrafish are relatively inexpensive to house and maintain compared to mammalian models, and only a small number of adult fish are required to produce thousands of embryos. Development of the iSpawn has reduced the amount of equipment needed to amass a large volume of embryos, as well as the input of time necessary for setup, collection, and breakdown [21]. Small embryo size—approximately 1 mm in diameter—conveniently allows molecules to be screened in small, multi-well plate format, economizing on both space and the cost of chemical compounds. Finally, the marriage of drug discovery and toxicological testing with use of an in vivo model is an efficient use of time and funds and provides a platform for rapid translation into the clinic.

Limitations and Future Directions of Small Molecule Screening in Zebrafish

The field of chemical genetics in zebrafish will continue to expand in the coming years due to the myriad of genetic and developmental characteristics that make the zebrafish system an excellent model for high-throughput chemical screening. While current technologies allow for a diversity of chemical screen designs, technological advancements in automated embryo handling and imaging will overcome current limitations and greatly expand the breadth and depth of the field.

Traditionally, the process of collecting, sorting, and arraying embryos into a multi-well format has been a time intensive task, representing a major bottleneck in high-throughput workflow. The introduction of robots that remove dead embryos, as well as selectively sort a specified number of embryos into multi-well plates has recently attenuated this obstacle. For example, the Complex Object Parametric Analyzer and Sorter, COPAS (Union Biometrica, Holliston, MA), can selectively sort a desired number of embryos into individual wells, while simultaneously eliminating dead embryos. Another automated system, the ZebraFactor, is capable of filling a 96-well plate within 11 min [41]. Systems that both dechorionate and sort embryos are also available for use [42], effectively reducing the burden of time placed on researchers. In addition to automated embryo sorting and arraying, more complex systems that integrate fluorescence-based technologies similar to fluorescent activated cell sorting (FACS) capabilities will be useful for future experiments (Union Biometrica).

Other limitations include constraints regarding chemical compounds. Screening of small molecules in vivo using zebrafish necessitates that compounds are readily soluble in water and easily absorbed by the skin and gills, thereby eliminating water insoluble chemicals as potential hits. For example, a study screening 23 drugs known to cause cardiotoxicity in humans resulted in 4 of 5 false negative results in zebrafish due to poor absorption. Drug efficacy of false negatives was confirmed by microinjection [24]. Thus, the hydrophilicity of compounds is an important consideration when designing a screen, and large-scale delivery of insoluble or impermeable compounds remains an obstacle. Further, compound permeability may be affected by the genetic background of the fish, and may vary from clutch to clutch. For this reason, it is beneficial to pool embryos prior to plate distribution.

The developmental stage of embryos can have a significant effect on the efficacy of a compound. For example, the epidermal layers of larvae and adults are less permeable to small molecules than the embryonic epidermal layer. Small molecules in larvae and adults are primarily taken up through the gills, rather than absorbed through the skin, and are thus subjected to different metabolic environments. Depending upon the developmental stage assayed, differences in drug distribution throughout the whole organism and in cell environment exposure can affect chemical activity and may lead to inconsistent results. Future chemical screens in adult zebrafish may require extensive characterization of pharmacodynamics and pharmacokinetics properties as a proof-of-principle.

Fluorescence imaging is a common endpoint for chemical screening, and the development of more powerful, automated, high-content imaging machines will provide a platform for improved embryo image acquisition. Fluorescent-based reporters rely on microscopes for signal detection, and a limitation includes the inability to visualize fluorescent expression in deep tissues. A benefit of high-powered confocal imaging is its ability to image deep tissues within the embryo. However, embryo orientation and positioning, time required for image acquisition, and the ability to uniformly analyze images remain challenges. Currently, researchers are developing microfluidics-based technologies, such as the Vertebrate Automated Screening Technology (VAST) Bioimager, that properly orient embryos such that imaging can be processed in a uniform, high-throughput manner [43–45]. The development of specific molds using 3-D printing technology can also aid in uniform positioning of embryos [46, 47]. Most importantly, the development of automated imaging methods will not only alleviate the burden of time that accompanies manual scoring, and but also improve upon the error rate attributed to human judgment.

Generally, high-throughput screening has remained challenging in juveniles and adults. The largest hurdle in using adult zebrafish in chemical studies is the difficulty of controlling drug delivery. Directly solubilizing chemicals into water for passive transport into the gills can be expensive and limits dosage control. Current techniques of drug administration in adults generally require anesthesia and include invasive approaches, such as retro-orbital injections [48], intraperitoneal injections [49], and oral gavage [50]. Therefore, current technologies render high-throughput and long-term drug administration a challenge in adult zebrafish.

The pigmentation of adult zebrafish also presents an obstacle, as phenotypes are less easily detected. The creation of a transparent adult zebrafish devoid of all pigment has addressed this issue to enable the visualization of internal phenotypes and organs [51]. The *casper* mutant, named for it's 'ghost-like' transparent phenotype, has laid the groundwork for future studies, including suppression screens utilizing transgenics and fluorescently-labeled tumor cells to identify anti-cancer and anti-metastatic drugs. From an economic standpoint, maintenance of adult zebrafish requires more resources. Increased size of adults notably necessitates a greater space for drug delivery—compared to the compact multi-well plates used for embryos—and thus larger volumes of compound. Unlike early-stage embryos, adult fish must be anesthetized prior to visualization and manipulation. The creation of a high-throughput method for imaging up to 30 unanesthetized adult fish in multiple fluorescent channels, however, is alleviating this burden [52].

Indeed, many researchers have opted to screen for embryonic markers, forgoing the availability of adult models, to circumvent the logistical obstacles associated with screening adults. However, a study in fin regeneration has revealed the feasibility of screening adults, albeit with a relatively limited number of compounds [33]. Similarly, Chen and colleagues performed a secondary transplant screen in adult zebrafish to identify compounds that inhibit embryonal rhabdomyosarcoma growth [53]. 12 hits from a large-scale screen that tested the effects of nearly 40,000 compounds on human embryonal rhabdomyosarcoma cell differentiation were combined with 83 compounds previously identified as having anti-RAS activity [14] for the adult screen [53]. The study indicates that adult screens may be more feasible as a follow-up to in vitro and in vivo embryonic screens performed on a larger scale.

Chemical Genetics and Cancer

Use of Chemical Genetics in Zebrafish to Identify Cancer Therapies

Cancer is a complex and heterogeneous disease that often involves the interaction of multiple signaling pathways. Chemical genetics offers an unbiased, broad-spectrum approach to identify molecules that interact with active oncogenic pathways, and to discover novel cancer therapies. The zebrafish is a valuable in vivo model for translational oncology [54–56] due to its adaptability in transgenesis [57–59], genome-editing [60–64], transplantation [65, 66], and imaging. Since the spontaneous development of cancer in the zebrafish is rare, transgenic cancer models are used to genetically and histopathologically mimic human cancers in the zebrafish. To date, a high-throughput screening approach has been used to identify targeted treatments for melanoma, leukemias (including AML and T-ALL), rhabdomyosarcoma, and prostate cancer (Table 1). Small molecule screens have also been employed to find inhibitors of events related to cancer metastasis, including angiogenesis and lymphangiogenesis; cell cycle aberrations; and oncogenic pathways, including Ras, Hedgehog, and Wnt/b-catenin (Table 1).

The reactivation of developmental programs that can accompany oncogenesis makes the zebrafish an attractive model for cancer studies. Embryos modeling cancer-initiating events have been used for high-throughput screens, economizing on

Table 1 Cancer-related chemical genetic screens in the zebrafish

Phenotype	Zebrafish	Scoring method	References
Angiogenesis	Tg(VEGFR:GRCFP)	Fluorescence imaging	[37]
Angiogenesis	*Tg(flk1:EGFP)*	Fluorescence imaging	[11, 67]
Angiogenesis	*Tg(fli-1a:EGFP)y1*	Fluorescence imaging	[10]
Cell cycle	*Crash and burn* (*crb*) mutant	IHC	[35]
Cell cycle	Wild type *and crb* mutant	IHC	[31]
Embryonal rhabdomyosarcoma	CG1-strain transplant recipient of ERMS cells from Tg(mylz-2mCherry; rag2:kRAS^{G12D})	Fluorescence imaging	[53]
Hedgehog signaling	*Tg(Gli-GFP)*	Fluorescence imaging	[26, 28]
Inflammation	Tg(BACmpx:GFP) and Tg(lysC:DsRed2)	Fluorescence imaging	[69]
Inflammation	Tg(zlyz:EGFP)	Fluorescence imaging	[70]
Leukemia (T-ALL)	Tg(rag2:MycER;rag2:dsRed2) in nacre background +4OHT	Fluorescence imaging	[15]
Leukemia (T-ALL)	*Tg(lck:EGFP)*	Fluorescence imaging	[13]
Acute myelogenous leukemia (AML)	*Tg(hsp:AML1-ETO)*	ISH	[71]
Lymphangiogenesis	Wild-type	ISH	[16]
Melanoma	Wild-type	ISH	[12]
Pancreatic differentiation/Notch signaling	Tg(Tp1:hmgb1-mCherry^{ih11}; pax6b:GFP^{ulg515})	Fluorescence imaging	[72]
Prostate cancer	*Tg(flk1:EGFP)*	Fluorescence imaging	[11]
Rhabdomyosarcoma/ RAS signaling	*Tg(hsp70-HRAS^{G12V})*	ISH	[14]
Wnt/β-catenin signaling	Wild-type	Morphology, dorsoventral patterning	[73]

time, space, and funds. For example, Le and colleagues ectopically expressed RAS during embryogenesis to investigate possible mechanisms at play in tumorigenesis. Using transgenic *hsp70-HRAS*^{G12V} embryos to screen for chemical suppressors of RAS signaling, two compounds were identified that subsequently had anti-tumor activity in both a zebrafish model of rhabdomyosarcoma and a human cell line of rhabdomyosarcoma [14]. More simply, small molecules can be used to inhibit oncogenic pathways already active during development. Hao et al. used wild-type embryos to identify small molecule inhibitors of the Wnt pathway [73]. Hits were determined by a straightforward morphological phenotype, specifically dorsoventral patterning [73]. The screen identified a small molecule, Windorphen, that selectively killed human cancer cells with aberrant Wnt signaling, including colon and prostate cancer cells [73].

The developing zebrafish embryo has been useful for identifying targeted therapies against cancer stem cells, a small population of quiescent, tumorigenic cells that contribute to metastasis and relapse. Screening in vivo avoids the difficulties of purifying these rare cells and culturing them in vitro. Using *Tg(hsp:AML1-ETO)* embryos, Yeh and colleagues screened a library of small molecules to identify novel compounds that inhibit the oncogenic effects of AML1-ETO (AE) on multipotent hematopoietic progenitor cells [74]. Nimesulide, a selective COX-2 inhibitor, effectively antagonized AE activity [74] and was subsequently shown to inhibit the initiation and progression of xenograft tumors [10].

Here, we highlight three chemical genetic screens in zebrafish with an emphasis in elucidating cancer mechanisms and identifying novel cancer therapies. The first study features a chemical suppressor screening strategy to identify novel cell cycle molecules and pathways involved in cancer progression [35]. The second study illustrates the power and promise of repurposing FDA-approved drugs for cancer therapy with a rapid translation into the clinic [12]. Finally, the third highlights the value of genetic manipulation in the zebrafish model, through the use of inducible oncogenes and fluorescent reporters, in identifying novel T-ALL therapies [15].

Chemical Screen to Identify Genes Involved in Oncogenic Pathways

Identifying and understanding genes involved in cancer formation can aid in the development of targeted therapies. In 2005, Shepard and colleagues performed a zebrafish forward genetic screen to identify novel genes involved in cancer formation and identified the crash&burn (*crb*) mutant. This *crb* mutant exhibited signs of mitotic arrest due to an observable increase in mitotic and apoptotic cells and further molecular characterization of the *crb* phenotype identified a splice-donor mutation in the *bymb* gene that resulted in decreased cyclin B1 expression, mitotic arrest, and genomic instability [8]. In a follow-up investigation, Stern and colleagues employed a zebrafish-based chemical suppressor screening strategy to identify novel molecules that interact with the *bmyb* pathway. In 16 weeks, thousands of zebrafish embryos were spawned from heterozygote *crb* mutant parents, arrayed into a 384-well microplate, and treated with chemicals from the DIVERSetE library of 16,320 compounds (ChemBridge Corp).

To identify chemicals that interact with the *bmyb* pathway, the authors used whole embryo immunohistochemistry to assay for mitotic cells. A serine-10-phosphorylated histone 3 (pH3) antibody was used, as histone H3 is phosphorylated on serine 10 from late G2 to early M phase and dephosphorylated during anaphase. The chemical screen identified a single chemical that suppressed the *crb* mutant phenotype of mitotic arrest without having any mitotic effect in WT embryos. Stern and colleagues resynthesized the compound; validated the chemical structure and potency based on liquid chromatography, mass spectrometry, and nuclear magnetic resonance spectra; and named the chemical suppressor persynthamide. In Fig. 3a, WT, heterozygote (Het), and mutant (Mut) *crb* embryos are

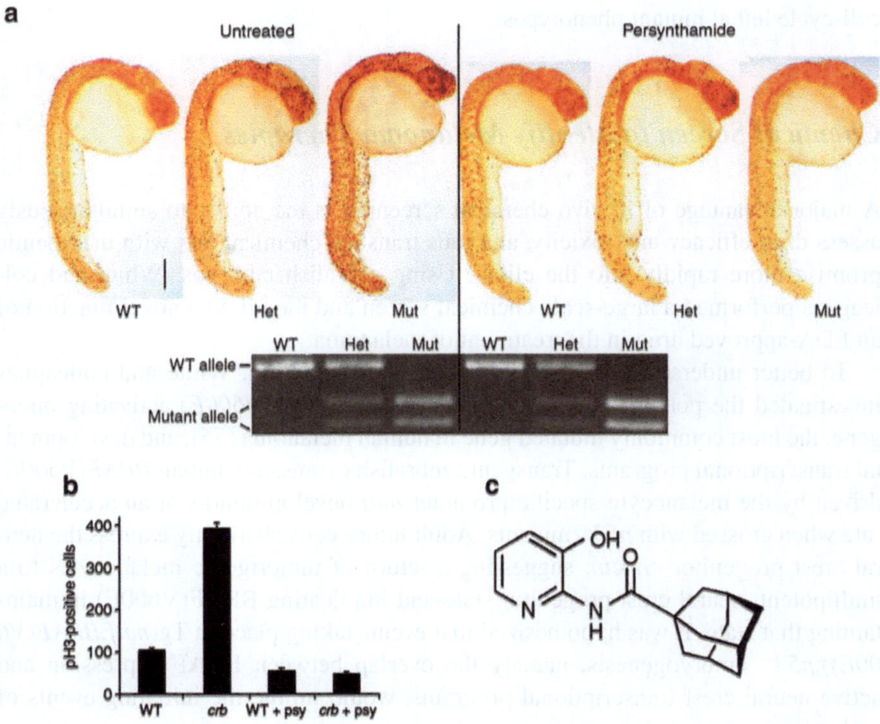

Fig. 3 Persynthamide suppresses mitotic accumulation in *crb* mutants. (**a**) Persynthamide abrogates mitotic accumulation in *crb* mutants as demonstrated by pH3 immunohistochemistry. (**b**) Quantified levels of persynthamide-mediated chemical suppression. (**c**) Chemical structure of synthesized chemical, persynthamide. Stern HM, Murphey RD, Shepard JL, Amatruda JF, Straub CT, et al. (2005) Small molecules that delay S phase suppress a zebrafish bmyb mutant. Nature chemical biology 1(7):366–370

stained for pH3 to quantify mitotic accumulation in the presence of the chemical hit persynthamide. The genotype was validated via restriction fragment length polymorphism. Figure 3b quantifies pH3-positive cells in the tails of 24 hpf embryos, and emphasizes the suppressive capacity of persynthamide, the chemical structure of which is depicted in Fig. 3c. Treatment with persynthamide restored additional *crb* mutant phenotypes, such as deformed head morphology, cell death, and polyploidy.

This screen highlights the power of chemical screening in zebrafish. The rapid development of zebrafish embryos, coupled with high fecundity, enabled researchers to screen 16,320 chemicals within 16 weeks. The chemical suppressor screen performed in *crb* heterozygous mutants identified a novel chemical compound that interacts with the *bmyb* pathway. The report demonstrates the advantages of using mutant zebrafish lines, chemical suppression of mutant phenotypes, and the elucidation of molecular characterization of cell-cycle mutants. The high degree of genetic conservation between zebrafish and humans, particularly the conservation of cell-cycle-related genes, allowed for a meaningful approach to study the mechanism of cell-cycle lethal mutant phenotypes.

Chemical Screen to Identify Melanoma Therapies

A major advantage of in vivo chemical screening is the ability to simultaneously assess drug efficacy and toxicity, and thus translate chemical hits with therapeutic promise more rapidly into the clinic. Using zebrafish embryos, White and colleagues performed a large-scale chemical screen and identified a novel function of an FDA-approved drug in the treatment of melanoma.

To better understand the events of melanoma initiation, White and colleagues investigated the potential interaction between the *BRAF*(*V600E*) activating oncogene, the most commonly mutated gene in human melanoma [75], and developmental transcriptional programs. Transgenic zebrafish expressing human *BRAF*(*V600E*) driven by the melanocyte-specific promoter *mitf* develop tumors at an accelerated rate when crossed with $p53^{-/-}$ mutants. Adult tumor cells aberrantly express the neural crest progenitor *crestin*, suggesting a return of tumorigenic melanocytes to a multipotent, neural crest progenitor state and implicating BRAF(V600E) in maintaining that state. It was hypothesized that events taking place in Tg(*mitf*:*BRAF*(*V600E*));$p53^{-/-}$ embryogenesis, namely the overlap between BRAF expression and active neural crest transcriptional programs, would mimic the initiating events of melanoma.

Based on the prediction that chemicals effective in suppressing neural crest progenitors would also be effective treatments for melanoma, a large-scale screen was performed using zebrafish embryos. 2000 compounds were tested in wild-type embryos. After treatment, embryos were fixed for a secondary ISH assay to evaluate *crestin* expression. One compound of unknown function, NSC210627, strongly reduced *crestin* expression. Using the chemoinformatic approach,

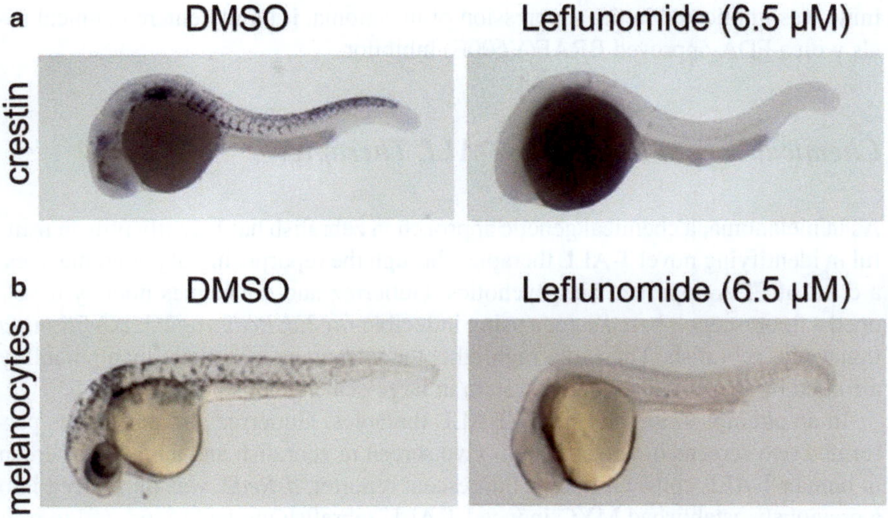

Fig. 4 Leflunomide suppresses neural crest development. (**a**) Leflunomide, a DHODH inhibitor, reduces *crestin* expression by in situ hybridization. (**b**) Leflunomide causes an absence of pigmented melanocytes. White RM, Cech J, Ratanasirintrawoot S, et al. (2011) DHODH modulates transcriptional elongation in the neural crest and melanoma. Nature 471:518–22. doi: 10.1038/nature09882

NSC210627 was found to have a similar structure to brequinar, a known dihydroorotate dehydrogenase (DHODH) inhibitor [76], and inhibited DHODH activity in vitro. Leflunomide, another DHODH inhibitor and FDA-approved anti-arthritis drug [40], phenocopied NSC210627 in vivo and was used in future studies.

The effects of leflunomide were further characterized in vivo and in vitro. Embryos treated with leflunomide lacked melanocytes at 36–48 hpf, and exhibited a nearly complete lack of melanocyte progenitors at 24 hpf (Fig. 4a, b). Treatment of a human melanoma cell line with A771726, the active metabolite of leflunomide, decreased proliferation in a dose-dependent manner. Jointly treating melanoma cells with A771726 and the BRAF(V600E) inhibitor PLX4720 had a synergistic effect in suppressing proliferation. A similar effect was observed in a mouse xenograft model, with joint treatment resulting in nearly complete tumor regression in 40 % of animals.

By selectively expressing oncogenic BRAF(V600E) in zebrafish melanocytes, White and colleagues were able to mimic melanoma initiation in embryos. Modeling melanoma onset in the zebrafish embryo provided a platform for a large-scale in vivo chemical screen aimed at identifying novel compounds that could suppress the neural crest. This study underscores zebrafish as a powerful model for drug discovery through large-scale in vivo chemical screening. By effectively combining drug discovery and animal testing and, further, by repurposing FDA-approved drugs, novel therapies can be rapidly translated into the clinic. Shortly after leflunomide was implicated in the suppression of melanoma, it jointly entered clinical trials with a FDA-approved BRAF(V600E) inhibitor.

Chemical Screen to Identify T-ALL Therapies

As in melanoma, a chemical genetic approach in zebrafish has recently proven fruitful in identifying novel T-ALL therapies through the repurposing of phenothiazines, a class of FDA-approved antipsychotics. Gutierrez and colleagues notably developed a fluorescence-based screen using inducible *rag2:dsRed2*;*rag2:MYC-ER*;*mitfa* transgenic zebrafish. The study highlights the advantage of genetic manipulability afforded by the zebrafish model system in large-scale chemical screening.

In an attempt to identify novel T-ALL therapies, Gutierrez and colleagues performed two screens in tandem: an in vivo screen in zebrafish and an in vitro screen in human T-ALL cells. Of note, a fluorescent reporter, *dsRed2*, was introduced into a previously established MYC-induced T-ALL zebrafish model [15, 65, 69] to provide a straightforward platform for phenotype detection. Four libraries—including 4880 FDA-approved drugs, drug-like small molecules, and natural products—were screened in *rag2:dsRed2*;*rag2:MYC-ER*;*mitfa* transgenic zebrafish larvae at 3 days post-fertilization (dpf). Larvae were simultaneously induced with 4-hydroxytamoxifen to activate the MYC-ER fusion protein. Chemicals were scored at 7 dpf for the ability to selectively kill MYC-overexpressing thymocytes, with either a complete loss of fluorescence or weak fluorescence constituting a hit (Fig. 5a). Perphenazine, an FDA-approved phenothiazine antipsychotic, qualified as a chemical hit (Fig. 5b) and was validated in a secondary screen by dose response.

Concurrently, the Broad Institute bioactives chemical library collection of 3194 compounds—including 2108 compounds from the in vivo screen—were assayed in KOPT-K1 cells, a NOTCH1-dependent T-ALL line. Compounds were evaluated for synergistic suppression of growth in combination with gamma-secretase inhibitors, which selectively target the NOTCH pathway. Perphenzine, again, scored as a top hit.

Follow-up studies revealed that perphenazine could suppress established tumors in zebrafish and T-ALL growth in culture in a Notch-independent manner by inducing mitochondrial apoptosis. Using ligand-affinity chromatography and mass spectrometry, protein phosphatase 2A (PP2A) was identified as a direct target of perphenazine. PP2A was proven to mediate the anti-leukemic effect of perphenazine through rapid dephosphorylation of substrates involved in oncogenic pathways, such as AKT, ERK, p70S6K, MYC, and BAD. Further, knockdown of PP2A subunits by shRNA was sufficient to impair the anti-leukemic effect of perphenazine.

The discovery of novel anti-leukemic activity by perphenazine that is Notch-independent holds great promise for the clinic, especially for patients in poor prognosis groups. While the high doses of perphenazine required to reach therapeutic benefit remain an obstacle in cancer treatment, the identification of PP2A as an important anti-leukemic agent will enable the development of new, targeted treat-

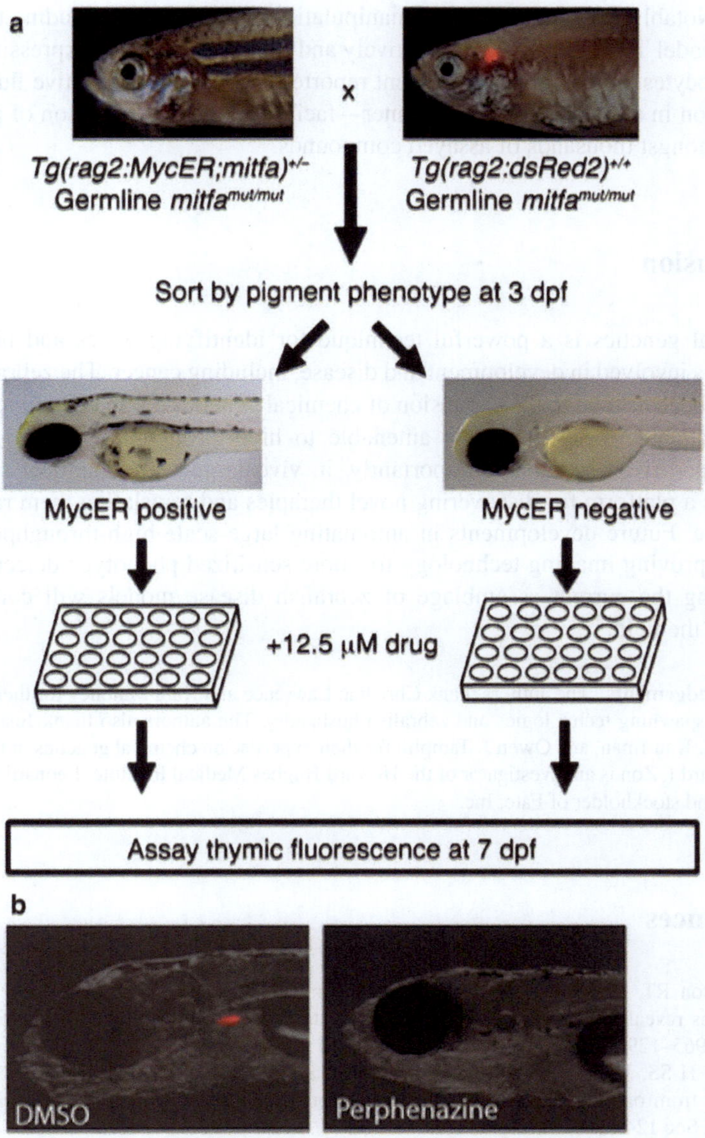

Fig. 5 Perphenazine has anti-T-ALL activity in vitro and in vivo. (**a**) The chemical screen design utilizes a 4-hydroxytamoxifen inducible fluorescent reporter in a transgenic zebrafish to visualize MYC-overexpressing thymocytes. (**b**) Perphenazine causes a decrease in MYC-overexpressing thymocytes in 7 dpf zebrafish larvae, as detected via fluorescence imaging. Gutierrez A, Pan L, Groen RW, et al. (2014) Phenothiazines induce PP2A-mediated apoptosis in T cell acute lymphoblastic leukemia. J Clin Invest 124:644–55. doi: 10.1172/JCI65093

ments. Notably, the ease of genetic manipulation in zebrafish—including the abilities to model T-ALL in vivo by selectively and conditionally over-expressing MYC in thymocytes, to introduce fluorescent reporters, and to assess relative fluorescent expression in a high-throughput manner—facilitated the identification of perphenazine amongst thousands of assayed compounds.

Conclusion

Chemical genetics is a powerful technique for identifying genes and biological pathways involved in development and disease, including cancer. The zebrafish system has been critical to the expansion of chemical genetics, as an economic, genetically malleablele model that is amenable to high-throughput, forward genetic, phenotype-driven screening. Importantly, in vivo large-scale chemical screening provides a platform for discovering novel therapies and translating them rapidly to the clinic. Future developments in automating large-scale high-throughput workflow, improving imaging technology for more sensitized phenotype detection, and expanding the current assemblage of zebrafish disease models will continue to advance the field.

Acknowledgements The authors thank Christian Lawrence and Kara Maloney for their input on zebrafish spawning technologies and zebrafish husbandry. The authors also thank Justin L. Tan, Charles K. Kauffman, and Owen J. Tamplin for their expertise on chemical genetics in the zebrafish. Leonard I. Zon is an investigator of the Howard Hughes Medical Institute. Leonard I. Zon is a founder and stockholder of Fate, Inc.

References

1. Peterson RT, Link BA, Dowling JE, Schreiber SL (2000) Small molecule developmental screens reveal the logic and timing of vertebrate development. Proc Natl Acad Sci U S A 97:12965–12969. doi:10.1073/pnas.97.24.12965
2. Moon H-SS, Jacobson EM, Khersonsky SM et al (2002) A novel microtubule destabilizing entity from orthogonal synthesis of triazine library and zebrafish embryo screening. J Am Chem Soc 124:11608–11609
3. Khersonsky SM, Jung D-WW, Kang T-WW et al (2003) Facilitated forward chemical genetics using a tagged triazine library and zebrafish embryo screening. J Am Chem Soc 125:11804–11805. doi:10.1021/ja035334d
4. Mathew LK, Sengupta S, Kawakami A et al (2007) Unraveling tissue regeneration pathways using chemical genetics. J Biol Chem 282:35202–35210. doi:10.1074/jbc.M706640200
5. Sachidanandan C, Yeh J-RJR, Peterson QP, Peterson RT (2008) Identification of a novel retinoid by small molecule screening with zebrafish embryos. PLoS One 3, e1947. doi:10.1371/journal.pone.0001947
6. Kokel D, Bryan J, Laggner C et al (2010) Rapid behavior-based identification of neuroactive small molecules in the zebrafish. Nat Chem Biol 6:231–237. doi:10.1038/nchembio.307

7. Peterson RT, Shaw SY, Peterson TA et al (2004) Chemical suppression of a genetic mutation in a zebrafish model of aortic coarctation. Nat Biotechnol 22:595–599. doi:10.1038/nbt963
8. Shepard JL, Amatruda JF, Stern HM et al (2005) A zebrafish bmyb mutation causes genome instability and increased cancer susceptibility. Proc Natl Acad Sci U S A 102:13194–13199. doi:10.1073/pnas.0506583102
9. Peal DS, Mills RW, Lynch SN et al (2011) Novel chemical suppressors of long QT syndrome identified by an in vivo functional screen. Circulation 123:23–30. doi:10.1161/CIRCULATIONAHA.110.003731
10. Zhang Y, Wang J, Wheat J et al (2013) AML1-ETO mediates hematopoietic self-renewal and leukemogenesis through a COX/β-catenin signaling pathway. Blood 121:4906–4916. doi:10.1182/blood-2012-08-447763
11. Wang C, Tao W, Wang Y et al (2010) Rosuvastatin, identified from a zebrafish chemical genetic screen for antiangiogenic compounds, suppresses the growth of prostate cancer. Eur Urol 58:418–426. doi:10.1016/j.eururo.2010.05.024
12. White RM, Cech J, Ratanasirintrawoot S et al (2011) DHODH modulates transcriptional elongation in the neural crest and melanoma. Nature 471:518–522. doi:10.1038/nature09882
13. Ridges S, Heaton WL, Joshi D et al (2012) Zebrafish screen identifies novel compound with selective toxicity against leukemia. Blood 119:5621–5631. doi:10.1182/blood-2011-12-398818
14. Le X, Pugach EK, Hettmer S et al (2013) A novel chemical screening strategy in zebrafish identifies common pathways in embryogenesis and rhabdomyosarcoma development. Development 140:2354–2364. doi:10.1242/dev.088427
15. Gutierrez A, Pan L, Groen RW et al (2014) Phenothiazines induce PP2A-mediated apoptosis in T cell acute lymphoblastic leukemia. J Clin Invest 124:644–655. doi:10.1172/JCI65093
16. Astin JW, Jamieson SM, Eng TC, Flores MV, Misa JP, Chien A, Crosier KE, Crosier PS (2014) An in vivo anti-lymphatic screen in zebrafish identifies novel inhibitors of mammalian lymphangiogenesis and lymphatic-mediated metastasis. Mol Cancer Ther 10:2450–2462
17. Rennekamp AJ, Peterson RT (2015) 15 years of zebrafish chemical screening. Curr Opin Chem Biol 24C:58–70. doi:10.1016/j.cbpa.2014.10.025
18. Kaufman CK, White RM, Zon L (2009) Chemical genetic screening in the zebrafish embryo. Nat Protoc 4:1422–1432. doi:10.1038/nprot.2009.144
19. Peterson RT, Fishman MC (2011) Designing zebrafish chemical screens. Methods Cell Biol 105:525–541. doi:10.1016/B978-0-12-381320-6.00023-0
20. Tan JL, Zon LI (2011) Chemical screening in zebrafish for novel biological and therapeutic discovery. Methods Cell Biol 105:493–516. doi:10.1016/B978-0-12-381320-6.00021-7
21. Adatto I, Lawrence C, Thompson M, Zon LI (2011) A new system for the rapid collection of large numbers of developmentally staged zebrafish embryos. PLoS One 6, e21715. doi:10.1371/journal.pone.0021715
22. Howe K, Clark MD, Torroja CF et al (2013) The zebrafish reference genome sequence and its relationship to the human genome. Nature 496:498–503. doi:10.1038/nature12111
23. Langheinrich U (2003) Zebrafish: a new model on the pharmaceutical catwalk. Bioessays 25:904–912. doi:10.1002/bies.10326
24. Milan DJ, Peterson TA, Ruskin JN et al (2003) Drugs that induce repolarization abnormalities cause bradycardia in zebrafish. Circulation 107:1355–1358
25. Van Leeuwen CJ, Grootelaar EM, Niebeek G (1990) Fish embryos as teratogenicity screens: a comparison of embryotoxicity between fish and birds. Ecotoxicol Environ Saf 20:42–52
26. Yang L, Ho NY, Alshut R et al (2009) Zebrafish embryos as models for embryotoxic and teratological effects of chemicals. Reprod Toxicol 28:245–253. doi:10.1016/j.reprotox.2009.04.013
27. Selderslaghs IW, Van Rompay AR, De Coen W, Witters HE (2009) Development of a screening assay to identify teratogenic and embryotoxic chemicals using the zebrafish embryo. Reprod Toxicol 28:308–320. doi:10.1016/j.reprotox.2009.05.004
28. Parng C, Seng WL, Semino C, McGrath P (2002) Zebrafish: a preclinical model for drug screening. Assay Drug Dev Technol 1(1):41–48

29. Spitsbergen JM, Tsai HW, Reddy A et al (2000) Neoplasia in zebrafish (Danio rerio) treated with 7,12-dimethylbenz[a]anthracene by two exposure routes at different developmental stages. Toxicol Pathol 28:705–715
30. Spitsbergen JM, Tsai HW, Reddy A et al (2000) Neoplasia in zebrafish (Danio rerio) treated with N-methyl-N'-nitro-N-nitrosoguanidine by three exposure routes at different developmental stages. Toxicol Pathol 28:716–725
31. Murphey RD, Stern HM, Straub CT, Zon LI (2006) A chemical genetic screen for cell cycle inhibitors in zebrafish embryos. Chem Biol Drug Des 68:213–219. doi:10.1111/j.1747-0285.2006.00439.x
32. Peal DS, Peterson RT, Milan D (2010) Small molecule screening in zebrafish. J Cardiovasc Transl Res 3:454–460. doi:10.1007/s12265-010-9212-8
33. Oppedal D, Goldsmith MI (2010) A chemical screen to identify novel inhibitors of fin regeneration in zebrafish. Zebrafish 7:53–60. doi:10.1089/zeb.2009.0633
34. Kokel D, Peterson RT (2011) Using the zebrafish photomotor response for psychotropic drug screening. Methods Cell Biol 105:517–524. doi:10.1016/B978-0-12-381320-6.00022-9
35. Stern HM, Murphey RD, Shepard JL, Amatruda JF, Straub CT et al (2005) Small molecules that delay S phase suppress a zebrafish bmyb mutant. Nat Chem Biol 1(7):366–370
36. Cao Y, Semanchik N, Lee SH et al (2009) Chemical modifier screen identifies HDAC inhibitors as suppressors of PKD models. Proc Natl Acad Sci U S A 106:21819–21824. doi:10.1073/pnas.0911987106
37. Tran TC, Sneed B, Haider J et al (2007) Automated, quantitative screening assay for antiangiogenic compounds using transgenic zebrafish. Cancer Res 67:11386–11392. doi:10.1158/0008-5472.CAN-07-3126
38. Kitambi SS, McCulloch KJ, Peterson RT, Malicki JJ (2009) Small molecule screen for compounds that affect vascular development in the zebrafish retina. Mech Dev 126:464–477. doi:10.1016/j.mod.2009.01.002
39. Leet JK, Lindberg CD, Bassett LA et al (2014) High-content screening in zebrafish embryos identifies butafenacil as a potent inducer of anemia. PLoS One 9, e104190. doi:10.1371/journal.pone.0104190
40. Schiff MH, Strand V, Oed C, Loew-Friedrich I (2000) Leflunomide: efficacy and safety in clinical trials for the treatment of rheumatoid arthritis. Drugs Today (Barc) 36(6): 383–394
41. Graf SF, Hötzel S, Liebel U et al (2011) Image-based fluidic sorting system for automated Zebrafish egg sorting into multiwell plates. J Lab Autom 16:105–111. doi:10.1016/j.jala.2010.11.002
42. Mandrell D, Truong L, Jephson C et al (2012) Automated zebrafish chorion removal and single embryo placement: optimizing throughput of zebrafish developmental toxicity screens. J Lab Autom 17:66–74. doi:10.1177/2211068211432197
43. Pardo-Martin C, Chang T-YY, Koo BK et al (2010) High-throughput in vivo vertebrate screening. Nat Methods 7:634–636. doi:10.1038/nmeth.1481
44. Wielhouwer EM, Ali S, Al-Afandi A et al (2011) Zebrafish embryo development in a microfluidic flow-through system. Lab Chip 11:1815–1824. doi:10.1039/C0LC00443J
45. Chang T-YY, Pardo-Martin C, Allalou A et al (2012) Fully automated cellular-resolution vertebrate screening platform with parallel animal processing. Lab Chip 12:711–716. doi:10.1039/C1LC20849G
46. Masselink W, Wong JC, Liu B et al (2014) Low-cost silicone imaging casts for zebrafish embryos and larvae. Zebrafish 11:26–31. doi:10.1089/zeb.2013.0897
47. Wittbrodt JN, Liebel U, Gehrig J (2014) Generation of orientation tools for automated zebrafish screening assays using desktop 3D printing. BMC Biotechnol 14:36. doi:10.1186/1472-6750-14-36
48. Pugach EK, Li P, White R, Zon L (2009) Retro-orbital injection in adult zebrafish. J Vis Exp. doi:10.3791/1645

49. Kinkel MD, Eames SC, Philipson LH, Prince VE (2010) Intraperitoneal injection into adult zebrafish. J Vis Exp. doi:10.3791/2126
50. Collymore C, Rasmussen S, Tolwani RJ (2013) Gavaging adult zebrafish. J Vis Exp. doi:10.3791/50691
51. White RM, Sessa A, Burke C et al (2008) Transparent adult zebrafish as a tool for in vivo transplantation analysis. Cell Stem Cell 2:183–189. doi:10.1016/j.stem.2007.11.002
52. Blackburn JS, Liu S, Raimondi AR et al (2011) High-throughput imaging of adult fluorescent zebrafish with an LED fluorescence macroscope. Nat Protoc 6:229–241. doi:10.1038/nprot.2010.170
53. Chen EY, DeRan MT, Ignatius MS et al (2014) Glycogen synthase kinase 3 inhibitors induce the canonical WNT/β-catenin pathway to suppress growth and self-renewal in embryonal rhabdomyosarcoma. Proc Natl Acad Sci U S A 111:5349–5354. doi:10.1073/pnas.1317731111
54. White R, Rose K, Zon L (2013) Zebrafish cancer: the state of the art and the path forward. Nat Rev Cancer 13:624–636. doi:10.1038/nrc3589
55. Yen J, White RM, Stemple DL (2014) Zebrafish models of cancer: progress and future challenges. Curr Opin Genet Dev 24:38–45. doi:10.1016/j.gde.2013.11.003
56. Barriuso J, Nagaraju R, Hurlstone A (2015) Zebrafish: a new companion for translational research in oncology. Clin Cancer Res. doi:10.1158/1078-0432.CCR-14-2921
57. Berghmans S, Murphey RD, Wienholds E et al (2005) tp53 mutant zebrafish develop malignant peripheral nerve sheath tumors. Proc Natl Acad Sci U S A 102:407–412. doi:10.1073/pnas.0406252102
58. Langenau DM, Feng H, Berghmans S et al (2005) Cre/lox-regulated transgenic zebrafish model with conditional myc-induced T cell acute lymphoblastic leukemia. Proc Natl Acad Sci U S A 102:6068–6073. doi:10.1073/pnas.0408708102
59. Langenau DM, Jette C, Berghmans S et al (2005) Suppression of apoptosis by bcl-2 overexpression in lymphoid cells of transgenic zebrafish. Blood 105:3278–3285. doi:10.1182/blood-2004-08-3073
60. Foley JE, Maeder ML, Pearlberg J et al (2009) Targeted mutagenesis in zebrafish using customized zinc-finger nucleases. Nat Protoc 4:1855–1867. doi:10.1038/nprot.2009.209
61. Foley JE, Yeh J-RJR, Maeder ML et al (2009) Rapid mutation of endogenous zebrafish genes using zinc finger nucleases made by Oligomerized Pool ENgineering (OPEN). PLoS One 4, e4348. doi:10.1371/journal.pone.0004348
62. Sander JD, Yeh J-RJR, Peterson RT, Joung JK (2011) Engineering zinc finger nucleases for targeted mutagenesis of zebrafish. Methods Cell Biol 104:51–58. doi:10.1016/B978-0-12-374814-0.00003-3
63. Hwang WY, Fu Y, Reyon D et al (2013) Heritable and precise zebrafish genome editing using a CRISPR-Cas system. PLoS One 8, e68708. doi:10.1371/journal.pone.0068708
64. Hwang WY, Fu Y, Reyon D et al (2013) Efficient genome editing in zebrafish using a CRISPR-Cas system. Nat Biotechnol 31:227–229. doi:10.1038/nbt.2501
65. Langenau DM, Traver D, Ferrando AA et al (2003) Myc-induced T cell leukemia in transgenic zebrafish. Science 299:887–890. doi:10.1126/science.1080280
66. Ignatius MS, Chen E, Elpek NM et al (2012) In vivo imaging of tumor-propagating cells, regional tumor heterogeneity, and dynamic cell movements in embryonal rhabdomyosarcoma. Cancer Cell 21:680–693. doi:10.1016/j.ccr.2012.03.043
67. Weng AP, Ferrando AA, Lee W et al (2004) Activating mutations of NOTCH1 in human T cell acute lymphoblastic leukemia. Science 306:269–71. doi:10.1126/science.1102160
68. Yang H, Xiang J, Wang N, Zhao Y, Hyman J et al (2009) Converse conformational control of smoothened activity by structurally related small molecules. Journal of Biological Chemistry 284(31):20876–20884
69. D'Alençon CA, Peña OA, Wittmann C et al (2010) A high-throughput chemically induced inflammation assay in zebrafish. BMC Biol 8:151. doi:10.1186/1741-7007-8-151
70. Liu Y-JJ, Fan H-BB, Jin Y et al (2013) Cannabinoid receptor 2 suppresses leukocyte inflammatory migration by modulating the JNK/c-Jun/Alox5 pathway. J Biol Chem 288:13551–62. doi:10.1074/jbc.M113.453811

71. Yeh JR, Munson KM, Elagib KE et al (2009) Discovering chemical modifiers of oncogene-regulated hematopoietic differentiation. Nat Chem Biol 5:236–43. doi:10.1038/nchembio.147
72. Rovira M, Huang W, Yusuff S et al (2011) Chemical screen identifies FDA-approved drugs and target pathways that induce precocious pancreatic endocrine differentiation. Proc Natl Acad Sci USA 108:19264–9. doi:10.1073/pnas.1113081108
73. Hao J, Ao A, Zhou L et al (2013) Selective small molecule targeting β-catenin function discovered by in vivo chemical genetic screen. Cell Rep 4:898–904. doi:10.1016/j.celrep.2013.07.047
74. Zhang Z-RR, Li J-HH, Li S et al (2014) In vivo angiogenesis screening and mechanism of action of novel tanshinone derivatives produced by one-pot combinatorial modification of natural tanshinone mixture from Salvia miltiorrhiza. PLoS One 9:e100416. doi:10.1371 journal.pone.0100416
75. Davies H, Bignell G, Cox C et al (2002) Mutations of the BRAF gene in human cancer. Nature 417:949–954. doi:10.1038/nature00766
76. McLean J, Neidhardt E, Grossman T, Hedstrom L (2001) Multiple inhibitor analysis of the brequinar and leflunomide binding sites on human dihydroorotate dehydrogenase. Biochemistry-us 40:2194–200. doi:10.1021/bi001810q

Genomic Approaches to Zebrafish Cancer

Richard M. White

Abstract The zebrafish has emerged as an important model for studying cancer biology. Identification of DNA, RNA and chromatin abnormalities can give profound insight into the mechanisms of tumorigenesis and the there are many techniques for analyzing the genomes of these tumors. Here, I present an overview of the available technologies for analyzing tumor genomes in the zebrafish, including array based methods as well as next-generation sequencing technologies. I also discuss the ways in which zebrafish tumor genomes can be compared to human genomes using cross-species oncogenomics, which act to filter genomic noise and ultimately uncover central drivers of malignancy. Finally, I discuss downstream analytic tools, including network analysis, that can help to organize the alterations into coherent biological frameworks that can then be investigated further.

Keywords Zebrafish • Cancer • Next-generation sequencing • RNA-seq • Oncogenomics

An Introduction to Zebrafish Cancer Models

The past decade has seen an explosion in the number of available zebrafish cancer models [1]. These range from transgenic overexpression of dominant acting oncogenes, to inactivation of tumor suppressor genes, and carcinogen induced tumors. One of the major advantages of performing cancer studies in zebrafish is that they can be easily manipulated with genetic tools, and the advent of CRISPR methods [2] will only continue to accelerate this process. In addition, the optical clarity of the developing larvae or adult *casper* strain [3] allows for in vivo imaging studies that would be prohibitive in typical murine models.

Regardless of the mode of oncogenesis, all of these tumors recapitulate certain aspects of human tumorigenesis in much the same way that mouse models do, albeit at a greater speed and with a larger number of available animals to study. For example, zebrafish models of BRAF-driven melanoma strongly resemble human melanoma at

R.M. White, M.D., Ph.D. (✉)
Memorial Sloan Kettering Cancer Center, 415 East 68th Street, New York, NY 10065, USA
e-mail: whiter@mskcc.org

D.M. Langenau (ed.), *Cancer and Zebrafish*, Advances in Experimental Medicine and Biology 910, DOI 10.1007/978-3-319-30654-4_6

the histological level [4], and in some ways, more so than mouse models do because zebrafish melanocytes are embedded in the dermal-epidermal junction like they are in humans. But in other respects, zebrafish models of cancer are divergent from human tumors, in the sense that they are generally less aggressive than many human cancers once metastasized. This may be in part due to the fact that most fish cancers are initiated with only a few genes, whereas most human cancers harbor hundreds to thousands of mutations, copy number changes, and structural rearrangements [5].

In order to continue improving the existing fish models, we must develop methods for interrogating the genomes of these tumors to discern where the similarities, and differences, occur compared to human tumors. Such information will be invaluable in taking full advantage of the genetic and optical strengths of the zebrafish system in a manner that complements what is available in murine, fly and human cancer models. The purpose of this review is to discuss methods for genomic analysis of zebrafish tumors, with a particular eye towards a comparison to human cancer genomics.

Cancer Genomics: A Primer

The term "cancer genomics" has evolved to mean any approach which uses large-scale methods to interrogate DNA, RNA, protein, chromatin or other molecules within tumor tissue. In its earliest iterations, technologies such as PCR and Sanger sequencing were used to analyze protein coding mutations found in the exons of pancreatic cancer [6]. With the advent of "next-generation" technologies such as the Illumina HiSeq platform (see below for more details), the ability to query large numbers of tumor genomes has become feasible from both a time and cost standpoint, and has led to a rapid increase in the number of such studies.

Because of the recognition that producing high quality genomic analyses of many tumors is complex and requires a great deal of expertise, several consortia have been formed to enable quality control and scalability. In the United States, this has taken the form of The Cancer Genome Atlas (TCGA, http://cancergenome.nih.gov/) which has aimed to characterize several thousand tumor-normal genome pairs from nearly every malignancy. The TCGA is strongly complemented, and in some cases subsumed, by international efforts such as the International Cancer Genome Consortium (ICGC, https://icgc.org/) and the Cancer Genome Project at the Wellcome-Trust Sanger Center (https://www.sanger.ac.uk/research/projects/cancergenome/). Together, these and other efforts have begun to yield conclusive genomic data for most cancers, and publications documenting their successes are beginning to appear in the literature.

The initial phase of the TCGA effort began in 2005, and focused on gliobastoma, lung and ovarian cancer, which rapidly moved into phase II in 2009 for other tumors. The vast majority of these early efforts centered on whole-exome sequencing, in order to identify recurrent or unique mutations in the protein coding regions. However, more recently, at least three other technologies have been incorporated into the TCGA effort: array CGH (typically using Affymetrix SNP 6.0 chips), RNA-seq (both mRNA and miRNA) as well as reverse phase protein array (RPPA). Together, these datasets for a given individual tumor are analyzed using so-called "integrative landscape" analyses, in order to try and give a picture of the entire array of molecular changes present in a patient's tumor.

The responsibility for each piece of the project is spread out across different centers, including those responsible for tumor acquisition, sequencing itself, informatics, analysis and downstream studies. For this reason, within each tumor type, as the data from each of these subprojects is collected, there is an embargo placed on the publication of the data until all members agree to a final, consortium publication from the TCGA. An illustrative example of such an approach is provided by the recent TCGA-sponsored publication for cutaneous melanoma [7]. In this manuscript, they describe an integrated DNA, RNA and protein analysis of 333 tumors, which were derived from 331 patients. The majority of these samples came from "regional" lymph node metastases, which are typically the most accessible to clinicians. Such accessibility is an important caveat for all large-scale sequencing studies in humans, since investigators are largely limited to what is practically available. For 199 of the samples, they were able to complete six types of analysis: whole-exome sequencing, DNA copy-number profiling, mRNA-sequencing, miRNA sequencing, DNA methylation profiling, and RPPA protein analysis. They also selectively used whole-genome sequencing and locus-specific PCR/Sanger sequencing of the TERT promoter for some samples. They then used an algorithm called iCluster [8], a platform for integrating disparate datasets to define the molecular subtypes of tumors, along with integrated pathway analysis to identify the major molecular alterations present in each subtype. From this, they could identify four major genomic subtypes of melanoma (BRAF, RAS, NF1, triple-negative), along with a transcriptomic subclassification into immune, keratin or MITF-low types. Interestingly, this analysis was also coupled with survival data on the patients, which implicated a new subtype of melanoma, based on immune markers, with a better prognosis. Perhaps most importantly, this landmark study also provides numerous new pathways and genes to explore in subsequent studies, some of which (e.g. KIT) may be therapeutically targetable with existing drugs. Outside of melanoma, the collection of TCGA "Pan-Cancer" datasets [9] allows for a broad sweep of mutational processes across tumor types, which revealed a core set of 127 genes that were altered across tumor types, spanning from well-known pathways like Wnt, to novel ones like histone modifications, splicing and metabolism. The number of such analyses is growing every month.

One particular challenge that many of these studies have highlighted are the remarkably diverse ways that tumors can genomically achieve the "hallmarks" of cancer such as avoidance of apoptosis. It is apparent that while major "drivers" such as RAS and BRAF are key for many tumors, there is an increasingly "long tail" of genomic alterations which may endow tumor cells with competitive advantages. In the classic definition of significant genes in cancer, most analyses have defined "driver" events as those that occur recurrently across tumors from different patients, and those falling below that cutoff as "passengers". Of course, this is somewhat arbitrary and based on statistical likelihood, taking into account the overall mutational burdens/patterns in a given tumor type. But it is also possible that many genes deemed a "passenger" at the population level may indeed be a driver in that individual patient. Adding to this complexity is the recent finding that subclones that make up a small percentage of a given tumor (e.g. those expressing IL11) can have dramatic cell non-autonomous effects on tumor growth [10]—in a typical genomic analysis, these would likely fall below thresholds for significance, yet they are essential for the overall success of the tumor. In addition, the extremely high mutational burden of some tumors makes it extremely difficult to assess which of those changes are truly important for tumor growth. Examples of these include UV-related melanomas, smoking-related lung cancers, as well as tumors deficient for DNA proofreading mechanisms such as those with POLD or POLE mutations [11].

It is in unraveling the complexity of human cancer genomics that the zebrafish may prove the most useful. In its simplest iteration, one can envision using a "cross-species" oncogenomics approach, in which the genomic landscape of a zebrafish cancer can be directly compared to that of the corresponding human tumor to find things in common between the two. In essence, this is a filtering approach, with the logic being that any alteration present in both species is likely to be a true driver event. It is designed to narrow down large potential genomic alterations into something manageable that can then be tested using focused downstream experiments. Other variations of this theme are cases in which lists of candidate alterations from human tumors can be tested, singly or in combination, using transgenic fish models [12]. All of these approaches will be discussed in more detail below, but it is first important to understand what has been done so far in zebrafish cancer genomics, and what tools are needed for such studies.

Zebrafish Cancer Genomics

A number of studies have now used a smattering of these technologies to interrogate zebrafish tumor models. Most of the published methods have relied upon chip-based technologies, but an increasing number are now using next-generation sequencing. It is likely that nearly all future studies will use these newer technologies.

Chip-Based Approaches

Array CGH

Array-based comparative genomic hybridization (aCGH) is a method where two different fluorescently labelled DNA samples can be hybridized to a chip that has a large array of complementary DNA fragments [13]. The labelled DNA will bind to its cognate target sequences present on the chip, and everything else washed away. Because it is generally done as a competition between two sources of DNA (i.e. tumor and normal DNA), each spot on the chip can then be scanned to calculate the relative fluorescence signal between the two samples. What is spotted onto the chip is up to each individual user, but usually it is a portion of the genome either from PCR arrays, BAC clones, cDNA clones, or in some cases small fragments of the entire genome. The level of resolution of aCGH depends entirely upon how much of the genome is spotted onto the chip, and how big each fragment is. aCGH is the gold standard for identify copy number alterations (CNAs) in tumor samples. More recently, investigators have achieved the same or better resolution that aCGH using SNP arrays, in which the fragments of DNA on the chip represent hotspots of single nucleotide polymorphisms—because those are common in the genome, they give an overall reasonable view of copy number changes, albeit not quite as a very dense aCGH chip. Although both aCGH and custom SNP arrays are available for the zebrafish genome, the majority of studies thus far have utilized the aCGH platform.

Using a BAC aCGH [14], three types of zebrafish tumors were analyzed for copy number changes: melanoma, rhabdomyosarcoma, and T-cell leukemia. Several areas of common, recurrent amplification and deletion were seen across the three tumor types, but several unique abnormalities, ranging from 1 to 28 copy number changes, were seen for individual tumors of each type. For the melanoma samples, five particular BACs were seen to be amplified in half the samples, suggesting some degree of positive selection for this region. Because this BAC array did not have very high resolution, it was not possible to discern which genes were specifically amplified in these samples, but several potentially important genes are contained within the region spanned, including EP300, PIM3, COL4A2, KIT, MITF and BRAF. Similar reports have utilized aCGH techniques for malignant peripheral nerve sheet tumors (MPNSTs) [15, 16], KRAS driven rhabdomyosarcoma [17], and T-cell ALL [18].

RNA Microarrays

Conceptually similar to aCGH, chip technologies have been also extensively applied to analysis of RNA. In this case, RNA is reverse transcribed to yield fluorescently labelled cDNA fragments, which are then hybridized to a chip containing complementary sequences. The intensity of the fluorescent spot can be inferred to be proportional to the amount of RNA species present in the original sample. Most studies tend to enrich the input RNA for mRNA, using either polyA priming or ribodepletion methods to eliminate ribosomal RNAs which make up the vast bulk of total RNA. A key aspect of this technology is what spots are placed on the chip. In the most widely used platform, the Affymetrix Zebrafish 1.0 chip, this primarily consisted of cDNAs that were derived from expressed sequence tag (EST) libraries. This chip has 14,900 transcripts, which covers much but certainly not all of the zebrafish transcriptome, which is estimated at over 25,000 genes. Other arrays are available, including an updated one from Affymetrix (although not commercially available) and an Agilent array containing 21,000 (v1) to 43,000 (v2–3) transcripts. One issue with all arrays, which is not unique to any particular technology, are the limitations in annotations of the transcripts. In part, this is related to the fact that some zebrafish transcripts do not have orthologues in other, especially mammalian, species. In other cases, the annotation difficulties are related to the genome-wide duplication that occurred in teleost evolution, making it difficult to map short oligonucleotide with confidence to a particular genome region.

Despite these limitations, microarrays have been the most widely used genomic technology in zebrafish cancer. One of the earliest attempts at this was using carcinogen-induced models of hepatocellular carcinoma in fish [19], in which they compared the tumor transcriptome to that of normal liver. This revealed a surprisingly large set of dysregulated transcripts—over 2300—which corresponded to 1920 human orthologs. The authors then used the technique of Gene Set Enrichment Analysis (GSEA) to compare the zebrafish HCC signature to human cancer [20]. GSEA is an extremely important technique for performing cross-species genomic comparisons. In short, it is a statistical method that takes in two sets of data. First, the user provides all of the expression data, for all genes and all samples, for a given genomic dataset (i.e. tumor versus normal RNA). Second, the user provides a list of genes that represents a state they might be interested in (i.e. genes associated with Wnt signaling). GSEA then uses a ranking algorithm to determine if members of a given gene set occur near the top or bottom ranking of a dataset, providing an Enrichment Score and associated p-values. GSEA is perhaps the most powerful statistical tool available to determine if fish genes are similarly enriched in human cancer genesets, especially when it is combined with the massive database of genesets available from the MSigDB database. MSigDB is essentially a collection of curated genesets representing thousands of phenotypic states, cancer and otherwise. Using GSEA, the authors were able to show that zebrafish HCC is not only enriched in human cancer in general, but most strongly in human HCC. The particular genes that formed this enrichment between the two species belonged to the Wnt/beta-catenin pathway and MAP kinase pathway. The implication of their finding is that these two pathways are so central to the biology of HCC that they are conserved across cancers that arise in species separated by millions of years of vertebrate evolution. As both of these pathways are under intense investigation for therapeutic targeting (i.e. IWR1 for Wnt, trametinib for MAP kinase), these types of studies can lead to meaningful translational outcomes.

A similar approach was taken for a KRAS-induced rhabdomyosarcoma (RMS) model [21]. Unexpectedly, expression of KRAS under the rag2 promoter led to muscle tumors in the fish. These tumors were then profiled using Affymetrix arrays (compared to normal muscle) and compared to human RMS subtypes, which revealed that the fish tumors were more similar to the embryonal, but not alveolar, types of human RMS. This is of key importance, since previous studies had not yet linked RMS to RAS activation, although this is now widely recognized to be the case in human disease. Interestingly their data also pointed out a core "RAS" signature that was not confined to RMS, but also found in pancreatic ductal adenocarcinoma, a truly RAS driven tumor. These data point out how the fish may yield unexpected pathway alterations that have strong relevance to the human counterpart, even in some cases before human genomics have made that apparent. Similar types of analyses have been performed for other induced models of HCC, including a KRASV12 [22], xmrk [23] and RAF [24].

Our group has used comparative profiling of zebrafish and human melanoma to identify specific developmental signatures common to the two species [25, 26]. Human melanoma sequencing has revealed two dominant genetic events: BRAF and NRAS mutations. Both of these have been used to create transgenic melanoma models, in which the mutated human gene (e.g. $BRAF^{V600E}$ or $NRAS^{Q61K}$) is driven under the melanocyte specific mitfa promoter [27]. In the context of a $p53^{-/-}$ background [28], all of the animals develop easily visible tumors, which were then profiled again using Affymetrix arrays. Two types of analysis were done with these datasets. First, we compared the fish tumors to human melanoma, nevi and normal skin, which showed a striking conservation of genes expressed in both species. Taking this a step further, we then asked which of the genes contained in the melanoma signature were enriched in genes normally expressed during neural crest or melanocyte development, since melanoma is known to be a "lineage addicted" cancer [29, 30]. We obtained the list of neural crest genes using the publically available ZFIN server (http://www.zfin.org), which provides a rich dataset of tissue specific expression during multiple stages of development. This analysis revealed a strong enrichment of the neural crest geneset in both fish and human melanomas, providing a rationale for a subsequent chemical screen we performed to identify small molecule suppressors of this neural crest signature. This screen ultimately identified leflunomide, a small molecule inhibitor of the metabolic enzyme dihydroorotate dehydrogenase (DHODH), which acts to suppress transcriptional elongation of neural crest genes and is now being tested as a therapeutic in human melanoma [25].

Next-Generation Sequencing Approaches

Although chip-based approaches will continue to play a role in all forms of cancer genomics, especially in regards to copy number changes, it is clear that the vast bulk of data for the foreseeable will be generated using 2nd or 3rd generation sequencing platforms. Until fairly recently, most DNA or RNA sequencing was confined to a relatively small number of samples, in which a PCR step or other form of DNA isolation was performed, followed by Sanger dideoxy chain termination sequencing. This technology generally produces short sequencing runs of 300–700 basepairs, and can be automated as found in instruments such as the ABI3730, a workhorse of many early sequencing projects. This and related technologies were used to perform the initial drafts of the human genome using shotgun approaches [31, 32]. A somewhat heroic effort using similar technology provided the first draft of a human cancer genome. Vogelstein and colleagues PCR amplified all the exons from a series of breast and colorectal cancers [33, 34] and then used the ABI3730 sequencers to delineate all of the coding mutations in these tumors.

By the early 2000s, it was clear that much higher throughput sequencing technologies were need not only for cancer genomes, but for genomics in general. A discussion of the evolution of sequencing technology is beyond the scope of this review, but summaries of this can be found elsewhere [35]. In short, most modern cancer genomic studies have begun to use so-called "massively parallel", short-fragment sequencing as typified by machines such as the Illumina GAII/HiSeq/MiSeq, SOLID and Roche 454 platforms. Although each individual piece of DNA sequenced is only 50–500 bp long, the machine can generate millions of these "reads" in parallel and relatively rapidly, which allows for near complete coverage of a genome in about a day. These technologies were brought to bear in cancer genomics by the Wellcome-Trust Sanger Center, who published the first "whole-genome sequence" of a human cancer in 2010 [36]. Using the Illumina GAII platform, Stratton and colleagues sequenced a human melanoma cell line along with matched normal cells (COLO829 and COLO829BL). In the melanoma, they identified over 33,000 somatic mutations, along with ~900 insertion/deletion events and 51 structural rearrangements. Only a small subset (292 of 33,345 mutations) were found in protein coding regions, highlighting that the exome sequencing approach will only capture a very limited landscape of cancer genomic changes. Since that time, thousands of human cancers have been sequenced either by exome sequencing or by whole-genome sequencing using the Illumina and related platforms. More recently, so-called "third generation" sequencing technologies have come on board, including platforms from PacBIO, Oxford Nanopore and IonTorrent. The major advantage of these newer systems are the dramatically longer read lengths, which can range up to 10,000 bases or more. This will massively improve the throughput and accuracy of genomic efforts not only in human cancer, but especially in model organisms such as zebrafish where genome alignment of small read fragments still remains a computational challenge.

Exome Sequencing of Zebrafish Cancers

In order to understand how zebrafish cancers compare to human cancers, our group undertook an effort to perform a large scale exome sequencing project in collaboration with the Sanger Institute [37]. We used melanoma as a prototype, because it had a particular advantage in terms of cross-species comparisons. In human melanoma, the mutational burden is very large, as mentioned above, primarily because of the high background rate of UV damage. It is believed that most of those mutations are of little functional consequence. In contrast, transgenically engineered zebrafish melanomas have essentially no UV exposure, allowing for a direct comparison between the two species to find the most likely true drivers. In this sense, the fish mutations act as a "filter" on the human mutations.

We performed whole-exome capture on a series of 53 transgenic zebrafish melanomas, along with matched normal tissue from that animal. Most of the tumors were of the mitf-BRAF^{V600E};p53$^{-/-}$ variety, with the rest being mitf-NRAS^{Q61K};p53$^{-/-}$. Several of the fish had additional candidate "driver" events built into the transgenic (e.g. SETDB1), in order to determine how increasing complexity of the transgenic affected the ultimate tumor genome. For each animal, tumor and normal DNA were isolated, and the exonic DNA was captured using the Zebrafish Agilent All Exon

SureSelect technology. The captured DNA was sequenced using a variety of next-generation technologies, including the Illumina GAIIx, HiSeq and Roche 454 platforms. The sequences reads were mapped to the Zv9 reference genome using the standard Burrows-Wheeler Algorithm. Several types of analyses were then performed: (1) mutations were called using the CaVEMan, SomaticSniper, and String Graph Assembler algorithms, (2) insertion/deletions (indels) identified through Pindel, (3) copy number variants (CNVs) were called using the ASCAT algorithm. Because the zebrafish genome has a relatively high number of germline SNPs (compared to humans), it is absolutely essential that normal DNA is sequenced alongside tumor for all zebrafish studies. Otherwise, simply using the zebrafish reference genome as the determinant of "normal" is fraught with problems and will give an exceptionally high false positive rate.

From the 53 tumors, a total of 403 point mutations were identified, along with 13 indels and 991 copy number variants. It is striking that, on average, a median of four exonic mutations per tumor was found. This is in stark contrast to UV-related cutaneous human melanoma, which has a median of 171 coding point mutations. However, the fish melanomas are much more in line with non-UV acral human melanomas, which have a median of 9 coding mutations per tumor. The predominant mutational signature were C>T transitions, the same that is found in UV-induced melanomas, which may indicate an underlying process favoring this substitution in melanoma, even in the absence of UV exposure. In at least one tumor, there were microclusters of mutations reminiscent of "kataegis", a phenomenon seen in human cancer yielding localized regions of hypermutation thought to arise from a single event [38]. Interestingly, few if any of the mutations in the fish were recurrent across individual fish, suggesting that either they are not important driver events, or that each fish harbors its own unique set of drivers. For the copy number changes, there were more consistent recurrent events, particularly a large 175 kb amplicon on chromosome 3 which occurred in 10/53 tumors. This region has several potentially interesting genes (i.e. prkascaa, samd1), several of which are being followed up as potentially important genes in melanoma.

One particularly striking outcome from this experiment was the tremendous degree of heterogeneity between the tumors. The vast genetic heterogeneity of human cancer is increasingly recognized [39–41], which confounds many analyses of which changes are important across populations but also within that individual patient. Conceptually, we had envisioned that transgenically engineered fish in reasonably homogeneous genetic backgrounds would harbor much less somatic variation between animals, but the exact opposite was found. For example, although the median mutation burden was four exonic mutations per tumor, the range of mutations varied from 0 to over 40. In fact, over half of the total mutations found in the entire study were identified in just 8 of the 53 tumors. Although copy number changes did show some degree of recurrence, here too there was tremendous heterogeneity. The reasons for this heterogeneity remain unclear, but a clue to this may be found in the observation that there is an inverse relationship between the number of "initiating" drivers and the subsequent number of somatic events. In other words—the more transgenes you start with, the fewer subsequent mutations you ultimately find in the tumor. It may also be due to the possibility that fish tumors are more driven by copy number, rather than mutational, events. Considering both the human and zebrafish data, it is clear that heterogeneity in cancer has a complex underpinning, the mechanisms of which remain to be identified.

From these results, it is fair to ask whether genomic sequencing of zebrafish cancer is justified given the expense and computational resources required. The answer to this depends in part on what the goals of such projects are. From a basic, mechanistic standpoint, it is likely that deeper interrogations of mutational process in tumors from fish and humans will help us to understand how specific mechanisms of genome integrity impact tumorigenesis. For example, a cross-species comparison of the relative impact of mutations versus copy number variants could be readily approached in the fish, and yield answers to why in humans, mutation and copy number variation seem to be somewhat mutually exclusive mechanisms of tumorigenesis [42]. Other questions that can be uniquely addressed in fish cancer genomics are mechanisms of processes like kataegis, which has been suggested to be due to AID-type events. Perhaps the most interesting way that the fish models can be used is to try and gain an understanding of where tumor heterogeneity comes from. Although this was initially a somewhat surprising finding from our study, it also points out that we truly do not understand the origins of genetic heterogeneity and how it relates to tumor progression. Transgenic and CRISPR models could help reveal the underlying mechanisms of these poorly understood events. Whether the fish can be used for more translational, actionable, "filtered" list of genes remains to be determined, but the data thus far indicates that for these more clinically relevant questions, the fish can be used to: (1) model potential lists of candidate DNA mutations/CNVs that arise from human TCGA data, as has been described using the miniCoopR system [12, 43], or (2) identify conserved pathways across species using RNA-seq approaches.

RNA-Sequencing of Zebrafish Cancer

As mentioned above, the vast majority of transcriptional profiling of zebrafish tumors has been done using microarray, chip based technologies. But as 2nd and 3rd generation sequencing technologies come down in price, and the informatics become more straightforward, it is likely that nearly all such studies will migrate to RNA-seq in the near future.

Our own group has used the zebrafish $BRAF^{V600E}$ melanomas to compare the performance of RNA-seq vs. the Affymetrix array platform (unpublished observations). In this experiment, we took total RNA and then split it to be used for either technology. The preparation for RNA-seq involves enrichment of mRNA, depletion of ribosomal RNA (very important for RNA experiments), reverse transcription into cDNA, and then fragmenting of the cDNA into small fragments. Adapters for a given sequencing platform are then ligated (in our case, the Illumina HiSeq2000 platform) in preparation for a sequencing "run". For most applications, read lengths of 50 bp is adequate, although the longer 100 bp or more reads will improve mapping (see below for more details on this). The sequences were then aligned using Burrows-Wheeler Algorithm, and transcript assembly and differential expression accomplished with a series of algorithms: TopHat, Cufflinks and Cuffdiff [44]. There are innumerable algorithms for doing similar types of analyses of RNA-seq data. The relative transcript abundances for each sample are calculated based on the number of "reads" aligning to a particular region of protein-coding genomic sequence, and are normalized to the total amount of RNA present in a given sample.

In general, we found that the melanoma data from the two approaches was very similar, especially for the more well-expressed genes. One major difference between the technologies is the vastly greater amount of information that can be extracted from the RNA-seq. In the chip-based approaches, the most you can generally learn about a gene is the expression of a fragment (usually a 3′EST) of the gene. In some more advanced chips, which contain all of the exons, you can discern information about alternative splicing/exon usage. But in RNA-seq, this data about splicing is immediately apparent and a standard part of the analytic pipelines, because in this case one will receive all of the information about which part of a transcript is actually being expressed. This can actually cause some degree of computational headaches, because it can make it challenging to come up with a single number that represents the "expression" level of a given gene. In other words, is it better to "average" the total number of reads for all exons in a given transcript, or is it better to "sum" the total number of reads for all exons in that transcript? For situations in which your two samples use the same exons, either approach is fine, but in cases where one sample strongly uses one exon over another, this can cause a misrepresentation of the actual biology. The analysis of alternative splicing is a major advantage of RNA-seq compared to microarrays. Other more recent RNA-seq advances have included the capacity for calling underlying DNA mutations (inferred from the RNA reads), RNA editing events, strand-specific transcription, microRNA profiling, and long noncoding RNA analyses (lncRNA).

A recent study from Gong [23] used RNA-seq to analyze hepatocellular carcinomas that arise as a result of transgenic expression of xmrk in the liver. They also profiled the tumors during regression phases. This revealed a similar expression pattern to that seen in a human subtype of HCC (S2) characterized by high Myc expression and enhanced proteasome, antigen processing, p53 and cell cycle pathways. The signature of the regressing tumors was characteristic of an immune response, which is typical of EGFR expression seen in human HCC treatment.

A Practical Guide to Zebrafish Cancer Genomics: The Nuts and Bolts

Pre-experiment Planning: The Most Important Step

Because of the rapidity with which many of the above discussed technologies have moved, it is important to determine when, if and how to use them for analyzing a zebrafish tumor model. In my experience, the single most important factor in a zebrafish cancer genomics project is planning the experiments before the money gets spent. This involves three people: (1) the person doing the lab work on the zebrafish tumor, who will (hopefully) know what genomic question they are interested in, (2) a person from the genomics facility who will be doing the actual sequencing, and (3) a bio-informatician capable of analyzing the resulting data. I cannot emphasize enough how important it is to have a conversation with #2 and #3 long before a pipette is picked up. The genomics facility will almost always have key information about a technology you are interested in, exactly what your samples need to look like, how to assess quality of your samples, what type/number of "reads" you will need to do, what machine should be used for the sequencing, should you multiplex your samples, the effect of batch variation, and the file formats they are capable of producing. Similarly, the informatician is essential for telling you how "deep" your coverage should be for your given question, is it better to do single end vs. paired end reads, what machines are going to be better than others, stranded vs. nonstranded reads, how to assess quality of the genomics facility output, how are they going to be able to access the giga or terabytes of data that you produce, what types of algorithms they feel comfortable running, are they willing to develop custom algorithms, do they feel comfortable working with zebrafish (as opposed to mammalian) datasets. Together, the genomics core and informatician will provide you with the most important framework for understanding if your experiment is likely to work, how much it will cost, and how long it will take. In general, the informatics takes longer than one might assume: informaticians are in extremely high demand at this time, and often spread across many projects, and very few have experience working with the zebrafish genome. Getting this right in the beginning will have tremendous benefit down the road, as developing a good relationship with a genomics facility and an informatician will then allow subsequent projects to move much more rapidly.

Sample Preparation/Requirements

Whether you are sequencing, DNA, RNA or chromatin, one of the first questions that always comes up is "how much do I need"? There is no straightforward answer to this, in part because it depends on what you are trying to achieve. For most techniques, somewhere around 1 μg of purified nucleic is plenty, but this is probably overkill for many projects. For DNA based approaches, 1 μg will allow you to do exome or whole-genome approaches with ease, but we have also had success with as little as 100 ng (or less in many cases). RNA and ChIP can be even more forgiving, as many facilities can handle sample inputs as low as 1 ng. One issue that will come up with these low input amounts is whether to perform a round of whole-genome amplification prior to library preparation. Some of the newer amplification methods (e.g. NuGEN Ovation) produce far less bias that older methods, so for samples with very low input (i.e. less than 1 ng) it is preferable to perform amplification rather than put in a very small amount to the sequencer that is unlikely to work. As long as all the samples in a given experiment are handled the same way (i.e. amplified or not) it is generally ok, because they will be internally controlled.

Another issue that often comes up regards library preparation. The purified DNA or RNA must be made compatible with the particular sequencer you plan to use (i.e. an Illumina library prep is not compatible with a IonTorrent prep, etc). This can involve the addition of adapters for that particular sequencer, as well as barcodes if one is multiplexing samples. Nearly all labs can prepare libraries themselves, since it involves a fairly straightforward set of molecular techniques and the protocols are well published. However, one thing we now consider is the cost-benefit ratio of doing the library preps ourselves versus within the genomics facility. Although at first glance, it may seem more cost efficient to just do it yourself (and then use for core facility to just run the sequencer), often this is not the case: many core facilities have automated equipment for library prep that reduces the per sample cost, and increases the likelihood of success. But this will clearly depend upon the facilities available at a given institution, so it is hard to generalize. One recent development has been the availability of large-scale private genomic facilities, allowing for "outsourcing" of a great deal of the work to these highly specialized groups. For example, we have worked with the New York Genome Center (a consortium amongst many different academic labs, http://www.nygenome.org/) and found this to be a very efficient and cost-effective way to complete projects. Innumerable other private companies (i.e. Genewiz, Axeq, Illumina themselves) allow you to send your samples to them and they are highly expert at library prep. In all cases, it is essential to perform some type of Quality Control (QC) step before and after preparing libraries. An initial first step is to analyze your sample on a Nanodrop or Qubit type of device, to get a sense of concentration and integrity. Beyond that, many facilities will also run devices like the Bioanalyzer (for RNA) to check things like ribosomal RNA bands (as a reflection of total RNA integrity).

How Much Sequencing?

An important consideration in these approaches is how "deep" do you want to sequence, meaning how much coverage of my sample do I want? One way of making this calculation is to determine what "X" coverage you want. This is a common nomenclature in the genomics field, which basically works out to "how many times do I want a given basepair of DNA to be read by the sequencer". So, if we state "this sample was sequenced to a depth of 40X", that simply means that, on average, each basepair of interest was covered by at least 40 "reads". This is clearly an average: because the DNA on the sequencer is generally randomly fragmented, a given read may cover a particular segment of the genome better or worse than another. Some regions will wind up covered at 100X, and some at 10X, and some at 0X. But the more sequencing you do, the deeper your average coverage will become.

An example is illustrative. Let's say you wanted to sequence the entire genome of a given cancer sample to 40X coverage. The fish genome is approximate 1.4Gb of DNA, so to cover every basepair at 40X, you would need $(1.4 \times 10^9 \times 40) = 56 \times 10^9$ bp of sequence. If your sequencing facility is going to generate 100 bp, paired-end reads (so each read generates 200 bp of usable sequence), then the total number of "reads" you would need, theoretically $= (56 \times 10^9$ bp)/(200 bp) $= 280 \times 10^6$ reads. But in reality you need probably 2–3X that amount, because many reads will not properly align to the genome (either because of errors or because of contamination with microbial constituents of most tissues) and not all reads will be of sufficient quality to be usable. In our own group, we recently performed whole-genome sequencing on a tumor:normal matched pair, and to generate ~40X coverage for each required about 750×10^6 reads per sample, or a total of 1500×10^6 reads for the two samples. Given the current capacities of Illumina 2500–4000 series machines, which generate somewhere between $200–350 \times 10^6$ reads per lane, this would require about six to eight Illumina lanes.

It is for these reasons that exome-seq or RNA-seq are much more common than whole-genome approaches. Because the protein coding genes only make up ~2 % of the genome [45], that means to get 40X coverage of the coding genes would only require 2 % as much sequencing. On a practical level, exome sequencing is typically done to much greater depths than 40X, often in the range of 100X to 1000X. This is because the main function of exome sequencing is to identify, with high confidence, mutations, and that often requires sequencing depths to greater than 40X. So, for things like mutation calling, there is a trade-off: whole-genome sequencing gives you a broader swath of the genome, but at lower depth, so you can only confidently call the most highly significant mutations. Exome sequencing gives you a small amount of the genome, but because you can go much more deeply, it is easier to call mutations that even occur at low allelic frequencies. For RNA-seq, the goal is usually not mutation calling—most experiments are done simply for differential expression. In that case, usually $20–40 \times 10^6$ reads per sample is adequate. Given a per lane capacity of 200–300 million reads, that makes multiplexing of RNA-seq samples extremely cost-effective compared to microarrays.

Basic Informatics: Raw Data to Primary Outputs

As mentioned above, most zebrafish cancer genomics projects will strongly benefit from an experienced bio-informatician, who can implement the algorithms typically used for analyzing next-generation data. Recently, there has been a community effort to make these tools more accessible to biomedical researchers through the Galaxy web portal (https://galaxyproject.org/) or Genepattern server (http://www.broadinstitute.org/cancer/software/genepattern/). Galaxy and Genepattern are powerful and relatively simple ways to run many of the algorithms discussed below, but it is still essential to have someone with experience act as a "check" on these analysis, since they will not give much information as to whether a given algorithm is appropriate—it will simply run it for you using an intuitive, web-based interface. A brief overview of some of the basic tools will be provided here.

The data output from the Illumina and other platforms is generally in FASTQ format, which is essentially a text file where each line contains the data from an individual "read" from the sequencer. An example is shown here:

```
@SEQ_ID
GATTTGGGGTTCAAAGCAGTATCGATCAAATAGTAAATCCATTTGTTCAA
  CTCACAGTTT
!"*((((***+))%%%++)(%%%%).1***-+*"))**55CCF>>>>>>CCCCCCC65
```

The top line contains the basepairs that were called by the machine, and the bottom line are the quality characters assigned to each of those, using a nomenclature specific to each platform. This format allows the subsequent algorithms to assign confidence to any given sequence using the quality metric, which is important for mapping them to the genome.

Once you have a collection of FASTQ files (each sample gets its own), these are then used as input into a "mapper" or "aligner", which attempts to align each read to its appropriate place in the genome. From an informatics standpoint this is the critical step, and in the zebrafish is especially challenging due to the less-well developed genome (compared to humans), an increased number of gene duplications (which make it hard to discern where a given 50 bp read "belongs") and an increased number of SNPs (which make it hard to discern if the read really belongs in that spot). There are many different mapping algorithms, but a common one for mapping DNA-seq reads is BWA (Burrows-Wheeler Algorithm). For mapping RNA-seq reads, a common one is the Bowtie/Tophat combination [46], since this allows not only for mapping of the sequence itself, but it also takes into account splice junctions where a given RNA read is likely to span two exons. These algorithms take in FASTQ files and produce files in the SAM/BAM format, which are essentially the sequence reads along with their genomic positions.

Following mapping of the reads, the next set of algorithms depends entirely upon what the question is. For detecting mutations in DNA, the TCGA has implemented a set of algorithms that are bundled into a pipeline called "Firehose" (http://gdac.broadinstitute.org/), primarily developed at the Broad Institute. Firehose itself is composed of several underlying algorithms including GATK [47] and MuTect [48], which try to determine whether a given basepair is different than the reference genome. Let us assume you have two samples—tumor and normal. You will generate FASTQ files for those, then map them to the genome to produce a BAM file of aligned reads. GATK will then take in the BAM files and determine if each basepair is different from the reference genome, and MuTect will then take in that data and determine if the tumor has a statistically different basepair than the matched normal sample. It will then generate a list of mutations in the VCF file format, which is essentially a list of what the basepair in the tumor is, compared to the normal sample along with the reference genome.

Mutation calling is not an either/or. It is a statistical argument as to whether a tumor sample has a higher likelihood of differing from the reference genome compared to the normal sample. An example: let us take a hypothetical chromosome position 3:1875678. At that position, let's assume that both the tumor and normal generated 100X coverage (e.g. 100 reads). Now let's say those reads are: tumor (G=99, A=1), normal (G=1, A=99) and the reference genome at that position=A. In that case, it is likely that the tumor has a mutation at that position, and the single "A" in the tumor sample is either a sequencing error or normal tissue contamination. But now let's say the reads are: tumor (G=80, A=20), normal (G=3, A=97) and the reference genome at that position=A. Is that position mutated in the tumor, or is it actually just heterozygous or contaminated with normal DNA? MuTect and related algorithms try to determine the statistical likelihood of mutation, taking these factors into account, but outside of very clear examples where the allelic frequencies are close to 100 %, it can be a difficult judgment to make.

For detecting differential gene expression from RNA, again there are many algorithms, but some commonly used pipelines include Cuffdiff (http://cole-trapnell-lab.github.io/cufflinks/cuffdiff/) and DeSeq (http://bioconductor.org/packages/release/bioc/html/DESeq.html). Here, the process is again to map all of the reads to the genome, and then generate a number which represents the level of expression of a given gene. The most commonly used one is FPKM, which stands for Fragments Per Kilobase of transcript per million mapped reads. Calculation of FPKM is essentially a normalized view of transcription for that given gene, and can be used to then compare across samples using statistical methods such as False Discovery Rate (FDR). Similar to a microarray analysis, the output from this analysis typically gives both a fold-change (tumor compared to normal) as well as a p-value and corrected q-value (typically from the FDR calculation). Genes can then be stratified by the user to determine what level of significance is most meaningful to them, i.e. a $q<0.05$, a $q<0.01$, etc. Similar to DNA-seq, there is no "either/or" for significance—it is simply a level of significance you feel comfortable with.

Downstream Analyses: Secondary Informatics Tools to Assign Meaning to the Data

All of the primary outputs from sequencing projects, whether they be DNA, RNA, or ChIP based, will ultimately generate a list of genomic fragments that are different between tumor and normal. Some of these differences will be in genes themselves, some will be in noncoding RNAs, promoters, enhancers, intergenic regions, UTRs, etc. It is tempting to initially just look at a list of these, especially ones in genes, and try to make a biological story from these lists. But this can be challenging to do when that list contains thousands of genes or regions, and so a large number of downstream tools have been developed to try and connect these findings to the broader biological literature. In fact, one can argue that individual investigators should be very cautious in looking at lists of genes and interpreting it without some type of secondary analysis, since the inherent biases we all have (based on our prior knowledge) can skew the meaning to be found in these large datasets. For this reason, it is important to use secondary tools with some statistical power to connect the data.

Pathway analysis is a common method to try and determine if your list of genes or regions are somehow connected to each other. Some commonly used ones are DAVID (http://david.abcc.ncifcrf.gov/), KEGG (http://www.genome.jp/kegg/), GO/Gene Ontology (http://geneontology.org/), GREAT (http://bejerano.stanford.edu/great/public/html/) and IPA (Ingenuity Pathway Analysis, http://www.ingenuity.com/products/ipa). IPA is particularly useful for zebrafish studies, since it will allow you to directly input zebrafish annotations, and then cross-reference that to data from other species on pathways. IPA is especially good at incorporating in not only the gene name, but also the level of significance for that particular gene in your dataset (i.e. q-value or fold-change). It will output a discrete set of "canonical" pathways altered in your dataset (i.e. Wnt/beta-catenin signaling or PI3K signaling, etc), and give you a corresponding p-value to determine how likely it is that your dataset is truly enriched in that pathway. IPA uses a combination of automated and manual curated pathways to connect genes to each other, and provides the level of evidence for those associations. It also includes a unique tool called "Upstream Regulatory Analysis", which attempts to statistically predict what upstream factors may have been responsible for a given gene expression signature (i.e. EGFR signaling might produce a given gene expression signature in lung cancer cells, etc.). The combination of pathway analysis plus upstream regulator analysis often leads to testable, discrete hypotheses that can then be tested in the laboratory. IPA a proprietary product that is continuously updated, and many academic institutions have subscriptions. Other tools like DAVID can provide somewhat similar data, and although not quite as comprehensive, it is free and very straightforward to implement.

As mentioned above, Gene Set Enrichment Analysis (GSEA) is another key tool for both pathway analysis as well as cross-species comparisons. It is a free tool provided by the Broad Institute (http://www.broadinstitute.org/gsea/index.jsp), and once one has mastered the unique file formats, it is very straightforward to run using its Java-based applet. The most powerful aspect of GSEA is that it is deeply connected to the MSigDB database (http://www.broadinstitute.org/gsea/msigdb/index.jsp), essentially a curated list of gene expression signatures encompassing nearly all biological states of interest (i.e. a signature of lung cancer, a signature of MAP kinase pathway, etc). For this reason, you can take a given dataset that emerges from a zebrafish study, and then use GSEA to compare it to the entire MSigDB database to figure out what your sample is most similar to. GSEA provides a statistically rich output that includes p-values, false discovery rates, and a unique score called an "Enrichment Score" which provides a convenient way to gauge how similar your dataset is to another one from the literature. The GSEA applet also produces publication quality visualization tools, especially Enrichment Plots that have become widely used in the literature and are easy to interpret.

In many cases, it will be important to compare a zebrafish cancer dataset to what is known about human cancer. GSEA can be used for this, but some specific tools focusing on human cancer can be especially powerful here. The cBIO Portal (http://www.cbioportal.org/) is a publically available tool for accessing all of the TCGA project data, which very cleanly shows all of the mutation, copy number, mRNA and phosphoproteomic data available for each tumor type. The input to the cBio Portal is human gene names, so zebrafish IDs will need to be converted to human orthologs using tools such as DRSC (http://www.flyrnai.org/cgi-bin/DRSC_orthologs.pl) or Ensembl/Biomart (http://www.ensembl.org/biomart/). Another useful tool for human cancer data is Oncomine (https://www.oncomine.org/), which has collected massive numbers of RNA expression profiles (either microarray or RNA-seq) along with a smattering of copy number and DNA-seq datasets. Oncomine is very good for querying one gene very deeply across human cancers, and determining how that gene is altered in those tumors. It is available as both a free and proprietary product (with enhanced features). A unique resource for protein data is the Human Protein Atlas (http://www.proteinatlas.org/), which is attempting to determine protein expression of every gene in the genome across both normal and cancerous tissues. They provide detailed data about each antibody they are using, along with validation status. They also provide photomicrographs of each sample, and a statistical measure of enrichment of a given protein in a given condition. Similar to Oncomine, it is very useful for deeply probing a particular gene for its role in human cancers.

Summary and Perspective

As the pace of studies using zebrafish as a cancer model has accelerated, so too has the pace of human cancer genomics. The major challenge over the next decade will be to determine how zebrafish cancer models can integrate with what is being done in human cancer such that the fish provides a truly unique tool in dissecting various aspects of tumorigenesis. It is nearly impossible to do this without some baseline interrogation of zebrafish cancer genomics, whether that be at the DNA, RNA, protein or epigenetic levels. The tools that have been developed for human cancer genomics can usually be applied without too much difficulty to zebrafish cancer genomics, and the number of tools that allow for cross-species oncogenomics continues to grow. The information presented here is meant to be a starting point for those interested in this interface between the two species, and what tools can be used to leverage the particular strengths of each system. These tools will continue to evolve, enabling fertile collaborations between the zebrafish and human cancer biologists.

Acknowledgments This work was supported by the NIH Directors New Innovator Award (DP2CA186572), K08AR055368, the Melanoma Research Alliance Young Investigator Award, an AACR/ASCO Young Investigator Award, and the Alan and Sandra Gerry Metastasis Research Initiative.

References

1. White R, Rose K, Zon L (2013) Zebrafish cancer: the state of the art and the path forward. Nat Rev Cancer 13:624–636
2. Ablain J, Durand EM, Yang S, Zhou Y, Zon LI (2015) A CRISPR/Cas9 vector system for tissue-specific gene disruption in zebrafish. Dev Cell 32:756–764
3. White RM, Sessa A, Burke C, Bowman T, LeBlanc J, Ceol C et al (2008) Transparent adult zebrafish as a tool for in vivo transplantation analysis. Cell Stem Cell 2:183–189
4. Patton EE, Widlund HR, Kutok JL, Kopani KR, Amatruda JF, Murphey RD et al (2005) BRAF mutations are sufficient to promote nevi formation and cooperate with p53 in the genesis of melanoma. Curr Biol 15:249–254
5. Vogelstein B, Papadopoulos N, Velculescu VE, Zhou S, Diaz LA Jr, Kinzler KW (2013) Cancer genome landscapes. Science 339:1546–1558
6. Jones S, Zhang X, Parsons DW, Lin JC, Leary RJ, Angenendt P et al (2008) Core signaling pathways in human pancreatic cancers revealed by global genomic analyses. Science 321:1801–1806
7. Cancer Genome Atlas Network (2015) Genomic Classification of Cutaneous Melanoma. Cell 161:1681–1696
8. Shen R, Olshen AB, Ladanyi M (2009) Integrative clustering of multiple genomic data types using a joint latent variable model with application to breast and lung cancer subtype analysis. Bioinformatics 25:2906–2912
9. Kandoth C, McLellan MD, Vandin F, Ye K, Niu B, Lu C et al (2013) Mutational landscape and significance across 12 major cancer types. Nature 502:333–339
10. Marusyk A, Tabassum DP, Altrock PM, Almendro V, Michor F, Polyak K (2014) Non-cell-autonomous driving of tumour growth supports sub-clonal heterogeneity. Nature 514:54–58
11. Heitzer E, Tomlinson I (2014) Replicative DNA polymerase mutations in cancer. Curr Opin Genet Dev 24:107–113
12. Ceol CJ, Houvras Y, Jane-Valbuena J, Bilodeau S, Orlando DA, Battisti V et al (2011) The histone methyltransferase SETDB1 is recurrently amplified in melanoma and accelerates its onset. Nature 471:513–517
13. Lee HJ, Lowdon RF, Maricque B, Zhang B, Stevens M, Li D et al (2015) Developmental enhancers revealed by extensive DNA methylome maps of zebrafish early embryos. Nat Commun 6:6315
14. Freeman JL, Ceol C, Feng H, Langenau DM, Belair C, Stern HM et al (2009) Construction and application of a zebrafish array comparative genomic hybridization platform. Genes Chromosomes Cancer 48:155–170
15. Zhang G, Hoersch S, Amsterdam A, Whittaker CA, Lees JA, Hopkins N (2010) Highly aneuploid zebrafish malignant peripheral nerve sheath tumors have genetic alterations similar to human cancers. Proc Natl Acad Sci U S A 107:16940–16945
16. Zhang G, Hoersch S, Amsterdam A, Whittaker CA, Beert E, Catchen JM et al (2013) Comparative oncogenomic analysis of copy number alterations in human and zebrafish tumors enables cancer driver discovery. PLoS Genet 9, e1003734
17. Chen EY, Dobrinski KP, Brown KH, Clagg R, Edelman E, Ignatius MS et al (2013) Cross-species array comparative genomic hybridization identifies novel oncogenic events in zebrafish and human embryonal rhabdomyosarcoma. PLoS Genet 9, e1003727
18. Rudner LA, Brown KH, Dobrinski KP, Bradley DF, Garcia MI, Smith AC et al (2011) Shared acquired genomic changes in zebrafish and human T-ALL. Oncogene 30:4289–4296
19. Lam SH, Wu YL, Vega VB, Miller LD, Spitsbergen J, Tong Y et al (2006) Conservation of gene expression signatures between zebrafish and human liver tumors and tumor progression. Nat Biotechnol 24:73–75
20. Subramanian A, Tamayo P, Mootha VK, Mukherjee S, Ebert BL, Gillette MA et al (2005) Gene set enrichment analysis: a knowledge-based approach for interpreting genome-wide expression profiles. Proc Natl Acad Sci U S A 102:15545–15550

21. Langenau DM, Keefe MD, Storer NY, Guyon JR, Kutok JL, Le X et al (2007) Effects of RAS on the genesis of embryonal rhabdomyosarcoma. Genes Dev 21:1382–1395
22. Nguyen AT, Emelyanov A, Koh CH, Spitsbergen JM, Lam SH, Mathavan S et al (2011) A high level of liver-specific expression of oncogenic Kras(V12) drives robust liver tumorigenesis in transgenic zebrafish. Dis Model Mech 4:801–813
23. Li Z, Luo H, Li C, Huo X, Yan C, Huang X et al (2014) Transcriptomic analysis of a transgenic zebrafish hepatocellular carcinoma model reveals a prominent role of immune responses in tumour progression and regression. Int J Cancer 135:1564–1573
24. He S, Krens SG, Zhan H, Gong Z, Hogendoorn PC, Spaink HP et al (2011) A DeltaRaf1-ER-inducible oncogenic zebrafish liver cell model identifies hepatocellular carcinoma signatures. J Pathol 225:19–28
25. White RM, Cech J, Ratanasirintrawoot S, Lin CY, Rahl PB, Burke CJ et al (2011) DHODH modulates transcriptional elongation in the neural crest and melanoma. Nature 471:518–522
26. Dovey M, White RM, Zon LI (2009) Oncogenic NRAS cooperates with p53 loss to generate melanoma in zebrafish. Zebrafish 6:397–404
27. Lister JA, Robertson CP, Lepage T, Johnson SL, Raible DW (1999) nacre encodes a zebrafish microphthalmia-related protein that regulates neural-crest-derived pigment cell fate. Development 126:3757–3767
28. Berghmans S, Murphey RD, Wienholds E, Neuberg D, Kutok JL, Fletcher CD et al (2005) tp53 mutant zebrafish develop malignant peripheral nerve sheath tumors. Proc Natl Acad Sci U S A 102:407–412
29. Garraway LA, Widlund HR, Rubin MA, Getz G, Berger AJ, Ramaswamy S et al (2005) Integrative genomic analyses identify MITF as a lineage survival oncogene amplified in malignant melanoma. Nature 436:117–122
30. Garraway LA, Weir BA, Zhao X, Widlund H, Beroukhim R, Berger A et al (2005) "Lineage addiction" in human cancer: lessons from integrated genomics. Cold Spring Harb Symp Quant Biol 70:25–34
31. Venter JC, Adams MD, Myers EW, Li PW, Mural RJ, Sutton GG et al (2001) The sequence of the human genome. Science 291:1304–1351
32. Lander ES, Linton LM, Birren B, Nusbaum C, Zody MC, Baldwin J et al (2001) Initial sequencing and analysis of the human genome. Nature 409:860–921
33. Sjoblom T, Jones S, Wood LD, Parsons DW, Lin J, Barber TD et al (2006) The consensus coding sequences of human breast and colorectal cancers. Science 314:268–274
34. Wood LD, Parsons DW, Jones S, Lin J, Sjoblom T, Leary RJ et al (2007) The genomic landscapes of human breast and colorectal cancers. Science 318:1108–1113
35. Liu L, Li Y, Li S, Hu N, He Y, Pong R et al (2012) Comparison of next-generation sequencing systems. J Biomed Biotechnol 2012:251364
36. Pleasance ED, Cheetham RK, Stephens PJ, McBride DJ, Humphray SJ, Greenman CD et al (2010) A comprehensive catalogue of somatic mutations from a human cancer genome. Nature 463:191–196
37. Yen J, White RM, Wedge DC, Van Loo P, de Ridder J, Capper A et al (2013) The genetic heterogeneity and mutational burden of engineered melanomas in zebrafish models. Genome Biol 14:R113
38. Nik-Zainal S, Alexandrov LB, Wedge DC, Van Loo P, Greenman CD, Raine K et al (2012) Mutational processes molding the genomes of 21 breast cancers. Cell 149:979–993
39. Gerlinger M, Rowan AJ, Horswell S, Larkin J, Endesfelder D, Gronroos E et al (2012) Intratumor heterogeneity and branched evolution revealed by multiregion sequencing. N Engl J Med 366:883–892
40. Navin N, Kendall J, Troge J, Andrews P, Rodgers L, McIndoo J et al (2011) Tumour evolution inferred by single-cell sequencing. Nature 472:90–94
41. Yachida S, Jones S, Bozic I, Antal T, Leary R, Fu B et al (2010) Distant metastasis occurs late during the genetic evolution of pancreatic cancer. Nature 467:1114–1117
42. Ciriello G, Miller ML, Aksoy BA, Senbabaoglu Y, Schultz N, Sander C (2013) Emerging landscape of oncogenic signatures across human cancers. Nat Genet 45:1127–1133

43. Iyengar S, Houvras Y, Ceol CJ (2012) Screening for melanoma modifiers using a zebrafish autochthonous tumor model. J Vis Exp. e50086
44. Kulahoglu C, Brautigam A (2014) Quantitative transcriptome analysis using RNA-seq. Methods Mol Biol 1158:71–91
45. Howe K, Clark MD, Torroja CF, Torrance J, Berthelot C, Muffato M et al (2013) The zebrafish reference genome sequence and its relationship to the human genome. Nature 496:498–503
46. Lindner R, Friedel CC (2012) A comprehensive evaluation of alignment algorithms in the context of RNA-seq. PLoS One 7, e52403
47. McKenna A, Hanna M, Banks E, Sivachenko A, Cibulskis K, Kernytsky A et al (2010) The Genome Analysis Toolkit: a MapReduce framework for analyzing next-generation DNA sequencing data. Genome Res 20:1297–1303
48. Cibulskis K, Lawrence MS, Carter SL, Sivachenko A, Jaffe D, Sougnez C et al (2013) Sensitive detection of somatic point mutations in impure and heterogeneous cancer samples. Nat Biotechnol 31:213–219

Transcriptomic Analyses in Zebrafish Cancer Models for Global Gene Expression and Pathway Discovery

Xiaoqian Huang, Ira Agrawal, Zhen Li, Weiling Zheng, Qingsong Lin, and Zhiyuan Gong

Abstract The past decade has witnessed a remarkable advancement of the zebrafish model in cancer research. With the rapid development of genomic tools, it is increasingly feasible to perform genome-wide analyses to identify changes associated with cancer in a wide array of model organisms. These genomic tools, particularly transcriptomic analyses using DNA microarray and RNA sequencing platforms, have now become widely used in zebrafish cancer models to uncover novel biology and common molecular pathways underlying hepatocellular carcinoma, intrahepatic cholangiocarcinoma, melanoma, embryonal rhabdomyosarcoma (ERMS), T cell acute lymphoblastic leukemia (T-ALL), Ewing's sarcoma and glioma. An important finding from these studies is the high similarity and conservation of molecular pathways that underlie cancer in complementary zebrafish models and their human counterparts. Finally, these transcriptomic tools have also proven effective in the development and the validation of specific assays for chemical compound screening. In the future, other genomic tools, such as epigenetic, proteomic and metabolomic tools will likely be incorporated into zebrafish cancer studies, further refining our understanding of cancer.

Keywords Zebrafish • Cancer • Transcriptome • RNA-Seq • Microarray • Pathway • Hepatocellular carcinoma • Leukemia • Melanoma • Embryonal rhabdomyosarcoma

---

X. Huang • I. Agrawal • W. Zheng • Q. Lin • Z. Gong (✉)
Department of Biological Sciences, National University of Singapore, Singapore, Singapore
e-mail: filippoinzaghi24@gmail.com; ira.agrawal@u.nus.edu; zhengweiling.echo@gmail.com; dbslinqs@nus.edu.sg; dbsgzy@nus.edu.sg

Z. Li
Genome Institute of Singapore, Singapore, Singapore
e-mail: liz10@gis.a-star.edu.sg

D.M. Langenau (ed.), *Cancer and Zebrafish*, Advances in Experimental Medicine and Biology 910, DOI 10.1007/978-3-319-30654-4_7

Introduction

The zebrafish has become an increasingly popular model in cancer research. Zebrafish develop cancer following chemical exposure to known carcinogens. Moreover, genetic models have now been developed to overexpress known human oncogenes and/or inactivate tumor suppressor genes [1, 2]. The zebrafish model has many advantages over mouse models of cancer, including reduced cost, optical clarity, and increased statistical power owing to the large number of fish that can be used for in vivo analyses. Zebrafish are also an easily accessible experimental model to functionally test "cancer drivers" discovered from human malignancy. Thus, it is not surprising that many studies used the zebrafish model for oncogene/pathway identification using systematic transgenic screening, cross-species genomic and transcriptomic comparison, tumor transplantation, in vivo live cell imaging, xenograft of human cancer cells, and forward genetic screening.

Zebrafish cancer were first generated using chemical carcinogens, with fish developing a wide array of cancers that were morphologically and histopathologically similarity to human cancer [3, 4]. Following on these studies, Langenau and Look developed the first zebrafish transgenic cancer model, describing a model of MYC-induced T cell leukemia in zebrafish [5]. Together, these studies nucleated interest in using the zebrafish as a cancer model that has expanded remarkably over the past decade. Molecular analyses of transcriptomes have been now been conducted in a wide range of cancers, uncovering remarkable molecular pathway conservation between zebrafish and human cancers. These cross-species transcriptomic analysis have now been completed in hepatocellular carcinoma (HCC), melanoma, rhabdomyosarcoma and other cancers, rendering the zebrafish model as a valid tool for modeling human cancer [6–8]. Such molecular profiling provides detailed insights into cancer mechanisms and has been applied for therapeutic drug discovery.

In this chapter, we will focus on the current status of transcriptomic analyses in zebrafish cancer research and its use in uncovering molecular and drug pathways in cancer.

Principles of Transcriptomic Data Analysis

Global transcriptomic can be generated using DNA microarrays or RNA Sequencing (RNA-seq). Microarray technology is based on designed probes which can hybridize labeled RNA or cDNA with high specificity. For microarray studies, one can use either in-house customized spotted arrays or commercial in situ synthesized arrays like the Affymetrix GeneChip®. The imaging signal of RNA hybridization to specific probes is then used to generate expression data following computational processing of scanned images and assigning intensity scores. This approach has been discussed in detail previously [9–11]. Microarrays are relatively inexpensive when compared to RNA-Seq. However, they are limited by a priori knowledge of which genes are to be assessed. Moreover, detection range is also limited by high background and saturation

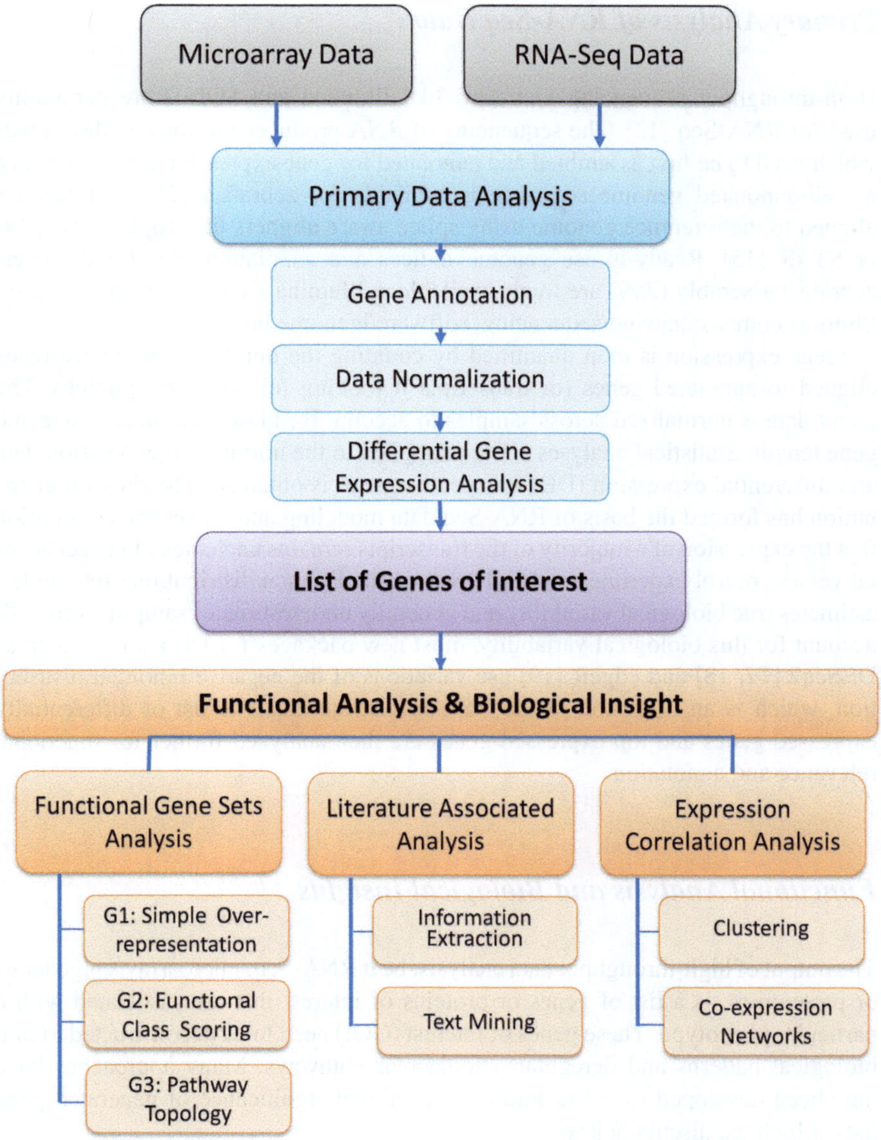

Fig. 1 Flow chart of transcriptomic analyses. Each step is described in section "Principles of Transcriptomic Data Analysis"

of signal at individual spotted probes. In contrast, RNA-Seq has no reliance on the prior knowledge of transcripts or genes to be assessed and RNA quantification has no arbitrary upper limit of detection. RNA-Seq also has the added advantage of identifying sequence variants, alternate gene splicing and post-transcriptional modifications [12]. Here, we provide a brief description of the principles and the tools commonly used for analysis of RNA-Seq data, including functional analyses and annotation. Figure 1 summarizes the workflow of these analyses.

Primary Analysis of RNA-Seq Data

High-throughput sequencing platforms like Illumina and SOLiD are commonly used for RNA-Seq [12]. The sequencing of RNA produces millions of short reads which need to be first assembled and annotated for gene expression analyses. Since a well-annotated genome sequence is available for zebrafish [2], the reads are aligned to the reference genome using splice aware aligners like TopHat2 [13, 14] or STAR [15]. Ready-to-use genome indices and annotation files for the latest zebrafish assembly (Zv9) are freely available at Illumina iGenomes (http://support.illumina.com/sequencing/sequencing_software/igenome.html).

Gene expression is then quantified by counting the number of sequence reads aligned to annotated genes (or transcripts if looking for alternate splicing). The count data is normalized across samples to account for biases like library size and gene length. Statistical analyses are then applied to the normalized expression data and differential expression (DE) for specific genes is obtained. The Poisson distribution has formed the basis of RNA-Seq data modeling and makes the assumption that the expression of a majority of the transcripts remains unchanged in experimental verses control experiments [16]. However, the Poisson distribution often underestimates true biological variability and generally underestimates sampling error. To account for this biological variability, most new packages for DE analysis such as DESeq2 [17, 18] and edgeR [19] use variations of the negative binomial distribution, which is an extension of the Poisson distribution. The list of differentially expressed genes and top expressed genes are then analyzed further for functional relevance and annotation.

Functional Analysis and Biological Insights

The output of high-throughput data analysis, be it RNA-Seq, microarray, epigenetics or proteomics, is a list of genes or proteins of interest that are associated with a particular phenotype. These genes of interest (GOI) need to be deconstructed to find biological patterns and deregulated molecular pathways. Many approaches have now been developed to aid in inferring functional significance of generated gene lists, which we discuss below.

Functional Gene Set Analysis

This approach breaks down the list of GOI into sets of genes sharing biological function. This strategy is dependent on previously defined biological classification of genes into functionally-annotated sets. For example, the Gene Ontology (GO) classification uses three attributes: cellular components, molecular function and biological process. Overrepresentation or enrichment of GOI in any one set suggests participation in that particular biological mechanism [20, 21]. The gene set information is taken from various sources such as GO annotation [22], pathway databases like MSigDb [23], KEGG [24] and Reactome [25]. These first generation tools take input of GOI which have met certain threshold values (e.g. FDR [false discovery rate] of 10 % and twofold change cutoff) and then tests for over- or under-representation of gene sets in the input list. Examples of overrepresentation analysis tools are GOstat [26], DAVID [27] and WebGestalt [28, 29].

The second generation of functional annotation tools include functional class scoring (FCS) approaches which rank GOI based on gene expression within a given sample. This approach removes the arbitrary threshold generally used by the first generation tools and takes into consideration that gene effects on the phenotype are often not only due to large gene expression changes [20, 21]. One popular approach is Gene Set Enrichment Analysis (GSEA), which yields an enrichment score for gene sets overrepresented at the top or bottom of a ranked gene list [30]. GSEA uses an easy graphical and command line interface and is one of the most popularly used functional analysis tools. Other FCS approaches include GAGE [31], GlobalANCOVA [32] and sigPathway [33].

Despite the utility of FCS approaches to associate gene expression changes with functional anotation, these methods often ignore pathway topology, such as the position of genes in a pathway and their interactions. For instance, genes may be involved in several pathways with different roles, and upstream genes are more likely to cause significant changes than downstream genes. Incorporating such information is a major challenge for the field and has resulted in the development of third generation pathway-topology based tools. These approaches treat pathways as graphs with genes forming the nodes and their interaction forming connections between different nodes. These analyses incorporate additional factors such as position of the gene in the pathway, types of interactions, number of interactions, and perturbation factors [20, 34]. SPIA [34] and Pathway-Express [35] are tools for these analyses.

Literature Associated Analysis
This is an information extraction based approach that identifies direct associations between genes which have been identified from the published literature. This method often extracts information based on both co-occurrence of genes/proteins mentioned in literature and sophisticated natural language processing algorithms that infer relationship between the genes/proteins [36]. These interactions and functional annotations are integrated and then curated by experts in the field. Examples include Ingenuity Pathway Analysis (IPA®, www.qiagen.com/ingenuity) and STRING [37], which integrates the gene associations with genomic and proteomic data.

Expression Correlation Analysis
Another approach to organizing data is by clustering genes with similar expression patterns. This clustering has been shown to be effective in grouping genes with similar function together [38]. Various algorithms are available for clustering [38, 39]. In hierarchical clustering, the genes are grouped based on similarity in expression, producing a dendrogram with hierarchical clusters. This can be time consuming and for a large list of genes k-means clustering or self-organizing maps (SOM) are often used. K-means clustering divides the genes into a fixed number of clusters

based on an average expression vector. SOM additionally arranges similar clusters as neighbors [40]. Some software which use clustering algorithms include TMeV, GeneSpring, and Eisen Cluster [41, 42].

Recent co-expression analysis makes use of network theory. Co-expression networks can reveal hubs in a network and evolutionary conserved gene interactions which are vital for survival and hence have not been perturbed across experimental conditions. This approach is useful to identify candidate biomarkers or even therapeutic targets. Weighted gene correlation network analysis (WGCNA) is an example of this approach and is popularly used tool [43].

Current Status of Zebrafish Cancer Transcriptome Research

There is a large body of published literature on using the zebrafish as a cancer model, but only a handful of these studies have incorporated transcriptomic analyses to uncover global transcriptional changes and molecular pathway perturbations found in cancer. Table 1 summarizes a subset of these studies that have been published to date.

Conservation of Tumor Transcriptomes Between Zebrafish and Human

It has been long recognized that zebrafish tumors have a great similarity with human cancers based on histology and many of the key genes involved in carcinogenesis are also conserved in both sequence and function [58]. However, the extent of molecular conservation between zebrafish and human tumors remained unclear until the first microarray-based transcriptomic analysis was performed on chemically-induced zebrafish liver tumors [7]. In this study, a gene set consisting of 2315 gene features (representing 1861 zebrafish Unigene clusters) was obtained through comparison of gene expression profiles of zebrafish liver tumor with normal liver. By mapping the zebrafish unigenes to human orthologs and using GO annotations, it was confirmed that differentially expressed genes in zebrafish liver cancer were similarly involved in the deregulation of cell cycle/proliferation, apoptosis, DNA replication and repair, metastasis, cytoskeletal organization, protein synthesis, and liver-specific functions, just like human liver cancers [44]. These zebrafish microarray data were further compared with human data from four cancer types (liver, gastric, prostate and lung) and it was found that zebrafish liver tumors were more similar to human liver cancers than to other tumor types. Additional comparisons also indicated a striking similarity between cancer progression in zebrafish and human. Altogether, these data provided the first comprehensive molecular evidence to support the modeling of human cancer using zebrafish and laid a foundation for transcriptomic studies in a wide range of zebrafish cancer models.

Table 1 Transcriptomic studies in zebrafish tumor models

Type of tumor	Platform	Significance	Reference
Chemical carcinogen induced liver tumor	Microarray	The first report for global conservation of gene expression between zebrafish and human liver cancers	[7, 44]
KRAS^{G12D} driven embryonal rhabdomyosarcoma (ERMS)	Microarray	Conservation of molecular signature of zebrafish and human ERMS and classification of different types of ERMS generated by different promoters to express KRAS^{G12D} from different stages of muscle development	[8, 45, 46]
Malignant peripheral nerve sheath tumors in *rp*$^{+/-}$ mutants	Microarray	Demonstration of MPNST tumor caused by *rp* mutation and *p53* mutation sharing similar signaling pathways	[47]
kras^{v12} driven hepatocellular carcinoma (HCC)	Microarray	Report of different pathways being turned on in different stages of *kras*-caused liver tumors	[48]
BRAF^{V600E} driven melanoma in *p53*$^{-/-}$ mutants	Microarray	Identification of conserved gene signature in *SETDB1*-overexpressing melanoma in zebrafish and human	[6]
BRAF^{V600E} driven melanoma in *p53*$^{-/-}$ mutants	Microarray	Discovery of a conserved set of genes in neural crest precursors and melanoma for drug screening	[49]
HBx and HCP induced intrahepatic cholangiocarcinoma	RNA-seq	Revealed the importance of TGF-β1 in HBx- and HCP-induced ICC development	[50]
EWS-FLI1 oncoprotein driven Ewing's sarcoma	Microarray	Identification of conserved genes in Ewing's sarcoma between zebrafish and human	[51]
RAS driven embryonal rhabdomyosarcoma (ERMS)	Microarray	Cross-species comparison of transcriptomes of ERMS for identification of readout genes for drug screening	[52]
Tet-on inducible *Myc* driven HCC	RNA-seq	Identification of common signature genes of Myc-induced HCCs from zebrafish, mice and human	[53]
Smoa1 driven glioma	Microarray	Transcriptomic profiling of retinal tumors	[54]
Tet-on inducible *xmrk* driven hepatocellular carcinoma	RNA-seq	Identified distinct immune responses in tumor progression and regression	[55]

(continued)

Table 1 (continued)

Type of tumor	Platform	Significance	Reference
UHRF1 driven HCC in *p53*$^{-/-}$ mutants	RNA-seq	Identification of deregulated genes in the *UHRF1* driven HCC model	[56]
Comparison of three types of HCCs induced by *kras*^{v12}, *xmrk* and *Myc*	RNA-seq	Demonstrating that each oncogene-driven zebrafish HCC model represents a subset of human HCCs	[57]

The molecular conservation of zebrafish tumors with human tumors has also been observed in oncogene-induced liver tumors using transgenic zebrafish. In the past few years, our group has generated several inducible liver tumor models by using the Tet-On or mifepristone-inducible transgenic approaches to target expression of specific oncogenes to hepatocytes including *kras*^{v12} [59, 60], *xmrk* (fish co-ortholog of human epidermal growth factor receptor [EGFR]) [61] or *Myc* [53]. HCC was induced in all of these oncogene expressing transgenic lines. Transcriptomic analyses of these different liver tumors has now been performed and compared with human HCC [57]. Our work uncovered that all of the three oncogenic models correlate highly with very advanced HCC. Interestingly, there were only 21 commonly up-regulated genes found in all three oncogenic models; and yet, these same 21 genes also showed the highest correlation with very advanced human HCC, suggesting their potential as novel molecular biomarkers of disease progression. We have also compared our zebrafish transcriptomic data with ten published human microarray data sets and found that *xmrk*, *kras* and *Myc* gene signatures significantly correlated with 30.8 %, 24.8 % and 25.6 % of HCC clinical samples respectively. Overall, 47.2 % of human HCCs share at least one of the three gene expression signatures identified in zebrafish; providing evidence that we have successfully modeled nearly half of human HCCs.

Transcriptomic Analyses of Zebrafish Liver Cancers for Biological Insights

Transcriptomic analysis is a powerful tool for delineating important biological pathways in tumorigenesis. In particular, cross-species comparison at the transcriptomic level frequently reveals important biological pathways and common biological events that underlie cancer. Using this approach, we have uncovered that carcinogen-induced liver tumors [7] deregulate the Wnt-β-catenin and Ras-MAPK pathways, which are also frequently upregulated in human liver cancers. This has been confirmed using the constitutive *kras*^{v12}-induced transgenic liver tumor model [48]. Further transcriptomic analyses of *kras*^{v12}-induced liver tumors also revealed important signaling pathways associated with different stages of liver tumors, e.g. p53 in hyperplastic liver (HL); TLR-NFκB, JAK-STAT, insulin-IGF and TGFβ in HCC; and Raf-MEK-ERK, PI3K-AKT, Wnt–β-catenin, VEGF and complement cascade in both HL and HCC. Further analysis of the transgenic *Myc*-induced tumor model using RNA-Seq analyses indicated that ribosome proteins are overwhelmingly upregulated in these liver tumors [53], consistent with previous reports that MYC is the key regulator of ribosome gene expression and protein translation [62]. Interestingly, *Myc*-induced zebrafish liver tumors also share a remarkable gene expression similarity with Myc-induced liver tumor models in mice [53].

The *Xmrk*-induced HCC zebrafish model has also been analyzed at the transcriptomic level [55], and was found to closely resemble the human HCC S2 proliferation subtype, which is characterized by activation of a Myc signature [63]. An interesting feature of this inducible transgenic line was tumor regression following oncogene inactivation in established tumors (upon removal of the chemical inducer for transgene expression). When the transcriptomic profiling was performed on regressing zebrafish liver tumors and compared with human HCC progression data sets, including cirrhotic, low grade dysplastic nodules (LGDN), high grade dysplastic nodules (HGDN), very early (veHCC), early (eHCC), advanced (aHCC) and very advanced HCC (vaHCC) [64], it was found that the regressing tumors had a reversed profile compared to human HCC progression. For example, the 1-week regressing tumors (R1) shared the highest resemblance with human LGDN/cirrhotic liver and the 2-week regressing tumors (R2) correlated with cirrhotic liver, while the *xmrk*-induced tumors had the highest resemblance with human vaHCC. The change in transcriptomic patterns was also consistent with histological observation that the liver phenotype gradually reverted to normal histology. GSEA analyses revealed that there are five significantly up-regulated pathways found in HCC: (1) antigen processing and presentation, (2) amino sugar metabolism, (3) cell cycle, (4) proteasome and (5) P53 pathway. Remarkably almost all proteasome genes were upregulated ($N=36$), including both the regulatory and core part of the 26S proteasome, immunoproteasome and chaperon regulators. The expression of these genes decreased in both R1 and R2 regressing tumors, indicating that this pathway played important roles in maintaining HCC state. In addition, R1 regressing tumors had up-regulation of cell cycle and antigen processing and presentation pathways, whereas R2 showed significant up-regulation of cell communication and antigen processing/presentation. It is interesting that antigen processing and presentation pathway was altered in all three stages, HCC, R1 and R2, and that there were different immune responses in established HCC and regressing tumor. These observations have now been validated histologically, with low levels of lymphocytes being observed in HCC samples, while infiltration of eosinophils was prominent in R2 samples. Both macrophages and neutrophils were also increased in HCC and regressing tumors, suggesting that immune cells actively participated in both progression and regression of liver tumors and might play different roles depending on tumor stage.

Hepatocarcinogenesis in human is quite diverse in terms of the genes and molecular pathways that drive tumor growth and progression [65]. In zebrafish, overexpression of many genes (alone or in combination) can induce liver tumors including *kras*, *Myc*, *xmrk*, *edn1*, *src*, *HBx*, *UHRF1* [48, 53, 55, 56, 59–61, 66]. RNA-Seq and transcriptomic comparison of the zebrafish liver tumors induced by *xmrk*, *Myc* and *kras* revealed quite distinct sets of deregulated biological pathways [57]. All tumors show significant deregulation of most of the cancer hallmarks elaborated by Hanahan and Weinberg [67]. *xmrk* mainly upregulated pathways involved in evading growth suppressors and avoiding immune destruction, which included activating cell cycle, promoting RNA transcription, up-regulating proteasome and altering immune properties. *Kras* provided tumor cells with the ability of self-sustaining proliferation by up-regulating EGFR, Raf-MEK-ERK, PI3K-AKT-mTOR and GSK3 signaling pathways, consistent earlier microarray gene expression studies using a different *kras* transgenic zebrafish HCC model [48]. *Myc* mainly up-regulated translation and proteolysis to assist tumor cells in evading growth suppressors, and also up-regulated the VEGF pathway to potentially induce angiogenesis. Pathways in reprogramming of energy metabolism as well as in normal liver function (e.g. blood factors, amino acid metabolism, detoxification and xenobiotic metabolism, fatty acid metabolism, bile synthesis, etc) were generally down-regulated in all three liver tumor models but the down-regulation in the *Myc* model was less apparent than the other two models (Fig. 2).

Recently, we have also completed transcriptomic comparisons of zebrafish liver tumors induced by a single oncogene (*xmrk* or *Myc*) or by expression of both oncogenes, with the latter tumors showing accelerated hepatocarcinogenesis and a more severe tumor phenotype when compared to either single oncogene-induced tumor models. Based on differentially expressed genes from these tumors, biological pathway analysis showed that pathways deregulated in the *Myc/xmrk*-induced liver tumors were also largely deregulated in the *Myc* and *xmrk* single transgenic tumors. Among the commonly up-regulated pathways, most of them showed more significant up-regulation in the double transgenic tumors than in single transgenic tumors (Fig. 3a). Interestingly, the pathways with opposite changes in the two single oncogene-induced liver tumors appeared to be counterbalanced in the double transgenic tumors (Fig. 3b).

Zebrafish models of HCC have also been generated by transgenic overexpression of human *UHRF1*, a gene required for DNA methylation and DNMT1 degradation. Transcriptomic gene expression studies uncovered a high molecular conservation of this model with human HCC [56]. In this model, high level expression of *UHRF1* caused DNA hypomethylation, triggered Tp53-mediated senescence in the liver, and resulted in a "microliver" phenotype in 5-dpf (day post-fertilization) larvae. RNA-Seq analysis uncovered down-regulation of proliferation associated genes including *ccnd1* and *myc*, and up-regulation of *tp53* and its target genes. However, by 8 dpf, senescence was significantly decreased in *UHRF1* transgenic fish with up to 70 % of the fish developing HCC by just 20 days of life. The reduction in senescence did not appear to be due to the re-establishment of DNA methylation or transgene silencing. Moreover, loss of one copy of *tp53* also resulted in increased tumor incidence, indicating that Tp53 had a role in both senescence and tumor onset. The same mechanism appeared to be present in human HCCs since there was a significant correlation between *UHRF1* overexpression, *TP53* mutation and genome integ-

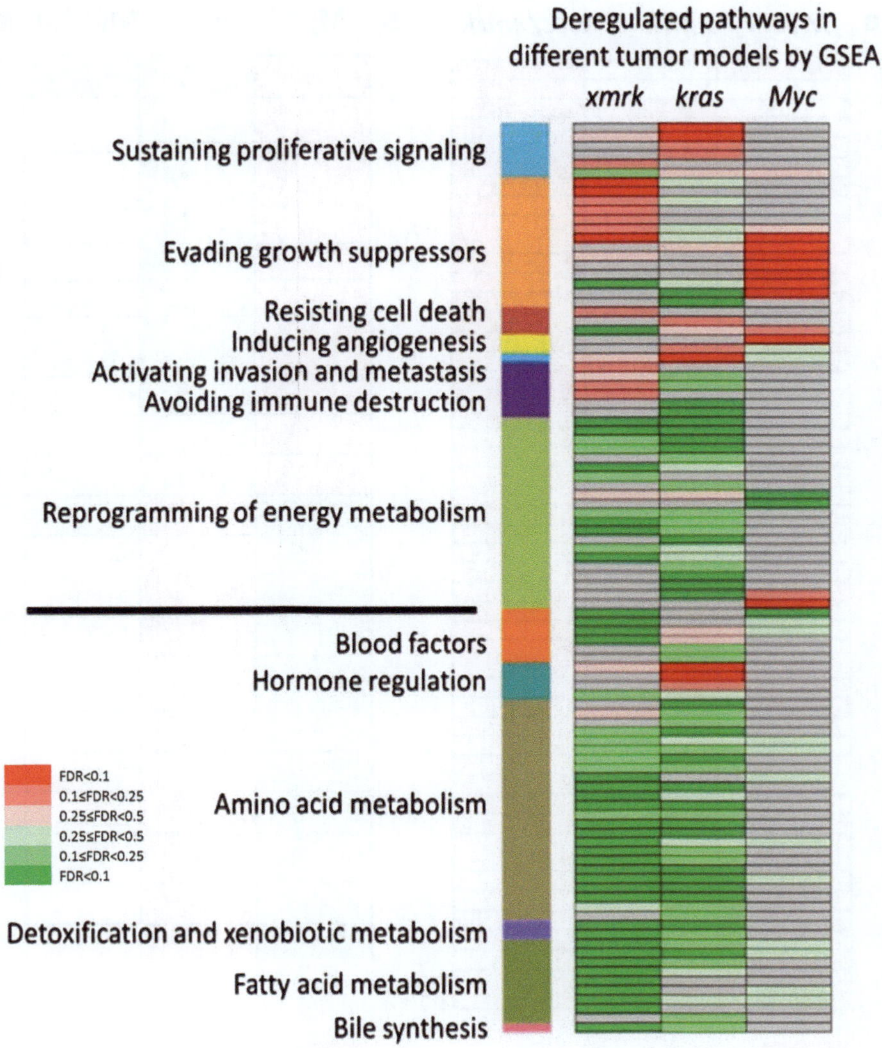

Fig. 2 Deregulated pathways in the three oncogene transgenic liver tumors. Up- and down-regulated pathways in each zebrafish transgenic HCC model were analyzed by GSEA. *Red* and *green* indicate up-and down-regulation, respectively. The color legend for assessing FDR is shown to the left. *Grey* denotes pathways that were unchanged. The up-regulated pathways were assigned into seven cancer hallmark categories according to Hanahan and Weinberg [67] and down-regulated pathways were classified based on liver metabolism. This figure was published previously in PLoS One [57]

rity, while genome-wide DNA hypomethylation was also present in most HCCs. A unique zebrafish liver cancer model has been recently reported that develops intrahepatic cholangiocarcinoma (ICC) following transgenic co-expression of hepatitis B virus X gene (Hbx) and hepatitis C virus core protein gene in hepatocytes [50].

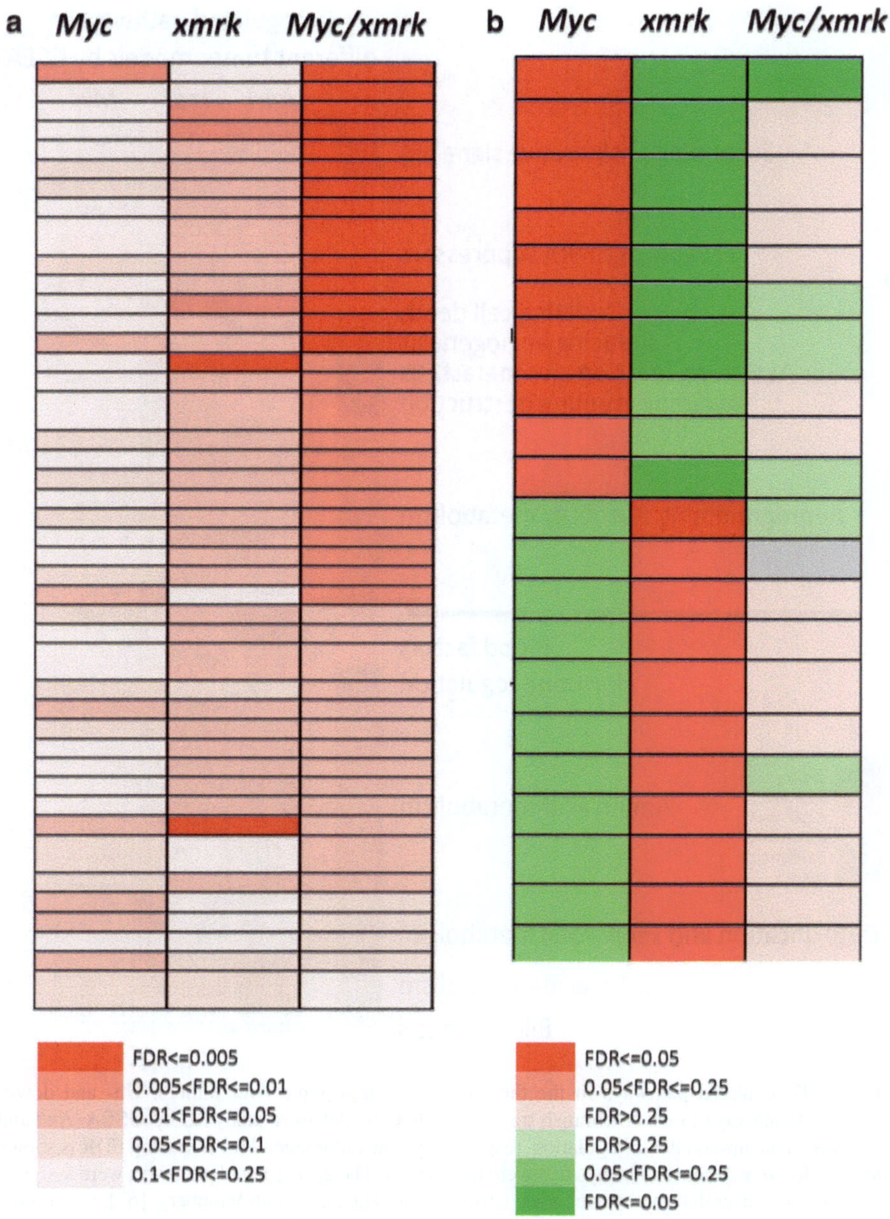

Fig. 3 Synergistic and counteractive pathway effects in the *Myc/xmrk* double transgenic zebrafish liver tumor model when compared to the single transgenic tumors. (**a**) Pathway synergy in genes co-regulated by *Myc* and *xmrk*. Forty-four canonical pathways were up-regulated in both *Myc*- and *xmrk*-induced zebrafish liver cancer. Thirty-two of them showed a more significant up-regulation in the double transgenic *Myc/xmrk* liver cancer model. (**b**) Counteractive effects of pathways oppositely regulated by *Myc* and *xmrk*. Pathways that were oppositely regulated in *Myc* and *xmrk* single transgenic models were counterbalanced in double transgenic tumors and thus, did not show any significant change

Transcriptomic analyses of the ICC model revealed a common set of genes up-regulated in both zebrafish and human ICCs. Further pathway analyses indicated a role of TGFβ pathway in tumorigenesis in this model, which was subsequently validated experimentally.

Transcriptomic Analyses of Zebrafish Cancer Models in Discovery of Novel Molecular Pathways

Transcriptomic analyses has now been widely applied to several other zebrafish tumors. In a mosaic transgenic zebrafish model of $KRAS^{G12D}$-induced embryonal rhabdomyosarcoma (ERMS) [8], two highly conserved gene signatures were identified following comparison of zebrafish and human ERMS. One gene signature was associated with tumor-specific and tissue-restricted gene expression in human ERMS and the other was specifically associated with RAS activity. This latter finding indicated that the RAS pathway was the dominant oncogenic driver in human ERMS. In subsequent studies, DNA constructs with different oncogenes, promoters and reporters (e.g. *rag2-KRASG12D*, *rag2-dsREDexpress*, *myogenin-H2B-RFP*, *myogenin-H2B-Amcyan*, and *mylz2-lyn-cyan*) were microinjected singly or in combination into zebrafish embryos at the one-cell stage and various ERMS subpopulations with different fluorescence labels were obtained [45]. Through fluorescence-activated cell sorting (FACS) and microarray gene expression studies, it was revealed that these different subpopulations had significant differences in gene expression and mimicked satellite cell and myoblast differentiation states found during normal muscle regeneration. One additional similarity identified between zebrafish and human ERMS was the compartmentalization of ERMS cell types based on myogenic transcription factor expression and differentiation status within the tumor. In vivo imaging studies went on to uncover that the myogenin positive ERMS cells were highly migratory. In a more recent study, zebrafish rhabdomyosarcoma (RMS) has been generated by expressing oncogenic $KRAS^{G12D}$ in different stages of muscle development by using different promoters, including *cdh15* and *rag2* promoters to drive gene expression in muscle progenitors, and *mylz2* promoter to drive expression in differentiating myoblasts [46]. Histopathological examination revealed that the oncogene driven by *cdh15* and *rag2* promoters resulted in tumors with a less differentiated phenotype than those driven by *mylz2* promoter. At molecular level, the *rag2*- and *cdh15*-driven tumors were more similar to each other than to the *mylz2* tumors. The zebrafish RMS gene signatures were then compared to human RMS tumors by GSEA and it was found that *mylz2* tumors best represent the well-differentiated human RMS while the *rag2* tumors recapitulate the undifferentiated subtypes. Ingenuity Pathway Analysis (IPA) further revealed differential regulation of glycolysis and Pten (phosphatase and tensin homolog) signaling pathways between *myzl2* and *rag2* tumors, indicating that changes in metabolism might be an underlying cause of distinct tumor subtypes.

A transgenic *Tg(mitfa:BRAF^{V600E})/p53(lf)* zebrafish melanoma model has been used to directly test human candidate genes for accelerating melanoma [6]. A novel human gene, *SETDB1*, was found to be amplified in human melanoma. When SETDB1 was overexpressed in the zebrafish melanoma model, tumors showed more invasive features. Microarray analyses revealed a gene signature consisting of 69 human orthologs down-regulated in *SETDB1* overexpressing melanomas and this gene signature was inversely correlated with *SETDB1* expression in 93 human melanoma short-term cultures and cell lines, suggesting the conserved role of this gene in both zebrafish and human melanomas. Chromatin immunoprecipitation sequencing (ChIP-Seq) was then performed in order to identify the direct targets of SETDB1. As a result, SETDB1 was found to methylate histone H3 on lysine 9 (H3K9) and its overexpression was likely to increase the activity of the H3K9 methyltransferase complex, causing changes in target specificity. In a related study using the same zebrafish transgenic model [49], microarray gene expression analysis was performed on human melanoma tissue and it was found that the majority of human melanomas expressed the neural crest marker *sox10*, thus, these tumor cells still possess progenitor-like features.

A zebrafish model for Ewing's sarcoma has been established using the human *EWS-FLI1* fusion oncogene under the control of either the zebrafish *hsp70* or the *β-actin* promoters [51]. In the *tp53* mutant background, a portion of transgenic fish develop small round blue cell tumors (SRBCTs). Despite the histopathological similarity of SRBCT to Ewing's sarcoma, gene expression profiling of this tumor revealed only a small set of overlapping genes with human Ewing's sarcoma. Among the six commonly up-regulated genes, *NKX2-2*, *MYC*, *MAPT*, *SALL2*, *PADI2* and *POU3F1*, *NKX2-2* were identified as important downstream targets of EWS-FLI1. These observations indicated that this zebrafish SRBCT model had biological relevance to human Ewing's sarcoma, thus providing a new way to study the function of *EWS-FLI1* fusion oncogene.

Transcriptomic analysis has also been used to aid the characterization of molecular pathways in other transgenic models. One example is the transgenic glioma model where tumors are induced by transgenic expression of *smoothened* in neural progenitor cells [54]. In this model, retinal dysplasia and retinal tumors developed in adults. Microarray gene expression comparison between gross eye tumors and age-matched control eyes identified 1901 differentially expressed genes, including upregulated tumor-associated genes that regulate cell cycle and the Hedgehog pathway genes including *gli1* and *gli2a*. KEGG (Kyoto Encyclopedia of Genes and Genomes) pathway analyses also revealed significant upregulation of cancer-related pathways such as those for carcinoma, glioma and small-cell lung cancer.

Transcriptomic analysis has also played an important role in classifying and identifying similar tumor types in the zebrafish [47]. One example was the confirmation of malignant peripheral nerve sheath tumors (MPNSTs) in ribosomal protein (*rp*) mutated zebrafish, which appeared histologically similar to MPNST that developed in *p53$^{M214K/M214K}$* fish and were confirmed to share similar gene expression as assessed by transcriptomic analyses. Gene expression profiles using microarray gene expression showed that MPNSTs arising from either *rp* mutation or *p53* mutation were closely related to one another. Further experiments revealed that although *p53* RNA remains intact in *rp* mutants, *p53* mRNAs fail to be translated into functional protein. Since p53 has been involved in most human cancers, this study raises the need for careful examination of p53 protein translation in evaluation of carcinogenesis.

In sum, these transcriptome-based molecular analyses have demonstrated the amazing conservation of global gene expression profiles and molecular mechanisms found in zebrafish and human cancer, thus providing further confidence for future application of the zebrafish model in cancer research. These studies have laid a solid foundation for future comparative transcriptomic studies between experimental zebrafish cancer models and human malignancy.

Transcriptomic Analyses in Chemical Screening for Drug Discovery

Transcriptomic analysis has become a powerful tool in drug discovery and development, particularly in the zebrafish model which has emerged as an excellent platform for high-throughput chemical screens. One prominent example used the zebrafish *Tg(mitfa:BRAF(V600E)* melanoma model to screen for potential anti-melanoma drugs [49]. In this study, microarray analysis was first used to compare gene expression profiles between transgenic zebrafish embryos and melanoma. This analysis identified a common gene signature that comprised 123 genes and included embryonic neural progenitor markers such as *crestin*, *sox10* and *ednrb*, and melanocyte markers such as *tyr* and *dct*. This suggested that melanoma cells are likely multipotent and similar to neural crest progenitors. Thus, a hypothesis was proposed that chemical suppressors of neural crest progenitors would identify common drugs that were also effective against melanoma. After screening 2000 chemicals, leflunomide, an inhibitor of dihydroorotate dehydrogenase (DHODH), was identified to reduce expression of *sox10* and *dct* in embryos. Microarray analyses further confirmed that this small molecule down-regulated 49 % of the genes up-regulated in the common 123-gene signature, and largely comprised genes found in neural crest cells. Further test using leflunomide and its derivative (A771726) on neural crest stem cells (NCSCs) indicated a prominent role in suppressing self-renewal of this target cell population. A771726 was able to decrease cell proliferation in a dose-dependent manner in human melanoma cell lines, similar to the results obtained with shRNA knock-down of DHODH. Combination treatment of leflunomide and PLX4720 ($BRAF^{V600E}$ inhibitor) resulted in suppressed human melanoma growth in xenotransplantation mouse models. In total, this study demonstrated the power of transcriptomic analyses to guide and to develop drug screening assays. Importantly, these findings have resulted in a Phase I/II clinical trial using Vemurafenib and Leflunomide for the treatment of human melanoma (http://clinicaltrials.gov/ct2/show/study/NCT01611675).

Another example of developing drug screening approaches based on transcriptomic analyses came from assessing the similarity of RAS signaling in embryonic development with transgenic zebrafish models of *KRAS*^{G12D}-induced embryonal rhabdomyosarcoma (ERMS). Specifically, heat shock inducible transgenic *Tg(hsp70-HRAS*^{G12V}*)* embryos were compared with wild-type embryos following heat shock [52]. Of the genes identified in this comparison, genes up-regulated in *HRAS*^{G12V} expressing animals were classified as "cancer" and were always ranked first in the "diseases and disorders" group, ahead of "developmental disorder" and "organismal injury and abnormalities", suggesting that the heat-shock induced RAS expression in embryos activated common oncogenic pathways. To identify downstream readout genes for use in chemical screening, commonly up-regulated genes are identified in comparing (1) *Tg(hsp70-HRAS*^{G12V}*)* embryos, (2) zebrafish *KRAS*^{G12D}-induced ERMS, (3) human ERMS, (4) human pancreatic cancer and (5) RAS-infected HMECs. From this analysis, *dusp6 (mkp3)*, a target of the FGF (fibroblast growth factor) signaling pathway and a negative regulator of Erk1/2 (Mapk3/1), was identified as the top readout gene for use in a chemical screen. With these experimental tools developed, among 2896 bioactive compounds screened, 31 exhibit complete suppression of *dusp6* expression. After further validation and assessment of the heat shock effect, 18 compounds are verified as specific inhibitors of the RAS pathway. The tumor inhibitory potential of these compounds was further tested in zebrafish ERMS and two chemicals, PD98059 (MEK inhibitor) and tosyl phenylalanyl chloromethyl ketone (TPCK, chymotrypsin-like serine protease inhibitor) were found to significantly delay tumor growth without affecting the gross morphology of the zebrafish. Further experiments demonstrated that these two compounds synergistically inhibited tumor progression in both zebrafish ERMS and a human rhabdomyosarcoma cell line that has a *NRAS*^{Q61H} mutation.

Future Perspectives

Thus far, global gene expression and pathway discovery in zebrafish cancer models have been entirely based on transcriptomic analyses using microarray and RNA-Seq platform technologies. With the completion of zebrafish genome sequencing and the ever improving gene annotation, comparative genomic analyses with human counterparts has become increasingly feasible and accurate. Now is the time to expand the frontier of genomic studies of cancer biology in zebrafish cancer models by incorporating other omic tools for epigenetic studies, proteomic and metabolomic profiling. These omic tools will enable a more holistic view of cancer and better utilization of zebrafish cancer models.

Cancer Transcriptomics for Alternative Splicing and microRNAs So far, transcriptomic studies have focused mainly on assessing RNA expression levels, with little information being provided on splice variant usage and/or micro-RNA expression in zebrafish cancer models. Up to 94 % of human genes are alternatively spliced and alternative splicing plays important roles in diverse cellular functions, including proliferation, apoptosis, metabolism and metastasis [68]. In the zebrafish, alternative splicing has important roles in regulating embryonic development [69, 70]; yet, systematic study of alternative splicing through RNA-Seq remains in its infancy, especially in the cancer setting. Thus, there will be an obvious need to systematically investigate alternative splicing in zebrafish cancer models. Another deficiency of current transcriptomic analyses in the zebrafish model is the lack of profiling of microRNAs, which have been discovered two decades ago and have been linked to many diseases including cancer [71]. Similar to the mRNA transcriptome, microRNA expression profiles have proven to be a valid cancer classifier as well as a cancer prognostic indicator [71]. Moreover, body fluids often contain microRNAs and can be used for biomarkers and cancer diagnosis [72]. Interestingly, microRNA could also function in cancer metastasis through both intra- [71] and inter-cellular mechanisms [73]. MicroRNAs play essential roles during zebrafish development [74], but their functions in zebrafish cancer is still poorly understood. Therefore, systematic study of microRNA profiling and function in zebrafish cancer models would greatly facilitate our understanding concerning the roles of microRNA's in human malignancy.

Cancer Epigenomics Although the roles of genomic changes in carcinogenesis have been well recognized, increasing evidence has indicated the importance of epigenetic changes as drivers of cancer growth and progression [75]. In the zebrafish, there are few reports on the role of epigenetics in cancer. To date, only one early report has examined CpG DNA methylation in promoter regions in carcinogen-induced zebrafish HCC [76], indicating conservation of altered DNA methylation between zebrafish and human HCCs. Now various epigenomic analyses have been applied to developmental analyses in zebrafish, including identification of histone modifications [77–79] and DNA methylation [80–82]. Therefore, these epigenetic tools will surely be applied to future studies in zebrafish cancer models.

Cancer Proteomics Proteomics in the zebrafish model is a relatively new research topic, but has been broadly applied to the study of embryonic development, tissue composition, and toxicological responses [83–87]. Proteomic analyses have been intensively applied in human cancer studies [88]. Proteomic technologies are generally classified as discovery-based and targeted-based. Discovery-based proteomics makes use of two-dimensional gel-electrophoresis (2-DE) and/or liquid chromatography–mass spectrometry (LC–MS) coupled with fluorescence-labeling (Difference Gel Electrophoresis [DIGE]), stable-isotope-labeling (e.g., Isobaric tag for relative and absolute quantitation [iTRAQ]) or stable-isotope labeling by amino acids in cell culture [SILAC]. Label-free approaches can also be used for the identification and quantification of proteins, variants, post-translational modifications, and protein-protein interactions. Targeted-based proteomics refers to multiple reaction monitoring-MS (MRM-MS) or similar technologies for the quantification of proteins of interest in complex samples, such as serum or plasma. It often couples with stable-isotope labeled internal standards and sample depletion/enrichment approaches to achieve higher accuracy, reproducibility and sensitivity, and has been widely used as a validation tool for discovery-based proteomics studies. In a recent report [89], sodium deoxycholate-based extraction buffer with heat denaturation was found to be the most effective approach for extracting proteins from complex tissues of the zebrafish and liver tumors. This new protocol has identified a total of 4790 proteins, the highest number by using shotgun proteomics approach with 2D liquid chromatography. With this discovery, proteomics approaches will likely be conducted in a wide array of zebrafish tumors and tissues to yield valuable data for further analyses.

Cancer Metabolomics Cancer metabolism is now widely recognized as a cancer hallmark [67]. Although cancers have different oncogenic drivers and use different molecular pathways for transformation, one common feature of all cancers is the rewiring of metabolic pathways to increase biomass and provide energy to rapidly dividing cells. Similar to "oncogene addiction", "metabolic addiction" is likely a major driver of cancer growth [83]. Many metabolic enzymes are targetable by small molecules for drug development [83]. However, there is an apparent limitation of RNA based analyses for metabolic pathways as many of the metabolic enzymes are not regulated at the transcriptional level. Thus, metabolomic tools will be critical for further validating findings from transcriptomic and proteomic analyses. Whole organism [90], tissue specific [86] and metabolite specific metabolomics [91] techniques have been developed in the zebrafish model. Several groups have identified novel aspects of disease metabolism regulation such as fatty acid-fuelled matrix metalloproteinase production in leukocyte migration during inflammation [87] as well as small molecule modulation of specific metabolic pathways including gluconeogenesis [92]. Application of metabolomics to zebrafish cancer models, along with comparison to human cancer data sets, will likely provide new insights into cancer metabolism and identification of conserved therapeutic targets in cancer.

Acknowledgement Our research is supported by National Medical Research Council of Singapore (R154000547511).

References

1. Yen J, White RM, Stemple DL (2014) Zebrafish models of cancer: progress and future challenges. Curr Opin Genet Dev 24:38–45
2. Howe K, Clark MD, Torroja CF, Torrance J, Berthelot C et al (2013) The zebrafish reference genome sequence and its relationship to the human genome. Nature 496:498–503
3. Spitsbergen JM, Tsai HW, Reddy A, Miller T, Arbogast D et al (2000) Neoplasia in zebrafish (*Danio rerio*) treated with *N*-methyl-*N′*-nitro-*N*-nitrosoguanidine by three exposure routes at different developmental stages. Toxicol Pathol 28:716–725
4. Spitsbergen JM, Tsai HW, Reddy A, Miller T, Arbogast D et al (2000) Neoplasia in zebrafish (*Danio rerio*) treated with 7,12-dimethylbenz[*a*]anthracene by two exposure routes at different developmental stages. Toxicol Pathol 28:705–715

5. Langenau DM, Traver D, Ferrando AA, Kutok JL, Aster JC et al (2003) Myc-induced T cell leukemia in transgenic zebrafish. Science 299:887–890
6. Ceol CJ, Houvras Y, Jane-Valbuena J, Bilodeau S, Orlando DA et al (2011) The histone methyltransferase SETDB1 is recurrently amplified in melanoma and accelerates its onset. Nature 471:513–517
7. Lam SH, Wu YL, Vega VB, Miller LD, Spitsbergen J et al (2006) Conservation of gene expression signatures between zebrafish and human liver tumors and tumor progression. Nat Biotechnol 24:73–75
8. Langenau DM, Keefe MD, Storer NY, Guyon JR, Kutok JL et al (2007) Effects of RAS on the genesis of embryonal rhabdomyosarcoma. Genes Dev 21:1382–1395
9. Hegde P, Qi R, Abernathy K, Gay C, Dharap S et al (2000) A concise guide to cDNA microarray analysis. Biotechniques 29:548–550, 552–544, 556 passim
10. Lam SH, Karuturi KMR, Gong Z (2009) Zebrafish spotted-microarray for genome-wide expression profiling experiments: data acquisition and analysis. Methods Mol Biol 546:197–226
11. Lam SH, Mathavan S, Gong Z (2009) Zebrafish spotted-microarray for genome-wide expression profiling experiments. Part I: array printing and hybridization. Methods Mol Biol 546:175–195
12. Wang Z, Gerstein M, Snyder M (2009) RNA-Seq: a revolutionary tool for transcriptomics. Nat Rev Genet 10:57–63
13. Trapnell C, Roberts A, Goff L, Pertea G, Kim D et al (2012) Differential gene and transcript expression analysis of RNA-seq experiments with TopHat and Cufflinks. Nat Protoc 7:562–578
14. Kim D, Pertea G, Trapnell C, Pimentel H, Kelley R et al (2013) TopHat2: accurate alignment of transcriptomes in the presence of insertions, deletions and gene fusions. Genome Biol 14:R36
15. Dobin A, Davis CA, Schlesinger F, Drenkow J, Zaleski C et al (2013) STAR: ultrafast universal RNA-seq aligner. Bioinformatics 29:15–21
16. Li J, Witten DM, Johnstone IM, Tibshirani R (2012) Normalization, testing, and false discovery rate estimation for RNA-sequencing data. Biostatistics 13:523–538
17. Anders S, Huber W (2010) Differential expression analysis for sequence count data. Genome Biol 11:R106
18. Anders S, McCarthy DJ, Chen Y, Okoniewski M, Smyth GK et al (2013) Count-based differential expression analysis of RNA sequencing data using R and Bioconductor. Nat Protoc 8:1765–1786
19. Robinson MD, McCarthy DJ, Smyth GK (2010) edgeR: a Bioconductor package for differential expression analysis of digital gene expression data. Bioinformatics 26:139–140
20. Khatri P, Sirota M, Butte AJ (2012) Ten years of pathway analysis: current approaches and outstanding challenges. PLoS Comput Biol 8, e1002375
21. Tarca AL, Bhatti G, Romero R (2013) A comparison of gene set analysis methods in terms of sensitivity, prioritization and specificity. PLoS One 8, e79217
22. Ashburner M, Ball CA, Blake JA, Botstein D, Butler H et al (2000) Gene ontology: tool for the unification of biology. The Gene Ontology Consortium. Nat Genet 25:25–29
23. Liberzon A, Subramanian A, Pinchback R, Thorvaldsdottir H, Tamayo P et al (2011) Molecular signatures database (MSigDB) 3.0. Bioinformatics 27:1739–1740
24. Kanehisa M, Goto S (2000) KEGG: kyoto encyclopedia of genes and genomes. Nucleic Acids Res 28:27–30
25. Croft D, Mundo AF, Haw R, Milacic M, Weiser J et al (2014) The Reactome pathway knowledgebase. Nucleic Acids Res 42:D472–D477
26. Beissbarth T, Speed TP (2004) GOstat: find statistically overrepresented Gene Ontologies within a group of genes. Bioinformatics 20:1464–1465
27. Dennis G Jr, Sherman BT, Hosack DA, Yang J, Gao W et al (2003) DAVID: database for annotation, visualization, and integrated discovery. Genome Biol 4:P3

28. Zhang B, Kirov S, Snoddy J (2005) WebGestalt: an integrated system for exploring gene sets in various biological contexts. Nucleic Acids Res 33:W741–W748
29. Wang J, Duncan D, Shi Z, Zhang B (2013) WEB-based GEne SeT AnaLysis Toolkit (WebGestalt): update 2013. Nucleic Acids Res 41:W77–W83
30. Subramanian A, Tamayo P, Mootha VK, Mukherjee S, Ebert BL et al (2005) Gene set enrichment analysis: a knowledge-based approach for interpreting genome-wide expression profiles. Proc Natl Acad Sci U S A 102:15545–15550
31. Luo W, Friedman MS, Shedden K, Hankenson KD, Woolf PJ (2009) GAGE: generally applicable gene set enrichment for pathway analysis. BMC Bioinf 10:161
32. Hummel M, Meister R, Mansmann U (2008) GlobalANCOVA: exploration and assessment of gene group effects. Bioinformatics 24:78–85
33. Tian L, Greenberg SA, Kong SW, Altschuler J, Kohane IS et al (2005) Discovering statistically significant pathways in expression profiling studies. Proc Natl Acad Sci U S A 102:13544–13549
34. Tarca AL, Draghici S, Khatri P, Hassan SS, Mittal P et al (2009) A novel signaling pathway impact analysis. Bioinformatics 25:75–82
35. Draghici S, Khatri P, Tarca AL, Amin K, Done A et al (2007) A systems biology approach for pathway level analysis. Genome Res 17:1537–1545
36. Jensen LJ, Saric J, Bork P (2006) Literature mining for the biologist: from information retrieval to biological discovery. Nat Rev Genet 7:119–129
37. von Mering C, Jensen LJ, Snel B, Hooper SD, Krupp M et al (2005) STRING: known and predicted protein-protein associations, integrated and transferred across organisms. Nucleic Acids Res 33:D433–D437
38. Eisen MB, Spellman PT, Brown PO, Botstein D (1998) Cluster analysis and display of genome-wide expression patterns. Proc Natl Acad Sci U S A 95:14863–14868
39. Quackenbush J (2001) Computational analysis of microarray data. Nat Rev Genet 2:418–427
40. Tamayo P, Slonim D, Mesirov J, Zhu Q, Kitareewan S et al (1999) Interpreting patterns of gene expression with self-organizing maps: methods and application to hematopoietic differentiation. Proc Natl Acad Sci U S A 96:2907–2912
41. Saeed AI, Sharov V, White J, Li J, Liang W et al (2003) TM4: a free, open-source system for microarray data management and analysis. Biotechniques 34:374–378
42. Chu L, Scharf E, Kondo T (2001) GeneSpringTM: tools for analyzing microarray expression data. Genome Inform 12:227–229
43. Langfelder P, Horvath S (2008) WGCNA: an R package for weighted correlation network analysis. BMC Bioinf 9, 559
44. Lam SH, Gong Z (2006) Modeling liver cancer using zebrafish: a comparative oncogenomics approach. Cell Cycle 5:573–577
45. Ignatius MS, Chen E, Elpek NM, Fuller AZ, Tenente IM et al (2012) In vivo imaging of tumor-propagating cells, regional tumor heterogeneity, and dynamic cell movements in embryonal rhabdomyosarcoma. Cancer Cell 21:680–693
46. Storer NY, White RM, Uong A, Price E, Nielsen GP et al (2013) Zebrafish rhabdomyosarcoma reflects the developmental stage of oncogene expression during myogenesis. Development 140:3040–3050
47. MacInnes AW, Amsterdam A, Whittaker CA, Hopkins N, Lees JA (2008) Loss of p53 synthesis in zebrafish tumors with ribosomal protein gene mutations. Proc Natl Acad Sci U S A 105:10408–10413
48. Nguyen AT, Emelyanov A, Koh CH, Spitsbergen JM, Lam SH et al (2011) A high level of liver-specific expression of oncogenic $Kras^{V12}$ drives robust liver tumorigenesis in transgenic zebrafish. Dis Model Mech 4:801–813
49. White RM, Cech J, Ratanasirintrawoot S, Lin CY, Rahl PB et al (2011) DHODH modulates transcriptional elongation in the neural crest and melanoma. Nature 471:518–522
50. Liu W, Chen JR, Hsu CH, Li YH, Chen YM et al (2012) A zebrafish model of intrahepatic cholangiocarcinoma by dual expression of hepatitis B virus X and hepatitis C virus core protein in liver. Hepatology 56:2268–2276

51. Leacock SW, Basse AN, Chandler GL, Kirk AM, Rakheja D et al (2012) A zebrafish transgenic model of Ewing's sarcoma reveals conserved mediators of EWS-FLI1 tumorigenesis. Dis Model Mech 5:95–106
52. Le X, Pugach EK, Hettmer S, Storer NY, Liu J et al (2013) A novel chemical screening strategy in zebrafish identifies common pathways in embryogenesis and rhabdomyosarcoma development. Development 140:2354–2364
53. Liao Y, Smyth GK, Shi W (2013) The Subread aligner: fast, accurate and scalable read mapping by seed-and-vote. Nucleic Acids Res 41, e108
54. Ju B, Chen W, Spitsbergen JM, Lu J, Vogel P et al (2014) Activation of Sonic hedgehog signaling in neural progenitor cells promotes glioma development in the zebrafish optic pathway. Oncogenesis 3, e96
55. Li Z, Luo H, Li C, Huo X, Yan C et al (2014) Transcriptomic analysis of a transgenic zebrafish hepatocellular carcinoma model reveals a prominent role of immune responses in tumour progression and regression. Int J Cancer 135:1564–1573
56. Mudbhary R, Hoshida Y, Chernyavskaya Y, Jacob V, Villanueva A et al (2014) UHRF1 overexpression drives DNA hypomethylation and hepatocellular carcinoma. Cancer Cell 25:196–209
57. Zheng W, Li Z, Nguyen AT, Li C, Emelyanov A et al (2014) Xmrk, kras and myc transgenic zebrafish liver cancer models share molecular signatures with subsets of human hepatocellular carcinoma. PLoS One 9, e91179
58. Amatruda JF, Shepard JL, Stern HM, Zon LI (2002) Zebrafish as a cancer model system. Cancer Cell 1:229–231
59. Nguyen AT, Emelyanov A, Koh CH, Spitsbergen JM, Parinov S et al (2012) An inducible *kras*^{V12} transgenic zebrafish model for liver tumorigenesis and chemical drug screening. Dis Model Mech 5:63–72
60. Chew TW, Liu XJ, Liu L, Spitsbergen JM, Gong Z et al (2014) Crosstalk of Ras and Rho: activation of RhoA abates Kras-induced liver tumorigenesis in transgenic zebrafish models. Oncogene 33:2717–2727
61. Li Z, Huang X, Zhan H, Zeng Z, Li C et al (2012) Inducible and repressable oncogene-addicted hepatocellular carcinoma in Tet-on xmrk transgenic zebrafish. J Hepatol 56:419–425
62. van Riggelen J, Yetil A, Felsher DW (2010) MYC as a regulator of ribosome biogenesis and protein synthesis. Nat Rev Cancer 10:301–309
63. Hoshida Y, Nijman SM, Kobayashi M, Chan JA, Brunet JP et al (2009) Integrative transcriptome analysis reveals common molecular subclasses of human hepatocellular carcinoma. Cancer Res 69:7385–7392
64. Wurmbach E, Chen YB, Khitrov G, Zhang W, Roayaie S et al (2007) Genome-wide molecular profiles of HCV-induced dysplasia and hepatocellular carcinoma. Hepatology 45:938–947
65. Santarius T, Shipley J, Brewer D, Stratton MR, Cooper CS (2010) A census of amplified and overexpressed human cancer genes. Nat Rev Cancer 10:59–64
66. Lu JW, Yang WY, Tsai SM, Lin YM, Chang PH et al (2013) Liver-specific expressions of HBx and src in the p53 mutant trigger hepatocarcinogenesis in zebrafish. PLoS One 8, e76951
67. Hanahan D, Weinberg RA (2011) Hallmarks of cancer: the next generation. Cell 144:646–674
68. Oltean S, Bates DO (2014) Hallmarks of alternative splicing in cancer. Oncogene 33:5311–5318
69. Markmiller S, Cloonan N, Lardelli RM, Doggett K, Keightley MC et al (2014) Minor class splicing shapes the zebrafish transcriptome during development. Proc Natl Acad Sci U S A 111:3062–3067
70. Ganassi M, Badodi S, Polacchini A, Baruffaldi F, Battini R et al (2014) Distinct functions of alternatively spliced isoforms encoded by zebrafish mef2ca and mef2cb. Biochim Biophys Acta 1839:559–570
71. Lujambio A, Lowe SW (2012) The microcosmos of cancer. Nature 482:347–355
72. Kong YW, Ferland-McCollough D, Jackson TJ, Bushell M (2012) microRNAs in cancer management. Lancet Oncol 13:e249–e258

73. Le MT, Hamar P, Guo C, Basar E, Perdigao-Henriques R et al (2014) miR-200-containing extracellular vesicles promote breast cancer cell metastasis. J Clin Invest 124:5109–5128
74. Mishima Y (2012) Widespread roles of microRNAs during zebrafish development and beyond. Develop Growth Differ 54:55–65
75. Suva ML, Riggi N, Bernstein BE (2013) Epigenetic reprogramming in cancer. Science 339:1567–1570
76. Mirbahai L, Williams TD, Zhan H, Gong Z, Chipman JK (2011) Comprehensive profiling of zebrafish hepatic proximal promoter CpG island methylation and its modification during chemical carcinogenesis. BMC Genomics 12:3
77. Andersen IS, Lindeman LC, Reiner AH, Ostrup O, Aanes H et al (2013) Epigenetic marking of the zebrafish developmental program. Curr Top Dev Biol 104:85–112
78. Haberle V, Li N, Hadzhiev Y, Plessy C, Previti C et al (2014) Two independent transcription initiation codes overlap on vertebrate core promoters. Nature 507:381–385
79. Lindeman LC, Andersen IS, Reiner AH, Li N, Aanes H et al (2011) Prepatterning of developmental gene expression by modified histones before zygotic genome activation. Dev Cell 21:993–1004
80. Potok ME, Nix DA, Parnell TJ, Cairns BR (2013) Reprogramming the maternal zebrafish genome after fertilization to match the paternal methylation pattern. Cell 153:759–772
81. Bogdanovic O, Fernandez-Minan A, Tena JJ, de la Calle-Mustienes E, Gomez-Skarmeta JL (2013) The developmental epigenomics toolbox: ChIP-seq and MethylCap-seq profiling of early zebrafish embryos. Methods 62:207–215
82. Chatterjee A, Ozaki Y, Stockwell PA, Horsfield JA, Morison IM et al (2013) Mapping the zebrafish brain methylome using reduced representation bisulfite sequencing. Epigenetics 8:979–989
83. Vander Heiden MG (2011) Targeting cancer metabolism: a therapeutic window opens. Nat Rev Drug Discov 10:671–684
84. Abramsson A, Westman-Brinkmalm A, Pannee J, Gustavsson M, von Otter M et al (2010) Proteomics profiling of single organs from individual adult zebrafish. Zebrafish 7:161–168
85. Carlson P, Smalley DM, Van Beneden RJ (2013) Proteomic analysis of arsenic-exposed zebrafish (*Danio rerio*) identifies altered expression in proteins involved in fibrosis and lipid uptake in a gender-specific manner. Toxicol Sci 134:83–91
86. Ong ES, Chor CF, Zou L, Ong CN (2009) A multi-analytical approach for metabolomic profiling of zebrafish (*Danio rerio*) livers. Mol Biosyst 5:288–298
87. Hall CJ, Boyle RH, Sun X, Wicker SM, Misa JP et al (2014) Epidermal cells help coordinate leukocyte migration during inflammation through fatty acid-fuelled matrix metalloproteinase production. Nat Commun 5:3880
88. Srinivas PR, Srivastava S, Hanash S, Wright GL Jr (2001) Proteomics in early detection of cancer. Clin Chem 47:1901–1911
89. Wang J, Lee YM, Li C, Li P, Li Z et al (2015) Dramatic improvement of proteomic analysis of zebrafish liver tumor by effective protein extraction with sodium deoxycholate and heat denaturation. Int J Anal Chem 2015, 763969
90. Huang SM, Xu F, Lam SH, Gong Z, Ong CN (2013) Metabolomics of developing zebrafish embryos using gas chromatography- and liquid chromatography-mass spectrometry. Mol Biosyst 9:1372–1380
91. Mugoni V, Medana C, Santoro MM (2013) 13C-isotope-based protocol for prenyl lipid metabolic analysis in zebrafish embryos. Nat Protoc 8:2337–2347
92. Gut P, Baeza-Raja B, Andersson O, Hasenkamp L, Hsiao J et al (2013) Whole-organism screening for gluconeogenesis identifies activators of fasting metabolism. Nat Chem Biol 9:97–104

Zebrafish Discoveries in Cancer Epigenetics

Yelena Chernyavskaya, Brandon Kent, and Kirsten C. Sadler

Abstract The cancer epigenome is fundamentally different than that of normal cells. How these differences arise in and contribute to carcinogenesis is not known, and studies using model organisms such as zebrafish provide an opportunity to address these important questions. Modifications of histones and DNA comprise the complex epigenome, and these influence chromatin structure, genome stability and gene expression, all of which are fundamental to the cellular changes that cause cancer. The cancer genome atlas covers the wide spectrum of genetic changes associated with nearly every cancer type, however, this catalog is currently unidimensional. As the pattern of epigenetic marks and chromatin structure in cancer cells is described and overlaid on the mutational landscape, the map of the cancer genome becomes multi-dimensional and highly complex. Two major questions

Y. Chernyavskaya
Department of Developmental and Regenerative Biology, Icahn School of Medicine at Mount Sinai, Box 1020, 1 Gustave L. Levy Place, New York, NY 10029, USA

Division of Liver Diseases, Department of Medicine, Icahn School of Medicine at Mount Sinai, Box 1020, 1 Gustave L. Levy Place, New York, NY 10029, USA

B. Kent
Department of Developmental and Regenerative Biology, Icahn School of Medicine at Mount Sinai, Box 1020, 1 Gustave L. Levy Place, New York, NY 10029, USA

Division of Liver Diseases, Department of Medicine, Icahn School of Medicine at Mount Sinai, Box 1020, 1 Gustave L. Levy Place, New York, NY 10029, USA

School of Biomedical Science, Icahn School of Medicine at Mount Sinai, Box 1020, 1 Gustave L. Levy Place, New York, NY 10029, USA

K.C. Sadler (✉)
Biology Program, New York University Abu Dhabi, Saadiyat Campus, P.O. Box 129188, Abu Dhabi, United Arab Emirates

Department of Developmental and Regenerative Biology, Icahn School of Medicine at Mount Sinai, Box 1020, 1 Gustave L. Levy Place, New York, NY 10029, USA

Division of Liver Diseases, Department of Medicine, Icahn School of Medicine at Mount Sinai, Box 1020, 1 Gustave L. Levy Place, New York, NY 10029, USA

School of Biomedical Science, Icahn School of Medicine at Mount Sinai, Box 1020, 1 Gustave L. Levy Place, New York, NY 10029, USA
e-mail: kirsten.edepli@nyu.edu

D.M. Langenau (ed.), *Cancer and Zebrafish*, Advances in Experimental Medicine and Biology 910, DOI 10.1007/978-3-319-30654-4_8

remain in the field: (1) how the epigenome becomes repatterned in cancer and (2) which of these changes are cancer-causing. Zebrafish provide a tractable in vivo system to monitor the epigenome during transformation and to identify epigenetic drivers of cancer. In this chapter, we review principles of cancer epigenetics and discuss recent work using zebrafish whereby epigenetic modifiers were established as cancer driver genes, thus providing novel insights into the mechanisms of epigenetic reprogramming in cancer.

Major Questions Surrounding the Cancer Epigenome: Answers from Zebrafish

The genetic landscape of cancer cells is dramatically different from normal cells. Mutations, chromosomal losses, gains and rearrangements have been described for virtually every type of cancer in humans [1, 2]. Advances in sequencing technology in the past decade have provided an exquisitely detailed view of the cancer genome in humans, and these have been used to identify candidate driver genes in nearly every type of cancer. Documenting the cancer cell epigenome is more complex, as epigenetics is influenced by DNA modifications, namely methylation and hydroxymethylation on cytosines, histone variants and histone modifications. Moreover, epigenetic modifications are clustered, and are influenced by underlying DNA sequence variations. Thus, integrating the mutational and epigenetic landscapes to generate a comprehensive genetic map of cancer to identify key regions that contribute to carcinogenesis remains a major goal of the field. In particular, sorting through these complex catalogs to identify features that cause or sustain malignancy requires tractable in vivo systems to allow functional assessment of candidates. Since epigenetic marks are well conserved in vertebrates, zebrafish represent an excellent system for such studies.

Those studying cancer epigenomics are addressing similar questions as those studying the cancer genome, including: (1) How does the epigenome get restructured in cancer cells? (2) Are there epigenetic signatures that can be used diagnostically or for identifying specific tumor sub-classes? (3) Which epigenetic changes contribute to tumorigenesis and, for those that are carcinogenic, what is the underlying mechanism by which these cause cancer? As with mutational analysis, epigenetic profiling of cancer cells has documented millions of differences compared to their normal counterparts. Importantly, many of these are conserved in zebrafish tumors [3–5]. The functional annotation of epigenetic differences between normal and cancer cells requires in vivo models. As many such epigenetic signatures are conserved in zebrafish, they has proven to be an excellent system to complement the more commonly use models [6].

Epigenetics is implicated in nearly every aspect of embryonic development and, as such, has been heavily investigated by developmental biologists using zebrafish. These studies are now being extended to the study of cancer. Work using zebrafish has recently identified epigenetic modifiers which impact melanoma [7], hepatocellular carcinoma (HCC) [5], myelodysplastic syndrome (MDS) [8] and rhabdomyosarcoma [9]. Here, the fundamental principles of epigenetic modifications and their relationship to cancer are discussed using zebrafish as a model organism.

The Complexity of the Epigenetic Code

Epigenetics is an umbrella term referring to factors that influence chromatin structure, which in turn regulate chromosome structure, permanent gene silencing and tissue-specific patterns of gene expression. The basic functional unit of chromatin structure is the nucleosome whereby active (euchromatin) and inactive (heterochromatin) domains dictate access of proteins that further modify the epigenome, mediate DNA replication and drive transcription. Many features of the cancer cell phenotype can arise from changes in the epigenome, and indeed, it is speculated that the marked differences in gene expression of malignant cells reflects the massive changes in their epigenome. As cancer cells are typically defined by the suppression of checkpoints that monitor DNA replication, DNA damage and cell cycle progression, it is possible that a monitoring system for epigenome integrity is another one of the checkpoints missing in cancer cells [10, 11].

Epigenetic modifications are added and removed by chromatin-modifying enzymes in a dynamic and tightly regulated fashion without impacting the underlying nucleotide sequence of the DNA (Fig. 1). Epigenetic control of gene activity and overall chromatin structure operates on three levels: DNA, histone proteins, and the nucleosome. Interplay between permissive and repressive domains dictate differential gene expression profiles and maintain a central, functional role during differentiation and development. Most notably, zebrafish research on the developmental functions of epigenetic modifiers has led to discoveries of the importance of epigenetics in fate decisions, gene expression patterning, and zygotic genome activation [12–18]. For instance, a large screen using morpholinos to knock down the expression of 425 chromatin modifiers in zebrafish embryos identified a distinct subset of modifiers that regulate erythroid cell formation and another that was important for hematopoietic stem and progenitor cells [19]. This demonstrates that multiple epigenetic modifiers are required for cell fate decisions and differentiation.

Historically, studies in cancer epigenetics have focused on differences of single epigenetic modifications at the gene regulatory region of a gene of interest—such as a tumor suppressor or a key oncogene—and have correlated these marks with gene expression. This has lead to the conclusions that there is a causative relationship between the mark under study and the expression level of the gene. However, as genome-wide techniques advance, it is clear that the relationship can work both ways (i.e. gene expression levels can influence the epigenetic landscape of the gene) and that the influence of single epigenetic marks is balanced by the conformation of the region as a whole. For instance, chromatin is not simply open or closed, but instead there are also intermediate states where epigenetic marks that are cataloged

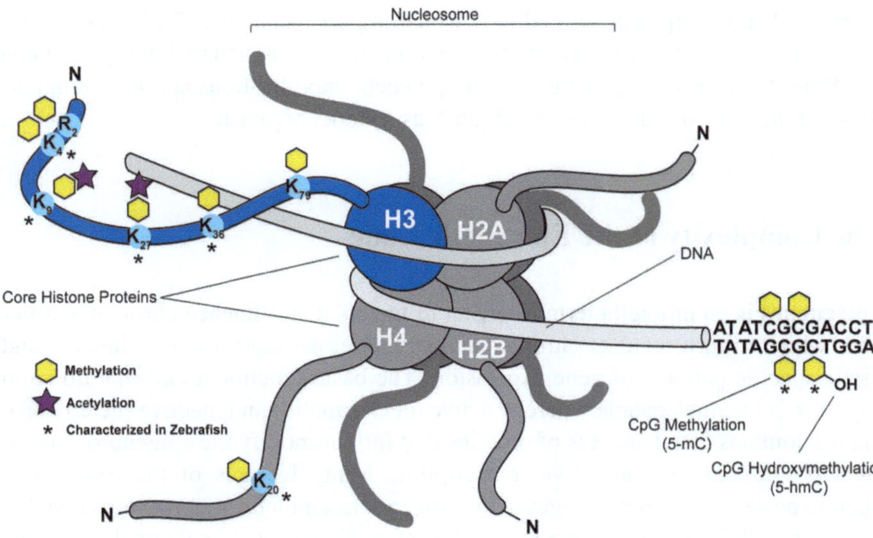

Fig. 1 Epigenetic modifications are conserved in zebrafish. Representation of the basic structure of the nucleosome composed of 147 base pairs of DNA wrapped around an octamer of histone H3/H4 and histone H2A/H2B dimmers. The core histone proteins have long N-terminal tails that extend out from the core particle that is rich in basic amino acid residues, lysine (K) and arginine (R), which can be extensively modified in a reversible, covalent manner. Lysine residues that are mono-, di-, or tri- methylated on histone H3 and H4 that are characterized in zebrafish have been indicated and are conserved across species including H3K4, H3K9, H3K27, H3K36, and H4K20. H3 lysine methylation elicits different transcriptional and structural responses depending on chromatin context and the residues that are modified. Histone acetyl marks have also been indicated and are associated with euchromatic regions amenable to gene transcription. Cytosine residues in DNA can be methylated in a CpG dinucleotide context throughout the genome by DNA methyltransferases (DNMT), which is a conserved process across vertebrate species and plants. DNA methylation is typically associated with irreversibly silenced regions in heterochromatin. Methylation of DNA can be reversed passively or actively through oxidation of the methyl mark. TET family enzymes carry out active de-methylation. Extensive DNA methylation profiling of the zebrafish has been performed in a number of studies

as repressive co-exist with marks that are associated with gene activation. Such "poised" genes are thus held in a repressed, but not completely closed, state that can rapidly be triggered when the signal arrives. Moreover, epigenetic marks have been shown to act at a distance [20]. Thus, the path forward for understanding how epigenetic modifications impact cancer gene expression should incorporate multiple epigenetic marks and a wide-angle view of the genomic region of interest.

Epigenetic readers and writers recognize and target DNA and histone modifications and the erasers remove these marks. The best-studied modifications of DNA and histones are predominantly mediated through writers such as methyltransferase enzymes (both DNA- and histone-methyltransferases; DNMTs and HMTs, respectively), histone demethylases, histone acetyltransferases (HATs), and histone deacetylases (HDACs). Furthermore, incorporating histone variants in place of the canonical histones can impact gene expression, chromatin structure, and can define specialized regions of the chromatin. Readers of the epigenetic code serve to target the writers and erasers and serve as a link between DNA and histone modification to generate a complex and overlapping set of modifications across the genome. DNA methylation is present in all vertebrates, yeast [21] and plants, but not in commonly used invertebrate animal models. Moreover, the factors that modify DNA are well conserved from humans to zebrafish, and make zebrafish an ideal model to study DNA methylation. Histones are over the most highly conserved proteins in vertebrates and conserved from human to zebrafish [22] and the epigenetic marks that regulate them are also well conserved. Thus far, all histone modifications described in humans that have been investigated in zebrafish have been identified, and the readers and writers that mediate these modifications are well conserved in zebrafish (Fig. 1 [23]).

Genome-wide expression studies have documented that the expression pattern of chromatin modifiers is dramatically altered in malignant cells (Fig. 2). Interestingly, while some of genes show similar expression changes across cancer types, others appear more specific. For instance, expression of Dnmt1 and Uhrf1, which are the key components of the DNA methylation machinery, are elevated across cancer types while TET2, a gene involved in cytosine de-methylation, is down-regulated primarily in leukemia (Fig. 2). Moreover, as chromatin modifiers are largely regulated by their interacting partners and pre-existing epigenetic modifications can direct whether the writers are able to make changes at each locus, altered expression of a single modifier may be less important than the combinatorial changes for multiple members of a complex.

DNA Methylation

Methylation of cytosine (5-mC) when it is paired with a guanine residue (i.e. CpG) is a critical mechanism of X-chromosome inactivation, imprinting, silencing repeats and transposons and heterochromatin formation [24–28]. Additionally, many studies implicate DNA methylation in transcriptional repression of differentially expressed genes, however whether DNA methylation is sufficient for fine-tuning gene expression is controversial [25]. The most heavily methylated regions of the genome are in intragenic regions, which are largely composed of repetitive sequences, transposable elements and the regions of the chromosome that give it structure (i.e. the centromeres and telomeres). Methylation of these regions is thought to provide an irreversible mechanism of repression and the ability to form higher order chromatin structure. Additionally, gene bodies are heavily methylated and this typically corresponds to actively transcribed genes [24], however, it the function of this region of DNA methylation is not clear [25]. An excellent graphical overview of DNA methylation has recently been published [29].

Methyl groups are transferred from S-adenosyl-L-methionine (SAM) to cytosine by DNMTs. DNMT1 has been extensively studied across species: it preferentially methylates hemimethylated CpG dinucleotide sequences generated during DNA

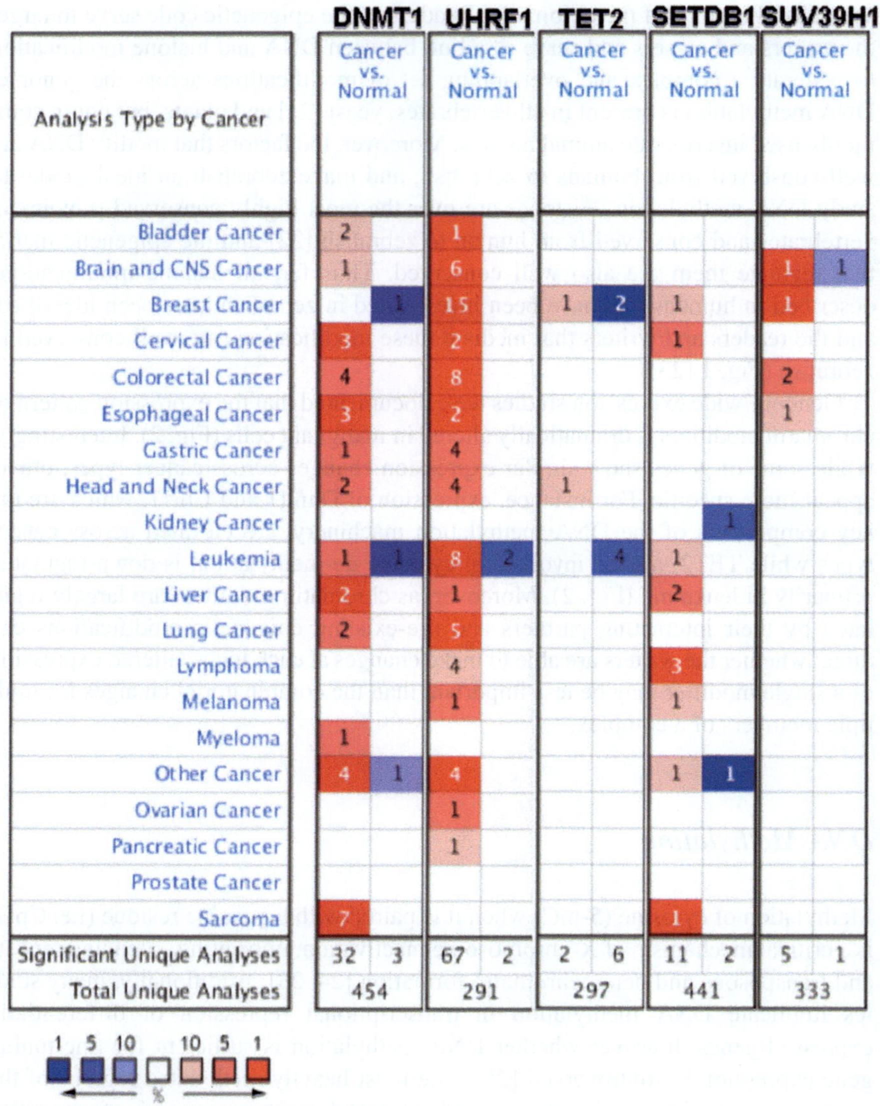

Fig. 2 Epigenetic modifiers discovered as cancer genes in zebrafish show unique expression patterns in human cancers. Oncomine expression analysis for each of the key DNA and histone modifying enzymes that have been discovered in zebrafish to play a role in cancer. Expression of each gene was monitored across a series of cancer samples compared to normal counterparts and their expression levels (high = *red*; low = blue) are indicated. Expression thresholds were set with a fold change of 2 or greater, p value <0.001; gene rank in the top 10 % of deregulated genes and

replication, and is thus considered the maintenance methyltransferase. This role has been confirmed in zebrafish through studying *dnmt1* mutants [11, 30, 31] and in mutants that cannot generate sufficient SAM [32]. DNMT3A and 3B, however, initiate methylation of those regions that were previously unmethylated (i.e. de novo DNA methylation), which is best studied in imprinted genes, which zebrafish do not appear to have. While DNMT1 can bind DNA [33], it is not very efficient at targeting CpG sites. This is significantly improved accomplished by the *U*biquitin-like with P*HD* and *R*ING *F*inger domains *1* (UHRF1) which recognizes hemimethylated CpG sequences during DNA replication and directly recruits DNMT1 to facilitate DNA methylation [34, 35]. 5mC is removed by the Ten-eleven translocation (TET) proteins, which convert 5mC to 5-hydroxymethyl cytosine [36].

Patterns of DNA methylation are dramatically different between cancer cells and their normal counterparts. In non-transformed cells, approximately 2–8 % of the genome is methylated. Certain regions of the genome contain high frequencies of CpG sites, which are termed CpG Islands (CGIs). CGIs occur in approximately 50 % of promoters of human genes and are largely un-methylated, while CpG sites in gene bodies and intergenic regions are mostly methylated. DNA methylation is thought to act as a repressive epigenetic mark and it was first described as a mechanism to assure that some regions of the genome are always maintained in a transcriptionally inactive state such retrotransposable elements [37] repeat sequences [38] imprinted genes [27] and centromeric and pericentromeric regions [39]. Most malignant cells have less DNA methylation than their normal counterparts, and this is largely attributed to loss of methylation at these regions which are typically heavily methylated.

Clusters of CpGs are found in some gene promoters (i.e. CpG islands), but methylation in these gene regulatory regions is largely absent in normal cells. Never-the-less, CpG island methylation patterns has received the most attention in the field. This is largely based on the hypothesis that since DNA methylation is required for silencing those regions that are typically heavily methylated, then DNA methylation in gene regulatory regions must also lead to formation of 'closed' chromatin structures rendering the underlying sequence inaccessible to transcription factors and RNA polymerase II. The finding that the promoters of some tumor suppressors are hypermethylated in cancer cells has fueled much of the study on DNA methylation as a mechanism of regulating gene expression in favor of promoting cell proliferation and transformation. However, although there are several clear examples where DNA methylation is inversely correlated with the expression of nearby genes, many studies also show that methylation in promoter regions can also be positively correlated with gene expression [40–43]. Moreover, since DNA methylation and other epigenetic modifications may act at a distance [20], it may be that the focus on differential methylation of CpG islands may be too narrow.

DNA methylation is frequently found to be co-localized with repressive histone modifications and histone variants, and thus it may also perform an instructive role for the recruitment of other chromatin modifiers to promote gene repression and higher order chromatin structure. Thus, in some cases, the correlation of DNA methylation in the regulatory regions of repressed genes may reflect the placement of DNA methylation side by side with other epigenetic marks that play a bigger role in the repression. In such cases, DNA methylation may serve a supportive role. Indeed, a recent study whereby the TET1 hydroxylase was targeted to methylated regulatory regions of specific genes to specifically demethylated these loci. In most cases tested, only modest demethylation was achieved and while demethylation of some sites significantly increased gene expression, this was not a universal finding [44]. While this exciting study is the first to show that reducing DNA methylation in specific promoters can increase gene expression, it is clear that other factors provide important and even dominant regulatory roles.

DNA Methylation in Zebrafish

Zebrafish mutants in the key genes regulating DNA methylation all have DNA hypomethylation, including S-adenosylhomocysteine hydrolase [32], *dnmt1* [11, 30, 31] and *uhrf1* [11, 45]. Additionally, *tet2* mutants show a loss of 5-HmC [8, 46], illustrating that the machinery that mediates the methylome is highly conserved.

Methylation patterning of the zebrafish genome is a highly dynamic process during development of the organism. In a very similar manner to mammals, zebrafish sperm is hypermethylated compared to its oocyte counterpart, which is then rapidly demethylated upon fertilization [47]. Following fertilization, the DNA methylation levels of the 1- to 2- cell stage of the zygote are lower then what is initially observed in the oocyte mediated by a wave of DNA demethylation [47]. De novo DNA methylation resets DNA methylation levels to levels seen in terminally differentiated tissues during the mid-blastula transition and in gastrula [47–49]. Next, a wave of DNA demethylation and remethylation occur during somitogenesis [49] and tissue specific DNA methylation patterns being established and maintained as cells differentiate. However, the evidence in support of DNA methylation as a biochemical system that regulates gene expression during early development through programmed methylation and demethylation patterns are correlative, at best [14, 50]. Therefore, the functional significance of the dynamic changes in DNA methylation remains unclear, and is under intense investigation.

In summary, the DNA methylome can be re-patterned during differentiation and transformation. The global loss of DNA methylation during carcinogenesis is seen across cancer types, but the mechanism leading to the demethylation of the cancer genome has not been established. Since DNA methylation impacts the interaction between DNA and histones within nucleosomes and promotes higher order chromatin formation, it clearly contributes to gene expression however, it does not appear to be the major mechanism that resets the transcriptome to favor cancer-promoting genes and suppress tumor suppressors. Instead, loss of DNA methylation can result in euchromatinization of the genome, which promotes chromosomal translocations, breaks, transposon activation and could thus serve as a major mechanism driving genomic instability.

Histone Modification

Histones can be post-translationally modified in a number of ways, including acetylation, methylation, ubiquitination, and phosphorylation. A large majority of these post-translational modifications occur on the unstructured N-terminal tails of the core histones, which are rich in basic amino acid residues lysine (K) and arginine (R), which confer a positively charged surface that extends out from the nucleosome and interacts with the negatively charged DNA (Fig. 1). The abundance of lysine and arginine residues on the N-terminal tails also make the core histone proteins amenable to post-translational modification by epigenetic enzymes and complexes to modify the local nucleosome structure. In turn, epigenetic modification of the core histone proteins can serve to recruit transcriptional repressors and activators to specific loci or maintain a baseline quiescent state. Here, we review the major types of histone modifications and focus on those that have been specifically shown to play a prominent role in cancer using zebrafish.

Histone Methylation

Histone methylation directly impacts chromatin structure and gene transcription. Lysine (K) and arginine (R) residues along the N-terminal tails of the core histones are methylated by histone methyl transferases (HMTs). The histone methylome is extremely complex, as amino acids can be bi or tri methylated, and the various combinations of the methylated residues ultimately dictate chromatin state. The well-established correlations between some histone methyl marks and gene expression allows the categorization of these marks as "activating" or "repressive". For instance, trimethylation of H3 on lysine 4 (H3K4Me3) is associated with actively transcribed genes and the di- and tri- methylation of lysine 9 of H3 (H3K9me2/3) is correlated with heterochromatin formation and suppression of gene transcription. Interestingly, a recent study found that SMYD3, a methyltransferase for K3K4Me3, promoted invasion of human tumor cells transplanted into zebrafish embryos. This was attributed to the induction of a metalloproteinase which aids tumor cell mobility [51]. H3K9 methylation is mediated by a number of proteins, including SETDB1 and SUV39H1, which are evolutionarily conserved and expressed in zebrafish [7, 9, 23, 52]. In addition to these, several other epigenetic marks also display a similar distribution across the zebrafish and mammalian genome [17, 45, 53]. Suv39h1 has been shown to cooperate with Dnmt1 to regulate the terminal differentiation of the intestine, exocrine pancreas, and the retina during zebrafish development [52]. Furthermore, Rai et al. demonstrated that H3K9me3 is also positively correlated with levels of DNA methylation in heterochromatic regions of the genome. It is thought that these two marks can collaborate to maintain genes in a repressed state [54]. Histone methyltransferases have been implicated in a wide range of cancers [55] and we discuss how zebrafish have been used to identify functional roles for SETDB1 and SUV39H1 in cancer.

Methylation of H3K27 is another well-studied epigenetic mark that is commonly incorporated into gene promoters and is mediated by Polycomb group proteins, namely the PRC2 complex [56]. High levels of H3K27Me3 were identified in zebrafish [3] and in humans [57]. On its own, H3K27me3 is typically associated with silenced genes and is thus seen as a repressive mark. However, H3K27 trimethylation is also associated with 'bivalent' transcriptional states. Promoter regions marked with H3K4me3 and H3K27me3 are thought to adopt these 'bivalent' states where transcription is primed and RNA polymerase II occupancy is permitted, yet the gene remains inactive.

Histone methylation plays a critical role during embryonic development. Patterning of histone methylation regulates zygotic genome activation in zebrafish [13, 14, 17, 58, 59]. Histone methylation directly influences nucleosome dynamics and stability and serves as a platform for the specific recruitment of transcription factors and remodeling complexes. Mounting evidence supports the crosstalk between DNA methylation and histone methylation. H3K4me3 is negatively correlated with DNA methylation, while H3K9me3 is significantly and positively correlated with DNA methylation. This direct cooperation indicates that a very fine tuned epigenetic signaling network regulates cell function.

Histone Phosphorylation, Acetylation and Histone Variants

Histone core particles can be phosphorylated and acetylated. Phosphorylation of serine, threonine, and tyrosine residues has been found to be critical in regulating chromatin condensation during mitosis, gene expression, and DNA repair. For instance, H3S10 phosphorylation in some contexts, promotes chromatin de-compaction to facilitate transcription, whereas during mitosis H3S10 phosphorylation is critical for chromatin condensation [60, 61]. Phosphorylation of the histone variant H2A.X marks areas of DNA damage and double-stranded breaks to facilitate the DNA repair pathway. This has been shown as a robust marker of DNA damage in mammalian and zebrafish cells [62]. It illustrates the complexity of the histone code, where the context can dictate the impact of the modification.

Lysine acetylation is a conserved, reversible, and highly regulated post-translational modification of core histone proteins that is mediated by HATs. The availability of multiple target sites enables stepwise regulation of acetylation allowing for the fine-tuning of chromatin remodeling mediating gene transcription. Histone deacetylation is mediated by HDACs. A large body of work has demonstrated that, most commonly, acetylated histones lead to an open chromatin configuration and thus HDAC recruitment serves to promote heterochromatin formation.

Histone acetylation is dynamically regulated during zebrafish embryogenesis [63–67]. H3K9ac levels were high at 24 h post fertilization (hpf), but rapidly diminished by 48 and 72 hpf [68], suggesting that this mark may play a regulatory role during early embryogenesis, further supported by findings of failed development in Hdac deficient embryos [64–67, 69]. Histone acetylation has also been linked to maintenance of genomic integrity and DNA repair and the acetylation and de-acetylation of histones has long been a targeted focus of cancer therapies [70] as disrupted histone acetylation dynamics have been associated with a number of cancers.

There are a large number of highly comparable forms of histones that are collectively defined as 'histone variants'. These variants can be incorporated into the nucleosome, replacing their canonical counterpart, resulting in changes in structural regions of the chromatin, gene expression, response to genotoxic events, genomic stability, and DNA repair. Furthermore, a growing body of evidence has demonstrated that histone variants can play critical roles during cancer development. Particularly in plants, the DNA wrapped in nucleosomes tends to be more methylated, suggesting that the position of the nucleosome influences the methylome pattern [71].

The concept of the 'histone code' indicates that the sum of histone modifications at a specific locus during a particular cellular process dictates nucleosome structure. Indeed, both developmental biologists and cancer biologists are now utilizing to a more complicated, but likely more accurate, model whereby the sum of the histone marks, DNA marks, associated co-factors, and transcription factors that creates a chromatin microenvironment which can either repress, induce or poise genes for expression. Studying the overlapping roles of covalent epigenetic marks with histone variant deposition to profile the epigenomic landscape of various cancers is a daunting, but important goal that has recently been tackled by the Roadmap Epigenomics Consortium who have generated a reference epigenome for 111 tumors [72]. The conservation of epigenetic marks and mechanisms of carcinogenesis from humans to zebrafish makes this an accessible and tractable model to study cancer epigenomics.

Common Epigenetic Changes in Cancer: Mechanisms and Questions

As the field of cancer epigenomics evolves, common themes are emerging. In nearly every cancer type, DNA methylation is reduced genome-wide (i.e. global DNA hypomethylation) yet in the same tumors that display global DNA hypomethylation, specific loci can be hypermethylated [73]. The use of ChIPSeq, histone array platforms and nucleosome positioning analysis (i.e. ENCODE) on tumor samples is yielding a rich encyclopedia of the integrated epigenome of over a 100 cancers [72]. Exciting discoveries combining both in silico analysis and in vivo functional studies using model organisms have begun to sift through the differences to uncover mechanisms by which key epigenetic modifiers contribute to cancer formation. In particular, experiments using the power of zebrafish genetics have demonstrated that two key epigenetic marks—DNA methylation and H3K9Me3—are functionally relevant to hepatocellular carcinoma (HCC), myelodysplastic syndrome (MDS), melanoma and rhabdomyosarcoma. These are reviewed below.

The Cancer Methylome: Causes, Consequences and Insight from Zebrafish

It has long been established that the global DNA methylation is significantly reduced in cancer cells [73]. This was first observed using techniques that monitor whole genome 5mC levels [74] which serves as an assay for methylation levels on the most heavily methylated and abundant regions of the genome: introns, gene bodies and intragenic regions containing repeats and transposons. However, those regions that are thought to serve regulate gene expression, such as those for imprinted genes [75] and CpG islands in differentially regulated genes do not influence the signal derived from global assessment of 5mC levels. More sophisticated approaches to monitor the cancer methylome has confirmed that it is the intragenic regions which are less methylated in cancer cells and, in some cases, the regulatory regions of specific genes within the same tumor become hypermethylated [73]. While it is clear that genome-wide approaches that allow locus specific resolution or base pair resolution provide the most comprehensive perspective of the cancer cell methylome, the mechanism by which and the significance of global DNA hypomethylation in cancer are not well understood.

Studies using zebrafish have both confirmed the loss of methylation in tumors [4] and have highlighted that proteins which modify the methylome are cancer genes [5]. Moreover, while it is well established that loss of DNA methylation can promote cancer-causing events, such as genome instability (Fig. 3), the mechanism of this is not yet clear. Three major questions in the field are: (1) How is DNA methylation lost during transformation (2) How does altering the mechanism impact cancer? (3) What is the mechanism of DNA methylation mediated carcinogenesis? Here, we describe how recent studies using zebrafish have provided the answers to these questions.

Loss of DNA methylation can occur via a passive mechanism, whereby methylation of a cytosine is not maintained after DNA replication, or via active demethylation either by the TET proteins which convert 5mC to 5-hydroxymethylcytosine (5hmC), or by spontaneous deamination followed by the repair of the deaminated cytosine [49]. TET2 mutation is found in patients with MDS [76] and it has been proposed that loss of TET2 reduces 5hmC which thereby promotes a more "stem-like" fate by suppressing the expression genes that promote differentiation. One study that used morpholinos to transiently knockdown *tet1*, *tet2* and *tet3* in zebrafish embryos found that this caused anemia associated with the loss of genes that promote erythropoiesis, such as *scl*, *gata1* and *cmyb*. This was associated with a moderate increase in the 5mC levels on a few CpG sites in the promoters of these genes [46], suggesting that the failure to convert 5mC to 5hmC resulted in the suppression of these genes. Another study using gene editing techniques to mutate *tet2* in zebrafish demonstrated reduced 5hmC levels in the kidney, the site of hematopoiesis in zebrafish. Similar to the phenotype of *tet1*, *2* and *3* morphant embryos [46], *tet2* mutation was associated with a marked decrease in the number of erythrocytes coupled with an increase in myelomonocytes by 11 months, leading to the development

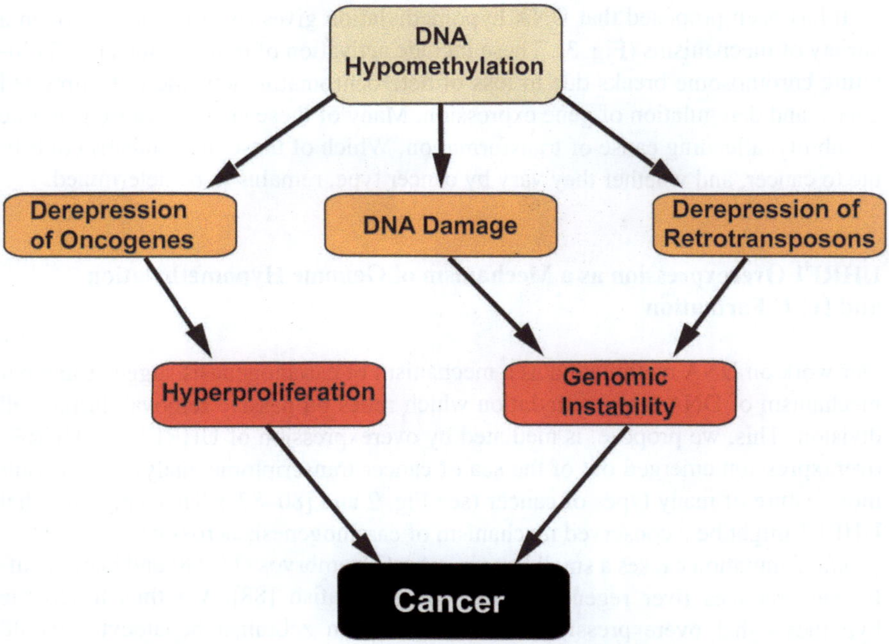

Fig. 3 Proposed mechanisms by which DNA hypomethylation contributes to cancer. Schematic depicting several pathways by which hypomethylation of the genome can contribute to cancer development. Loss of CpG methylation can lead to general genomic instability and mutation or de-repress typically silenced oncogenes and tumor promoting factors

of myelodysplasia in 2 year old fish [8]. While these interesting and clinically relevant studies suggest that blocking Tet activity can reduce the expression of pro-differentiation genes due to retention of DNA methylation marks in their promoters, it is possible that the loss of 5hmC could affect erythrocyte development and MDS by other mechanisms, as 5hmC also can directly affect gene expression [77].

Work in zebrafish has identified a second mechanism by which DNA methylation is actively removed. In an elegant study that capitalized on the power of zebrafish genetics, Rai et al. demonstrated that Aid/Apobec deaminates 5mC, and then Mbp and Gadd45a recognize and excise this aberrant nucleotide [49]. Inappropriate activation of this demethylation program occurs in zebrafish intestinal epithelial cells that lack Apc, a key tumor suppressor in colon cancer. Loss of Apc leads to hypomethylation and upregulation of genes controlling intestinal cell differentiation, including *tcf* family members, and the downregulation of the pathway that generates retinoic acid, a key driving factor for intestinal cell differentiation. This was shown to be mediated by the failure of Apc mutant cells to activate the Aid/Apobec demethylase program. They conclude that DNA methylation is retained in the regulatory region of genes that keep intestinal cells undifferentiated [78], which could lead to cancer. It is interesting to speculate that demethylation by Aid/Apobec could be one mechanism contributing to DNA hypomethylation in colon and gastric cancer [79].

It has been proposed that DNA hypomethylation gives rise to cancer through a variety of mechanisms (Fig. 3). These include activation of retrotransposons, facilitating chromosome breaks due to loss of heterochromatin, activation of imprinted genes, and deregulation of gene expression. Many of these changes cause genomic instability, a leading cause of transformation. Which of these mechanisms contribute to cancer, and whether they vary by cancer type, remains to be determined.

UHRF1 Overexpression as a Mechanism of Genome Hypomethylation and HCC Formation

Our work on DNA methylation as a mechanism of carcinogenesis suggests another mechanism of DNA hypomethylation which relies on passive removal during cell division. This, we propose, is mediated by overexpression of UHRF1 [5]. UHRF1 overexpression emerged out of the sea of cancer transcriptome analyses as a common feature of many types of cancer (see Fig. 2 and [80–87]). This suggested that UHRF1 might be a conserved mechanism of carcinogenesis across cancer types.

uhrf1 mutation causes a small liver in zebrafish embryos [11, 88] and haploinsufficiency reduces liver regeneration in adult zebrafish [88]. We thus tested the hypothesis that overexpressing human UHRF1 in zebrafish hepatocytes would cause HCC. We found that high UHRF1 levels increased destabilization of Dnmt1, potentially due to its E3 ubiquitin ligase activity [89, 90], and delocalized Dnmt1 away from the chromatin, resulting in global DNA hypomethylation. This phenotype was accompanied by p53 mediated senescence of hepatocytes, a tumor suppressive mechanism that, when overcome, leads to HCC in nearly all fish younger than 20 days old. Analysis of human HCC samples revealed elevated levels of UHRF1 and a general downregulation of the p53 senescence program, as predicted by findings in zebrafish. Moreover, in a classical transformation assay using NIH-3 T3 cells, UHRF1 overexpression was shown to cooperate with another senescence-inducing oncogene, Ras, to mediate transformation. These data indicate that UHRF1 is an epigenetic modifier that acts as an oncogene in HCC and, given its widespread overexpression in a range of cancers types, it may also function as an oncogene in other cancers. Moreover, since high levels of UHRF1 paradoxically caused DNA hypomethylation, it is possible that this is a mechanism by which DNA methylation is lost in cancer cells.

Histone Methyltransferases Discovered as Cancer Genes Using Zebrafish

There are multiple histone methyltransferases which target the same residue for methylation, for instance, both SETDB1 and SUV39H1 mediate H3K9 trimethylation, and it is unclear how these two proteins that perform a similar function

achieve specificity. Interestingly, each of these enzymes have a unique expression profile across cancer types (Fig. 2) and have been implicated in distinct cancers. For instance, down regulation of SETDB1 is observed in metastatic lung cancer [91], yet overexpression of SUV39H1 has been reported in HCC [92]. This suggests that the functional similarity between these two HMTs may become more distinct depending on cancer type.

SETDB1

SETDB1 was identified as the first epi-oncogene in zebrafish through a screen for genes that cooperate with the most common mutations in melanoma, activating BRAF and loss of p53 [7]. These mutations are found in nearly 25–60 % of melanomas and a previous study demonstrated that they cooperate to cause melanoma in zebrafish [93]. Since melanomas carry more mutations than virtually any other tumor type likely due to the life-long exposure of melanocytes to UV induced DNA damage and there is a high variability in tumor latency, even after these mutations occur, it has been proposed that cooperating mutations are important for tumor onset. A common hypothesis in cancer genomics is that amplicons which are conserved across tumors harbor oncogenes or other genes required for tumor survival. Chromosome 1q21 is commonly amplified in melanoma [94], hence the authors selected 17 genes in the 1q21 region to screen for their capacity to increase tumor-induced mortality in zebrafish engineered to have melanocytes with deleted p53 and overexpressed activated BRAF. Of these, only SETDB1 overexpression increased the incidence of melanoma from 53 to 94 % and decreased survival by nearly half. Importantly, mutant versions of SETDB1 that lack MT activity were not cancer promoters in this system, implicating H3K9Me3 in this phenotype. Since human melanoma cells transfected with SETDB1 show significant enrichment of SETDB1 on the regulatory regions of genes that are expressed at low levels, but not on those that are upregulated in this system, it is assumed that SETDB1 mediates gene repression through H3K9Me3 deposition. However, they report the unanticipated finding that a subset of the presumed SETDB1 target genes in human cells were not marked by H3K9Me3 in cells overexpressing SETDB1, suggesting that this epigenetic mark alone cannot account for the mechanism by which the genes that are bound by SETDB1 become repressed. Since SETDB1 has recently been reported to exist in a complex with SuV39H1 and other factors [95], there is a possibility that SETDB1 is primarily required for increased targeting of the HMT complex to H3K9. Moreover, catalytically inactive SETDB1 was still able to associate with the HMT complex without any reduction in H3K9me3 and drive the onset of melanoma [7]. This raises the interesting possibility that all of the cancer causing effects of SETDB1 overexpression in melanoma may not be all mediated through its function as an H3K9 methyltransferase.

SUV39H1

Another of the H3K9 methyltransferases, SUV39H1, was found to repress rhabdomyosarcoma formation in the zebrafish [9]. Furthermore, SUV39H1 has been implicated in promoting a number of other cancers including HCC [92], however, the study in zebrafish was the first to demonstrate the oncogenic potential of this important epigenetic regulator. In humans, activating mutation in a ras-pathway gene is common in rhabdomyosarcomas, and overexpressing activated KRAS in zebrafish muscle stem cells causes a high incidence of tumors with an onset prior to 20 days post fertilization (dpf) with over 50 % lethality by 2 months of age [96]. These tumors in zebrafish share many common genetic features with human rhabdomyosarcomas [97], making this a useful model for translational studies on this cancer type. Albacker et al. evaluated human RMS samples to identify chromatin modifiers that were commonly overexpressed, and then used zebrafish to assay the impact of overexpressing 19 of these modifiers on the survival of zebrafish overexpressing $KRAS^{G12D}$ in muscle cells. Only SUV39H1 overexpression imparted a survival advantage compared to controls expressing $KRAS^{G12D}$ plus GFP. Gene expression analysis revealed that cell cycle regulator genes were suppressed in the samples expressing SUV39H1 and $KRAS^{G12D}$ compared to $KRAS^{G12D}$ alone [9]. Interestingly, the effects of SUV39H1 were observed prior to any tumor formation, as early as 7 dpf. It remains unclear whether the suppression of these genes is mediated by enhanced H3K9Me3 deposition in their regulatory regions, and the effect of SUV39H1 overexpression on the epigenetic landscape has yet to be determined.

All of the H3K9 MT proteins—SETDB1 SUV39H1, G9a and GLP—have been identified in a complex [95] raising the possibility that they may work together or co-regulate to control H3K9Me3 domains. However, the mechanism of loci selection for H3K9 methylation by individual methyltransferases has not been identified. Interestingly, overexpression of both SETDB1 and human SUV39H1 was also capable of promoting melanoma in the zebrafish system, and a mutant of SETDB1 lacking methyltransferase activity was not a pro-cancer gene [7]. This indicates that H3K9 methylation is pro-tumorigenic in this cancer type, however, since only SETDB1 is overexpressed in human melanoma (Fig. 2), it suggests that some feature of SETDB1 specifically provides a selective advantage for melanoma cells (Table 1).

With the tools in hand to manipulate the expression of these and other MTs in several zebrafish cancer models, it will be possible to assess their impact on gene expression and on other epigenetic marks. The ability to dissect both their individual and combinatorial effect on chromatin structure will further contribute to understanding the formation of melanoma [7] and rhabdomyosarcoma [9], in addition to other cancers where these genes are overexpressed.

Table 1 Frequent epigenetic marks which are deregulated in cancers

Deregulated mark	Change in cancer	Modifier	Associated with	References
DNA methylation	Regional hypermethylation	DNMT1, DNMT3, UHRF1[a]	Gene repression	Herman et al. (1995) [98], Ai et al. (2006) [40], Bae et al. (2004) [41], Kornegoo et al. (2012) [99], Roll et al. (2013) [43], Zou et al. (2009) [100]
	Global hypomethylation		Genomic instability	Bedford and van Helden (1987) [101], Cadieux et al. (2006) [102], Feinberg and Vogelstein (1983) [74], Fraga et al. (2005) [103], Mudbhary et al. (2014) [5], Rauch et al. (2008) [104]
			Activation of retrotransposons	Jackson-Grusby et al. (2001) [105], Howard et al. (2007) [106], Minoguchi and Iba (2008) [107]
H3K9me3	Hypermethylation	G9a, SETDB1, SUV39H1	Increased gene repression	Frigola et al. (2006) [108], Nguyen et al. (2002) [109]
H3k27me2–me3	Hypermethylation	EZH1, EZH2	DNA hypermethylation	Chinaranagari et al. (2014) [110], Ohm et al. (2007) [111], Popovic et al. (2014) [112], Schlesinger et al. (2007) [113]
H3K4me3	Hypomethylation	MLL4	Gene repression	Kang et al. (2015) [114], Nguyen et al. (2002) [109], Lee et al. (2013) [115]

List of common epigenetic marks and their analogous regulators are shown in relation to reported changes in chromatin or gene expression resulting from their misregulation

[a]Although it does not modify the DNA methylome, UHRF1 is an essential necessary component of the DNMT1 complex

Dissecting the Cancer Epigenome: Benefits and Limitations of Zebrafish

The ability to interrogate the epigenome has been transformed over the past decade with the advent of next generation sequencing (NGS) technologies. Additionally, using histone modifications and DNA methylation patterns as biomarkers for diagnosing diseases has generated new opportunities for the use of epigenome data. Typically, increasing resolution requires a trade off in genome coverage. An array of techniques for both genome-wide and locus specific analysis of specific epigenetic marks as well as for the accessibility of the DNA, reflecting open chromatin have been optimized and are used routinely in mammalian systems. Most of these approaches have been established in early embryos [4, 116] however, using samples composed of heterogeneous cells, such as whole embryos at post-cleavage stages of development or of tissues containing both cancer cells and the normal cells that surround the tumors can complicate analysis, because it is difficulty to assign the cellular origin of peaks in the sequencing data. To circumvent this, separating different tissue types requires the ability to label and isolate cells of interest, which is feasible using transgenic zebrafish expressing fluorescent proteins in the cell type of interest (Table 2).

Since the structure of epigenetic marks are not species specific, many of the reagents to probe these marks can be used across phyla. Indeed, several histone marks [3, 4, 58, 128] and 5mC [14, 17, 48] have been profiled in zebrafish embryos and in tumors [3, 5]. However, antibody resources in zebrafish are a limitation in the field, finding antibodies that recognize endogenous chromatin modifiers in zebrafish to perform chromatin immunoprecipitation (ChIP) and other studies are a challenge. There are alternative approaches that can be applied to zebrafish samples to yield exciting and informative perspectives on the epigenome. Here, we review some of the common approaches in epigenomic analysis and discuss their applicability to zebrafish samples.

Methylome Analysis

Approaches for analyzing the methylome are numerous, and selection is based on cost, sample volume and downstream application, all of which are important considerations when using zebrafish. Genome-wide base pair resolution is the most informative of all the approaches, but it is also the most costly and often the most difficult to obtain, since mapping reads that are comprised entirely of repeat sequences is an informatic challenge [129], and this is confounded when mapping to the zebrafish genome, which is not as well assembled as the human or mouse. An alternative is the use of arrays, which provide a quick and relatively affordable means to investigate a subset of the genomic regions that are pre-selected, and the read out is as simple as a spreadsheet with relative—but not actual—values in

Table 2 Chromatin modifiers implicated in cancer and their analogous zebrafish models

Chromatin modifier	Human cancers	Zebrafish cancer model	Mechanism of carcinogenesis in fish	References
UHRF1	Multiple	HCC	Senescence escape, genomic instability	Cui et al. (2014) [117], Mudbhary et al. (2014) [5]
DNMT1	Multiple	No	Unknown	Li et al. (2010) [118], Chen et al. (2006) [119], Xu et al. (2010) [120]
DNMT3a/b	AML, melanoma	No	Unknown	Im et al. (2014) [121], Nguyen et al. (2002) [109]
TET2	Leukemia	Myelodysplasia	Unknown	Gjini et al. (2015) [8], Scourzic et al. (2015) [122]
EZH2	Prostate, HCC, myeloma	No	Unknown	Chinaranagari et at. (2014) [110], Hung et al. (2014) [123]
SETDB1	Melanoma	Melanoma	Cooperation with BRAFV^{600E}	Ceol et al. (2011) [7]
SUV39H1	HCC, gastric colorectal, breast	Rhabdo-myosarcoma	Cyclin B1 repression	Albacker et al. (2013) [9], Cai et al. (2014) [124], Chiba et al. (2013) [92], Khanal et al. (2013) [125]
G9a	Squamous cell carcinoma, glioma	No	Unknown	Tao et al. (2014) [126], Zhong et al. (2015) [127]

methylation levels for each locus. On the other end of the spectrum is the use of methylation-sensitive restriction enzymes. The range of approaches used to investigate DNA methylation and hydroxymethylation and their applicability to zebrafish samples is presented in Table 3.

Briefly, DNA methylation analyses can be subdivided into those that focus on locus-specific vs. genome-wide assessment (with varying resolutions for each) and the ability to obtain base pair resolution compared to global genomic DNA methylation status. For a list of techniques for assessing DNA methylation status please refer to detailed reviews [129, 132, 133] and Table 3.

Identifying whether a cytosine is methylated or unmethylated in the genome relies on the mutagen, bisulfite, which converts any unmethylated cytosine to an uracil, but 5mC is protected and remains a C. Thus, sequencing bisulfite converted DNA will provide information about the methylation status of each cytosine. PCR-based DNA methylation analyses are the most common way of assessing locus

Table 3 Approaches to interrogating the epigenome and their utility in zebrafish samples

Approach	Application	Level	Resolution	Genome coverage	Enriched	Cost	Zebrafish reference
IHC/IF/immuno-blot	DNA and Histone	Global	Lowest	High	No	$	Mudhary et al. (2014) [5], Jacoh et al. (2015) [11]
Methylation sensitive restriction digest	DNA methylation	Global	Low	High	No	$	Mudbhary et al. (2011) [130]
MNase assay	Nucleosome position	Global	Low	High	No	$	
Methylation array	DNA methylation	Locus specific	Moderate	Moderate	No	$$	Mirbahai et al. (2011) [4]
MSP	DNA methylation	Locus specific	High	Lowest	No	$$	
BSP	DNA methylation	Locus specific	Highest	Lowest	No	$$	
Chip-seq	Histone modification	Global/specific	Moderate	Moderate	Yes	$$$	Lindeman et al. (2009) [116]
RRBS	DNA methylation	Global/Specific	Moderate	Moderate	Yes	$$$	Chatterjee et al. (2013) [131]
WG-BS	DNA methylation	Global/Specific	Highest	Highest	No	$$$$	–

Different techniques used for analyzing the prevalence and distribution of different epigenetic marks are ranked based on genomic coverage, resolution and cost specific methylation patterns and are routinely used in research using zebrafish. They fall into two categories; methylation specific PCR (MSP) and bisulfite sequencing PCR (BSP). Methylation status of specific loci can be presented as total methylation at that locus or as percent methylation for a particular CpG dinucleotide at a specific position. The utility and advantage of locus specific assessment of DNA methylation patterns is counterbalanced by the fact that methylation changes are not isolated events but phenomenon that occur across the entire genome.

Genome-wide analysis of specific base pairs or loci is provided by array and sequencing platforms. Affinity enriched microarrays provide a detailed, genome-wide look at certain parts of the genome such as promoters, gene bodies, CpG islands and shores, and have the advantage of being customizable. MeDIP uses an antibody to 5mC to bind the methylated regions of the genome and then precipitate them using the standard approaches for chromatin immunoprecipitation (ChIP). The precipitated DNA can then be used for microarray or sequencing. The advantages of using MeDIP for arrays are cut short by the inherent limitation of the array design. Although high-throughput sequencing can eliminate array based design biases it still faces the shortcomings associated with MeDIP such as mapping difficulty (see the section "Overcoming Limitations of the Zebrafish Model"), low coverage and the fact that ChIP only identifies those regions that are enriched for 5mC, but does not indicate the actual methylation status of CpGs in this region.

The most recent development in DNA methylation analysis, sequence-based profiling, couples bisulfite conversion with NGS. Since no enrichment is required for this method, all regions of the genome are equally represented and single base pair resolution can be achieved. This technology, however, is still costly and often such depth is unnecessary if one wishes to focus primarily on CpG rich regions. RRBS is one method that utilizes NGS with methylation enrichment and bisulfite conversion. Since RRBS features a digestion step with MspI, which recognizes CpGs, it ensures that the sample is enriched for those dinucleotides after size selection and PCR amplification. The amount of required input DNA is also a lot less for RRBS than some of the other methods, which makes this ideal for working with small sized samples such as zebrafish tissues or tumors. However, it is important to note that even though the frequency of CpG dinucleotides is more common in zebrafish genome relative to human or mouse [131], there are fewer MspI restriction sites and thus less enrichment of CpG high regions for these samples.

WGS of bisulfite converted DNA provides the most amount of information, however, this is a challenge, as it requires 1–5 μg of starting genomic material [73]. Amounts such as these can be difficult to obtain from zebrafish tissues or tumors unless samples are pooled. In addition, NGS based approaches often take many weeks to months to carry out, produce many gigabytes of data and require sophisticated analytic ability that is not available to all laboratories. Finally, the relative cost is an important consideration: ranging from a few 100 dollars to analyze hundreds of samples by methylation restriction digestion to several 1000 dollars for NGS of a single WGS sample. Therefore, while these approaches are powerful, it is important to balance the drive to obtain an omics-level perspective of all data with the ability to generate large sample sizes quickly using zebrafish.

Histone Modifications, Variants and Nucleosome Positioning

Detecting the presence and absence of certain histone marks and positions of nucleosomes on chromatin can lend insight into the development and nature of different cancers. Immunofluorescence and Western blotting can give a general overview of histone variant abundance and epigenetic marks since these are well conserved in zebrafish and most commercially developed antibodies for histone and DNA modifications cross react with zebrafish. Thus, analyzing global levels of histone marks or variants are good first approach methods. However, often times it is necessary to know how certain histone marks change across the genome, or how they change at specific loci, where these techniques fall short.

ChIP can be used to analyze changes in histone modifications as they relate to certain regions of the genome, such as promoters or enhancers. ChIP relies on the idea that histones and DNA are close enough in proximity to become covalently linked. Briefly, cells undergo formaldehyde cross-linking and sonication to fragment their DNA. An antibody recognizing the PMT of choice is used to immuno-precipitate the sample followed by cross-link reversal. The freed DNA can then undergo NGS (ChIP-seq) or microarray analysis (ChIP-chip) to determine which genomic regions were enriched for that PMT. ChIP-seq can be easily applied to zebrafish and a methods paper—"Fish and ChIPs"—has been recently published [116].

Nucleosome positioning is another useful epigenetic marker when looking at the cancer genome. Nucleosome occupancy is a critical factor in determining chromatin structure and density and is thus strongly linked to the balance between hetero- and euchromatin and DNA methylation [71]. Methods for studying nucleosome positioning, such as MNAse digestion or DNAse hypersensitivity assay, take advantage of enzymes that are able to cut DNA on nucleosome barren regions thus generating short fragments that can be subjected to sequencing and mapped to specific loci.

Additional insights can be gained by combining epigenomics data with previously reported mRNA expression profiles. Databases, such as ENCODE, allow researchers to overlay genome-wide epigenetic interrogations with expression profiles to gain a better understanding of gene regulation. However, such large-scale, genomic collections are still in their early phases, with limited sample data.

Overcoming Limitations of the Zebrafish model

Although many epigenetic techniques have been successfully coupled with the zebrafish model, there are still several limitations that must be overcome before the fish system can rival the current mammalian standards. The small size of zebrafish, which enables high-throughput genetic screens, becomes problematic when trying to collect a sufficient amount of sample to carry out genome-wide epigenome interrogations. Obtaining sufficient genomic DNA from a larval liver is impossible and thus calls for pooling of samples. This can limit research applications—for example, pooling samples is acceptable for assessing liver development, where it is presumed that there is low animal-to-animal variability in hepatoblasts, but becomes unacceptable for assessing tumors, which are inherently heterogeneous even within a single animal. Since it is unlikely that the size of the zebrafish will increase in the near future, resolution to this problem depend on enhancing the sensitivity of techniques used to generate epigenetic data.

A second problem stems from the limited annotation of the zebrafish genome. This, coupled with the generation of short DNA fragments from NGS makes it difficult to map repetitive elements to correct positions in the genome. Only ~30 % of the zebrafish library generated from RRBS maps to unique sites in the genome—the rest maps to multiple regions [131]. Incomplete annotation results in "orphaned" data that could contain potentially useful information but cannot be accessed without knowing its genomic address. It is anticipated that the continual efforts to improve the annotation and assembly of the zebrafish genome—the tenth version of the assembled genome was just released (GRCz10; http://www.sanger.ac.uk/resources/zebrafish/genomeproject.html_)—and advances in bioinformatics approaches and tools for mapping will overcome this limitation.

Clinical Impacts and Future Perspectives

The cancer genomic landscape is complex, and dissecting the functional relevance of complex genetic and epigenetic changes in cancer cells requires in vivo systems. Zebrafish provide an excellent system to carry these out, as all epigenetic marks investigated to date are conserved in zebrafish, and the mechanism for genome editing and tissue specific over-expression is straightforward, as demonstrated by many of the studies reviewed here. Moreover, the ability to carry out epigenetic analysis is improving in this system, which provides a unique opportunity to sort the genome-wide data obtained from human tumors into a relevant framework. As the ability to screen drugs for efficacy in zebrafish tumor models advances [6, 134], this provides a tractable system to not only identify important cancer genes and pathways, but also to screen for drugs that will halt cancer.

References

1. Futreal PA, Coin L, Marshall M et al (2004) A census of human cancer genes. Nat Rev Cancer 4(3):177–183
2. Cancer Genome Atlas Research Network, Weinstein JN, Collisson EA et al (2013) The cancer genome atlas pan-cancer analysis project. Nat Genet 45(10):1113–1120
3. Anelli V, Santoriello C, Distel M et al (2009) Global repression of cancer gene expression in a zebrafish model of melanoma is linked to epigenetic regulation. Zebrafish 6(4):417–424
4. Mirbahai L, Williams TD, Zhan H et al (2011) Comprehensive profiling of zebrafish hepatic proximal promoter CpG island methylation and its modification during chemical carcinogenesis. BMC Genomics 12:3

5. Mudbhary R, Hoshida Y, Chernyavskaya Y et al (2014) Overexpression of UHRF1 drives DNA hypomethylation and hepatocellular carcinoma. Cancer Cell 25:1–14
6. White R, Rose K, Zon L (2013) Zebrafish cancer: the state of the art and the path forward. Nat Rev Cancer 13(9):624–636
7. Ceol CJ, Houvras Y, Jane-Valbuena J et al (2011) The histone methyltransferase SETDB1 is recurrently amplified in melanoma and accelerates its onset. Nature 471(7339):513–517
8. Gjini E, Mansour MR, Sander JD et al (2015) A zebrafish model of myelodysplastic syndrome produced through tet2 genomic editing. Mol Cell Biol 35(5):789–804
9. Albacker CE, Storer NY, Langdon EM et al (2013) The histone methyltransferase SUV39H1 suppresses embryonal rhabdomyosarcoma formation in zebrafish. PLoS One 8(5), e64969
10. Milutinovic S, Zhuang Q, Niveleau A et al (2003) Epigenomic stress response. Knockdown of DNA methyltransferase 1 triggers an intra-S-phase arrest of DNA replication and induction of stress response genes. J Biol Chem 278(17):14985–14995
11. Jacob V, Chernyavskaya Y, Chen X et al (2015) DNA hypomethylation induces a DNA replication-associated cell cycle arrest to block hepatic outgrowth in uhrf1 mutant zebrafish embryos. Development 142(3):510–521
12. Almeida RD, Loose M, Sottile V et al (2012) 5-hydroxymethyl-cytosine enrichment of non-committed cells is not a universal feature of vertebrate development. Epigenetics 7(4): 383–389
13. Andersen IS, Ostrup O, Lindeman LC et al (2012) Epigenetic complexity during the zebrafish mid-blastula transition. Biochem Biophys Res Commun 417(4):1139–1144
14. Andersen IS, Lindeman LC, Reiner AH et al (2013) Epigenetic marking of the zebrafish developmental program. Curr Top Dev Biol 104:85–112
15. Bogdanovic O, Fernandez-Minan A, Tena JJ et al (2013) The developmental epigenomics toolbox: ChIP-seq and MethylCap-seq profiling of early zebrafish embryos. Methods 62(3):207–215
16. Hirose K, Shimoda N, Kikuchi Y (2013) Transient reduction of 5-methylcytosine and 5-hydroxymethylcytosine is associated with active DNA demethylation during regeneration of zebrafish fin. Epigenetics 8(9):899–906
17. Lindeman LC, Winata CL, Aanes H et al (2010) Chromatin states of developmentally-regulated genes revealed by DNA and histone methylation patterns in zebrafish embryos. Int J Dev Biol 54(5):803–813
18. Long HK, Sims D, Heger A et al (2013) Epigenetic conservation at gene regulatory elements revealed by non-methylated DNA profiling in seven vertebrates. Elife 2, e00348
19. Huang HT, Kathrein KL, Barton A et al (2013) A network of epigenetic regulators guides developmental haematopoiesis in vivo. Nat Cell Biol 15(12):1516–1525
20. Bert SA, Robinson MD, Strbenac D et al (2013) Regional activation of the cancer genome by long-range epigenetic remodeling. Cancer Cell 23(1):9–22
21. Tang Y, Gao XD, Wang Y et al (2012) Widespread existence of cytosine methylation in yeast DNA measured by gas chromatography/mass spectrometry. Anal Chem 84(16):7249–7255
22. McGhee JD, Felsenfeld G (1980) Nucleosome structure. Annu Rev Biochem 49:1115–1156
23. Sun XJ, Xu PF, Zhou T et al (2008) Genome-wide survey and developmental expression mapping of zebrafish SET domain-containing genes. PLoS One 3(1), e1499
24. Suzuki MM, Bird A (2008) DNA methylation landscapes: provocative insights from epigenomics. Nat Rev Genet 9(6):465–476
25. Bestor TH, Edwards JR, Boulard M (2014) Notes on the role of dynamic DNA methylation in mammalian development. Proc Natl Acad Sci U S A 112:6796–6799
26. Mohandas T, Sparkes RS, Shapiro LJ (1981) Reactivation of an inactive human X chromosome: evidence for X inactivation by DNA methylation. Science 211(4480):393–396
27. Li E, Beard C, Jaenisch R (1993) Role for DNA methylation in genomic imprinting. Nature 366(6453):362–365
28. Onodera Y, Haag JR, Ream T et al (2005) Plant nuclear RNA polymerase IV mediates siRNA and DNA methylation-dependent heterochromatin formation. Cell 120(5):613–622

29. Ooi SK, O'Donnell AH, Bestor TH (2009) Mammalian cytosine methylation at a glance. J Cell Sci 122(Pt 16):2787–2791
30. Anderson RM, Bosch JA, Goll MG et al (2009) Loss of Dnmt1 catalytic activity reveals multiple roles for DNA methylation during pancreas development and regeneration. Dev Biol 334:213–223
31. Goll MG, Anderson R, Stainier DY et al (2009) Transcriptional silencing and reactivation in transgenic zebrafish. Genetics 182(3):747–755
32. Matthews RP, Lorent K, Manoral-Mobias R et al (2009) TNFα-dependent hepatic steatosis and liver degeneration caused by mutation of zebrafish s-adenosylhomocysteine hydrolase. Development 136(5):865–875
33. Song J, Rechkoblit O, Bestor TH et al (2011) Structure of DNMT1-DNA complex reveals a role for autoinhibition in maintenance DNA methylation. Science 331(6020):1036–1040
34. Bostick M, Kim JK, Esteve PO et al (2007) UHRF1 plays a role in maintaining DNA methylation in mammalian cells. Science 317(5845):1760–1764
35. Sharif J, Muto M, Takebayashi S et al (2007) The SRA protein Np95 mediates epigenetic inheritance by recruiting Dnmt1 to methylated DNA. Nature 450(7171):908–912
36. Hill PW, Amouroux R, Hajkova P (2014) DNA demethylation, Tet proteins and 5-hydroxymethylcytosine in epigenetic reprogramming: an emerging complex story. Genomics 104(5):324–333
37. Swindle CS, Kim HG, Klug CA (2004) Mutation of CpGs in the murine stem cell virus retroviral vector long terminal repeat represses silencing in embryonic stem cells. J Biol Chem 279(1):34–41
38. Yu W, McIntosh C, Lister R et al (2014) Genome-wide DNA methylation patterns in LSH mutant reveals de-repression of repeat elements and redundant epigenetic silencing pathways. Genome Res 24(10):1613–1623
39. John RM, Lefebvre L (2011) Developmental regulation of somatic imprints. Differentiation 81(5):270–280
40. Ai L, Kim WJ, Kim TY et al (2006) Epigenetic silencing of the tumor suppressor cystatin M occurs during breast cancer progression. Cancer Res 66(16):7899–7909
41. Bae YK, Brown A, Garrett E et al (2004) Hypermethylation in histologically distinct classes of breast cancer. Clin Cancer Res 10(18 Pt 1):5998–6005
42. Gutierrez-Arcelus M, Lappalainen T, Montgomery SB et al (2013) Passive and active DNA methylation and the interplay with genetic variation in gene regulation. Elife 2, e00523
43. Roll JD, Rivenbark AG, Sandhu R et al (2013) Dysregulation of the epigenome in triple-negative breast cancers: basal-like and claudin-low breast cancers express aberrant DNA hypermethylation. Exp Mol Pathol 95(3):276–287
44. Maeder ML, Angstman JF, Richardson ME et al (2013) Targeted DNA demethylation and activation of endogenous genes using programmable TALE-TET1 fusion proteins. Nat Biotechnol 31(12):1137–1142
45. Feng S, Cokus SJ, Zhang X et al (2010) Conservation and divergence of methylation patterning in plants and animals. Proc Natl Acad Sci U S A 107(19):8689–8694
46. Ge L, Zhang RP, Wan F et al (2014) TET2 plays an essential role in erythropoiesis by regulating lineage-specific genes via DNA oxidative demethylation in a zebrafish model. Mol Cell Biol 34(6):989–1002
47. Mhanni AA, McGowan RA (2004) Global changes in genomic methylation levels during early development of the zebrafish embryo. Dev Genes Evol 214(8):412–417
48. Andersen IS, Reiner AH, Aanes H et al (2012) Developmental features of DNA methylation during activation of the embryonic zebrafish genome. Genome Biol 13(7), R65
49. Rai K, Huggins IJ, James SR et al (2008) DNA demethylation in zebrafish involves the coupling of a deaminase, a glycosylase, and gadd45. Cell 135(7):1201–1212
50. McGaughey DM, Abaan HO, Miller RM et al (2014) Genomics of CpG methylation in developing and developed zebrafish. G3 (Bethesda) 4(5):861–869
51. Cock-Rada AM, Medjkane S, Janski N et al (2012) SMYD3 promotes cancer invasion by epigenetic upregulation of the metalloproteinase MMP-9. Cancer Res 72(3):810–820

52. Rai K, Nadauld LD, Chidester S et al (2006) Zebra fish Dnmt1 and Suv39h1 regulate organ-specific terminal differentiation during development. Mol Cell Biol 26(19):7077–7085
53. Tena JJ, Gonzalez-Aguilera C, Fernandez-Minan A et al (2014) Comparative epigenomics in distantly related teleost species identifies conserved *cis*-regulatory nodes active during the vertebrate phylotypic period. Genome Res 24(7):1075–1085
54. Hashimoto H, Vertino PM, Cheng X (2010) Molecular coupling of DNA methylation and histone methylation. Epigenomics 2(5):657–669
55. Albert M, Helin K (2010) Histone methyltransferases in cancer. Semin Cell Dev Biol 21(2):209–220
56. Margueron R, Reinberg D (2011) The Polycomb complex PRC2 and its mark in life. Nature 469(7330):343–349
57. Kampilafkos P, Melachrinou M, Kefalopoulou Z et al (2015) Epigenetic modifications in cutaneous malignant melanoma: EZH2, H3K4me2, and H3K27me3 immunohistochemical expression is enhanced at the invasion front of the tumor. Am J Dermatopathol 37(2):138–144
58. Vastenhouw NL, Zhang Y, Woods IG et al (2010) Chromatin signature of embryonic pluripotency is established during genome activation. Nature 464(7290):922–926
59. Lindeman LC, Andersen IS, Reiner AH et al (2011) Prepatterning of developmental gene expression by modified histones before zygotic genome activation. Dev Cell 21(6):993–1004
60. de la Barre AE, Gerson V, Gout S et al (2000) Core histone N-termini play an essential role in mitotic chromosome condensation. EMBO J 19(3):379–391
61. Van Hooser A, Goodrich DW, Allis CD et al (1998) Histone H3 phosphorylation is required for the initiation, but not maintenance, of mammalian chromosome condensation. J Cell Sci 111(Pt 23):3497–3506
62. Pereira S, Bourrachot S, Cavalie I et al (2011) Genotoxicity of acute and chronic gamma-irradiation on zebrafish cells and consequences for embryo development. Environ Toxicol Chem 30(12):2831–2837
63. Karmodiya K, Anamika K, Muley V et al (2014) Camello, a novel family of Histone Acetyltransferases that acetylate histone H4 and is essential for zebrafish development. Sci Rep 4:6076
64. Stadler JA, Shkumatava A, Norton WH et al (2005) Histone deacetylase 1 is required for cell cycle exit and differentiation in the zebrafish retina. Dev Dyn 233(3):883–889
65. Cunliffe VT, Casaccia-Bonnefil P (2006) Histone deacetylase 1 is essential for oligodendrocyte specification in the zebrafish CNS. Mech Dev 123(1):24–30
66. Farooq M, Sulochana KN, Pan X et al (2008) Histone deacetylase 3 (hdac3) is specifically required for liver development in zebrafish. Dev Biol 317:336–353
67. Noel ES, Casal-Sueiro A, Busch-Nentwich E et al (2008) Organ-specific requirements for Hdac1 in liver and pancreas formation. Dev Biol 322(2):237–250
68. Li Y, Wang J, Xie Y et al (2014) Pattern of change in histone 3 lysine 9 acetylation and histone deacetylases in development of zebrafish embryo. J Genet 93(2):539–544
69. Yamaguchi M, Tonou-Fujimori N, Komori A et al (2005) Histone deacetylase 1 regulates retinal neurogenesis in zebrafish by suppressing Wnt and Notch signaling pathways. Development 132(13):3027–3043
70. Shabason JE, Tofilon PJ, Camphausen K (2010) HDAC inhibitors in cancer care. Oncology (Williston Park) 24(2):180–185
71. Chodavarapu RK, Feng S, Bernatavichute YV et al (2010) Relationship between nucleosome positioning and DNA methylation. Nature 466(7304):388–392
72. Roadmap Epigenomics C, Kundaje A, Meuleman W et al (2015) Integrative analysis of 111 reference human epigenomes. Nature 518(7539):317–330
73. Stirzaker C, Taberlay PC, Statham AL et al (2014) Mining cancer methylomes: prospects and challenges. Trends Genet 30(2):75–84
74. Feinberg AP, Vogelstein B (1987) Alterations in DNA methylation in human colon neoplasia. Semin Surg Oncol 3(3):149–151

75. Rainier S, Johnson LA, Dobry CJ et al (1993) Relaxation of imprinted genes in human cancer. Nature 362(6422):747–749
76. Jadersten M, Hellstrom-Lindberg E (2010) New clues to the molecular pathogenesis of myelodysplastic syndromes. Exp Cell Res 316(8):1390–1396
77. Colquitt BM, Allen WE, Barnea G et al (2013) Alteration of genic 5-hydroxymethylcytosine patterning in olfactory neurons correlates with changes in gene expression and cell identity. Proc Natl Acad Sci U S A 110(36):14682–14687
78. Rai K, Sarkar S, Broadbent TJ et al (2010) DNA demethylase activity maintains intestinal cells in an undifferentiated state following loss of APC. Cell 142(6):930–942
79. Lao VV, Grady WM (2011) Epigenetics and colorectal cancer. Nat Rev Gastroenterol Hepatol 8(12):686–700
80. Zhou L, Zhao X, Han Y et al (2013) Regulation of UHRF1 by miR-146a/b modulates gastric cancer invasion and metastasis. FASEB J 27(12):4929–4939
81. Wang F, Yang YZ, Shi CZ et al (2012) UHRF1 promotes cell growth and metastasis through repression of $p16^{ink4a}$ in colorectal cancer. Ann Surg Oncol 19:2753–2762
82. Babbio F, Pistore C, Curti L et al (2012) The SRA protein UHRF1 promotes epigenetic crosstalks and is involved in prostate cancer progression. Oncogene 31:4878–4887
83. Daskalos A, Oleksiewicz U, Filia A et al (2011) UHRF1-mediated tumor suppressor gene inactivation in nonsmall cell lung cancer. Cancer 117(5):1027–1037
84. Unoki M, Daigo Y, Koinuma J et al (2010) UHRF1 is a novel diagnostic marker of lung cancer. Br J Cancer 103(2):217–222
85. Unoki M, Kelly JD, Neal DE et al (2009) UHRF1 is a novel molecular marker for diagnosis and the prognosis of bladder cancer. Br J Cancer 101(1):98–105
86. Jin W, Chen L, Chen Y et al (2009) UHRF1 is associated with epigenetic silencing of BRCA1 in sporadic breast cancer. Breast Cancer Res Treat 123:359–373
87. Jenkins Y, Markovtsov V, Lang W et al (2005) Critical role of the ubiquitin ligase activity of UHRF1, a nuclear RING finger protein, in tumor cell growth. Mol Biol Cell 16(12):5621–5629
88. Sadler KC, Krahn KN, Gaur NA et al (2007) Liver growth in the embryo and during liver regeneration in zebrafish requires the cell cycle regulator, uhrf1. Proc Natl Acad Sci U S A 104(5):1570–1575
89. Du Z, Song J, Wang Y et al (2010) DNMT1 stability is regulated by proteins coordinating deubiquitination and acetylation-driven ubiquitination. Sci Signal 3(146):ra80
90. Qin W, Leonhardt H, Spada F (2011) Usp7 and Uhrf1 control ubiquitination and stability of the maintenance DNA methyltransferase Dnmt1. J Cell Biochem 112(2):439–444
91. Wu PC, Lu JW, Yang JY et al (2014) H3K9 histone methyltransferase, KMT1E/SETDB1, cooperates with the SMAD2/3 pathway to suppress lung cancer metastasis. Cancer Res 74(24):7333–7343
92. Chiba T, Saito T, Yuki K et al (2015) Histone lysine methyltransferase SUV39H1 is a potent target for epigenetic therapy of hepatocellular carcinoma. Int J Cancer 136(2):289–298
93. Patton EE, Widlund HR, Kutok JL et al (2005) BRAF mutations are sufficient to promote nevi formation and cooperate with p53 in the genesis of melanoma. Curr Biol 15(3):249–254
94. Macgregor S, Montgomery GW, Liu JZ et al (2011) Genome-wide association study identifies a new melanoma susceptibility locus at 1q21.3.. Nat Genet 43(11):1114–1118
95. Fritsch L, Robin P, Mathieu JR et al (2010) A subset of the histone H3 lysine 9 methyltransferases Suv39h1, G9a, GLP, and SETDB1 participate in a multimeric complex. Mol Cell 37(1):46–56
96. Langenau DM, Keefe MD, Storer NY et al (2007) Effects of RAS on the genesis of embryonal rhabdomyosarcoma. Genes Dev 21(11):1382–1395
97. Chen EY, Dobrinski KP, Brown KH et al (2013) Cross-species array comparative genomic hybridization identifies novel oncogenic events in zebrafish and human embryonal rhabdomyosarcoma. PLoS Genet 9(8), e1003727

98. Herman JG, Merlo A, Mao L, Lapidus RG, Issa JP, Davidson NE, Sidransky D, Baylin SB (1995) Inactivation of the CDKN2/p16/MTS1 gene is frequently associated with aberrant DNA methylation in all common human cancers. Cancer Res 55:4525–4530
99. Kornegoor R, Moelans CB, Verschuur-Maes AH, Hogenes MC, de Bruin PC, Oudejans JJ, van Diest PJ (2012) Promoter hypermethylation in male breast cancer: analysis by multiplex ligation-dependent probe amplification. Breast Cancer Res 14:1–9
100. Zou X-P, Zhang B, Zhang X-Q, Chen M, Cao J, Liu W-J (2009) Promoter hypermethylation of multiple genes in early gastric adenocarcinoma and precancerous lesions. Hum Pathol 40:1534–1542
101. Bedford MT, van Helden PD (1987) Hypomethylation of DNA in pathological conditions of the human prostate. Cancer Res 47:5274–5276
102. Cadieux B, Ching T-T, VandenBerg SR, Costello JF (2006) Genome-wide hypomethylation in human glioblastomas associated with specific copy number alteration, methylenetetrahydrofolate reductase allele status, and increased proliferation. Cancer Res 66:8469–8476
103. Fraga MF, Ballestar E, Villar-Garea A, Boix-Chornet M, Espada J, Schotta G, Bonaldi T, Haydon C, Ropero S, Petrie K et al (2005) Loss of acetylation at Lys16 and trimethylation at Lys20 of histone H4 is a common hallmark of human cancer. Nat Genet 37:391–400
104. Rauch TA, Zhong X, Wu X, Wang M, Kernstine KH, Wang Z, Riggs AD, Pfeifer GP (2008) High-resolution mapping of DNA hypermethylation and hypomethylation in lung cancer. Proc Natl Acad Sci U S A 105:252–257
105. Jackson-Grusby L, Beard C, Possemato R, Tudor M, Fambrough D, Csankovszki G, Dausman J, Lee P, Wilson C, Lander E et al (2001) Loss of genomic methylation causes p53-dependent apoptosis and epigenetic deregulation. Nat Genet 27:31–39
106. Howard G, Eiges R, Gaudet F, Jaenisch R, Eden A (2007) Activation and transposition of endogenous retroviral elements in hypomethylation induced tumors in mice. Oncogene 27:404–408
107. Minoguchi S, Iba H (2008) Instability of retroviral DNA methylation in embryonic stem cells. Stem Cells 26:1166–1173
108. Frigola J, Song J, Stirzaker C, Hinshelwood RA, Peinado MA, Clark SJ (2006) Epigenetic remodeling in colorectal cancer results in coordinate gene suppression across an entire chromosome band. Nat Genet 38:540–549
109. Nguyen CT, Weisenberger DJ, Velicescu M, Gonzales FA, Lin JCY, Liang G, Jones PA (2002) Histone H3-lysine 9 methylation is associated with aberrant gene silencing in cancer cells and is rapidly reversed by 5-Aza-2′-deoxycytidine. Cancer Res 62:6456–6461
110. Chinaranagari S, Sharma P, Chaudhary J (2014) EZH2 dependent H3K27me3 is involved in epigenetic silencing of ID4 in prostate cancer. Oncotarget 5:7172–7182
111. Ohm JE, McGarvey KM, Yu X, Cheng L, Schuebel KE, Cope L, Mohammad HP, Chen W, Daniel VC, Yu W et al (2007) A stem cell–like chromatin pattern may predispose tumor suppressor genes to DNA hypermethylation and heritable silencing. Nat Genet 39:237–242
112. Popovic R, Martinez-Garcia E, Giannopoulou EG, Zhang Q, Zhang Q, Ezponda T, Shah MY, Zheng Y, Will CM, Small EC et al (2014) Histone methyltransferase MMSET/NSD2 alters EZH2 binding and reprograms the myeloma epigenome through global and focal changes in H3K36 and H3K27 methylation. PLoS Genet 10, e1004566
113. Schlesinger Y, Straussman R, Keshet I, Farkash S, Hecht M, Zimmerman J, Eden E, Yakhini Z, Ben-Shushan E, Reubinoff BE et al (2007) Polycomb-mediated methylation on Lys27 of histone H3 pre-marks genes for de novo methylation in cancer. Nat Genet 39:232–236
114. Kang J-Y, Song S-H, Yun J, Jeon M-S, Cha Y, Lee S-H, Kim H-P, Jeong E-G, Han S-W, Cho N-Y et al (2015) Identification of long-range epigenetic silencing on chromosome 15q25 and its clinical implication in gastric cancer. Am J Pathol 185(3):666–678
115. Lee J-E, Wang C, Xu S, Cho Y-W, Wang L, Feng X, Baldridge A, Sartorelli V, Zhuang L, Peng W et al (2013) H3K4 mono- and di-methyltransferase MLL4 is required for enhancer activation during cell differentiation. eLife 2, e01503

116. Lindeman LC, Vogt-Kielland LT, Alestrom P et al (2009) Fish'n ChIPs: chromatin immunoprecipitation in the zebrafish embryo. Methods Mol Biol 567:75–86
117. Cui L, Chen J, Zhang Q, Wang X, Qu J, Zhang J, Dang S (2015) Up-regulation of UHRF1 by oncogenic Ras promoted the growth, migration, and metastasis of pancreatic cancer cells. Mol Cell Biochem 400:223–232
118. Li A, Omura N, Hong S-M, Goggins M (2010) Pancreatic cancer DNMT1 expression and sensitivity to DNMT1 inhibitors. Cancer Biol Ther 9:321–329
119. Chen D, Wang P, Huang A, Zhang B (2006) Effects of pshRNA-DNMT1 on the proliferation and apoptosis of gastric cancer: experiments in vitro and in vivo. Zhonghua Yi Xue Za Zhi 86:1534–1539
120. Xu H, Wang Z, Liu D, Li Q, Dai S, Wang E (2010) High expression of DNMT1 was correlated with beta-catenin accumulation and malignant phynotype of lung squamous cell carcinoma and adenocarcinoma. Chin J Lung Cancer 13:856–860
121. Im AP, Sehgal AR, Carroll MP, Smith BD, Tefferi A, Johnson DE, Boyiadzis M (2014) DNMT3A and IDH mutations in acute myeloid leukemia and other myeloid malignancies: associations with prognosis and potential treatment strategies. Leukemia 28:1774–1783
122. Scourzic L, Mouly E, Bernard OA (2015) TET proteins and the control of cytosine demethylation in cancer. Genome Med 7:1–16
123. Hung S-Y, Lin H-H, Yeh K-T, Chang J-G (2014) Histone-modifying genes as biomarkers in hepatocellular carcinoma. Int J Clin Exp Pathol 7:2496–2507
124. Cai L, Ma X, Huang Y, Zou Y, Chen X (2014) Aberrant histone methylation and the effect of Suv39H1 siRNA on gastric carcinoma. Oncol Rep 31:2593–2600
125. Khanal P, Kim G, Lim S-C, Yun H-J, Lee KY, Choi H-K, Choi HS (2013) Prolyl isomerase Pin1 negatively regulates the stability of SUV39H1 to promote tumorigenesis in breast cancer. FASEB J 27:4606–4618
126. Tao H, Li H, Su Y, Feng D, Wang X, Zhang C, Ma H, Hu Q (2014) Histone methyltransferase G9a and H3K9 dimethylation inhibit the self-renewal of glioma cancer stem cells. Mol Cell Biochem 394:23–30
127. Zhong X, Chen X, Guan X, Zhang H, Ma Y, Zhang S, Wang E, Zhang L, Han Y (2015) Overexpression of G9a and MCM7 in oesophageal squamous cell carcinoma is associated with poor prognosis. Histopathology 66:192–200
128. Aday AW, Zhu LJ, Lakshmanan A et al (2011) Identification of *cis* regulatory features in the embryonic zebrafish genome through large-scale profiling of H3K4me1 and H3K4me3 binding sites. Dev Biol 357(2):450–462
129. Bock C, Tomazou EM, Brinkman AB et al (2010) Quantitative comparison of genome-wide DNA methylation mapping technologies. Nat Biotechnol 28(10):1106–1114
130. Mudbhary R, Sadler KC (2011) Epigenetics, development, and cancer: zebrafish make their mark. Birth Defects Res Part C Embryo Today Rev 93:194–203
131. Chatterjee A, Ozaki Y, Stockwell PA et al (2013) Mapping the zebrafish brain methylome using reduced representation bisulfite sequencing. Epigenetics 8(9):979–989
132. Bock C (2012) Analysing and interpreting DNA methylation data. Nat Rev Genet 13(10):705–719
133. Gupta R, Nagarajan A, Wajapeyee N (2010) Advances in genome-wide DNA methylation analysis. Biotechniques 49(4):3–11
134. Rennekamp AJ, Peterson RT (2014) 15 years of zebrafish chemical screening. Curr Opin Chem Biol 24C:58–70

Lymphatics, Cancer and Zebrafish

Jonathan W. Astin and Philip S. Crosier

Abstract Many solid tumors are known to metastasize through the lymphatic vasculature. This process is facilitated by the generation of new lymphatic vessels (tumor lymphangiogenesis) and also by the remodelling of existing lymphatics. Together these processes enable the spread of tumor cells to distant sites. Currently our understanding of tumor lymphangiogenesis has been informed from mouse tumor models and from studies of developmental lymphangiogenesis. Since the discovery of *bona fide* lymphatic vessels in zebrafish in 2006, zebrafish have become a well-established model of developmental lymphangiogenesis. The attributes that make zebrafish such an important model of blood vessel development—the ability to live image developing vessels, genetic tractability and the conserved nature of development—also make fish an attractive model of lymphatic vessel development. In particular, zebrafish have made important contributions to our understanding of the processes of lymphatic vessel sprouting from veins and the mechanisms by which lymphatic precursors remodel into mature vessels. To date, zebrafish have not been used to directly model tumor lymphangiogenesis. In this chapter we will summarise the contributions zebrafish have made to our understanding of lymphangiogenesis and investigate the possibilities of combining zebrafish transgenic cancer lines or tumor transplantation models with existing lymphatic reporter lines, which could provide valuable insights into the process of tumor-induced lymphangiogenesis. In addition the utility of using the zebrafish lymphatic model as a platform to screen and develop novel anti-lymphatic therapeutics will also be discussed.

Keywords Zebrafish • Lymphatic • Cancer • Lymphangiogenesis • Vascular • Development • Tumor lymphatics

J.W. Astin (✉) • P.S. Crosier (✉)
Department of Molecular Medicine and Pathology, School of Medical Sciences,
University of Auckland, Private Bag 92019 Auckland Mail Centre,
Auckland 1142, New Zealand
e-mail: j.astin@auckland.ac.nz; ps.crosier@auckland.ac.nz

D.M. Langenau (ed.), *Cancer and Zebrafish*, Advances in Experimental Medicine and Biology 910, DOI 10.1007/978-3-319-30654-4_9

Introduction to the Lymphatic System

The lymphatic vasculature is a hierarchical set of vessels that functions to regulate tissue fluid homeostasis within the body. Interstitial fluid and protein (lymph) that escapes from blood capillaries is absorbed by specialised blind-ended lymphatic capillaries; these capillaries are highly permeable vessels, with little to no basement membrane or support cells, which are attached to the extracellular matrix by specialised elastic filaments called anchoring filaments. In contrast to the blood vasculature, the lymphatic system transports fluid unidirectionally with lymph being transported from the lymphatic capillaries, through progressively larger precollecting and collecting lymphatic vessels before being returned to the blood circulation through lymphovenous connections. Unlike lymphatic capillaries, the collecting vessels are sheathed with contractile smooth muscle cells, which help to drive the flow of lymph. Collecting vessels also contain specialised valves that ensure that lymph flows back to the veins. In mammals all the lymph from the legs and the left side of the body is returned to the largest collecting vessel, the thoracic duct, while the right lymphatic duct collects the lymph from the rest of the body [1].

In addition to maintaining fluid homeostasis, the mammalian lymphatics also play key roles in immune responses with leukocytes, antigens and antigen presenting cells all being transported through lymphatic vessels. Lymph nodes are wired into lymphatic collecting vessels and enable the interaction between antigens contained in the lymph and lymphocytes present in the lymph nodes. Finally, specialised lymphatic capillaries in the intestinal villi, called lacteals, are key for the absorption of long-chain fatty acids released by enterocytes.

The lymphatic system develops from the blood vasculature in a process termed lymphangiogenesis. Florence Sabin first proposed the currently accepted mechanism of lymphangiogenesis in 1902, in which she suggested that the lymphatic vessels arise from lymph sacs that bud from the veins [2]. More recent lineage tracing experiments in mice have confirmed the venous origin of lymphatic vessels [3]. The first lymphatic endothelial cells (LEC) bud from the veins at embryonic day 10 in mice (embryonic week 6–7 in humans) and migrate away to form lymph sacs. By E14-15 these lymph sacs have remodelled and the first lymphatic vessel networks are observed.

Lymphatic vessels are present in all vertebrates, including birds, amphibians and fish and are present in most vascularised tissues in the body, the exception being the central nervous system. Unlike blood vessels, there appears to be a large amount of morphological divergence in different vertebrate lymphatic networks [4]. Historical studies suggest that fish lack both lymph nodes and lymphatic valves [5] and recent studies of the zebrafish lymphatic system have so far supported these earlier findings, although more investigation into the structure of adult fish lymphatics needs to be conducted to confirm this. It is possible that the differences in adapting to an aquatic compared with a terrestrial environment might account for the divergence in the morphology between mammalian and fish lymphatics. Nevertheless, despite these differences, it is clear that the process of lymphangiogenesis is largely conserved between mouse and zebrafish.

Zebrafish Lymphatic Anatomy

Studies in the eighteenth century demonstrated that fish contain a lymphatic network [5], however, the first evidence of a lymphatic system in zebrafish was not provided until 2006. In these studies the zebrafish thoracic duct was described using the *fli1a:egfp* transgenic line which expresses GFP in all endothelial cells [6–8]. Functional assays confirmed that zebrafish lymphatics are able to absorb and transport interstitial fluid back to the veins while electron microscopy revealed that the developing thoracic duct consists of a thin layer of endothelial cells with anchoring filament-like structures [8]. Since then, more zebrafish lymphatic reporter lines have been developed (Table 1). Most of these lines also label venous endothelial cells as well as other non-vascular tissues; because of this, lymphatic reporters are often combined with additional transgenic reporters for blood endothelial cells to allow the lymphatics to be identified (for example, *fli1a:egfp;kdrl:mcherry* or *lyve1b:dsred;kdrl:egfp*). Using these compound transgenics, the entire lymphatic vessel network has been mapped in embryonic zebrafish up to 15 days post fertilisation (dpf). Three lymphatic vascular beds have been described: the trunk lymphatics (which include the thoracic duct), the facial lymphatics and the intestinal lymphatics [9].

Table 1 Zebrafish transgenic lines with expression in lymphatic vessels

Lymphatic reporter line	Expression pattern	Reference
Tg(fli1a:egfp)^{y1}	Highly expressed pan-endothelial line—expressed in both lymphatic and blood vessels. Non-vascular expression in neural crest derived tissues in the head	[7]
Tg(stab1^{BAC}:yfp)	Expressed in veins and lymphatics. Expression relatively weak	[34]
Tg(sagff27c;uas:egfp)	Expressed in trunk lymphatics, weak expression in PCV. Non-vascular expression in many tissues including the intestine, lens and cardiac muscle	[10]
Tg(-5.2lyve1b:egfp)^{nz150} *Tg(-5.2lyve1b:dsred)^{nz101}*	Expressed in major axial veins and lymphatics. Weak non-vascular expression in fins. GFP first observed in PCV and CV from 30 hpf, DsRed expression weak until 48 hpf	[9]
Tg(-6.6flt4:YFP)^{hu4881}	Expression in DA, PCV and ISVs at 24 hpf. Expression becomes weak from 48 hpf	[29]
Tg(flt4^{BAC}:YFP)		[31]
Tg(flt4^{BAC}:mCitrine)^{hu7135}	Expressed in blood vessels initially, becomes increasingly venous from 26 hpf and by 5 dpf has a similar expression pattern to *lyve1b*	[11]
Tg(prox1a^{BAC}:KalTA4; *4xUASE1b:unctagRFP)^{nim5}*	Expressed in a number of non-vascular tissues such as the myotome, liver, neuromasts, lens and retina. Expressed in both arterial and venous vessels before 24 hpf. Vascular expression becomes restricted to venous cells by 32 hpf and is lymphatic-specific by 5 dpf	[22]

Currently the trunk lymphatic network is by far the most extensively studied [4] (Fig. 1). Here, lymphatic vessel development is linked to the process of secondary (venous) sprouting (SS) from the posterior cardinal vein (PCV). At this stage of development the dorsal aorta (DA), intersomitic vessels (ISVs) and the PCV have already formed by vasculogenesis and primary angiogenesis. At 36 h post fertilisation (hpf) SSs from the PCV migrate dorsally—half of these will anastomose with ISVs to form a venous connection but the other half will become the precursors of the trunk lymphatic network [10]. These lymphatic precursors continue migrating dorsally before turning to align with the horizontal myoseptum. Here they become completely detached from the PCV and are termed parachordal lymphangioblasts (PLs). They subsequently migrate either dorsally or ventrally along the arterial ISVs to form the trunk lymphatic network, which by 5 dpf consists of the thoracic duct (TD), intersomitic lymphatic vessels (ISLVs) and the dorsal longitudinal lymphatic vessels (DLLVs). Both the TD and the DLLV form from the bidirectional fusion of vessel fragments along their lengths. In the posterior part of the trunk, the thoracic duct is a single vessel but splits into a left and right branch towards the anterior, mimicking the bifurcation observed in the DA and PCV [11].

The facial lymphatics develop initially from a single lymphatic sprout that forms at 36 hpf from the CCV and migrates anteriorly along the primary head sinus (PHS). Here, other lymphangioblast populations delaminate from the PHS, and later from an unidentified vein(s) near the ventral aorta called the ventral aorta lymphangioblast—VAL (Fig. 1). These lymphangioblasts add to the tip of the growing FLS and drive migration towards the jaw. By 4 dpf the facial lymphatics consist of three vessels. The FLS matures into the lateral facial lymphatic (LFL); the major facial lymphatic vessel that grows along the PHS before turning under the eye, while the medial facial lymphatic (MFL) and the otolithic lymphatic vessel (OLV) sprout from the LFL at approximately 3 dpf. By 5 dpf additional lymphatics form along the brachial arteries, known as the brachial lymphatics (LAAs). At this stage the initial connection to the CCV is lost and the facial lymphatics form a connection to the left and right thoracic duct via a new vessel termed the jugular lymphatic vessel (JLV) [9]. The connection between the lymphatic and venous networks has not been described in zebrafish but is thought to be located at the anterior end of the TD, near where the JLV joins, as anatomical examination of lymphatic vessels in other teleosts suggest that this is the site of lymphovenous connection in fish [12].

Fig. 1 Development of the zebrafish facial and trunk lymphatics. Schematic diagram highlighting the development of the zebrafish facial (*left*) and trunk (*right*) lymphatic networks. *CCV* common cardinal vein, *LDA* lateral dorsal artery, *DA* dorsal aorta, *PCV* posterior cardinal vein, *ISV* intersomitic vessel, *DLAV* dorsal longitudinal anastomotic vessel, *FLS* facial lymphatic sprout, *PHS* primary head sinus, *SS* secondary sprout, *PCeV* posterior cerebral vein, *VAL* ventral aorta lymphangioblast, *PL* paracordal lymphangioblast, *LFL* lateral facial lymphatic, *AA* brachial arteries, *vISV* venous intersomitic vessel, *MFL* medial facial lymphatic, *OLV* otolithic lymphatic vessel, *LAA* brachial lymphatic vessels, *JLV* jugular lymphatic vessel, *TD* thoracic duct, *ISLV* intersomitic lymphatic vessel, *DLLV* dorsal longitudinal lymphatic vessel

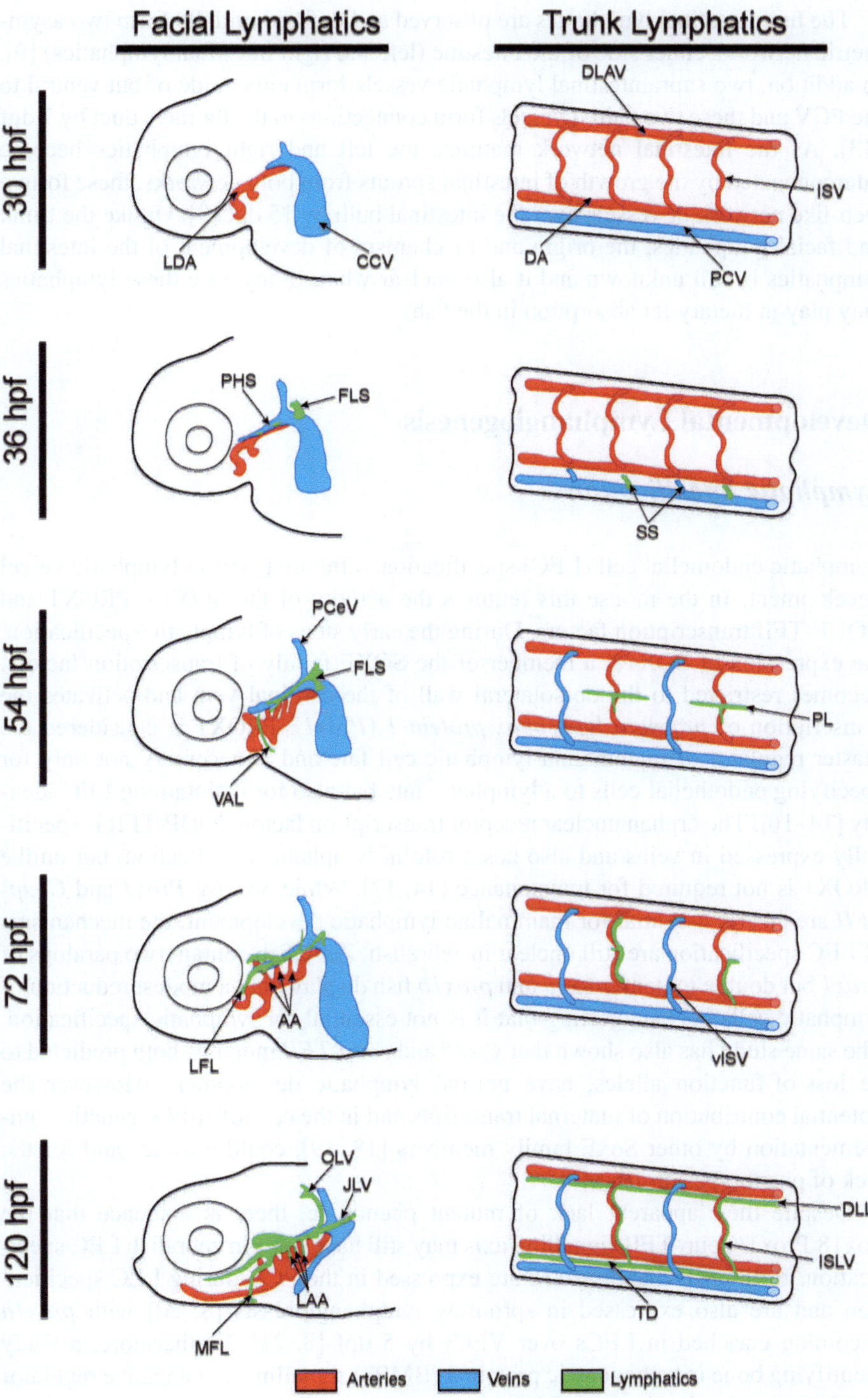
Facial Lymphatics
Trunk Lymphatics
30 hpf
LDA
CCV
DLAV
ISV
DA
PCV
36 hpf
PHS
FLS
SS
54 hpf
PCeV
FLS
VAL
PL
72 hpf
AA
LFL
vISV
120 hpf
OLV
JLV
LAA
MFL
DLLV
ISLV
TD
Arteries
Veins
Lymphatics

The first intestinal lymphatics are observed at 4 dpf and initially form two asymmetric networks either side of the intestine (left and right intestinal lymphatics) [9]. In addition, two supraintestinal lymphatic vessels form either side of but ventral to the PCV and these two paired vessels form connections to the thoracic duct by 7 dpf [13]. As the intestinal network matures, the left and right lymphatics become interconnected by the growth of intestinal sprouts from both networks, these form a web-like network of vessels over the intestinal bulb by 15 dpf [9]. Unlike the trunk and facial lymphatics, the origin and mechanism of development of the intestinal lymphatics is still unknown and it also unclear what, if any, role these lymphatics may play in dietary fat absorption in the fish.

Developmental Lymphangiogenesis

Lymphatic Specification

Lymphatic endothelial cell (LEC) specification is the first step in lymphatic vessel development. In the mouse this requires the activity of the SOX18, PROX1 and COUP-TFII transcription factors. During the early steps of lymphatic specification, the expression of SOX18, a member of the SOXF family of transcription factors, becomes restricted to the dorsolateral wall of the cardinal vein and activates the transcription of *prospero homeobox protein 1* (*Prox1*). PROX1 is considered the master regulator of mammalian lymphatic cell fate and is necessary not only for specifying endothelial cells to a lymphatic fate but also for maintaining LEC identity [14–16]. The orphan nuclear receptor transcription factor COUP-TFII is specifically expressed in veins and also has a role in lymphatic specification, but unlike PROX1 is not required for maintenance [14, 17]. While *Sox 18, Prox1* and *Coup-TFII* are clearly essential for mammalian lymphatic development, the mechanisms of LEC specification are still unclear in zebrafish. Zebrafish contain two paralogs of *Prox1* but double mutant *prox1a* and *prox1b* fish display only a modest reduction of lymphatic cells [11] suggesting that it is not essential for lymphatic specification. The same study has also shown that *sox18* and *coup-TFII* mutants, both predicted to be loss-of-function alleles, have normal lymphatic development. However the potential contribution of maternal transcripts and in the case of *sox18,* genetic complementation by other SoxF family members [18, 19], could also account for the lack of phenotypes observed.

Despite their apparent lack of mutant phenotype, there is evidence that the Sox18/Prox1/Coup-TFII signalling axis may still have roles in zebrafish LEC specification. Both *prox1a* and *prox1b* are expressed in the PCV during LEC specification and are also expressed in sprouting lymphangioblasts [8, 20] with *prox1a* becoming enriched in LECs over VECs by 5 dpf [8, 21]. Furthermore, a study identifying bone morphogenetic protein 2 (BMP2) signalling as a negative regulator of LEC specification in zebrafish and mouse stem cells, has implicated *prox1a* as the key gene in this process, with the expression of *prox1a* negatively regulated by

Bmp2 signalling [22]. Expression of *coup-TFII* in the PCV and PLs is regulated by *sox18* and *sox7* while morpholino silencing of *coup-TFII* results in a reduction of both *lyve1* and *prox1a* transcripts in the trunk [23, 24].

Induction of LEC fate in zebrafish appears to require Notch signalling, as loss of either *delta-like ligand 4* (*dll4*) or *notch 1b* causes almost all SS to become vISVs, suggesting they have adopted a blood vessel fate. However an alternative explanation for the increase in vISVs is that Notch signalling is necessary for the migration of LECs to the horizontal myoseptum and in the absence of Notch activity, SSs that are unable to migrate and form PLs instead anastomose with ISVs [25]. In mammals, the role of Notch signalling in lymphatic fate specification is less clear with several conflicting studies being reported. A mouse knockout of canonical Notch signalling (*Rbpj*) in *Tie2*-expressing lymphatic precursors had no effect on LEC induction [17], whereas loss of *Notch1* in *Prox*-1 expressing lymphatic precursors revealed that it was a negative regulator of lymphatic cell fate and lymphangiogenesis [26, 27]. Taken together, these data reveal a complex picture underlying LEC specification and more research is required to fully elucidate the signalling pathways responsible for determining lymphatic fate decisions in zebrafish.

Lymphatic Sprouting and Migration

Despite the uncertainty surrounding LEC fate specification in fish, the process of lymphatic sprouting from veins is highly conserved with vascular endothelial growth factor receptor 3 (VEGFR3) signalling being essential for all lymphatic sprouting in both zebrafish and mice [4]. In mice, *Vegfr3* expression is upregulated in LECs through PROX1 activity [28] while the ligand, VEGFC is expressed in the surrounding mesenchyme. The binding of the extracellular VEGFC protein to VEGFR3 activates downstream signalling, resulting in migration and proliferation of LECs from the vein. In fish this process is conserved as Vegfr3 (also known as Flt4) is expressed in lymphatic cells and is essential for all lymphatic vessel development [9, 29]. Similarly, zebrafish Vegfc is expressed in the mesenchyme and blood vessels adjacent to Vegfr3-expressing cells and is essential for trunk and intestinal lymphatic development [6, 30–33], while knockdown of the Vegfr3 signalling axis prevents secondary sprouting from the PCV and the formation of the FLS from the CCV [9]. Interestingly, the facial lymphatics appear to be able to grow in the absence of Vegfc, due to the expression of another Vegfr3 ligand, Vegfd in the head [30].

A forward genetic screen, looking for lymphatic null mutants in the zebrafish, identified collagen and calcium binding EGF domains 1 (Ccbe1), a novel component of the Vegfr3 signalling axis [34]. Ccbe1 mutant zebrafish display no lymphatic sprouting and consequently lack lymphatic vessels. Subsequently it was shown that mutations in human CCBE1 result in Hennekam syndrome, a rare autosomal recessive primary lymphedema disorder [35]. The molecular identification of a human lymphatic disorder through zebrafish research was instrumental in establishing the fish as a model of mammalian lymphatic development. Genetic epistasis studies in the fish have shown that *ccbe1* and *vegfc* genetically interact [36] and an in vitro study has shown that CCBE1 is indirectly required for the proteolytic cleavage of VEGFC to its active form [37]. It is thought that *ccbe1* expression provides additional spatiotemporal control of Vegfr3 activity during lymphatic sprouting and in support of this, the spatiotemporal expression pattern of *ccbe1* in the zebrafish trunk correlates with the path of migrating lymphangioblasts [34].

In mice, VEGFR3 signalling activates a number of downstream pathways that lead to cell migration and proliferation. These include the RAS/MAPK/ERK pathway and the PI3K/AKT/mTOR pathway. In fish, the role of these pathways in lymphatic vessel development has not yet been determined. However, chemical inhibition of mTOR or MAPK does block lymphatic vessel development suggesting conservation in VEGFR3 signalling between fish and mammals [38, 39].

Genetic studies in the fish have led to the identification of two LEC-expressed genes with roles in lymphangioblast migration. The first, *arap3*—a GTPse activating protein with known roles in cell migration, was shown to mediate LEC migration in both zebrafish and mice and is thought to act downstream of the VEGFC/VEGFR3 pathway [40]. The second, *polycystic kidney disease 1a* (*pkd1a*), a zebrafish orthologue of PKD1 that is commonly mutated in autosomal dominant polycystic kidney disease, was shown to be required for the migration of lymphangioblasts following sprouting from the veins. Loss of PKD1 in mouse LECs results in defective dermal lymphatic morphogenesis and branching [21, 41]. Further work is required to elucidate the pathways modulated by PKD1 during lymphangioblast migration but it appears to affect the polarity and adherens junctions of migrating LECs.

Guidance of Lymphatic Vessels

As LECs migrate from the veins to form mature vessels they are known to interact with surrounding tissues and use them as "guides" to help them migrate correctly. The majority of this work has been conducted in the zebrafish and has implicated both blood vessels and neurons as key tissues involved in the guidance of lymphatic vessel precursors.

Studies have shown that many mammalian lymphatic vessels align closely to blood vessels [42, 43]. Initial observations of lymphatic anatomy in the zebrafish trunk noted that there was a high degree of congruence between ISLVs and arterial ISVs (aISVs). Live imaging confirmed that migrating parachordal lymphangioblasts associate with arterial ISVs and that loss of this arterial substrate in the *kdrl*^{hu5088} mutant causes defective PL migration [10]. Further studies have linked chemokine signalling to this process where a chemokine ligand (*cxcl12b)* expressed in arterial cells helps to drive the migration PLs along aISVs through the signalling of chemokine receptors Cxcr4a and 4b expressed in the lymphangioblasts [44]. Later in development, *cxcl12b* expression in the dorsal aorta (DA) as well as *cxcl12a* expression in the PCV is thought to guide lymphangioblasts between these two blood vessels to mediate the formation of the TD [44]. However, the trunk lymphatic defects observed in chemokine mutants are not completely penetrant, suggesting that other signalling pathways are also involved in the cellular interplay between lymphatics and blood vessels. One additional candidate is Notch signalling as knockdown of the Notch ligand *dll4* not only results in reduced numbers of LECs but also disrupts the migration of lymphangioblasts along aISVs [25]. Surprisingly, chemokine signalling does not appear to have a role in the formation of the facial lymphatics and the guidance cues utilised in the formation of these lymphatics remain to be determined [30].

It has been noted that while many zebrafish lymphatic vessels align with arterial vessels, some intestinal lymphatic vessels align with veins (the SIL with the PCV), while some facial lymphatic vessels (OLV, anterior-LFL, MFL) do not appear to align with any blood vessels at all [9]. One example of a non-vascular lymphatic guidance tissue is motoneurons. As PLs migrate into the horizontal myoseptum they migrate along a set of rostral primary motoneuron axons and associated secondary motoneuron axons that extend along the horizontal myoseptum. Genetic or physical ablation of the neurons prevents the migration of PLs and leads to defects in trunk lymphatic vessel development [45, 46] but further work is required to uncover the molecular signals that drive lymphangioblast migration along motoneurons.

Taken together, these observations suggest that there is still much to learn about the signalling pathways and tissues underlying the guidance of developing lymphatic vessels. However, zebrafish are perhaps the best model to investigate this and have already shown us that developing lymphatic vessels utilise different guidance tissues depending on their location and that some lymphangioblasts (the PLs for example) may interact with different tissues during vessel formation.

Tumor Lymphangiogenesis

The importance of the lymphatic vasculature in metastasis is underlined by cancer staging and treatment, with the dissemination of tumor cells to regional lymph nodes a major criterion for staging the progression of many solid tumors such as carcinomas of the breast, colon, prostate and also melanoma [47, 48]. It has been estimated that over 80 % of solid tumors metastasize, at least partially, through the lymphatic vasculature [49].

The growth of new lymphatic vessels into and around a tumor, a process termed tumor-induced lymphangiogenesis, is thought to be an important step in cancer metastasis. The increase in lymphatic vessel density in the tumor has been proposed to facilitate the metastatic spread of cancer cells, as it has been correlated with an increased incidence of lymph node metastases and a consequential decrease in patient survival in many cancers [49, 50]. In addition, more recent studies indicate that existing lymphatic vessels both proximal and distal to the tumor are remodelled to facilitate the metastatic spread of tumor cells. Such remodelling includes the enlargement of lymphatic vessels [51, 52] as well as lymphangiogenesis in the tumor draining lymph nodes [53, 54].

Tumor-Induced Lymphangiogenesis

During tumorigenesis, both the tumor cells and the various cells that make up the tumor microenvironment produce pro-lymphatic growth factors that can promote lymphatic vessel growth from existing lymphatic vessels. This growth includes both the proliferation and migration of LECs. In general two types of lymphatic vessels are observed in tumors, lymphatic vessels within tumors (intratumoral) and lymphatic vessels that surround the tumor (peritumoral) [55]. It is thought that intratumoral vessels do not greatly contribute to metastasis as they appear to be collapsed [56]. Instead the peritumoral lymphatics are thought to facilitate metastasis.

As with developmental lymphangiogenesis, the major driver of tumor-induced lymphangiogenesis is the VEGFR3 signalling pathway [57]. Both VEGFC and VEGFD are known to promote tumor lymphangiogenesis and lymphatic-mediated metastases in mouse models [53, 57, 58] and blockade of VEGFR3 signalling inhibits these processes [59–62]. VEGFC and VEGFD expression is observed in many human tumors [63–65] and can be secreted by the tumor cells or by the immune cells and fibroblasts in the tumor stroma [66]. Furthermore, it has been shown that both ligands can induce the enlargement of existing lymphatic vessels that surround the tumor [51, 52, 67] and have also been linked to structural changes in the tumor-draining lymph nodes that help provide a niche for metastatic cancer cells [53, 68].

While the role of VEGFR3 signalling in tumor lymphangiogenesis is relatively well defined, there are other pathways involved in this process. VEGFA is known to induce tumor lymphatics and increase lymph node metastases in some mouse models [54] but is not able to consistently induce tumor lymphatics, suggesting its role is somewhat context dependent [57]. In addition other non-VEGF signalling pathways can also contribute towards tumor lymphatic growth and remodelling. For example, fibroblast growth factor 2 (FGF2) is known to complement VEGFC-induced lymphatic vessel growth and lymphatic mediated metastasis [69]. In addition, the observation that platelet derived growth factor (PDGF) receptor is expressed on tumor lymphatics has led to the suggestion that PDGF signalling may help to drive proliferation and specification of LECs [70]. Other growth factors such as epidermal growth factor (EGF) [71], erythropoietin [72], angiopoietin [73, 74] and adrenomedullin [75] have also been shown to promote tumor lymphatic vessel growth.

Insights from Zebrafish Developmental Lymphangiogenesis

To date, zebrafish have been primarily used to study developmental lymphangiogenesis, with little focus on tumor lymphatics. Nevertheless, by contributing to our understanding of how lymphatic vessels grow normally, zebrafish research has indirectly aided our understanding of tumor lymphangiogenesis. In the future, more lymphatic-focused genetic and chemical zebrafish-based screens may help to reveal novel therapeutic strategies to prevent lymphatic-mediated metastasis.

Genetic Screens

A zebrafish forward genetic screen has already led to the identification of a number of novel mammalian lymphatic genes, including *Ccbe1*, a modulator of VEGFR3 signalling that could be a new target with which to inhibit tumor lymphangiogenesis [34]. There is the potential for further genetic screens to isolate other genes and pathways involved in lymphatic vessel growth and remodelling that could be potentially targeted in tumorigenesis.

Drug Screens

As lymphangiogenesis does not normally occur in adult tissues, it has been proposed that targeting tumor lymphangiogenesis in patients with solid tumors should be well tolerated. In support of this, loss of VEGFR3 signalling in adult mice does not appear to affect normal lymphatic vessel function [76]. Current strategies to block tumor lymphangiogenesis are focused around the VEGFR3 pathway with various neutralising antibodies used against VEGFR3, VEGFC and VEGFD as well as soluble dominant negative forms of VEGFR3. Many of these have shown promise in preclinical studies and have progressed to Phase I clinical trials [57]. In addition, a number of small molecule inhibitors of VEGFR signalling have been developed, primarily against VEGFR2, to prevent neo-angiogenesis. Many of these inhibitors also display activity against VEGFR3 and could therefore be used as anti-lymphatic agents. However, while many of these approaches have shown promise in pre-clinical trials there is increasing evidence that blockade of the VEGFR3 pathway is not able to prevent all tumor lymphangiogenesis and metastasis and there is a need to identify compounds that can act outside of this pathway [57].

Zebrafish present a well established biological system for conducting small molecule screens. As an in vivo model, zebrafish have clear advantages to existing cell-based lymphatic sprouting assays [77] and there is an opportunity to develop the lymphatic zebrafish model as a platform to identify and develop novel anti-lymphatic compounds [78]. We recently conducted the first zebrafish-based anti-lymphatic drug screen in which the Prestwick Chemical library, containing 1120 mainly FDA-approved compounds, was assessed for the ability to inhibit developmental lymphangiogenesis in zebrafish embryos [79]. Three classes of anti-lymphatic compounds were isolated from this screen which include kaempferol, a natural product found in plants, leflunomide, an inhibitor of pyrimidine biosynthesis, and cinnarizine and flunarizine, members of the type IV class of calcium channel antagonists. Of note, a murine in vivo lymphangiogenesis Matrigel plug assay showed that all three compounds were able to inhibit lymphatic vessel growth in mammals, showing that zebrafish are a viable platform for the identification and development of mammalian anti-lymphatic compounds.

One advantage of using zebrafish in chemical screens is that analysis of the lymphatic defect induced by the drug can help determine which pathway it is targeting. For example, kaempferol was found to completely inhibit secondary sprouting in zebrafish embryos, a phenotype identical to when the Vegfr3/Vegfc/Ccbe1 pathway is inhibited, suggesting it could be a novel inhibitor of this pathway. Further analysis showed that kaempferol was able to inhibit human VEGFR2 and VEGFR3 with an IC_{50} in the same range as the doses used in zebrafish embryos. Kaempferol was also able to reduce the density of tumor-associated lymphatic vessels as well as the incidence of lymph node metastases in a metastatic breast cancer xenograft model. Clearly there is an opportunity for more chemical screens to be conducted either with other chemical libraries to identify more compounds with anti-lymphatic activity or as a tool to develop and refine existing anti-vascular compounds.

Zebrafish Genetic Cancer Models

While the process of tumor lymphangiogenesis is clearly an important step in the metastatic spread of many types of cancer, our understanding of this process is currently limited to analysis of fixed tissue sections from mouse and human tumors. While transgenic lymphatic reporter lines have been developed in mice [80], due to the technical difficulty of long-term imaging mouse tumors, there is very little in vivo data on the process of tumor lymphangiogenesis. This presents a knowledge gap which zebrafish lymphatic models may be able to address. One approach would be to combine a lymphatic reporter line with an existing zebrafish transgenic model of cancer. Perhaps the area with best potential is melanoma, as human melanoma is well known to metastasize through the lymphatic system, with tumoral lymphatic density prognostic for lymph node metastasis and overall survival [50, 81–83]. Genetic models of melanoma exist in the fish, which use either the $BRAF^{V600E}$ [84], $NRAS^{Q61K}$ [85] or $HRAS^{G12V}$ [86, 87] oncogenes driven by a melanocyte specific promoter. When the $BRAF^{V600E}$ or $NRAS^{Q61K}$ lines were crossed with a *p53* mutant, both oncogenes were able to induce melanoma by early adulthood. Importantly these zebrafish tumors were not only histologically similar to their human counterparts but they also display considerable overlap in their gene expression profile to human melanomas [85].

By crossing a lymphatic reporter (see Table 1) into one of these melanoma models it may be possible to observe tumoral lymphatic growth in a zebrafish tumor. One difficulty with this model is that many melanoma tumors are highly pigmented which may make visualisation of the tumor vasculature difficult. Another consideration is that with all of these melanoma models, distant metastases were not observed, with local invasion the predominant form of cancer cell spread. This observation, coupled with the lack of lymph nodes in adult fish means that zebrafish are probably only suitable for modelling the effects of the primary tumor on the lymphatic vasculature.

Outside of melanoma, the human cancers with the best association with tumoral lymphatics and metastasis are lung, breast and colon cancer; obviously breast and lung cancer are difficult to model in fish and as yet a colorectal cancer model has yet to be developed. However there are a number of pancreatic [88, 89], neuroblastoma [90] and liver cancer [91, 92] models in fish which may be useful to observe tumoral lymphatic growth, as advanced human forms of these cancers are known to spread to lymph nodes, suggesting a lymphatic component to their invasion.

Tumor Transplantation

Zebrafish have increasingly been utilised as a host for xenografting mammalian cancer cells in order to understand the processes of tumor angiogenesis [93, 94] tumor cell extravasation [95, 96] and migration, and metastasis [97]. One advantage of tumor transplantation techniques over genetic models is the high penetrance and rapid progression of tumor formation. In addition, the embryonic grafts enable high-resolution analysis of tumor-endothelial cell interactions that are currently not possible in mouse xenografts and are technically challenging in adult fish models. This technique has the potential to be extended into the field of tumor lymphangiogenesis and could provide the ability to visualise the remodelling of host lymphatics to the tumor.

The majority of xenograft studies have been undertaken in zebrafish embryos, which do not have a fully intact immune system, allowing the transplantation of mammalian cells. Xenograft experiments have also been conducted in juvenile fish using dexamethasone as an immunosuppressant [96]. Techniques for allografting zebrafish tumors back into host animals have been developed for many zebrafish cancers including melanoma [84–86, 98]. The advantage of these tumors is that they are able to grow in their preferred temperature as opposed to mammalian xenografts which must be grown in sub-optimal temperatures, as zebrafish embryos are not viable at 37°. In addition, two recent advances allow the generation of zebrafish transplanted tumors without the need for immunosuppression; the first is the development of isogenic zebrafish lines which allow the allografting of clonal cells into syngenic hosts [99–101], the second is the creation of an immunocompromised *rag2* mutant that can tolerate the long term engraftment of tumor cells isolated from genetically diverse zebrafish [102].

We have shown that injection of human breast cancer cells (MDA-MB-231) into the perivitelline space of 2 dpf embryos was able to induce ectopic *lyve1*-expressing vessels into the xenograft (Fig. 2), suggesting that at least the process of lymphatic sprouting into tumors could be modelled in this assay [30]. The majority of these vessels appear to sprout from existing veins including the CCV and sub intestinal vein (SIV). These vessels were responsive to Vegfc and Vegfd, with xenografts expressing either of these pro-lymphatic ligands, displaying a higher degree of *lyve1*-positive vascularisation than xenografts expressing a vector control. However these neo-vessels also express the blood vessel marker *kdrl* and it remains to be determined whether these vessels are venous blood vessels or immature lymphatic vessels

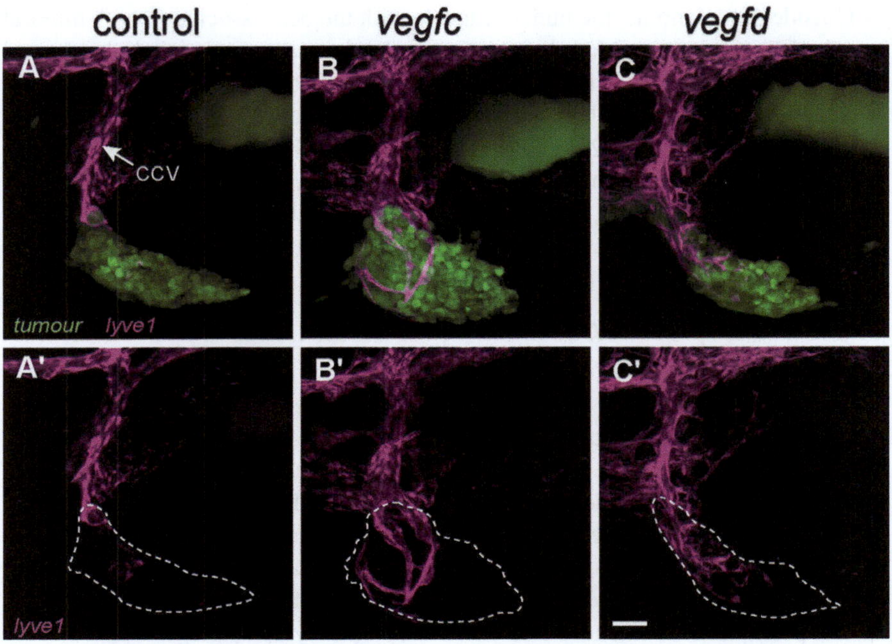

Fig. 2 Human breast cancer xenografts expressing *vegfc* or *vegfd* can induce ectopic *lyve1*-positive vessels in zebrafish embryos. (**A–C**) Confocal images of human breast cancer cells (MDA-MD-231) labelled with cell tracker green xenografted into *lyve1:dsRed* embryos and imaged at 5 dpf. (**A**) MDA-MD-231 cells transfected with vector only (**B**) MDA-MD-231 cells transfected with zebrafish *vegfc* (**C**) MDA-MD-231 cells transfected with zebrafish *vegfd.* (**A′–C′**) Confocal images of the *lyve1:dsred* vessels only. Cancer cells in *green*, *lyve1* in *magenta. Dotted lines* represent the boundary of the xenograft. The xenografts expressing either *vegfc* or *vegfd* are able to induce more ectopic *lyve1*-positive sprouts from the CCV than the control xenografts. *CCV* common cardinal vein. Image reproduced from Astin et al., (2014) Development 141: 2680–2690

that have residual *kdrl* expression following their migration from the vein (Fig. 2). While this assay shows promise as a model of tumor lymphangiogenesis, it is limited in that mammalian tumor grafts only persist in embryos for 3–4 days which may insufficient time for the formation of *bone fide* tumor lymphatic vessels. Because of this, other techniques such as the transplantation of cancer cells into immunocompromised fish, or conducting allograft studies using fish cancer cells, may be better suited for longer-term engraftment and the modelling of tumor lymphangiogenesis.

Future Directions

Zebrafish are increasingly becoming a valuable model of developmental lymphangiogenesis. The ability to live image lymphatic vessel growth makes zebrafish an important tool in identifying the molecular pathways that drive lymphatic vessel development. There are still many areas of lymphangiogenesis that remain unclear, such as how lymphatic precursors are specified in the fish, how lymphangioblasts remodel to form an entire network and what guides lymphatic vessel growth in different tissues. Importantly this research may lead to the identification of novel therapeutic targets of tumor lymphangiogenesis.

Currently zebrafish are underutilised as a model of tumor lymphangiogenesis. However, there is potential in combining existing transgenic and tumor transplantation models with lymphatic reporter lines, which could provide valuable insights into the process of tumor-induced lymphangiogenesis. Proof-of-principle studies have shown that zebrafish are a viable model for the identification of novel mammalian anti-lymphatic compounds and there is opportunity for more refined zebrafish drug screens to be used to identify and develop more compounds to help prevent tumor lymphangiogenesis and metastasis.

Acknowledgements We thank Ben Hogan and Kazuhide Okuda for advice on lymphatic reporter lines and Kathryn Crosier for helpful comments on the manuscript.

Funding
The authors were supported by a Ministry of Business, Innovation and Employment grant [UOAX0813] awarded P.S.C, a Health Research Council of New Zealand project grant (14/105) awarded to P.S.C and J.W.A and an Auckland Medical Research Foundation project grant awarded to J.W.A.

References

1. Tammela T, Alitalo K (2010) Lymphangiogenesis: molecular mechanisms and future promise. Cell 140:460–476
2. Sabin FR (1902) On the origin of the lymphatic system from the veins and the development of lymph hearts and the thoracic duct in the pig. Am J Anat 1:367–389
3. Srinivasan RS, Dillard ME, Lagutin OV, Lin FJ, Tsai S et al (2007) Lineage tracing demonstrates the venous origin of the mammalian lymphatic vasculature. Genes Dev 21:2422–2432
4. Koltowska K, Betterman KL, Harvey NL, Hogan BM (2013) Getting out and about: the emergence and morphogenesis of the vertebrate lymphatic vasculature. Development 140:1857–1870
5. Hewson W, Hunter W (1769) An account of the lymphatic system in fish. By the same. Philos Trans (1683–1775) 59:204–215
6. Kuchler AM, Gjini E, Peterson-Maduro J, Cancilla B, Wolburg H et al (2006) Development of the zebrafish lymphatic system requires VEGFC signaling. Curr Biol 16:1244–1248
7. Lawson ND, Weinstein BM (2002) In vivo imaging of embryonic vascular development using transgenic zebrafish. Dev Biol 248:307–318
8. Yaniv K, Isogai S, Castranova D, Dye L, Hitomi J et al (2006) Live imaging of lymphatic development in the zebrafish. Nat Med 12:711–716
9. Okuda KS, Astin JW, Misa JP, Flores MV, Crosier KE et al (2012) lyve1 expression reveals novel lymphatic vessels and new mechanisms for lymphatic vessel development in zebrafish. Development 139:2381–2391
10. Bussmann J, Bos FL, Urasaki A, Kawakami K, Duckers HJ et al (2010) Arteries provide essential guidance cues for lymphatic endothelial cells in the zebrafish trunk. Development 137:2653–2657

11. van Impel A, Zhao Z, Hermkens DM, Roukens MG, Fischer JC et al (2014) Divergence of zebrafish and mouse lymphatic cell fate specification pathways. Development 141:1228–1238
12. Kampmeier OT (1969) Evolution and comparative morphology of the lymphatic system. Thomas, London
13. Coffindaffer-Wilson M, Craig MP, Hove JR (2011) Determination of lymphatic vascular identity and developmental timecourse in zebrafish (Danio rerio). Lymphology 44:1–12
14. Johnson NC, Dillard ME, Baluk P, McDonald DM, Harvey NL et al (2008) Lymphatic endothelial cell identity is reversible and its maintenance requires Prox1 activity. Genes Dev 22:3282–3291
15. Wigle JT, Harvey N, Detmar M, Lagutina I, Grosveld G et al (2002) An essential role for Prox1 in the induction of the lymphatic endothelial cell phenotype. EMBO J 21:1505–1513
16. Wigle JT, Oliver G (1999) Prox1 function is required for the development of the murine lymphatic system. Cell 98:769–778
17. Srinivasan RS, Geng X, Yang Y, Wang Y, Mukatira S et al (2010) The nuclear hormone receptor Coup-TFII is required for the initiation and early maintenance of Prox1 expression in lymphatic endothelial cells. Genes Dev 24:696–707
18. Duong T, Koltowska K, Pichol-Thievend C, Le Guen L, Fontaine F et al (2013) VEGFD regulates blood vascular development by modulating SOX18 activity. Blood 123:1102–1112
19. Pendeville H, Winandy M, Manfroid I, Nivelles O, Motte P et al (2008) Zebrafish Sox7 and Sox18 function together to control arterial-venous identity. Dev Biol 317:405–416
20. Del Giacco L, Pistocchi A, Ghilardi A (2010) prox1b Activity is essential in zebrafish lymphangiogenesis. PLoS One 5:e13170
21. Coxam B, Sabine A, Bower NI, Smith KA, Pichol-Thievend C et al (2014) Pkd1 regulates lymphatic vascular morphogenesis during development. Cell Rep 7:623–633
22. Dunworth WP, Cardona-Costa J, Bozkulak EC, Kim JD, Meadows S et al (2014) Bone morphogenetic protein 2 signaling negatively modulates lymphatic development in vertebrate embryos. Circ Res 114:56–66
23. Aranguren XL, Beerens M, Vandevelde W, Dewerchin M, Carmeliet P et al (2011) Transcription factor COUP-TFII is indispensable for venous and lymphatic development in zebrafish and Xenopus laevis. Biochem Biophys Res Commun 410:121–126
24. Swift MR, Pham VN, Castranova D, Bell K, Poole RJ et al (2014) SoxF factors and Notch regulate nr2f2 gene expression during venous differentiation in zebrafish. Dev Biol 390:116–125
25. Geudens I, Herpers R, Hermans K, Segura I, Ruiz de Almodovar C et al (2010) Role of delta-like-4/Notch in the formation and wiring of the lymphatic network in zebrafish. Arterioscler Thromb Vasc Biol 30:1695–1702
26. Fatima A, Culver A, Culver F, Liu T, Dietz WH et al (2014) Murine Notch1 is required for lymphatic vascular morphogenesis during development. Dev Dyn 243:957–964
27. Murtomaki A, Uh MK, Choi YK, Kitajewski C, Borisenko V et al (2013) Notch1 functions as a negative regulator of lymphatic endothelial cell differentiation in the venous endothelium. Development 140:2365–2376
28. Srinivasan RS, Escobedo N, Yang Y, Interiano A, Dillard ME et al (2014) The Prox1-Vegfr3 feedback loop maintains the identity and the number of lymphatic endothelial cell progenitors. Genes Dev 28:2175–2187
29. Hogan BM, Herpers R, Witte M, Helotera H, Alitalo K et al (2009) Vegfc/Flt4 signalling is suppressed by Dll4 in developing zebrafish intersegmental arteries. Development 136:4001–4009
30. Astin JW, Haggerty MJ, Okuda KS, Le Guen L, Misa JP et al (2014) Vegfd can compensate for loss of Vegfc in zebrafish facial lymphatic sprouting. Development 141:2680–2690
31. Gordon K, Schulte D, Brice G, Simpson MA, Roukens MG et al (2013) Mutation in vascular endothelial growth factor-C, a ligand for vascular endothelial growth factor receptor-3, is associated with autosomal dominant milroy-like primary lymphedema. Circ Res 112:956–960
32. Ober EA, Olofsson B, Makinen T, Jin SW, Shoji W et al (2004) Vegfc is required for vascular development and endoderm morphogenesis in zebrafish. EMBO Rep 5:78–84

33. Villefranc JA, Nicoli S, Bentley K, Jeltsch M, Zarkada G et al (2013) A truncation allele in vascular endothelial growth factor c reveals distinct modes of signaling during lymphatic and vascular development. Development 140:1497–1506
34. Hogan BM, Bos FL, Bussmann J, Witte M, Chi NC et al (2009) Ccbe1 is required for embryonic lymphangiogenesis and venous sprouting. Nat Genet 41:396–398
35. Alders M, Hogan BM, Gjini E, Salehi F, Al-Gazali L et al (2009) Mutations in CCBE1 cause generalized lymph vessel dysplasia in humans. Nat Genet 41:1272–1274
36. Le Guen L, Karpanen T, Schulte D, Harris NC, Koltowska K et al (2014) Ccbe1 regulates Vegfc-mediated induction of Vegfr3 signaling during embryonic lymphangiogenesis. Development 141:1239–1249
37. Jeltsch M, Jha SK, Tvorogov D, Anisimov A, Leppanen VM et al (2014) CCBE1 Enhances lymphangiogenesis via ADAMTS3-mediated VEGF-C activation. Circulation 129:1962–1971
38. Fevurly RD, Hasso S, Fye A, Fishman SJ, Chan J (2012) Novel zebrafish model reveals a critical role for MAPK in lymphangiogenesis. J Pediatr Surg 47:177–182
39. Flores MV, Hall CJ, Crosier KE, Crosier PS (2010) Visualization of embryonic lymphangiogenesis advances the use of the zebrafish model for research in cancer and lymphatic pathologies. Dev Dyn 239:2128–2135
40. Kartopawiro J, Bower NI, Karnezis T, Kazenwadel J, Betterman KL et al (2014) Arap3 is dysregulated in a mouse model of hypotrichosis-lymphedema-telangiectasia and regulates lymphatic vascular development. Hum Mol Genet 23:1286–1297
41. Outeda P, Huso DL, Fisher SA, Halushka MK, Kim H et al (2014) Polycystin signaling is required for directed endothelial cell migration and lymphatic development. Cell Rep 7:634–644
42. Alitalo K, Tammela T, Petrova TV (2005) Lymphangiogenesis in development and human disease. Nature 438:946–953
43. Clark ER, Clark EL (1937) Observations on living mammalian lymphatic capillaries–their relation to the blood vessels. Am J Anat 60:253–298
44. Cha YR, Fujita M, Butler M, Isogai S, Kochhan E et al (2012) Chemokine signaling directs trunk lymphatic network formation along the preexisting blood vasculature. Dev Cell 22:824–836
45. Lim AH, Suli A, Yaniv K, Weinstein B, Li DY et al (2011) Motoneurons are essential for vascular pathfinding. Development 138:3847–3857
46. Navankasattusas S, Whitehead KJ, Suli A, Sorensen LK, Lim AH et al (2008) The netrin receptor UNC5B promotes angiogenesis in specific vascular beds. Development 135:659–667
47. Sundlisaeter E, Dicko A, Sakariassen PO, Sondenaa K, Enger PO et al (2007) Lymphangiogenesis in colorectal cancer—prognostic and therapeutic aspects. Int J Cancer 121:1401–1409
48. Nathanson SD (2003) Insights into the mechanisms of lymph node metastasis. Cancer 98:413–423
49. Alitalo A, Detmar M (2012) Interaction of tumor cells and lymphatic vessels in cancer progression. Oncogene 31:4499–4508
50. Dadras SS, Paul T, Bertoncini J, Brown LF, Muzikansky A et al (2003) Tumor lymphangiogenesis: a novel prognostic indicator for cutaneous melanoma metastasis and survival. Am J Pathol 162:1951–1960
51. Hoshida T, Isaka N, Hagendoorn J, di Tomaso E, Chen YL et al (2006) Imaging steps of lymphatic metastasis reveals that vascular endothelial growth factor-C increases metastasis by increasing delivery of cancer cells to lymph nodes: therapeutic implications. Cancer Res 66:8065–8075
52. Karnezis T, Shayan R, Caesar C, Roufail S, Harris NC et al (2012) VEGF-D promotes tumor metastasis by regulating prostaglandins produced by the collecting lymphatic endothelium. Cancer Cell 21:181–195
53. Hirakawa S, Brown LF, Kodama S, Paavonen K, Alitalo K et al (2007) VEGF-C-induced lymphangiogenesis in sentinel lymph nodes promotes tumor metastasis to distant sites. Blood 109:1010–1017

54. Hirakawa S, Kodama S, Kunstfeld R, Kajiya K, Brown LF et al (2005) VEGF-A induces tumor and sentinel lymph node lymphangiogenesis and promotes lymphatic metastasis. J Exp Med 201:1089–1099
55. Fukumura D, Duda DG, Munn LL, Jain RK (2010) Tumor microvasculature and microenvironment: novel insights through intravital imaging in pre-clinical models. Microcirculation 17:206–225
56. Padera TP, Kadambi A, di Tomaso E, Carreira CM, Brown EB et al (2002) Lymphatic metastasis in the absence of functional intratumor lymphatics. Science 296:1883–1886
57. Stacker SA, Williams SP, Karnezis T, Shayan R, Fox SB et al (2014) Lymphangiogenesis and lymphatic vessel remodelling in cancer. Nat Rev Cancer 14:159–172
58. Karpanen T, Egeblad M, Karkkainen MJ, Kubo H, Yla-Herttuala S et al (2001) Vascular endothelial growth factor C promotes tumor lymphangiogenesis and intralymphatic tumor growth. Cancer Res 61:1786–1790
59. He Y, Kozaki K, Karpanen T, Koshikawa K, Yla-Herttuala S et al (2002) Suppression of tumor lymphangiogenesis and lymph node metastasis by blocking vascular endothelial growth factor receptor 3 signaling. J Natl Cancer Inst 94:819–825
60. Lin J, Lalani AS, Harding TC, Gonzalez M, Wu WW et al (2005) Inhibition of lymphogenous metastasis using adeno-associated virus-mediated gene transfer of a soluble VEGFR-3 decoy receptor. Cancer Res 65:6901–6909
61. Pytowski B, Goldman J, Persaud K, Wu Y, Witte L et al (2005) Complete and specific inhibition of adult lymphatic regeneration by a novel VEGFR-3 neutralizing antibody. J Natl Cancer Inst 97:14–21
62. Roberts N, Kloos B, Cassella M, Podgrabinska S, Persaud K et al (2006) Inhibition of VEGFR-3 activation with the antagonistic antibody more potently suppresses lymph node and distant metastases than inactivation of VEGFR-2. Cancer Res 66:2650–2657
63. Debinski W, Slagle-Webb B, Achen MG, Stacker SA, Tulchinsky E et al (2001) VEGF-D is an X-linked/AP-1 regulated putative onco-angiogen in human glioblastoma multiforme. Mol Med 7:598–608
64. Onogawa S, Kitadai Y, Tanaka S, Kuwai T, Kimura S et al (2004) Expression of VEGF-C and VEGF-D at the invasive edge correlates with lymph node metastasis and prognosis of patients with colorectal carcinoma. Cancer Sci 95:32–39
65. Schietroma C, Cianfarani F, Lacal PM, Odorisio T, Orecchia A et al (2003) Vascular endothelial growth factor-C expression correlates with lymph node localization of human melanoma metastases. Cancer 98:789–797
66. Schoppmann SF, Birner P, Stockl J, Kalt R, Ullrich R et al (2002) Tumor-associated macrophages express lymphatic endothelial growth factors and are related to peritumoral lymphangiogenesis. Am J Pathol 161:947–956
67. He Y, Rajantie I, Pajusola K, Jeltsch M, Holopainen T et al (2005) Vascular endothelial cell growth factor receptor 3-mediated activation of lymphatic endothelium is crucial for tumor cell entry and spread via lymphatic vessels. Cancer Res 65:4739–4746
68. Farnsworth RH, Karnezis T, Shayan R, Matsumoto M, Nowell CJ et al (2011) A role for bone morphogenetic protein-4 in lymph node vascular remodeling and primary tumor growth. Cancer Res 71:6547–6557
69. Cao R, Ji H, Feng N, Zhang Y, Yang X et al (2012) Collaborative interplay between FGF-2 and VEGF-C promotes lymphangiogenesis and metastasis. Proc Natl Acad Sci U S A 109:15894–15899
70. Cao R, Bjorndahl MA, Religa P, Clasper S, Garvin S et al (2004) PDGF-BB induces intratumoral lymphangiogenesis and promotes lymphatic metastasis. Cancer Cell 6:333–345
71. Bracher A, Cardona AS, Tauber S, Fink AM, Steiner A et al (2013) Epidermal growth factor facilitates melanoma lymph node metastasis by influencing tumor lymphangiogenesis. J Invest Dermatol 133:230–238
72. Lee AS, Kim DH, Lee JE, Jung YJ, Kang KP et al (2011) Erythropoietin induces lymph node lymphangiogenesis and lymph node tumor metastasis. Cancer Res 71:4506–4517

73. Fagiani E, Lorentz P, Kopfstein L, Christofori G (2011) Angiopoietin-1 and -2 exert antagonistic functions in tumor angiogenesis, yet both induce lymphangiogenesis. Cancer Res 71:5717–5727
74. Holopainen T, Saharinen P, D'Amico G, Lampinen A, Eklund L et al (2012) Effects of angiopoietin-2-blocking antibody on endothelial cell-cell junctions and lung metastasis. J Natl Cancer Inst 104:461–475
75. Karpinich NO, Kechele DO, Espenschied ST, Willcockson HH, Fedoriw Y et al (2013) Adrenomedullin gene dosage correlates with tumor and lymph node lymphangiogenesis. FASEB J 27:590–600
76. Karpanen T, Wirzenius M, Makinen T, Veikkola T, Haisma HJ et al (2006) Lymphangiogenic growth factor responsiveness is modulated by postnatal lymphatic vessel maturation. Am J Pathol 169:708–718
77. Schulz MM, Reisen F, Zgraggen S, Fischer S, Yuen D et al (2012) Phenotype-based high-content chemical library screening identifies statins as inhibitors of in vivo lymphangiogenesis. Proc Natl Acad Sci U S A 109:E2665–E2674
78. Zon LI, Peterson RT (2005) In vivo drug discovery in the zebrafish. Nat Rev Drug Discov 4:35–44
79. Astin JW, Jamieson SM, Eng TC, Flores MV, Misa JP et al (2014) An in vivo antilymphatic screen in zebrafish identifies novel inhibitors of Mammalian lymphangiogenesis and lymphatic-mediated metastasis. Mol Cancer Ther 13:2450–2462
80. Choi I, Chung HK, Ramu S, Lee HN, Kim KE et al (2011) Visualization of lymphatic vessels by Prox1-promoter directed GFP reporter in a bacterial artificial chromosome-based transgenic mouse. Blood 117:362–365
81. Balch CM, Gershenwald JE, Soong SJ, Thompson JF, Atkins MB et al (2009) Final version of 2009 AJCC melanoma staging and classification. J Clin Oncol 27:6199–6206
82. Rinderknecht M, Detmar M (2008) Tumor lymphangiogenesis and melanoma metastasis. J Cell Physiol 216:347–354
83. Shayan R, Karnezis T, Murali R, Wilmott JS, Ashton MW et al (2012) Lymphatic vessel density in primary melanomas predicts sentinel lymph node status and risk of metastasis. Histopathology 61:702–710
84. Patton EE, Widlund HR, Kutok JL, Kopani KR, Amatruda JF et al (2005) BRAF mutations are sufficient to promote nevi formation and cooperate with p53 in the genesis of melanoma. Curr Biol 15:249–254
85. Dovey M, White RM, Zon LI (2009) Oncogenic NRAS cooperates with p53 loss to generate melanoma in zebrafish. Zebrafish 6:397–404
86. Santoriello C, Gennaro E, Anelli V, Distel M, Kelly A et al (2010) Kita driven expression of oncogenic HRAS leads to early onset and highly penetrant melanoma in zebrafish. PLoS One 5, e15170
87. Anelli V, Santoriello C, Distel M, Koster RW, Ciccarelli FD et al (2009) Global repression of cancer gene expression in a zebrafish model of melanoma is linked to epigenetic regulation. Zebrafish 6:417–424
88. Park SW, Davison JM, Rhee J, Hruban RH, Maitra A et al (2008) Oncogenic KRAS induces progenitor cell expansion and malignant transformation in zebrafish exocrine pancreas. Gastroenterology 134:2080–2090
89. Liu S, Leach SD (2011) Screening pancreatic oncogenes in zebrafish using the Gal4/UAS system. Methods Cell Biol 105:367–381
90. Zhu S, Lee JS, Guo F, Shin J, Perez-Atayde AR et al (2012) Activated ALK collaborates with MYCN in neuroblastoma pathogenesis. Cancer Cell 21:362–373
91. Li Z, Huang X, Zhan H, Zeng Z, Li C et al (2012) Inducible and repressable oncogene-addicted hepatocellular carcinoma in Tet-on xmrk transgenic zebrafish. J Hepatol 56:419–425
92. Nguyen AT, Emelyanov A, Koh CH, Spitsbergen JM, Parinov S et al (2012) An inducible kras(V12) transgenic zebrafish model for liver tumorigenesis and chemical drug screening. Dis Model Mech 5:63–72

93. Haldi M, Ton C, Seng WL, McGrath P (2006) Human melanoma cells transplanted into zebrafish proliferate, migrate, produce melanin, form masses and stimulate angiogenesis in zebrafish. Angiogenesis 9:139–151
94. Nicoli S, Ribatti D, Cotelli F, Presta M (2007) Mammalian tumor xenografts induce neovascularization in zebrafish embryos. Cancer Res 67:2927–2931
95. Stoletov K, Kato H, Zardouzian E, Kelber J, Yang J et al (2010) Visualizing extravasation dynamics of metastatic tumor cells. J Cell Sci 123:2332–2341
96. Stoletov K, Montel V, Lester RD, Gonias SL, Klemke R (2007) High-resolution imaging of the dynamic tumor cell vascular interface in transparent zebrafish. Proc Natl Acad Sci U S A 104:17406–17411
97. Marques IJ, Weiss FU, Vlecken DH, Nitsche C, Bakkers J et al (2009) Metastatic behaviour of primary human tumours in a zebrafish xenotransplantation model. BMC Cancer 9:128
98. Langenau DM, Traver D, Ferrando AA, Kutok JL, Aster JC et al (2003) Myc-induced T cell leukemia in transgenic zebrafish. Science 299:887–890
99. Mizgireuv IV, Revskoy SY (2006) Transplantable tumor lines generated in clonal zebrafish. Cancer Res 66:3120–3125
100. Mizgirev I, Revskoy S (2010) Generation of clonal zebrafish lines and transplantable hepatic tumors. Nat Protoc 5:383–394
101. Smith AC, Raimondi AR, Salthouse CD, Ignatius MS, Blackburn JS et al (2010) High-throughput cell transplantation establishes that tumor-initiating cells are abundant in zebrafish T-cell acute lymphoblastic leukemia. Blood 115:3296–3303
102. Tang Q, Abdelfattah NS, Blackburn JS, Moore JC, Martinez SA et al (2014) Optimized cell transplantation using adult rag2 mutant zebrafish. Nat Methods 11:821–824

In Vivo Imaging of Cancer in Zebrafish

Myron S. Ignatius, Madeline Hayes, and David M. Langenau

Abstract Zebrafish cancer models have greatly advanced our understanding of malignancy in humans. This is made possible due to the unique advantages of the zebrafish model including ex vivo development and large clutch sizes, which enable large-scale genetic and chemical screens. Transparency of the embryo and the creation of adult zebrafish devoid of pigmentation (*casper*) have permitted unprecedented ability to dynamically visualize cancer progression in live animals. When coupled with fluorescent reporters and transgenic approaches that drive oncogenesis, it is now possible to label entire or subpopulations of cancer cells and follow cancer growth in near real-time. Here, we will highlight aspects of in vivo imaging using the zebrafish and how it has enhanced our understanding of the fundamental aspects of tumor initiation, self-renewal, neovascularization, tumor cell heterogeneity, invasion and metastasis. Importantly, we will highlight the contribution of cancer imaging in zebrafish for drug discovery.

Keywords Zebrafish • Cancer • Imaging • Self-renewal • Cancer stem cells

Introduction

Over a decade ago, preeminent cancer researchers Dr. Douglas Hanahan and Dr. Robert Weinberg proposed a set of features termed "The Hallmarks of Cancer" that drive the progression of normal cells to malignant transformation [1]. This originally included six traits, including limitless replicative potential, sustained angiogenesis, tissue invasion and metastasis, apoptosis evasion, and growth factor independence. Subsequently, it became apparent that malignant transformation enables additional capabilities including the ability to evade the immune system and

M.S. Ignatius • M. Hayes • D.M. Langenau (✉)
Molecular Pathology Unit, Department of Pathology, Massachusetts General Hospital, 149 13th Street, #6012, Charlestown, MA 02129, USA

Center of Cancer Research, Massachusetts General Hospital, Charlestown, MA 02129, USA

Harvard Stem Cell Institute, Boston, MA 02129, USA
e-mail: mignatius@mgh.harvard.edu; mnhayes@mgh.harvard.edu; dlangenau@mgh.harvard.edu

D.M. Langenau (ed.), *Cancer and Zebrafish*, Advances in Experimental Medicine and Biology 910, DOI 10.1007/978-3-319-30654-4_10

the deregulation of metabolic pathways to promote tumor growth [2]. Cancer imaging in live animals has provided a unique window to directly visualize many of these aspects of tumor biology. Although mouse models of cancer remain the mainstay for modeling human tumors, they are not optimal for in vivo imaging mainly due the fact that special imaging windows are required to visualize cancers through the opaque skin of the mouse.

Imaging individual tumor cells in transgenic zebrafish models has significantly enhanced our understanding of cancer biology. A number of unique characteristics make zebrafish particularly suitable for in vivo imaging. Embryonic, larval and *casper* adult zebrafish are transparent facilitating imaging at single cell resolution [3, 4]. Large clutch sizes and substantially lower husbandry costs provide a platform for chemical-genetic screening of primary tumors and xenograft transplantation into zebrafish embryos [5, 6]. Moreover, when coupled with genome editing using TALEN and CRISPR nuclease technologies [7, 8], the mutational and genetic heterogeneity present in individual patients can now be modeled in live animals. Finally, imaging protocols are well established, for example confocal, spinning disc, and two-photon microscopy have been elegantly used to image developmental processes including cell movements in embryogenesis [9, 10], vascular development [11, 12], and the emergence of hematopoietic stem cells from the dorsal aorta [13–15]. Similar techniques are now being widely applied by zebrafish cancer researchers to directly observe cancer initiation and progression.

Importantly, zebrafish models of cancer closely mimic their human counterparts, making studies in zebrafish highly relevant. The utility of zebrafish in research was initially highlighted in 1996 by the publication of an entire edition of the journal Development, which systematically catalogued mutant phenotypes obtained from forward genetic screens [16]. Over the next decade, careful characterization of these mutants identified key genes regulating a large spectrum of biological process, most of which are conserved in humans. Similar mutagenesis approaches were also used to generate cancer in zebrafish and in the early 2000s the ability to rapidly generate transgenic zebrafish allowed oncogenes to be expressed in specific cell types along with fluorescent transgenes. In 2003, Langenau, Look and colleagues used a transgenic approach to express mouse c-Myc fused with GFP in lymphocyte precursors and generated a model of myc-induced T-cell acute lymphoblastic leukemia (T-ALL) [17]. Subsequently, transgenic oncogene expression and/or tumor suppressor inactivation have been widely used to generate a variety of cancer types in the zebrafish [24, 25, 26, 30, 31, 44, 58–64]. For example, p53-deficient fish develop malignant peripheral nerve sheath tumors (MPNSTs) while PTEN-inactivation leads to ocular tumors [18, 19].

A wide variety of zebrafish tumor models have been generated including embryonal rhabdomyosarcoma (ERMS), melanoma, liver cancer, malignant peripheral nerve sheath tumors (MPNSTs), Ewing's Sarcoma, T-cell acute lymphoblastic leukemia (T-ALL), neuroblastoma, pancreatic cancer and chordoma. Transplantation of fluorescent labeled human and zebrafish tumors into recipient zebrafish have been performed, followed by imaging of transplanted recipients for a wide variety of cancer-associated phenotypes. The generation of both clonal and immune-compromised zebrafish that can accept transplanted tumor cells has greatly enhanced the ability to study tumor biology in vivo [20, 21].

In the following sections, we will discuss the significance of imaging zebrafish tumors to assess tumor initiation, self-renewal, neovascularization, invasion, metastasis, interaction with the microenvironment, and tumor cell heterogeneity. Importantly, we will highlight how these approaches have formed the basis of drug screens to identify compounds that reduce tumor incidence and growth.

Imaging Tumor Initiation

Zebrafish cancer models have been used extensively to study tumor initiation, with direct observation providing unique insight into the cell of origin, the tissue microenvironment, as well as the dynamics of tumor onset and growth. The first fluorescent zebrafish cancer was generated by expressing mouse *cMYC* fused to the green fluorescent protein (GFP) in developing lymphocytes using the *rag2* promoter [17]. Serial imaging of leukemic animals revealed that T-ALLs initiate in the thymus by 21 days of life, form thymic masses and expand locally into the gills, olfactory bulb, and structures in the head. Cells eventually infiltrate the kidney marrow and disseminate widely, causing morbidity by 90 days of life. Additional work went on to show that this model accurately recapitulates the molecular underpinnings of human T-ALL, mimicking a common and treatment resistant variant of human T-ALL that express SCL and LMO2. Importantly, this model has now been widely used to uncover important biology of human T-ALL, much of which has centered on fluorescent imaging platforms to assess leukemia growth in live animals [22, 23].

Since the initial publication of GFP-labeled zebrafish T-ALL, it has now become common practice to label cancer cells in zebrafish models using fluorescent transgenic approaches. For example, in a mifepristone-inducible transgenic model of RAS driven liver cancer, malignancy can be visualized by the co-expression of GFP. Activation of RAS results in hyperplasia after a week of induction and tumor formation after approximately 4 weeks [24]. Similarly, cancer initiation can be visualized in a RAS driven zebrafish model of pancreatic cancer where oncogenic *kRAS*^{G12V} is expressed by the *ptf1a* promoter [25]. In this model, oncogenic *kRAS*^{G12V} does not affect the specification of pancreatic cells but rather prevents cells from undergoing differentiation [25]. Live imaging of cancer has also been reported in zebrafish models of neuroblastoma, where the zebrafish MYCN oncogene is driven by the *dopamine-β-hydroxylase* promoter in sympathetic neuronal precursors [26]. Tumor initiation can be visualized in vivo when MYCN expression is coupled with a fluorescent GFP reporter. Tumors in this model arise at approximately 8 months of age and tumor initiation is first detected when MYCN-expressing sympathetic neuron precursors fail to undergo apoptosis. By tracking early stages of tumor initiation, researchers have discovered that mutant anaplastic lymphoma kinase (ALK) can accelerate time to tumor onset in this model [26], revealing the utility of imaging tumor onset for pathway discovery in cancer initiation. Similar approaches have also been used in zebrafish models of melanoma, malignant brain tumors, glioma, and chordoma [4, 27–30]. Thus, the zebrafish is particularly well suited for imaging tumor initiation at a macroscopic level, which has been further bolstered through use of optically-clear *casper* adult zebrafish [4].

It is also possible to directly visualize initiation and stages of tumor growth at single cell resolution. This perhaps has been best described using the transgenic zebrafish model of kRASG12D-driven embryonal rhabdomyosarcoma (ERMS) where the *rag2* promoter drives expression of activated human kRAS*G12D* (Fig. 1). The truncated 6.5 kb *rag2* promoter leads to transgene expression within the mononuclear compartment of developing muscle due to uncovering of a myoD binding site. Furthermore, it has been shown that co-injection of multiple linearized DNA plasmids into one-cell zebrafish embryos results in co-integration and co-expression of transgenes [22]. This technique enables the co-expression of oncogenes and/or fluorescent reporters to label ERMS cells based on their differentiation status [31–33]. To image initiation in ERMS, our group created tumors that express a combination of three fluorescent reporter proteins, labeling the tumor propagating cell component with myf5-GFP, the mid-differentiated tumor cells with myogenin-H2B-RFP, and the more terminally differentiated cells with mylz2-lyn-cyan [32] (Fig. 1). Live animal imaging revealed that early stage tumors consist of an expansion of single myf5-GFP+/myogenin-H2B-RFP-negative/mylz2-lyncyan-negative cells that occupy the normal muscle cell niche [32]. These cells proliferate and give rise to more differentiated cell types that eventually take over the myotome compartment [32]. We visualized ERMS cells invading the neighboring muscle, reorganizing the normal muscle architecture and replacing it with malignant cells [32]. Interestingly, we found that the ERMS cell subtypes are not uniformly distributed, but rather localized in distinct niches occupied by either tumor propagating cells or more differentiated cell types [32] (Fig. 1).

In the future, zebrafish models will commonly be used to follow the birth of cancer and the earliest events in tumor formation. Outstanding questions in the field about how pre-tumorigenic fields of cells acquire changes to become fully transformed will be imaged at single cell resolution greatly enhancing our understating of human cancer.

Imaging Neovascularization and Extravasation

The ability to recruit vasculature for tumor growth is one of the hallmarks of cancer formation and can be dynamically imaged in zebrafish. Zebrafish develop a structurally characteristic network of blood vessels by 2-days post-fertilization and several fluorescent transgenic lines are available that mark the vasculature including *fli1-EGFP* [11], *mtie2-GFP* [34], and *flk1-EGFP* [35]. The use of zebrafish to follow angiogenic recruitment and remodeling by tumor cells is well established. First studied in tumor xenograft studies performed in early embryos, Nicoli et al. demonstrated that fluorescent-labeled tumorigenic mouse aortic endothelial cells that overexpress FGF2 could engraft into 2-day-old transgenic *VEGFR2:G-RCFP* embryos. Remarkably, FGF2 stimulated new vessel growth which could be dynamically visualized over time [36]. Lee et al. also used imaging approaches to study the

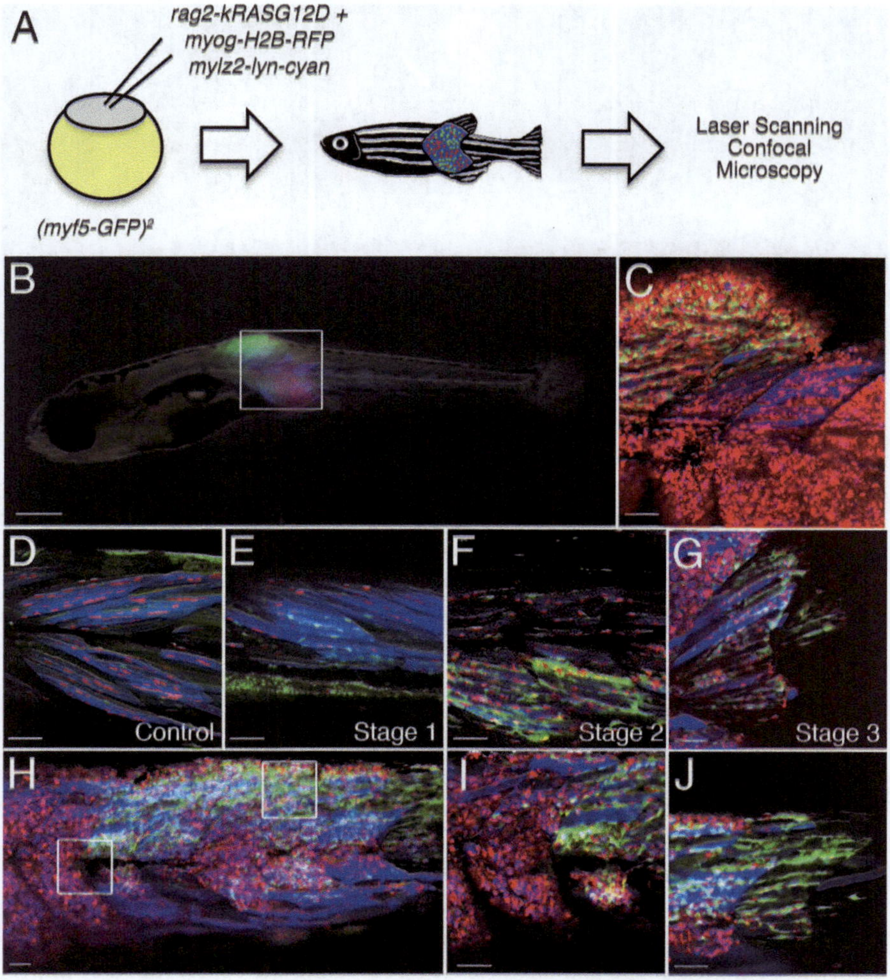

Fig. 1 Imaging tumor initiation in embryonal rhabdomyosarcoma. (**A**) Schematic of the experimental design showing the fluorescent color combination used for fluorescently labeling transgenic ERMS. (**B**) A 15 day old larval zebrafish that developed triple fluorescent-labeled ERMS. (**C**) Merged confocal image of the *boxed region* shown in B at 250X magnification showing heterogeneity at cellular magnifications. (**D**) Control *Tg(myf5-GFP)* transgenic animals injected with *myogenin-H2B-RFP* and *mylz2-lyn-cyan* have few *myf5-GFP+* satellite cells adjacent muscle fibers. (**E–G**) Representative images of ERMS-affected zebrafish labeled with *myf5-GFP, myogenin-H2B-RFP* and *mylz2-lyn-cyan* at tumor initiation (stage 1) and progression (stage 2 and 3). (**H**) Late stage ERMS from a triple fluorescent-labeled zebrafish imaged at 100x magnification. (**I, J**) *Boxed regions* in H imaged at higher magnification showing distinct tumor regions that contain either differentiated tumor cells or ERMS-propagating cells. Scale bars are 500 μm in (**B**); and 50 μm in (**C–J**) (Ignatius et. al., Cancer Cell 2012)

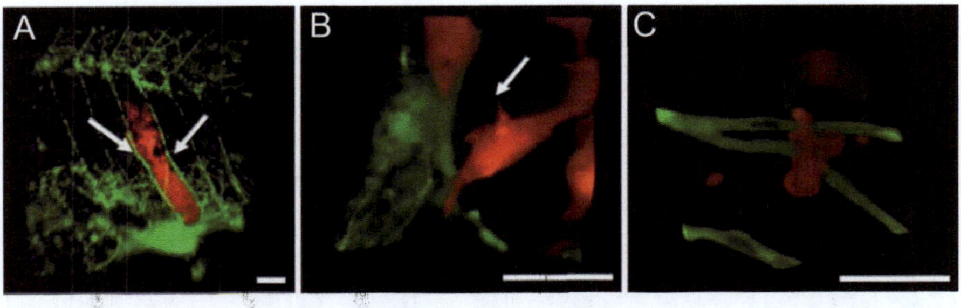

Fig. 2 Imaging neovascularization and extravasation in larval zebrafish. (**A**) Merged confocal reconstruction of engrafted human MDA-435 dsRed-labeled cells in the body wall between the intersegmental vessels (*arrows*) of *Tg(fli1:EGFP)* fish, 4 dpi. (**B**) Three-dimensional reconstruction of single invading MDA-435 cell. (**C**) Three-dimensional reconstruction of single invading HT1080 cell. Scale bars **A** = 200 μm, and **B**, **C** = 20 μm (Stoletov et al., PNAS 2007)

effects of hypoxia on neo-vascularization using murine T241 fibrosarcoma cells [37]. Tumor cells migrated from the primary site of transplantation only under hypoxic conditions, which was associated with increased vascularization adjacent engrafted human cancer cells [37]. This hypoxia-induced migration could be mimicked by overexpression of a hypoxia-regulated angiogenic factor, while inhibition of VEGF-receptor signaling abrogated the dissemination of tumor cells [37].

The Klemke laboratory has performed incredibly detailed analyses of neovascularization at single cell resolution to map the cellular behavior of metastatic human cancer cells. Using xenotransplantation of fluorescently labeled human cancer cells into 30-day-old dexamethasone-treated *fli1-EGFP* transgenic zebrafish [38–40], they showed that adenocarcinoma MDA-435 cells could induce vessel remodeling 4 days following transplantation [38] (Fig. 2). Remarkably, imaging showed that new vessels arose from preexisting vessel surfaces [38]. Importantly, MDA-435 cells engineered to secrete human VEGF displayed an increased propensity to induce vessel "sprouting" [38]. These remodeled vessels displayed unusual variability in both their shape and thickness [38], revealing the conservation of important angiogenic factors between zebrafish and mammals. In this study, MDA-435 cells were also engineered to stably over-express RhoC and time-lapse imaging revealed that RhoC-expressing cells were more invasive and displayed increased membrane protrusions [38]. This activity contributed to local invasion of MDA-435 cells into surrounding tissue of the host animal and together with VEGF, could induce cellular extrusions and extravasation into the lumen of fli1-EGFP labeled vessels [38]. This data suggested that RhoC amplification in certain cancer cells contributes to tumor cell extravasation by increasing membrane protrusions capable of penetrating VEGF-induced vessels. The same model was subsequently used for the characterization of other genes linked to the metastatic process including *Twist* and *Integrin β1 (ITGB1)* [40]. Overexpression of Twist significantly increased the percentage of cells that could extravasate into fli1-EGFP labeled vessels, whereas silencing of *ITGB1* reduced the number of tumor cells entering vessels [40]. Like RhoC, integrin-dependent extravasation could be dynamically visualized and was associated with increased membrane protrusions and remodeling of vascular endothelium and cytoskeleton [40].

Imaging neovascularization has now been extended to both primary and transplanted tumors including T-ALL and ERMS [32, 45]. These studies have provided important insights into the dynamic nature of angiogenesis and extravasation, developing valuable tools for the identification and understanding of clinical markers to predict invasive potential in human cancer.

Assessing Self Renewal and Functional Heterogeneity in Cancer

Cancer cells can make more of themselves through the process of self-renewal. Self-renewal is often confined to a small subset of molecularly-defined tumor propagating cells (TPCs) that drive continued tumor growth and ultimately initiate relapse. Traditionally, TPCs have been isolated using antibodies to cell surface proteins and fluorescent activated cell sorting; however, antibodies for zebrafish proteins, especially cell surface proteins, are severely limited, making it difficult to identify functionally important tumor cell subpopulations using traditional techniques. Rather, fluorescent reporters when expressed in a tissue and lineage restricted manner, can often reliably label tumor cell subpopulations, analogous to cell surface antibody markers. For example, to study cancer cell self-renewal in zebrafish ERMS, kRASG12D-induced ERMS were generated where all tumor cells were labeled with dsRED while differentiated cells expressed *alpha-actin-GFP* [31]. Using this approach, early progenitor populations were confined to the dsRED+/GFP-negative fraction [31]. These cells expressed early muscle stem cell markers including *myf5*, *cmet*, and *mcadherin* and could engraft robustly into irradiated recipient fish [31]. By contrast, the dsRED+/GFP+ mature ERMS cell subfractions had significantly reduced tumor propagating potential [31]. Building on these observations, we and others have refined the cancer stem cell population in ERMS using syngeneic *myf5-GFP/mylz2-mCherry* transgenic zebrafish [32, 46]. Moreover, use of limiting dilution cell transplantation in syngeneic zebrafish identified a *myf5*-positive cell population that was unique in its ability to engraft when transplanted into syngeneic host animals, with an average of 1 in 146 cells capable of reinitiating tumors in recipient animals [32]. In contrast, more differentiated *myosin*-positive cells were unable to engraft [32], suggesting that ERMS follows the traditional cancer stem cell model.

The ability to differentially label tumor-propagating cells in zebrafish ERMS has also been used to determine the effects of chemical-genetic perturbations on self-renewal. For example, Chen et al. showed that GSK3 inhibitors decreased the numbers of myf5-GFP+/mylz2-mherry-negative TPCs through activating the WNT/β-catenin pathway to ultimately induce differentiation [33]. GSK3-inhibitors also suppressed ERMS growth and diminished self-renewal capacity in vivo [33]. GSK3-inhibitors dramatically reduced self-renewal of human ERMS in vitro and curbed ERMS cell growth [33], indicating a conserved function for this pathway in modulating ERMS self-renewal. Thus, this approach reveals an intriguing potential for GSK3 inhibitors as therapies for treating ERMS.

Blackburn et al. also recently defined new ways to assess tumor propagating potential in fluorescent-labeled T cell leukemia using an unbiased cell transplantation assay [41] (Fig. 3A-C). Specifically, large-scale cell transplantation experiments were completed using fluorescently-labeled T-ALL cells engrafted into recipient fish as single cells [41]. This approach allowed for assessment of functional differences between clones and when coupled with large-scale limiting dilution cell transplantation experiments, leukemia propagating cell (LPC) number could be directly assessed [41]. By performing serial transplantation of leukemias, the authors were also able to follow how mutations are amassed with time and impact functional changes in leukemic cells [41]. These experiments allowed direct visualization of the effects of clonal evolution on growth, LPC frequency, and therapy responses [41]. This work uncovered that primary T-ALLs are heterogeneous and contain clones with different self-renewal and aggression properties [41]. Rare clones contained in the leukemia continued to amass new genetic and epigenetic changes that increased self-renewal and growth potential [41]. 50 % of aggressive clones acquired Akt pathway activation, confirming previous observations linking the AKT pathway to outcome in human patients [42]. Importantly, by comparing pre and post evolved leukemias, we are now identifying new drivers of progression, self-renewal and growth.

Tumor cell heterogeneity can also be assessed using fluorescently labeled human cells and engrafting cells into embryos prior to maturation of the acquired immune system at 2 days of life. For example, Chapman and colleagues transplanted invasive and non-invasive human melanoma cell lines and dynamically visualized tumor growth in larval fish [43]. This work uncovered that non-invasive melanoma cells can co-invade new areas of tumor growth only when grown with invasive melanoma cells [43] (Fig. 3D-E). This work also revealed that the invasive melanoma altered their mode of invasion when co-cultured with non-invasive cancer cells, switching from protease independent to metaloprotease dependent migration [43], suggesting that heterogeneous tumor cell populations are not isolated entities but can interact with each other to promote growth. Importantly, using differential labeling of tumor cell fractions, this work opens new and exciting methods to compete tumor cells

Fig. 3 (continued) Representative fluorescent images of animals following 28 days of engraftment are shown. (**B**) Animals were transplanted with 25 LPCs of dsRED-positive clone 11.2 (1:78 LPC frequency, 88 days latency) and 25 LPCs of GFP-positive clone 4.3 (1:246 LPC frequency, 28 days latency). Representative images of whole fish and confocal images of T-ALL cells harvested at 15 days and 45 days post-transplantation. The percentages of dsRED-positive and GFP-positive cells at 15 days and 45 days were analyzed by flow cytometry. Data are represented as ± SE (n = 4–7 transplant recipients per time-point). (**C**) In an independent competition assay, 25 LPCs from clone 1.3 (zsYellow-positive, 1:13 LPC frequency, 58 days latency) were competed with 25 LPCs from clone 5.1 (amCyan-positive, 1:184 LPC frequency, 30 days latency), and analyzed as in (**B**). Scale bars, whole-mount zebrafish = 5 mm and confocal images = 40 μm. (**D, E**) Assessing functional heterogeneity of human melanoma cell cells engrafted into embryos alone or in combination. (**D**) Homogeneous melanoma cell line xenografts into zebrafish embryos imaged at 1 (*upper*) and 4 days (*lower*) post transplantation. WM266-4 cells (*green*) are invasive and 501mel cells (*red*) are non-invasive (**E**) Engraftment of both WM266-4 and 501mel results in invasion of both cell types. Xenografts were imaged at 1 (*upper*) and 4 dpi (*lower*). *Arrows* indicate directions of invasion (Blackburn et al., 2014; Chapman et al., Cell Reports 2014)

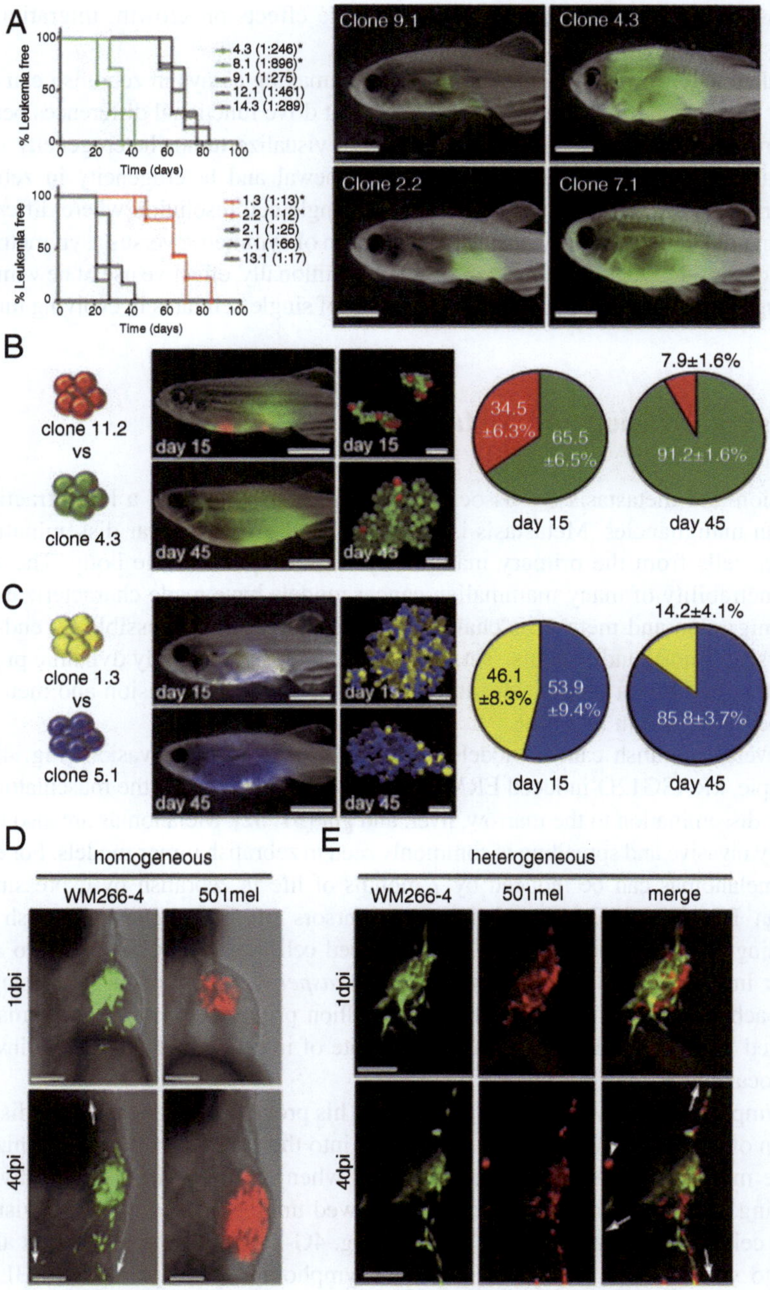

Fig. 3 Assessing functional heterogeneity in cancer. (**A–C**) Individual leukemic cells arising in Myc-induced T-ALL have marked differences in LPC frequency and latency. (**A**) Syngeneic CG1 fish were transplanted with 25 LPCs from various leukemic clones and assessed for time to leukemia onset (n = 8–10 animals transplanted per individual clone). *denotes significant differences in latency between clones that have low (*upper left panel*) or high (*lower left panel*) LPC frequencies (<0.001).

against one another and dynamically visualize effects on growth, migration, and metastatic progression in live fish.

Taken together, these studies highlight how imaging assays in zebrafish can effectively be used to dissect important pathways that drive functional differences between tumor cells and provides assessable models to visualize tumor heterogeneity in live animals. In the future, we expect that self-renewal and heterogeneity in zebrafish tumors will become increasingly observed at single cell resolution where differential labeling may allow for the real-time visualization of symmetric versus asymmetric cell divisions in TPCs that sustain tumor growth. Additionally, effective use of new lineage-tracing technologies will allow for the analysis of single cell fates in evolving tumors.

Imaging Invasion and Metastasis

Invasion and metastasis are associated with poor prognosis in a large fraction of human malignancies. Metastasis is characterized by the vascular dissemination of cancer cells from the primary malignancy to other parts of the body. The visual impenetrability of many mammalian cancer models have made characterization of cell migration and metastasis challenging and often, only accessible by end-point analysis of moribund animals. Yet, cancer progression is a highly dynamic process, and it is therefore a unique advantage of the zebrafish that invasion and metastasis can be visualized in real time.

Several zebrafish cancer models have been used to study invasion (Fig. 4). For example, kRASG12D-induced ERMS exhibit local invasion into the musculature and wider dissemination to the marrow, liver, and gut [31, 32]. Melanomas are also highly locally invasive and spreading is commonly seen in zebrafish tumor models. For example, melanomas can be induced by 7 months of life in zebrafish by expressing the human BRAF(V600E) in melanocyte precursors of *tp53* mutant zebrafish [44]. Building on this model, White et al. developed cell transplantation assays to assess tumor invasion in irradiated, optically clear, *casper* strain zebrafish [4]. Using this approach, allograft transplantation of five million primary zebrafish melanoma cells resulted in robust tumor engraftment at the site of injection and subsequent invasion into local tissues [4].

Lymphomas often progress to leukemia. This process is associated with dissemination of transformed cells from the thymus into the blood stream and colonization of the marrow. Zebrafish transgenic models when coupled with dynamic live cell imaging of fluorescent cancer cells has allowed unprecedented access to visualize these cell state transitions in live animals (Fig. 4G-L). For example, Feng et al. was able to visualize the differences in T-cell lymphoblastic lymphoma (T-LBL) and leukemia in live fluorescent transgenic zebrafish [45]. Feng et al. found that *bcl2* collaborated with *myc* to accelerate time to T-LBL, but also paradoxically inhibited intravasation and dissemination of tumor cells throughout the animal [45]. Intravasation was directly assessed by confocal microscopy in live animals 6 days post engraftment of dsRED labeled malignant cells in *fli-EGFP casper* transgenic fish [45]. These authors went on to show that *bcl2* blocks intravasation through

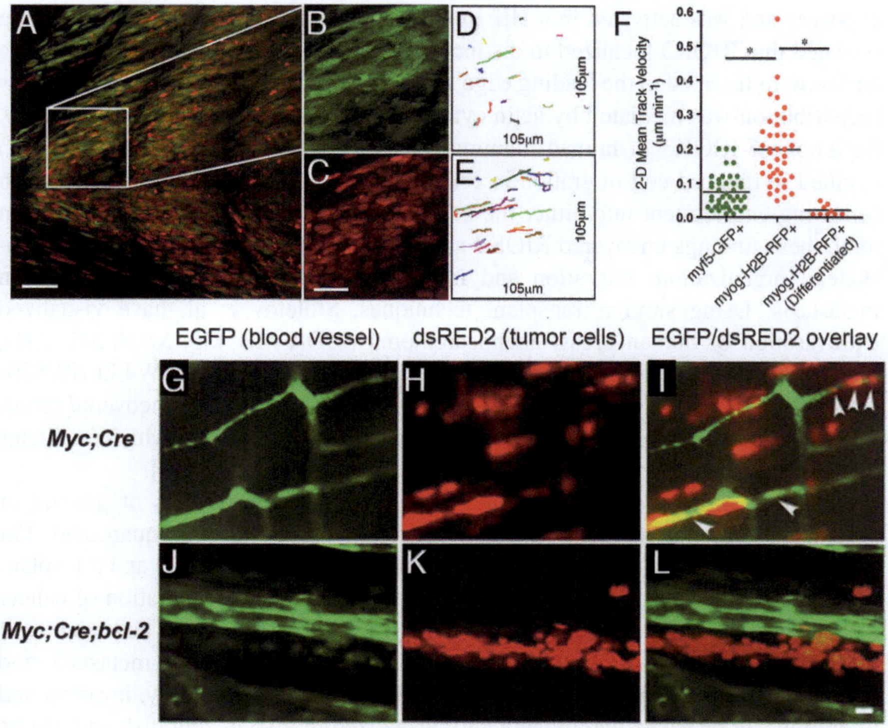

Fig. 4 Imaging invasion and metastasis in embryonal rhabdomyosarcoma and T cell lymphoma. (**A–E**) Still image of a multiphoton movie recording of a stage 3 ERMS arising in *myf5-GFP/myogenin-H2B-RFP* transgenic zebrafish revealing that myf5-GFP+ TPCs are less migratory than differentiated ERMS cells that express myogenin. (**B**, **C**) Magnified view of the *boxed region* showing *myf5-GFP+* (**B**) or *myogenin-H2B-RFP+* ERMS cells (**C**). (**D**, **E**) Tracks of cell movements over the 6.7 h observation period. The same areas are shown as in panels **B** and **C**, respectively. (**F**) Mean track velocities of representative myf5-GFP+ and myogenin-H2B-RFP+ cell types contained within the tumor mass. (**G–L**) Assessing a role for Bcl2 in T cell lymphoma. (**G–I**) dsRED2-expressing lymphoma cells (**H**) from the *Myc;Cre* fish intravasate into *fli1*-EGFP-labeled vasculature (**M**) of the transplant host (*fli1-EGFP;Casper*) by 6 days post-transplantation (see *arrowheads* in **I**); (**J–L**) In contrast, dsRED2-expressing lymphoma cells (**K**) from the *Myc;Cre;bcl-2* fish fail to intravasate vasculature (**L**) of the transplant hosts by 6 days post-transplantation (compare panel **I** with **L**). Scale bar for **A** = 50 μm **B**, **C** = 15 μm and **G–L** = 10 μm (Ignatius et al., Cancer Cell 2012; Feng et al., Cancer Cell 2010)

modulating *S1P1/ICAM1*-mediated cell adhesion [45], suggesting a role for cell-cell interactions in limiting spread away from the site of origin. Finally, using tumor cell transplantation and co-injection of an SP1 inhibitor, the authors identified that SP1 inhibition restored invasive potential to T-LBL cells [45].

Invasion assays have also been performed using human and murine tumor xenografts into 2-day-old fish. For example, Rouhi and colleagues have engrafted fluorescent-labeled human cells into *fli1:EGFP* transgenic zebrafish to assess the effects of hypoxia on angiogenesis and cell migration [47]. Using this approach, RIOK3 (Right Open reading frame kinase 3) was identified as a target of hypoxic exposure and was activated in a HIF1α-dependent manner [48]. Live cell imaging revealed that RIOK3 localized to distinct cytoplasmic aggregates in normoxic cells and were redistribed to the leading edge following hypoxia in cancer cells [48]. This redistribution was mediated by actin cytoskeleton reorganization [48]. Importantly, depletion of RIOK3 in human mammary adenocarcinoma cells (MDA-MB-231) resulted in reduced cell migration in cell culture and blunted cancer dissemination following engraftment into either the zebrafish and mouse animal models [48]. In total, these findings uncovered RIOK3 as a critical factor in maintaining actin cytoskeletal organization, migration and invasion in the context of hypoxia-driven metastasis. Using similar transplant techniques, Stoletov et al. have visualized extravasation of human breast adenocarcinoma (MDA-MB-435, MDA-MB-231), human fibrosarcoma (HT1080), and human colon adenocarcinoma (SW480, SW620) cells in *fli1-EGFP* transgenic embryos [38, 40]. These studies have uncovered prominent roles for *β-1-integrin* in local remodeling of the blood vessels, which facilitates intravasation of cancer cells into new sites of tumor colonization [40].

Coupling invasion assays with automated imaging, the effects of genetic or chemical manipulations on tumor cell migration can be also be quantified. For example, Ghotra et al. have used DiL-labeling of human cancer cells and transplantation into 2-day-old *casper* fli1-EGFP embryos to assess dissemination of cancer cells from the initial site of injection [49, 50]. Using this system, the authors identified SYK kinase and SIRT1 as critical factors in prostrate cancer cell metastasis and tumor growth in Ewing Sarcoma, respectively [50, 51]. In summary, invasion and migration can be easily and dynamically visualized in vivo in zebrafish and can be used to uncover novel biology underlying these processes.

Imaging Tumor, Stromal Cell and Immune Cell Interactions

Transformed tumor cells grow in an environment replete with immune and stromal cell populations. These non-cancerous cells are not passive bystanders but actively promote or repress tumor growth. Immune cell based treatments are an exciting new therapeutic approach in the clinic, however there remains a need to develop in vivo systems to directly visualize tumor and immune cell interactions. Zebrafish share many of the same innate and adaptive immune cells including T, B, and likely NK cells that are found in mice and humans, suggesting the fish will provide many new insights into immune modulation of cancer growth in vivo.

The innate immune system is well characterized in zebrafish and multiple groups have contributed to our understanding of the origin and properties of these important cell types [52]. For example, Feng et al. used fluorescent transgenic reporter lines to label different innate immune cells and study tumor cell interactions [53, 54]. They showed that HRAS transformed melanocytes activate tumor-associated neutrophils and macrophages in a hydrogen peroxide–dependent manner. Blocking this interaction reduced tumor growth, suggesting that innate immune cells provide trophic factors (for example, PGE2) to promote tumor growth [54] (Fig. 5).

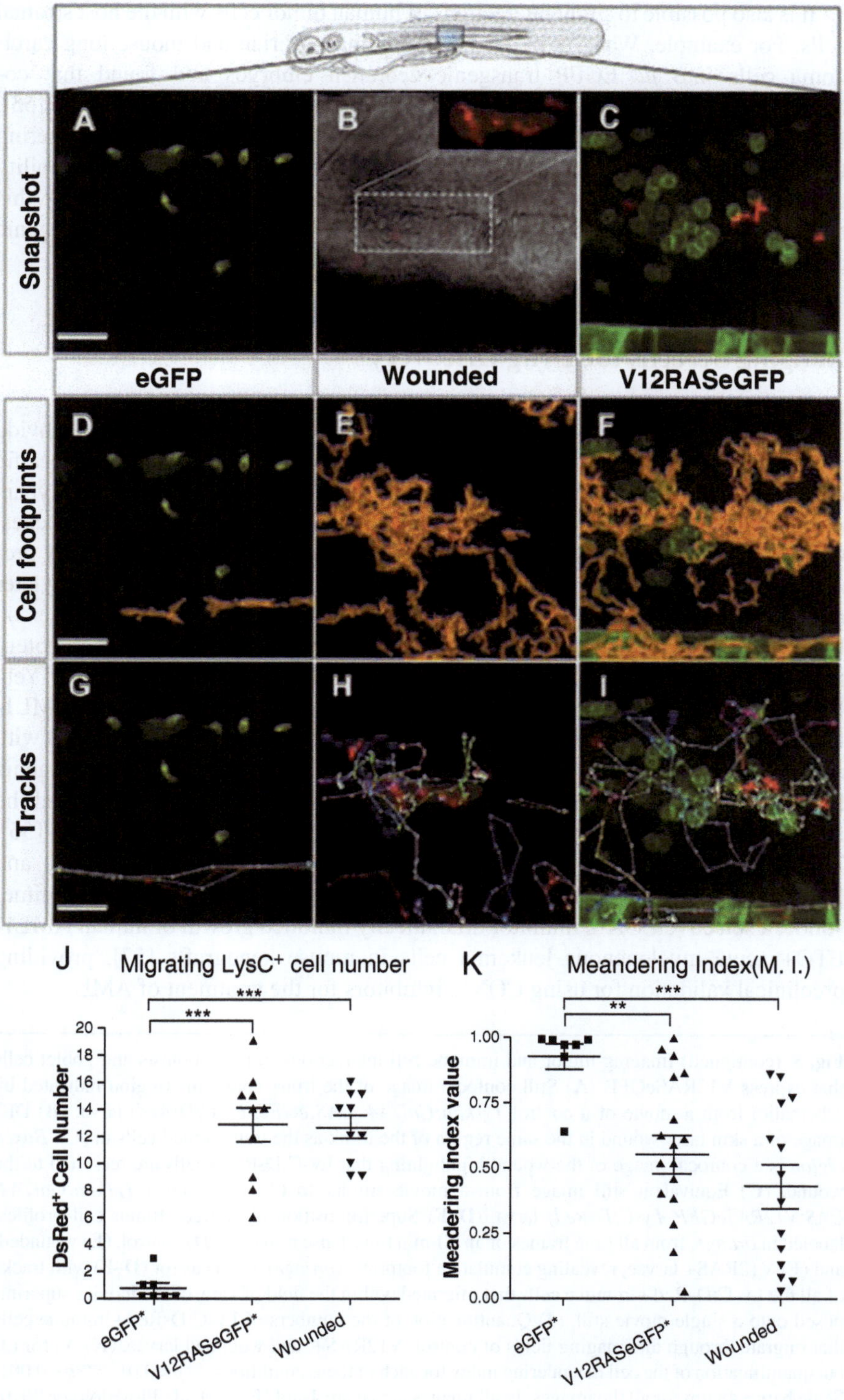

Fig. 5 (continued)

It is also possible to study interactions of human tumor cells with the host stromal cells. For example, Wang et al. introduced human ovarian and mouse lung carcinoma cells into *fli1*:EGFP transgenic zebrafish embryos and found that co-transplantation of tumor-associated macrophages (TAMs) enhanced metastasis [55]. In this report, the authors were able to visually capture TAMs binding and entering the blood stream together with tumor cells [55], again highlighting the unique ability to visualize a role for the immune system in cancer using zebrafish. Importantly, this type of analysis could have only been achieved using fluorescent transgenic approaches, optically transparent zebrafish, and live animal imaging.

Imaging Screens for Drug Discovery

Zebrafish imaging screens are particularly useful for drug discovery, as they provide a robust in vivo platform for fast and cost effective analysis of drugs on specific cellular phenotypes. In this regard, it is important to note that over the past 20 years phenotypic screening has far outperformed targeted drug discovery approaches. One of the major reasons for this is that screening in animal models can easily predict drug toxicity and side effects early on in the discovery process, preventing later and more expensive phases of drug development. Using zebrafish, the Peterson laboratory has become a leader in using chemical genetic screens to identify potential therapeutics for the treatment disease, including cancer. For example, Yeh, Peterson and colleagues were the first to describe a transgenic model of AML1-ETO (AE)—the most common chromosomal translocation product associated with acute myelogenous leukemia (AML). Using these heatshock-inducible transgenic lines, they found that the histone deacetylase inhibitor Trichostatin A reduced the accumulation of granulocytic cells associated with AE in zebrafish embryos [56]. The Yeh laboratory also found that AE induced cyclooxygenase (COX)-2 and β-catenin signaling in zebrafish embryos and human cells [57]. Remarkably, nimesulide, a selective COX-2 inhibitor dramatically inhibited growth of human AML1-ETO+ acute myelogenous leukemia cells in mouse xenografts [57], providing preclinical validation for using COX-2 inhibitors for the treatment of AML.

Fig. 5 (continued) Imaging tumor and immune cell interactions in melanoblasts and goblet cells that express V12RASeGFP. (**A**) Still confocal image of the trunk epidermis (region indicated by schematic) from a movie of a control *Tg(kita:GalTA4, UAS:eGFP, LysC:DsRed)* larva. (**B**) DIC image of a skin laser wound in the same region of the trunk as the transformed cells in (**C**). *Boxed region* is a confocal image of the wound highlighting that LysC:DsRed+cells are recruited to the wound. (**C**) Equivalent still image from a movie similar to (**A**) but from a *Tg(kita:GalTA4, UAS:V12RASeGFP, LysC:DsRed)* larva. (**D–F**) Superimposition of DsRed+tumor cell profiles, labeled in *orange*, from all time frames of an 80-min time-lapse movie of (**D**) control, (**E**) wounded, and (**F**) V12RAS+ larvae, revealing cumulative footprints (*orange*). (**G–I**) as for (**D–F**) with tracks of all the LysC:DsRed+immune cells that migrated within the field of view of the movies superimposed onto a single movie still. (**J**) Quantification of the numbers of LysC:DsRed+immune cells that migrate through the imaging fields of control, V12RAS+, and wounded larvae. (**K**) As for (**J**) but quantification of the cell meandering index for each of these conditions. $**p<0.01$; $***p<0.001$. Scale bars = 48 μm for all the images. In all images larvae are 4-dpf (Feng et. al., PlosBiology, 2010)

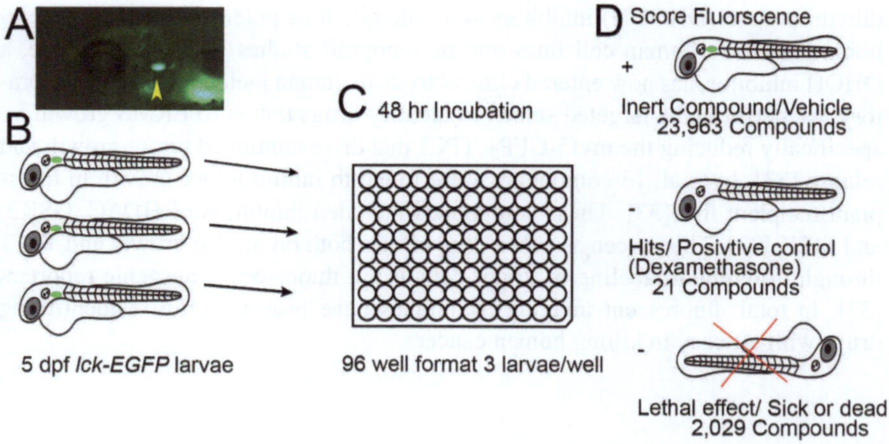

Fig. 6 Zebrafish larval drug screen identifies anti-leukemia compounds. *Tg(lck:EGFP)* transgenic zebrafish have T cells labeled with EGFP. Loss of EGFP expression was used to screen a small molecule library. (**A**) Lateral view of 5 day-post-fertilization (dpf) normal *Tg/lck:EGFP* larva (*yellow arrowhead* lck-EGFP+ thymus). (**B–D**) Schematic of experimental design; (**B**, **C**) Three larvae were treated in 96-well plates, DMSO (vehicle) and dexamethasone (Dex, positive control). (**C**) Fluorescence emission was evaluated after 48 h of drug treatment (**D**) no effect/normal fluorescence (inert compounds) and toxic effect (2029 compounds). Of 26,400 compounds screened, 21 compounds potently killed immature T cells including, Lenaldekar (LDK), a compound that eliminates immature T cells in developing zebrafish without affecting the cell cycle in other cell types (Ridges et al., Blood 2012). LDK killed human T-ALL cells in vitro and in xenograft studies not shown

Ridges et al. also used a chemical screen to identify drugs that kill fluorescent-labeled *lck*:EGFP+ thymocytes, with the hope of identifying drugs that kill T-ALL (Fig. 6). 4 day old- zebrafish were incubated in drugs for 2 days and then assessed for effects on T cell growth [23]. In total, this screen assessed the efficacy of 26,000 different compounds for killing T cells in live fish [23]. From this work, Lenaldekar (LDK) emerged as the top hit and was subsequently tested in the myc-induced model of T-ALL, leading to durable reductions in leukemia growth in vivo [23]. LDK was also found to kill human T-ALL cells in vitro and in vivo in mouse xenograft studies. LDK exhibited exquisite selectivity for killing only thymocytes and T-ALL cells, and had no impact on normal HSCs [23]. Mechanistically, LDK led to dephosphorylation of members of the PI3 kinase/AKT/mTOR pathway and affected cell cycle progression [23]. Using a similar approach, Gutierrez et al. screened chemical libraries for compounds that kill fluorescently labeled T cells that co-express both dsRED and cMyc. Perphenazine was found to potently kill MYC-overexpressing thymocytes and proved to be a potent agonist of protein phosphatase 2A (PP2A) [58]. Human T-ALL cells treated with perphenazine exhibited suppressed cell growth and dephosphylation of PP2A targets both in vitro and in vivo [58], indicating a therapeutic potential for the phenothiazine class of drugs in treating patients with T-ALL.

Visual screening approaches have also been used to identify chemical modifiers of solid tumors. For example, White et al. performed a chemical screen to identify drugs that kill neural crest cells and melanoma [28]. From this work, dihydroorotate (DHODH) inhibitors were identified as potent inhibitors of melanoma growth in human cell lines and in xenograft studies [28]. Leflunomide, a DHOH inhibitor, has now entered clinical trials in human melanoma. In our laboratory, we performed a targeted screen to identify drugs that curb ERMS growth by specifically reducing the myf5-GFP+ TPCs that drive continued tumor growth and relapse [33]. In total, 13 compounds were found to inhibit tumor growth in transplant recipient fish [33]. These compounds included inhibitors of HDAC, GSK3, and MEK [33]. This screen assessed drug effects both on in vivo growth and TPCs through differential labeling of tumor cells using fluorescent transgenic reporters [33]. In total, fluorescent imaging techniques have proven useful in identifying drugs with efficacy in killing human cancers.

Future of Imaging in Cancer in Zebrafish

Given the rapid advances in imaging cancer in zebrafish, we believe that research using this model will be used both to discover fundamental aspects involved in tumor initiation and growth, while providing an in vivo platform for early phases of drug discovery. Many of the hallmarks of cancer proposed by Hanahan and Weinberg can be visualized in live zebrafish including tumor initiation, self-renewal, invasion and metastasis, neovascularization and heterogeneity [60]. Interactions between tumor cells and their microenvironment, including stromal and immune cell components, can also now be readily imaged using zebrafish cancer models. As highlighted in this chapter, several aspects make imaging in zebrafish particularly useful to modeling human disease. Cancer cells grow in a specific niche or microenvironment and important cellular interactions between tumor and stromal cells can be visualized directly in vivo. It is thought that cancer cells can modify their microenvironment to promote survival and growth; however, the nature of many of these interactions including the self-renewing tumor propagating cells and the stromal and/or immune cells, are currently unknown. Thus, zebrafish primary and transplant tumor models will be critical for understanding many environmentally specific cellular and molecular mechanisms regulating tumor growth.

It is becoming increasing clear that human tumors are exceedingly heterogeneous. While many tumor models in mice and zebrafish are generated by single oncogene expression and/or tumor suppressor loss, whole genome and exome sequencing of human tumors has identified large combinations of co-occurring mutations. Therefore, the coupling of imaging in zebrafish with genome editing technologies including TALENs and CRISPR/CAS tools should enable the creation of tumors that better reflect combinatory mutations found in patients. These precision models of human cancer will be valuable for use in chemical-genetic screens, while at the same time aid in understand the consequences of combinations of both genetic and/or epigenetic changes on tumor biology.

Importantly, imaging zebrafish embryos and larvae is increasingly becoming automated, which promises to improve the reproducibility of data collection and to increase the number of assays that can be performed for chemical-genetic screening. For this reason, imaging-based drug screening in zebrafish is an area of expected growth. Drug screening with zebrafish is substantially cheaper and can predict drug toxicity early on during the discovery process. In conclusion, imaging cancer in zebrafish will continue to play an important role in defining mechanistic insights into cancer and identifying novel therapies to treat human malignancy for years to come.

Acknowledgments This work was supported by Alex's Lemonade Stand Foundation (MSI-A award and DML), the Live Like Bella Foundation for Childhood Cancer, American Cancer Society, St. Baldrick's Foundation, the MGH Howard Goodman Fellowship, and NIH grants K99CA175184, R24OD016761, U54CA168512, and 1R01CA154923.

References

1. Hanahan D (2000) The hallmarks of cancer. Cell 100:57–70
2. Hanahan D, Weinberg RA (2011) Hallmarks of cancer: the next generation. Cell 144:646–674
3. Megason SG, Fraser SE (2003) Digitizing life at the level of the cell: high-performance laser-scanning microscopy and image analysis for in toto imaging of development. Mech Dev 120:1407–1420
4. White RM et al (2008) Transparent adult zebrafish as a tool for in vivo transplantation analysis. Cell Stem Cell 2:183–189
5. Zon LI, Peterson RT (2005) In vivo drug discovery in the zebrafish. Nat Rev Drug Discov 4:35–44
6. Peal DS, Peterson RT, Milan D (2010) Small molecule screening in zebrafish. J Cardiovasc Transl Res 3:454–460
7. Sander JD et al (2011) Targeted gene disruption in somatic zebrafish cells using engineered TALENs. Nat Biotechnol 29:697–698
8. Hwang WY et al (2013) Efficient genome editing in zebrafish using a CRISPR-Cas system. Nat Biotechnol 31:227–229
9. Gong Y, Mo C, Fraser SE (2004) Planar cell polarity signalling controls cell division orientation during zebrafish gastrulation. Nature 430:689–693
10. Keller PJ, Schmidt AD, Wittbrodt J, Stelzer EHK (2008) Reconstruction of zebrafish early embryonic development by scanned light sheet microscopy. Science 322:1065–1069
11. Lawson N, Weinstein B (2002) In vivo imaging of embryonic vascular development using transgenic zebrafish. Dev Biol 248:307–318
12. Nicoli S, Presta M (2007) The zebrafish/tumor xenograft angiogenesis assay. Nat Protoc 2:2918–2923
13. Nguyen PD et al (2014) Haematopoietic stem cell induction by somite-derived endothelial cells controlled by meox1. Nature 512:314–318
14. Bertrand JY, Kim AD, Teng S, Traver D (2008) CD41+ cmyb + precursors colonize the zebrafish pronephros by a novel migration route to initiate adult hematopoiesis. Development 135:1853–1862
15. Kissa K, Herbomel P (2010) Blood stem cells emerge from aortic endothelium by a novel type of cell transition. Nature 464:112–115
16. Haffter P et al (1996) The identification of genes with unique and essential functions in the development of the zebrafish, Danio rerio. Development 123:1–36

17. Langenau DM et al (2003) Myc-induced T cell leukemia in transgenic zebrafish. Science 299:887–890
18. Berghmans S et al (2005) tp53 mutant zebrafish develop malignant peripheral nerve sheath tumors. Proc Natl Acad Sci U S A 102:407–412
19. Faucherre A, Taylor GS, Overvoorde J, Dixon JE, den Hertog J (2008) Zebrafish pten genes have overlapping and non-redundant functions in tumorigenesis and embryonic development. Oncogene 27:1079–1086
20. Smith ACH et al (2010) High-throughput cell transplantation establishes that tumor-initiating cells are abundant in zebrafish T-cell acute lymphoblastic leukemia. Blood 115:3296–3303
21. Tang Q et al (2014) Optimized cell transplantation using adult rag2 mutant zebrafish. Nat Methods 11:821–824
22. Langenau DM et al (2008) Co-injection strategies to modify radiation sensitivity and tumor initiation in transgenic zebrafish. Oncogene 27:4242–4248
23. Ridges S et al (2012) Zebrafish screen identifies novel compound with selective toxicity against leukemia. Blood 119:5621–5631
24. Nguyen AT et al (2012) An inducible krasV12 transgenic zebrafish model for liver tumorigenesis and chemical drug screening. Dis Model Mech 5:63–72
25. Park SW et al (2008) Oncogenic KRAS induces progenitor cell expansion and malignant transformation in zebrafish exocrine pancreas. Gastroenterology 134:2080–2090
26. Zhu S et al (2012) Activated ALK collaborates with MYCN in neuroblastoma pathogenesis. Cancer Cell 21:362–373
27. Ceol CJ et al (2011) The histone methyltransferase SETDB1 is recurrently amplified in melanoma and accelerates its onset. Nature 471:513–517
28. White RM et al (2011) DHODH modulates transcriptional elongation in the neural crest and melanoma. Nature 471:518–522
29. Jung IH et al (2013) Glioma is formed by active Akt1 alone and promoted by active Rac1 in transgenic zebrafish. Neuro Oncol 15:290–304
30. Burger A et al (2013) A zebrafish model of chordoma initiated by notochord-driven expression of HRASV12. Dis Model Mech. doi:10.1242/dmm.013128
31. Langenau DM et al (2007) Effects of RAS on the genesis of embryonal rhabdomyosarcoma. Genes Dev 21:1382–1395
32. Ignatius MS et al (2012) In vivo imaging of tumor-propagating cells, regional tumor heterogeneity, and dynamic cell movements in embryonal rhabdomyosarcoma. Cancer Cell 21:680–693
33. Chen EY et al (2014) Glycogen synthase kinase 3 inhibitors induce the canonical WNT/β--catenin pathway to suppress growth and self-renewal in embryonal rhabdomyosarcoma. Proc Natl Acad Sci U S A 111:5349–5354
34. Motoike T et al (2000) Universal GFP reporter for the study of vascular development. Genesis 28:75–81
35. Cross LM, Cook MA, Lin S, Chen JN, Rubinstein AL (2003) Rapid analysis of angiogenesis drugs in a live fluorescent zebrafish assay. Arterioscler Thromb Vasc Biol 23:911–912
36. Nicoli S, Ribatti D, Cotelli F, Presta M (2007) Mammalian tumor xenografts induce neovascularization in zebrafish embryos. Cancer Res 67:2927–2931
37. Lee SLC et al (2009) Hypoxia-induced pathological angiogenesis mediates tumor cell dissemination, invasion, and metastasis in a zebrafish tumor model. Proc Natl Acad Sci U S A 106:19485–19490
38. Stoletov K, Montel V, Lester RD, Gonias SL, Klemke R (2007) High-resolution imaging of the dynamic tumor cell–vascular interface in transparent zebrafish. Proc Natl Acad Sci U S A 104:17406–17411
39. Stoletov K et al (2009) Vascular lipid accumulation, lipoprotein oxidation, and macrophage lipid uptake in hypercholesterolemic zebrafish. Circ Res 104:952–960
40. Stoletov K et al (2010) Visualizing extravasation dynamics of metastatic tumor cells. J Cell Sci 123:2332–2341
41. Blackburn JS et al (2014) Clonal evolution enhances leukemia-propagating cell frequency in T cell acute lymphoblastic leukemia through Akt/mTORC1 pathway activation. Cancer Cell 25:366–378

42. Gutierrez A et al (2009) High frequency of PTEN, PI3K, and AKT abnormalities in T-cell acute lymphoblastic leukemia. Blood 114:647–650
43. Chapman A et al (2013) Heterogeneous tumor subpopulations cooperate to drive invasion. Cell Rep. doi:10.1016/j.celrep.2014.06.045
44. Patton EE et al (2005) BRAF mutations are sufficient to promote nevi formation and cooperate with p53 in the genesis of melanoma. Curr Biol 15:249–254
45. Feng H et al (2010) T-lymphoblastic lymphoma cells express high levels of BCL2, S1P1, and ICAM1, leading to a blockade of tumor cell intravasation. Cancer Cell 18:353–366
46. Storer NY, White RM, Uong A, Price E, Nielsen GP, Langenau DM, Zon LI (2013) Zebrafish rhabdomyosarcoma reflects the developmental stage of oncogene expression during myogenesis. Development. 140(14):3040–50
47. Rouhi P et al (2010) Hypoxia-induced metastasis model in embryonic zebrafish. Nat Protoc 5:1911–1918
48. Singleton DC et al (2015) Hypoxic regulation of RIOK3 is a major mechanism for cancer cell invasion and metastasis. Oncogene 34:4713–4722. doi:10.1038/onc.2014.396
49. Ghotra VPS et al (2012) Automated whole animal bio-imaging assay for human cancer dissemination. PLoS One 7
50. Ghotra VPS et al (2014) SYK is a candidate kinase target for the treatment of advanced prostate cancer. Cancer Res 75:230–240
51. Ban J et al (2014) Suppression of deacetylase SIRT1 mediates tumor-suppressive NOTCH response and offers a novel treatment option in metastatic Ewing sarcoma. Cancer Res 74:6578–6588
52. Iwanami N (2014) Zebrafish as a model for understanding the evolution of the vertebrate immune system and human primary immunodeficiency. Exp Hematol 42:697–706
53. Feng Y, Santoriello C, Mione M, Hurlstone A, Martin P (2010) Live imaging of innate immune cell sensing of transformed cells in zebrafish larvae: parallels between tumor initiation and wound inflammation. PLoS Biol 8
54. Feng Y, Renshaw S, Martin P (2012) Live imaging of tumor initiation in zebrafish larvae reveals a trophic role for leukocyte-derived PGE2. Curr Biol 22:1253–1259
55. Wang J et al (2015) Novel mechanism of macrophage-mediated metastasis revealed in a zebrafish model of tumor development. Cancer Res 75:306–315
56. Yeh J-RJ et al (2008) AML1-ETO reprograms hematopoietic cell fate by downregulating scl expression. Development 135:401–410
57. Zhang Y et al (2013) AML1-ETO mediates hematopoietic self-renewal and leukemogenesis through a COX/β-catenin signaling pathway. Blood 121:4906–4916
58. Gutierrez A et al (2014) Phenothiazines induce PP2A-mediated apoptosis in T cell acute lymphoblastic leukemia. J Clin Invest 124:644–655
59. Santoriello C et al (2010) Kita driven expression of oncogenic HRAS leads to early onset and highly penetrant melanoma in zebrafish. PLoS One 5:1–11
60. Tang Q, Moore JC, Ignatius MS, Tenente IM, Hayes MN, Garcia EG, Torres Yordán N, Bourque C, He S, Blackburn JS, Look AT, Houvras Y, Langenau DM (2016) Imaging tumour cell heterogeneity following cell transplantation into optically clear immune-deficient zebrafish. Nat Commun 7:10358
61. Shin J et al (2012) Zebrafish neurofibromatosis type 1 genes have redundant functions in tumorigenesis and embryonic development. Dis Model Mech 5:881–894
62. Zheng W et al (2014) Xmrk, Kras and Myc transgenic zebrafish liver cancer models share molecular signatures with subsets of human hepatocellular carcinoma. PLoS One 9
63. Li Z et al (2013) A transgenic zebrafish liver tumor model with inducible Myc expression reveals conserved Myc signatures with mammalian liver tumors. Dis Model Mech 6:414–423
64. Mudbhary R et al (2014) UHRF1 overexpression drives DNA hypomethylation and hepatocellular carcinoma. Cancer Cell 25:196–209
65. Ju B et al (2015) Oncogenic KRAS promotes malignant brain tumors in zebrafish. Mol Cancer 14:18. doi:10.1186/s12943-015-0288-2
66. Ju B et al (2014) Activation of Sonic hedgehog signaling in neural progenitor cells promotes glioma development in the zebrafish optic pathway. Oncogenesis 3, e96

Imaging Cancer Angiogenesis and Metastasis in a Zebrafish Embryo Model

C. Tulotta, S. He, W. van der Ent, L. Chen, A. Groenewoud, H.P. Spaink, and B.E. Snaar-Jagalska

Abstract Tumor angiogenesis and metastasis are key steps of cancer progression. In vitro and animal model studies have contributed to partially elucidating the mechanisms involved in these processes and in developing therapies. Besides the improvements in fundamental research and the optimization of therapeutic regimes, cancer still remains a major health threatening condition and therefore the development of new models is needed. The zebrafish is a powerful tool to study tumor angiogenesis and metastasis, because it allows the visualization of fluorescently labelled tumor cells inducing vessel remodeling, disseminating and invading surrounding tissues in a whole transparent embryo. The embryo model has also been used to address the contribution of the tumor stroma in sustaining tumor angiogenesis and spreading. Simultaneously, new anti-angiogenic drugs and compounds affecting malignant cell survival and migration can be tested by simply adding the compound into the water of living embryos. Therefore the zebrafish model offers the opportunity to gain more knowledge on cancer angiogenesis and metastasis in vivo with the final aim of providing new translational insights into therapeutic approaches to help patients.

Keywords Angiogenesis • Extravasation • Metastasis • Tumor inflammation • In vivo imaging

Introduction

Tumor angiogenesis and metastasis are hallmarks of cancer and often associate with poor patient prognosis [1]. Current therapeutic regimes are based on anti-angiogenic drugs and inhibitors designed to target highly proliferative cancer cells. In vitro and in vivo models have been crucial to understand tumor induced vessel growth and cancer cell dissemination, however, although progress has been made, metastatic cancer remains a leading cause of death. Therefore it is necessary to establish new

C. Tulotta • S. He • W. van der Ent • L. Chen • A. Groenewoud • H.P. Spaink
B.E. Snaar-Jagalska (✉)
Institute of Biology, Leiden University, Einsteinweg 55, 2333 CC Leiden, The Netherlands
e-mail: b.e.snaar-jagalska@biology.leidenuniv.nl

D.M. Langenau (ed.), *Cancer and Zebrafish*, Advances in Experimental Medicine and Biology 910, DOI 10.1007/978-3-319-30654-4_11

models in order to further elucidate molecular mechanisms of tumor angiogenesis and metastatic spreading, with the ultimate goal of developing more efficient cures.

The zebrafish has emerged as a powerful tool to study tumor biology and several models have been established [2, 3]. Carcinogen induced neoplasia has been used to study liver and pancreatic tumors, intestine adenocarcinoma, chondrosarcoma, spindle cell sarcoma [4–7] and germ cell tumors upon ENU-induced mutagenesis [8]. Transgenic zebrafish of specific tumor types have been developed for T cell acute lymphoblastic leukemia (T-ALL) [9], kRAS-induced Rhabdomyosarcoma [10], pancreatic adenocarcinoma [11], RAS-induced liver cancer [12] and melanoma [13]. Moreover, the visualization of tumor progression and transplants in zebrafish adults is greatly facilitated by the casper mutant, where the double mutation in *nacre* and *roy* result in the absence of melanocyte and iridophores giving adult zebrafish transparent skin [13]. Younger zebrafish have also contributed in better understanding of cancer, mainly using xenotransplantation techniques [14]. The transparency of the embryo, quickly developing outside of the mother, and the availability of reporter and mutant lines, make the zebrafish embryo an excellent in vivo model for phenotype assessment of engrafted human tumors. The implantation of human cancer cells is possible without rejection of the graft, due to the absence of the adaptive immunity in the early stages of development (a fully mature immune system is complete in 3–4 weeks-old-juveniles) [15]. The temporal separation between the innate and adaptive arms of the immune system also bears the advantage of following the interaction between tumor cells and myeloid cells and to understand how their cross-talk might be supportive for tumor progression. Orthotopic transplantation has also been recently reported, allowing analysis of tumors in their environment of origin [16, 17]. Findings in the embryo have been confirmed in chemically immune-suppressed 30-day-old juveniles, and could be extended to the immunocompromised Rag2$^{-/-}$ mutants [18], currently suitable for allograft engraftment of adult skeletal muscle, blood cells and transgenic models of cancer [19].

The use of fluorescence stereo and confocal microscopy, two-photon and second harmonic generation (SHG) has provided qualitative and quantitative information on tumor behavior on a single cell and whole organism level. High and low resolution imaging of tumor and stroma cells allow the visualization of cell-cell interaction and analysis of tumor burden, also upon drug treatment.

In this chapter we focus on the use of the zebrafish xenotransplantation model to study tumor angiogenesis and metastasis, additionally evaluating the contribution of the immune system. The role of the microenvironment is also highlighted in transgenic models of early malignant transformation.

Tumor Angiogenesis

The Process, the Models and the Therapies

Angiogenesis is the process of new blood vessel formation, starting from preexisting vasculature and occurring in physiological as well as pathological conditions [20]. It is tightly regulated by the balance between anti- and pro-angiogenic factors, especially during embryogenesis, wound healing and the female reproductive cycle [1]. When the balance is skewed towards pro-angiogenic events the so called "angiogenic switch" occurs, leading to vessel growth and consequent enrichment of the vascular network. Angiogenesis is regulated by several signaling molecules such as vascular endothelial growth factor (VEGF), basic fibroblast growth factor (bFGF), platelet-derived growth factor (PDGF), placenta growth factor (PGF), hepatocyte growth factor or scatter factor (HGF or SF), pro-angiogenic chemokines and cytokines, as well as angiopoietins [21]. Until now, VEGF has been the major target of investigation. VEGF family members, VEGF-A,-B,-C,-D,-E and -F, together with PGF, are differentially coupled to correspondent receptors like VEGFR-1 (Flt-1), -2 (Kdr) and -3 [22]. Generally, endothelial cells express VEGF receptors and start the angiogenic process upon VEGF sensing, through filopodia extension. Accordingly, endothelial cells at the front edge, called "tip" cells, begin to migrate, after degradation of the surrounding extracellular matrix (ECM) by metalloproteases (MMPs). The tip cells are consequently followed by the "stalk" cells, which keep contact with the mother vascular trunk and undergo duplication to further create the lumen of the forming vessel [23, 24]. First reported by Caduff et al., 1986 and then renamed by Burri and Tarek in 1990, intussusceptive angiogenesis is an alternative form of angiogenesis, where endothelial cell proliferation is substituted by vessel intraluminal growth. The process is fast (within minutes to hours) and it is believed to occur in tumors simultaneously to canonical vessel sprouting, leading to enhanced vascular perfusion [25–27].

Angiogenesis is one of the hallmarks of cancer and takes place to sustain uncontrolled growth, which often leads to a hypoxic state and consequent higher need of oxygen and nutrients. The VEGFR signaling is initiated during hypoxia, a condition that primarily results in the activation of HIF1α (Hypoxia inducible factor 1α) [28, 29]. VEGF can be secreted by either tumor cells, or stroma cells, such as platelets, erythrocytes and leukocytes. In particular, tumor associated macrophages (TAMs), which are often recruited to the tumor mass through the CSF-1/CSF1R axis, play a tumor-supportive role, by producing MMPs and VEGF, to facilitate endothelial cell attraction and proliferation, together with a plethora of chemokines for the activation of tumor invasion [30]. Furthermore, so called "resident" macrophages localized along blood vessels have been reported to promote tumor cell intravasation [31]. TNFα secreted by these myeloid cells has been associated with higher permeability of the vascular wall, in a 3D microfluidic model [32, 33]. The formation of a vascular network infiltrating a tumor mass can support tumor growth and facilitate the metastatic cascade. The link between angiogenesis and metastasis and the possibility of an anti-angiogenic approach for tumor suppression was first proposed by Folkman in 1971 [34]; since then, several models to study (tumor) angiogenesis have been established and anti-angiogenesis drugs have been developed and tested for anti-cancer therapies.

In vitro models to study endothelial cell proliferation, differentiation and migration, together with co-culture systems, offer the advantage of high reproducibility, easy handling and quick experimental assessment. However, endothelial cells are often quiescent in vivo and the immortalization in culture likely alters physiological responses. The use of primary lines is possible, but long-term analysis of these cultures is not possible since physiological characteristics are lost after passage in culture [35].

Microvessel modifications, including the function of the neighboring non-endothelial cells, can be studied using ex vivo organ cultures, embedded in the matrix. These are practically simple and low cost systems, where angiogenesis is examined outside the context of inflammation. Yet, standardized quantification is limited with this method [35].

In vivo experimental models offer the opportunity to study mechanisms, kinetics and dynamics in the context of a complex organism. In the case of angiogenesis, the chorioallantoic membrane (CAM) of a chicken embryo is used without rejection of the graft [36, 37]. As this is a relatively easy and low cost assay that allows drug testing to be completed. The high complexity of vascularization in this model can make it difficult to identify vessel formation. As developmental angiogenesis goes on until day 11, it is also not always easy to distinguish whether vessel sprouts are a treatment response or alteration of a physiological process [35].

The use of rodent models allows assessment of tumor angiogenesis in mammals. Although it is relatively easy to visualize subcutaneous xenograft induced angiogenesis, orthotopic transplantations are preferred, due to known differences in microenvironment contributions to tumor behavior. Although this resembles a more natural environment, transplantations in the same organ of tumor origin have the disadvantage that the experimental readout is assessed mainly on fixed material and application of live imaging techniques to monitor angiogenesis are limited. Moreover, other systems have been developed such as implantation of sponges and polymers, corneal angiogenesis assays, dorsal air sac model and chamber assay, all reviewed in detail in Staton et al. [35].

Currently tumor inhibition via angiogenesis targeting has been achieved with the development of drugs that belong to the classes of monoclonal antibodies (bevacizumab, cetuximab, trastuzumab, panitumumab), tyrosine kinase inhibitors (axitinib, cabonzantinib, pazopanib, regorafenib, sorafenib, sunitinib, vandetanib) and mTOR targeting compounds. Anti-angiogenic therapies are often prescribed in combination with other anti-cancer medicines, in order to limit the onset of drug resistance. Even though targeting angiogenesis has registered positive outcomes, there are many limitations. Recurrence, for example, occurs when the therapy is suspended, in spite of the fact that an initial inhibition was manifested. Increased incidence of metastasis has also been reported, together with other side effects, such as bleeding, thrombosis, hypertension, proteinuria, leukopenia and hyperthyroidism [38]. As previously mentioned, drug resistance is another limiting factor for successful anti-angiogenic therapies, which has been related to both VEGF dependent and independent mechanisms, as well as linked to the influence of the surrounding tumor microenvironment. Therefore the application of combination therapies is a valid option and is currently in use clinically. Interestingly, development of drug delivery systems using nanotechnologies are in development and might allow circumvention of side effects, improving patient prognosis [39]. Considering the limitations of anti-angiogenic approaches, there is a need to identify new bio-markers of angiogenesis as well as models to assess drug efficacy and development of resistance. Additionally, verifying the effect of the anti-angiogenic therapy on different stages of tumor progression will be critically important [21].

Imaging Tumor Angiogenesis in Zebrafish Embryo

The zebrafish embryo model has the potential to be a translational system to study tumor angiogenesis. The transparency of zebrafish embryos and the availability of reporter lines with fluorescent vascular system [40] allow easy intravital imaging of blood vessels sprouting towards engrafted tumor cells of human or mouse origin. Other supporting cells, like pericytes [41] fibroblasts, circulating erythrocytes [42] and immune cells [43, 44] are present and it is possible to visualize these cells using tissue-specific fluorescent transgenic fish.

Originally described by Nicoli and colleagues in 2007, FGF2-T-MAE cells, an immortalized murine aortic endothelial cell line, were engrafted into the perivitellin space of 48 h post fertilization (hpf) zebrafish embryo. The perivitellin space is between the yolk syncytial layer and the periderm (Fig. 1a, b). Remarkably, engraftment resulted in tumor angiogenesis that could be dynamically imaged using fluorescent reporters that label the vasculature [45]. Specifically, cells were mixed with Matrigel and then engrafted in VEGFR2:G-RCFP. Growth of zebrafish blood vessels from the sub-intestinal vein plexus (SIV) was observed after 24–48 h of engraftment. The assay allowed for the assessment of tumor-induced angiogenesis over a short time window. It is a more rapid approach compared to other systems, such as the CAM assay, the murine and rabbit cornea methods and the xenografts in rodents. Moreover, anti-angiogenic compounds could be easily tested by either premixing the inhibitors with the cell suspension or adding it directly into the water. Importantly, blood vessels surrounding the tumor graft expressed endothelial markers and their sprouting was inhibited by morpholino-mediated VEGFR knockdown [46]. Using this assay, the angiogenic properties of FGF2 were demonstrated. Injection of the recombinant growth factor stimulated SIV complex growth while inhibition using SU5402 antagonized growth [47].

A wide range of tumor cell types can be used in embryo assays. Zhao et al. engrafted the melanoma line B16 and the murine colon carcinoma CT26, in the abdominal perivitellin space, close to the opening into the heart cavity. Tumor growth was accompanied by clear vessel sprouting that was visualized to 6 days post implantation (dpi). While the overexpression of VEGF165 in the B16 line resulted in longer vessel length, increased branching, and larger diameter of neo-vessels, the use of a specific inhibitor against VEGFR2 abolished their growth [48]. In these experiments, sprouting occurred from the surrounding vascular network. By contrast van der Ent et al. showed that Ewing's sarcoma cell lines EW3, EW7, L1062, SKNMC, TC32 were able to induce growth directly from the SIV complex when engrafted into 2 days-post-fertilization (dpf) zebrafish embryo [49]. A similar response was observed for another Ewing sarcoma cell line, A673 (Fig. 1c). Neuroendocrine tumors have been shown to induce vessel attraction from the SIV and common cardinal vein (CCV) towards cancer cells engrafted in the perivitellin space [50].

Tumor induced angiogenesis has also been observed when human cancer cells were engrafted in the hindbrain ventricle and in the yolk sac of zebrafish embryos [51].

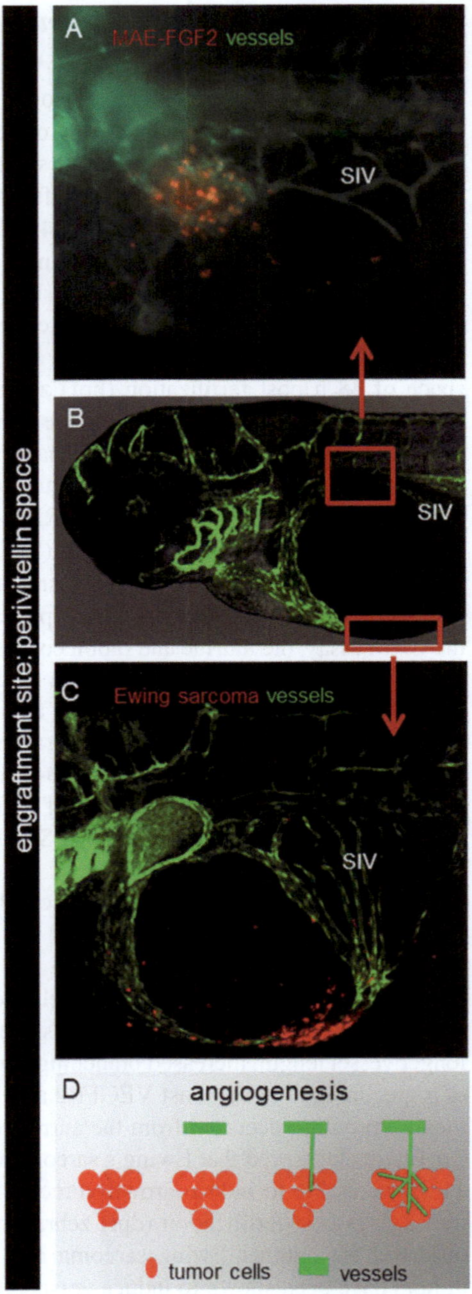

Fig. 1 Induction of angiogenesis by murine and human tumor cells. In the zebrafish embryo, the reorganization of the host vasculature towards cancer cells allows the study of tumor induced angiogenesis. Xenoengraftment of tumor cells is performed by implantation into the perivitellin space of a 2-day-old zebrafish embryo (**a**, **c**), with fluorescently traceable blood vessels (Tg(fli1a:EGFP)^{y1} (**a**, **c**); Tg(kdrl:EGFP)^{s843} [96] (**b**)). As the perivitellin layer surrounds the yolk, the implantation at this site was done in the cavity between the yolk and the trunk for MAE-FGF2 (**a**) or at the bottom of the yolk for Ewing sarcoma, A673 (**c**). SIV response occurred upon both engraftments. Sites of implantation are indicated in (**b**). In (**d**) a schematic representation of the angiogenic process is shown

As previously described, angiogenesis can occur as a result of local low oxygen tension, which results in the activation of HIF signaling to induce the transcription of downstream targets, including VEGF. Systemic hypoxia enhanced local invasion, distant migration from the engraftment site and increased infiltration of blood vessels, when zebrafish embryos, inoculated with murine fibrosarcoma, were exposed to 7.5 % oxygen level for up to 72 h. The same aggressive phenotype was present upon VEGF overexpression. On the other hand, inhibition of tumor volume, minimization of tumor foci and reduction of vascularization in the primary tumor mass were obtained via anti-angiogenic treatment, using Sunitinib, a multi-targeted tyrosine kinase inhibitor, and VEGFR2 morpholino knockdown [52].

Cancer cells have been successfully engrafted into the perivitellin space, yolk sac and hindbrain. Engraftment into each of these sites has resulted in dynamic visualization of angiogenesis, especially in the context of engrafted tumor cells that aggregate, mimicking a primary tumor mass. Additionally, the growth of blood vessels has been reported in the case of tumor aggregates distantly localized from the implantation site. These distal growths arise following inoculation of tumor cells directly in the blood circulation. Tobia et al. have shown that murine melanoma B16-BL6, stably expressing DsRed protein, induced the formation of a microvascular network both at the metastatic and engraftment sites [53].

Vasculogenesis, the formation of blood vessels starting from endothelial cell precursors, was induced by human tumor cells inoculated in the Duct of Cuvier (Fig. 2a). The process started 6 h after engraftment, while a complex vascular network formed during the following days. As the primary tumor mass expanded, different stages of vascularization could be visualized and distinguished. After initiation, vessel remodeling occurred between 12 and 18 h post implantation (hpi), followed by connection formation at 1 dpi, network establishment at 2 dpi and finally lumenization, with initiation of blood flow, at 3 dpi (Fig. 2b, c). A detailed microscopic analysis, using confocal laser scanning microscopy (CLSM), revealed altered and tortuous structural composition of the newly induced vessels, compared to the physiological architecture of zebrafish vessels. Treatments with VEGFR inhibitors (KRN633 and Sunitinib) resulted in the reduction of microvessel density and tumor size. A possible effect of VEGFR inhibition on vessel integrity was excluded via microangiography and immunohistochemistry, using an antibody against the tight junction protein ZO-1 [54].

The ability to study tumor-induced angiogenesis using intravital imaging has allowed testing of alternative anti-angiogenic compounds. Most studies can be completed by simply adding the drug into the water and changing drug daily. Non-water-soluble drugs can also be used, but require a carrier. For instance, the packaging of the flavonoid Quercetin or the fungal metabolite Chetomin in polymeric micelles permitted the delivery of drug to engrafted embryos and exhibited potent antitumor effects [55, 56]. Furthermore, inhibition of MDA-MB-231 or B16-F10 induced angiogenesis could be suppressed by packaging of LY294002 in nanoparticles which were delivered via injection in the embryos [57]. The distribution of nanoparticles or other carrier system to tumor cell masses often relies on the enhanced permeability and retention (EPR) effect on cancer vessels [58, 59]. Current research aims to design targeted delivery systems of anti-cancer drugs. These methods will allow targeted delivery based on recognizing tumor antigens. Moreover, slow chemical release mechanisms will likely reduce toxicity and increase efficacy in target cell killing [60].

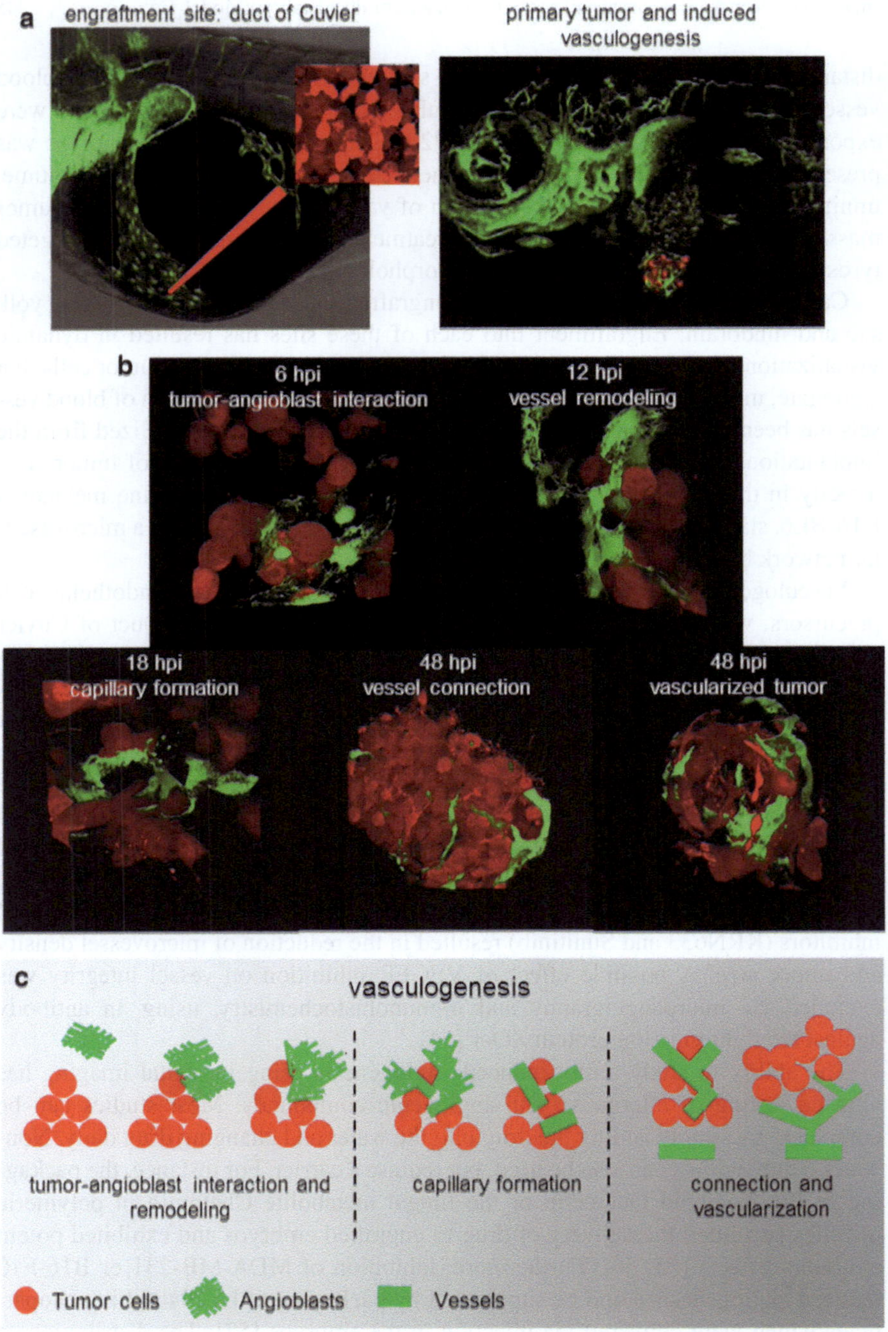

Fig. 2 Vessel network formation following tumor engraftment. Implantation of red fluorescent tumor cells into the duct of Cuvier leads to tumor cell circulating with the blood of a 2-day-old Tg(fli1a:EGFP)[y1] zebrafish embryo. Tumor masses are supported by blood vessels (**a**). The formation of the tumor vascular network starts with the recruitment of endothelial cell precursors (angioblasts) which, upon interaction with the tumor cells, are oriented and form capillaries. Vessel connection subsequently occurs until a complex network has formed (**b**). 3D reconstruction of figures in He et al. [54] are shown in the panel in (**b**). In (**c**) the steps of vasculogenesis are illustrated

Extravasation Dynamics Occur Upon Tumor-Endothelial Cell Interactions

The interaction between tumor cells and endothelium is not restricted to angiogenesis and vasculogenesis. These interactions are also crucial during the metastatic process where tumor cells traveling in the blood extravasate and colonize secondary organs. It is likely that endothelial cells of the vascular wall exert roles in tumor cell survival as cells transit through the vessel and subsequently extravasate at distant sites. Intravital imaging in transparent transgenic zebrafish embryos, has revealed key aspects of the extravasation dynamics [61]. Tumor cells engrafted in the pericardium were found in the blood circulation, the head and the tail regions at 3–5 hpi. The comparison of different cell lines revealed the correlation between extravasation and high metastatic potential. Tumor cells in the intersegmental vessels (ISVs) required adaptation inside the narrow compartment of the capillaries. Contrary to trapped beads, cancer cells clearly showed a more active behavior, characterized by extension of protrusions, anatomical adjustment to vessel diameter and intravascular migration upon interaction with endothelial cells. Pressure on the endothelium and concomitant flattening allowed tumor cell migration against the blood flow, leading to the hypothesis that cancer cells "scan" the endothelium, and extravasate at a favorable site. Therefore the capability of tumor cells to extravasate into surrounding tissues is greatly influenced by their adhesive features. When contacted by tumor cells, the endothelium actively responded by increasing the thickening of the vessel wall, culminating in a single endothelial cell clustered around an arrested tumor cell. The migration outside the blood vessels was found to correlate to the expression of candidate genes like Twist, VEGFA and integrin β1. The process could be either dependent or independent on the integrin. For example, blocking integrin β1 resulted in reduction of extravasation in MDA cells expressing VEGF, but not in MDA cells which had higher levels of Twist. A central role for the integrin β1 was also elucidated by assessing active clustering of endothelial cells around tumor cells. VEGFA expressing cancer cells induced quick movements of endothelial cells, in an integrin β1 dependent fashion. On the other hand, MDA-Twist cells, able to interact with the endothelium in a integrin β1 independent manner, did not cause thickening of the vessel wall or rapid aggregation of endothelial cells. Hence, the authors conclude that malignant cells arrested in the vasculature, create contact with vessels by integrin β1. This signaling results in endothelium changes that facilitate the extravasation process. While no increase in number of endothelial cells was found, the number of endothelial junctions increased in the ISVs. Extravasation of tumor cells could also occur independently of integrin β1 and was associated with Twist over expression. This processes was mediated in part by ROCK kinase modulation and associated changes in the cytoskeleton. Inhibition of ROCK by Y27632 lead to a dramatic decrease in Twist-dependent extravasation [61].

When tumor cells extravasate and invade the tissues, different types of movement can take place. Tumor cells can move directionally as single cells through use of invadopodia. However, cells can also exhibit amoeboid cell movement either in single cells or collective cells. Ameboid cells have a characteristic round shape, are small size and form numerous blebs and commonly overexpress RhoC [62]. In this study, co-implantation of tumor cell lines with and without overexpression of RhoC were used to assess whether the small GTPase could play a role in the secretion of specific cytokines and soluble factors to alter cell movements and influence invasion phenotypes. RhoC enabled tumor cells to intravasate by extending long filopodia into the vessel lumen. This process only occurred after VEGF secretion and vascular opening [62].

Metastasis

Dynamics, Models and Drug Treatments

Metastasis is a leading cause of death in cancer patients. Metastasis formation is a multistep process. Tumor cells initially disseminate from the primary malignant lesion by invading surrounding tissue and then entering the blood or the lymphatic circulation through a process called intravasation. Secondary organ sites are colonized following extravasation of tumor cells from the vessels into the target site. It is not surprising that these metastatic steps lead to dynamic changes in tumor cells including modulation of tumor cell survival and growth, ultimately culminating in the formation of micro- and macrometastasis. In order to initiate the first-step of local invasion, cancer cells minimize cell-to-cell adhesion. This usually occurs by loss of E-cadherin and increased expression of N-cadherin and vimentin. This shift in use of adhesion molecules favors cell migration and is a hallmark of the epithelial-to-mesenchymal transition (EMT). EMT is characterized by alterations in cell morphology, from epithelial to fibroblastic shape [63]. The TGFβ signaling pathway is a key regulator of this process and the transcription factors Snail, Slug, Zeb 1/2 and Twist are often overexpressed [64]. The EMT has been well established for epithelial tumors, but a similar process that promotes cell migration and cell state changes in mesenchymal tumors is unknown [1]. Invasion is usually initiated in malignant cells localized at the edges of a tumor mass and is associated with the degradation of the surrounding matrix by secreted enzymes. These proteolytic enzymes are produced by both cancer cells and stromal cells, including recruited blood cells. This process not only allows break down of cell-cell adhesions required for migration, but also leads to activation of cytokines and growth factors that are trapped in the ECM [1].

The colonization step is carried out by tumor cells that have survived the hostile blood circulation and extravasate into distal sites and tissues. Whilst in the vascular bed, tumor cells are shielded by aggregating platelets [65] and can proliferate in the lumen [1]. Platelets contribute to the interaction between tumor cells and endothelium, favoring adhesion and extravasation [66]. When compared to diapedesis, the multistep mechanism that leads to recruitment of immune cells to inflamed tissues from the blood circulation, cancer cell extravasation might be instead a "clumsy" process, that facilitates in part the breakdown of the endothelium [67]. Following extravasation, adaptation to the new microenvironment is necessary for successful metastatic growth. In the case of epithelial tumors, the mesenchymal to epithelial transition (MET) precedes the establishment of micrometastasis. Interestingly, tumors that have metastasized are able to undergo the reseeding process, sometimes localizing back to the original organ (self-seeding) [68]. Metastasis can arise years after the primary lesion has been diagnosed and surgically removed or pharmacologically treated. Mechanisms of dormancy seem to be primarily responsible. Preferential location of metastasis associates with the organization of the cardiocirculatory system in 66 % of tumors, while in 20 % of the circumstances the local microenvironment plays a major role; in the residual 14 %, secondary tissues hijack the colonization, even though highly perfused [67].

In 1889 physician Stephen Paget formulated the "seed and soil" theory as explanation for tropism of metastasis. He postulated that tumor cells disseminate widely throughout the body, but can only metastasize to supportive environments [69]. The main caveat in the theory is the difficulty in explaining the very low incidence of bilateral tumors (e.g. breast and kidney cancers). Hostile tissues might accommodate malignancy in specific circumstances, such as inflammation. Moreover, secondary locations can be colonized, upon release of chemoattractants. Chemokine signaling seems to play a cardinal role; particularly the CXCL12/SDF1 gradient facilitates the seeding of tumor cells expressing the correspondent receptor CXCR4, through a directional cell migration event, known as chemotaxis [70]. In addition, tumors release exosomes, that might facilitate organ tropism for metastasis, have an effect on immunoediting and EMT [71].

The heterogeneity of cancer has made it thus far difficult to unravel the biological mechanisms underlying metastasis in different tumor types. It has been shown for breast and prostate cancers that the tendency to home to the bone relates to their interactions with osteoblasts and osteoclasts respectively. In particular, once in the bone, metastatic breast cancer cells re-utilize a mechanism that is activated during lactation. Production of the parathyroid hormone-related peptide (PTHrP) induces the exposure on the osteoblast cell membrane of the substrate RANKL (receptor activator for NF-κB ligand); this molecule binds RANK on osteoclasts, resulting in their activation. Functional osteoclasts demineralize the bone, which leads to release of calcium, needed during lactation, degradation of ECM and release of growth factors, supporting tumor growth [67].

Current understanding of metastatic spread has been achieved with the use of animal models. A common method to monitor metastasis formation in rodents uses bioluminescence imaging, which is useful for assessing tumor spread to secondary sites, but is limited by resolution. It is also possible to introduce imaging windows on organs in the abdomen [72] and perform live imaging up to a month, as has been done to study tumor angiogenesis, by injecting FITC-dextran and using dorsal skinfold windows [73]. Moreover cranial windows allow imaging of metastasis to the brain and have been used to follow angiogenesis [74]. Imaging windows have also been placed on lungs, but imaging has been technically difficult due to the respiratory rhythm [75]. Liver exteriorization showed angiogenesis in colorectal cancer metastasis [76] and ex vivo lung imaging has highlighted the role of the myeloid cells in tumor cell extravasation and metastasis [77]. These methods are limited by the challenge of balancing the imaging of a complex and possibly long process with the survival of the animal and does not allow simultaneous visualization of colonization in different sites. All applications are more difficult when tumors arise or metastasize in deeper tissues [33].

Current therapies to treat cancer are mainly surgery, chemotherapy, radiotherapy, hormone-based treatments for specific tumor types and targeted therapies, when available. As previously discussed, combination therapies are often better approaches to limit resistance development, which normally occurs due to the presence of resistant subclones or alterations in the tumor microenvironment (TME). Most drugs currently in use work by disrupting the cell cycle and the genome integrity, hence causing systemic long-term side effects. By contrast, targeted therapy is now widely applied in the clinic, where the consideration of patient specific mutations allows optimal treatment. Therapies that target the tumor microenvironment (TME) are also under active development. Reprogramming of TAMs is a promising future therapeutic strategy [78].

Although the cellular steps leading to metastasis are known (invasion, intravasation, dissemination through circulation, extravasation, micrometastasis, colonization and macrometastasis), the molecular mechanisms responsible for each of them are poorly understood and new approaches as well as new therapies are needed.

Imaging Tumor Invasion, Foci Formation and Dissemination in the Zebrafish Embryo Model

Xenograft engraftment of tumor cells into optically clear zebrafish has allowed dynamic visualization and new mechanistic insights into the processes by which cancer cells migrate, colonize local tissues, and disseminate throughout the rest of the body. Several implantation sites of tumor cells have been described in zebrafish embryo [14]. Engraftment in the yolk sac of 2-day-old zebrafish embryo avoids disruption of organ development. In 2006, Haldi and colleagues showed that human metastatic melanoma cells WM-266-4 migrated from the yolk to other tissues in the embryo and formed masses in neighboring organs, like intestine, liver and pancreas [51].

An automated method has now been developed to quantify tumor cell migration and dissemination in the whole body. Using confocal imaging, very similar patterns of tumor cell dissemination were revealed in zebrafish as in the mouse model. Importantly, this work showed that similar metastatic processes were observed in a wide range of cancers including prostate, breast and colorectal cancers [79]. Using the xenograft zebrafish experimental platform, a role for the non-tyrosine kinase (SYK) in advanced prostate cancer was discovered. Inoculation of prostate cell line PC3-M-Pro4 luc transduced with shRNA against SYK, resulted in less bone metastasis and less disseminated tumor foci both in mouse and zebrafish. Pharmacological inhibition of SYK also resulted in a suppressed tumor growth both in vitro and when grown in the zebrafish xenograft model [80].

Quantification of tumor burden in the whole organism has also led to important findings concerning Ewing sarcoma. A diverse migration pattern was identified following engraftment of several sarcoma lines into embryos and juvenile zebrafish. Using chemical treatment with Nutlin-3a and YK-4-279, p53 could be stabilized and inhibit tumor growth in vivo [49]. Furthermore, p53 acetylation, required to destabilize the interaction with the ubiquitinase MDM2, was preserved. Cell cycle was also arrested when the activity of the deacetylases SIRT 1/2 was impaired, using tnv-6. Thus, the TC-252 Ewing sarcoma cell line, with high levels of SIRT 1/2, failed to migrate and to proliferate upon tnv-6 treatment in vivo [81].

Another advantage of the zebrafish embryo model is the ability to apply compounds directly to the water and assess drug efficacy on tumor burden, while also monitoring possible side effects on embryo development. As an example, a pharmacological approach was designed to target uveal melanoma. A set of cell lines derived from primary tumor and metastasis from the same patients was used. The employment of experimental drugs affecting Src signaling (dasatinib), histone deacetylase activity (quisinostat) and neddylation pathway (MLN-4924) resulted in reduction of tumor proliferation and migration [82].

As described so far, there are many pathways and mechanisms that contribute to the invasive properties of tumor cells. Additionally, using zebrafish as a model for cancer biology has uncovered the importance of communication between cells to facilitate invasion. For example, the simultaneous implantation of two melanoma cell lines, highly invasive WM-2664 and less invasive 501mel has shown that co-culture will lead to increased invasion of less migratory 501mel cells in larval zebrafish. Metalloproteases (MMP) and deposition of extracellular matrix (ECM) and fibronectin fibers had roles in mediating this cooperative invasive mechanism. WM-2664 expressed higher levels of MMPs, compared to the less aggressive 501mel cell line. Inhibition of MMP function in WM-2664 cells by siRNA resulted in reduced cooperative invasion but had no effect on migration of these cells when implanted singly. These data suggest that the most aggressive cell line normally invades through mechanisms which are independent from the MMP function. Interestingly, deposition of collagen and fibronectin fibers increased in the graft derived of both cell types but was not seen in tumors arising from only WM-2664 cells. These data again strongly suggest cross-talk between the two cell populations [83].

Currently one of the main challenges for cancer biologist is to understand the mechanisms of cancer recurrence following therapy. Mechanisms of drug resistance are believed to be mainly associated with the survival of often quiescent tumor initiating cells (TICs). These cells are also commonly called tumor propagating cells and have stem cell-like properties that drive self-renewal and continued tumor growth. The zebrafish embryo model has been used to evaluate TICs in prostate cancer. TICs were isolated from both cell lines and primary patient tumors based on their preferential high adherence to plating on collagen and laminin. The highly adherent cells had increased levels of the stem cell markers CD44 and $\alpha 2\beta 1$ integrin and showed augmented expression of CD133 when grown as spheroids. TICs were labelled with quantum dots (QD) and inoculated into zebrafish embryos. Remarkably, TICs formed localized tumor masses at the site of injection and could migrate to other tissues including the brain. Tumor cells engrafted in the zebrafish maintained their TIC features, as their isolation and re-implantation lead to the formation of new tumors when serially engrafted into embryos or juvenile zebrafish [84].

The role of the stem cell pool in glioblastoma invasion has also been investigated in the zebrafish xenograft model. Glioblastoma is the most common tumor of the central nervous system and is a very aggressive type of cancer. U87-RFP cells could engraft into 2 dpf embryos and appeared to intravasate and then widely disseminate through the vasculature. The more invasive glioblastoma cells were positive for the stem cell marker CD133. Moreover, pre-sorted $CD133^+$ glioblastoma U87 cells showed higher migratory properties, compared to $CD133^-$ cells. Interestingly, the presence of $CD133^+$ cells correlated with an elevated expression of zebrafish MMP9 and MMP2. MMP9 inhibitor treatment resulted in reduced U87 aggressiveness, suggesting a key role of the host ECM in influencing tumor behavior [85]. Rampazzo and colleagues also evaluated roles for stem cells in glioblastoma multiforme (GBM). For example, they showed the interplay between Hif1α, Wnt and Notch signaling could lead to neuronal differentiation of glioblastoma multiforme (GBM) stem-like cells following engraftment into the brain of zebrafish embryos [86]. Further xenograft studies revealed a novel mechanism by which GBM undergoes cell death. This work defined that GBM cells die by a catastrophic vacuolization with formation of macropinosomes that was not linked to apoptosis or autophagy. The delivery of Vacquinol-1, belonging to a class of small molecules originally designed for malaria, resulted in successfully induction of cell death in GBM cells and prolongation of rodent survival. Similar results were seen in xenografts grown in larval zebrafish where Vacquinol-1 inhibited growth of tumor-like masses and suppressed infiltration into the brain. These findings may lead to new therapies that target quiescent tumor cells with stem cell like features [87].

Imaging Micrometastasis Formation in a Zebrafish Xenograft Model

The formation of micrometastasis can be studied upon tumor cell injection directly into the blood circulation following injection into the Duct of Cuvier or CCV [54]. Circulating tumor cells travel in the blood and later localize at the end of the circulatory loop, where the dorsal aorta (DA) turns into the caudal vein (CV). Depending on the cell line, single cell or collective invasion of the surrounding tissues can occur. Tumor cells often accumulate at the end of the tail and proliferate, both inside and outside the blood vessels. In some cases, the linear structure of the endothelium (Fig. 3a) is lost and a visible enlargement of the vein (Fig. 3b) occurs due to vessel obstruction by accumulating and proliferating tumor cells (Fig. 3c, d). The ability of tumor cells to leave the blood flow and colonize the tissue of the local tail fin is an active process and is not observed in all cell lines [54, 88].

As previously noted, the epithelial-to-mesenchymal transition (EMT) is an important process, responsible for a more invasive and aggressive behavior of cancers.

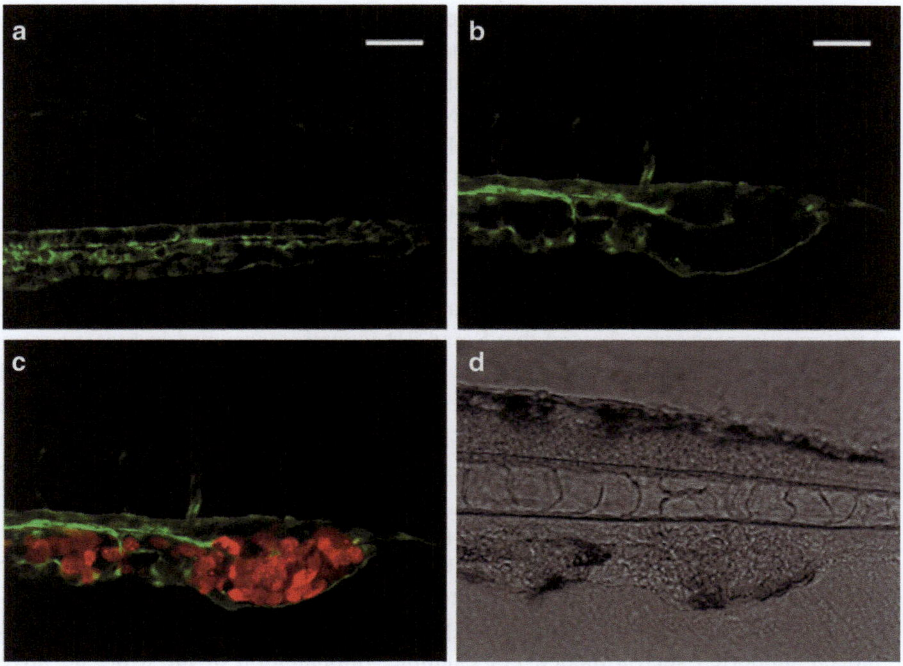

Fig. 3 Tumor cells modify the vascular lumen. Circulating tumor cells form a tumor clog and interact with the inner endothelial layer of the vessel, in proximity of the circulatory loop at the end of the tail. The regular vessel structure in a Tg(kdrl:EGFP)^{s843} embryo (**a**) is compared with a thickened vein in (**b**), where MDA-231-B DsRed tumor cells pressure the endothelium (**c**). Tumor cell aggregate can be visualized by bright field imaging (**d**). Scale bar: 50 μm

Evidence from the zebrafish model have contributed to our understanding of how the TGFβ pathway participates in promoting the invasive phenotype of human tumor cells. Ectopic expression of murine mSnail and mSlug in H-Ras transformed MCF10AT (M-II) led to a shift from collective to single cell invasion motility pattern, in in vitro 3D spheroid assay as well as in zebrafish embryo. The invasive phenotype appeared at 3 dpi and was fully developed at 5 dpi where >55 % of the larvae showed signs of disseminated tumor cells. This represents more than tenfold increase in number of embryos developing tail metastasis (>30 cells) as has been observed for tumor cells overexpressing one of the two TGFβ related transcription factors. Therefore Snail and Slug are involved in the metastatic process, promoting single cell invasion [89]. Inhibition of invasion and metastasis was also addressed using chemical inhibitors of TGFβ type I receptor kinase, namely SB-431542 and LY-294002. Breast cancer cell lines were exposed to these drugs for 24 h prior to implantation into embryos. Following engraftment, animals were administered drug every other day until 6 dpi. Upon activation of the TGFβ signaling pathway, phosphorylation of Smad2 and dimerization with Smad4 occurred. This was consequent with migration of the Smad complex into the nucleus. The pharmacological treatment with the TGFβ type I receptor kinase inhibitor impaired the phosphorylation of

Smad2 and at the same time the invasive phenotype of the breast cancer cell lines. Moreover, reduction of the metastatic phenotype was seen upon genetic and chemical inhibition of Smad4 by either RNA interference or siRNA. The link between the TGFβ signaling and metalloprotease (MMP) activity was also considered, as inhibition of MMP2/9 has been found to correlate with diminished breast cancer invasion and was dependent on TGFβ signaling. The use of a MMP2/9 inhibitors and general MMP inhibitors like Galardin or Ilomastat led to reduction in cell migration in the zebrafish embryo model. By contrast, activation of the TGFβ signaling pathway led to a remarkable increase in invasion most likely by effecting paracrine communication. In vitro assays demonstrated human tumor cells respond to zebrafish TGFβ. Luciferase experiments showed that Smad 3/4 induction occurred in breast cancer cell lines upon stimulation with zTGFβ. Zebrafish Smad2 was phosphorylated and translocated into the nucleus of zebrafish cells, upon addition of human TGFβ in the culture medium [88]. These data suggest a cross talk of signaling between zebrafish host cells and xenografted human cells is likely in live animals and important for regulating cell motility.

The zebrafish embryo represents a powerful tool to visualize and scrutinize micrometastasis formation for different tumor types. It is also possible to compare the behavior of tumor lines in vivo and relate that to their classified level of malignancy. The xenogeneic engraftment of a panel of human breast cancer cell lines has indeed supported such a theory. For example, implantation of the triple negative breast carcinoma line MDA-MB-231 resulted in invasion of the tail fin after 12 h after hematologic cell inoculation. Metastasis was found in 54 % of the larvae engrafted with MDA-MB-231 cells A similar phenotype was found upon implantation of high grade carcinoma MCF10CA1a.cl1 or M4 breast cancer cells which have been shown to be high metastatic in other cancer models. By contrast, normal breast epithelial cells MCF10A or M1 and pre-malignant cells MCF10AT1k.cl2 or M2 exhibited limited invasive ability [88]. Interestingly, aggressive MDA-MB-231 mainly colonized the caudal hematopoietic tissue (CHT) (Fig. 4a) as single cells (Fig. 4a, left and c). Similar behavior was observed for MAE-T-FGF2 (Fig. 4b) and melanoma lines (Fig. 4d). M4 breast lines disseminated as collective cells (Fig. 4a, right and e) and were localized between the blood vessels. We also observed collective cell invasion and extravasation using the prostate cell line PC3-Pro-4 (Fig. 4f) and the Ewing sarcoma line A673 (Fig. 4g). Diversity in malignant behavior was also shown by He et al., 2012 where tumor cells differently formed micrometastasis in the tail fin and were able to induce vascularization and localized tumor growth. FGF2-T-MAE cells induced angiogenesis in 97 % of the larvae and tumor invasion in 32.8 % of animals. 4T1 mouse breast cancer cells were highly angiogenic, but less invasive, while human MDA-MB-231 did not lead to new vessel formation but displayed tumor metastasis in almost 50 % of the larvae. The opposite behavior was recorded for human prostate cancer cell line PC3, with 25 % of angiogenesis and no invasion in the tail region. Finally, mouse endothelial cells MAE and immortalized zebrafish lines ZF4 and PAC2 did not have effects on vascularization, localized tumor growth or tumor invasion and micrometastasis [54]. Taken together, these studies suggest that invasion is an active process and can be studied in zebrafish, which represent a powerful tool to gain more knowledge on tumor metastasis.

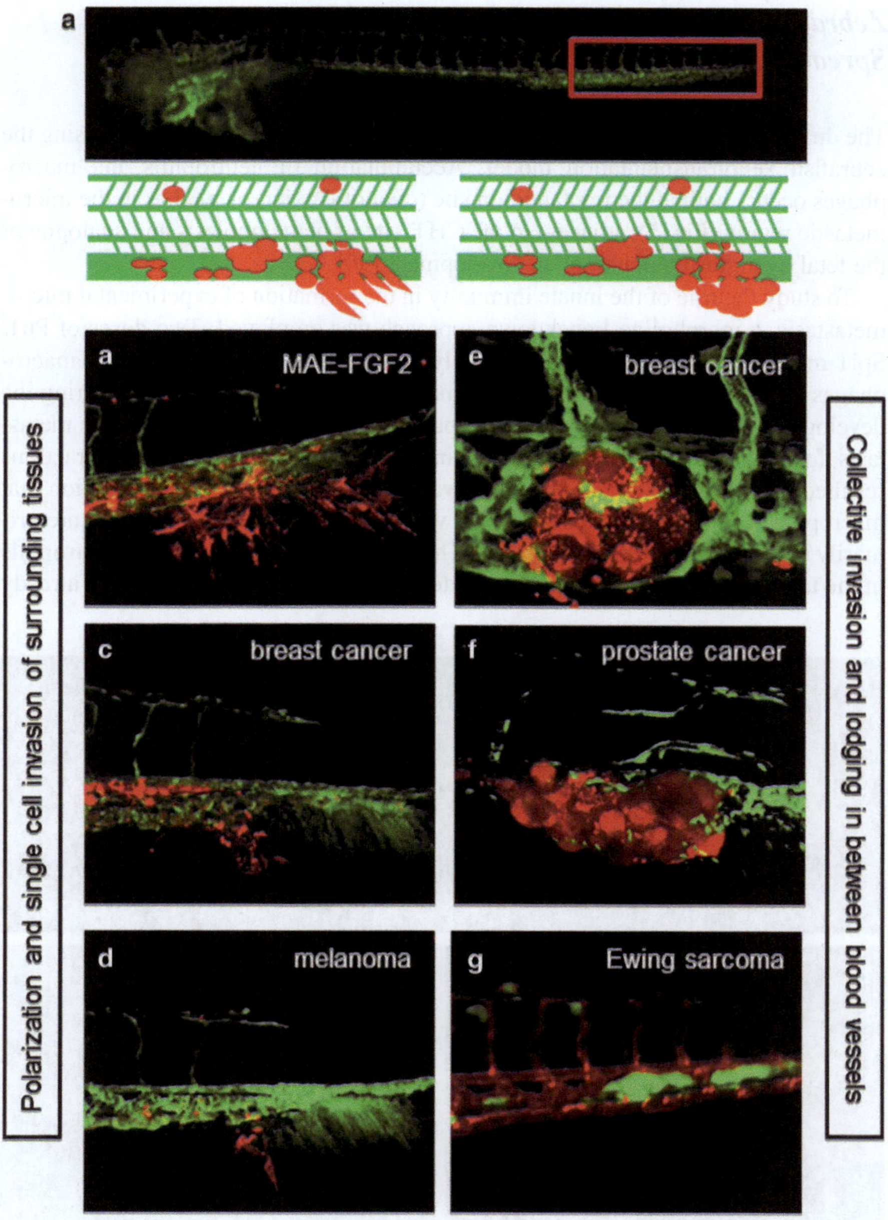

Fig. 4 Single cell and collective invasion lead to tumor micrometastasis formation. Human tumor cell lines were engrafted in the blood circulation of 2-day-old zebrafish embryo and micrometastasis occurred in the CHT region. Schematic in (**a**) show single (*left*) and collective (*right*) tumor invasion in the CHT (*red box*). Tissue colonization occurred upon single cell invasion in the case of MAE-FGF2 (**b**), MDA-MB-231 (**c**) and melanoma (**d**) or collective invasion in PC3-Pro-4 (**f**). Growth of tumor cells in between blood vessels was also observed for M4 (**e**) and Ewing sarcoma A673 (**g**). Tumor cells were engrafted in Tg(fli1a:EGFP)^{y1} (**b–f**) or Tg(kdrl:mCherry) [97] (**g**)

Zebrafish Neutrophils and Macrophages Support Metastatic Spreading

The dual role of the immune system in cancer biology has been addressed using the zebrafish xenotransplantation model. Accumulation of neutrophils and macrophages occurs both at the implantation site (duct of Cuvier) as well as in the micrometastic region (Fig. 5), adjacent to the CHT, site of hematopoiesis and analogue of the fetal liver, during mammalian development [90].

To study the role of the innate immunity in the formation of experimental micrometastasis, a morpholino knockdown approach was employed. Two doses of Pu1/Spi1 morpholino were used to selectively block the development of only macrophages (1 mM) or both neutrophils and macrophages (2 mM). While impairing the development of the only macrophage population did not have any effects on metastasis formation by FGF2-T-MAE, the inhibition of both myeloid cell fractions resulted in a clear inhibition of tumor invasion. Our data led to the conclusion that macrophages play a major role in tumor vascularization, while neutrophils are primarily responsible of tumor metastasis. The physiological movement of neutrophils in the tail fin was studied using the reporter line *Tg(mpx:GFP)*^{il114} [43] and already

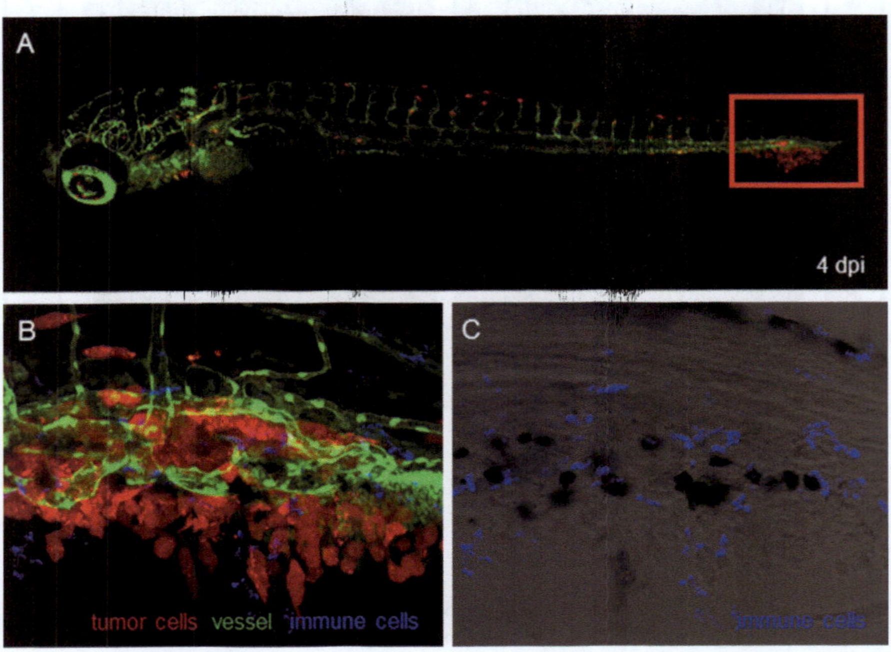

Fig. 5 Tissue colonization by tumor cells is accompanied by immune cell migration. Metastatic breast cancer cell line MDA-231-B stably expressing DsRed protein is implanted in the blood circulation of 2-day-old Tg(fli1a:EGFP)^{y1} zebrafish embryo and showed single cell invasion of the tail fin at 4 dpi (**a, b**). The site of invasion is infiltrated by L-Plastin^{+} immune cells (**b**). (**c**) is a bright field image of (**b**)

in absence of cancer cells it was found that these immune cells have tendency to randomly migrate in and out of the caudal hematopoietic tissue. During their transmigration, neutrophils create paths in ECM, modifying the collagen structure. Neutrophils trajectories were visualized with video time-lapse and the deformation of the collagen fibers was analyzed using two-photon confocal microscopy and second harmonic generation (SHG) microscopy. The presence of tumor cells in the region of transmigration did not affect neutrophil motility, instead the random migration created an ideal environment for tumor cell invasion. It was therefore possible to conclude that neutrophils are involved in the preparation of the metastatic niche. Inhibition of neutrophil migration was achieved by immunosuppression through beclomethasone and resulted in reduced tumor invasion, while increased metastatic potential was observed upon β-APN (β-aminopropionitrile) treatment, which diminished the collagen fibers and led to an enhanced neutrophil migration. No effects on macrophage behavior were observed, supporting the hypothesis that neutrophils are the main players in the metastatic process. Interestingly, the formation of experimental micrometastasis was found to be enhanced when VEGFR inhibitor treatment was applied in order to reduce angiogenesis at the primary tumor site. A diminished tumor vasculature was accompanied by lower macrophage number and change in cell shape, with a blockage of migration. On the other hand, the number of neutrophils was the same, but the VEGFR inhibition caused increased neutrophil motility from the CHT to the tail fin and vice versa, hence promoting tumor cell invasiveness. Restoration of normal neutrophil motility after VEGFR signaling alteration was reached by beclomethasone treatment and normal tumor metastasis was re-established. The role of inflammation in tumor invasion was also analyzed and the metastatic process was compared to wound-healing mechanism. Wounding of the tail fin showed a decrease in tumor metastasis, perhaps due to higher phagocytic activity of immune cells. Parallels between tumor and wound-associated inflammation have been made, but immune cells might activate a different response towards tumor cells, supportive of tumor angiogenesis and metastasis, in the absence of wound-related inflammation and characterized by anti-tumor properties in case of tissue damage [54].

The interaction between tumor cells and associated macrophages and how the cross talk influences tumor behavior in terms of dissemination and metastatic potential have been visualized in transparent zebrafish embryos. Tumor cells and macrophages from mouse and human origin have been labelled with transient cell tracker dyes (DiI and DiD respectively) and co-injected in the perivitellin space of 2-day-old zebrafish embryos, with fluorescent blood vessels. Only upon co-injection with tumor associated macrophages (TAMs), cancer cells displayed an aggressive phenotype, characterized by intravasation and formation of cell aggregates, distantly from the engraftment site, at 4 dpi. Development of tumor foci from T241 cells occurred in presence of macrophages isolated from Lewis lung carcinoma (LLC) cells, engineered to overexpress IL6 and TNFα cytokines. Exclusively in the latter case, cancer cells co-localized with macrophages. The same results were obtained upon injection of macrophages, differentiated from bone-marrow derived monocytes, exposed in vitro to the same cytokines. It was found that only M2 and not M1 macrophages were able to support tumor dissemination. First T241 fibrosarcoma cells were engrafted together with F4/80⁺PDL1⁺M1 or F4/80⁺CD206⁺M2 macrophages isolated from mouse LLC tumors, then the co-injection was performed with primary bone marrow monocytes differentiated towards M1, via LPS and IFNy stimulation or skewed towards M2 phenotype using IL4, IL10 and TGFβ. Only in presence of isolated M2 macrophages or M2 differentiated monocytes, intravasation and formation of single tumor foci was observed. Moreover, T241 cells were associated to the macrophages with M2 characteristics. The clinical relevance of the finding was underlined by isolation of macrophages from different human tumor tissues and co-injection with human cancer cells. The ovarian tumor cell line OVCAR8 displayed a more aggressive behavior, when co-injected with macrophages derived from metastatic human endometrial cancer. Macrophages associated to a non-metastatic form of cancer did not contribute to the OVCAR8 spreading in zebrafish.

In this way, the zebrafish embryo model has allowed the visualization on a whole organism level of important dynamics of the early steps of tumor metastasis and the role of macrophages, specifically with a M2 phenotype, in worsening cancer spreading and consequent patient prognosis [91].

Myeloid Cells Provide Trophic Signals and Sustain Early Phases of Malignant Transformation in Zebrafish Transgenic Models

The role of the microenvironment in cancer biology, particularly myeloid cells, has been also addressed in the context of zebrafish transgenic models to study tumor initiation. Parallels in the response of the innate immune system towards wounds and transformed cells have been made. Neutrophils and macrophages often associate with transformed cells through filopodia extension and cytoplasmic tethering. Similar interactions are found in stromal cells at wound sites. Immune cells appeared to display an anti-tumor behavior, as neutrophils broke off membrane parts belonging to transformed cells, while macrophages were able to activate phagocytic mechanisms. On the other hand, immune cell recruitment showed important contributions in transformed cell proliferation, which was prevented in Pu1 morphants. Hydrogen peroxide (H_2O_2) was identified as a trigger that regulated leukocyte accumulation to transformed cell as well as damaged tissues [92]. The recruitment of immune cells and their accumulation around transformed cells resemble the acute and chronic phase of inflammation at wounds, suggesting a comparison between cancer and never-healing-lesions. Leukocytes sustained tumor cell growth, providing trophic signals like prostaglandin PGE_2. Interestingly, inhibition of the COX-2 enzyme, involved in the cascade synthesis of PGE_2, limited transformed cell growth, linking the administration of aspirin, belonging to the non-steroidal anti-inflammatory class of drugs (NSAID), to reduced tumor incidence [93].

The recruitment of immune cells, mainly neutrophils, to transformed cells has been found to be regulated also by the chemokine signaling, particularly the IL8-CXCR2 axis. The outcome of the interaction between neutrophils and transformed cells, expressing activated HRas ($HRas^{V12}$), was the induction of epithelial to mesenchymal transition (EMT). Snail, slug and vimentin were up-regulated in $HRas^{V12+}$ cells, while mmp9 and slug were down-regulated in a *Tg (mpx:mcherry, Rac2*^{D57N}*)* background, characterized by impaired neutrophil motility towards wounds [94]. Macrophages were also detected in the transformed cell area, but no effect on EMT was detected, when their development was impaired as a result of irf8 morpholino knockdown.

The CXCL8-CXCR2 axis has been proposed as signaling pathway that mediates neutrophil attraction towards transformed cells and regulates the EMT process. $HRas^{V12+}$ cells were found to express IL8 and were surrounded by a lower number of neutrophils when CXCR2 signaling was impaired in the host via receptor knockdown. Moreover CXCR2 signaling inhibition led to a reduced expression of the EMT markers. Interestingly, neutrophil recruitment and the EMT activation were also affected when CXCR2 was inhibited exclusively in the transformed cells [95].

Conclusions

The zebrafish embryo model greatly facilitates the study of key mechanisms of cancer progression and therefore complements already established in vivo systems to study tumor angiogenesis and metastasis. It contributes to the field of tumor biology by revealing live tumor-stroma cell interactions and dynamic live cell imaging of tumor development in a whole organism. Imaging of single cell interactions is relatively easy and allows the visualization of the angiogenic process as well as early and late steps of metastasis formation. Zebrafish represent a promising model to unravel molecular mechanisms that orchestrate metastatic spreading and offers the opportunity to develop new therapeutic strategies.

Acknowledgments The authors would like to thank A.H. Meijer for scientific discussion and P.M. Elks for critical reading of the chapter. The work was supported by the Netherlands Organization for Scientific Research (TOP GO Grant: 854.10.012).

References

1. Hanahan D, Weinberg RA (2011) Hallmarks of cancer: the next generation. Cell 144(5):646–674
2. Amatruda JF, Shepard JL, Stern HM et al (2002) Zebrafish as a cancer model system. Cancer Cell 1(3):229–231
3. Amatruda JF, Patton EE (2008) Genetic models of cancer in zebrafish. Int Rev Cell Mol Biol 271:1–34

4. Lam SH, Wu YL, Vega VB et al (2006) Conservation of gene expression signatures between zebrafish and human liver tumors and tumor progression. Nat Biotechnol 24(1):73–75
5. Mizgireuv IV, Revskoy SY (2006) Transplantable tumor lines generated in clonal zebrafish. Cancer Res 66(6):3120–3125
6. Spitsbergen JM, Tsai HW, Reddy A et al (2000) Neoplasia in zebrafish (Danio rerio) treated with 7,12-dimethylbenz[a]anthracene by two exposure routes at different developmental stages. Toxicol Pathol 28(5):705–715
7. Spitsbergen JM, Tsai HW, Reddy A et al (2000) Neoplasia in zebrafish (Danio rerio) treated with N-methyl-N′-nitro-N-nitrosoguanidine by three exposure routes at different developmental stages. Toxicol Pathol 28(5):716–725
8. Neumann JC, Dovey JS, Chandler GL et al (2009) Identification of a heritable model of testicular germ cell tumor in the zebrafish. Zebrafish 6(4):319–327
9. Langenau DM, Traver D, Ferrando AA et al (2003) Myc-induced T cell leukemia in transgenic zebrafish. Science 299(5608):887–890
10. Langenau DM, Keefe MD, Storer NY et al (2007) Effects of RAS on the genesis of embryonal rhabdomyosarcoma. Genes Dev 21(11):1382–1395
11. Park SW, Davison JM, Rhee J et al (2008) Oncogenic KRAS induces progenitor cell expansion and malignant transformation in zebrafish exocrine pancreas. Gastroenterology 134(7):2080–2090
12. Santoriello C, Deflorian G, Pezzimenti F et al (2009) Expression of H-RASV12 in a zebrafish model of Costello syndrome causes cellular senescence in adult proliferating cells. Dis Model Mech 2(1-2):56–67
13. White RM, Sessa A, Burke C et al (2008) Transparent adult zebrafish as a tool for in vivo transplantation analysis. Cell Stem Cell 2(2):183–189
14. Konantz M, Balci TB, Hartwig UF et al (2012) Zebrafish xenografts as a tool for in vivo studies on human cancer. Ann N Y Acad Sci 1266:124–137
15. Lam SH, Chua HL, Gong Z et al (2004) Development and maturation of the immune system in zebrafish, Danio rerio: a gene expression profiling, in situ hybridization and immunological study. Dev Comp Immunol 28(1):9–28
16. Jo DH, Son D, Na Y et al (2013) Orthotopic transplantation of retinoblastoma cells into vitreous cavity of zebrafish for screening of anticancer drugs. Mol Cancer 12
17. Eden CJ, Ju B, Murugesan M et al (2014) Orthotopic models of pediatric brain tumors in zebrafish. Oncogene
18. Tang Q, Abdelfattah NS, Blackburn JS et al (2014) Optimized cell transplantation using adult rag2 mutant zebrafish. Nat Methods 11(8):821–824
19. Tenente IM, Tang Q, Moore JC et al (2014) Normal and malignant muscle cell transplantation into immune compromised adult zebrafish. J Vis Exp 94
20. Carmeliet P, Jain RK (2000) Angiogenesis in cancer and other diseases. Nature 407(6801):249–257
21. Gacche RN, Meshram RJ (2014) Angiogenic factors as potential drug target: efficacy and limitations of anti-angiogenic therapy. Biochim Biophys Acta 1846(1):161–179
22. Ferrara N, Gerber HP, LeCouter J (2003) The biology of VEGF and its receptors. Nat Med 9(6):669–676
23. Gerhardt H, Golding M, Fruttiger M et al (2003) VEGF guides angiogenic sprouting utilizing endothelial tip cell filopodia. J Cell Biol 161(6):1163–1177
24. Wilkinson RN, van Eeden FJM (2014) The zebrafish as a model of vascular development and disease. Prog Mol Biol Transl Sci 124:93–122
25. Caduff JH, Fischer LC, Burri PH (1986) Scanning electron-microscope study of the developing microvasculature in the postnatal rat lung. Anat Rec 216(2):154–164
26. Burri PH, Tarek MR (1990) A novel mechanism of capillary growth in the rat pulmonary microcirculation. Anat Rec 228(1):35–45
27. Hlushchuk R, Makanya AN, Djonov V (2011) Escape mechanisms after antiangiogenic treatment, or why are the tumors growing again? Int J Dev Biol 55(4-5):563–567

28. Germain S, Monnot C, Muller L et al (2010) Hypoxia-driven angiogenesis: role of tip cells and extracellular matrix scaffolding. Curr Opin Hematol 17(3):245–251
29. Bertout JA, Patel SA, Simon MC (2008) The impact of O2 availability on human cancer. Nat Rev Cancer 8(12):967–975
30. Goswami S, Sahai E, Wyckoff JB et al (2005) Macrophages promote the invasion of breast carcinoma cells via a colony-stimulating factor-1/epidermal growth factor paracrine loop. Cancer Res 65(12):5278–5283
31. Wyckoff JB, Wang Y, Lin EY et al (2007) Direct visualization of macrophage-assisted tumor cell intravasation in mammary tumors. Cancer Res 67(6):2649–2656
32. Zervantonakis IK, Hughes-Alford SK, Charest JL et al (2012) Three-dimensional microfluidic model for tumor cell intravasation and endothelial barrier function. Proc Natl Acad Sci U S A 109(34):13515–13520
33. Fein MR, Egeblad M (2013) Caught in the act: revealing the metastatic process by live imaging. Dis Model Mech 6(3):580–593
34. Folkman J (1971) Tumor angiogenesis: therapeutic implications. N Engl J Med 285(21):1182–1186
35. Staton CA, Reed MW, Brown NJ (2009) A critical analysis of current in vitro and in vivo angiogenesis assays. Int J Exp Pathol 90(3):195–221
36. Vogel HB, Berry RG (1975) Chorioallantoic membrane heterotransplantation of human brain tumors. Int J Cancer 15(3):401–408
37. Ribatti D (2014) The chick embryo chorioallantoic membrane as a model for tumor biology. Exp Cell Res 328(2):314–324
38. Verheul HM, Pinedo HM (2007) Possible molecular mechanisms involved in the toxicity of angiogenesis inhibition. Nat Rev Cancer 7(6):475–485
39. Estanqueiro M, Amaral MH, Conceicao J et al (2015) Nanotechnological carriers for cancer chemotherapy: the state of the art. Colloids Surf B Biointerfaces 126:631–648
40. Lawson ND, Weinstein BM (2002) In vivo imaging of embryonic vascular development using transgenic zebrafish. Dev Biol 248(2):307–318
41. Wang Y, Pan L, Moens CB et al (2014) Notch3 establishes brain vascular integrity by regulating pericyte number. Development 141(2):307–317
42. Traver D, Paw BH, Poss KD et al (2003) Transplantation and in vivo imaging of multilineage engraftment in zebrafish bloodless mutants. Nat Immunol 4(12):1238–1246
43. Renshaw SA, Loynes CA, Trushell DM et al (2006) A transgenic zebrafish model of neutrophilic inflammation. Blood 108(13):3976–3978
44. Ellett F, Pase L, Hayman JW et al (2011) mpegl promoter transgenes direct macrophage-lineage expression in zebrafish. Blood 117(4):e49–e56
45. Nicoli S, Presta M (2007) The zebrafish/tumor xenograft angiogenesis assay. Nat Protoc 2(11):2918–2923
46. Nicoli S, Ribatti D, Cotelli F et al (2007) Mammalian tumor xenografts induce neovascularization in zebrafish embryos. Cancer Res 67(7):2927–2931
47. Nicoli S, De Sena G, Presta M (2009) Fibroblast growth factor 2-induced angiogenesis in zebrafish: the zebrafish yolk membrane (ZFYM) angiogenesis assay. J Cell Mol Med 13(8B):2061–2068
48. Zhao C, Wang X, Zhao Y et al (2011) A novel xenograft model in zebrafish for high-resolution investigating dynamics of neovascularization in tumors. PLoS One 6(7), e21768
49. van der Ent W, Jochemsen AG, Teunisse AF et al (2014) Ewing sarcoma inhibition by disruption of EWSR1-FLI1 transcriptional activity and reactivation of p53. J Pathol 233(4):415–424
50. Vitale G, Gaudenzi G, Dicitore A et al (2014) Zebrafish as an innovative model for neuroendocrine tumors. Endocr Relat Cancer 21(1):R67–R83
51. Haldi M, Ton C, Seng WL et al (2006) Human melanoma cells transplanted into zebrafish proliferate, migrate, produce melanin, form masses and stimulate angiogenesis in zebrafish. Angiogenesis 9(3):139–151

52. Lee SL, Rouhi P, Dahl Jensen L et al (2009) Hypoxia-induced pathological angiogenesis mediates tumor cell dissemination, invasion, and metastasis in a zebrafish tumor model. Proc Natl Acad Sci U S A 106(46):19485–19490
53. Tobia C, Gariano G, De Sena G et al (2013) Zebrafish embryo as a tool to study tumor/endothelial cell cross-talk. Biochim Biophys Acta 1832(9):1371–1377
54. He S, Lamers GE, Beenakker JW et al (2012) Neutrophil-mediated experimental metastasis is enhanced by VEGFR inhibition in a zebrafish xenograft model. J Pathol 227(4):431–445
55. Wu Q, Deng S, Li L et al (2013) Biodegradable polymeric micelle-encapsulated quercetin suppresses tumor growth and metastasis in both transgenic zebrafish and mouse models. Nanoscale 5(24):12480–12493
56. Wu Q, Li G, Deng S et al (2014) Enhanced antitumor activity and mechanism of biodegradable polymeric micelles-encapsulated chetomin in both transgenic zebrafish and mouse models. Nanoscale 6(20):11940–11952
57. Harfouche R, Basu S, Soni S et al (2009) Nanoparticle-mediated targeting of phosphatidylinositol-3-kinase signaling inhibits angiogenesis. Angiogenesis 12(4):325–338
58. Matsumura Y, Maeda H (1986) A new concept for macromolecular therapeutics in cancer chemotherapy: mechanism of tumoritropic accumulation of proteins and the antitumor agent smancs. Cancer Res 46(12 Pt 1):6387–6392
59. Torchilin V (2011) Tumor delivery of macromolecular drugs based on the EPR effect. Adv Drug Deliv Rev 63(3):131–135
60. Brambilla D, Luciani P, Leroux JC (2014) Breakthrough discoveries in drug delivery technologies: the next 30 years. J Control Release 190:9–14
61. Stoletov K, Kato H, Zardouzian E et al (2010) Visualizing extravasation dynamics of metastatic tumor cells. J Cell Sci 123(Pt 13):2332–2341
62. Stoletov K, Montel V, Lester RD et al (2007) High-resolution imaging of the dynamic tumor cell vascular interface in transparent zebrafish. Proc Natl Acad Sci U S A 104(44): 17406–17411
63. Wells A, Chao YL, Grahovac J et al (2011) Epithelial and mesenchymal phenotypic switchings modulate cell motility in metastasis. Front Biosci (Landmark Ed) 16:815–837
64. Yang J, Weinberg RA (2008) Epithelial-mesenchymal transition: at the crossroads of development and tumor metastasis. Dev Cell 14(6):818–829
65. Quail DF, Joyce JA (2013) Microenvironmental regulation of tumor progression and metastasis. Nat Med 19(11):1423–1437
66. Gay LJ, Felding-Habermann B (2011) Contribution of platelets to tumour metastasis. Nat Rev Cancer 11(2):123–134
67. Weinberg RA (2007) The biology of cancer. Garland Science, New York
68. Comen E, Norton L, Massague J (2011) Clinical implications of cancer self-seeding. Nat Rev Clin Oncol 8(6):369–377
69. Paget S (1989) The distribution of secondary growths in cancer of the breast. 1889. Cancer Metastasis Rev 8(2):98–101
70. Vandercappellen J, Van Damme J, Struyf S (2008) The role of CXC chemokines and their receptors in cancer. Cancer Lett 267(2):226–244
71. Vella LJ (2014) The emerging role of exosomes in epithelial-mesenchymal-transition in cancer. Front Oncol 4:361
72. Ritsma L, Steller EJA, Beerling E et al (2012) Intravital microscopy through an abdominal imaging window reveals a pre-micrometastasis stage during liver metastasis. Sci Transl Med 4(158)
73. Zhou ZN, Boimel PJ, Segall JE (2011) Tumor-stroma: in vivo assays and intravital imaging to study cell migration and metastasis. Drug Discov Today Dis Models 8(2-3):95–112
74. Kienast Y, von Baumgarten L, Fuhrmann M et al (2010) Real-time imaging reveals the single steps of brain metastasis formation. Nat Med 16(1):116–122
75. Wagner WW Jr (1969) Pulmonary microcirculatory observations in vivo under physiological conditions. J Appl Physiol 26(3):375–377

76. Tanaka K, Morimoto Y, Toiyama Y et al (2012) In vivo time-course imaging of tumor angiogenesis in colorectal liver metastases in the same living mice using two-photon laser scanning microscopy. J Oncol 2012:265487
77. Qian BZ, Li J, Zhang H et al (2011) CCL2 recruits inflammatory monocytes to facilitate breast-tumour metastasis. Nature 475(7355):222–225
78. Klemm F, Joyce JA (2015) Microenvironmental regulation of therapeutic response in cancer. Trends Cell Biol 25:198–213
79. Ghotra VP, He S, de Bont H et al (2012) Automated whole animal bio-imaging assay for human cancer dissemination. PLoS One 7(2), e31281
80. Ghotra VP, He S, van der Horst G et al (2015) SYK is a candidate kinase target for the treatment of advanced prostate cancer. Cancer Res 75(1):230–240
81. Ban J, Aryee DN, Fourtouna A et al (2014) Suppression of deacetylase SIRT1 mediates tumor-suppressive NOTCH response and offers a novel treatment option in metastatic Ewing sarcoma. Cancer Res 74(22):6578–6588
82. van der Ent W, Burrello C, Teunisse AF et al (2014) Modeling of human uveal melanoma in zebrafish xenograft embryos. Invest Ophthalmol Vis Sci 55(10):6612–6622
83. Chapman A, Fernandez del Ama L, Ferguson J et al (2014) Heterogeneous tumor subpopulations cooperate to drive invasion. Cell Rep 8(3):688–695
84. Bansal N, Davis S, Tereshchenko I et al (2014) Enrichment of human prostate cancer cells with tumor initiating properties in mouse and zebrafish xenografts by differential adhesion. Prostate 74(2):187–200
85. Yang XJ, Cui W, Gu A et al (2013) A novel zebrafish xenotransplantation model for study of glioma stem cell invasion. PLoS One 8(4), e61801
86. Rampazzo E, Persano L, Pistollato F et al (2013) Wnt activation promotes neuronal differentiation of glioblastoma. Cell Death Dis 4, e500
87. Kitambi SS, Toledo EM, Usoskin D et al (2014) Vulnerability of glioblastoma cells to catastrophic vacuolization and death induced by a small molecule. Cell 157(2):313–328
88. Drabsch Y, He S, Zhang L et al (2013) Transforming growth factor-beta signalling controls human breast cancer metastasis in a zebrafish xenograft model. Breast Cancer Res 15(6):R106
89. Naber HP, Drabsch Y, Snaar-Jagalska BE et al (2013) Snail and Slug, key regulators of TGF-beta-induced EMT, are sufficient for the induction of single-cell invasion. Biochem Biophys Res Commun 435(1):58–63
90. Chen AT, Zon LI (2009) Zebrafish blood stem cells. J Cell Biochem 108(1):35–42
91. Wang J, Cao Z, Zhang XM et al (2015) Novel mechanism of macrophage-mediated metastasis revealed in a zebrafish model of tumor development. Cancer Res 75(2):306–315
92. Feng Y, Santoriello C, Mione M et al (2010) Live imaging of innate immune cell sensing of transformed cells in zebrafish larvae: parallels between tumor initiation and wound inflammation. PLoS Biol 8(12), e1000562
93. Feng Y, Renshaw S, Martin P (2012) Live imaging of tumor initiation in zebrafish larvae reveals a trophic role for leukocyte-derived PGE(2). Curr Biol 22(13):1253–1259
94. Deng Q, Yoo SK, Cavnar PJ et al (2011) Dual roles for Rac2 in neutrophil motility and active retention in zebrafish hematopoietic tissue. Dev Cell 21(4):735–745
95. Freisinger CM, Huttenlocher A (2014) Live imaging and gene expression analysis in zebrafish identifies a link between neutrophils and epithelial to mesenchymal transition. PLoS One 9(11), e112183
96. Jin SW, Beis D, Mitchell T et al (2005) Cellular and molecular analyses of vascular tube and lumen formation in zebrafish. Development 132(23):5199–5209
97. Wang Y, Kaiser MS, Larson JD et al (2010) Moesin1 and Ve-cadherin are required in endothelial cells during in vivo tubulogenesis. Development 137(18):3119–3128

Allograft Cancer Cell Transplantation in Zebrafish

John C. Moore and David M. Langenau

Abstract Allogeneic cell transplantation is the transfer of cells from one individual into another of the same species and has become an indispensable technique for studying development, immunology, regeneration and cancer biology. In experimental settings, tumor cell engraftment into immunologically competent recipients has greatly increased our understanding of the mechanisms that drive self-renewal, progression and metastasis in vivo. Zebrafish have quickly emerged as a powerful genetic model of cancer that has benefited greatly from allogeneic transplantation. Efficient engraftment can be achieved by transplanting cells into either early larval stage zebrafish that have not yet developed a functional acquired immune system or adult zebrafish following radiation or chemical ablation of the immune system. Alternatively, transplantation can be completed in adult fish using either clonal syngeneic strains or newly-generated immune compromised zebrafish models that have mutations in genes required for proper immune cell function. Here, we discuss the current state of cell transplantation as it pertains to zebrafish cancer and the available models used for dissecting important processes underlying cancer. We will also use the zebrafish model to highlight the power of cell transplantation, including its capacity to dynamically assess functional heterogeneity within individual cancer cells, visualize cancer progression and evolution, assess tumor-propagating potential and self-renewal, image cancer cell invasion and dissemination and identify novel therapies for treating cancer.

Keywords Zebrafish • Allogeneic transplantation • Immune compromised • Syngeneic • Engraftment

---

J.C. Moore
Molecular Pathology Unit, Cancer Center and Regenerative Medicine, Massachusetts General Hospital, 149 13th Street, office #6133, Charlestown, MA 02129, USA

Harvard Stem Cell Institute, Cambridge, MA, USA
e-mail: moore.john@mgh.harvard.edu

D.M. Langenau (✉)
Molecular Pathology and Cancer Center, Massachusetts General Hospital, Charlestown, MA, USA
e-mail: dlangenau@mgh.harvard.edu

D.M. Langenau (ed.), *Cancer and Zebrafish*, Advances in Experimental Medicine and Biology 910, DOI 10.1007/978-3-319-30654-4_12

Introduction

The ability to perform allogeneic cell transplantation has profoundly impacted our understanding of stem cell biology and cancer. Allogeneic transplantation is the transfer of cells, organs or tissues from one individual into another of the same species. Nearly a century ago, Spemann and Mangold pioneered the use of allogeneic transplantation in embryonic salamanders to demonstrate the inductive properties of the vertebrate organizer [1]. Around the same time, Little developed mouse strains that allowed efficient engraftment of tumors between animals, thereby initiating an entirely new approach to understanding cancer processes and identifying malignant cells [2]. The culmination of allogeneic transplantation occurred with the first successful kidney transplant between human twins [3], which proved the procedure could have not only profound effects on our understanding of disease but also the potential for clinical applications. Following successful organ transplantation, hematopoietic stem cell transplantation was shown to be an effective treatment for radiation-induced immune cell ablation and a durable cure for hematological malignancies [4, 5]. Today, cellular transplantation remains an indispensable research tool in the fields of development, immunology, regeneration and cancer biology. Transplantation studies have become the gold standard for identifying stem and progenitor cells in various tissues and assessing regenerative capacity. More specifically, tumor cell engraftment has become a powerful experimental platform for revealing the mechanistic intricacies of self-renewal, homing, migration and malignancy in vivo.

The zebrafish has a number of qualities that make it an excellent transplantation model for understanding cancer's cellular and molecular mechanisms. Zebrafish produce externally fertilized, transparent embryos that greatly facilitate high-throughput chemical and genetic screens. These screens, when coupled with high-content imaging, can be used to study developmental and/or oncogenic processes. Additionally, zebrafish are highly fecund, yielding large numbers of donor and recipient fish for transplantation experiments. The ability to genetically manipulate zebrafish through knockdowns, knockouts and transgenesis has greatly enhanced our ability to create and characterize new cancer models. To date, there are a myriad of chemically- and genetically-induced zebrafish cancer models, including T and B cell leukemia [6–8], Acute Myeloid Leukemia [9], melanoma [10–12], neuroblastoma [13], malignant peripheral nerve sheath tumors [14–16], liver tumors [17, 18], pancreatic tumors [17, 19–21], germ cell tumors [22] and sarcoma [23]. Importantly, these models are derived by manipulating oncogenes and tumor suppressors found in human cancers. Zebrafish tumors share histopathologic and molecular similarity to human cancers [10, 23–25], highlighting the evolutionarily conserved programs that drive malignancy in vertebrates. The transplantability of zebrafish cancers has enabled this model to serve as a robust platform for probing mechanisms of initiation, clonal evolution, metastasis, relapse and self-renewal in a variety of tumor sub-types.

In this chapter, we will discuss the models and techniques currently being used for allogeneic cancer cell transplantation in the zebrafish. We will also explore how tumor cell transplantation allows us to dissect certain cancer processes that would be impossible to study with other model organisms. Lastly, we will look to the future of transplantation techniques in the zebrafish and outline ongoing challenges.

Zebrafish Allogeneic Transplantation Models

Because it actively scans for and neutralizes foreign cells, the vertebrate immune system is the biggest challenge to transplantation approaches in the zebrafish. The zebrafish has fully functional innate and adaptive immune systems by 21 days of life [26, 27]. Thus, long-term transplantation studies spanning into adulthood require immune-suppression and/or histocompatible tolerance. Research in the mouse has pioneered allogeneic transplantation techniques, including inbreeding to genetic homozygosity resulting in syngeneic mouse strains, chemical immunosuppression, radiation-induced immunosuppression and genetic ablation to modify immune cell function. Allogeneic transplantation techniques in the zebrafish have also taken advantage of these and other approaches, each of which has its own strengths and disadvantages, which are discussed below and highlighted in Fig. 1.

Embryonic Transplantation

Cellular transplantation into embryonic and larval-stage zebrafish is a useful way to study cancer because the larval fish are relatively transparent and lack competent adaptive immune systems until 3 weeks post fertilization [28]. Larval fish are most often used for xenotransplanting human and mouse cancer cell lines and provide an accessible platform for visualizing complex behaviors including engraftment, neovascularization, proliferation, migration and therapeutic response. Embryonic transplantation into zebrafish has several benefits beyond immune competency. The embryos are small in size, thereby facilitating high-throughput chemical screens using automated bio-imaging and allows the direct visualization of cellular processes [29, 30]. However, growth kinetics and tumor stage must be considered when transplanting into larval zebrafish. For example, the human melanoma line C8161-RFP engrafts when injected into blastula-stage zebrafish embryos but does not form tumors [31]. In contrast, the WM-266-4 human melanoma cell line forms visible tumors by 7 days post injection when transplanted into the 2-day-old yolk sack of zebrafish embryos [32].

One of the first reported allogeneic transplantation into larval zebrafish was a Myc-transgenic T-cell acute lymphoblastic leukemia (T-ALL) [6]. T cells were generated to transgenically express GFP fused in frame with the mouse *c-Myc* gene under control of the zebrafish *rag2* promoter. T-ALL developed after 30 days of life and could be transplanted into the sinus venous of non-immune matched, 2-day-old embryos. Of the 25 fish that engrafted T-ALL by 5 dpf, only one developed leukemia by 44 days post injection, a result that indicated the fish were not likely fully tolerized to engrafted cells. Subsequent experiments engrafting solid tumors into 2-day old fish also suggested innate immunity plays an important role in clearing engrafted tumor cells (Ignatius and Langenau, unpublished observation). In addition to low engraftment rates, transplantation into larval zebrafish has other distinct disadvantages. The small size of larval-stage zebrafish prevents the engraftment of large numbers of cancer cells, with most studies engrafting only 25–100 cells/fish [33].

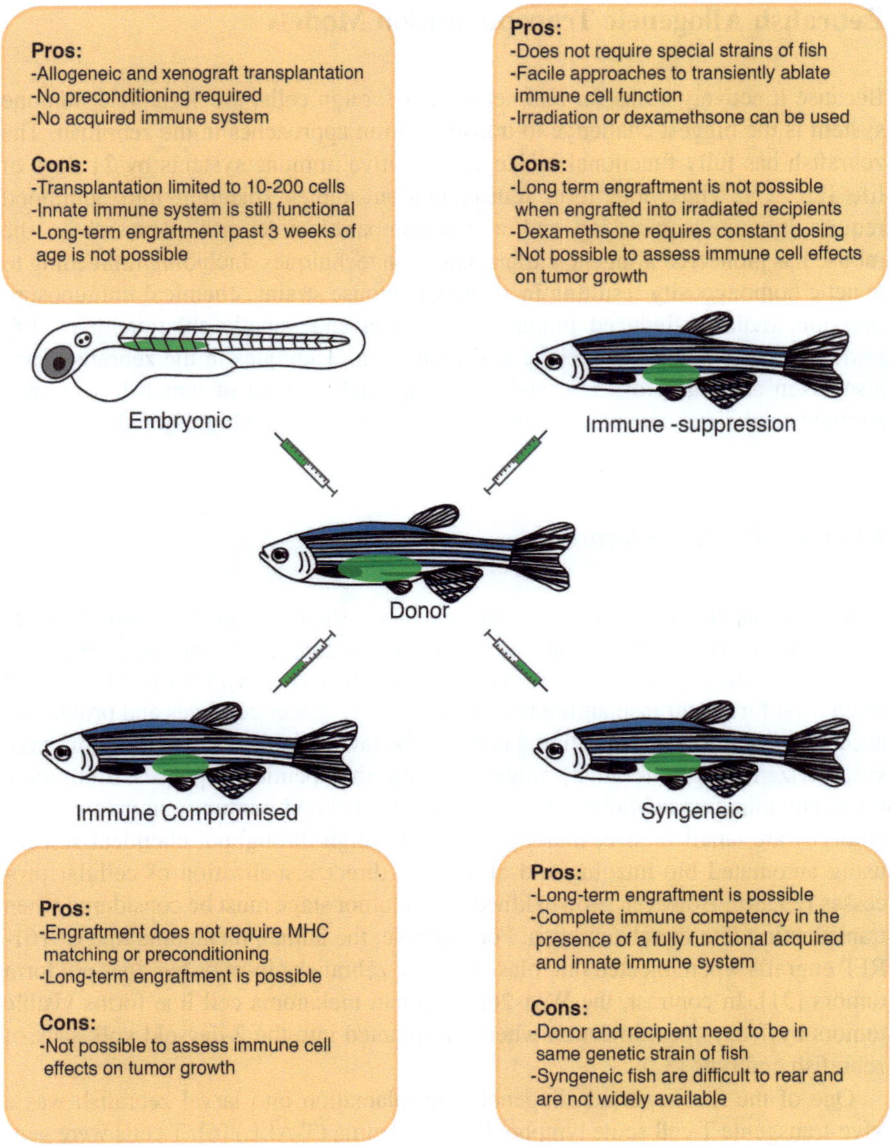

Fig. 1 Allogeneic cell transplantation approaches utilized in zebrafish

In mice, >1 × 10^6 cells are usually required to generate palpable tumors [34], which raises concerns that embryo transplantation experiments may not directly study long-term engrafting tumor propagating cells. Moreover, analyzing the full complement of immune cell interactions within the tumor microenvironment is impossible with larval-stage engraftment because fish lack functional T and B cells. Fortunately, some of these problems have been overcome by the development of immunologically tolerant adult transplantation models.

Cell Transplantation Following Immunosuppression

Allogeneic cancer cell engraftment into adult zebrafish can be completed following immune suppression using either γ-irradiation or dexamethasone treatment. In a technique pioneered by Traver et al., zebrafish were treated with graded doses of γ-irradiation to identify a single sub-lethal radiation dose of 20–25 Gy, which results in immune suppression and 90 % survival [35]. These experiments went on to demonstrate that hematopoiesis is reinitiated by 12 days post-irradiation, and the marrow is fully restored by 20 days post irradiation. These studies were the first to show that irradiation can lead to immune suppression in the zebrafish and made possible engraftment of both hematopoietic and cancerous tissues [6, 35, 36]. One major pitfall of this approach is that once the immune system recovers, by 20 days post transplantation, it kills engrafted cells, thereby preventing long-term analysis of cancer cell engraftment [36]. Alternatively, the immune system can be chemically ablated using the steroidal compound dexamethasone before cancer cell engraftment. For example, the Klemke group has shown that human tumor cells engraft into 25–35 day old zebrafish continually dosed with dexamethasone [37]. Dexamethasone treatment ablates T and B cells and can effectively kill T cells in larval zebrafish [38, 39]. Immune ablation with dexamethasone is effective for solid tumor transplantation approaches but not lymphoid leukemias [39, 40].

The strategies detailed above utilize cancer cell engraftment following transient ablation of the immune system, which presents several problems. Firstly, the immune system recovers following γ-irradiation and causes rejection. Thus, long-term analyses of tumor growth in engrafted fish is not possible [35, 36]. Secondly, inefficient immune system ablation and subsequent recovery kills tumor cells, resulting in vast underestimation of tumor-initiating cell frequency [36]. Further, outbred and polymorphic zebrafish strains are differentially sensitivity to γ-irradiation, adding variability and complicating data analysis. For example, *casper* strain zebrafish lack γ-irradiation absorbing melanophores and are much more sensitive to γ-irradiation. Adult caper-strain zebrafish do not survive following γ-irradiation doses of 25 Gy. The absence of an immune cell response following γ-irradiation also precludes the study of tumor/immune cell interactions and prevents the screening of immunologically-mediated therapeutic compounds. Despite their numerous disadvantages, irradiation and chemical ablation of the immune system allow efficient engraftment of cancers from non-immune-matched fish. This is crucial because many zebrafish tumor models are not derived from clonal, syngeneic zebrafish lines, and engraftment into immune-matched zebrafish is neither widely used nor feasible for most laboratories at this time [41].

Clonal (Syngeneic) Zebrafish Models

Streisinger et al. reported the first vertebrate clone through parthenogenesis, forever solidifying the zebrafish as a genetic model three decades ago [42]. The Revskoy laboratory has since applied similar in vitro fertilization and heat shock techniques to create clonal, syngeneic zebrafish [17, 43, 44]. Cancers arising in syngeneic donors can be directly transplanted into a sibling recipient without eliciting an immune response or regression. Importantly, transplanting into syngeneic donors precludes the need for irradiation or dexamethasone pre-conditioning [36]. Developing a clonal zebrafish line starts with in vitro fertilization using UV-inactivated sperm. The eggs are then subjected to heat shock prior to the first cell division in order to prevent cytokinesis and produce gynogenic diploid animals that can be raised to adulthood. Later eggs are harvested from adult gynogenic diploid females and in vitro fertilized with UV-irradiated sperm followed by heatshock. The resulting progeny are genetically identical and can be maintained by incrossing siblings or backcrossing to founder females [17]. Remarkably, a subset of clonal fish lines are viable and fertile, and genetically similar enough to allow high-throughput serial transplantation of various tumors [36, 43, 44]. The Revskoy laboratory, for example, initially demonstrated the utility of using clonal zebrafish lines by serially transplanting *N*-nitrosodiethylamine-induced primary hepatic and pancreatic carcinomas into syngeneic recipient fish [44]. These lines have been invaluable for quantifying self-renewal potential and long-term engraftment [24, 36, 40, 45–48].

Because the donor and recipient have fully competent, matched immune systems, tumor transplantation using syngeneic zebrafish facilitates long-term tumor engraftment and allows researchers to analyze interactions between immune cells and tumor microenvironment. In addition, large-scale high-throughput transplantation into adults or embryos affords a convenient and robust platform for in vivo chemical and genetic screens. Moreover, the successful transplantation of fluorescently-labeling cancers allows unparalleled access to visualize tumor processes in vivo. Nevertheless, the approach is far from ideal: only tumors made in clonal zebrafish lines will readily transplant into related recipients, and the limited number of lines prevents the research community from adopting the approach more broadly [17, 24, 36, 40, 43–48].

Immune Compromised Transplantation Models

Immune compromised mice are commonly used to engraft normal and malignant cells. The first model utilized for cell transplantation was the nude mouse, which harbors a point mutation in the *Forkhead box N1* (*Foxn1*) gene [49, 50]. Foxn1 is required for thymic epithelium function and is critical for T cell maturation. This mutation caused concomitant loss of T-cell differentiation and consequently inhibited the adaptive immune response [51]. However, because nude mice retain limited T cell function, especially as they mature, this model is only partially immune compromised. To create mice that completely lack T and B cell-mediated adaptive immunity required the removal of functional *Recombination activation gene 1/2* (*Rag1* and *Rag2*) or *Protein Kinase, DNA-Activated, Catalytic Polypeptide* (*Prkdc*, *SCID*) [52–54].

These genes are required to V(D)J recombination and the production of functional T and B cell receptors. To date, no one genetic mutation has resulted in ablation of both the adaptive and innate immune system. Thus, compound mutants have been created that lack T, B, and NK cells. Currently, the most immune deficient mouse is the *NSG* (*N*OD.Cg-prkdc^{scid}Il2rg^{tm1Wjl}) model with mutations in three genes [55]: the Non-Obese Diabetic (NOD) mutation exhibits multiple immunophenotypes including defects in the hemolytic complement system; the *Prkdc* mutation effectively abrogates T and B cells by stopping non-homologous end joining repair required for V(D) J recombination; the *interleukin 2 receptor*, *gamma* (*Il2rg*) mutation prevents NK-cell maturation. Combining these mutations within a single animal eliminates both innate and adaptive immune response. Allogeneic transplantation and xenotransplantation into the NSG model has quickly become the preferred model to test tumor malignancy and assess new therapeutic strategies in human cancer, now widely considered an obligatory step in pre-clinical modeling studies. Transplantation into genetically immune compromised mice has reached a maturity only now being realized in the zebrafish. For example, new models of immune compromised zebrafish (ICZ) are fast becoming available, largely due to the development of facile genome editing approaches [56–66].

There are several zebrafish mutants with impaired immune cell function, but few have been used for cell transplantation [27]. In fact, the first targeted mutation created in zebrafish was the *recombinant activating gene 1* (*rag1*) [67]. This mutation was identified from ENU-mutagenized progeny and Targeting Induced Local Lesions in Genomes (TILING). The zebrafish *rag1* mutant contains a frame shift mutation causing a premature stop codon that impairs receptor recombination and T and B cell function. However, the mutant line proved difficult to maintain and consequently is not widely used as a transplantation model. Another mutation involves the *v-myb avian myeloblastosis viral oncogene homolog* (myb) gene, which is a transcription factor required for definitive hematopoiesis and maintenance of hematopoietic stem cells [68]. Zebrafish *myb* mutants are anemic and die 2–3 months post-fertilization, but they can be readily transplanted with allogeneic hematopoietic stem cells without the need for pre-conditioning [69]. Immune system reconstitution with wild type hematopoietic stem and progenitor cells can rescue the early lethality phenotype. Although these fish are clearly immune deficient, their usefulness in cancer cell transplantation as adults is limited by death resulting from marrow failure and anemia. Cancer engraftment studies using early embryonic or larval stage transplantation in this line have yet to be reported.

Our group recently used Zinc-Finger nucleases to create mutants in the *rag2* loci [70]. We engineered mutations into the plant homeodomain (PHD), causing a frame shift mutation that resulted in a premature stop codon. Mutations in the PHD domain disrupt the ability of Rag2 to gain access to the chromatin and partially impair V(D) J recombination in vivo [71], thus preventing T and B cell maturation. The zebrafish $rag2^{E450fs}$ mutation produces a hypomorphic protein that results in loss of T cells and a variable reduction in differentiated B cells, yet $rag2^{E450fs}$ mutant zebrafish are still

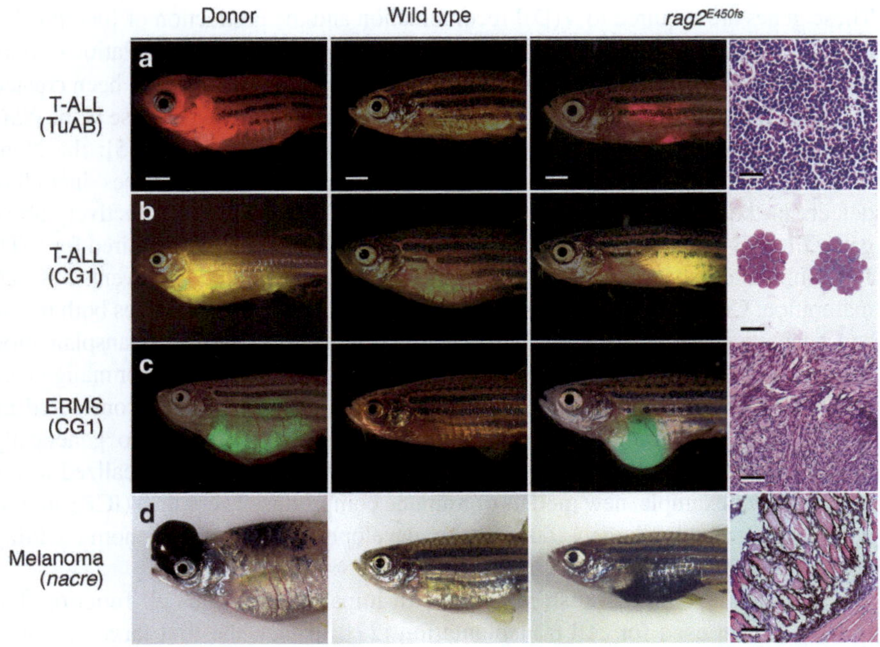

Fig. 2 Engraftment of zebrafish tumors into immune-compromised *rag2*^{E450fs} mutant zebrafish. Donor tumors used for cell transplantation (*left panels*). Wild-type and *rag2*^{E450fs} mutant recipient animals 30 days after cell transplantation (*three right panels*). (**a**) dsRED-labeled *Myc*-induced T-ALL arising in TuAB strain fish, (**b**) zsYellow-labeled T-ALL from CG1-strain fish, (**c**) GFP-labeled embryonal rhabdomyosarcoma (ERMS) from CG1-strain fish and (**d**) *mitfa* and *BRAF*^{V600E} induced melanoma arising in *p53*-deficient *nacre* strain fish. Transplanted tumor histology (**a**–**d**, *extreme right panels*), images of whole kidney marrow sections (**a**), leukemia cytospins (**b**) and tumor cell infiltrates into skeletal muscle (**c, d**). Scale bars in fish images are 2 mm, and histopathology images are 25 μm (**a**), 20 μm (**b**) and 50 μm (**c, d**). Figure modified from [70]

viable and fertile. The *rag2*^{E450fs} mutant zebrafish robustly engraft both normal and malignant cells without the need for immunosuppressive preconditioning or MHC matching [70, 72]. For example, primary T-ALL, rhabdomyosarcoma (RMS) and melanoma engraft following injection into the peritoneal cavity (see Fig. 2a–d). Importantly, homozygous *rag2*^{E450fs} mutants can be reared under normal laboratory conditions, and engrafted tumors do not regress over time. Despite these lines utility, we also clearly need to continue refining transplantation models. For example, combining NK deficient lines with animals that lack an adaptive immune system will likely provide even better models for engrafting both zebrafish and mammalian cells. Moreover, our initial work created *rag2*^{E450fs} mutant fish in the pigmented, AB strain, which mostly precludes tracking single cancer cells in vivo. Creating the second generation of optically clear immune compromised lines will likely be invaluable for tracking invasion, metastasis, stem cell function and therapy responses in vivo at single cell resolution [73].

Understanding Cancer Processes Through Allogeneic Cell Transplantation

Numerous zebrafish allogeneic transplantation studies have helped increase our understanding of the cellular and genetic mechanisms driving tumor progression while also providing a platform for drug discovery. Allogeneic cell transplantation experiments have been predominantly performed in adult transplant recipient fish. Broadly speaking, carefully executed transplantation experiments can increase our knowledge of the cellular and molecular mechanisms of malignant disease. Experimental considerations and detailed protocols for processing and transplanting normal and malignant cells into adult fish have been published elsewhere [45, 72, 74, 75], but an abridged workflow can be found in Fig. 3a–e.

Allogeneic Cell Transplantation to Establish Tumorigenicity

The most common use for zebrafish cell transplantation is to study malignancy, which is defined in part by tumorigenic cells ability to grow in secondary recipients The first published transgenic zebrafish cancer model used the lymphocyte-specific *rag2* promoter to express a GFP fusion with the mouse *c-myc* gene [6]. These transgenic animals developed T-cell acute lymphoblastic leukemia (T-ALL) that initiated in thymic T cells, spread to the marrow and extended to the periphery. Importantly, leukemic cells isolated from these fish readily engrafted into irradiated recipients. T-ALL cells were first visualized in engrafted fish at the site of injection and were then observed spreading throughout the peritoneal cavity. Transplanted leukemia cells also homed to the thymus while retaining the phenotypic characteristics of the donor cells. Similarly, a transgenic zebrafish expressing TEL-AML1 oncogenic fusion resulted in B cell acute lymphoblastic leukemia, which also demonstrated serial transplantability [8]. Melanomas, rhabdomyosarcoma, neuroblastoma, hepatocellular carcinomas and pancreatic carcinoma have all been engrafted into irradiated donor animals, as well as into syngeneic and immune compromised fish [44, 48, 70, 76]. These experiments validate that tumors are fully transformed rather than premalignant lesions or hyperproliferative disorders. These assays have shown that myeloproliferative disorders are unable to serially transplant into recipient fish [77]. Moreover, zebrafish Notch-induced leukemias contain multiple pre-malignant clones that cannot be serially transplanted into recipient fish [24]. In total, cell transplantation has been an important tool for assessing tumorigenicity.

Progression and Evolution

Cancer progression from a benign hyperplasia to a malignant growth is a complex and stochastic process involving genomic instability and/or epigenetic changes. Heterogeneous cancer cell populations continue to amass mutations and undergo

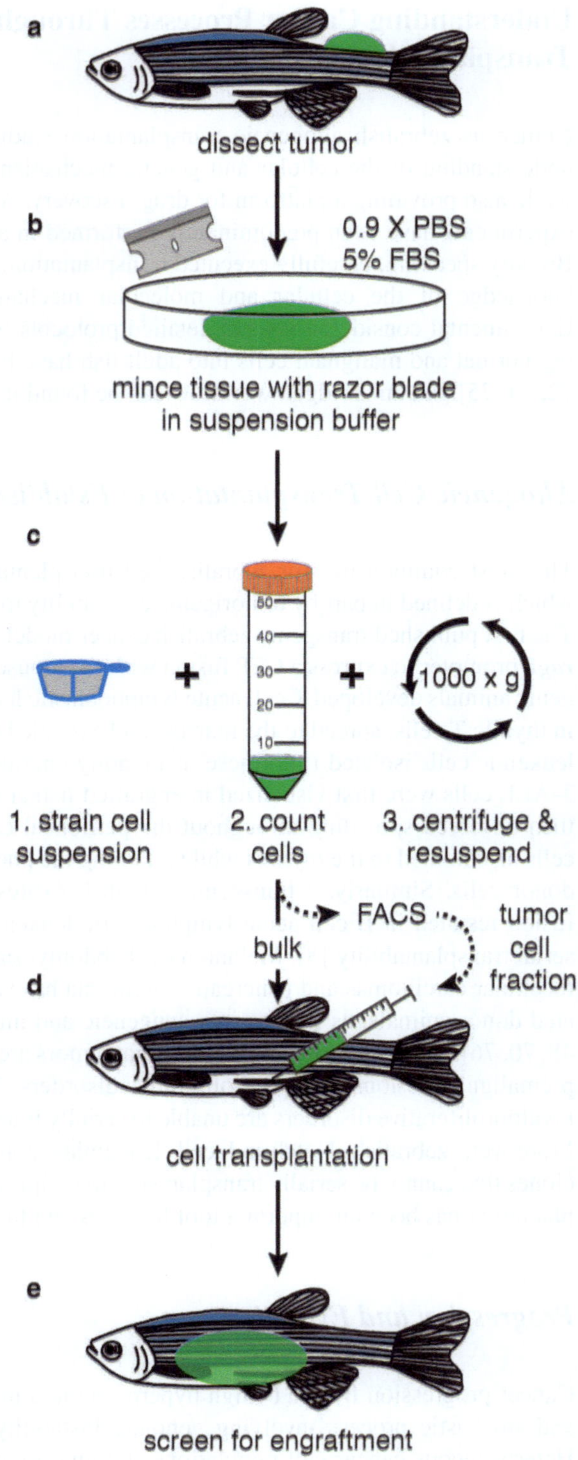

Fig. 3 Protocol for allogeneic cancer cell transplantation into adult recipient zebrafish. (**a**) Remove tumor from donor zebrafish. (**b**) Mechanically homogenize the tumor using repeated mincing with a razorblade in the presence of 5 % FBS in 0.9 X PBS cell suspension buffer. (**c**) Filter the cell suspension through a 45 μM cell strainer, and then perform cell counts. Centrifuge the cell suspension at 1000×g for 10 min and re-suspend in suspension buffer to the desired cell concentration. (**d**) Transplant tumor cells using a 26-gauge Hamilton syringe. Cells can be implanted from bulk tumor or isolated by fluorescent activated cell sorting (FACS). (**e**) Recipient animals are screened for engraftment (**e**). Figure modified from [72]

Darwinian selection in a process termed clonal evolution [78, 79]. Genetic lesions that enhance fitness are ultimately propagated, culminating in a tumor comprised of genetically distinct clones. Evolution happens throughout tumor progression and selects for distinct traits depending on tumor stage and treatment [80]. For example, in early stages mutations that confer proliferative or anti-apoptotic advantage would provide tumor cell survival advantage, whereas in later stages lesions that provide increased invasiveness or therapy resistance would lead to continued survival and progression.

Using a syngeneic zebrafish model of Myc-induced T-ALL and high-throughput cell transplantation, our group engrafted zebrafish T-ALLs from a single T-ALL cell [36, 45]. In concert with large scale-limiting dilution cell transplantation approaches, we estimated the frequency of leukemia propagating cells (LPCs) both in bulk leukemia and leukemias generated in recipient fish from single cells (see Fig. 4a–c). These studies proved that the percentage of leukemia propagating cells was much higher than previously predicted [36]. Expanding on these initial studies, we sought to test functional heterogeneity and evolution in the transplant setting. Using 16 different primary T-ALL, we generated 47 monoclonal T-ALL by transplanting single cells into 1200 adult syngeneic animals. Engrafted leukemia where derived from a single leukemia cell, confirmed by TCRβ clonality assays and array CGH studies. Leukemias that developed from the same primary malignancy were functionally distinct, having different LPC frequency, latency and in vivo drug responses. Remarkably, continued serial transplantation of the clones into over 5000 adult zebrafish allowed us to demonstrate that a subset of cells acquired additional driver mutations culminating in more aggressive T-ALLs. Subsequent studies demonstrated that 50 % of the evolved clones had acquired AKT signaling pathway activation, which had not previously been implicated in expanding leukemia propagating cell numbers but is a signaling pathway with known prognostic value [81]. Using a similar approach, Frazer et al. also demonstrated heterogeneity in transplanted T-ALL, and they identified amplified and deleted genes associated with increased aggression [82]. The ability to cost-effectively perform high-throughput zebrafish cell transplantation studies will help uncover novel oncogenic drivers. Moreover, functionally annotating tumors generated from single cells complexed with whole-genome deep sequencing will likely uncover genetic changes that impart novel functions to individual cancer cells.

Tumor Propagating Cells and Self-Renewal

A subset of cancer cells have the ability to self-renew and drive continued tumor growth, whereas other cell types neither proliferate nor sustain tumor growth. These cancer-driving tumor cells are commonly called tumor-propagating cells (TPCS) and are responsible for initiating metastasis and relapse. TPCs have the unique capacity to produce more of themselves, akin to self-renewal found in normal stem cells, but also generate more specialized cell types that often make up the significant bulk of the tumor mass. Prevailing theories on tumor cell heterogeneity describe a

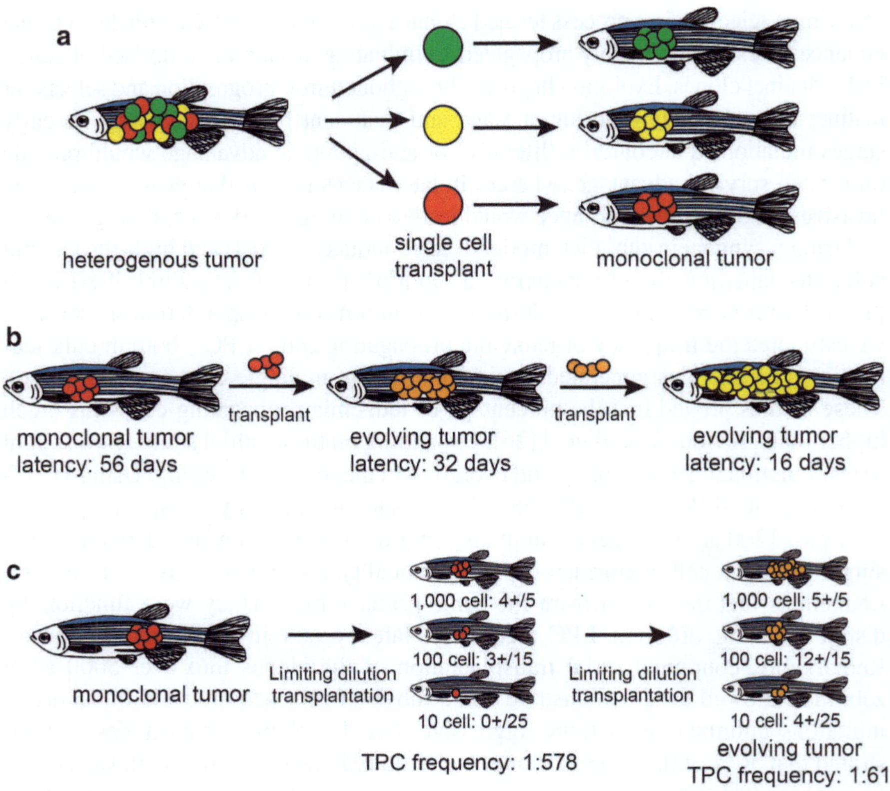

Fig. 4 Assessing intratumoral heterogeneity and clonal evolution using single-cell tumor transplantation. (**a**) Single cells from a heterogeneous primary tumor can be isolated by FACS and transplanted by intra-peritoneal injection into immune competent adult recipients. Each color in this figure represents a different clone within the primary tumor. Monoclonal tumors derived from single-cell engraftment can be assessed for changes in latency (**b**) or tumor-propagating cell numbers following serial passaging (**c**). Tumor-propagating cells frequency can be assessed by limiting-dilution cell transplantation, in which varying doses of cells, such as 10, 100 and 1000 cells, are transplanted into recipient animals. The frequency of tumor-propagating cells is calculated based on the ratio of animals that develop a tumor compared to the total number transplanted at each dose. Tumor-propagating cell (TPC). Figure modified from [97]

sequential differentiation program that starts with a self-renewing stem cell-like population called the cancer stem cell (CSC). The CSC model implies that targeting the stem cell population alone will have therapeutic efficacy. The stochastic model, on the other hand, suggests that all cells within a tumor have the capacity to self-renew, meaning all tumor cells must be killed to prevent reoccurrence. How cancers generate cancer cell self-renewal likely varies by cancer and cell type. Thus, experimental models are necessary to define which models apply to specific cancers and more importantly, what molecular programs drive self-renewal of TPCs [83]. Limiting dilution cell transplantation is the gold standard for identifying and quantifying TPCs found within a malignancy; in the zebrafish, this method has helped reveal important cellular and molecular mechanisms driving self-renewal and, ultimately, cancer relapse.

The Langenau lab developed a well-characterized model of KRASG12D induced embryonal rhabdomyosarcoma (ERMS), a muscle malignancy that is both histologically and molecularly similar to human disease [23]. Building on this work, Ignatius et al. used well-known muscle-specific transgenic reporter lines that faithfully recapitulate endogenous gene expression and fluorescently-labeled ERMS cells based on muscle differentiation status to identify a molecularly defined TPC that exclusively drove continued tumor growth when engrafted into syngeneic recipient fish [48]. The *myf5-GFP+/mylz2-mCherry-negative* TPCs express cMet, mCadherin and myf5, all markers of activated satellite cells. The *myf5-GFP$^+$/mylz2-mCherry$^-$* ERMS cell population is more proliferative than the differentiated *mylz2-mCherry+* ERMS cell types. Furthermore, when ERMS cell subpopulations were FACS purified by flow cytometry into four populations (GFP$^+$/mCherry$^-$, GFP$^+$/mCherry$^+$, GFP$^-$/mCherry$^+$ and GFP$^-$/mCherry$^-$), only the *myf5-GFP$^+$/mylz-mCherry$^-$* single positive cell population could propagate disease when serially transplanted (see Fig. 5a–l). These studies indicate that ERMS follows the cancer stem cell model and highlights the utility of allogeneic transplantation studies to functionally differentiate between heterogeneous tumor cell populations. Additional work by our group used this model to identify novel drugs that curb growth by specifically killing and/or differentiating TPCs [47].

Imaging Cancer Cell Invasion and Dissemination

Metastatic cancer is the hardest to treat and has exceedingly poor prognosis. Understanding the intrinsic cellular and micro-environmental mechanisms that contribute to tumor dissemination throughout the body continues to remain a challenge that necessitates new models. The ability to make transplantable fluorescent transgenic zebrafish models has given researchers an important avenue to study metastatic processes. Developing the transparent *casper* adult zebrafish overcame one of the key technical hurdles for imaging cancer invasion and metastasis [76]. For example, White et al. transplanted pigmented melanoma cells into irradiated recipient *casper* fish and were able to assess engraftment and proliferation without sacrificing the animal (see Fig. 6a) [76]. They were also able to track local tumor invasion by serially imaging the same animal over time (see Fig. 6a). Reports from the White lab have since described new genetic models of GFP-labeled melanoma that suggest serially transplanted melanomas demonstrate distinct distal growth akin to metastasis and may provide a new experimental platform for understanding cancer cell invasion, spread and dissemination in the zebrafish [84].

One of the first steps involved in tumor dissemination is entrance into the blood stream, which the tumor completes by passing through the basal membrane of the vasculature in a process called intravasation. Using *casper*, *Tg(fli:egfp)^{Y1}* transgenic fish that have fluorescent labeled vasculature, Feng et al. were able to identify a

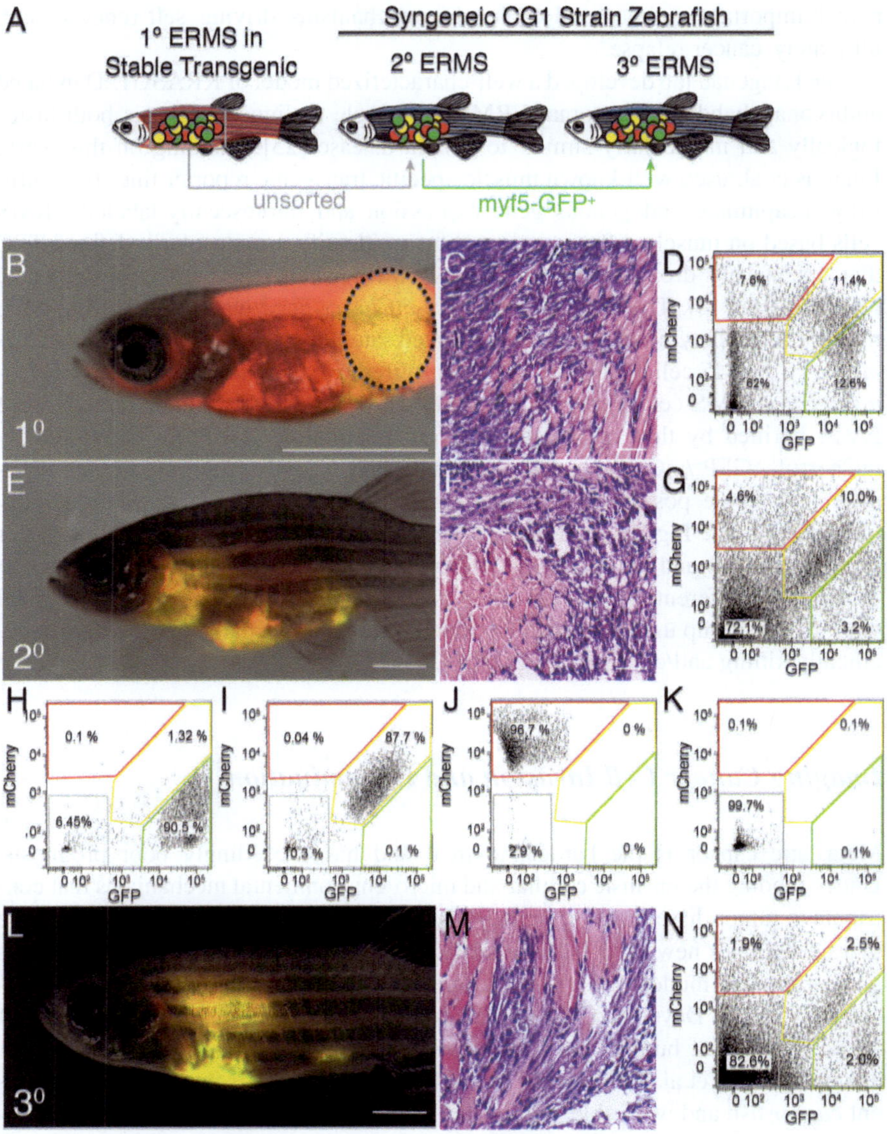

Fig. 5 Embryonal rhabdomyosarcoma contains a molecularly defined tumor propagating cell that expresses myf5-GFP. (**a**) Schematic of cell transplantation approach. Briefly, kRASG12D-induced ERMS were created in myf5-GFP/mylz2-mCherry CG1-strain transgenic fish. Unsorted cells were implanted into secondary recipients to expand the tumor cell numbers, and fluorescent ERMS subpopulations were isolated by FACs and introduced into tertiary recipient fish. (**b–d**) A primary ERMS arising in a clonal syngeneic myf5-GFP/mylz2-mCherry transgenic zebrafish (35 dpf). *Broken black line* denotes tumor area. (**e–g**) Engraftment of unsorted ERMS cells into syngeneic secondary recipient animal. (**h–k**) FACS plots of fluorescent-labeled ERMS cells isolated from secondary recipient fish following two rounds of FACS. (**l–n**) Transplantation of myf5-GFP$^+$/mylz2-mCherry$^-$ FACs sorted ERMS cells into tertiary transplant animals leads to robust tumor engraftment (**n**). Hematoxylin- and eosin-stained sections (**c**, **f** and **m**) and FACS (**d**, **g** and **n**) of primary and serially passaged ERMS. Scale bars, 2 mm (for **b**, **e** and **l**) and 100 mm (for **c**, **f** and **m**). Figure modified from [48]

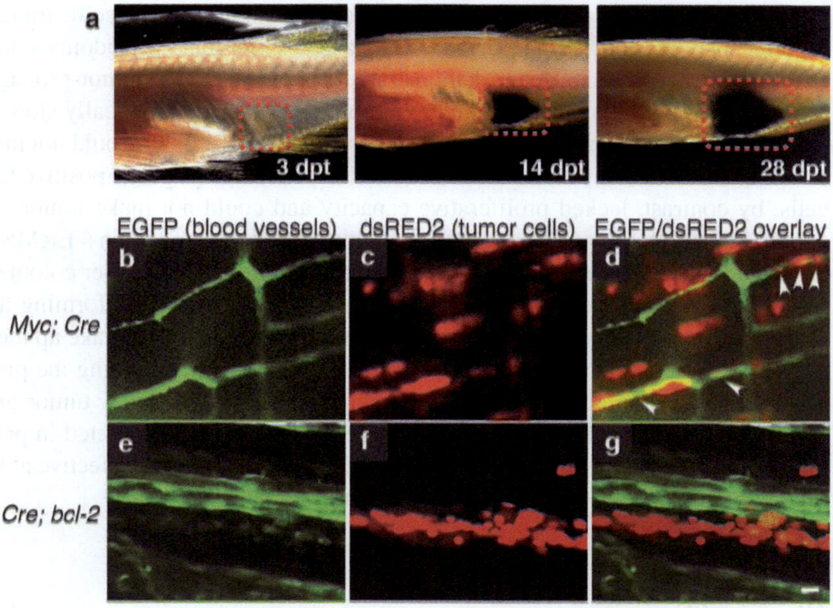

Fig. 6 Imaging cancer dissemination and intravasation. (**a**) mitf-NRAS-GFP;p53$^{-/-}$ driven melanoma cells were transplanted into irradiated *casper* adults. Repeated imaging of the same fish over a 1 month period showed gradually increased tumor volume without needing to sacrifice the recipient animals. (**b–d**) dsRED2-expressing lymphoma cells from *Myc*;*Cre expressing* fish (**c**) intravasate into EGFP-labeled vasculature (**b**) of the transplant recipient (*fli1-EGFP*; *casper*) by 6 day post-transplantation (see *arrowheads* in **d**). (**e, f**) In contrast, dsRED2-expressing lymphoma cells (**f**) from the *Myc*; *Cre*; *bcl-2* fish fail to intravasate into the vasculature (**e**) of the transplant recipients by 6 days post-transplantation (compare panel **g** with **d**). Note aggregates of the *Myc*; *Cre*; *bcl-2* lymphoma cells in panels **f** and **g**. Scale bar for panels **b–g** = 10 μm. Panel (**a**) was originally published in [76]. Panels **b–g** were originally published in [85]

mechanism by which T-lymphoblastic lymphoma (T-LBL) progresses to the more aggressive and malignant T cell acute lymphoblastic leukemia (T-ALL) [85]. These studies showed that overexpression of the *b-cell lymphoma 2* (*bcl2*) gene, when co-expressed with c-myc, accelerated malignant thymocyte transformation by suppressing MYC-induced apoptosis. However, lymphocytes remained trapped in an expanded thymus and were unable to undergo intravasation (see Fig. 6b–g). In this case intravasation was blocked by elevated expression of Sphingosine-1-phosphate receptor 1(S1P1) and Intercellular Adhesion Molecule 1(ICAM1), which together led to leukemia cells increasing homeotypic adhesion and failing to transit into the vasculature. These observations were extended to patient samples and helped define the cellular and molecular differences between human T-LBL and T-ALL. Additionally, this research revealed that progression from T-LBL to T-ALL requires weakened intercellular adhesion prior to intravasation and dissemination.

Tumor invasion and metastasis require seeding new areas and establishing niches that will sustain tumor growth, but the mechanisms controlling these processes are poorly understood. Using the kRASG12D-indcued ERMS model described above,

Ignatius et al. were able to image niche partitioning using lineage-specific transgenic reporter fish in both primary and transplanted embryonal rhabdomyosarcoma (ERMS) [48]. The same team demonstrated that myf5-GFP+ tumor-propagating cells are regionally confined within the tumor and are characteristically slow moving yet highly proliferative. Remarkably, these myf5-GFP+ TPCs could not intravasate and circulate in the blood stream. The differentiated, myogenin-positive ERMS cells, by contrast, lacked proliferative capacity and could not make tumors when engrafted into recipient fish. Yet, these highly migratory myogenin+ ERMS cells were the first cell types to seed new areas of tumor growth. Only after colonization by the differentiated cells did myf5+ TPCs migrate into the newly forming tumor. These data indicate that non-tumor-propagating cells, which often make up the bulk of a tumor, likely have important supportive roles in both maintaining the primary tumor microenvironment and serving as pioneer cells that seed new tumor growth areas. Although, the above experiments were predominantly completed in primary tumor bearing fish, the same imaging approaches would likely be effective at tracking tumor cell behavior in the transplant setting.

Chemical Screens

In 2000, Peterson et al. demonstrated for first time that zebrafish could be used in chemical screens by simply adding chemicals to the water and treating fish in 96-well plates [29], cementing the zebrafish as an important tool for drug discovery. Since then, over 60 different chemicals screens have been performed in the zebrafish with developmental, behavioral, metabolic, proliferative, regenerative, and cancer-based endpoints [86]. Here, we highlight a number of important chemical-based screening experiments relevant to cancer and the use allogeneic transplantation as a fundamental assay for assessing drug efficacy (highlighted in Fig. 7a, b).

Mizgirev and Revskoy were the first to establish that zebrafish and human tumors respond similarly to clinically relevant cancer drugs. Using the Myc-induced T-ALL model, the researchers generated T-ALL in clonal CG2-strain zebrafish [6, 43]. These fish developed fluorescently-labeled leukemia that could robustly engraft into 5-day-old CG2 larvae. Transplanted larvae developed T-ALL rapidly and become moribund just 5 days after transplantation, yet larval fish treated with vincristine and cyclophosphamide exhibited increased life span. This work showed that current clinical chemotherapies effectively curb tumor growth in zebrafish models. Moreover, this line of experiments showed that high-throughput transplantation in the zebrafish is feasible and can be coupled with in vivo drug screens to identify anti-cancer compounds.

Our group recently performed two independent drug screens to identify drugs that suppress kRASG12D-induced ERMS. First, embryos expressing the *Tg(hsp70:HRAS*^{G12D}) transgene were heat shocked in the presence of 2896 individual small molecules that were assessed for their ability to inhibit Ras pathway

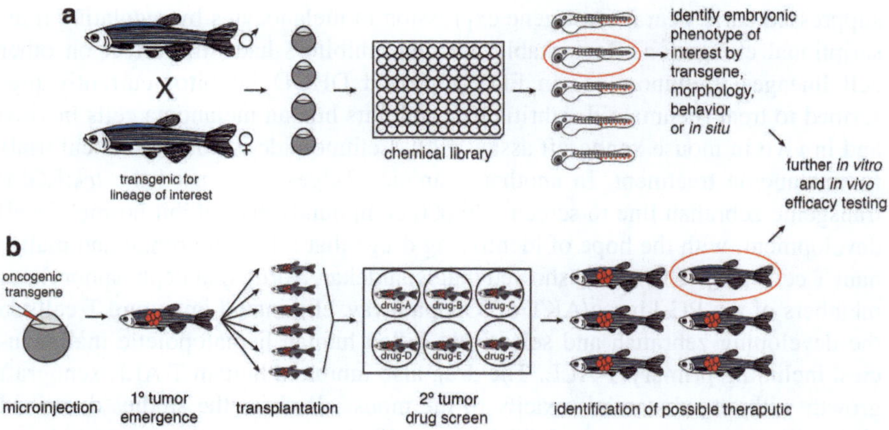

Fig. 7 Chemical genetic approaches to identify anti-cancer drugs in zebrafish. (**a**) Schematic of embryonic/larval based chemical screen. Progeny from adult transgenic zebrafish are placed in a library of chemical compounds. Embryos are screened for loss of a specific cellular lineage demarcated by fluorescence transgene. An alternative approach is to screen for specific morphological defects representing known signaling pathway perturbation or for changes in gene expression following whole-mount by in situ hybridization. Compounds that inhibit embryonic phenotypes are then assessed for efficacy in curbing tumor growth in vitro and in vivo. (**b**) Chemical screening for cancer phenotypes in transplant recipient fish. Single cell zebrafish are injected with oncogenic transgenes to induce a primary tumor, which is then isolated and transplanted into hundreds of immune competent recipient fish. These secondary tumor-bearing fish are then treated with a library of compounds to test anti-tumor efficacy. Hit compounds are then assessed for additional therapeutic value in human cell lines and patient-derived xenografts in mice

activation [87]. In a separate high-content imaging screen of 38,000 novel compounds and 47% of the FDA-approved chemotoxic drugs (~2500 drugs), we identified compounds that induce terminal differentiation of human ERMS cells [47] Compounds identified in these screens were then evaluated for suppressing tumor growth in zebrafish transplanted with RAS-induced ERMS. Our analysis identified several classes of compounds, the most promising being Glycogen synthase kinase 3 (GSK3) inhibitors. Treating ERMS-bearing zebrafish with GSK3 inhibitors activated the WNT/β-catenin pathway and resulted in suppressed ERMS growth, depleted tumor propagating cells, reduced self-renewal and terminal tumor cell differentiation. These findings were surprising because WNT/β-catenin signaling was largely considered an oncogenic driver in cancer [88, 89].

Cancer is a disease that commonly reinitiates developmental signaling pathways to drive malignant transformation and sustained tumor growth. Thus, it is not surprising that a number of studies have used embryonic phenotypes to screen for drugs that alter cell states or modulate cancer/developmental pathways. For example, the Zon group performed a chemical screen to identify suppressors of neural crest progenitor specification in larval zebrafish [90]. They identified inhibitors of the metabolic enzyme dihydroorotate dehydrogenase (DHOD) that suppressed early neural crest gene expression in melanocytes by regulating transcriptional elongation. Remarkably, DHOD inhibitors had little effect on other cell lineages. Leflunomide, an FDA-approved DHOD inhibitor currently prescribed to treat rheumatoid arthritis [91], inhibits human melanoma cells in vitro and in vivo in mouse xenograft assays [90]. Leflunomide is now in clinical trials for melanoma treatment. In another example, Ridges et al. used the *lck*:EGFP transgenic zebrafish line to screen ~26,000 compounds that inhibit normal T cell development, with the hope of identifying drugs that kill both normal and malignant T cells [92]. The screen showed that lenaldekar, a drug that dephosphorylates members of the PI3 kinase/AKT/mTOR pathway, eliminated immature T cells in the developing zebrafish and selectively killed human hematopoietic malignancies, including primary T-ALL. The drug also inhibited human T-ALL xenograft growth without substantial toxicity to the mouse. Both of the studies described above illustrate the potential of finding effective therapeutic compounds by screening early embryonic phenotypes. Using complementary strategies, the Yeh group showed that PGE2 inhibitors potently suppress AML-ETO-driven leukemia and suppress the growth of xenografted AML when introduced into mice [93]. Although these examples did not utilize allogeneic transplantation in zebrafish to test chemical efficacy in suppressing cancer growth, one could easily devise a primary or secondary chemical screen that uses allogeneic transplantation to study chemical effects directly on engrafted cancer.

Current Challenges, Future Directions, and Conclusions

One major hurdle to cell transplantation being more widely adopted is the lack of experimental models. Clonal syngeneic lines are currently only being used by a handful of labs and are notoriously hard to maintain. With support from the Office of the Director at the NIH, our group has utilized genome-editing approaches to develop and characterize a library of immune compromised zebrafish models. These models are now being methodically assessed for immune cell dysfunction and superior allogeneic engraftment potential. The publication of the *rag2*^{E450fs} mutant lines demonstrated that these recipient fish can engraft a wide range of cancer types. In order to expand this approach's accessibility and utility, we are currently characterizing mutants in *DNA-dependent protein kinase* (*prkdc*), *janus kinase 3* (*jak3*), *interleukin 2 receptor gamma* (*Il2rg*), *zeta-chain* (*TCR*) *associated protein kinase 70* (*zap70*) and *forkhead box N1* (*foxn1*/*nude*). Mutations in these genes are known to have immune deficient phenotypes in both humans and mouse. An additional benefit of having multiple genetically-defined transplantation models is the ability to test how the lack of a particular immune cell type affects cancer initiation, progression and metastasis. For example, *prkdc* mutants have a complete loss of differentiated T and B cells, whereas the *rag2* mutant is hypomorphic with complete T cell loss and variable B cell loss. It is possible that comparing tumor growth between these two recipient models may reveal how B cells combat tumorigenicity. Developing casper-strain, immunocompromised zebrafish will provide unprecedented capacity for dynamically visualizing the hallmarks of cancer in optically clear engrafted zebrafish. We recently created *casper* strain, *rag2*^{E450fs} mutant zebrafish that engraft fluorescently-labelled zebrafish cancer cells, thereby providing optical access to visualize cancer cell heterogeneity in embryonal rhabdomyosarcoma (ERMS), the emergence of clonal dominance in T-cell acute lymphoblastic leukemia (T-ALL) and the evolution of metastatic potential in zebrafish melanoma models. Remarkably, these approaches provide single cell resolution of fluorescent-labelled cancer cells in vivo [73]. Finally, the experimental attributes and innovative tools available to zebrafish researchers will likely establish ICZ models as the next-generation, low-cost, high-throughput transplantation platform for engraftment of both allogeneic and mammalian cells. We envision that a subset of ICZ models will efficiently engraft patient derived cancers, and zebrafish avatars will be used to evaluate therapeutic efficacy of new drugs and will help stratify patients into open current clinical trials. These screening strategies have the potential to revolutionize personal medicine and clinical trial design.

Marrying new technologies available for developmental zebrafish studies with cell transplantation approaches outlined above is sure to provide new insights into cancer and its progression. For example, the zebrabow cell lineage tracing system will likely allow exquisite dissection of tumor cell heterogeneity in both primary and transplanted tumors at single cell resolution [94]. Additionally, dynamic live cell imaging of blood cell types in cancer will likely help define how immune cells regulate tumor growth. These approaches will necessarily require both engraftment into syngeneic zebrafish lines that have fully functional immune systems and blood cell labeling using fluorescent transgenic reporters. For example, blood lineage-specific transgenics like *tg(mpx:eGFP)*, *tg(mpeg1:eGFP)*, and Tg(lck:eGFP) could be used to clarify neutrophil, monocyte/macrophage and T lymphocyte behaviors in cancer growth in the transplantation setting [39, 95, 96]. Finally, the CRISPR/Cas9 genome editing technologies can now be employed for tissue modification to create tumors that even more accurately mimic human cancer [66]. These models will greatly benefit from the cell transplantation approaches outlined here.

Acknowledgments We would like to thank Nora Torres Yordán, Qin Tang and Kaley Joyes for their critical review and helpful feedback. We would also like to acknowledge members of the Langenau Lab for helpful discussions related to transplantation techniques, future technologies and practical applications. Special thanks to Richard White, Hui Feng, Myron Ignatius, Jessica Blackburn, Qin Tang, Ines Tenente, Len Zon and Tom Look for allowing us to reproduce their work in figures. This chapter is dedicated to Nicole and Kellen. Support was provided by Alex's Lemonade Stand Foundation, Live Like Bella Foundation for Childhood Cancer, American Cancer Society, St. Baldrick's Foundation, the MGH Howard Goodman Fellowship and NIH grants R24OD016761, U54CA168512, and 1R01CA154923.

Disclosure Statement We have neither competing nor financial conflicts of interests to declare.

References

1. Spemann H, Mangold H (2001) Induction of embryonic primordia by implantation of organizers from a different species. 1923. Int J Dev Biol 45(1):13–38
2. Little CC (1920) The Heredity of Susceptibility to a Transplantable Sarcoma (J. W. B.) of the Japanese Waltzing Mouse. Science 51(1323):467–468
3. Merrill JP et al (1956) Successful homotransplantation of the human kidney between identical twins. J Am Med Assoc 160(4):277–282
4. Thomas ED et al (1957) Intravenous infusion of bone marrow in patients receiving radiation and chemotherapy. N Engl J Med 257(11):491–496
5. Kim TH et al (1977) Total-body irradiation with a high-dose-rate linear accelerator for bone-marrow transplantation in aplastic anemia and neoplastic disease. Radiology 122(2):523–525
6. Langenau DM et al (2003) Myc-induced T cell leukemia in transgenic zebrafish. Science 299(5608):887–890
7. Langenau DM et al (2005) Cre/lox-regulated transgenic zebrafish model with conditional myc-induced T cell acute lymphoblastic leukemia. Proc Natl Acad Sci U S A 102(17):6068–6073
8. Sabaawy HE et al (2006) TEL-AML1 transgenic zebrafish model of precursor B cell acute lymphoblastic leukemia. Proc Natl Acad Sci U S A 103(41):15166–15171
9. Zhuravleva J et al (2008) MOZ/TIF2-induced acute myeloid leukaemia in transgenic fish. Br J Haematol 143(3):378–382
10. Dovey M, White RM, Zon LI (2009) Oncogenic NRAS cooperates with p53 loss to generate melanoma in zebrafish. Zebrafish 6(4):397–404
11. Patton EE et al (2005) BRAF mutations are sufficient to promote nevi formation and cooperate with p53 in the genesis of melanoma. Curr Biol 15(3):249–254
12. Santoriello C et al (2010) Kita driven expression of oncogenic HRAS leads to early onset and highly penetrant melanoma in zebrafish. PLoS One 5(12), e15170
13. Zhu S et al (2012) Activated ALK collaborates with MYCN in neuroblastoma pathogenesis. Cancer Cell 21(3):362–373
14. Berghmans S et al (2005) tp53 mutant zebrafish develop malignant peripheral nerve sheath tumors. Proc Natl Acad Sci U S A 102(2):407–412
15. Amsterdam A et al (2004) Many ribosomal protein genes are cancer genes in zebrafish. PLoS Biol 2(5), E139
16. Lai K et al (2009) Many ribosomal protein mutations are associated with growth impairment and tumor predisposition in zebrafish. Dev Dyn 238(1):76–85
17. Mizgirev I, Revskoy S (2010) Generation of clonal zebrafish lines and transplantable hepatic tumors. Nat Protoc 5(3):383–394
18. Nguyen AT et al (2012) An inducible kras(V12) transgenic zebrafish model for liver tumorigenesis and chemical drug screening. Dis Model Mech 5(1):63–72
19. Liu S, Leach SD (2011) Screening pancreatic oncogenes in zebrafish using the Gal4/UAS system. Methods Cell Biol 105:367–381
20. Park SW et al (2008) Oncogenic KRAS induces progenitor cell expansion and malignant transformation in zebrafish exocrine pancreas. Gastroenterology 134(7):2080–2090
21. Haramis AP et al (2006) Adenomatous polyposis coli-deficient zebrafish are susceptible to digestive tract neoplasia. EMBO Rep 7(4):444–449
22. Gill JA et al (2010) Enforced expression of Simian virus 40 large T-antigen leads to testicular germ cell tumors in zebrafish. Zebrafish 7(4):333–341
23. Langenau DM et al (2007) Effects of RAS on the genesis of embryonal rhabdomyosarcoma. Genes Dev 21(11):1382–1395
24. Blackburn JS et al (2012) Notch signaling expands a pre-malignant pool of T-cell acute lymphoblastic leukemia clones without affecting leukemia-propagating cell frequency. Leukemia 26(9):2069–2078
25. Lam SH et al (2006) Conservation of gene expression signatures between zebrafish and human liver tumors and tumor progression. Nat Biotechnol 24(1):73–75

26. Trede NS et al (2004) The use of zebrafish to understand immunity. Immunity 20(4):367–379
27. Iwanami N (2014) Zebrafish as a model for understanding the evolution of the vertebrate immune system and human primary immunodeficiency. Exp Hematol 42(8):697–706
28. Lam SH et al (2004) Development and maturation of the immune system in zebrafish, Danio rerio: a gene expression profiling, in situ hybridization and immunological study. Dev Comp Immunol 28(1):9–28
29. Peterson RT et al (2000) Small molecule developmental screens reveal the logic and timing of vertebrate development. Proc Natl Acad Sci U S A 97(24):12965–12969
30. Ordas A et al (2015) Testing tuberculosis drug efficacy in a zebrafish high-throughput translational medicine screen. Antimicrob Agents Chemother 59(2):753–762
31. Topczewska JM et al (2006) Embryonic and tumorigenic pathways converge via Nodal signaling: role in melanoma aggressiveness. Nat Med 12(8):925–932
32. Haldi M et al (2006) Human melanoma cells transplanted into zebrafish proliferate, migrate, produce melanin, form masses and stimulate angiogenesis in zebrafish. Angiogenesis 9(3):139–151
33. Veinotte CJ, Dellaire G, Berman JN (2014) Hooking the big one: the potential of zebrafish xenotransplantation to reform cancer drug screening in the genomic era. Dis Model Mech 7(7):745–754
34. Morton CL, Houghton PJ (2007) Establishment of human tumor xenografts in immunodeficient mice. Nat Protoc 2(2):247–250
35. Traver D et al (2004) Effects of lethal irradiation in zebrafish and rescue by hematopoietic cell transplantation. Blood 104(5):1298–1305
36. Smith AC et al (2010) High-throughput cell transplantation establishes that tumor-initiating cells are abundant in zebrafish T-cell acute lymphoblastic leukemia. Blood 115(16):3296–3303
37. Stoletov K et al (2007) High-resolution imaging of the dynamic tumor cell vascular interface in transparent zebrafish. Proc Natl Acad Sci U S A 104(44):17406–17411
38. Baschant U, Tuckermann J (2010) The role of the glucocorticoid receptor in inflammation and immunity. J Steroid Biochem Mol Biol 120(2–3):69–75
39. Langenau DM et al (2004) In vivo tracking of T cell development, ablation, and engraftment in transgenic zebrafish. Proc Natl Acad Sci U S A 101(19):7369–7374
40. Blackburn JS et al (2014) Clonal evolution enhances leukemia-propagating cell frequency in T cell acute lymphoblastic leukemia through Akt/mTORC1 pathway activation. Cancer Cell 25(3):366–378
41. de Jong JL, Zon LI (2012) Histocompatibility and hematopoietic transplantation in the zebrafish. Adv Hematol 2012:282318
42. Streisinger G et al (1981) Production of clones of homozygous diploid zebra fish (Brachydanio rerio). Nature 291(5813):293–296
43. Mizgirev IV, Revskoy S (2010) A new zebrafish model for experimental leukemia therapy. Cancer Biol Ther 9(11):895–902
44. Mizgireuv IV, Revskoy SY (2006) Transplantable tumor lines generated in clonal zebrafish. Cancer Res 66(6):3120–3125
45. Blackburn JS, Liu S, Langenau DM (2011) Quantifying the frequency of tumor-propagating cells using limiting dilution cell transplantation in syngeneic zebrafish. J Vis Exp 53, e2790
46. Blackburn JS et al (2011) High-throughput imaging of adult fluorescent zebrafish with an LED fluorescence macroscope. Nat Protoc 6(2):229–241
47. Chen EY et al (2014) Glycogen synthase kinase 3 inhibitors induce the canonical WNT/beta-catenin pathway to suppress growth and self-renewal in embryonal rhabdomyosarcoma. Proc Natl Acad Sci U S A 111(14):5349–5354
48. Ignatius MS et al (2012) In vivo imaging of tumor-propagating cells, regional tumor heterogeneity, and dynamic cell movements in embryonal rhabdomyosarcoma. Cancer Cell 21(5):680–693
49. Giovanella BC et al (1972) Development of invasive tumors in the "nude" mouse after injection of cultured human melanoma cells. J Natl Cancer Inst 48(5):1531–1533
50. Flanagan SP (1966) 'Nude', a new hairless gene with pleiotropic effects in the mouse. Genet Res 8(3):295–309

51. Nehls M et al (1996) Two genetically separable steps in the differentiation of thymic epithelium. Science 272(5263):886–889
52. Mombaerts P et al (1992) RAG-1-deficient mice have no mature B and T lymphocytes. Cell 68(5):869–877
53. Shinkai Y et al (1992) RAG-2-deficient mice lack mature lymphocytes owing to inability to initiate V(D)J rearrangement. Cell 68(5):855–867
54. McCune JM et al (1988) The SCID-hu mouse: murine model for the analysis of human hematolymphoid differentiation and function. Science 241(4873):1632–1639
55. Shultz LD et al (2005) Human lymphoid and myeloid cell development in NOD/LtSz-scid IL2R gamma null mice engrafted with mobilized human hemopoietic stem cells. J Immunol 174(10):6477–6489
56. Doyon Y et al (2008) Heritable targeted gene disruption in zebrafish using designed zinc-finger nucleases. Nat Biotechnol 26(6):702–708
57. Meng X et al (2008) Targeted gene inactivation in zebrafish using engineered zinc-finger nucleases. Nat Biotechnol 26(6):695–701
58. Moore FE et al (2012) Improved somatic mutagenesis in zebrafish using transcription activator-like effector nucleases (TALENs). PLoS One 7(5), e37877
59. Huang P et al (2011) Heritable gene targeting in zebrafish using customized TALENs. Nat Biotechnol 29(8):699–700
60. Sander JD et al (2011) Targeted gene disruption in somatic zebrafish cells using engineered TALENs. Nat Biotechnol 29(8):697–698
61. Sander JD et al (2011) Selection-free zinc-finger-nuclease engineering by context-dependent assembly (CoDA). Nat Methods 8(1):67–69
62. Foley JE et al (2009) Rapid mutation of endogenous zebrafish genes using zinc finger nucleases made by Oligomerized Pool ENgineering (OPEN). PLoS One 4(2), e4348
63. Hwang WY et al (2013) Efficient genome editing in zebrafish using a CRISPR-Cas system. Nat Biotechnol 31(3):227–229
64. Hwang WY et al (2013) Heritable and precise zebrafish genome editing using a CRISPR-Cas system. PLoS One 8(7), e68708
65. Jao LE, Wente SR, Chen W (2013) Efficient multiplex biallelic zebrafish genome editing using a CRISPR nuclease system. Proc Natl Acad Sci U S A 110(34):13904–13909
66. Ablain J et al (2015) A CRISPR/Cas9 vector system for tissue-specific gene disruption in zebrafish. Dev Cell 32(6):756–764
67. Wienholds E et al (2002) Target-selected inactivation of the zebrafish rag1 gene. Science 297(5578):99–102
68. Soza-Ried C et al (2010) Essential role of c-myb in definitive hematopoiesis is evolutionarily conserved. Proc Natl Acad Sci U S A 107(40):17304–17308
69. Hess I et al (2013) Zebrafish model for allogeneic hematopoietic cell transplantation not requiring preconditioning. Proc Natl Acad Sci U S A 110(11):4327–4332
70. Tang Q et al (2014) Optimized cell transplantation using adult rag2 mutant zebrafish. Nat Methods 11(8):821–824
71. Matthews AG et al (2007) RAG2 PHD finger couples histone H3 lysine 4 trimethylation with V(D)J recombination. Nature 450(7172):1106–1110
72. Tenente IM et al (2014) Normal and malignant muscle cell transplantation into immune compromised adult zebrafish. J Vis Exp 94
73. Tang Q, Moore JC, Ignatius MS, Tenente IM, Hayes MN, Garcia EG, Torres Yordán N, Bourque C, He S, Blackburn JS, Look AT, Houvras Y, Langenau DM (2016) Imaging tumour cell heterogeneity following cell transplantation into optically clear immune-deficient zebrafish. Nat Commun 7:10358
74. Diep CQ, Davidson AJ (2011) Transplantation of cells directly into the kidney of adult zebrafish. J Vis Exp 51
75. LeBlanc J, Bowman TV, Zon L (2007) Transplantation of whole kidney marrow in adult zebrafish. J Vis Exp 2:159

76. White RM et al (2008) Transparent adult zebrafish as a tool for in vivo transplantation analysis. Cell Stem Cell 2(2):183–189
77. Le X et al (2007) Heat shock-inducible Cre/Lox approaches to induce diverse types of tumors and hyperplasia in transgenic zebrafish. Proc Natl Acad Sci U S A 104(22):9410–9415
78. Swanton C (2012) Intratumor heterogeneity: evolution through space and time. Cancer Res 72(19):4875–4882
79. Almendro V, Marusyk A, Polyak K (2013) Cellular heterogeneity and molecular evolution in cancer. Annu Rev Pathol 8:277–302
80. Marusyk A, Polyak K (2010) Tumor heterogeneity: causes and consequences. Biochim Biophys Acta 1805(1):105–117
81. Gutierrez A et al (2009) High frequency of PTEN, PI3K, and AKT abnormalities in T-cell acute lymphoblastic leukemia. Blood 114(3):647–650
82. Frazer JK et al (2009) Heritable T-cell malignancy models established in a zebrafish phenotypic screen. Leukemia 23:1825–1835
83. Ignatius MS, Langenau DM (2009) Zebrafish as a model for cancer self-renewal. Zebrafish 6(4):377–387
84. White R, Rose K, Zon L (2013) Zebrafish cancer: the state of the art and the path forward. Nat Rev Cancer 13(9):624–636
85. Feng H et al (2010) T-lymphoblastic lymphoma cells express high levels of BCL2, S1P1, and ICAM1, leading to a blockade of tumor cell intravasation. Cancer Cell 18(4):353–366
86. Rennekamp AJ, Peterson RT (2015) 15 years of zebrafish chemical screening. Curr Opin Chem Biol 24:58–70
87. Le X et al (2013) A novel chemical screening strategy in zebrafish identifies common pathways in embryogenesis and rhabdomyosarcoma development. Development 140(11):2354–2364
88. Korinek V et al (1997) Constitutive transcriptional activation by a beta-catenin-Tcf complex in APC−/− colon carcinoma. Science 275(5307):1784–1787
89. Anastas JN, Moon RT (2013) WNT signalling pathways as therapeutic targets in cancer. Nat Rev Cancer 13(1):11–26
90. White RM et al (2011) DHODH modulates transcriptional elongation in the neural crest and melanoma. Nature 471(7339):518–522
91. Golicki D et al (2012) Leflunomide in monotherapy of rheumatoid arthritis: meta-analysis of randomized trials. Pol Arch Med Wewn 122(1–2):22–32
92. Ridges S et al (2012) Zebrafish screen identifies novel compound with selective toxicity against leukemia. Blood 119(24):5621–5631
93. Zhang Y et al (2013) AML1-ETO mediates hematopoietic self-renewal and leukemogenesis through a COX/beta-catenin signaling pathway. Blood 121(24):4906–4916
94. Pan YA et al (2013) Zebrabow: multispectral cell labeling for cell tracing and lineage analysis in zebrafish. Development 140(13):2835–2846
95. Renshaw SA et al (2006) A transgenic zebrafish model of neutrophilic inflammation. Blood 108(13):3976–3978
96. Ellett F et al (2011) mpegl promoter transgenes direct macrophage-lineage expression in zebrafish. Blood 117(4):e49–e56
97. Blackburn JS, Langenau DM (2014) Zebrafish as a model to assess cancer heterogeneity, progression and relapse. Dis Model Mech 7(7):755–762

The Zebrafish Xenograft Platform: Evolution of a Novel Cancer Model and Preclinical Screening Tool

Jaime Wertman, Chansey J. Veinotte, Graham Dellaire, and Jason N. Berman

Abstract Animal xenografts of human cancers represent a key preclinical tool in the field of cancer research. While mouse xenografts have long been the gold standard, investigators have begun to use zebrafish (*Danio rerio*) xenotransplantation as a relatively rapid, robust and cost-effective in vivo model of human cancers. There are several important methodological considerations in the design of an informative and efficient zebrafish xenotransplantation experiment. Various transgenic fish strains have been created that facilitate microscopic observation, ranging from the completely transparent *casper* fish to the Tg(*fli1:eGFP*) fish that expresses fluorescent GFP protein in its vascular tissue. While human cancer cell lines have been used extensively in zebrafish xenotransplantation studies, several reports have also used primary patient samples as the donor material. The zebrafish is ideally suited for transplanting primary patient material by virtue of the relatively low number of cells required for each embryo (between 50 and 300 cells), the absence of an adaptive immune system in the early zebrafish embryo, and the short experimental timeframe (5–7 days). Following xenotransplantation into the fish, cells can be tracked using in vivo or ex vivo measures of cell proliferation and migration, facilitated by fluorescence or human-specific protein expression. Importantly, assays have been

J. Wertman
Department of Microbiology and Immunology, Dalhousie University, Halifax, NS, Canada, B3H 4R2

C.J. Veinotte
Department of Pediatrics, IWK Health Centre, Halifax, NS, Canada

G. Dellaire
Department of Pathology, Dalhousie University, Halifax, NS, Canada
e-mail: dellaire@dal.ca

J.N. Berman (✉)
Department of Microbiology and Immunology, Dalhousie University, Halifax, NS, Canada, B3H 4R2

Department of Pediatrics, IWK Health Centre, Halifax, NS, Canada
e-mail: Jason.berman@iwk.nshealth.ca

D.M. Langenau (ed.), *Cancer and Zebrafish*, Advances in Experimental Medicine and Biology 910, DOI 10.1007/978-3-319-30654-4_13

developed that allow for the reliable detection of in vivo human cancer cell growth or inhibition following administration of drugs of interest. The zebrafish xenotransplantation model is a unique and effective tool for the study of cancer cell biology.

Keywords Zebrafish • Xenotransplant • Xenograft • Cancer • Model organism

Introduction to Xenotransplantation

Human cancer research is inherently impacted by the obvious ethical and technical constraints concerning the use of human or other large animals in a research setting. It is for this reason that cancer research relies heavily on the use of model organisms for experimental purposes. One classical example of an experiment that necessitates the use of a model organism is xenotransplantation. Xenotransplantation is defined as the transfer of material isolated from one species into another [1–3]. In a cancer research or drug discovery context, this normally translates into the transplantation of human cancer cells into a smaller organism. It is important to note, then that xenotransplant models are not an animal model of disease, but rather a human-in-animal model of disease that offers both unique advantages and challenges in comparison with other model types. The following chapter will summarize the current state of the field concerning human cancer xenotransplantation in the zebrafish (*Danio rerio*).

Historically, the mouse has been viewed as the "gold standard" for xenotransplantation experiments [1, 2] for a number of reasons. Firstly, the mouse is a relatively small and easy to care for animal, with moderate husbandry requirements. Secondly, there is a high degree of genetic similarity between the human and the mouse, at approximately 85 % [4]. Mouse transplant experiments also benefit from the fact that the mouse shares the presence of most anatomical features with the human that may be relevant to cancer studies, such as the breast pad or the testes [1, 2]. In typical xenograft experiments, 0.5–1 million human tumor cells, either derived from a primary patient tissue sample or from a cell line, are injected subcutaneously into a mouse. The mice are then monitored for health and visible or palpable signs of tumor burden. This process is lengthy, requires a relatively large number of human tumor cells, and is hindered by the fact that it is can be difficult to detect or image small numbers of cells within a mouse (e.g. micrometastases).

Mouse xenograft experiments also require immunocompromised animals, as normally the mouse immune system would reject the human cells, obviating tumor studies when the adaptive immune system is intact. To avoid immune rejection, typically murine xenograft experiments employ severely immunodeficient (severe combined immunodeficient (SCID) or nude) mice [1]. Since the advent of these mice, other transgenic mice have been created that may represent a superior xenograft recipient than the traditional NOD/*scid*-related mice [5–8]. For example, the NOD/

LtSz-*scid IL2R*γ^{null} (NSG) mouse have a deficiency in the common cytokine receptor γ-chain, causing a lack of B, T and natural killer (NK) cells [5]. It has been shown that human acute myeloid leukemia (AML) cells transplanted into NSG mice have more efficient engraftment and more rapid disease progression than in other immunocompromised strains [6]. Since human cancer is treated within the context of a generally intact immune system and in particular, leukemia engraftment may require cytokines not provided by the host mouse, the relevance of murine xenograft experiments in immunocompromised mice can be called into question. Therefore, some investigators have begun to supplement the xenograft recipients with constitutively expressed cytokines, to more closely mimic the human environment, or "humanize" the mice in leukemia studies [9, 10]. Similarly, for solid tumor studies, transfer of cancer-associated fibroblasts or stroma along with tumor cells can functionally "reconstruct" a species-specific tumor niche [11].

Despite the applicability of the mouse in human cancer xenograft being challenged, its utility has been demonstrated through its moderate ability to predict cancer drug effectiveness in Phase II clinical trials [3, 12]. Nonetheless, the fact remains that the majority of drugs that show a chemotherapeutic effect in a pre-clinical setting ultimately fail once they enter human trials. Improvements to the current murine xenograft model have been suggested, such as drastically increasing the number of cell lines tested, fully understanding the pharmacology of the compound tested, and using primary patient tumors as transplantable material [1]. Unfortunately, due to the finite budget and working capacity of most laboratories and pharmaceutical companies, coupled with the challenges associated with imaging small numbers of human cells within a living mouse, these recommendations may be difficult to put into practice [13].

While great strides have been realized in both the screening for and treatment of human cancer, improvements need to be made in order to facilitate the transfer of knowledge from basic science to the clinical setting. It is clear that xenograft experiments will likely remain a fixture in cancer research for the foreseeable future. It is for this reason that the classical models used in pre-clinical research constantly need to be reassessed and revamped in order to increase the relevance and utility of xenograft experiments. Ultimately, the sobering reality is that no animal model will be able to substitute for actual human trials. However, by employing multiple animal models such as zebrafish in addition to mice, the likelihood of success in human trials could be much greater if evolutionary conserved drug-tumor interactions between species can be identified.

Why Zebrafish?

The zebrafish, *Danio rerio,* was first used in a formal research setting in the early 1980s. The zebrafish emerged as a model organism primarily in the field of developmental biology, largely due to its external fertilization, rapid and external development and optical clarity [8, 14–16]. The use of the zebrafish in cancer research is a more recent phenomenon that evolved following a 1982 study that demonstrated the development of neoplasms in zebrafish exposed to carcinogens, such as N-nitramines [17]. In retrospect, the use of an animal model typically employed for developmental studies in cancer research now seems intuitive, as cancer is recognized as a pathology resulting from hijacked developmental pathways.

There are several important features of the zebrafish that lend it to cancer research, and more specifically, to xenotransplantation experiments. First and foremost, the zebrafish possesses striking genetic conservation to humans, estimated at approximately 70 % similarity [4]. To put this in context, zebrafish are roughly as similar to mice as they are to humans [8] and in fact, even epigenetic marks regulating conserved genes between these species are also highly conserved [18]. Additionally, there are several characteristics of the zebrafish that render its husbandry costs lower than that of mice or other small rodents, including lower space requirements, more offspring per breeding pair, rapid maturation, and lower maintenance requirements [13–15]. Zebrafish are known to develop spontaneous neoplasms at a low incidence rate, for example, pancreatic ductal adenocarcinomas [15, 19]. Neoplasms can also be induced intentionally in an experimental setting through administration of various compounds, including ethylnitrosourea [20]. Of importance and speaking to the fidelity of the zebrafish for modeling human disease, the histology of these tumors is quite similar to those found in humans [19]. Relevant for the use of fish for xenotransplantation studies, the zebrafish does not develop a functional adaptive immune system until approximately 30 days post fertilization (dpf), making immunosuppression in young embryos unnecessary, while also enabling the opportunity to examine innate and adaptive waves of the immune system separately [21]. Zebrafish are also optically transparent for the first portion of their life, as they begin developing visible pigment cells at 48 hpf, facilitating microscopic imaging. Finally, multiple transgenic and mutant zebrafish lines have been created that greatly facilitate fixed and live microscopy, including the *casper* mutant line that is optically transparent by virtue of lacking melanophores and iridiphores [38]. The use of these transgenic lines in xenotransplantation experiments will be discussed in more detail later in this chapter.

Early Reports of Xenografts in the Zebrafish: The Birth of a Model

The first study of transplantation of human cells into the zebrafish was published in 2005 by Lee et al. who injected human melanoma cells into blastula stage zebrafish embryos [22]. This study was prompted by the realization that metastatic melanoma cells possess many molecular markers of de-differentiated, plastic cell types [22, 23]. It was thought that an unknown signal could allow these cells to transition into epithelial cells, possibly transiently suppressing their cancerous phenotype, as had been previously demonstrated in mice [24]. Thus, in this study, between 1 and 100 cells from the human C8161-RFP metastatic cutaneous melanoma cell line were transplanted into the blastodisc region of wild type zebrafish embryos ranging from 3.5 to 4.5 h post-fertilization (hpf). For comparison, the same number of fibroblasts and melanocytes were transplanted in a similar manner. The embryos were maintained at 31 °C (an intermediate temperature between the 28 °C ideal for the zebrafish and the 37 °C that is optimal for human cells) and tracked for several days by fluorescence microscopy. The cells survived in the embryos, and by 5 dpf, were found in the head, trunk, and sometimes in the tail region of the fish. Notably, there were no visible tumors in these fish, even at 3 months old. Both the melanocytes and fibroblasts were also able to live in the embryo, albeit they had significantly less migration and dispersal throughout the animal. This is an important observation for two reasons. The first is that this study proves that both tumor stroma and the tumor itself can be xenografted, opening up the possibility of exploring tumor-stromal interactions using the fish. Secondly, these results led Lee et al. to conclude that there was an environmental signal present in the zebrafish embryo that suppressed the malignant phenotype of the melanoma cells, and that this would be an ideal system to study the effects of the extracellular environment on cancer cell behavior [22].

Once this innovative model was published, other groups began xenotransplanting multiple cancer cell types of various tissue origin into zebrafish embryos, often without a clear idea of what host or tissue factors might be required for engraftment or if suitable orthopic sites existed in the fish (e.g. no suitable orthotopic sites exist in the fish for lung, mammary and prostate cancer xenografts). As researchers soon found out, there are many experimental variables that can influence this type of experiment. Haldi et al. were the first group to publish several aspects of the troubleshooting process for zebrafish xenotransplantation [25]. This study was also the first group to use 48 hpf embryos as the host for injection. This group believed that this time point was less likely to allow for passive diffusion of xenotransplanted cells, as gastrulation is complete and the main body plan of the animal has been formed. They went on to label WM-266-4 human melanoma cells and CCD-1092Sk normal human fibroblasts with Cm-DiI, a lipophilic fluorescent cell stain, and inject them into either the yolk sac, hindbrain ventricle, or the circulation of the zebrafish embryos. This group determined a set of parameters that they believed to be optimal for zebrafish xenograft experiments, namely: the use of albino zebrafish embryos to facilitate microscopy, 48 hpf embryos as the host, the yolk sac as the optimal site for injection of 50–100 cancer cells, and 35 °C as the optimal temperature to maintain engrafted embryos. Additionally, this group was the first to attempt quantification of the proliferation of xenografted cells within the fish. They were able to dissociate the engrafted fish to approximate single cell suspensions, fix the cells, and visually count the number of stained human cancer cells at 1 day and 4 days post injection (dpi). The comparison of number of cells on day 1 with the number of cancer cells counted on day 4 provided a rough estimate of cell proliferation, which could be used to determine the approximate in vivo doubling time [25]. Our group later refined this technique for use in drug-tumor response studies by employing semi-automated quantification, cytospins and the use of human-specific antigens for immunocytochemical detection and enumeration of tumor cells [30, 35].

Another pivotal development in the zebrafish xenotransplantation experiment field was carried out by Marques and colleagues [26]. This study was the first to use primary patient material as the transplanted tissue. They labeled small samples from pancreas, colon and stomach adenocarcinomas, and non-cancerous tissue from the same resected sample with Cm-DiI, and used a glass transplantation needle to force a section of tissue ~1/5 the size of the yolk sac into the yolk. The successfully xenografted fish were then monitored by live laser scanning confocal microscopy. Human cancer cells were seen migrating out of the injection area at 12 h post injection (hpi), and some micrometastases appeared 24 hpi. This study defined a micrometastasis as the presence of daughter cells in the zebrafish at 3 dpi, and invasion/migration as at least 5 cells outside the yolk sac and within the zebrafish. This study also employed several different transgenic zebrafish lines as the host organism for xenograft. The homozygous *cloche* mutant fish (cloche $^{-/-}$, described in [27]), that lacks a functional vascular system was used to demonstrate that the observed invasion of PaTu-8988-T pancreatic cancer cells into wild-type host tissue was reliant on an intact circulatory system, as cloche $^{-/-}$ fish did not show any invasion of the same cells. Transgenic *Tg(fli1::eGFP)* zebrafish that expresses GFP within the endothelial cells of the vasculature [28], were exploited to demonstrate the dissemination of fluorescently labeled Ras-transformed epithelial cells throughout the zebrafish embryo. The *Tg(fli1::eGFP)* zebrafish has since been used to demonstrate angiogenic responses to human tumor xenografts in the zebrafish embryo and to study interactions between tumors and the existing vasculature [29], and will be discussed in detail later.

Subsequent studies have used hosts from different genetic backgrounds, developmental stages, diverse locations of injection, variable numbers of and type of human cancer cells, and variable methods of post-xenograft observation [25, 26, 29–35]. While the reasons for some of these procedural divergences will be made clear in later sections of this chapter, a summary of procedural details from published xenograft studies can be found in Table 1.

Xenotransplantation Methodological Considerations

Host Selection

Choice of the zebrafish strain to serve as the xenograft recipient is a key initial consideration. While the earliest transplantation experiments employed wild-type or albino zebrafish, this was likely due to the restricted availability or existence of the multiple transgenic and mutant zebrafish lines available today [22, 25].

While zebrafish embryos are transparent, older zebrafish are opaque, due to their characteristic stripes. These stripes are made up of pigment cells: black melanophores, reflective iridophores and yellow xanthophores [36]. In 2008, White et al. crossed two established mutant fish lines, *nacre*, which exhibits a total loss of melanocytes [37] and the *roy orbison* line, exhibiting a loss of iridophores to create the *casper* zebrafish line [38]. The *casper* fish is completely transparent throughout adulthood, with macroscopically visible organs, large black eyes, and is functionally and reproductively normal. Indeed, observations of single and groups of transplanted

Table 1 Summary of major peer-reviewed zebrafish xenograft studies

Donor material	Fluorescence labeling method	Recipient animal	Number of cells injected	Developmental stage at injection	Location of injection	Reference
Malignant cutaneous melanoma cell line (C8161) Human dermal fibroblasts Human melanocytes	Stable transfection with RFP Transient transfection with RFP	WT	1–100	High stage—Oblong sphere stage (3.5–4.5 hpf)	Blastodisc, halftway between animal and margin	[22]
Metastatic melanoma cell line (WM-266-4) Human fibroblast cell line (CCD-1092Sk)	Stained with Cm-DiI	WT; albino	50	48 hpf	Yolk sac; hindbrain ventricle; circulation	[25]
Human cutaneous melanoma cell lines (C8161, C81-61, WM793, 1205Lu, A375P) Uveal melanoma cell lines (MUM-2B and MUM-2C)	Transient transfection with GFP	WT (derived from AB line)	50–100	Blastula stage (~6 hpf)	Near ventral margin of blastidisc	[34]
FGF-2 overexpressing murine aortic cell line (FGF2-MAE-T) Breast cancer cell line (MDA-MB-435) Ovarial carcinoma cell line (A2780) Murine melanoma cell line (B16-BL16)	Unspecified	WT (AB); *VEGFR2:G-RCFP*	1000–2000, imbedded in matrigel	48 hpf	Yolk sac, close to duct of Cuvier	[32]
Adenocarcinoma cell line (MDA-435) Fibrosarcoma cell line (HT1080) Murine melanoma cell line (B16)	Transient transfection with GFP, CFP, DsRed	Chemically immunosuppressed WT (AB); tg(*fli1:eGFP*)	0–300	25–35 dpf	Peritoneal cavity	[29]
Malignant glioblastoma cell line (U251)	Stably transfected with RFP	WT	1–100	High stage—Oblong sphere stage (3.5–4.5 hpf)	Blastidisc, between the margin and the animal pole or the center of the embryonic yolksac	[73]
Malignant glioblastoma cell line (U251)	Stably transfected with RFP	WT, *Tg(fli1a::eGFP)*	50–200	High stage—Oblong sphere stage (3.5–4.5 hpf), 48 hpf	Centre of yolk sac	[60]

(continued)

Table 1 (continued)

Donor material	Fluorescence labeling method	Recipient animal	Number of cells injected	Developmental stage at injection	Location of injection	Reference
Mouse mammary epithelial cell line transformed with oncogenic Ras (EpRas) Pancreatic cancer cell lines (PaTu8988T and PaTu8898S) Tissue sections and dissociated cells from primary patient stomach, colon and pancreatic carcinoma	Stained with Cm-DiI and DiO	Albino; tg(*fli1:eGFP*);*cloche*$^{-/-}$	200 Piece of tissue approximately 1/2–1/5 size of yolk	48 hpf	Yolk sac	[26, 71]
Malignant glioma cell line (U87-MG)	Stably transfecting with firefly luciferase gene (designated U87-L)	WT	100	Blastula stage (~6 hpf)	Yolk sac	[33]
Human pancreatic cancer cell lines used were (Pa- Tu8988T, Panc-125)	Stained with Cm-DiI	tg(*fli1:eGFP*)	Embedded in matrigel	48 hpf	Yolk sac, close to perivitelline space	[43]
Metastatic malignant breast adenocarcinoma cell line (MDA-MB-231) Murine melanoma (B16-F10)	Stably transfected with GFP Labeled with QTracker Red kit	WT (AB)	500 cells, embedded in matrigel	48 hpf	Yolk sac, close to subintestinal vessels	[72]
Ovarian carcinoma cell line (OVCA 433)	Stained with Cm-DiI	Unspecified	100	48 hpf	Yolk sac	[74]
Ovarian carcinoma cell line (SKOV3) Fibrosarcoma cell line (HT1080)	Stably transfected with eGFP	WT; tg(*fli1a:eGFP*)	1000–2000 cells in matrigel	48 hpf	Yolk sac, close to subintestinal vessels	[42]
Leukemic cell lines (K562, Jurkat and NB4) Primary patient CD4^{+} and CD4^{-} cells isolated from peripheral blood of patients with acute myeloblastic leukemia (AML)	Stained with Cm-DiI	WT (AB)	50–200	48 hpf	Through yolk sac, into posterior cardinal vein (PCV)	[54]
Leukemia cell lines (K562 and NB-4)	Stained with Cm-DiI	*Casper*	25–50	48 hpf	Yolk sac	[30]
Murine melanoma cells line (B16) Murine colon cancer cell line (CT26) Human embryonic kidney cells (HEK293)	Stably transfected with dsRed	Tg(*flk1:eGFP*)	50–100	48 hpf	Perivitelline space	[75]
Prostate cancer cell line (PC3) Breast cancer cell line (MDA-MB-231) Mouse endothelial cell line (MAE, FGF-T-MAE)	Stably transfected with mCherry or dsRed Stained with Cm-DiI	Tg(*fli1a:eGFP*), tg(*mpx:GFP*)	50–400	48 hpf	Circulation via duct of Cuvier	[49]

Prostate cancer cell lines (PC3, LNCaP) Murine breast carcinoma cell line (4 T1)	Stably transfected with mCherry Stained with Cm-DiI	WT and *casper* background Tg(*fli1a:eGFP*)	100	48 hpf	Yolk sac	[76]
Oral cancer cell lines (YD10B, HSC-2) Colon cancer cell lines (HCT116, DLD-1) Primary patient carcinoma associated fibroblasts (CAF)	Stained with Cm-DiI	Unspecified	100 (tested 25, 100, 200)	48 hpf	Yolk sac	[53]
Glioblastoma cell line (U87 sphere cells)	Stably transfected with RFP	Tg(*fli1:eGFP*)	200 (tested 0–1000)	48 hpf	Middle of yolk sac	[77]
Breast cancer cell line (MDA-MB-231) MCF10A-derived breast epithelial cell lines (MCF10A, MCF10AT1k.cl2, MCF10CA1a.cl1) Human embryonic kidney cells (Hek293T)	Stably transfected with mCherry Stained with Cm-DiI	tg(*fli1:GFP*)	400	48 hpf	Circulation via duct of Cuvier	[78]
Breast cancer cell lines (MDA-231, T47D) Prostate cancer cell lines (DU145, LNCaP) Pancreatic ductal adenocarcinoma cell lines (AsPC-1 and BxPC3) Colon cancer cell lines (SW620 and HT29)	Stained with Cm-DiI	Tg(*kdrl:eGFP*), WT (AB)	300	48 hpf	Pervitelline space	[79]
Uveal melanoma cell lines (92.1, Mel270) Metastatic uveal melanoma cell lines (OMM1, OMM2.3, OMM2.5)	Stably transfected with mCherry	Tg(*fli1:eGFP*)	400–500	48 hpf	Yolk sac	[80]
T-cell acute lymphoblastic leukemia (T-ALL) cell lines (Jurkat, TALL1, Karpas45) Primary patient samples derived from bone marrow collected at diagnosis with T-ALL	Stained with Cm-DiI Immunohistochemical detection of promyelocytic leukemia (PML) bodies Stably transfected with GFP	*Casper*	50–100 for cell lines, 500 for patient samples	48 hpf	Yolk sac	[35]
TC32 Ewing sarcoma cell line	Stained with Cm-DiI	*Casper*	150–200	48 hpf	Yolk sac	[55]

hematopoietic stem/progenitor cells (HSPCs) were made with a resolution of 5 μm at a maximum depth of approximately 88 μm with a Zeiss confocal microscope [38]. Since its creation, the *casper* strain has been used extensively for xenotransplantation work, as it is an organism ideally suited to microscopy, optically transparent with little autofluorescence, allowing for the visualization of fluorescently tagged cancer cells in vivo, without needing to sacrifice the fish [29, 30, 35, 39, 40].

Fluorescent reporter fish lines can be crossed to the *casper* fish, creating a virtually limitless combination of fish strains, some examples of which can be seen in Fig. 1. *Tg(fli1::eGFP)*, with fluorescent vasculature have incredible potential for imaging the interactions between the fish vasculature and the engrafted tumor, or possible angiogenesis, and will be discussed further later [25, 29, 31, 32, 41–44]. Other reporter lines label components of the immune system, which in this context facilitates the real-time visualization of the interaction between the immune system and transplanted tumors, an active and complex area of cancer research [45]. Some examples of these reporter lines include, but are not limited to *Tg(mpx::eGFP)*, expressing fluorescent neutrophils [46]; Tg(*mpeg::eGFP*), with GFP expression in monocytes and macrophages [47]; and *Tg(cd41::eGFP)*, with fluorescent platelets [48]. Some studies have used tail wounding, and subsequent quantification of neutrophilic infiltration as a measure of inflammatory immune response, a model that could lend itself to investigations of neutrophil-tumor interactions [46]. *Tg(mpx::eGFP)* zebrafish were used to show that neutrophil migration within the fish caused collagen matrix deformation, which the authors argue subsequently allowed for the infiltration of tumor cells at this location [49].

In addition to the imaging potential offered by fluorescently tagged transgenic fish, there are other transgenic zebrafish lines that can be considered as the xenograft host. Advances in the field of genomic editing, both for knock-in and knockout of genes of interest, have facilitated the investigation into gene function in both cell-based and in vivo models. Recently, clustered regularly interspaced palindromic repeats (CRISPR)/Cas9 technology has significantly simplified genome editing in the zebrafish, and has been reviewed at length elsewhere [50, 51]. The potential impact of this technology on xenograft experiments should not be understated, making it possible to investigate the effects of a certain gene on the engraftment, proliferation and migration of human cancer cells. With this technology, it will be possible to modify the microenvironment of the tumor, by creating a CRISPR fish, or the cell itself, with the creation of a CRISPR-modified cell line. Additionally, knock-in technology would allow researchers to "humanize" the zebrafish, by adding in requisite factors, such as cytokines, receptors or other growth factors, as has been done in mouse models [10, 52].

Donor Material Type and Pre-treatment

Successful xenotransplantation experiments require careful selection of both the host and the donor materials. The choice of donor material, normally human cancer cells or tumor tissue, is usually predetermined by the goals of the study.

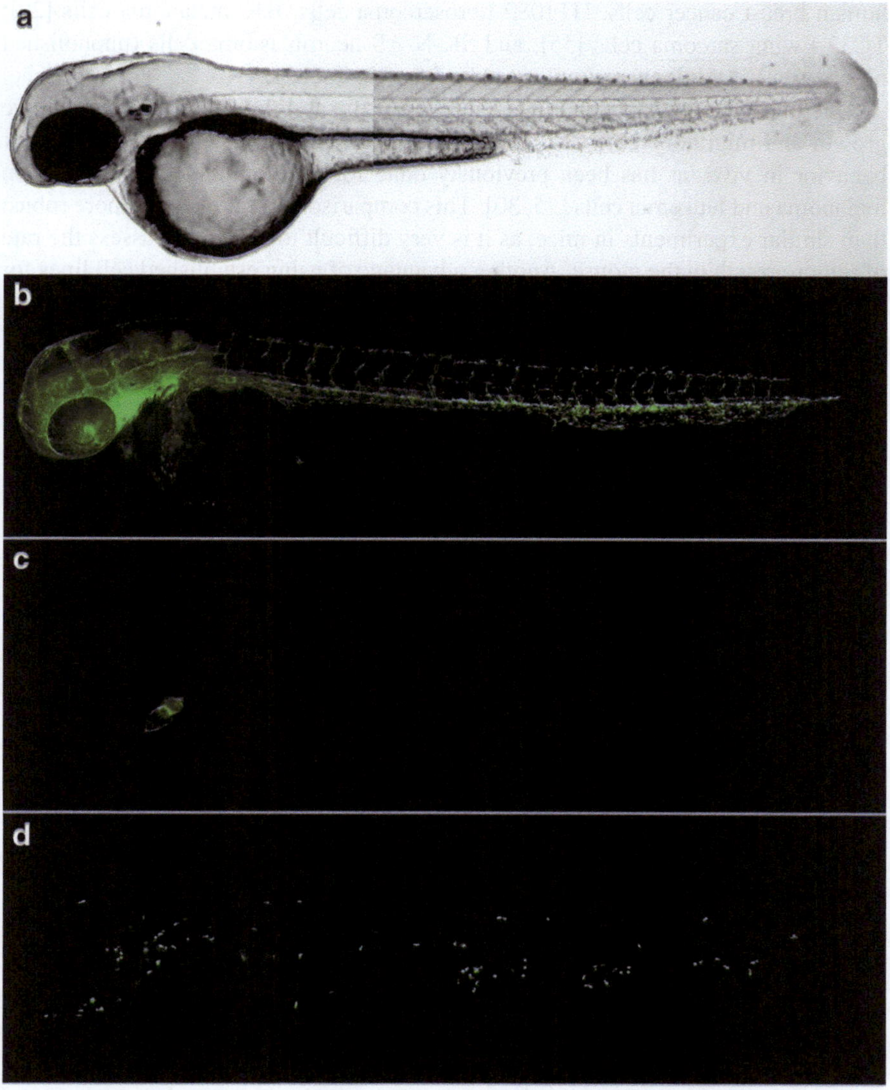

Fig. 1 Fluorescently labeled transgenic fish strains facilitate in vivo microscopy. (**a**) Bright field image of a 48 h post-fertilization (hpf) *casper* embryo, representative of brightfield images of embryos depicted in **b–d**, (**b**) Fluorescent image of *Tg(fli::egfp/casper)* embryo at 48 hpf, exhibiting fluorescent vasculature, (**c**) Fluorescent image of *Tg(myl7::egfp/casper)* embryo at 48 hpf, depicting specific fluorescence of the cardiomyocytes in the heart, (**d**) Fluorescent image of *Tg(mpx::egfp/casper)* embryo at 48 hpf, demonstrating fluorescence in neutrophils. 5X magnification

Many xenograft studies employ commercially available cell lines, derived from a multitude of cancer types [8]. Some of the many cell lines that have been successfully xenografted into zebrafish include: PaTu8988-S and PaTu8988-T pancreatic tumor cells [26]; YD10B and HSC-2 human oral squamous cell carcinoma cells [53]; various leukemia cell lines including K562, Jurkat and NB4 [30, 35, 54]; MDA-MB-435 human breast cancer cells, HT1080 fibrosarcoma cells, B16 melanoma cells [29]; TC32 Ewing sarcoma cells [55]; and SK-N-AS neuroblastoma cells (unpublished observation). Additional examples can be found in Table 1.

One of the benefits of xenotransplantation is the ability to compare the relative growth and migratory potential of the human cells in the in vitro system to their behavior in vivo, as has been previously done for the doubling time for human melanoma and leukemia cells [25, 30]. This comparison is significantly more robust than similar experiments in mice, as it is very difficult to accurately assess the rate of tumor growth in the mouse. Another advantage of using established cell lines for xenograft donor material is the relative ease with which they can be obtained, cultured, and manipulated. Cell lines can easily be modified with either transient or stable transfection of gene of interest prior to engraftment. Conversely, cells can be treated with either targeted small-interfering RNA (siRNA) or small hairpin RNA (shRNA) in order to transiently or more permanently knockdown a particular gene prior to transplantation. For example, an siRNA screen originally identified ribosomal s6 kinase 1 (RSK1) as a potential regulator of lung cancer metastasis, a finding that was supported when siRNA-mediated knockdown of RSK1 significantly enhanced the migration of lung cancer cells in a zebrafish xenograft model [56]. A similar approach was also used to demonstrate the knockdown of the LIM kinases LIMK1 and LIMK2 could significantly inhibit the invasion and metastatic potential of pancreatic cell lines in the zebrafish [57]. A recent example exemplifies the advantages of this technique. Ewing family tumor cells (TC-32) were xenografted into the yolk sac of 48 hpf *casper* embryos. Knockdown of the Y-box binding protein 1 (YB-1) prior to their injection resulted in significantly impaired cell migration, in comparison with the parental TC-32 cells, findings that were recapitulated in a murine sub-renal capsule xenograft model [55, 58].

Primary Patient Samples

Xenotransplantation experiments may also employ primary patient-derived samples as the donor material [26, 35, 54]. The use of primary samples from cancer patients is an important step for xenograft experiments. Throughout the field of cancer research, studies are critiqued for their use of commercially available cells lines, as results gleaned from experiments from these cell lines do not always correlate with those found in primary cancers [8]. There are several main reasons proposed for these inconsistencies. First, the long-term maintenance of cell lines in a set of conditions may inadvertently select for a subset of cells best suited to the constant environment, while cells in a tumor would be exposed to a microenvironment that varies both spatially and temporally. Second, cell lines do not capture the heterogeneity that exists within a given cancer. Lastly, genetic evolution within a given tumor or cancer will not be replicated by the possible genetic drift of a cell line over time [8, 59]. While cells from cell lines can easily be expanded, primary patient material is exceptionally difficult to obtain, and expensive to store. In this respect, the zebrafish presents a distinct advantage over murine xenograft experiments, as zebrafish xenografts require only a fraction of the number of cells that would be needed for a similar mouse experiment (approximately 50–200 cells/zebrafish vs. 0.5–1 million cells/mouse). Additionally, if the zebrafish embryo is used as a host, immune suppression is unnecessary, as the embryo does not have a fully developed adaptive immune system until approximately 30 dpf, as will be discussed further below [21].

Several leukemia studies have used primary patient material in zebrafish xenograft experiments. Leukemia is a relatively easily accessible cancer to obtain for xenograft studies. Blood and bone marrow are collected from patients at the time of diagnosis and at certain intervals during treatment for clinical purposes. With the appropriate ethical approval, some of this material can then be transferred to the laboratory and stored in liquid nitrogen until the time of the experiment. For example, blast cells that were separated from peripheral blood from patients with acute myeloid leukemia (AML) and injected into the posterior cardinal vein (PCV) of the 48 hpf embryo survive until 6 dpi, whereas CD34+hematopoietic cells derived from human cord blood rapidly disappear [54]. In a study from our research group, cells isolated from bone marrow aspirate samples from patients with T-cell acute lymphoblastic leukemia (T-ALL) proliferated in zebrafish embryos over several days [35].

Solid tumors are more challenging to model in a xenograft experiment, as they are normally more difficult to harvest or obtain and need to be dissociated into smaller fragments to physically implant in the host. Diagnostic biopsy samples are small, with little residual tissue for research purposes following necessary diagnostic studies. Furthermore, the biopsy sample may be fairly heterogeneous from a tissue point of view containing both malignant and surrounding normal cellular elements. Nonetheless, human pancreas, colon and stomach carcinomas have been successfully xenografted into zebrafish embryos [26]. In this particular study, tissue fragments were implanted whole by manually forcing them into the yolk sac using a transplantation needle, and following dissociation to a single cell suspension. Similar results were obtained with each technique, with evidence of micrometastases at 24 hpi in the liver and intestines at 3 dpi [26]. Interestingly, this group also compared non-invasive pancreatitis tissue xenografts with those from infiltrating pancreatic adenocarcinoma, and found that only the latter invaded the embryo [26].

Distinguishing Human from Zebrafish Cell Populations

An additional factor to consider when xenografting human cancer cells into zebrafish is the method by which the human cells will be distinguished from cells belonging to the fish host. Cell membrane stains, such as Cm-DiI, PKH, or CMTMR have been used extensively due to their bright fluorescence, ease of use and applicability for primary patient samples [25, 26, 30, 35, 57, 58]. While these stains technically have in vivo staining persistence for up to 100 days, they invariably become more dilute as cells divide and proliferate in the fish. Thus, some investigators have favored the use of stably-transfected cell lines for xenotransplantation [22, 29, 32, 34, 60]. However, a recent report from our group confirms that the measured proliferation rates of human leukemia cells xenografted in the zebrafish are equivalent regardless of whether the cells were stained with Cm-DiI or stably expressing GFP [35]. Another way to identify human cells in the fish is to take advantage of proteins and cellular properties not conserved between humans and zebrafish. For example, the gene encoding the promyelocytic leukemia (PML) tumor suppressor does not exist in zebrafish but is present in all human cells [35, 39, 61]. As a result, the subnuclear domains formed by the PML protein, known as PML bodies [61, 62] are only present in human cells from the xenografted tumor. Thus, following dissociation of the xenografted embryo to single cells, immunohistochemistry against PML bodies, or any other non-conserved protein, allows for specific identification of human cells [35]. Cell surface molecules such as glycoproteins, unique to human cells may also be utilized for tumor cell identification and isolation. Using commercially available antibodies for specific human cell surface antigens, column based and/or magnetic field based separation techniques can be used to sequester human tumor cells from host cells post tissue dissociation. SEP systems such as the EasySEP™, offer the ability to pass antibody labeled samples through a magnetic field where positively selected cells are retained in the field while negative cells can be removed from the liquid suspension. Both cell enrichment and depletion kits can be used to isolate select cell populations from mixed samples. Once separated, many properties of the tumor cells can be examined post-transplant including apoptosis, proliferation rate, DNA damage and other cell modifications, without physical interference from the host tissue/cells. Our group has begun using these enrichment and depletion targeted magnet based kits to isolate human sarcoma and leukemia cells from dissociated human cell and zebrafish cell sample mix and to date have observed partial recovery of these injected cells. Further optimization is needed to increase the efficiency of this technique.

Fluorescent activated cell sorting (FACS) may also be applied to separate human cells from fish tissue post-dissociation as in most xenotransplantation experiments human cells have been pre-labeled with a temporary lipophilic dye or permanently labeled by fluorescent protein transfection before transplantation. Like magnetic-based cell sorting, the use of FACS to sort and isolate tumor cells from fish cells populations may require large volumes of sample for adequate recovery. Tumor xenograft volume corresponds to a very small portion of the total cell number achieved through embryonic dissociation and increasing the sample size of xenotransplanted fish may overcome this hurdle.

Developmental Time Point

As previously mentioned, past studies have used embryos at various developmental stages for xenograft experiments, some of which are summarized in Table 1. The first reported zebrafish xenograft study transplanted approximately 100 metastatic cutaneous melanoma cells into the blastodisc region of a 3.5–4.5 hpf embryo [22]. The authors reported migration and survival of the cancer cells up to 5 dpf. In comparison, injected fibroblasts and melanocytes survived, but did not migrate to the same extent as the cancer cells. The unexpected lack of tumor formation led the authors to conclude that there were environmental signals present in the zebrafish embryo that effectively suppressed the metastatic phenotype of the melanoma cells. Several studies that use zebrafish embryos at this time point comment on the lack of developmental defects that occurred in fish injected with cancer cells [22, 60]. By contrast, Topczewska et al. reported ectopic outgrowths or axial duplication when 3 hpf embryos were injected with the aggressive human melanoma cell line C8161 [34]. These authors concluded that the observed malformations were caused by Nodal secretion of the xenotransplanted cancer cells. Thus, there is somewhat conflicting evidence on the ability of injected cancer cells to induce developmental defects. The location of injection and cancer cell type seem to influence the capability of human cancer cells to induce abnormalities, as the less aggressive C81-61 melanoma cells did not induce ectopic outgrowths in this same study [34]. While this early embryonic time point obviously benefits from the ability to observe possible developmental effects from cancer cell xenotransplantation, the possibility of confounding phenotypic features as a result of this poorly understood phenomenon also hampers the experimental design of studies using blastula-stage embryos.

The vast majority of available zebrafish xenotransplantation studies use 48 hpf embryos. The first advantage of using 48 hpf embryos in xenograft experiments is that at this point, the body plan of the fish is relatively established. The vasculature has largely been mapped and the organ systems are somewhat developed. Therefore, at this stage there is less of a chance of inducing developmental defects in the fish that could confound the experiment. At this time point, the embryos are still young enough that they have not yet developed an adaptive immune system, which avoids problems of xenograft rejection [21]. Two day old embryos also possess a large yolk sac, facilitating the safe injection and accommodation of a larger number of cancer cells [39]. Another benefit of using embryos at this time point is that most of their nutrition is derived from the yolk sac, meaning that the embryos do not require feeding. This is important for two reasons, namely that it reduces husbandry requirements and also eliminates a possible variable of food variety and quantity. Embryos at this age are also more resistant to drug treatments than younger fish, enabling drug treatments at an increased range of doses.

Adult fish are larger and more mobile than their embryonic counterparts, meaning that they require more space. They also require more husbandry, including regular feeding and breeding. Another obstacle in the use of adult zebrafish for xenotransplantation studies is the presence of an adaptive immune system. Similar to challenges faced in mouse xenograft experiments, zebrafish that are >30 dpf will have an intact adaptive immune system that results in transplant rejection [21]. Various methods can be used to immunosuppress zebrafish. For example, 1 month old zebrafish were treated with 10 μg/ml dexamethasone for 2 days prior to experimentation [29]. These methods result in an immune suppressed fish that permit the survival and proliferation of human MDA-435 cells for ~13.5 days following injection into the peritoneal cavity of the fish, after which the fish succumb to their tumor burden. Another technique that can be used to impede the zebrafish immune system is the sublethal irradiation of the fish prior to injection. Studies suggest that 20–25 Gy is tolerated by approximately 90 % of the fish, and tumors can engraft 2 days post irradiation at a dose of 20 Gy [63]. This method can also be used on embryos as young as 6 dpf. Stoletov et al. discovered that a single dose of irradiation at 15 Gy would ablate the T cells of fish aged 6–30 dpf [29]. However, it should be noted that the immune system of these fish recovers approximately 20 days post-irradiation, so long-term studies cannot rely on this method of immune ablation alone [63, 64].

There have been several attempts to create transgenic zebrafish strains that would facilitate transplantation. For allograft transplantation techniques, Mizgireuv and Revskoy created two strains of homozygous diploid clonal zebrafish that can be used to transplant any tissue type, including an induced tumor, between fish [65]. However, this method does not enable xenograft cell engraftment. Recently, genome editing approaches have aided in the creation of immunocompromised fish, akin to the SCID or NSG mouse strains discussed previously [5, 6, 64]. An example of this approach is the use of the zinc-finger nuclease genetic engineering technology to create a $RAG2^{E450fs}$ mutant zebrafish line. We contributed to studies demonstrating that these mutant zebrafish have low levels of B and T cells, allowing animals to accept allografts [64]. Yet, to date, these lines have not been assessed for the ability to engraft human tissues or cancer. It is likely that more immunocompromised fish strains will become available in the coming years, as zebrafish genetic engineering technology begins to catch up to the technology available for mice.

Injection Location

The location of injection within the embryo is a fundamental aspect of xenograft experiments. Orthotopic injection is defined as the transplantation of cells or a tissue into its native anatomic location. However, due to some of the anatomical differences between zebrafish and humans, orthotopic xenotransplantation is not always possible, such as with breast, lung, or prostate cancer cells. Thus, xenotransplantation experiments in zebrafish are frequently heterotopic, where the donor tissue location is not the same as the host's recipient location. However, studies have injected cancer cells into a variety of locations in the zebrafish embryos, including the blastodisc region, the yolk sac, the hindbrain ventricle, and into the circulation via the duct of Cuvier (studies summarized in Table 1). The choice of location for injection is intrinsically linked to the developmental stage of the fish due to the rapid development of the embryos, which can result in dramatic changes in body plan and organ structure within the first few hours and days post-fertilization. For example, the yolk sac is completely reabsorbed by the embryo within 6–7 days post-fertilization, preventing injection to this site in embryos older than 7 days.

The yolk sac is the most common location for injection. There are several main reasons for this, many of which were highlighted by Haldi et al. The yolk sac was deemed the ideal injection site as it allowed a higher burden of cancer cells than the hindbrain ventricle, is an acellular and partly avascular environment that is unlikely to support passive cell transport, and the nutrient-rich setting is supportive of cell proliferation [25], findings we recently confirmed for sarcoma cells [55].

Another benefit of the yolk sac location is the relatively large site for injection. This makes it significantly easier to manually transplant a sufficient number of cells for engraftment. In comparison, the hindbrain ventricle is small. Additionally, several studies have reported difficulties surrounding the survivability of fish following hindbrain ventricle injection [25, 41]. Our group has developed a novel transpharangeal hindbrain ventricle injection technique that facilitates engraftment and survivability of the embryos following injection (unpublished observation). Further work on the refinement of this technique will likely lead to the use of this technique for the xenotransplantation of gliomas and other central nervous system tumors.

Post-Xenotransplant Observation and Applications

Drug Screening

The mouse is typically considered to be the "gold standard" for preclinical drug testing, however, the zebrafish represents a valuable complementary tool for the preclinical testing of cancer compounds, as the use of fish will likely shorten the experimental timeline, reduce cost, and decrease the quantity of compound required. Several small-molecule drug screens have yielded important findings that reach beyond the cancer field. An example was the discovery by Peterson et al. that several compounds have an effect on the development of the central nervous system of the fish, causing increases in the size of the hindbrain ventricle, and other unusual features [66]. Another example from the cancer field involved a small-molecule screen using zebrafish embryos that identified several compounds that previously were not known to possess cell cycle modifying activity [67].

The zebrafish xenograft model has significant potential as a tool in the process of drug discovery and pre-clinical evaluation [30, 35, 39, 53, 54]. In comparison with mice, the small size of the zebrafish means that experiments require significantly less drug per animal, which may be especially important given the difficulty and costs associated with obtaining large quantities of experimental chemicals [39]. Additionally, the aquatic environment of the zebrafish means that many drugs (depending on solubility) can be added directly to the embryo water and absorbed through the fish, avoiding the burden of administering drug to each individual animal [39, 68]. If the drug in question is not soluble in water, it can either be diluted with a vehicle such as dimethyl sulfoxide (DMSO), and added to the water, or injected directly into the fish [39]. Another advantage is the ease with which drug toxicity can be evaluated in the zebrafish. Embryos can be arrayed in 96-well plates and treated with increasing concentrations of the drug in question and observed to monitor toxicity [30, 35]. In our group, as a starting point we typically keep the concentration of the drug used for experiments below 50 % of the maximum tolerated dose determined in toxicity assays [30, 35, 39]. It is absolutely essential to determine the toxicity curve for each compound before embarking on xenograft studies, as the effective dose to inhibit cancer growth within the zebrafish can range from 1 to 20X the concentration required in vitro. Following completion of toxicity curves, xenografted zebrafish can be exposed to the appropriate concentration of drug and observed over time for cancer progression, in terms of cell proliferation, migration and angiogenesis.

In order for drug treatment to be a realistic option, long-term and severe side effects need to be minimized as much as possible. Recently, in collaboration with the Peterson laboratory, we established a zebrafish model of doxorubicin-induced cardiotoxicity in order to screen for cardioprotective agents [69]. The malate dehydrogenase inhibiting agent, visnagin was identified as a potential cardioprotectant and administered in the water at a dose of 20 μM. Human T-immunophenotype acute leukemia cells (Jurkat) were fluorescently labeled with Cm-DiI, injected into the yolk sac of 48 hpf embryos, and analyzed for cell proliferation following treatment with doxorubicin and visnagin. The chemotherapeutic effect in both fish and subsequently in mouse breast cancer xenografts was unaffected by the co-administration of visagnin with doxorubicin [69]. The zebrafish xenograft model served as a critical in vivo tool to determine that reduced cardiotoxic adverse effects of current therapies could be achieved while preserving the cytotoxic effect of the drug treatment on the cancer cells in question. This study demonstrated for the first time the utility of zebrafish xenotransplantation studies for examining drug efficacy and safety.

Proliferation

There have been several attempts to quantify the in vivo proliferation of human cancer cells within the zebrafish [25, 30, 35, 60]. Our group has developed a highly reproducible and robust ex vivo method for quantification of tumor proliferation that has been verified in drug treatment experiments [30]. Briefly, cancer cells are labeled with a membrane dye and injected into the yolk sac of 48 hpf *casper* embryos, an example of which can be seen in Fig. 2. On the first and fourth days following injection, groups of 15–20 fish are dissociated into a single cell suspension using collagenase, dropped onto slides, imaged, and an in silico program like ImageJ (NIH, Bethesda, MD) is used to count the average number of fluorescent cancer cells per embryo (Schematic in Fig. 3; [30]). This method has been used successfully with a number of different cancer types. Notably, the proliferation experiments in this study were able to demonstrate molecularly targeted selectivity, whereby xenografted chronic myelogenous leukemia cells responded to imatinib mesylate, while acute promyelocytic leukemia cells responded to all-trans retinoic acid [30]. Additionally, an adaptation of this ex vivo proliferation assay has been used to evaluate the response of primary patient T-ALL samples to several drugs. Remarkably, this approach allowed the rapid identification of γ-secretase inhibitor susceptibility of one patient's leukemia [35]. This response to treatment observed in the zebrafish xenotransplantation assay suggested a potential mutation in the NOTCH signaling pathway, which was later confirmed to be a gain-of-function mutation in *NOTCH* by Sanger sequencing [35]. Due to the short one-week timeframe of the zebrafish xenotransplantation assay, it has the potential to inform treatment decisions for patients, possibly bringing the field closer to the reality of personalized medicine.

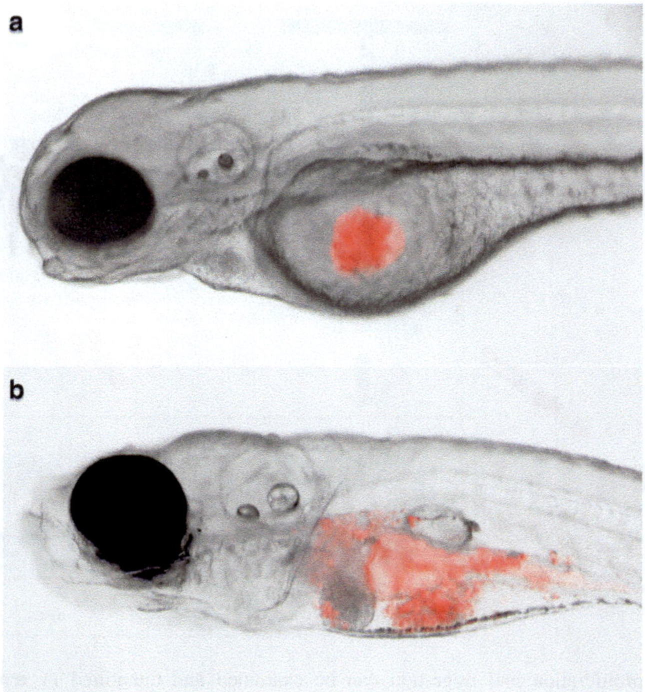

Fig. 2 Lung cancer cells survive in 48 hpf embryos, proliferate, and spread throughout the fish over 4 days. A549 lung carcinoma cells are fluorescently labelled with CMTMR-Orange CellTracker, then manually injected into the yolk sac of 48 h post-fertilization (hpf) *casper* embryos. Approximately 150–200 cells are injected per embryo. One day post injection (dpi), embryos are screened for the presence of a consistently sized bolus of cells in the yolk sac. Selected fish are maintained at 35 °C incubator for the duration of the experiment. (**a**) Brightfield-RFP fluorescent overlay image of a representative xenotransplanted fish at 1 dpi (72 hpf). Note the presence of a round group of fluorescent human cancer cells in the centre of the yolk sac. (**b**) Corresponding brightfield-RFP fluorescent overlay image of the same fish at 4 dpi (6 days post fertilization (dpf)). Note that the fluorescent cancer cells have disseminated throughout the fish, and appear to have proliferated

Migration

In addition to the growth and division of cancer cells, their dissemination through the body is a subject of interest for further investigation, as up to 90 % of cancer-related deaths are due to cancer spread, or metastasis [70]. In addition to the proliferation of human cancer cells in the zebrafish, many studies report the migration of cells from the site of injection to a secondary location [25, 26, 43, 71]. Our group has established a xenograft cell migration experiment that requires few supplementary materials or technology. In short, 48 hpf *casper* embryos are injected with 50–200 Cm-DiI-stained cancer cells in the yolk sac. The embryos are placed at 28 °C to recover for 1 h, then transfered to 35 °C for the remainder of the experiment. The embryos are examined using a fluorescent microscope at 24 hpi, and

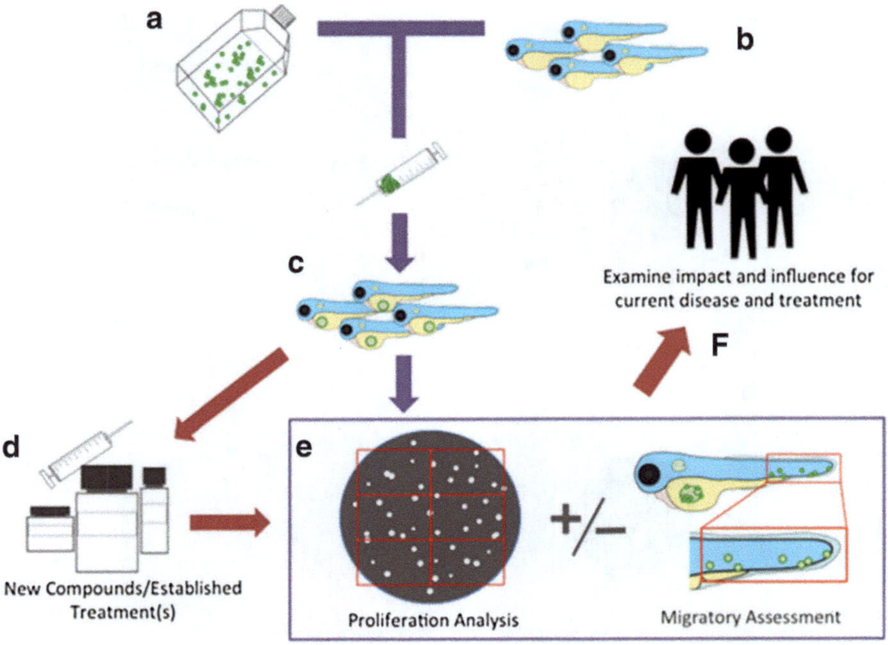

Fig. 3 Cell proliferation and migration can be examined and quantified in xenotransplanted embryos in either the presence or absence of drug treatment. (**a**) Human cells are fluorescently labeled and transplanted into the yolk sacs of (**b**) 48 h post-fertilization (hpf) zebrafish embryos. (**c**) Positive transplants are confirmed before (**d**) the initiation of drug evaluation. (**e**) Engrafted cells are assessed in vivo for effects on cell proliferation and migration. (**f**) Findings in zebrafish xenografts may influence prioritization of further drug testing and/or inform choice of patient treatments

embryos are only selected if they have a consistent-sized bolus in the middle of the yolk sac and they do not have any visible cancer cells in the tail of the embryo. Fish are observed and imaged daily, and counted as "positive" for migration when there are more than 6 cancer cells observed beyond the cloaca of the embryo, extending to the tip of the tail (Fig. 3). Using this method, we were able to show that the YB-1 transcription/translation factor is an important player in the migration of sarcomas, data that was then recapitulated in murine xenografts [55].

Angiogenesis

The optical clarity and predictable development of the vasculature of the zebrafish, coupled with the availability of transgenic fish with fluorescent blood vessels, can significantly enhance the study of real-time tumor-vasculature interactions [31]. Angiogenesis is a normal part of development, when blood vessels form to deliver oxygen and nutrients to developing tissues. It has been well-documented that tumors hijack the process of angiogenesis to deliver nutrients and to enable cancer cells to extravasate cells from the primary tumor and metastasize [41, 44]. Several reports have demonstrated that the zebrafish xenograft model is valuable in the study of angiogenesis and/or neovascularization [25, 29, 31, 32, 41, 43, 72]. Xenotransplantation of several cancer types into zebrafish, including melanoma, breast adenocarcinoma, and pancreatic cancer has been shown to induce recruitment of nearby endothelial cells and form vessels [25, 29, 32]. This experimental approach can be used to identify molecular factors involved in the angiogenic response. For example, knockdown of LIMK1 and LIMK2 in pancreatic cancer cell lines significantly reduced angiogenesis following xenotransplantation [43]. Several reports have also highlighted the potential for this high-throughput, relatively inexpensive, pre-clinical xenograft model to investigate the potential anti-angiogenic effects of certain drugs [7, 39, 44, 72].

Looking Forward: A Place for Zebrafish Xenografts in Cancer Research

Although there are advantages and disadvantages to the use of xenotransplantation studies, they will likely remain a fixture in cancer research for the foreseeable future. Zebrafish xenograft experiments are an inexpensive, relatively high-throughput, and extremely rapid counterpart to similar studies in other model organisms. We argue that the use of zebrafish and other vertebrate animal models such as the mouse, increases the likelihood that drug-treatments found to be effective across evolutionarily distant models would have a strong possibility of being conserved in humans. The use of human tumor xenotransplantation in zebrafish has proven to be a versatile and highly reproducible technique being adopted by numerous research groups. More enticingly, zebrafish xenotransplantation is potentially applicable as a personalized medicine strategy by allowing the drug sensitivity of a patient's tumor to be determined in a clinically actionable timeframe. The discovery of a NOTCH pathway mutation in a young patient with T-ALL was made possible through the response of the primary patient sample in vivo in the zebrafish embryo to γ-secretase inhibitor [35]. This is just one example highlighting the potential for these xenotransplantation experiments to inform patient stratification and treatment. Collection of these types of samples occurs routinely for clinical purposes, so within a week, this entire experiment could be completed with a selection of possible chemotherapy drugs. This result, in particular, highlights the main benefits that the zebrafish xenograft model possesses over similar experiments in the mouse model, namely the small number of donor cells required for tumor engraftment and the optical clarity of the fish, facilitating microscopic observations.

Zebrafish xenotransplantation of human cancer cells continues to gain acceptance by the wider cancer community as a robust platform for in vivo validation of in vitro findings related to the contribution of oncogenic factors and drug responses. As the cost of clinical trials continues to increase, mandating the judicious selection of validated biological targets and selective therapies, the zebrafish xenograft model will find its place in the preclinical pipeline, by providing a rapid, reliable and cost-effective means to prioritize next steps in cancer therapy evaluation.

Acknowledgments JW is funded by a Killam Predoctoral Award, a Nova Scotia Health Research Foundation Scotia Scholar award, and the Colleen Elliott Award for Excellence in Cancer Research. This work was funded by a Collaborative Health Research Project (CHRP) grant funded by the Canadian Institutes of Health (CIHR) and the Natural Sciences and Engineering Research Council (NSERC) to GD and JNB, and an NSERC Discovery Grant to GD.

References

1. Sharpless NE, Depinho RA (2006) The mighty mouse: genetically engineered mouse models in cancer drug development. Nat Rev Drug Discov 5:741–754
2. Cariati M, Marlow R, Dontu G (2011) Cancer cell culture. Methods Mol Biol 731:471–482
3. Voskoglou-Nomikos T, Pater JL, Seymour L (2003) Clinical predictive value of the in vitro cell line, human xenograft, and mouse allograft preclinical cancer models. Clin Cancer Res 9:4227–4239
4. Howe K, Clark MD, Torroja CF, Torrance J, Berthelot C, Muffato M, Collins JE, Humphray S, McLaren K, Matthews L, McLaren S, Sealy I, Caccamo M, Churcher C, Scott C, Barrett JC, Koch R, Rauch G-J, White S, Chow W, Kilian B, Quintais LT, Guerra-Assunção JA, Zhou Y, Gu Y, Yen J, Vogel J-H, Eyre T, Redmond S, Banerjee R et al (2013) The zebrafish reference genome sequence and its relationship to the human genome. Nature 496:498–503
5. Agliano A, Martin-Padura I, Mancuso P, Marighetti P, Rabascio C, Pruneri G, Shultz LD, Bertolini F (2008) Human acute leukemia cells injected in NOD/LtSz-scid/IL-2R?? null mice generate a faster and more efficient disease compared to other NOD/scid-related strains. Int J Cancer 123(August):2222–2227
6. Shultz LD, Lyons BL, Burzenski LM, Gott B, Chen X, Chaleff S, Kotb M, Gillies SD, King M, Mangada J, Greiner DL, Handgretinger R (2005) Human lymphoid and myeloid cell development in NOD/LtSz-scid IL2R gamma null mice engrafted with mobilized human hemopoietic stem cells. J Immunol 174:6477–6489
7. Konantz M, Balci TB, Hartwig UF, Dellaire G, André MC, Berman JN, Lengerke C (2012) Zebrafish xenografts as a tool for in vivo studies on human cancer. Ann N Y Acad Sci 1266:124–137
8. Berman JN, Chiu PPL, Dellaire G (2014) Preclinical animal models for cancer genomics. In Dellaire G, Berman J, Arceci R (eds) Cancer genomics from bench to bedside, 1st edn. Elsevier, pp 109–131
9. Rathinam C, Poueymirou WT, Rojas J, Murphy AJ, Valenzuela DM, Yancopoulos GD, Rongvaux A, Eynon EE, Manz MG, Flavell RA (2011) Efficient differentiation and function of human macrophages in humanized CSF-1 mice. Blood 118:3119–3128
10. Wunderlich M, Chou F-S, Link KA, Mizukawa B, Perry RL, Carroll M, Mulloy JC (2010) AML xenograft efficiency is significantly improved in NOD/SCID-IL2RG mice constitutively expressing human SCF, GM-CSF and IL-3. Leukemia 24:1785–1788
11. Kuperwasser C, Chavarria T, Wu M, Magrane G, Gray JW, Carey L, Richardson A, Weinberg RA (2004) Reconstruction of functionally normal and malignant human breast tissues in mice. Proc Natl Acad Sci U S A 101:4966–4971
12. Johnson JI, Decker S, Zaharevitz D, Rubinstein LV, Venditti JM, Schepartz S, Kalyandrug S, Christian M (2001) Relationships between drug activity in NCI preclinical in vitro and in vivo models and early clinical trials. Br J Cancer 84:1424–1431
13. White R, Rose K, Zon L (2013) Zebrafish cancer: the state of the art and the path forward. Nat Rev Cancer 13:624–636
14. Lieschke GJ, Currie PD (2007) Animal models of human disease: zebrafish swim into view. Nat Rev Genet 8:353–367
15. Stern HM, Zon LI (2003) Cancer genetics and drug discovery in the zebrafish. Nat Rev Cancer 3:533–539

16. Streisinger G, Walker C, Dower N, Knauber D, Singer F (1981) Production of clones of homozygous diploid zebra fish (Brachydanio rerio). Nature 291:293–296
17. Pliss GB, Zabezhinski MA, Petrov AS, Khudoley VV (1982) Peculiarities of N-nitramines carcinogenic action. Arch Geschwulstforsch 52:629–634
18. Long HK, Sims D, Heger A, Blackledge NP, Kutter C, Wright ML, Grützner F, Odom DT, Patient R, Ponting CP, Klose RJ (2013) Epigenetic conservation at gene regulatory elements revealed by non-methylated DNA profiling in seven vertebrates. Elife 2013:1–19
19. Spitsbergen J, Kent M (2003) The state of the art of the zebrafish model for toxicology and toxicologic pathology research – advantages and current limitations. Toxicol Pathol 31:62–87
20. Beckwith LG, Moore JL, Tsao-Wu GS, Harshbarger JC, Cheng KC (2000) Ethylnitrosourea induces neoplasia in zebrafish (Danio rerio). Lab Invest 80:379–385
21. Lam S, Chua H, Gong Z, Lam T, Sin Y (2004) Development and maturation of the immune system in zebrafish, Danio rerio: a gene expression profiling, in situ hybridization and immunological study. Dev Comp Immunol 28:9–28
22. Lee LMJ, Seftor EA, Bonde G, Cornell RA, Hendrix MJC (2005) The fate of human malignant melanoma cells transplanted into zebrafish embryos: assessment of migration and cell division in the absence of tumor formation. Dev Dyn 233:1560–1570
23. Hendrix MJC, Seftor EA, Hess AR, Seftor REB (2003) Vasculogenic mimicry and tumour-cell plasticity: lessons from melanoma. Nat Rev Cancer 3:411–421
24. Mintz B, Illmensee K (1975) Normal genetically mosaic mice produced from malignant teratocarcinoma cells. Proc Natl Acad Sci U S A 72:3585–3589
25. Haldi M, Ton C, Seng WL, McGrath P (2006) Human melanoma cells transplanted into zebrafish proliferate, migrate, produce melanin, form masses and stimulate angiogenesis in zebrafish. Angiogenesis 9:139–151
26. Marques IJ, Weiss FU, Vlecken DH, Nitsche C, Bakkers J, Lagendijk AK, Partecke LI, Heidecke C-D, Lerch MM, Bagowski CP (2009) Metastatic behaviour of primary human tumours in a zebrafish xenotransplantation model. BMC Cancer 9:128
27. Herpers R, Van De Kamp E, Duckers HJ, Schulte-Merker S (2008) Redundant roles for sox7 and sox18 in arteriovenous specification in Zebrafish. Circ Res 102:12–15
28. Lawson ND, Weinstein BM (2002) In vivo imaging of embryonic vascular development using transgenic zebrafish. Dev Biol 248:307–318
29. Stoletov K, Montel V, Lester RD, Gonias SL, Klemke R (2007) High-resolution imaging of the dynamic tumor cell vascular interface in transparent zebrafish. Proc Natl Acad Sci U S A 104:17406–17411
30. Corkery DP, Dellaire G, Berman JN (2011) Leukaemia xenotransplantation in zebrafish—chemotherapy response assay in vivo. Br J Haematol 153:786–789
31. Nicoli S, Presta M (2007) The zebrafish/tumor xenograft angiogenesis assay. Nat Protoc 2:2918–2923
32. Nicoli S, Ribatti D, Cotelli F, Presta M (2007) Mammalian tumor xenografts induce neovascularization in zebrafish embryos. Cancer Res 67:2927–2931
33. Zhao H, Tang C, Cui K, Ang B-T, Wong STC (2009) A screening platform for glioma growth and invasion using bioluminescence imaging. Laboratory investigation. J Neurosurg 111:238–246
34. Topczewska JM, Postovit L-M, Margaryan NV, Sam A, Hess AR, Wheaton WW, Nickoloff BJ, Topczewski J, Hendrix MJC (2006) Embryonic and tumorigenic pathways converge via Nodal signaling: role in melanoma aggressiveness. Nat Med 12:925–932
35. Bentley VL, Veinotte CJ, Corkery DP, Pinder JB, LeBlanc MA, Bedard K, Weng AP, Berman JN, Dellaire G (2015) Focused chemical genomics using zebrafish xenotransplantation as a pre-clinical therapeutic platform for T-cell acute lymphoblastic leukemia. Haematologica 100:70–76
36. Rawls JF, Mellgren EM, Johnson SL (2001) How the zebrafish gets its stripes. Dev Biol 240:301–314
37. Lister JA, Robertson CP, Lepage T, Johnson SL, Raible DW (1999) Nacre encodes a zebrafish microphthalmia-related protein that regulates neural-crest-derived pigment cell fate. Development 126:3757–3767

38. White RM, Sessa A, Burke C, Bowman T, LeBlanc J, Ceol C, Bourque C, Dovey M, Goessling W, Burns CE, Zon LI (2008) Transparent adult zebrafish as a tool for in vivo transplantation analysis. Cell Stem Cell 2:183–189
39. Veinotte CJ, Dellaire G, Berman JN (2014) Hooking the big one: the potential of zebrafish xenotransplantation to reform cancer drug screening in the genomic era. Dis Model Mech 7:745–754
40. Taylor AM, Zon LI (2009) Zebrafish tumor assays: the state of transplantation. Zebrafish 6:339–346
41. Tobia C, De Sena G, Presta M (2011) Zebrafish embryo, a tool to study tumor angiogenesis. Int J Dev Biol 55:505–509
42. Cheng J, Yan-Juan G, Wang Y, Cheng SH, Wong W-T (2011) Nanotherapeutics in angiogenesis: synthesis and in vivo assessment of drug efficacy and biocompatibility in zebrafish embryos. Int J Nanomedicine 6:2007–2021
43. Vlecken DH, Bagowski CP (2009) LIMK1 and LIMK2 are important for metastatic behavior and tumor cell-induced angiogenesis of pancreatic cancer cells. Zebrafish 6:433–439
44. Camus S, Quevedo C, Menéndez S, Paramonov I, Stouten PFW, Janssen RA, Rueb S, He S, Snaar-Jagalska BE, Laricchia-Robbio L, Izpisua Belmonte JC (2012) Identification of phosphorylase kinase as a novel therapeutic target through high-throughput screening for anti-angiogenesis compounds in zebrafish. Oncogene 31:4333–4342
45. De Visser KE, Eichten A, Coussens LM (2006) Paradoxical roles of the immune system during cancer development. Nat Rev Cancer 6:24–37
46. Renshaw SA, Loynes CA, Trushell DMI, Elworthy S, Ingham PW, Whyte MKB (2006) A transgenic zebrafish model of neutrophilic inflammation. Blood 108:3976–3978
47. Ellett F, Pase L, Hayman JW, Andrianopoulos A, Lieschke GJ (2011) Mpeg1 promoter transgenes direct macrophage-lineage expression in zebrafish. Blood 117:49–56
48. Lin H-F, Traver D, Zhu H, Dooley K, Paw BH, Zon LI, Handin RI (2005) Analysis of thrombocyte development in CD41-GFP transgenic zebrafish. Blood 106:3803–3810
49. He S, Lamers GE, Beenakker J-WM, Cui C, Ghotra VP, Danen EH, Meijer AH, Spaink HP, Snaar-Jagalska BE (2012) Neutrophil-mediated experimental metastasis is enhanced by VEGFR inhibition in a zebrafish xenograft model. J Pathol 227:431–445
50. Auer TO, Del Bene F (2014) CRISPR/Cas9 and TALEN-mediated knock-in approaches in zebrafish. Methods 69:142–150
51. Prykhozhij S, Rajan V, Gaston D, Berman JN (2015) CRISPR MultiTargeter: a web tool to find common and unique CRISPR single guide RNA targets in a set of similar sequences. PLoS One 10, e0119372
52. Willinger T, Rongvaux A, Strowig T, Manz MG, Flavell RA (2011) Improving human hemato-lymphoid-system mice by cytokine knock-in gene replacement. Trends Immunol 32:321–327
53. Jung D-W, Oh E-S, Park S-H, Chang Y-T, Kim C-H, Choi S-Y, Williams DR (2012) A novel zebrafish human tumor xenograft model validated for anti-cancer drug screening. Mol Biosyst 8:1930–1939
54. Pruvot B, Jacquel A, Droin N, Auberger P, Bouscary D, Tamburini J, Muller M, Fontenay M, Chluba J, Solary E (2011) Leukemic cell xenograft in zebrafish embryo for investigating drug efficacy. Haematologica 96:612–616
55. El-Naggar A, Veinotte C, Tognon C, Corkery D, Cheng H, Tirode F, Grunewald T, Kyle A, Baker J, Mathers J, Somasekharad S, LePard N, McKinney S, Bennewith K, Minchinton A, Delattre O, Wang Y, Dellaire G, Berman J, Sorensen P (2015) Translational activation of HIF1a by YB-1 promotes sarcoma metastasis. Cancer Cell 27:682–697
56. Lara R, Mauri FA, Taylor H, Derua R, Shia A, Gray C, Nicols A, Shiner RJ, Schofield E, Bates PA, Waelkens E, Dallman M, Lamb J, Zicha D, Downward J, Seckl MJ, Pardo OE (2011) An siRNA screen identifies RSK1 as a key modulator of lung cancer metastasis. Oncogene 30:3513–3521
57. Vlecken DH, Bagowski CP (2009) LIMK1 and LIMK2 are important for metastatic behavior of pancreatic. Cancer Cells 6

58. Veinotte CJ, Corkery D, Dellaire G, El-Naggar A, Sinclair K, Bernstein ML, Sorensen PB, Berman JN (2012) Using zebrafish xenotransplantation to study the role of Y-Box binding protein (YB-1) in the metastasis of Ewing family tumors. In Proceedings of the 103rd Annual Meeting of the American Association for Cancer Research, 2012, Abstract no 1398
59. Greaves M, Maley CC (2012) Clonal evolution in cancer. Nature 481:306–313
60. Geiger GA, Fu W, Kao GD (2008) Temozolomide-mediated radiosensitization of human glioma cells in a zebrafish embryonic system. Cancer Res 68:3396–3404
61. Dellaire G, Bazett-Jones DP (2004) PML nuclear bodies: dynamic sensors of DNA damage and cellular stress. Bioessays 26:963–977
62. Corkery D, Cann K, Dellaire G (2011) Nuclear subdomains. In Nabi I (ed) Cell domains. Wiley, pp 393–414
63. Traver D, Paw BH, Poss KD, Penberthy WT, Lin S, Zon LI (2003) Transplantation and in vivo imaging of multilineage engraftment in zebrafish bloodless mutants. Nat Immunol 4:1238–1246
64. Tang Q, Abdelfattah NS, Blackburn JS, Moore JC, Martinez SA, Moore FE, Lobbardi R, Tenente IM, Ignatius MS, Berman JN, Liwski RS, Houvras Y, Langenau DM (2014) Optimized cell transplantation using adult rag2 mutant zebrafish. Nat Methods 11:6–11
65. Mizgireuv IV, Revskoy SY (2006) Transplantable tumor lines generated in clonal zebrafish. Cancer Res 66:3120–3125
66. Peterson RT, Link BA, Dowling JE, Schreiber SL (2000) Small molecule developmental screens reveal the logic and timing of vertebrate development. Proc Natl Acad Sci U S A 97:12965–12969
67. Murphey RD, Stern HM, Straub CT, Zon LI (2006) A chemical genetic screen for cell cycle inhibitors in zebrafish embryos. Chem Biol Drug Des 68:213–219
68. Zon LI, Peterson RT (2005) In vivo drug discovery in the zebrafish. Nat Rev Drug Discov 4:35–44
69. Liu Y, Asnani A, Zou L, Bentley VL, Yu M, Wang Y, Dellaire G, Sarkar KS, Dai M, Chen HH, Sosnovik DE, Shin JT, Haber DA, Berman JN, Chao E, Peterson RT (2014) Visnagin protects against doxorubicin-induced cardiomyopathy through inhibition of mitochondrial malate dehydrogenase. Sci Transl Med 6:266ra170
70. Jemal A, Bray F, Ferlay J (2011) Global cancer Statistics. CA Cancer J Clin 61:69–90
71. Weiss FU, Marques IJ, Woltering JM, Vlecken DH, Aghdassi A, Partecke LI, Heidecke C-D, Lerch MM, Bagowski CP (2009) Retinoic acid receptor antagonists inhibit miR-10a expression and block metastatic behavior of pancreatic cancer. Gastroenterology 137:2136–2145.e1–7
72. Harfouche R, Basu S, Soni S, Hentschel DM, Mashelkar RA, Sengupta S (2009) Nanoparticle-mediated targeting of phosphatidylinositol-3-kinase signaling inhibits angiogenesis. Angiogenesis 12:325–338
73. Lally BE, Geiger GA, Kridel S, Arcury-Quandt AE, Robbins ME, Kock ND, Wheeler K, Peddi P, Georgakilas A, Kao GD, Koumenis C (2007) Identification and biological evaluation of a novel and potent small molecule radiation sensitizer via an unbiased screen of a chemical library. Cancer Res 67:8791–8799
74. Latifi A, Abubaker K, Castrechini N, Ward AC, Liongue C, Dobill F, Kumar J, Thompson EW, Quinn MA, Findlay JK, Ahmed N (2011) Cisplatin treatment of primary and metastatic epithelial ovarian carcinomas generates residual cells with mesenchymal stem cell-like profile. J Cell Biochem 112:2850–2864
75. Zhao C, Wang X, Zhao Y, Li Z, Lin S, Wei Y, Yang H (2011) A novel xenograft model in zebrafish for high-resolution investigating dynamics of neovascularization in tumors. PLoS One 6:1–9
76. Ghotra VPS, He S, de Bont H, van der Ent W, Spaink HP, van de Water B, Snaar-Jagalska BE, Danen EHJ (2012) Automated whole animal bio-imaging assay for human cancer dissemination. PLoS One 7
77. Yang X-J, Cui W, Gu A, Xu C, Yu S-C, Li T-T, Cui Y-H, Zhang X, Bian X-W (2013) A novel zebrafish xenotransplantation model for study of glioma stem cell invasion. PLoS One 8, e61801

78. Drabsch Y, He S, Zhang L, Snaar-Jagalska BE, Ten Dijke P (2013) Transforming growth factor-β signalling controls human breast cancer metastasis in a zebrafish xenograft model. Breast Cancer Res 15:R106
79. Teng Y, Xie X, Walker S, White DT, Mumm JS, Cowell JK (2013) Evaluating human cancer cell metastasis in zebrafish. BMC Cancer 13:453
80. Van der Ent W, Burrello C, Teunisse AF, Ksander BR, van der Velden PA, Jager MJ, Jochemsen AG, Snaar-Jagalska BE (2014) Modeling of human uveal melanoma in zebrafish xenograft embryos. Invest Ophthalmol Vis Sci 55:6612–6622

Automation of Technology for Cancer Research

Wietske van der Ent, Wouter J. Veneman, Arwin Groenewoud, Lanpeng Chen, Claudia Tulotta, Pancras C.W. Hogendoorn, Herman. P. Spaink, and B. Ewa Snaar-Jagalska

Abstract Zebrafish embryos can be obtained for research purposes in large numbers at low cost and embryos develop externally in limited space, making them highly suitable for high-throughput cancer studies and drug screens. Non-invasive live imaging of various processes within the larvae is possible due to their transparency during development, and a multitude of available fluorescent transgenic reporter lines. To perform high-throughput studies, handling large amounts of embryos and larvae is required. With such high number of individuals, even minute tasks may become time-consuming and arduous. In this chapter, an overview is given of the developments in the automation of various steps of large scale zebrafish cancer research for discovering important cancer pathways and drugs for the treatment of human disease. The focus lies on various tools developed for cancer cell implantation, embryo handling and sorting, microfluidic systems for imaging and drug treatment, and image acquisition and analysis. Examples will be given of employment of these technologies within the fields of toxicology research and cancer research.

Keywords Zebrafish • Cancer • Automation • High-throughput • Robotics • Microfluidics • Image analysis

W. van der Ent
Institute of Biology, Leiden University, Einsteinweg 55, 2333 CC Leiden, The Netherlands

Department of Pathology, Leiden University Medical Center, Albinusdreef 2, 2333 ZA Leiden, The Netherlands

W.J. Veneman • A. Groenewoud • L. Chen • C. Tulotta • H.P. Spaink (✉)
B.E. Snaar-Jagalska (✉)
Institute of Biology, Leiden University, Einsteinweg 55, 2333 CC Leiden, The Netherlands
e-mail: h.p.spaink@biology.leidenuniv.nl; b.e.snaar-jagalska@biology.leidenuniv.nl

P.C. Hogendoorn
Department of Pathology, Leiden University Medical Center, Albinusdreef 2, 2333 ZA Leiden, The Netherlands

D.M. Langenau (ed.), *Cancer and Zebrafish*, Advances in Experimental Medicine and Biology 910, DOI 10.1007/978-3-319-30654-4_14

Introduction

The use of zebrafish in cancer research has become increasingly widespread, and different models have been developed for a variety of cancer types. In initial models, tumor development was induced by the exposure of embryos, larvae or adult fish to carcinogens, giving rise to hepatic, mesenchymal, neural and epithelial neoplasms [1–3]. Then, with improvement of techniques to generate transgenic animals, Langenau et al. developed the first transgenic cancer model in zebrafish, in which expression of murine *mMyc* in lymphoid cells drove the onset of leukemia [4]. Other transgenic cancer models followed [5–15], such as activated human $BRAF^{V600E}$ leading to invasive melanoma formation in p53-deficient fish [7], or an embryonic model for rhabdomyosarcoma induced by activated human *RAS* [9]. In addition to transgenic models, xenotransplantation models were developed, and tumors cells from a range of cancer types and species were shown to be able to proliferate, migrate and induce neovascularization [16–24]. In these models, cells can be implanted at different stages, from blastula and later embryonic stages to (immunosuppressed) adults, as well as in different sites, like the yolk, Duct of Cuvier [25], perivitelline space [17] or brain cavity [26]. In addition to xenotransplantation with cancer cells of human or murine origin, allotransplantation with cells from transgenic zebrafish cancer models or zebrafish transformed cells have been performed [27, 28].

Many papers highlight the opportunity provided by these animals to perform large-scale chemical screens in aid of finding novel anti-cancer drugs. Before gaining ground as a model for cancer, zebrafish embryos were used in screens identifying small molecules affecting development [29], and immersion of embryos in compounds is now an established technique for treatment [30, 31]. With the development of various cancer models in zebrafish, performing screens with large libraries of compounds to find improved treatment strategies for patients is a logical next step. However, performing large scale screens can be a labor-intensive, monotonous task, and automation of different steps of the process would both increase speed, precision and reproducibility of results.

Automated systems for injection, compound treatment, imaging and data analysis are being developed. Many of these new systems are designed with zebrafish embryos in mind, not adults, so the main application will be for engraftment models, or transgenic models where there is a distinct phenotype in larval stages. Here, we will provide an overview of the advances in automation regarding each stage of cancer research in zebrafish embyos and larvae.

Automated Microinjection Systems

Microinjection is an indispensable technique in zebrafish research, with many applications. Microinjection is used for generation of transgenic lines, mutant lines (using the TALEN or CRISPR/Cas9 system), transient gene knock-down (using morpholinos, siRNA or antibodies), transient gene overexpression (by mRNA injection), infection studies (by injection of microbes) and cancer cell engraftment. Most of these injections are performed in embryos up to 16 cell stage, and currently available automated injection systems are designed with this stage in mind. At these early stages, the yolk cell is the largest cell in the embryo, and compounds injected there will be taken up by neighboring cells. However, with the rapid development that embryos are undergoing, injections have to be performed at a high rate, to ensure proper uptake by all the cells of the embryo.

Two automated injection systems for zebrafish were reported in 2007. In the first system, published by Wang et al., embryos were positioned using a 5×5 vacuum-based holding grid [32]. Using image recognition software, different structures in the embryos could be recognized, and the site of injection was determined based on this information. In this first injection system, 25 embryos were injected per 2-minute run, with an accuracy of 99 %. The second injection system, published by Hogg et al., was primarily designed for injection of south-African clawed frog *Xenopus laevis* oocytes, but the application for zebrafish microinjections was also shown [33]. Here, embryos were placed in commercially available 96-well microplates with conical wells, and the site of injection was based on the spacing between wells. This negated the need for image capturing and recognition software. The setup had a plunger-driven injection system, which allowed automated cleaning and refilling of the injection needle, making it possible to inject up to seven different solutions. Up to 600 injections could be performed per hour.

In 2011 Carvalho et al. published an automated injection system that, in addition to being suitable for gene disruption injection in early-stage embryos, could also be applied to inject pathogens such as *Mycobacterium marinum* into the yolk of embryos of up to 1024-cell stage [34]. In this injection system, embryos were positioned in an agarose grid with a honeycomb pattern of hemispherical wells, and the site of injection was based on the consistent spacing between the wells (Fig. 1a). The freshly cast agarose grid could be designed according to the experiment in mind, with a variety in number of wells for small or large scale experiments, or multiple small grids of wells to distinguish groups of embryos injected with different compounds or parameters (Fig. 1b). With a built-in camera, the volume of the injected droplet could be calibrated on-screen, and easily adapted during the run of the experiment, if desired. With this robotic injection method, the authors showed that embryos could be infected with *M. marinum* bacteria at a rate of 2000 embryos per hour. This high rate of infection makes it an attractive tool for high-throughput screens. Furthermore, successful morpholino injection at one to two cell stage has been described in this system, as well as the possibility to perform gene knockdown by injection of antibodies or CRISPR/CAS constructs, and generating transgenic lines by DNA injection [35]. For additional details, we refer to publications by Veneman et al. that provide for an overview of the setup [36] and its application in *Staphylococcus epidermidis* infection studies [37].

The possibilities for performing cancer research in zebrafish by manual xenotransplantation has been shown in a large number of papers. Therefore the applicability of an automated injection system for cancer cell xenografts was also investigated. An overview paper by Spaink et al. reports that cells of a number of

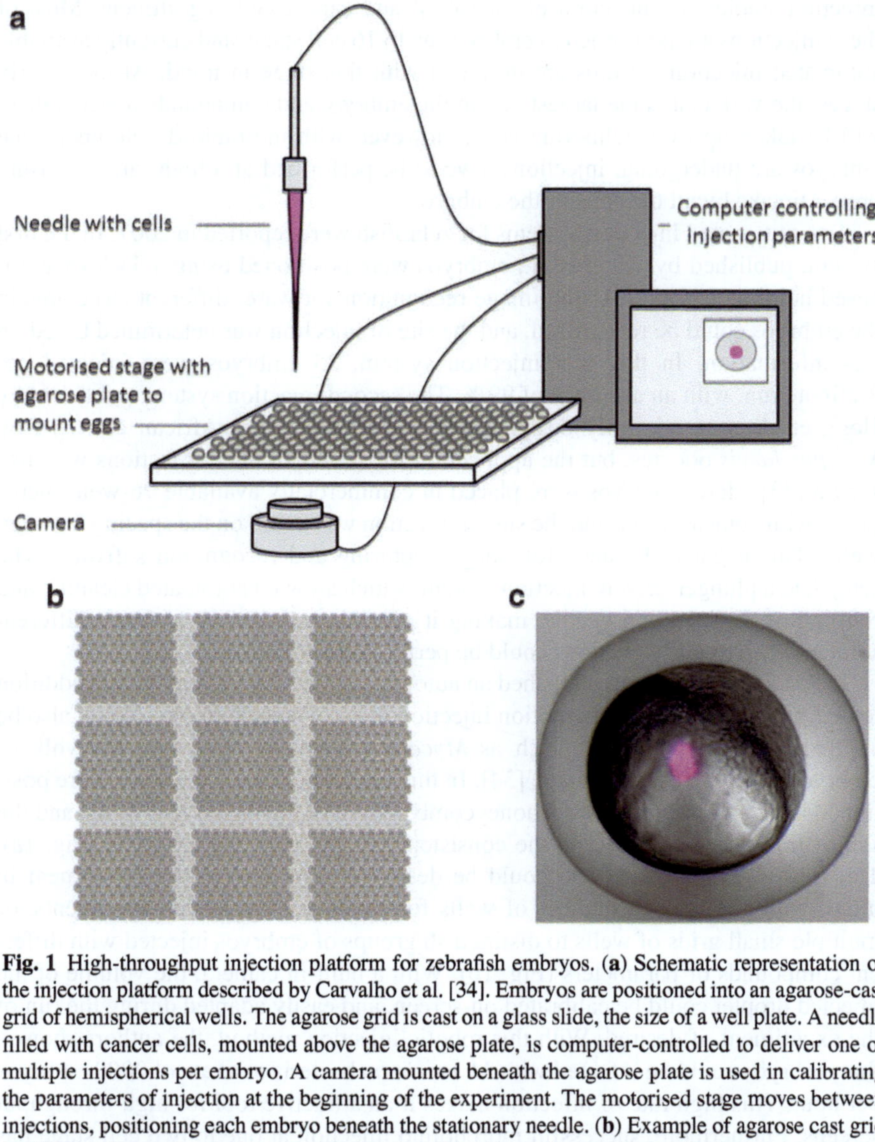

Fig. 1 High-throughput injection platform for zebrafish embryos. (**a**) Schematic representation of the injection platform described by Carvalho et al. [34]. Embryos are positioned into an agarose-cast grid of hemispherical wells. The agarose grid is cast on a glass slide, the size of a well plate. A needle filled with cancer cells, mounted above the agarose plate, is computer-controlled to deliver one or multiple injections per embryo. A camera mounted beneath the agarose plate is used in calibrating the parameters of injection at the beginning of the experiment. The motorised stage moves between injections, positioning each embryo beneath the stationary needle. (**b**) Example of agarose cast grid in different forms: the *dark gray areas* show a grid-pattern when multiple injection parameters are used, or different compounds are injected. *Each dark gray square* holds 100 embryos. The *light and dark gray areas* combined can hold up to 1024 embryos. Grid-molds for holding up to 2580 embryos are available. (**c**) Embryo injected with cancer cells (*magenta*), 30 min after implantation

different cancer types could be successfully injected into the yolk of embryos between 2 and 4 h post-fertilisation (hpf) (Fig. 1c) [35]. Osteosarcoma cells (from the SJSA-1 cell line), cutaneous melanoma (Mel57), and prostate cancer cells (PC3 and LNCap) were all found to proliferate and in several cases shown to disseminate at 6 days post-injection (dpi) and onwards. The injection system operated largely in the same manner as described by Carvalho et al., with only a few adaptations to injection parameters. Due to the larger size of tumor cells when compared to bacteria, needles with a larger opening were used for these injections. Additionally, the larger tumor cells tended to sediment in the needle more rapidly, and clump together, which could cause needle clogging. A higher concentration of 14 % polyvinylpyrrolidone-40 (PVP-40) carrier solution PVP was used to prevent this from happening. With these two changes, successful implantation rates were 80–90 %, and showcases the suitability of the system for high-throughput applications. Some optimization of injection parameters may have to be performed for each cancer cell line to be injected. In addition, it is recommended to exclude the possible effects of the embryonic developmental program on the cancer cells injected at this stage [38]. Furthermore, it was shown that some cancer cell types do not exit the yolk after implantation at early embryonic stages although the proliferate very rapidly.

Thus far, the described automated injection systems are primarily used to achieve injections into the yolk cell of early stage embryos. However, when looking at cancer cell engraftment models established in zebrafish, often engraftment takes place in the yolk of older embryos from two days post fertilization. At this point, the embryos are less fragile, and thus more likely to survive higher numbers of engrafted cells. As well as tolerating larger volumes, another key difference of older stages is that it is also possible to engraft in compartments other than the yolksac. Engraftment into the bloodstream is achieved via injection into the Duct of Cuvier [25] or the heart cavity, and enables following extravasation processes and micrometastases formation. For studying angiogenesis, an engraftment model of cancer cells into the perivitelline space (PVS) close to the subintestinal vessel complex has been described [17]. The PVS was also used as an implantation site in a recent publication by the group of Prof. Y. Cao, which studies the effect of tumor associated macrophages in the tumor microenvironment on intravasation and metastasis formation [39]. The effect of the zebrafish microenvironment on Glioblastoma multiforme cells was investigated by performing injections to the midbrain/hindbrain boundary [26]. In infection studies, injection of bacteria in the hindbrain have been performed to study macrophage recruitment [40–42], and injection of bacteria into the otic vesicle induce macrophage and neutrophil recruitment [43]. Similar injections could be of use in investigation of leukocyte recruitment in cancer research. However, a limiting factor into achieving successful automation of these types of injections is the inter-individual variations in body shape of the embryos. As the injections mentioned in this paragraph are largely high-precision maneuvers concerning minute structures, the slightest deviation in injection site or depth will result in a failed engraftment. Furthermore, with the loss of radial symmetry during embryonic development, positioning of the embryos is also a more time-consuming process.

A system for performing these kinds of injections was recently published by the group of Prof. M.F. Yanik [44]. This systems positions zebrafish larvae in an array of hydrogel droplets, using a microfluidic dispensing system. Each droplet contains one larva, which can be orientated in either a dorsal orientation by several pulses of mechanical vibration (eliciting the 'startle' response), or in the lateral orientation by addition of an anesthetic to the hydrogel. After orientation of the embryos, the hydrogel will solidify after a brief period of cooling. Using a high-speed camera and image-recognition software, the position of each embryo within a droplet is identified and can be zoomed in on. The eyes and posterior-anterior axis serve as reference coordinates to determine the site of implantation with a front-loaded micropipette. This system achieved a success rate between 84 % and 93 % of implanting 4 dpf larvae in different orientations in a variety of organs such as forebrain, midbrain, ventricles, eyes, heart and liver.

Another aspect to consider when aiming to perform high-throughput screens, is the amount of embryos needed. Variation of response, loss of individual embryos during the experiment, and the intention to use multiple concentrations per compound means that tens of thousands of reproducibly injected embryos are needed to screen a typical compound library. While the robotic injection systems described above will allow these reproducible injections, acquiring such numbers of uniformly staged embryos is the first necessary step. Commonly, to acquire embryos at the same developmental stage, single couples of zebrafish are placed in small tanks in which a spacer is placed between the female and male. This spacer prevents the fish from spawning, and is only removed once spawning is desired. Setting up these types of crosses when very large amounts of eggs are required, is laborious and takes up large amounts of time and space. A solution for these problems was presented in a paper by Addato et al., which describes development of a large breeding vessel, in which 180 zebrafish can be place at a time [45]. In the breeding vessel, females and males are separated until the desired spawning time, when the researcher removes the barrier separating the fish. Subsequently, eggs can be harvested at multiple time intervals from the bottom of the vessel. In this way, in a relatively short time, around 8000 eggs were collected on average.

Embryo-to-Plate Dispensing Systems

Commonly, drug testing in zebrafish is performed in 96- or 384- well plates. In these plates, individual embryos can be treated with small volumes of compound to be taken up from the water in which they are immersed. Benefits of this set up are that the required volumes are relatively small, and liquid handling robots for these standardized plates are commercially available, having been previously developed for cell culture systems. Furthermore, embryos may be imaged directly in the plates in which they receive compound treatment. Confocal microscopy is possible in plates with an optical bottom, or sideview plates in which prisms allow viewing embryos from two different angles [46].

Manually filling these plates with embryos is a laborious undertaking in large drug screens. Various systems have been developed to automate the dispensing of embryos to well plates, and are described in the following paragraphs.

Pfriem et al. developed a fish sorting system intended for dispensing embryos from a Petridish to a well plate that works by taking a photograph of a plate with anaesthetized embryos (hatched or unhatched), and analyzing this image to determine coordinates of each embryo [47]. They are then subsequently taken up from the Petridish via a pipette tip, whilst a built-in sensor detects if the embryos are indeed aspirated. The same sensor is able to detect if the aspirated embryo is living or coagulated. Living embryos are transferred to the well plate, either of 96- or 384-well format. As no fluorescent screening step is incorporated, the simple design may be cheaper than other available embryo sorters. Notably, this fish sorting system is compatible with other robots performing a variety of different tasks, in a 'modular cube' system. This modular design allows the addition or removal of parts, dependent on the design and needs of the researcher.

A similar system is described by Mandrell et al., where a 4-axis *S*elective *C*ompliant *A*ssembly *R*obot *A*rm (SCARA, Denso Inc.) picks up 5–6 hpf embryos from a Petridish based on coordinates obtained from a photograph [48]. Here, the distinction between living and dead embryos is made based on rapid analysis of the same photographic record. Living, semi-transparent embryos are taken up with the pipette mounted on the SCARA. Dead embryos will appear bright white and will not be aspirated.

In addition to such 'pick-and-place' devices, other flow-based sorting systems are available. One such system is the COPAS from Union Biometrica [49]. This '*C*omplex *O*bject *P*arametric *A*nalyzer and *S*orter' takes up embryos from a reservoir in the system, and leads them through tubing past a sensor measuring the time-of-flight (to get an indication of length), the optical density (to get an indication of thickness), and the presence and intensity of fluorescent signal within the embryos. By gating the parameters, the system can deposit the embryos in well plates of various formats, or discard them if they do not confirm to defined factors. Since it is possible to analyze the intensity and presence of fluorescent signal, this system can be employed to perform selection based on the presence and amount of fluorescent tumor cells. This can be particularly useful in combination with an automated injection system, to separate the embryos implanted with sufficient amounts of cancerous cells from those with little to no cells.

Another flow-based system, called the ZebraFactor, was described by Graf et al. [50]. The ZebraFactor consists of two devices working in sync. The CellSorter unit uses a static and a sliding ring to create a circular fluidic channel in which suspended embryos are caught via drag and friction forces. Cameras placed to visualize a part of the channel can be used to observe and sort the embryos. Single embryos are pushed, by redirection of the buffer, into the WellPlateFeeder. This second unit will dispense the embryos in wells of a 96-well plate. This setup makes use of light barriers to control opening and closing of various valves, to ensure correct embryo placement.

Microfluidic Systems

The ability to automate the dispense of embryos in micro titer plates is a great boon to zebrafish research. But the well plate format is not always the ideal experimental set up. When analysis of the embryos requires a staining procedure, a multitude of washing steps are involved. Such steps are not easily carried out in the well plate format. An alternative are microfluidic systems, or *L*ab-*O*n-*C*hip (LOC) devices.

These devices have been designed to perform rapid fluid perfusion, and to allow in-device imaging. A number of devices are designed to generate a continuous flowthrough of fresh medium, which aids in the survival of embryos when they are kept in low volumes of medium.

An in-depth explanation of the rationale of design and mechanics of LOC devices for zebrafish handling can be read in the papers by Khoshmanesh et al. [51] and Akagi et al. [52], from the group of Dr D. Wlodkovic. In short, PDMS chips bonded to microscope slides are designed to have a small fluidic channel in which embryos are loaded. The channel goes past an array of interconnected embryo traps. When embryos (within their chorions) are introduced into this channel via the inlet of the LOC device, hydrodynamic forces cause the docking and immobilization of embryos at these traps, whilst allowing remaining embryos to pass. After loading the chip with embryos, drugs or dyes can be completely perfused through the system via the inlet and outlet in a matter of minutes, without disrupting the positioned embryos. The internal volume of the described LOC device in these papers was under 1 ml, highlighting the small amounts of fluids necessary. As the devices are made with microscopic glass slides, in-device microscopy can be performed for easy imaging and analyzing of the embryos.

As a proof of concept, Akagi et al. performed an on-chip angiogenesis assay [52]. After loading transgenic *TG(fli1:EGFP)* embryos with fluorescent vasculature into the chip, they are perfused by eggwater with either vehicle control, or Tivozanib, a VEGFR inhibitor effectively inhibiting angiogenesis. The development of intersegmental bloodvessels could be monitored in the array for a period of 48 h. In follow-up papers, the group of Wlodkovic describe a further developed LOC device (Fig. 2a) [53, 54]. The new design includes a small suction channel connected to each well, to increase immobilization efficiency via combined gravitational sedimentation and low-pressure suction forces. Additionally, the chips are fitted out with an integrated electronic automation interface, and include an automated stage and fluorescent microscope. This integrated LOC device allows automated loading of embryos, liquid perfusion control, microenvironment maintenance, and fluorescent imaging of embryos over time.

Other presented LOC devices have their own unique features. The device described by Zheng et al. [35, 55] (Fig. 2b) is a device where the embryos are loaded manually into the open wells of the chip. What makes this an interesting system is that actuator-regulated monolithic valves are present in the channels leading into and out of each individual well of the chip. This allows for rapid, automated aspiration and reperfusion of the wells. This LOC device was used to demonstrate the ability to monitor effects of drug treatment on the cancer-associated hedgehog pathway and vasculature development [55]. Both this study and that of Akagi et al. [52] show that it is possible to investigate cancer-related processes in these LOC devices, with only very small amounts of compound necessary.

In the device presented by Wielhouwer et al. [56], embryos are manually loaded into the wells of a chip, where a constant flow-through of medium is attained via the presence of multiple in- and out-let channels. Furthermore, this chip has integrated heating channels, making it possible to maintain stable temperature gradients on a small scale.

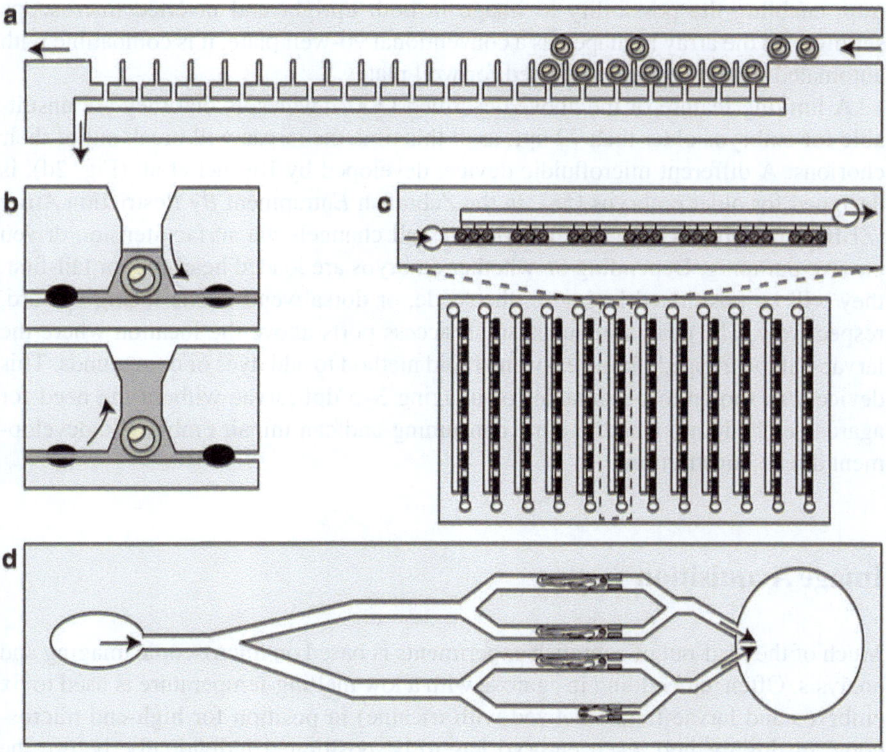

Fig. 2 Microfluidic devices for imaging or compound treatment purposes *Arrows* indicate the direction of flowthrough. (**a**) Schematic representation of the cross-section a microfluidic device described by Wang et al. [54]. Embryos are loaded at the inlet through the main channel, along which embryo traps are present. Each embryo trap is connected via a small channel to a suction channel underneath. Via combined gravitational sedimentation and low-pressure suction forces, embryos are immobilized in the embryo traps. In this LOC device, embryos can be imaged from above. The array was integrated in a platform with an electronic interface regulating automated embryo immobilization, culture, treatment and time-resolved image acquisition. (**b**) Schematic representation of the cross-section of one well of a microfluidic device described by Zheng et al. [55]. In this device, consisting of 24 wells of 40 μl, each well is open and embryos are loaded from above. Channels feeding into the wells have actuator-controlled monolithic valves, which open and close independently from eachother as old medium is removed (*upper panel*) and fresh medium is added (*lower panel*). (**c**) Schematic representation of an microfluidic array, described by Zhu et al. [57]. In this device, embryos are immobilized via combined gravitational sedimentation and low-pressure suction forces. As suction channels are positioned on the side of the embryo traps, imaging can be performed in both upright- and inverted imaging setups. The whole array is the size of a 96-well plate, and can contain 252 embryos. (**d**) Schematic representation of the ZEBRA device described by Bischel et al. [58]. The device employs passive pumping to drive embryos through the device. Embryos are trapped in individual channels. Dependant on the manner of loading embryos (head-first or tail-first), embryos are trapped allowing side-view or in dorsal/ventral view

Another device, described by Zhu et al. [57] (Fig. 2c) is an array of multiple identical microfluidic segments, the size of a 96-well plate. It employs combined gravitational and suction forces to trap the embryos, and can contain up to 252 individual embryos. The suction channels are positioned on the side of each embryo trap, enabling the possibility to image in both upright and inverted microscope settings. As the array is shaped as a conventional 96-well plate, it is compatible with automated imaging setups designed for well plates.

A limiting feature of the above described LOC devices, is that they are unsuitable for embryos older than 72 hpf, as at this time the larvae will break out of their chorions. A different microfluidic device, developed by Bischel et al. (Fig. 2d), is designed for older embryos [58]. In the *Z*ebrafish *E*ntrapment *B*y *R*estriction *A*rray (ZEBRA), embryos are guided through small channels via surface-tension driven passive pumping. Depending on whether embryos are loaded head-first or tail-first, they will be positioned laying on their side, or dorsal/ventral side facing upward, respectively. The design includes small access ports above the location where the larvae will be trapped, for an easy and rapid method to add dyes or compounds. This device was shown to be suitable for imaging 3–5 dpf larvae without the need for agarose embedding, which is time consuming and can impair embryonic development due to constriction.

Image Acquisition

Much of the read-out of zebrafish experiments is based on microscopic imaging and analysis. Often, embedding in agarose with a low melting-temperature is used to fix embryos and larvae (anaesthetized with tricaine) in position for high-end microscopy. In this method, each embryo has to be positioned individually, before the agarose solidifies. This technique is not suited for large-scale experiments. In the previous section on microfluidic devices, the possibility of doing image acquisition in the LOC devices has already been discussed. In microtiter plate format, optical glass bottom plates allow confocal imaging [59, 60]. As embryos are still quite small relative to one well of a 96-well plate, the use of predefined imaging coordinates is hampered. Plates can be modified with an array of agarose molds, to restrict the space the embryo will occupy [61]. Additionally, Physical Sciences, Inc. (Andover, MA, USA) has manufactured the Sideview Microplate. In this plate, the wells are designed as narrow rectangles, so that embryos are limited in their orientation. Prisms placed adjacent to the wells allow imaging the embryos from the side of the well as well as the bottom. In a microscope with an automated stage, complete 96-well plates of consistently oriented embryos can be imaged with minimal effort of the investigator. Alternatively, if embryo orientation within the well is not fixed, microscopes are developed with integrated detection software to locate and recognize the orientation of the embryo [62].

A flow-based device, called the *V*ertebrate *A*utomated *S*creening *T*echnology (VAST, Fig. 3a) [63, 64], similarly takes up anesthetized embryos from a suspension cup or plate and leads them through tubing past a sensor. When an embryo is detected, the water flow is adjusted and eventually stopped so that the embryo is positioned in a glass capillary mounted under a microscope. This capillary is immersed in water and has a similar refractive index as water, making microscopy with high numeric aperture water dipping objectives possible. Motors drive the

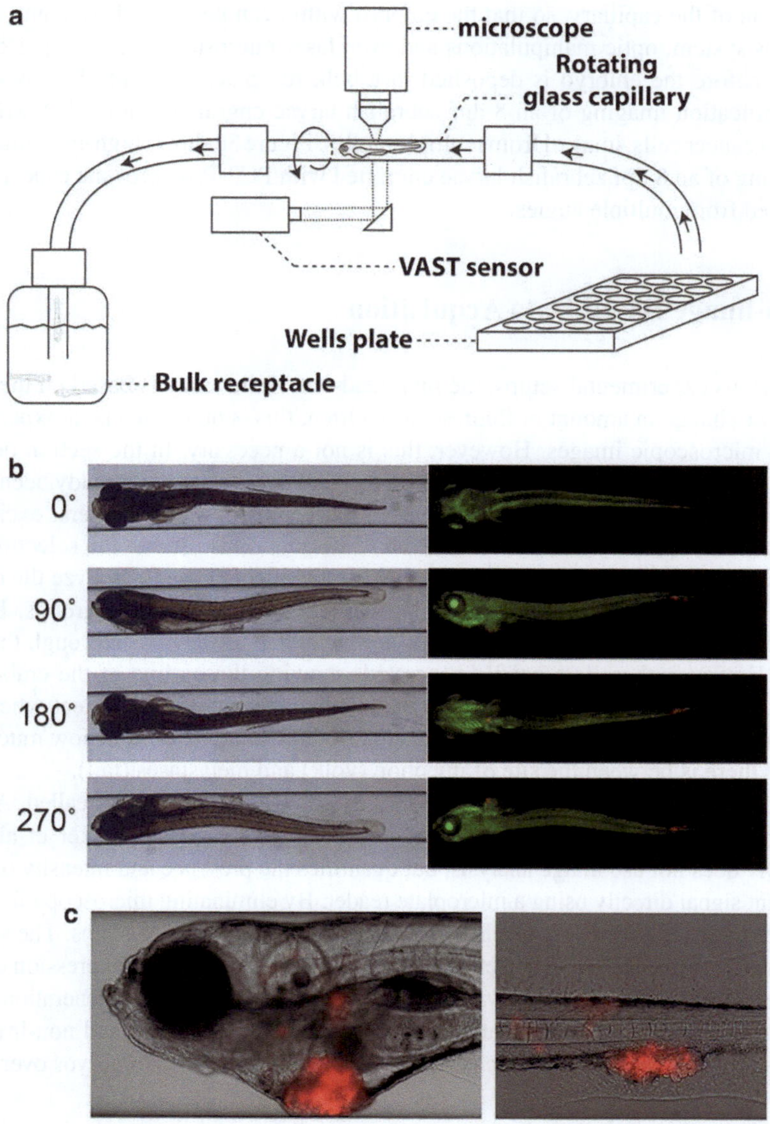

Fig. 3 Automated imaging using Vertebrate Automated Screening Technology (VAST). (**a**) Schematic representation of the *V*ertebrate *A*utomated *S*creening *T*echnology (VAST) setup. Briefly, embryos are taken up from a well plate (or bulk receptacle) and guided through tubing past a sensor. As an embryo is detected by the sensor, fluid flowthrough stops or is reverted, until the embryo is positioned in a glass capillary mounted beneath a microscope. Two stepper motors on either end of the capillary can cause it to rotate, allowing the embryo to be imaged from multiple sides. After imaging, fluid flowthrough is reinitiated, and the embryo is guided to a bulk receptacle. (**b**) Example of images taken in the VAST setup, using a 2x objective. Bright-field and fluorescent overlays of fixed 6 dpi *TG(fli1:EGFP)* embryo (vasculature in *green*) implanted with breast-cancer cells (MDA-MB-157, *red*) are shown at multiple angles. (**c**) Example of images taken in the VAST setup, using a 4x objective. Bright-field and fluorescent overlays of fixed 6 dpi embryo implanted with prostate-cancer cells (PC3-Pro4, *red*) are shown at multiple angles

rotation of the capillary, so that the embryo within can be imaged from any angle. In this system, optic manipulations and even laser microsurgery can be performed [63], before the embryo is deposited in a bulk receptacle. Figure 3b shows low-magnification imaging of an 8 dpf zebrafish larvae engrafted with MDA-MB-157 breast cancer cells, imaged from multiple angles. Figure 3c shows high-magnification imaging of an 8 dpf zebrafish larvae engrafted with PC3-Pro4 prostate cancer cells, imaged from multiple angles.

Non-image Based Data Acquisition

In various experimental setups, the final read-out of the assay is based on the presence or change in amount of fluorescence. Often, this is quantified post-experiment from microscopic images. However, this is not a necessity. In the section dealing with embryo-to-plate dispensing devices, the COPAS system has already been mentioned. As each individual embryo passes the beams of various lasers, excitation levels can be measured and recorded. This information is used in the selection and sorting of embryos during dispensing, but can be of equal use to analyze the difference in presence of various fluorescent markers in post-treatment groups. In this setup, no actual images of the embryos are generated as they pass through the system. However, there is a profile generated, showing the outline of the embryo in combination with the fluorescence signal. This could be used to detect where the fluorescent signal is located within the embryo body, and determine how much distance there is between the site of injection (yolk) and metastases (tail).

For high-throughput reporter-based assays in zebrafish, a tool called ARQiv (*A*utomated *R*eporter Quantification *i*n *v*ivo) was presented by Walker et al. [65] ARQiv does not use image analysis, but quantifies the presence and intensity of fluorescent signal directly using a microplate reader. By eliminating microscopy from the process, higher throughput levels can be achieved than in other systems. The system was demonstrated to allow detection of inter-individual variation of expression of several reporters. ARQiv was shown to detect cell loss, although cell regeneration could not be as robustly measured. Furthermore, as the process is rapid and non-invasive, alterations of expression levels can be followed within individual embryos over time.

Image Analysis

In development and toxicology research, alterations in phenotype of the embryo is an important readout. For this purpose, automated image analysis software packages designed for recognizing various structures within the embryo are now available [61, 66–69].

To assess fluorescent tumor cell burden in zebrafish, several different analysis programs are available, each with unique attributes. Pixel-counting programs are available and are a useful tool to quickly determine differences in fluorescent (tumor cell) burden. With this type of analysis, care needs to be taken that the detected fluorescent signal indeed comes from the tumor cells within the fish, and not from debris which may be visible in the background of an image. A way to reduce the interference of non-relevant signal, is to use recognition software to find the zebrafish body, and only count fluorescent signal within this area of the image [70]. Pixel-counting is a rapid analysis tool, but provides no information on the migration capabilities of cancer cells within the zebrafish.

An automated image analysis tool specifically designed for analyzing cancer cell engraftment models in zebrafish, was presented by Ghotra et al. [60] Here, a macro is able to detect the body of the larvae, no matter in which orientation the image was taken, and the fluorescent tumor cell burden within. Based on the body plan of the larva and its tumor cell burden, the site of implantation is determined. For each individual tumor cell cluster, the size and migration distance away from the site of implantation is determined (Fig. 4).

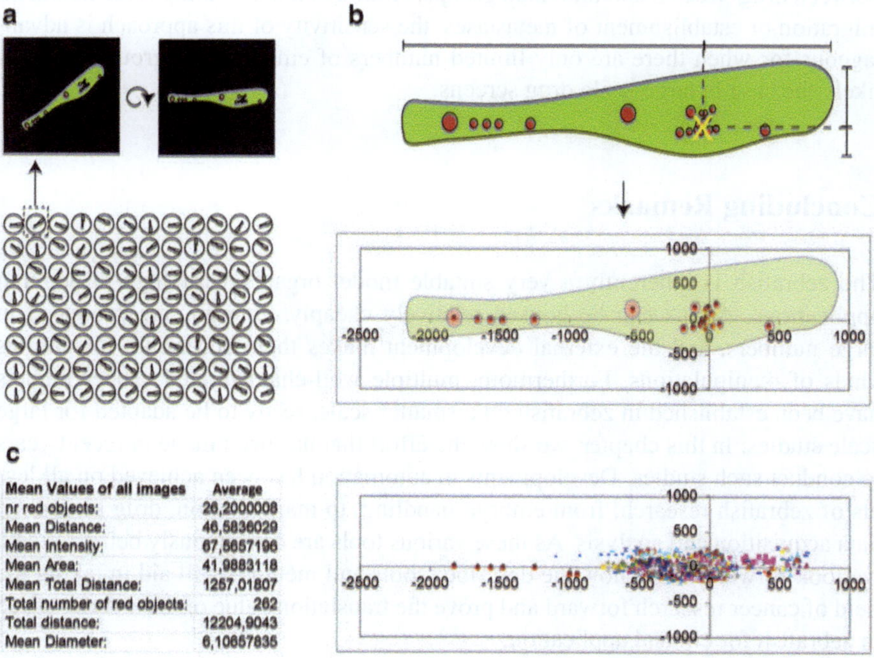

Fig. 4 Automated image analysis tool. (**a**) Schematic of embryos imaged in glass-bottom 96-well plate. Based on a macro written for Image-Pro Analyzer software (Media Cybernetics, USA) which determines the body axis of the larva based on signal of the green fluorescent channel, all images are rotated so that the body axes of all larvae align. (**b**) In the correctly re-orientated images, the same macro uses the red fluorescent signal to identify all tumor cell foci. Simultaneously, based on the larval body axes, the site of implantation (SOI) is determined (indicated by a *yellow X*, *top panel*). Of each larva, the number of tumor cell foci, the size and intensity, and the distance of migration away from the SOI is recorded. The coordinates of each tumor cell cluster is plotted in a dot plot, where the SOI corresponds to the origin (*lower panel*). (**c**) A Microsoft Excel (Microsoft Corporation, USA) based macro is used to summarize data required from all embryos in one group (*left panel*), and generates a dot plot (*right panel*)

In another program called zebIAT, developed by Annila et al., a body plan is mapped to the image of engrafted embryos, and segregated into 12 different tissues/structures [71]. ZebIAT can then assess the presence of tumor cells per region. The program is not fully automated as of yet, as there is a requirement for manually identifications of part of the landmarks needed to map all structures. However, this software provides useful information on the seeding amount and preference of cells to home to certain organs or structures and warrants further development to allow analysis of large data sets generated in engraftment screens.

One method to quantify the tumor burden, presented by Corkery et al. [72] does not make use of images of whole embryos containing engrafted cells, but images fluorescent cancer cells ex vivo. For this purpose, engrafted embryos are dissociated, the cells pelleted and then resuspended in PBS. Images of fluorescent cells are analysed in silico using a semi-automated ImageJ (National Institutes of Health, USA) macro, and was shown to allow detecting of difference in proliferation between drug- treated and untreated groups. Although not providing information on migration or establishment of metastases, the sensitivity of this approach is advantageous for when there are only limited numbers of embryos per group, which is likely the case in large-scale drug screens.

Concluding Remarks

The zebrafish is inherently a very suitable model organism for high-throughput applications. Adults can be housed relatively cheaply, embryos are produced in large numbers, and the external development makes them accessible for various kinds of manipulations. Furthermore, multiple well-characterized cancer models have been established in zebrafish on a smaller scale, ready to be adapted for large scale studies. In this chapter, we show the effort that has been made in recent years to conduct such studies. Developments in automation has been achieved on all levels of zebrafish research, from embryo handling, to manipulation, drug screening, data acquisition and analysis. As these various tools are continuously being refined, we look forward to see how the described tools and methods will aid in taking the field of cancer research forward and prove the translation value of discoveries made in zebrafish for clinical application.

Acknowledgements This work was supported by Stichting Kinderen Kankervrij, Project 30677

References

1. Beckwith LG, Moore JL, Tsao-Wu GS, Harshbarger JC, Cheng KC (2000) Ethylnitrosourea induces neoplasia in zebrafish (Danio rerio). Lab Invest 80(3):379–385
2. Spitsbergen JM, Tsai HW, Reddy A, Miller T, Arbogast D, Hendricks JD, Bailey GS (2000) Neoplasia in zebrafish (Danio rerio) treated with N-methyl-N′-nitro-N-nitrosoguanidine by three exposure routes at different developmental stages. Toxicol Pathol 28(5):716–725

3. Spitsbergen JM, Tsai HW, Reddy A, Miller T, Arbogast D, Hendricks JD, Bailey GS (2000) Neoplasia in zebrafish (Danio rerio) treated with 7,12-dimethylbenz[a]anthracene by two exposure routes at different developmental stages. Toxicol Pathol 28(5):705–715
4. Langenau DM, Traver D, Ferrando AA, Kutok JL, Aster JC, Kanki JP, Lin S, Prochownik E, Trede NS, Zon LI, Look AT (2003) Myc-induced T cell leukemia in transgenic zebrafish. Science 299(5608):887–890. doi:10.1126/science.1080280
5. Yang HW, Kutok JL, Lee NH, Piao HY, Fletcher CD, Kanki JP, Look AT (2004) Targeted expression of human MYCN selectively causes pancreatic neuroendocrine tumors in transgenic zebrafish. Cancer Res 64(20):7256–7262. doi:10.1158/0008-5472.CAN-04-0931
6. Langenau DM, Jette C, Berghmans S, Palomero T, Kanki JP, Kutok JL, Look AT (2005) Suppression of apoptosis by bcl-2 overexpression in lymphoid cells of transgenic zebrafish. Blood 105(8):3278–3285. doi:10.1182/blood-2004-08-3073
7. Patton EE, Widlund HR, Kutok JL, Kopani KR, Amatruda JF, Murphey RD, Berghmans S, Mayhall EA, Traver D, Fletcher CD, Aster JC, Granter SR, Look AT, Lee C, Fisher DE, Zon LI (2005) BRAF mutations are sufficient to promote nevi formation and cooperate with p53 in the genesis of melanoma. Curr Biol 15(3):249–254. doi:10.1016/j.cub.2005.01.031
8. Feng H, Langenau DM, Madge JA, Quinkertz A, Gutierrez A, Neuberg DS, Kanki JP, Look AT (2007) Heat-shock induction of T-cell lymphoma/leukaemia in conditional Cre/lox-regulated transgenic zebrafish. Br J Haematol 138(2):169–175. doi:10.1111/j.1365-2141.2007.06625.x
9. Langenau DM, Keefe MD, Storer NY, Guyon JR, Kutok JL, Le X, Goessling W, Neuberg DS, Kunkel LM, Zon LI (2007) Effects of RAS on the genesis of embryonal rhabdomyosarcoma. Genes Dev 21(11):1382–1395. doi:10.1101/gad.1545007
10. Le X, Langenau DM, Keefe MD, Kutok JL, Neuberg DS, Zon LI (2007) Heat shock-inducible Cre/Lox approaches to induce diverse types of tumors and hyperplasia in transgenic zebrafish. Proc Natl Acad Sci U S A 104(22):9410–9415. doi:10.1073/pnas.0611302104
11. Chen YH, Wang YH, Yu TH, Wu HJ, Pai CW (2009) Transgenic zebrafish line with overexpression of Hedgehog on the skin: a useful tool to screen Hedgehog-inhibiting compounds. Transgenic Res 18(6):855–864. doi:10.1007/s11248-009-9275-y
12. Dovey M, White RM, Zon LI (2009) Oncogenic NRAS cooperates with p53 loss to generate melanoma in zebrafish. Zebrafish 6(4):397–404. doi:10.1089/zeb.2009.0606
13. Ju B, Spitsbergen J, Eden CJ, Taylor MR, Chen W (2009) Co-activation of hedgehog and AKT pathways promote tumorigenesis in zebrafish. Mol Cancer 8:40. doi:10.1186/1476-4598-8-40
14. Santoriello C, Gennaro E, Anelli V, Distel M, Kelly A, Koster RW, Hurlstone A, Mione M (2010) Kita driven expression of oncogenic HRAS leads to early onset and highly penetrant melanoma in zebrafish. PLoS One 5(12), e15170. doi:10.1371/journal.pone.0015170
15. Leacock SW, Basse AN, Chandler GL, Kirk AM, Rakheja D, Amatruda JF (2012) A zebrafish transgenic model of Ewing's sarcoma reveals conserved mediators of EWS-FLI1 tumorigenesis. Dis Model Mech 5(1):95–106. doi:10.1242/dmm.007401
16. Lee LM, Seftor EA, Bonde G, Cornell RA, Hendrix MJ (2005) The fate of human malignant melanoma cells transplanted into zebrafish embryos: assessment of migration and cell division in the absence of tumor formation. Dev Dyn 233(4):1560–1570. doi:10.1002/dvdy.20471
17. Nicoli S, Ribatti D, Cotelli F, Presta M (2007) Mammalian tumor xenografts induce neovascularization in zebrafish embryos. Cancer Res 67(7):2927–2931. doi:10.1158/0008-5472.CAN-06-4268
18. Stoletov K, Montel V, Lester RD, Gonias SL, Klemke R (2007) High-resolution imaging of the dynamic tumor cell vascular interface in transparent zebrafish. Proc Natl Acad Sci U S A 104(44):17406–17411. doi:10.1073/pnas.0703446104
19. Marques IJ, Weiss FU, Vlecken DH, Nitsche C, Bakkers J, Lagendijk AK, Partecke LI, Heidecke CD, Lerch MM, Bagowski CP (2009) Metastatic behaviour of primary human tumours in a zebrafish xenotransplantation model. BMC Cancer 9:128. doi:10.1186/1471-2407-9-128
20. Eguiara A, Holgado O, Beloqui I, Abalde L, Sanchez Y, Callol C, Martin AG (2011) Xenografts in zebrafish embryos as a rapid functional assay for breast cancer stem-like cell identification. Cell Cycle 10(21):3751–3757. doi:10.4161/cc.10.21.17921

21. Zhao C, Wang X, Zhao Y, Li Z, Lin S, Wei Y, Yang H (2011) A novel xenograft model in zebrafish for high-resolution investigating dynamics of neovascularization in tumors. PLoS One 6(7), e21768. doi:10.1371/journal.pone.0021768
22. Drabsch Y, He S, Zhang L, Snaar-Jagalska BE, ten Dijke P (2013) Transforming growth factor-beta signalling controls human breast cancer metastasis in a zebrafish xenograft model. Breast Cancer Res 15(6):R106. doi:10.1186/bcr3573
23. van der Ent W, Burrello C, Teunisse AF, Ksander BR, Van der Velden PA, Jager MJ, Jochemsen AG, Snaar-Jagalska BE (2014) Modelling of human uveal melanoma in zebrafish xenograft embryos. Invest Ophthalmol Vis Sci. doi:10.1167/iovs.14-15202
24. van der Ent W, Jochemsen AG, Teunisse AF, Krens SF, Szuhai K, Spaink HP, Hogendoorn PC, Snaar-Jagalska BE (2014) Ewing sarcoma inhibition by disruption of EWSR1-FLI1 transcriptional activity and reactivation of p53. J Pathol 233(4):415–424. doi:10.1002/path.4378
25. He S, Lamers GE, Beenakker JW, Cui C, Ghotra VP, Danen EH, Meijer AH, Spaink HP, Snaar-Jagalska BE (2012) Neutrophil-mediated experimental metastasis is enhanced by VEGFR inhibition in a zebrafish xenograft model. J Pathol 227(4):431–445. doi:10.1002/path.4013
26. Rampazzo E, Persano L, Pistollato F, Moro E, Frasson C, Porazzi P, Della Puppa A, Bresolin S, Battilana G, Indraccolo S, Te Kronnie G, Argenton F, Tiso N, Basso G (2013) Wnt activation promotes neuronal differentiation of glioblastoma. Cell Death Dis 4:e500. doi:10.1038/cddis.2013.32
27. Tang Q, Abdelfattah NS, Blackburn JS, Moore JC, Martinez SA, Moore FE, Lobbardi R, Tenente IM, Ignatius MS, Berman JN, Liwski RS, Houvras Y, Langenau DM (2014) Optimized cell transplantation using adult rag2 mutant zebrafish. Nat Methods 11(8):821–824. doi:10.1038/nmeth.3031
28. He S, Krens SG, Zhan H, Gong Z, Hogendoorn PC, Spaink HP, Snaar-Jagalska BE (2011) A DeltaRaf1-ER-inducible oncogenic zebrafish liver cell model identifies hepatocellular carcinoma signatures. J Pathol 225(1):19–28. doi:10.1002/path.2936
29. Peterson RT, Link BA, Dowling JE, Schreiber SL (2000) Small molecule developmental screens reveal the logic and timing of vertebrate development. Proc Natl Acad Sci U S A 97(24):12965–12969. doi:10.1073/pnas.97.24.12965
30. Stern HM, Zon LI (2003) Cancer genetics and drug discovery in the zebrafish. Nat Rev Cancer 3(7):533–539. doi:10.1038/nrc1126
31. Goessling W, North TE, Zon LI (2007) New waves of discovery: modeling cancer in zebrafish. J Clin Oncol 25(17):2473–2479. doi:10.1200/JCO.2006.08.9821
32. Wang W, Liu X, Gelinas D, Ciruna B, Sun Y (2007) A fully automated robotic system for microinjection of zebrafish embryos. PLoS One 2(9), e862. doi:10.1371/journal.pone.0000862
33. Hogg RC, Bandelier F, Benoit A, Dosch R, Bertrand D (2008) An automated system for intracellular and intranuclear injection. J Neurosci Methods 169(1):65–75. doi:10.1016/j.jneumeth.2007.11.028
34. Carvalho R, de Sonneville J, Stockhammer OW, Savage ND, Veneman WJ, Ottenhoff TH, Dirks RP, Meijer AH, Spaink HP (2011) A high-throughput screen for tuberculosis progression. PLoS One 6(2), e16779. doi:10.1371/journal.pone.0016779
35. Spaink HP, Cui C, Wiweger MI, Jansen HJ, Veneman WJ, Marin-Juez R, de Sonneville J, Ordas A, Torraca V, van der Ent W, Leenders WP, Meijer AH, Snaar-Jagalska BE, Dirks RP (2013) Robotic injection of zebrafish embryos for high-throughput screening in disease models. Methods 62(3):246–254. doi:10.1016/j.ymeth.2013.06.002
36. Veneman WJ, Marin-Juez R, de Sonneville J, Ordas A, Jong-Raadsen S, Meijer AH, Spaink HP (2014) Establishment and optimization of a high throughput setup to study Staphylococcus epidermidis and Mycobacterium marinum infection as a model for drug discovery. J Vis Exp 88:e51649. doi:10.3791/51649
37. Veneman WJ, Stockhammer OW, de Boer L, Zaat SA, Meijer AH, Spaink HP (2013) A zebrafish high throughput screening system used for Staphylococcus epidermidis infection marker discovery. BMC Genomics 14:255. doi:10.1186/1471-2164-14-255
38. Hendrix MJ, Seftor EA, Seftor RE, Kasemeier-Kulesa J, Kulesa PM, Postovit LM (2007) Reprogramming metastatic tumour cells with embryonic microenvironments. Nat Rev Cancer 7(4):246–255. doi:10.1038/nrc2108

39. Wang J, Cao Z, Zhang XM, Nakamura M, Sun M, Hartman J, Harris RA, Sun Y, Cao Y (2015) Novel mechanism of macrophage-mediated metastasis revealed in a zebrafish model of tumor development. Cancer Res 75(2):306–315. doi:10.1158/0008-5472.CAN-14-2819
40. Herbomel P, Thisse B, Thisse C (1999) Ontogeny and behaviour of early macrophages in the zebrafish embryo. Development 126(17):3735–3745
41. Davis JM, Clay H, Lewis JL, Ghori N, Herbomel P, Ramakrishnan L (2002) Real-time visualization of mycobacterium-macrophage interactions leading to initiation of granuloma formation in zebrafish embryos. Immunity 17(6):693–702
42. Clay H, Davis JM, Beery D, Huttenlocher A, Lyons SE, Ramakrishnan L (2007) Dichotomous role of the macrophage in early Mycobacterium marinum infection of the zebrafish. Cell Host Microbe 2(1):29–39. doi:10.1016/j.chom.2007.06.004
43. Le Guyader D, Redd MJ, Colucci-Guyon E, Murayama E, Kissa K, Briolat V, Mordelet E, Zapata A, Shinomiya H, Herbomel P (2008) Origins and unconventional behavior of neutrophils in developing zebrafish. Blood 111(1):132–141. doi:10.1182/blood-2007-06-095398
44. Chang TY, Shi P, Steinmeyer JD, Chatnuntawech I, Tillberg P, Love KT, Eimon PM, Anderson DG, Yanik MF (2014) Organ-targeted high-throughput in vivo biologics screen identifies materials for RNA delivery. Integr Biol 6(10):926–934. doi:10.1039/c4ib00150h
45. Adatto I, Lawrence C, Thompson M, Zon LI (2011) A new system for the rapid collection of large numbers of developmentally staged zebrafish embryos. PLoS One 6(6), e21715. doi:10.1371/journal.pone.0021715
46. Peterson RT, Fishman MC (2011) Designing zebrafish chemical screens. Methods Cell Biol 105:525–541. doi:10.1016/B978-0-12-381320-6.00023-0
47. Pfriem A, Pylatiuk C, Alshut R, Ziegener B, Schulz S, Bretthauer G (2012) A modular, low-cost robot for zebrafish handling. Conf Proc IEEE Eng Med Biol Soc 2012:980–983. doi:10.1109/EMBC.2012.6346097
48. Mandrell D, Truong L, Jephson C, Sarker MR, Moore A, Lang C, Simonich MT, Tanguay RL (2012) Automated zebrafish chorion removal and single embryo placement: optimizing throughput of zebrafish developmental toxicity screens. J Lab Autom 17(1):66–74. doi:10.1177/2211068211432197
49. Pulak R (2006) Techniques for analysis, sorting, and dispensing of C. elegans on the COPAS flow-sorting system. Methods Mol Biol 351:275–286. doi:10.1385/1-59745-151-7:275
50. Graf SF, Hotzel S, Liebel U, Stemmer A, Knapp HF (2011) Image-based fluidic sorting system for automated Zebrafish egg sorting into multiwell plates. J Lab Autom 16(2):105–111. doi:10.1016/j.jala.2010.11.002
51. Khoshmanesh K, Akagi J, Hall CJ, Crosier KE, Crosier PS, Cooper JM, Wlodkowic D (2012) New rationale for large metazoan embryo manipulations on chip-based devices. Biomicrofluidics 6(2):24102–2410214. doi:10.1063/1.3699971
52. Akagi J, Khoshmanesh K, Evans B, Hall CJ, Crosier KE, Cooper JM, Crosier PS, Wlodkowic D (2012) Miniaturized embryo array for automated trapping, immobilization and microperfusion of zebrafish embryos. PLoS One 7(5), e36630. doi:10.1371/journal.pone.0036630
53. Akagi J, Zhu F, Hall CJ, Crosier KE, Crosier PS, Wlodkowic D (2014) Integrated chip-based physiometer for automated fish embryo toxicity biotests in pharmaceutical screening and ecotoxicology. Cytometry A 85(6):537–547. doi:10.1002/cyto.a.22464
54. Wang KI, Salcic Z, Yeh J, Akagi J, Zhu F, Hall CJ, Crosier KE, Crosier PS, Wlodkowic D (2013) Toward embedded laboratory automation for smart Lab-on-a-Chip embryo arrays. Biosens Bioelectron 48:188–196. doi:10.1016/j.bios.2013.04.033
55. Zheng C, Zhou H, Liu X, Pang Y, Zhang B, Huang Y (2014) Fish in chips: an automated microfluidic device to study drug dynamics in vivo using zebrafish embryos. Chem Commun (Camb) 50(8):981–984. doi:10.1039/c3cc47285j
56. Wielhouwer EM, Ali S, Al-Afandi A, Blom MT, Riekerink MB, Poelma C, Westerweel J, Oonk J, Vrouwe EX, Buesink W, vanMil HG, Chicken J, van't Oever R, Richardson MK (2011) Zebrafish embryo development in a microfluidic flow-through system. Lab Chip 11(10):1815–1824. doi:10.1039/c0lc00443j
57. Zhu F, Akagi J, Hall CJ, Crosier KE, Crosier PS, Delaage P, Wlodkowic D (2013) A high-throughput lab-on-a-chip interface for zebrafish embryo tests in drug discovery and ecotoxicology. pp 892345–892349

58. Bischel LL, Mader BR, Green JM, Huttenlocher A, Beebe DJ (2013) Zebrafish Entrapment By Restriction Array (ZEBRA) device: a low-cost, agarose-free zebrafish mounting technique for automated imaging. Lab Chip 13(9):1732–1736. doi:10.1039/c3lc50099c
59. Takaki K, Cosma CL, Troll MA, Ramakrishnan L (2012) An in vivo platform for rapid high-throughput antitubercular drug discovery. Cell Rep 2(1):175–184. doi:10.1016/j.celrep.2012.06.008
60. Ghotra VP, He S, de Bont H, van der Ent W, Spaink HP, van de Water B, Snaar-Jagalska BE, Danen EH (2012) Automated whole animal bio-imaging assay for human cancer dissemination. PLoS One 7(2), e31281. doi:10.1371/journal.pone.0031281
61. Westhoff JH, Giselbrecht S, Schmidts M, Schindler S, Beales PL, Tonshoff B, Liebel U, Gehrig J (2013) Development of an automated imaging pipeline for the analysis of the zebrafish larval kidney. PLoS One 8(12), e82137. doi:10.1371/journal.pone.0082137
62. Spomer W, Pfriem A, Alshut R, Just S, Pylatiuk C (2012) High-throughput screening of zebrafish embryos using automated heart detection and imaging. J Lab Autom 17(6):435–442. doi:10.1177/2211068212464223
63. Pardo-Martin C, Chang TY, Koo BK, Gilleland CL, Wasserman SC, Yanik MF (2010) High-throughput in vivo vertebrate screening. Nat Methods 7(8):634–636. doi:10.1038/nmeth.1481
64. Chang TY, Pardo-Martin C, Allalou A, Wahlby C, Yanik MF (2012) Fully automated cellular-resolution vertebrate screening platform with parallel animal processing. Lab Chip 12(4):711–716. doi:10.1039/c1lc20849g
65. Walker SL, Ariga J, Mathias JR, Coothankandaswamy V, Xie X, Distel M, Koster RW, Parsons MJ, Bhalla KN, Saxena MT, Mumm JS (2012) Automated reporter quantification in vivo: high-throughput screening method for reporter-based assays in zebrafish. PLoS One 7(1), e29916. doi:10.1371/journal.pone.0029916
66. Liu T, Li G, Nie J, Tarokh A, Zhou X, Guo L, Malicki J, Xia W, Wong ST (2008) An automated method for cell detection in zebrafish. Neuroinformatics 6(1):5–21. doi:10.1007/s12021-007-9005-7
67. Kamali M, Day LJ, Brooks DH, Zhou X, O'Malley DM (2009) Automated identification of neurons in 3D confocal datasets from zebrafish brainstem. J Microsc 233(1):114–131. doi:10.1111/j.1365-2818.2008.03102.x
68. Peravali R, Gehrig J, Giselbrecht S, Lutjohann DS, Hadzhiev Y, Muller F, Liebel U (2011) Automated feature detection and imaging for high-resolution screening of zebrafish embryos. Biotechniques 50(5):319–324. doi:10.2144/000113669
69. Eames BF, DeLaurier A, Ullmann B, Huycke TR, Nichols JT, Dowd J, McFadden M, Sasaki MM, Kimmel CB (2013) FishFace: interactive atlas of zebrafish craniofacial development at cellular resolution. BMC Dev Biol 13:23. doi:10.1186/1471-213X-13-23
70. Nezhinsky A, Verbeek F (2010) Pattern recognition for high throughput zebrafish imaging using genetic algorithm optimization. In: Dijkstra TH, Tsivtsivadze E, Marchiori E, Heskes T (eds) Pattern recognition in bioinformatics, vol 6282, Lecture Notes in Computer Science. Springer, Heidelberg, pp 301–312. doi:10.1007/978-3-642-16001-1_26
71. Annila T, Lihavainen E, Marques IJ, Williams DR, Yli-Harja O, Ribeiro A (2013) ZebIAT, an image analysis tool for registering zebrafish embryos and quantifying cancer metastasis. BMC Bioinformatics 14(Suppl 10):S5. doi:10.1186/1471-2105-14-S10-S5
72. Corkery DP, Dellaire G, Berman JN (2011) Leukaemia xenotransplantation in zebrafish—chemotherapy response assay in vivo. Br J Haematol 153(6):786–789. doi:10.1111/j.1365-2141.2011.08661.x

Part II
Cancer Models in Fish

Zebrafish Models of Human Leukemia: Technological Advances and Mechanistic Insights

Nicholas R. Harrison, Fabrice J.F. Laroche, Alejandro Gutierrez, and Hui Feng

Abstract Insights concerning leukemic pathophysiology have been acquired in various animal models and further efforts to understand the mechanisms underlying leukemic treatment resistance and disease relapse promise to improve therapeutic strategies. The zebrafish (*Danio rerio*) is a vertebrate organism with a conserved hematopoietic program and unique experimental strengths suiting it for the investigation of human leukemia. Recent technological advances in zebrafish research including efficient transgenesis, precise genome editing, and straightforward transplantation techniques have led to the generation of a number of leukemia models. The transparency of the zebrafish when coupled with improved lineage-tracing and imaging techniques has revealed exquisite details of leukemic initiation, progression, and regression. With these advantages, the zebrafish represents a unique experimental system for leukemic research and additionally, advances in zebrafish-based high-throughput drug screening promise to hasten the discovery of novel leukemia therapeutics. To date, investigators have accumulated knowledge of the genetic underpinnings critical to leukemic transformation and treatment resistance and without doubt, zebrafish are rapidly expanding our understanding of disease mechanisms and helping to shape therapeutic strategies for improved outcomes in leukemic patients.

Keywords Zebrafish • T-ALL • B-ALL • AML • CLL • CML • Xenograft • Leukemia • Transplantation

---

Nicholas R. Harrison and Fabrice J.F. Laroche contributed equally to the chapter writing.

N.R. Harrison • F.J. Laroche • H. Feng (✉)
Department of Pharmacology and Experimental Therapeutics, Cancer Research Center, Section of Hematology and Medical Oncology, Boston University School of Medicine, Boston, MA, USA
e-mail: nichar@bu.edu; laroche@bu.edu; huifeng@bu.edu

A. Gutierrez
Division of Hematology, Boston Children's Hospital, Boston, MA, USA
e-mail: Alejandro.Gutierrez@childrens.harvard.edu

D.M. Langenau (ed.), *Cancer and Zebrafish*, Advances in Experimental Medicine and Biology 910, DOI 10.1007/978-3-319-30654-4_15

Abbreviations

4HT	4-hydroxytamoxifen
ALL	Acute lymphoblastic leukemia
AML	Acute myeloid leukemia
B-ALL	B-cell acute lymphoblastic leukemia
CLL	Chronic lymphocytic leukemia
CML	Chronic myeloid leukemia
CRISPR/Cas9	Clustered regularly interspaced short palindromic repeats/ CRISPR-associated protein 9
dpf	Days post fertilization
EGFP	Enhanced green fluorescent protein
ENU	*N*-ethyl-*N*-nitrosourea
GFP	Green fluorescent protein
GMP	Granulocyte/macrophage progenitors
HOXA9	Homeobox A9
HSCs	Hematopoietic stem cells
HSCT	Hematopoietic stem cell transplantation
ICN1	Intracellular portion of human *NOTCH1*
LED	Light emitting diode
LP	Lymphoid progenitor (cells)
LPC	Leukemia-propagating cells
LSCM	Laser scanning fluorescent confocal microscopy
MDS	Myelodysplastic syndrome
MEP	Megakaryocyte-erythrocyte progenitors
MP	Myeloid progenitor (cells)
MPN	Myeloproliferative neoplasm
NUP98	Nucleoporin 98 kDa
PP2A	Protein phosphatase 2
Pro-B	Pro-B-lymphocytes
Pro-T	Pro-T-lymphocytes
RFP	Red fluorescent protein
S1P1	Singhosine-1 phosphate receptor 1
SDCM	Spinning disc confocal microscopy
TALENs	Transcription activator-like effector nucleases
T-ALL	T-cell acute lymphoblastic leukemia
TET2	Tet methylcytosine dioxygenase 2
TILLING	Targeting induced local lesions IN Genomes
T-LBL	T-cell acute lymphoblastic lymphoma
ZFNs	Zinc finger nucleases

Introduction

Leukemias are hematopoietic malignancies commonly subdivided into four major types—acute lymphoblastic leukemia (ALL), chronic lymphocytic leukemia (CLL), acute myeloid leukemia (AML), and chronic myeloid leukemia (CML)—and other less common types, such as hairy cell leukemia and chronic myelomonocytic leukemia [1]. The incidence (number of new cases per year) of leukemias differs from their prevalence (total cases of disease). AML has the highest incidence, but CLL is the most prevalent. Although leukemia occurs more frequently in older adults, ALL is the most common form of cancer in children and adolescents. Leukemogenic processes are generally induced by multiple genetic changes including chromosomal translocations, mutations, and deletions [2, 3].

The Genetic Basis of Human Leukemia

T-cell acute lymphoblastic leukemia (T-ALL) is characterized by constitutively activated NOTCH1 signaling accompanied by deletions of the tumor suppressor genes *p16/INK4A* and *p14/ARF* in more than 70 % of cases (Table 1). This NOTCH1 activation is the result of mutations in the *NOTCH1* gene, involving the extracellular heterodimerization and/or the C-terminal PEST domains [13], and about 15 % of cases are associated with *FBXW7* mutations that reduce the degradation of activated NOTCH1 [14, 15]. Additionally, aberrant expression of certain transcription factor genes, such as *TAL1/SCL, TAL2, LYL1, BHLHB1, MYC, LMO1,* and *LMO2*, contributes to T-ALL pathogenesis (Table 1) [4]. These transcription factors normally remain quiescent during T cell development, but become aberrantly activated during T-ALL pathogenesis. Their aberrant expression results from either recurrent chromosomal translocations or, in some cases, mutations affecting upstream regulatory pathways that can inactivate transcriptional repressor networks or induce aberrant transcriptional activation. B-cell acute lymphoblastic leukemia (B-ALL) is characterized by aneuploidy or gross chromosomal rearrangements in approximately 75 % of cases (Table 1). Chromosomal rearrangements affecting genes related to hematopoiesis, tumor suppression, and tyrosine kinase activity are common and in combination with further genetic aberrations such as aberrant oncogene activation, lead to the leukemic phenotype [16].

AML is a genetically complex disease that has historically been sub-classified based on both large-scale cytogenetic and individual gene abnormalities [17]. Importantly, knowledge about the mutational status of genes for *FLT3*, *NPM1*, *CEBPA*, and *KIT* is helping to refine treatment protocols; whereas other mutations—including *DNMT3A*, *IDH1/2*, and *TET2*—are poised to improve prognostic prediction (Table 1) [18]. Though the biological functions of many AML-related genes have been detailed [17], it remains an ongoing effort to assign driver vs. passenger mutations (i.e., to distinguish pathogenic driver mutations from coincident alterations). Such assignments will allow better definition of clinically relevant patient populations.

Table 1 Summary of genetic alterations in human ALL, AML, CLL, and CML

Disease	Gene activation, fusion, inactivation or mutations	Translocations	Deletions
ALL	*RB1, WT1, GATA3, CCND2, ETV6, RUNX1, EZH2, SUZ12, EED, PHF6, NOTCH1, FBXW7, PTEN, NRAS, NF1, JAK1, JAK3, FLT3, IL7R, CDKN2A/2B, BCL11B, LEF1, MYC, CDKN1B, NUP214-ABL1, TEL-AML1, EML1-ABL1, ETV6-ABL1, ETV6-JAK2, BCR-ABL1*	t(14;21)(q11.2;q22) t(11;14)(p13;q11) t(1;14)(p32;q11) t(1;7)(p32;q34) t(7;7)(p15;q34) t(10;11)(p13;q14) t(11;19)(q23;p13) t(7;14)(q34;q13) t(14;20)(q11;p11) t(6;7)(q23;q34) t(7;19)(q34;p13) t(7;11)(q34;p15) t(7;12)(q34;p12) t(11;14)(p15;q11) t(7;9)(q34;q32) t(7;11)(q34;p13) inv(7)(p15q34) inv(14)(q11.2q13) inv(14)(q13q32.33)	del(1p32) del(11p13) del(9q34) del(9p21)
AML	*IDH2, NRAS, TP53, KRAS, RUNX1, NPM1, FLT3, DNMT3A, IDH1, TET2, CEBPA, WT1, PTPN11, KIT, MYST3-NCOA2, AML1-ETO, NUP98-HOXA9*	t(15;17) t(8;21)(q22;q22) inv(16)(11q23) inv(8)(p11q13)	del(5q) del(7q)
CLL	*ZMYM3, CHD2, SF3B1, U2AF2, SFRS1, XPO1, ATM, TP53, LRP1B, MAPK1, MYD88, IGHV, NOTCH1, FBXW7, DDX3X, TLR2, POT1*	t(14;18)(q32;q21) t(14;19)(q32;q13) t(11;14)(q13;q32) t(17;18)(p11;q11)	del(11q22-23) del(17p13) del(13q14)
CML	*BCR-ABL1, EVI1, RB1, TP53, CDKN2A, ETV6-JAK2*	t(9;22)(q34;q11) inv(3)(q21q26) t(3;21)(q26;q22) t(15;17) (q22;q12-21)	*IKZF1* Δ3–6

Information summarized from Van Vlierberghe and Ferrando [4], Takeuchi et al. [5] (ALL); Cancer Genome Atlas Research [6] (AML); Wang et al. [7], Martin-Subero et al. [8], Baliakas et al. [9] (CLL); Johansson et al. [10], Rumpold and Webersinke [11], Mulligan et al. [12] (CML)

CLL is a leukemia subtype with high genetic heterogeneity, where the prevalence of any given mutation totals only 10–15 % of cases (Table 1). However, associated gene mutations affect distinct pathways including those of NOTCH signaling, mRNA splicing, processing, and transport, and the DNA damage response (Table 1) [8]. Despite the quantity of genetic mutations in CLL, their roles in disease initiation and progression remain elusive. For instance, Quesada et al. defined 78 recurrently mutated CLL genes, with particular focus on the splicing factor *SF3B1* [19]. As in T-ALL, *NOTCH1* mutations are found in CLL cases, but it remains unclear whether these mutations drive leukemogenesis. Larger sample sizes may facilitate the detection of putative driver mutations affecting smaller patient subsets (2–5 % of cases) [20]. The genetic heterogeneity in CLL has led to the investigation of numerous candidate genes and a handful of commonly affected pathways.

Unlike CLL, greater than 90 % of CML cases are associated with a unifying genetic abnormality known as the Philadelphia chromosome—a specific chromosomal translocation between the long arms of chromosomes 9 & 22:t(9;22)(q34;q11) (Table 1) [21]. This translocation causes the expression and constitutive action of the oncogenic tyrosine kinase BCR-ABL1 [22]. The most common BCR-ABL1 fusion protein—p210$^{BCR\text{-}ABL}$—results from translocation within the break point cluster region (BCR) between exons 12 and 16 on chromosome 22 (Table 1). This fusion protein activates a number of downstream signaling pathways, including: (a) RAS-MAPK signaling that transcriptionally upregulates *BCL-2*; (b) AKT signaling that stabilizes MYC and thus increases transcriptional activation of MYC downstream targets; and (c) STAT5 signaling that leads to the transcription of *BCL-xL* [23]. Although targeting the BCR-ABL1 fusion protein with the tyrosine kinase inhibitors (e.g., Imatinib) is an effective therapeutic strategy for the management of CML during the chronic phase, the disease remains incurable upon progression [24].

Remaining Challenges in Leukemia Treatment

Intensification of standard therapeutic agents has improved the clinical outcome in many leukemia subtypes. However, 5-year survival rates for leukemia remain low: 68.8 % for ALL, 24.9 % for AML, 83.1 % for CLL, and 58.6 % for CML [1]. Among adult patients, only 20–30 % with ALL and 30–35 % (younger than 60 years) with AML achieve complete remission and are considered cured [2, 3]. These high mortalities are resulted from induction failure, disease progression and relapse, and toxicities associated with current treatment regimens [25]. For example, following disease relapse, patients with acute leukemia face an extremely poor 5-year survival rate (~10 %) where treatment intensification often increases toxicity without improving outcome [26]. Direct and indirect toxicological complications include anthracycline-related cardiotoxicities, hepatotoxicity, peripheral neuropathy, central neurotoxicity, nephrotoxicity, cutaneous toxicity, myelosuppression—leading to anemia, thrombocytopenia, neutropenia, and infection—capillary leak syndrome, cytokine release syndrome, hypogammaglobulinemia, graft-versus-host disease, and veno-occlusive disease [26–30]. The need for more tolerable and targeted therapies is clear.

The development of monoclonal antibodies, immunotherapies, cancer vaccines, and improvements in hematopoietic stem cell transplantation (HSCT) are contributing to the more effective management and treatment of human leukemia. Antibody therapies are generally more tolerable than traditional chemotherapeutics and remain an area of developmental focus despite limited monotherapeutic efficacy [31]. Very recently, the bi-specific T-cell engager monoclonal antibody blinatumomab received approval from the U.S. Food and Drug Administration. Other antibody therapeutics in clinical trial includes traditional naked antibodies, immunoconjugates, and immunotoxins [26]. As compared to traditional small molecule drugs, these next-generation antibody therapeutics are characterized by improved malignant-cell-specific cytotoxicity [31].

Immune-based strategies for leukemia treatment represent an area of growing research. Cancer vaccines and immunotherapies are currently in clinical trial and provide potential means of reducing relapse following chemotherapy and HSCT [25, 32]. Small clinical trials for ALL and CLL have demonstrated that patients' own T cells can be genetically engineered (e.g. chimeric antigen receptor-modified T cells) to potently eradicate cancer cells by targeting tumor antigens [32–34]. Additionally, drugs modifying cancer-related immunosuppression or enhancing innate immunity are likely to improve the efficacy and durability of next-generation leukemia treatments [1].

These advances, both as monotherapies and potentially as combination therapies with current standard protocols, promise to significantly advance the treatment of leukemia. To fully exploit this potential, research to identify and develop biomarkers that can predict drug efficacies in individual patients is necessary [35]. Predictive biomarkers will be valuable for assessing immunotherapies, particularly as applied to cancer vaccinations, where discordant patterns of antibody responses arise from even identical treatment regimens [36]. Unsurprisingly, even targeted therapies are not completely effective and are complicated by therapeutic resistance, characterized by the expansion of cancer clones lacking expression of the target gene [26]. Therefore, in vivo research using animal models of human leukemia will allow us to better understand mechanisms of treatment resistance, advanced disease, and treatment toxicities, along with immunopathologies.

Conservation of Zebrafish Hematopoietic Program

The zebrafish (*Danio rerio*) is a freshwater fish with unique characteristics, including high fecundity, rapid development, transparency, and closer homology to humans than invertebrates. Zebrafish hematopoiesis is highly conserved, making it well suited to study both normal hematopoiesis and malignant transformation [37–39]. As in mammals, the cellular components of zebrafish blood are produced and replenished by distinct yet overlapping waves of primitive and definitive hematopoiesis [37, 40, 41]. Beginning 12 h post-fertilization (hpf), primitive erythropoiesis occurs in the intermediate cell mass and myelopoiesis initiates in the anterior lateral mesoderm. Definitive hematopoiesis begins around 26 hpf when hemogenic endothelial cells give rise to hematopoietic stem cells (HSCs) in the ventral wall of the dorsal aorta—a structure within the aorta-gonad-mesonephros [42, 43]. These HSCs then migrate to secondary hematopoietic compartments: the caudal hematopoietic tissue (an intermediate site of hematopoiesis comparable to the liver of

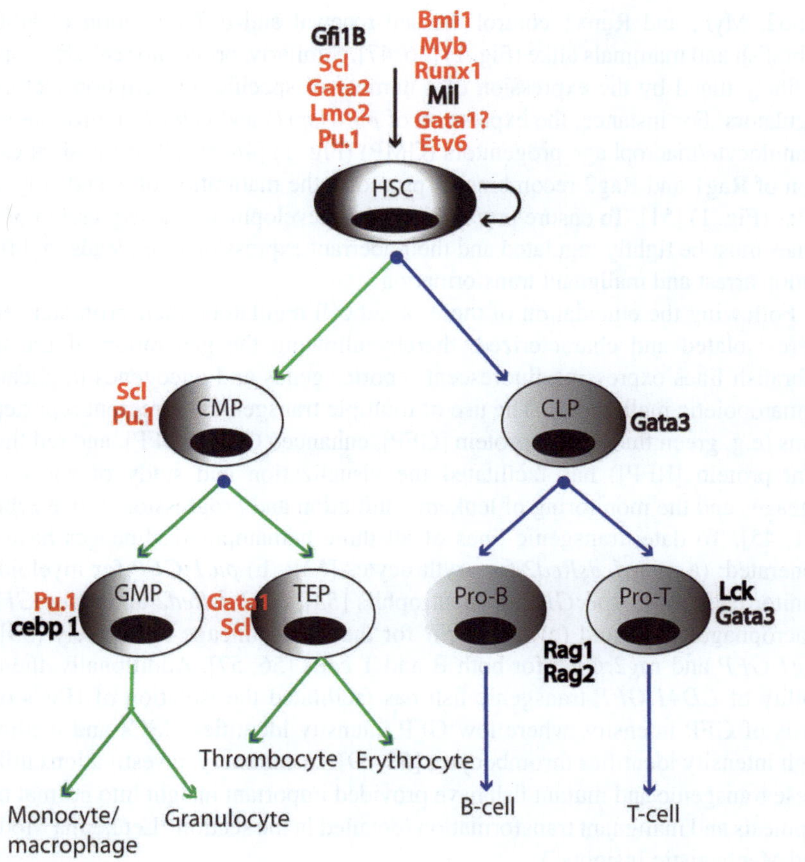

Fig. 1 Schematic overview of zebrafish definitive hematopoiesis. *Green arrow lines* indicate the myeloid lineage and *blue arrow lines* indicate the lymphoid lineage. Transcription factors implicated in hematopoiesis are shown next to each cell type and those associated with mutations, chromosomal abnormalities, or misexpression in leukemia are shown in *red. HSC* hematopoietic stem cells, *CMP* common myeloid progenitor, *CLP* common lymphoid progenitor, *GMP* granulocyte/macrophage progenitors, *TEP* thrombocyte and erythroid progenitor, *Pro-B* pro-B-lymphocytes, *Pro-T* pro-T-lymphocytes

mammals), the thymus (an organ where T lymphocytes mature), and the kidney marrow (equivalent to mammalian bone marrow). HSCs in the kidney subsequently give rise to all hematopoietic lineages (erythroid, myeloid, and lymphoid) throughout the life of the fish (Fig. 1).

Since 1963, and especially during the past two decades [44], significant knowledge of the molecular mechanisms underlying blood cell development has been acquired through investigations of gene inactivation and/or heritable mutations in zebrafish [45, 46]. Indeed, the genetic programs governing zebrafish blood cell development are highly conserved. Critical transcription factors such as Gata2, Scl (Tal1),

Lmo2, Myb, and Runx1 control the self-renewal and differentiation of HSCs in zebrafish and mammals alike (Fig. 1) [46, 47]. Similarly, progenitor cell development is finely tuned by the expression of a number of specific transcription factors and regulators. For instance, the expression of *pu.1* (*spi1*) and *cebp 1* control the fate of granulocyte/macrophage progenitors (GMP) (Fig. 1) [48–50]. The transient expression of Rag1 and Rag2 recombinases promotes the maturation of T and B lymphocytes (Fig. 1) [51]. To ensure proper blood cell development, the expression of these genes must be tightly regulated and their aberrant expression often leads to differentiation arrest and malignant transformation.

Following the elucidation of these blood cell regulators, their promoter regions were isolated and characterized, thereby allowing the generation of transgenic zebrafish lines expressing fluorescent reporter genes and oncogenes implicated in hematopoietic malignancy. The use of multiple transgenic fluorescent reporter systems (e.g. green fluorescent protein [GFP], enhanced GFP [EGFP], and red fluorescent protein [RFP]) has facilitated the visualization and study of specific cell lineages and the monitoring of leukemic initiation and progression in live zebrafish [41, 45]. To date, transgenic lines of all three hematopoietic lineages have been generated: (a) *gata1:dsRed2* for erythrocytes [52]; (b) *pu.1:GFP* for myeloid progenitor cells [50], *mpo:GFP* for neutrophils [53], *lyzC:dsRed2* and *mpeg:GFP* for macrophages [54]; and (c) *lck:EGFP* for the T cell lineage exclusively [55], and *rag1:GFP* and *rag2:GFP* for both B and T cells [56, 57]. Additionally, the availability of *CD41:GFP* transgenic fish has facilitated the isolation of HSCs on the basis of GFP intensity, where low GFP intensity identifies HSCs and medium or high intensity identifies thrombocytes [58, 59]. In summary, investigations utilizing these transgenic and mutant fish have provided important insight into normal hematopoiesis and malignant transformation (detailed in the section "Leukemia Modeling and Mechanistic Insights").

Technological Advances Enabling the Experimental Tractability of Zebrafish

In the last decade, researchers have developed and optimized multiple methodologies for the zebrafish, making this system particularly amenable to experimental manipulations. Improvements in transgenic techniques have enabled the misexpression of various oncogenes in blood lineages, while ongoing mutagenesis efforts within the zebrafish community have generated a large collection of mutants, facilitating forward genetic screens and identification of mutations modifying leukemic phenotypes. Moreover, the ability to precisely edit the zebrafish genome has facilitated the study of tumor suppressor inactivation in relation to leukemogenesis. Finally, the transparency and optical clarity of zebrafish embryos and the adult *Casper* strain—so called for its ghost-like appearance [60]—allow the tracking of leukemic cell behavior at single-cell resolution, in real time, without surgical intervention.

Transgenesis and Mutagenesis

The application of meganuclease I-SceI- or Tol2-mediated transgenesis has been shown to significantly improve rates of germ-line integration as compared to conventional transgenesis with linear-DNA injection (16–50 % vs. 1–5 %) [61, 62]. The ability to efficiently generate transgenic zebrafish in the injected (F0) generation represents a major advantage over other animal models, where the generation of stable transgenic animals typically requires two to three generations before these new lines can be used experimentally. The development of conditional transgenesis (e.g. Cre/lox-regulated, tamoxifen-regulated, GAL4-UAS-regulated, and tetracycline-inducible systems) has provided further capacity to enable the expression of oncogenes and tumor suppressors with spatial and temporal specificity [63–66]. Moreover, the ability to deliver two to three transgenic constructs into the same cell through co-injection is particularly useful. Co-injection facilitates the identification of transgenic fish by the activity of a fluorescent reporter construct alongside the transgene of interest, enabling the selection of fish with positive transgene expression [67]. Additionally, the co-expression of multiple transgenes of interest in the same cells enables the dissection of multiple genetic pathways in the leukemic context [67, 68]. These improvements in transgenic techniques have led to the generation of various zebrafish leukemia models and the functional study of candidate leukemogenic genes (detailed in the section "Leukemia Modeling and Mechanistic Insights").

While zebrafish transgenesis enables the misexpression of exogenous genes in a specific cell lineage, mutagenesis can effect enhanced expression, dominant-negative effects, and/or the inactivation of an endogenous gene throughout the entire organism. With UV irradiation, chemicals (e.g. *N*-ethyl-*N*-nitrosourea [ENU]), viruses, and transposons (e.g. Tol2) as mutagens, high-throughput mutagenesis projects have generated large collections of zebrafish mutants, including blood mutants, over the past two decades [69, 70]. These mutant fish represent a rich resource for investigators working to identify mutations capable of modifying leukemic onset and progression. A recent genetic suppressor screen utilizing a collection of insertional mutants led to the identification of dihydrolipoamide S-succinyltransferase as an important contributor to Myc-induced T-ALL pathogenesis (Anderson and Feng, 2015, personal communications). Targeting Induced Local Lesions in Genomes (TILLING) combines the use of a chemical mutagen (e.g. ENU) with a single-base-mutation screening technique. With advances in whole-genome sequencing, TILLING-by-sequencing (Illumina) has been developed to rapidly identify single-nucleotide changes and thus significantly speed the screening process. This technique has led to the identification of multiple mutants affecting blood development, including the *rag1*^{R797stop} and *pu.1*^{G242D} mutants (Fig. 1) [71, 72]. The *rag1*^{R797stop} mutant is defective in V(D)J recombination and fails to generate mature T and B cells [72]. The *pu.1* ^{G242D} mutant results in reduced Pu.1 activity and increased granulocytes with patterns comparable to human AML [71].

Gene Inactivation and Genome Editing

Gene inactivation is the preferred method for the functional study of tumor suppressors in the leukemic context. Although antisense morpholino oligonucleotides have been used for gene knockdown in zebrafish, they are susceptible to undesirable off-target effects and their transience is unsuited to studies in larval or adult fish [73]. Small hairpin RNA gene inactivation is effective in embryos [74]; however, its feasibility in adult fish remains to be proven. The zebrafish community has successfully adapted multiple genome editing techniques, including zinc finger nucleases (ZFNs), transcription activator-like effector nucleases (TALENs), and clustered regularly interspaced short palindromic repeats/CRISPR associated protein 9 (CRISPR/Cas9) technologies, enabling site-directed mutagenesis and precise single-base genome editing [75–77]. The Langenau group successfully applied TALENs to generate *rag2**E450fs*** mutants with reduced numbers of mature T and B cells [78, 79]. Both *rag1*^{R797stop} and *rag2**E450fs*** mutant fish are immunocompromised and provide excellent tools for the long-term engraftment of various tissues and leukemic cells [72, 78]. With efficiencies comparable to ZFNs and TALENs, the CRISPR/Cas9 system provides an adaptable, simple, and rapid method to generate loss-of-function alleles by targeting open reading frames [76]. Because leukemia is a heterogeneous disease, often involving the simultaneous misexpression of oncogenes and the inactivation of tumor suppressors, the ability to genetically manipulate the zebrafish genome easily and precisely makes zebrafish particularly well suited for modeling human leukemia.

Improved Imaging Techniques

Given the optical clarity of embryos and the transparency of adult *Casper* fish, the zebrafish is an excellent model for tracking the movement of normal or malignant blood cells with live-cell imaging [80]. Fluorochrome labeling of each blood cell lineage enables researchers to monitor leukemic initiation, dissemination, and regression with a fluorescent microscope in real time. The most commonly used fluorochrome is GFP, though blue fluorescent protein (e.g., lyn-Cyan), zsYellow, mCherry, and dsRed2 are available as well. To track fluorescent tumor cells, researchers fuse GFP or EGFP to an oncoprotein of choice; examples include EGFP-mMYC [81], EGFP-TEL-AML1 [82], EGFP-ICN1 [83], and MYST3/NCOA2-EGFP [84]. However, this strategy is limited by the fact that, in some cases, fast degradation of the oncogene can weaken the GFP signal. Additionally, the fluorochrome tag may adversely affect oncogene functionality and/or subcellular localization [80]. To avoid these pitfalls, co-injection is used to independently drive the expression of both the oncogene and the fluorescent protein with separate transgenic constructs within the same cells [67, 68].

Stereomicroscopy is broadly applied for the imaging and dissection of fluorescently labeled tissues in zebrafish. However, stereomicroscopes are limited to imaging fish individually. Moreover, it is difficult to acquire an image of the entire adult fish, even at low magnification. To facilitate high-throughput screening and scoring of zebrafish, the Langenau group developed a light emitting diode (LED) fluorescent macroscope system [85, 86]. This LED fluorescent macroscope is cost effective and capable of simultaneously imaging up to 30 fish within a petri dish. Additionally, LED macroscopy can differentiate five different fluorochromes and is suited to the real-time imaging and time-lapse video recording of live fish. A drawback of LED macroscopy is its reduced sensitivity, limiting detailed imaging of small tumors and single circulating cells.

Laser scanning confocal microscopy (LSCM) is a powerful technology with increased optical resolution and contrast. Laser light is focused on a single point within a defined focal plane, eliminating noise from other planes and reducing image distortions. Fluorescent scanning of multiple focal planes can be reconstructed as a z-series overlay and mapped in three dimensions onto a zebrafish anatomical atlas. Sabaawy et al. used LSCM to image *EGFP-TEL1-RUNX1* expression in adult fish [82]. Zhang et al. demonstrated the ability to track and count circulating blood cells in adult zebrafish via the combination of LSCM and in vivo flow cytometry [87]. Disadvantages of LSCM scanning include its rate of acquisition (each pixel is acquired individually) and high levels of laser exposure, often causing tissue photobleaching and phototoxicity. By contrast, spinning disc confocal microscopy (SDCM) utilizes a disc with different pinhole apertures, allowing the acquisition of multiple pixels simultaneously, thereby decreasing laser exposure. Although the resolution of SDCM is not as good as that of LSCM, its imaging is significantly faster, thereby limiting photobleaching and phototoxicity.

Significant progress has been made in improving the tractability of zebrafish for the modeling and mechanistic elucidation of leukemogenesis. With the increasing sophistication and expansion of genetic techniques and imaging modalities, the zebrafish has proven its experimental utility in disease research alongside both murine models and cell culture systems.

Leukemia Modeling and Mechanistic Insights

As previously described, zebrafish have unique advantages that make them especially suited to the investigation of hematopoietic malignancies. To date, investigators have generated zebrafish models of T-ALL, B-ALL, and AML that can be experimentally manipulated to elucidate both the molecular and genetic components of leukemic transformation, progression, and maintenance.

Disease Modeling

T-ALL

Multiple zebrafish T-ALL models have been generated during the past 12 years. The first model was developed via transgenesis in the Look laboratory. In this T-ALL model, the murine *c-Myc* gene (*mMyc*) was fused to *EGFP* to drive the expression of the EGFP-mMyc fusion protein under the lymphocyte-specific *rag2* promoter (Tables 1 and 2) [81]. This model provides the capacity to monitor the dissemination of tumor cells in a live vertebrate organism under a fluorescent microscope. However, high mortality rates among these heavily diseased leukemic fish preclude their fecundity. To overcome this, Langenau et al. developed a Cre/lox-regulated conditional system in which *EGFP-mMyc* expression is controlled by Cre-mediated recombination of the *loxP-dsRed2-loxP* cassette upon *Cre* mRNA injection (Table 2) [64]. This technique allows the maintenance of tumor-free *rag2:loxP-dsRed2-loxP-EGFP-mMyc* transgenic fish in the absence of *Cre* expression.

This conditional T-ALL model was further improved by outbreeding it with *hsp70:Cre* transgenic fish, generating *rag2:loxP-dsRed2-loxP-EGFP-mMyc;hsp70:Cre* double-transgenic fish (Fig. 2) [88]. Following heat-shock treatment at 3 days post fertilization (dpf), T-ALL developed in over 81 % of double-transgenic fish around 120 dpf. In addition to this Cre/lox-regulated conditional T-ALL model, Gutierrez et al. developed a tamoxifen-regulated zebrafish model of T-ALL [63]. In this model, 4-hydroxytamoxifen (4HT) treatment induces human MYC activation and T-ALL development in lymphocytes, and withdrawal of 4HT results in apoptosis and tumor regression (Fig. 3a, c). In addition to these *MYC*-induced T-ALL models, Chen et al. developed a zebrafish model of T-ALL by overexpressing the intracellular portion of human *NOTCH1* (*ICN1*) under the *rag2* promoter (Table 2) [83]. Constitutive murine Akt2 activation in lymphocytes also promotes T-ALL development with low penetrance in zebrafish (Table 2) [63]. In addition to transgenic T-ALL models mentioned above, Frazer et al. used ENU-mediated mutagenesis to identify three mutants—*hrk*, *slk*, and *otg*—developing genetic heritable T-ALL in the background of *lck:EGFP* transgenic fish [92] (Table 2). The gene identity of these mutations is currently under investigation.

B-ALL

Modeling B-ALL in zebrafish has proven challenging. Mysteriously, oncogene overexpression driven by the *rag2* promoter in both T and B lineages only leads to T-ALL development [63, 64, 81, 83, 88]. Despite multiple attempts in different laboratories using various oncogenes and promoters, the only zebrafish model of B-ALL generated to date is by overexpressing the *TEL-AML1* fusion gene under ubiquitous promoters. The t(12, 21)(p13;q22) *TEL-AML1 (ETV6-RUNX1)* chromosomal translocation is the most common genetic rearrangement in childhood cancer, presenting in 25 % of pediatric pre-B ALL (Table 1) [102]. Sabaawy et al. overexpressed the human *TEL-AML1* gene under the *elongation factor*, *β-actin*, or *rag2* promoters in an attempt to determine the cellular origin and molecular pathogenesis of *TEL-AML1*-induced B-ALL (Table 2) [82]. Interestingly, B-ALL arises in transgenic fish only when *TEL-AML1* is expressed under ubiquitous promoters, but not when expressed under the lymphocyte-specific *rag2* promoter. Leukemia development in these fish is associated with B cell differentiation arrest, loss of *TEL* expression, and an elevated Bcl-2/Bax ratio. Long disease latency (8–12 months) and low

Table 2 Summary of zebrafish models of leukemia

Tumor	Method	Transgenes or genes disrupted	Earliest disease onset	Characteristics/application	References
T-ALL	Stable transgenesis	*rag2:EGFP-mMyc*	22 days	The first zebrafish transgenic model of cancer	[81]
T-ALL	Cre/lox-regulated	*rag2:loxP-dsRed2-loxP-EGFP-mMyc* *Cre* mRNA injection	4 months	Similar to human T-ALL subtype with SCL and LMO2/LMO1 activation	[64]
T-ALL	Cre/lox-regulated	*rag2:loxP-dsRed2-loxP-EGFP-mMyc;hsp70:Cre*	2 months	Heat-shock inducible; Elucidation of genetic basis of T-LBL dissemination to T-ALL; Determined the effect of p53 inactivation on Myc-induced T-ALL onset	[88–90]
T-ALL	Tamoxifen-regulated	*rag2:MYC-ER;mitfa*	4 weeks	Mechanistic dissection of MYC-PTEN-AKT-BIM regulation; Small molecule screens to identify perphenazine with anti-leukemic activity	[63, 90, 91]
T-ALL	Stable transgenesis	*rag2:EGFP-ICN1*	5 months	Demonstrated the effect of bcl-2 to accelerate ICN1-induced T-ALL onset	[83]
T-ALL	Transient transgenesis	*rag2:Myr-Akt2*	6 weeks	Determined the role of Akt2 in T-ALL maintenance and progression	[63, 89]
T-ALL	Random ENU-mediated mutagenesis	*srk* *hlk* *otg*	3–6 months	Discovery of potential new genetic players for T-ALL	[92]
B-ALL	Stable transgenesis	*β-actin:EGFP-TEL-AML1* *ef1:EGFP-TEL-AML1*	8 months	The first B-ALL model in zebrafish; Provided a tool for enhancer screens	[82]
AML	Stable transgenesis	*pu.1:MYST3-NCOA2-EGFP*	14 months	Demonstrated the oncogenic potential of the MYST3-NCOA2 fusion protein in vivo	[84]

(continued)

Table 2 (continued)

Tumor	Method	Transgenes or genes disrupted	Earliest disease onset	Characteristics/application	References
MPN	Cre/lox-regulated	*β-actin:loxP-dsRed2-loxP-KRASG12D;* *hsp70-Cre*	3–6 months	The first zebrafish MPN model; Studied the role of MAP/ERK pathway in MPN	[93]
MPN	Stable transgenesis	*hsp70:AML1-ETO*	1-dpf embryos	Expanded myelopoiesis in embryos with gene expression resembling human AML; Small molecule screens to identify the Cox1 inhibitor	[94, 95]
MPN	Cre/lox-regulated	*pu.1:loxP-EGFP-loxP-NUP98-HOXA9;* *hsp70-Cre*	19–23 months	Zebrafish MPN model induced by NUP98-HOXA9; Applicable for small molecule screens	[96]
MPN	GAL4-UAS-regulated	*fli1:GAL4-FF;* *UAS-GFP-HRASG12V*	Embryos	Discovery of repression of NOTCH signaling by RAS expression	[97]
MPN	Transient mRNA injection	Cytoplasmic *NPM1* mutant	Embryos	Increased primitive myeloid cells	[98]
MPN	Transient expression of mRNA	*RUNX1-CBF2T1*	Embryos	Perturbation of normal hematopoiesis	[99]
MPN	Heat-shock inducible Stable transgenesis	*EGFP:HSE:mMycN*	2 months	Myeloid blasts infiltration in zebrafish organs	[100]
MDS	TALENs	*tet-2 inactivation*	11 months	Utilized genome editing technology to generate a MDS model in zebrafish	[101]
MDS	TILLING	$pu.1^{G242D}$	18 months	Expanded myelopoiesis; Tested current standard therapeutics on these fish	[71]

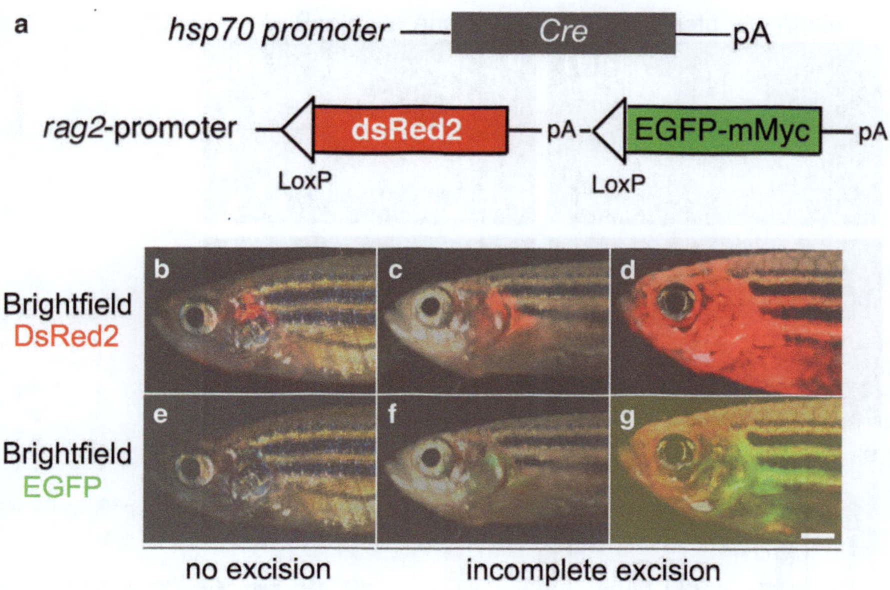

Fig. 2 The Cre/lox-regulated T-ALL model in zebrafish. (**a**) Schematic of the *hsp70:Cre* and *rag2:loxP-dsRed2-loxP-EGFP-mMyc* constructs. (**b–g**) Overlay of brightfield and fluorescence image of T cells expressing either dsRed2 and/or EGFP-mMyc shows T-LBL/T-ALL development in double transgenic *rag2:loxP-dsRed2-loxP-EGFP-mMyc;hsp70:Cre* fish, with no (*red* only) or incomplete (*red+green*) excision of *loxP-dsRed2-loxP* cassettes. (**a–g**). (Republished with permission from Wiley-Blackwell) [88]

tumor penetrance (3 %) in this model indicate the necessity of compound genetic alterations for B cell transformation. Thus, *TEL-AML1* transgenic fish are useful for genetic screens to identify cooperating mutations that may accelerate B-ALL development in the context of *TEL-AML1* expression.

AML and Other Myeloid Malignancies

The only AML model currently available was developed by the Delva group and is based on the inv(8)(p11q13) chromosomal abnormality (Tables 1 and 2) [84]. This chromosomal translocation fuses two histone acetyltransferase genes—*MYST3* and *NCOA2*—together. Expression of the *MYST3-NCOA2* fusion gene under the *pu.1* (*spi1*) promoter led to AML development in 2 of 180 F0 mosaic fish at 14 and 26 months of life, respectively, with characteristics including an extensive invasion of myeloid blasts in the kidney. This study demonstrated the oncogenic potential of the *MYST3-NCOA2* fusion gene resulting from the inv(8)(p11q13) chromosomal abnormality within an in vivo animal model.

In addition to the AML model described above, a number of zebrafish transgenic lines have been generated via the exploitation of additional genetic abnormalities in AML. Interestingly, instead of developing AML, these transgenic lines have

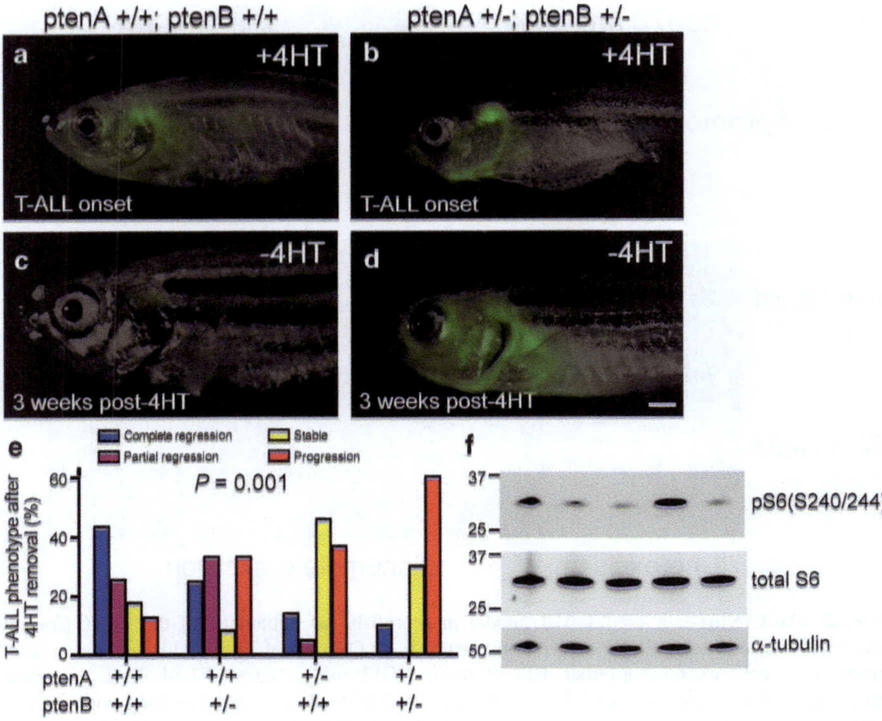

Fig. 3 *pten* haploinsufficiency promotes loss of MYC transgene dependence. (**a**, **c**) One representative *rag2:MYC-ER; rag2:GFP* double-transgenic *pten*–wild-type zebrafish, shown at time of T-ALL onset (**a**) and 3 weeks after removal from 4HT (**c**). (**b**, **d**) Representative *rag2:MYC-ER; rag2:GFP* double-transgenic zebrafish that harbored heterozygous mutations of both *ptenA* and *ptenB*, shown at time of T-ALL onset (**b**) and 3 weeks after removal from 4HT (**d**). Bar, 1 mm. (**e**) Quantitation of T-ALL phenotypes after 4HT removal, based on *pten* genotype. Number of fish with T-ALL analyzed per genotype: *ptenA* +/+, *ptenB* +/+, $n=39$; *ptenA* +/+, *ptenB* +/−, $n=12$; *ptenA* +/−, *ptenB* +/+, $n=22$; *ptenA* +/−, *pten*B +/−, $n=10$. (**f**) Western blot analysis for phosphorylation of S6 ribosomal protein, a marker of Akt pathway activation, in sorted T-ALL cells from five different *rag2:MYC-ER pten*–wild-type zebrafish in which T-ALL progressed despite MYC downregulation. Units for the molecular mass markers shown are in kD. (Republished with permission from The Rockefeller University Press) [63]

developed myeloproliferative neoplasm (MPN) or myelodysplastic syndrome (MDS), disorders with potential to progress to AML. The Look laboratory recently generated a zebrafish model of MDS through inactivation of the tumor suppressor Tet2 via TALENs [101]. *TET2* encodes a DNA methylcytosine oxidase that is essential for demethylating DNA within genomic CpG islands [103]. *TET2* loss-of-function mutations were found in multiple myeloid malignancies, including MDS, MPN, AML, and CML (Table 1) [104]. Notably, hematopoiesis in *tet2* homozygous mutants is normal throughout embryogenesis and larval development, whereas progressive clonal myelodysplasia arises in aged adult fish.

One example of modeling MPN is the *AML1-ETO* transgenic fish generated by the Peterson group (Table 2) [94]. The AML1-ETO fusion protein results from the t(8;21)(q22;q22) chromosomal translocation in AML (Table 1). Patients expressing the *AML1-ETO* fusion gene exhibit an accumulation of granulocytic precursors in the bone marrow and peripheral blood [105]. Similarly, expressing the human *AML1-ETO* gene under the *hsp70* promoter redirects myeloerythroid progenitor cells to the granulocytic cell fate, leading to the expansion of granulocyte precursors in transgenic *AML1-ETO* embryos. This change is accompanied by a loss of *gata1* expression and an elevated *pu.1* expression in myeloerythroid progenitor cells. This zebrafish model has been utilized to discover *AML1-ETO*-induced oncogenic signaling and to screen for compounds that exert anti-leukemic effects (see the section "Drug Discovery & Small Molecule Screens").

An MPN model in *NUP98-HOXA9* transgenic fish was developed in the Berman laboratory [96]. The NUP98-HOXA9 fusion protein is formed by translation of the 5′ region of nucleoporin 98 kDa (*NUP98*) on human chromosome 11 in frame to the 3′ coding sequence of homeobox A9 (*HOXA9*) on human chromosome 7 (Table 1) [96, 106]. Forrester et al. utilized Cre/lox recombination to conditionally express the human *NUP98-HOXA9* gene under the *pu.1* promoter (Table 2) [96]. Their study demonstrated that *NUP98-HOXA9* expression impaired hematopoiesis, leading to an expansion of myeloid precursors [96]. Within 19–23 months, 23 % of fish developed MPN. The *NUP98-HOXA9* transgenic fish represents yet another tool for drug discovery and investigation of the pathogenesis of MPN. See Table 2 for details on other zebrafish MPN models [71, 93, 97, 100].

Mechanistic Insights

The availability of these models has allowed investigators to study the genetic basis of leukemia, from disease initiation to progression and drug resistance. Importantly, the molecular mechanisms underlying zebrafish leukemogenesis are strikingly similar to those in humans, providing a unique opportunity to utilize the imaging and genetic advantages intrinsic to zebrafish for continued insights into human disease. In this subsection, we will discuss several advances made by investigations of zebrafish leukemia models in conjunction with other experimental systems.

Genetic Similarities Between Human Disease and Zebrafish Models of Leukemia

As studies utilizing genome-wide approaches in zebrafish models emerged over the past few years, it became clear that gene expression signatures and genetic alterations in zebrafish models of leukemia are comparable to those in humans. Langenau et al. analyzed Myc-induced T-ALL cells in zebrafish and found that these cells upregulate *scl* and *lmo2* oncogenes, resembling a major subtype of human T-ALL [64, 107]. Rudner et al. performed comparative genomic hybridization with Myc-induced zebrafish T-ALL cells obtained from transplanted fish, aiming to identify the recurrent genomic changes underlying disease relapse [108]. When zebrafish data were compared to a cohort of 75 published human T-ALL databases, 893 genes (67 %) were found to overlap between fish and human T-ALL. Blackburn et al. performed gene expression profiling analysis on zebrafish T-ALL cells induced by activated Notch1 signaling [109]. A cross-comparison between zebrafish and human T-ALL microarray data revealed a conserved T-ALL signature shared between fish and human leukemia. Together, these studies support that zebrafish and human leukemia—particularly T-ALL—are driven and controlled by shared oncogenic pathways.

The NOTCH-MYC Axis

The NOTCH signaling pathway plays an indispensable role in hematopoiesis and T cell development [110]. Activating NOTCH1 mutations were discovered in both human and murine T-ALL samples [13, 111]. In zebrafish, overexpression of human *ICN1* in the lymphocyte lineage leads to the activation of NOTCH1 downstream genes (e.g., *her6* and *her9*) and T-ALL development [83]. Long latency and low penetrance of tumor development in this model indicates the requirement of additional cooperating mutations for leukemic transformation together with the activated NOTCH1 signaling. The oncogenic role of *MYC* in the lymphoid lineage was first suggested by clinical observation of the translocation of *MYC* into the immunoglobulin heavy chain locus (Eμ) in human Burkitt lymphoma cells [112, 113]. Subsequently, researchers generated a transgenic *Eμ:Myc* murine model in which B cell lymphoma rapidly arose, demonstrating MYC's oncogenic effect in vivo [114]. In zebrafish, overexpression of the human *MYC* or murine *Myc* gene under the control of the *rag2* promoter leads to the rapid development of T-ALL but not B-ALL (Table 2) [63, 64, 81, 88]. This rapid MYC-driven T-ALL development was later connected to the activation of NOTCH1 signaling as MYC was identified to be a direct transcriptional target of NOTCH1 in mammals [13, 115–119]. These studies established the important role of NOTCH and MYC in T-ALL pathogenesis. Different from its effect in mammals, Notch activation in zebrafish does not transcriptionally induce *myc*, thus providing a unique system to study NOTCH or MYC signaling independent from each other [109].

The PTEN-PI3K-AKT Pathway

Initial investigations of PTEN-PI3K-AKT signaling in T-ALL cells began after discovering that NOTCH1 inhibition by γ-secretase is only sensitive in a subset of T-ALL cell lines [120]. Palomero et al. elucidated the mechanisms that confer resistance of T-ALL cells to γ-secretase inhibition, leading to the discovery of NOTCH1's regulation of *PTEN* expression [116]. Loss-of-function *PTEN* mutations confer the resistance of T-ALL cells to the inhibition of NOTCH1 signaling due to activation of PI3K-AKT signaling.

Gutierrez et al. employed their zebrafish *rag2:MYC-ER* transgenic fish to investigate genetic factors that are required for T-ALL maintenance [63]. In this model, tamoxifen treatment activates MYC signaling to encourage leukemogenesis. Subsequent removal of tamoxifen inactivates MYC, after which T-ALL cells undergo apoptosis rapidly in the absence of MYC (Fig. 3a, c). Importantly, when *pten* loss-of-function mutations are introduced into these fish, T-ALL cells no longer undergo apoptosis; instead, the disease progresses independently of MYC activation (Fig. 3b, d, e). This result can be explained by MYC's ability to transcriptionally repress *PTEN,* so that PTEN upregulation following MYC inactivation is a key mediator of T-ALL regression in this context. Notably, AKT pathway activation downstream of PTEN signaling is sufficient for leukemic maintenance and progression despite inactivation of the *MYC* transgene (Fig. 3f). This work highlights the importance of the PTEN-PI3K-AKT pathway in MYC-independent disease maintenance, identifying mechanisms used by drug resistant T-ALL cells in the context of PTEN loss or AKT activation.

Escape of Leukemia Cells from Apoptosis

On the road to malignant transformation, cells must dampen the apoptotic response triggered by oncogenic stress. Acquired mutations, chromosomal deletions, and gene transcriptional repression or overexpression are a few means that cancer cells utilize to escape apoptosis (Table 1) [121]. Decades ago, using the murine *Eμ:Myc* model noted above, researchers learned the importance of *Bcl-2* overexpression and suppression of the Arf-P53 apoptotic pathway in Myc-mediated transformation [122, 123]. As in mammals, *bcl-2* overexpression in zebrafish dramatically accelerates Myc- or NOTCH1-induced tumor onset in the T cell lineage [83, 89]. Moreover, in a zebrafish model of B-ALL, an elevated Bcl-2/Bax ratio was observed in B-ALL cells overexpressing the *TEL-AML1* fusion gene [82]. This evidence underscores the importance of escaping apoptosis during zebrafish T and B cell transformation. Surprisingly, the zebrafish does not appear to have an *arf* locus as evidenced by a recent study demonstrating that P53 inactivation does not impact T-ALL onset [90].

Combining the analysis of a tamoxifen-regulated T-ALL model with human T-ALL cell lines, Reynolds et al. identified repressed expression of *BIM* (a pro-apoptotic BCL2 family member) in both human and zebrafish T-ALL cells, allowing them to escape apoptosis [91]. The repressed *BIM* expression is mediated by *MYC* overexpression, as MYC downregulation upon tamoxifen removal restores *bim* expression in zebrafish T-ALL cells, promoting apoptosis and tumor regression in vivo. Alternatively, AKT activation can lead to the suppression of *BIM* expression. When human T-ALL cells were treated with MYC or PI3K-AKT pathway inhibitors, *BIM* expression was upregulated and subsequently promoted T-ALL cell apoptosis. These results indicate that the identifying molecular mechanisms that leukemic cells use to escape apoptosis, and restoration of these apoptotic pathways (e.g. BIM-mediated apoptosis), represents a promising therapeutic approach.

Genetic Basis of T-LBL Dissemination to T-ALL

T-ALL and T-cell acute lymphoblastic lymphoma (T-LBL) have distinct clinical presentations. T-ALL is defined by a bone marrow biopsy with a composition of ≥30 % T-lymphoblasts, while T-LBL is defined by a mediastinal mass where <30 % of blasts are in circulation. Feng et al. utilized a Cre/lox-regulated T-ALL model in combination with clinical patient sample analyses to investigate the molecular and cellular mechanisms underlying the progression of T-LBL to T-ALL (Figs. 2 and 4) [88, 89]. Interestingly, when *bcl-2* was overexpressed in this Myc-induced T-ALL model, despite the rapid onset of the malignant transformation, the disease remained localized and failed to disseminate, compared to fish expressing *Myc* alone. They demonstrated that both human and zebrafish T-LBL cells with increased BCL2 levels possess a distinct cellular phenotype including impaired vascular invasion, metabolic stress, and autophagy. This T-LBL phenotype results from elevated levels of singhosine-1 phosphate receptor 1 (S1P1) and intercellular adhesion molecule 1, promoting homotypic cell-cell adhesion and blocking intravasation of tumor cells. Treatment with the S1P1-antagonist W146 promotes intravasation of *bcl-2*-overexpressing T-LBL cells in zebrafish (Fig. 4). Moreover, AKT activation

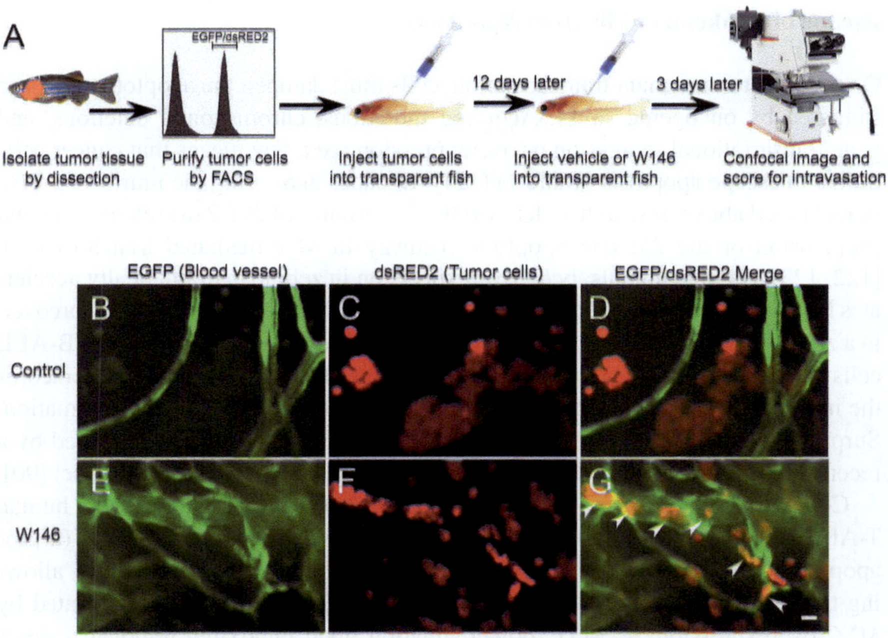

Fig. 4 The selective S1P1 antagonist W146 promotes intravasation of *bcl-2*-overexpressing T-LBL cells in vivo. (**a**) Schematic drawing of the experimental strategy. (**b–g**) Confocal images of EGFP-labeled blood vessels (**b**, **e**), dsRed2-labeled lymphoma cells (**c**, **f**), and the merged images of a vehicle-treated (**d**; n = 29) and a W146-treated transplanted animal (**g**; n = 18) demonstrate that W146 treatment promotes intravasation of *bcl-2*-overexpressing lymphoma cells (*arrowheads*) in vivo (cf. **g** to **d**). Note that W146 treatment also inhibited the in vivo formation of lymphoma cell aggregates (cf. **f** to **c**). Scale bar for (**b**)–(**g**), 10 mM. (Republished with permission from Elsevier) [89]

overcomes this intravasation blockade and promotes the rapid dissemination of T-LBL to T-ALL. This work elucidated the molecular events underlying the progression of T-LBL to T-ALL, and helped to explain clinical disparities between these two disease entities.

Lineage-Specific TEL-JAK2 Action

The human *TEL* (*ETV6*)-*JAK2* fusion gene results from translocations between *JAK2* (9p24) and *TEL* (12p13) in cases of T-ALL and CML (Table 1) [124]. In T-ALL cells, this translocation fuses exon 5 of *TEL* with exon 9 of *JAK2*, whereas the translocation in CML cells fuses exon 5 of *TEL* with exon 12 of *JAK2* [124]. To better understand the role of *TEL-JAK2* translocation in leukemogenesis, Onnebo et al. generated transgenic fish expressing the zebrafish *tel-jak2* fusion genes, mimicking different leukemic variants of human *TEL-JAK2* fusion genes [125]. Interestingly, T-ALL-derived *tel-jak2* significantly perturbed lymphopoiesis. Conversely, CML-derived *tel-jak2* led to more myelopoiesis-associated defects. This experimental evidence supports the idea that TEL-JAK2 exerts lineage-specific oncogenic effects.

Leukemia Transplantations

Tumor transplantations are important techniques for understanding the pathophysiology of cancer. Given their high fecundity and unique imaging advantages, zebrafish make excellent transplantation models. Zebrafish tumor cells have been transplanted (i.e. allotransplantation, or transplanted from one zebrafish to another) to quantify leukemia-propagating cells (LPC) and dissect the molecular mechanisms underlying LPC self-renewal and drug resistance [86, 126]. Moreover, human leukemia cell lines or patient cells can be transplanted into zebrafish embryos to generate various xenograft models (i.e., transplanted from one species to another), providing efficient in vivo tools for screening novel anti-leukemic therapeutics and determining compound efficacy and toxicity.

Syngeneic Transplantation and Assessing LPC Function in Zebrafish T-ALL

Although zebrafish transplantation experiments started as early as the 1990s [127], a major hurdle in their broader application was the need for immunosuppression. Recipient fish are often subjected to γ-irradiation to ablate immune cells, experiencing high rates of mortality due to immunosuppression-related complications. To overcome this limitation, Smith et al. generated a Myc-induced T-ALL model in syngeneic zebrafish (CG1 strain), allowing the transplantation of T-ALL cells into immune-matched CG1 sibling fish without the need for immunosuppression [86]. This method enabled investigators to quantify the frequency of LPC using limiting-dilution cell transplantation in zebrafish. Their research demonstrated that T-ALL can be initiated from a single cell, and such LPCs are abundant in T-ALL despite differences in their individual tumor-initiating potential. While this research was based on T-ALL, the same strategy can be applied to other zebrafish cancer models. Identification of cancer stem cells such as LPCs and elucidation of molecular mechanisms underlying their self-renewal may reveal therapeutic strategies directed against disease relapse.

The NOTCH1 and AKT Signaling in LPC

To rapidly assess the functional role of two to three genes in F0 fish without the need for stable transgenic fish, Langenau et al. established a co-injection strategy to introduce multiple transgenes into one-cell-stage embryos, permitting transgenes to synergize for tumorigenesis [68]. Combining the co-injection strategy with limiting-dilution cell transplantation, Blackburn et al. determined the effects of NOTCH1 signaling on the self-renewal of LPCs [109]. Utilizing a zebrafish Notch1 defective in its ability to transcriptionally activate Myc, they demonstrated that NOTCH1 can expand pre-leukemic clones; however, *NOTCH* signaling does not increase the overall frequency of LPCs, either alone or in collaboration with *MYC*. These results suggest that the main function of *NOTCH* signaling in T-ALL is to promote the proliferation of premalignant T cells, of which only a small number acquire additional mutations and become LPCs.

To identify mutations driving leukemic clonal evolution and intratumoral heterogeneity, Blackburn et al. performed a cell transplantation screen based on phenotypic differences among leukemic clones (Fig. 5) [126]. Using this approach, after serial clonal selection, they identified AKT activation in leukemic cells with elevated LPC frequency. Indeed, enforced AKT activation increased LPC frequency by enhancing mTORC1 activity and shortened tumor latency by promoting MYC stability. However, MYC stabilization alone was not sufficient to promote LPC frequency. AKT activation promoted treatment resistance, and this resistance could be overcome by the combined treatment of an AKT inhibitor and dexamethasone. These results indicate that continuous selection in T-ALL cells promotes clonal evolution and disease progression, and that clonal selection can occur before therapeutic intervention.

Zebrafish Xenograft Models of Human Leukemia

Xenograft models are widely applied in rodents to investigate tumor proliferation and drug sensitivity. Zebrafish embryos are particularly suited for xenografting given the following: (a) zebrafish lack a mature adaptive immune system until

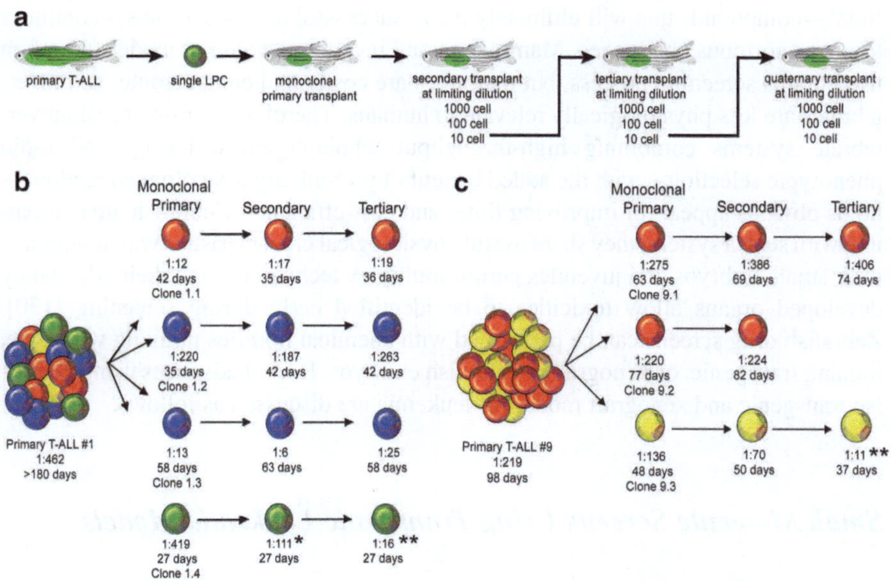

Fig. 5 Clonal evolution drives intratumoral heterogeneity and can lead to increased LPC Frequency. (**a**) A schematic of the cell transplantation screen designed to identify phenotypic differences between single leukemic clones. (**b**, **c**) Schematic of results from primary T-ALL #1 (**b**) and T-ALL #9 (**c**). *Denotes a significant increase in LPC frequency from monoclonal primary to secondary transplant ($p = 0.02$). **Denotes a significant increase in LPC frequency from monoclonal primary transplant T-ALL compared with tertiary transplanted leukemia ($p < 0.0001$). Clones are color coded based on Tcrβ-rearrangements. (Republished with permission from Elsevier) [126]

4 weeks post-fertilization [128], allowing transplantation without the need for immunosuppression; (b) zebrafish are optically transparent and, as with injected cells, can be specifically labeled with various fluorochromes, allowing visualization of in vivo cell-cell interactions; and (c) zebrafish are highly fecund and inexpensive to maintain, permitting high-throughput drug discovery screening. Since 2011, researchers have been transplanting human leukemia cell lines and primary leukemic cells into zebrafish embryos (48 hpf) to screen novel compounds and to assay drug efficacy and toxicity for preclinical validation (see the section "Drug Discovery & Small Molecule Screens").

Drug Discovery & Small Molecule Screens

Drug discovery and development is a time-consuming and expensive process, in some instances costing upwards of $1 billion over 10–15 years as a drug progresses from early target identification through clinical trials and market approval [129]. Current techniques in drug discovery allow researchers to generate an enormous number of potential targets for further development, but correctly identifying 'hits'—compounds that will ultimately prove successful in clinical trials—continues to pose enormous challenges. Mammalian and invertebrate animal models are often used in this screening process, but mammals are costly and cumbersome, and invertebrates are less physiologically relevant to humans. Therefore, non-mammalian vertebrate systems combining high-throughput whole-organism testing and rapid phenotypic selection—with the added benefit of preliminary toxicological readout—holds obvious appeal for improving time- and cost-efficiency. Zebrafish are compatible with such a system; they share useful physiological characteristics with mammals; their small embryos and juveniles permit multiplex techniques; and their adequately developed organs allow toxicities to be identified early during screening [130]. Zebrafish drug screens can be performed with chemical libraries utilizing wild-type, mutant, transgenic, or xenografted zebrafish embryos. Recent advances using zebrafish transgenic and xenograft models of leukemia are discussed as follows.

Small Molecule Screens Using Transgenic Leukemia Models

Chemical library screening for leukemic research began in 2010 when a zebrafish Myc-induced T-ALL model demonstrated its capacity to recapitulate clinical responses to approved therapeutics (e.g. cyclophosphamide, vincristine, and prednisolone) as observed in humans [131]. Since then chemical library screening has been applied to zebrafish leukemia models including an AML model and multiple T-ALL models [95, 132, 133].

Yeh et al. performed one of the first chemical library screens with a transgenic line aiming to model AML [95]. Their model expressed an oncogenic AML driver of hematopoietic differentiation termed AML1-ETO, a fusion protein resulting from the t(8;21) translocation (Table 1). The investigators hypothesized that inhibition of the self-renewal ability of hematopoietic stem cells could provide a complementary therapy to the inhibition of cell proliferation (a common approach of AML therapy). As described earlier (Table 2), the oncogenic AML1-ETO fusion protein was expressed under the *hsp70* promoter in homozygous fish outcrossed with wild-type fish. Their screen utilized five heterozygous embryos per well in 96-well plates collectively treated with the SPECTRUM compound library—comprised of drugs (60 %), natural products (25 %), and other bioactive components (15 %). Plates were heat shocked to induce AE expression within embryos, thereafter assayed for phenotypic change. Notably, use of the *hsp70* promoter resulted in false-positive hits when select screening agents interfered with proteins related to the heat shock response; the authors suggest using a different promoter to avoid this pitfall. Follow-up study with their lead hit—the COX-2 inhibitor nimesulide—identified a previously unknown role for COX-2 and β-catenin in AE-mediated hematopoietic differentiation. Specifically, nimesulide antagonizes the effects of AML1-ETO on hematopoietic differentiation by interrupting the COX-2 dependent β-catenin signaling pathway [95]. Therefore, future studies might aim to manipulate COX-2 or β-catenin signaling in AML patients with AE-associated pathology.

Whereas Yeh et al. performed chemical screening with a library of characterized compounds, Ridges et al. performed the first embryonic zebrafish screen using a library of compounds with unknown activity [95, 133]. Lacking a model manifesting T-ALL during embryogenesis, Ridges et al. used a sequential approach by identifying compounds with activity against immature zebrafish T cells—with the idea that these cells are most similar to leukemic T lymphoblasts—and thereafter validated hits for anti-neoplastic effect in an adult zebrafish Myc-induced T-ALL model. Their lead hit—lenaldekar—affects the PI3 kinase/AKT/mTOR pathway and shows selective activity against malignant hematopoietic cell lines and primary leukemias [133].

Motivated by the need for NOTCH-independent therapeutics, Gutierrez et al. modified a tamoxifen-regulated MYC-induced T-ALL model in a fluorescence-based screen using dsRed2 expression in thymocytes as a screen read-out (Fig. 6) [81, 132]. With the ability to select for embryos expressing MYC and to image the intensity of their thymic dsRed2 fluorescence, they identified perphenazine in a screen of FDA-approved compounds. Interestingly, in a complementary cell line screen, perphenazine was identified for its synergy with NOTCH inhibitors [132]. Perphenazine is best-known as an inhibitor of dopamine receptor signaling, but the biologic target mediating its anti-leukemic effect appeared to be unrelated to any of

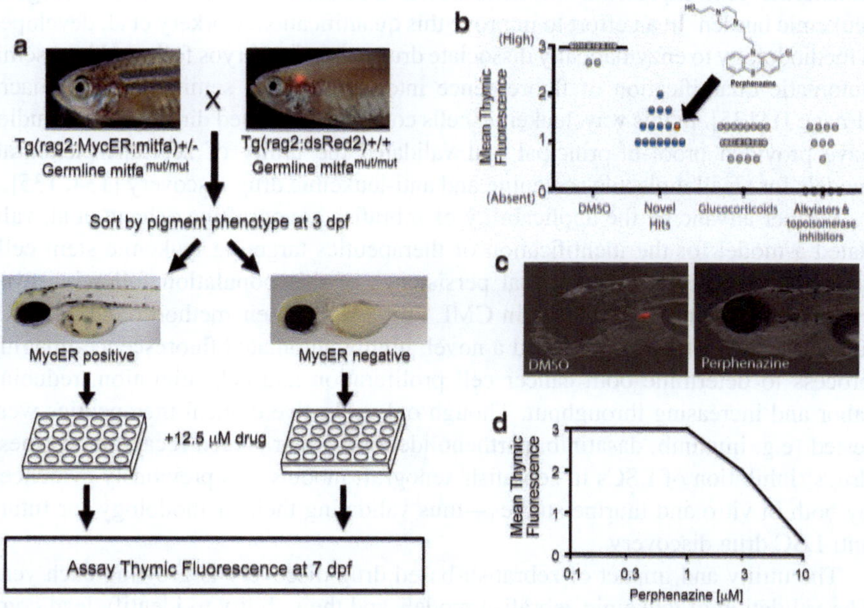

Fig. 6 Zebrafish screen for small molecules that are toxic to MYC-overexpressing thymocytes. (**a**) Primary screen design. (**b**) Hits from the primary screen. *Arrow* denotes the result obtained with perphenazine (**c**). Representative images of DMSO (control) or perphenazine treated zebrafish larvae. (**d**) Dose response curve from secondary screen of perphenazine, with six zebrafish larvae treated per concentration. Drug doses higher than 10 µM induced general toxicity (not shown). Error bars = SD. (Modified and republished with permission from American Society for Clinical Investigation) [132]

its known molecular targets. Using a novel proteomics approach, the tumor suppressor protein phosphatase 2A (PP2A) was found to be a target of perphenazine that is activated by perphenazine binding. Their data provided evidence that activators of PP2A (e.g. perphenazine) in combination with NOTCH-dependent therapeutics might provide an effective T-ALL therapy [132].

Leukemic Xenograft Models in Drug Discovery

When transgenic zebrafish models for particular leukemias are unavailable for screening, zebrafish xenograft models are an alternative and have moved into the mainstream since early studies in 2011 [134]. Creating these models involves a labor-intensive xenografting procedure, requiring the manual injection of cells prior to experimental manipulation. Pruvot et al. were the first to establish zebrafish xenograft models of human leukemia for drug discovery [134]. After performing zebrafish xenografting, the investigators aimed to identify drug-induced decreases in leukemic burden. When flow cytometry could not provide sufficient sensitivity to quantify cell proliferation, human *L32* and hypoxanthine-guanine phosphoribosyltransferase gene expression was measured as a correlate of drug-induced change in leukemic burden. In an effort to improve this quantification, Corkery et al. developed a methodology to enzymatically dissociate drug-treated embryos followed by a semi-automatic quantification of fluorescence intensity using a semi-automated macro (Image J) [135]. In this way, leukemic cells could be quantified directly. Such studies have provided proof-of-principal and validated the utility of zebrafish xenograft models for small molecule screening and anti-leukemic drug discovery [134, 135].

Further advancing the applicability of zebrafish xenografting, Zhang et al. validated a model for the identification of therapeutics targeting leukemic stem cells (LSCs), motivated by the clinical persistence of this population following even complete cytogenetic responses in CML [136, 137]. Their method required LSC-xenografted embryos and utilized a novel, highly automated fluorescence-imaging process to determine both cancer cell proliferation and cell migration, reducing labor and increasing throughput. Though only selective clinical therapeutics were tested (e.g. imatinib, dasatinib, parthenolide, etc.), their screen recapitulated these drugs' inhibition of LSCs in zebrafish xenograft models—as previously evidenced by both in vitro and murine studies—thus validating their methodology for future anti-LSC drug discovery.

The utility and impact of zebrafish-based drug discovery is growing each year. The validation of leukemic zebrafish models and their ability to identify lead compounds like lenaldekar promote further study and methodological development. Although thousands of compounds were screened in the studies discussed above, their setups were largely manual and their execution required years of work. High-throughput, fully automated robotic methods are in development and promise to automate the selection, placement, and injection of zebrafish embryos with DNA and cancer cells, among other desired materials [138–140].

Summary & Future Perspectives

Since the birth of the first zebrafish model of leukemia (T-ALL) in 2003 [81], zebrafish have contributed greatly to leukemia research. Multiple different zebrafish lines have been generated to model various human leukemias and pre-leukemic diseases, including T-ALL, B-ALL, AML, MPN, and MDS. The ease with which the zebrafish can be genetically modified and subsequently imaged has allowed investigators to elucidate the molecular mechanisms of leukemogenesis, as well as leukemic progression and maintenance. In fact, investigation of zebrafish models of leukemia has significantly improved our understanding of leukemogenesis. Examples of the advances include: (a) knowledge of a shared genetic signature between human and zebrafish T-ALL; (b) the importance of NOTCH and MYC in T-ALL pathogenesis [13, 63, 64, 81, 83, 88, 115–119]; (c) the contribution of the PTEN-PI3K-AKT pathways to T-ALL drug resistance in the context of MYC-independent disease [63]; (d) BIM-mediated apoptosis as an avenue for therapy [91]; (e) genetic factors underlying T-LBL to T-ALL dissemination [89]; and (f) the expression of disease-specific fusion proteins [125]. The development of syngeneic transplantation has allowed functional assessment of LPCs in zebrafish, and a transplantation-based screen has elucidated the role of AKT in LPC clonal evolution and treatment resistance [86, 126]. Exploiting high-throughput capabilities of the zebrafish system, investigators have performed small molecule screens and identified multiple agents with anti-leukemic activity, including nimesulide, lenaldekar, and perphenazine [95, 132, 133]. These advances in zebrafish modeling and disease studies have helped to overcome initial skepticism concerning their utility in cancer research, establishing an important biomedical role for this humble organism.

Moving forward, much additional research can be done using the zebrafish. Most zebrafish models of leukemia recapitulate human T-ALL, and although models of pre-B-ALL and AML exist, they have only limited utility. Multiple models of MPN and MDS have been generated, but these fish fail to develop AML. This is likely due to single genetic manipulation in these zebrafish lines, whereas multiple genetic alterations are required for leukemogenesis in humans. Given technological advances, it is now possible to simultaneously manipulate multiple genes within the same cell lineage. We anticipate new zebrafish models of leukemia—including T-ALL, B-ALL, AML, CLL, and CML—will be generated through the use of both transgenic and genome editing technologies, resulting in the overexpression of oncogenes and the inactivation of tumor suppressors. These new models will provide important mechanistic insight into leukemic initiation and progression. Despite scientists' current capacity to conduct successful small molecule screening with zebrafish embryos, the process remains limited by the necessity to manually deposit embryos into multiple wells, as well as the number of embryos that can be manually injected when xenograft models are required. Future automation and standardization of these procedures will enable truly high-throughput small molecule screening, which could eventually contribute to personalized medicine. For example, ex vivo screening of an individual patient's leukemia samples could be applied in a zebrafish xenograft model for the testing of multiple drugs and drug combinations, and the readout would inform clinicians on measures of both drug potency and toxicity, improving therapy selection on a case-by-case basis.

Zebrafish have been welcomed into the oncological research community for their usefulness in delineating factors contributing to tumorigenesis and treatment resistance, as well as for identifying novel agents for experimental therapeutics. However, due to their evolutionary distance from humans, results from zebrafish models require further validation in mammalian models, particularly as candidate therapeutics approach clinical development. Moreover, gene duplication and functional redundancy within the zebrafish genome pose challenges for gene inactivation studies, and antibody production in zebrafish remains daunting due to surface glycosylation modification of proteins. Despite these limitations, maximizing their strengths in combination with the use of mammalian systems, zebrafish will make waves in an ocean of unexploited oncological knowledge, especially in leukemic research.

Acknowledgements The authors thank Joseph Hirsch for editorial assistance, and Leah Huiting and Dr. Nicole M. Anderson for helpful comments and suggestions. N.R.H is supported by a training grant (T32GM008541) from the National Institutes of Health. H.F. is supported by a grant (R00CA134743) from the National Institutes of Health, a career development grant from the St. Baldrick's Foundation, a Karin Grunebaum Faculty Fellowship from the Karin Grunebaum Cancer Foundation, a Ralph Edwards Career Development Professorship from Boston University, and an institutional grant (IRG –72-001-36-IRG) from the American Cancer Society.

References

1. Cancer Facts & Figures (2014). American Cancer Society, Atlanta, Georgia
2. Seiter K (2014) ALL. Acute Lymphoblastic Leukemia – Medscape Reference, vol 2014
3. Seiter K (2014) AML. Acute Myelogenous Leukemia – Medscape Reference, vol 2014
4. Van Vlierberghe P, Ferrando A (2012) The molecular basis of T cell acute lymphoblastic leukemia. J Clin Invest 122(10):3398–3406. doi:10.1172/JCI61269
5. Takeuchi S, Koike M, Seriu T, Bartram CR, Slater J, Park S, Miyoshi I, Koeffler HP (1997) Homozygous deletions at 9p21 in childhood acute lymphoblastic leukemia detected by microsatellite analysis. Leukemia 11(10):1636–1640
6. Cancer Genome Atlas Research Network (2013) Genomic and epigenomic landscapes of adult de novo acute myeloid leukemia. N Engl J Med 368(22):2059–2074. doi:10.1056/NEJMoa1301689
7. Wang L, Lawrence MS, Wan Y, Stojanov P, Sougnez C, Stevenson K, Werner L, Sivachenko A, DeLuca DS, Zhang L, Zhang W, Vartanov AR, Fernandes SM, Goldstein NR, Folco EG, Cibulskis K, Tesar B, Sievers QL, Shefler E, Gabriel S, Hacohen N, Reed R, Meyerson M, Golub TR, Lander ES, Neuberg D, Brown JR, Getz G, Wu CJ (2011) SF3B1 and other novel cancer genes in chronic lymphocytic leukemia. N Engl J Med 365(26):2497–2506. doi:10.1056/NEJMoa1109016
8. Martin-Subero JI, Lopez-Otin C, Campo E (2013) Genetic and epigenetic basis of chronic lymphocytic leukemia. Curr Opin Hematol 20(4):362–368. doi:10.1097/MOH.0b013e32836235dc
9. Baliakas P, Iskas M, Gardiner A, Davis Z, Plevova K, Nguyen-Khac F, Malcikova J, Anagnostopoulos A, Glide S, Mould S, Stepanovska K, Brejcha M, Belessi C, Davi F, Pospisilova S, Athanasiadou A, Stamatopoulos K, Oscier D (2014) Chromosomal translocations and karyotype complexity in chronic lymphocytic leukemia: a systematic reappraisal of classic cytogenetic data. Am J Hematol 89(3):249–255. doi:10.1002/ajh.23618

10. Johansson B, Fioretos T, Mitelman F (2002) Cytogenetic and molecular genetic evolution of chronic myeloid leukemia. Acta Haematol 107(2):76–94
11. Rumpold H, Webersinke G (2011) Molecular pathogenesis of Philadelphia-positive chronic myeloid leukemia – is it all BCR-ABL? Curr Cancer Drug Targets 11(1):3–19
12. Mullighan CG, Miller CB, Radtke I, Phillips LA, Dalton J, Ma J, White D, Hughes TP, Le Beau MM, Pui CH, Relling MV, Shurtleff SA, Downing JR (2008) BCR-ABL1 lymphoblastic leukaemia is characterized by the deletion of Ikaros. Nature 453(7191):110–114. doi:10.1038/nature06866
13. Weng AP, Ferrando AA, Lee W, Morris JP, Silverman LB, Sanchez-Irizarry C, Blacklow SC, Look AT, Aster JC (2004) Activating mutations of NOTCH1 in human T cell acute lymphoblastic leukemia. Science 306(5694):269–271. doi:10.1126/science.1102160
14. O'Neil J, Grim J, Strack P, Rao S, Tibbitts D, Winter C, Hardwick J, Welcker M, Meijerink JP, Pieters R, Draetta G, Sears R, Clurman BE, Look AT (2007) FBW7 mutations in leukemic cells mediate NOTCH pathway activation and resistance to gamma-secretase inhibitors. J Exp Med 204(8):1813–1824. doi:10.1084/jem.20070876
15. Thompson BJ, Buonamici S, Sulis ML, Palomero T, Vilimas T, Basso G, Ferrando A, Aifantis I (2007) The SCFFBW7 ubiquitin ligase complex as a tumor suppressor in T cell leukemia. J Exp Med 204(8):1825–1835. doi:10.1084/jem.20070872
16. Mullighan CG (2012) Molecular genetics of B-precursor acute lymphoblastic leukemia. J Clin Invest 122(10):3407–3415. doi:10.1172/JCI61203
17. Marcucci G, Haferlach T, Dohner H (2011) Molecular genetics of adult acute myeloid leukemia: prognostic and therapeutic implications. J Clin Oncol 29(5):475–486. doi:10.1200/JCO.2010.30.2554
18. Chung SS (2014) Genetic mutations in acute myeloid leukemia that influence clinical decisions. Curr Opin Hematol 21(2):87–94. doi:10.1097/MOH.0000000000000024
19. Quesada V, Ramsay AJ, Lopez-Otin C (2012) Chronic lymphocytic leukemia with SF3B1 mutation. N Engl J Med 366(26):2530. doi:10.1056/NEJMc1204033
20. Landau DA, Stewart C, Reiter JG, Lawrence M, Sougnez C, Brown JR, Lopez-Guillermo A, Gabriel S, Lander E, Neuberg DS, Lopez-Otin C, Campo E, Getz G, Wu CJ (2014) Novel putative driver gene mutations in chronic lymphocytic leukemia (CLL): results from a combined analysis of whole-exome sequencing of 262 primary CLL samples. Blood 124(21):1952–1952
21. Sawyers CL (1999) Chronic myeloid leukemia. N Engl J Med 340(17):1330–1340. doi:10.1056/NEJM199904293401706
22. Faderl S, Talpaz M, Estrov Z, O'Brien S, Kurzrock R, Kantarjian HM (1999) The biology of chronic myeloid leukemia. N Engl J Med 341(3):164–172. doi:10.1056/NEJM199907153410306
23. Quintas-Cardama A, Cortes J (2009) Molecular biology of bcr-abl1-positive chronic myeloid leukemia. Blood 113(8):1619–1630. doi:10.1182/blood-2008-03-144790
24. Besa EC (2014) CML. Chronic Myelogenous Leukemia – Medscape Reference, vol 2014
25. Hosen N, Maeda T, Hashii Y, Tsuboi A, Nishida S, Nakata J, Nakae Y, Takashima S, Oji Y, Oka Y, Kumanogoh A, Sugiyama H (2014) Vaccination strategies to improve outcome of hematopoietic stem cell transplant in leukemia patients: early evidence and future prospects. Expert Rev Hematol 7(5):671–681. doi:10.1586/17474086.2014.953925
26. Ai J, Advani A (2014) Current status of antibody therapy in ALL. Br J Haematol 168:471–480. doi:10.1111/bjh.13205
27. Balducci L (2001) The geriatric cancer patient: equal benefit from equal treatment. Cancer Control 8(2 Suppl):1–25, quiz 27–28
28. Breslin S (2007) Cytokine-release syndrome: overview and nursing implications. Clin J Oncol Nurs 11(1 Suppl):37–42. doi:10.1188/07.CJON.S1.37-42
29. Mavroudis D, Barrett J (1996) The graft-versus-leukemia effect. Curr Opin Hematol 3(6):423–429
30. Rezvani AR, Storb RF (2008) Separation of graft-vs.-tumor effects from graft-vs.-host disease in allogeneic hematopoietic cell transplantation. J Autoimmun 30(3):172–179. doi:10.1016/j.jaut.2007.12.002
31. Wayne AS, Fitzgerald DJ, Kreitman RJ, Pastan I (2014) Immunotoxins for leukemia. Blood 123(16):2470–2477. doi:10.1182/blood-2014-01-492256

32. Burkhardt UE, Hainz U, Stevenson K, Goldstein NR, Pasek M, Naito M, Wu D, Ho VT, Alonso A, Hammond NN, Wong J, Sievers QL, Brusic A, McDonough SM, Zeng W, Perrin A, Brown JR, Canning CM, Koreth J, Cutler C, Armand P, Neuberg D, Lee JS, Antin JH, Mulligan RC, Sasada T, Ritz J, Soiffer RJ, Dranoff G, Alyea EP, Wu CJ (2013) Autologous CLL cell vaccination early after transplant induces leukemia-specific T cells. J Clin Invest 123(9):3756–3765. doi:10.1172/JCI69098
33. CAR T-Cell Therapy: Engineering Patients' Immune Cells to Treat Their Cancers. (2014) National Cancer Institute. Accessed Web Page 2014
34. Porter DL, Levine BL, Kalos M, Bagg A, June CH (2011) Chimeric antigen receptor-modified T cells in chronic lymphoid leukemia. N Engl J Med 365(8):725–733. doi:10.1056/NEJMoa1103849
35. Carella AM, Branford S, Deininger M, Mahon FX, Saglio G, Eiring A, Khorashad J, O'Hare T, Goldman JM (2013) What challenges remain in chronic myeloid leukemia research? Haematologica 98(8):1168–1172. doi:10.3324/haematol.2013.090381
36. Qin L, Smith BD, Tsai HL, Yaghi NK, Neela PH, Moake M, Fu J, Kasamon YL, Prince GT, Goswami M, Rosner GL, Levitsky HI, Hourigan CS (2013) Induction of high-titer IgG antibodies against multiple leukemia-associated antigens in CML patients with clinical responses to K562/GVAX immunotherapy. Blood Cancer J 3:e145. doi:10.1038/bcj.2013.44
37. Davidson AJ, Zon LI (2004) The 'definitive' (and 'primitive') guide to zebrafish hematopoiesis. Oncogene 23(43):7233–7246. doi:10.1038/sj.onc.1207943
38. Detrich HW 3rd, Kieran MW, Chan FY, Barone LM, Yee K, Rundstadler JA, Pratt S, Ransom D, Zon LI (1995) Intraembryonic hematopoietic cell migration during vertebrate development. Proc Natl Acad Sci U S A 92(23):10713–10717
39. Liu S, Leach SD (2011) Zebrafish models for cancer. Annu Rev Pathol 6:71–93. doi:10.1146/annurev-pathol-011110-130330
40. Bertrand JY, Traver D (2009) Hematopoietic cell development in the zebrafish embryo. Curr Opin Hematol 16(4):243–248. doi:10.1097/MOH.0b013e32832c05e4
41. Ellett F, Lieschke GJ (2010) Zebrafish as a model for vertebrate hematopoiesis. Curr Opin Pharmacol 10(5):563–570. doi:10.1016/j.coph.2010.05.004
42. Bertrand JY, Chi NC, Santoso B, Teng S, Stainier DY, Traver D (2010) Haematopoietic stem cells derive directly from aortic endothelium during development. Nature 464(7285):108–111. doi:10.1038/nature08738
43. Kissa K, Herbomel P (2010) Blood stem cells emerge from aortic endothelium by a novel type of cell transition. Nature 464(7285):112–115. doi:10.1038/nature08761
44. Colle-Vandevelde A (1963) Blood anlage in teleostei. Nature 198:1223
45. Carroll KJ, North TE (2014) Oceans of opportunity: exploring vertebrate hematopoiesis in zebrafish. Exp Hematol 42(8):684–696. doi:10.1016/j.exphem.2014.05.002
46. Paik EJ, Zon LI (2010) Hematopoietic development in the zebrafish. Int J Dev Biol 54(6–7):1127–1137. doi:10.1387/ijdb.093042ep
47. Zhang C, Patient R, Liu F (2013) Hematopoietic stem cell development and regulatory signaling in zebrafish. Biochim Biophys Acta 1830(2):2370–2374. doi:10.1016/j.bbagen.2012.06.008
48. Crowhurst MO, Layton JE, Lieschke GJ (2002) Developmental biology of zebrafish myeloid cells. Int J Dev Biol 46(4):483–492
49. Hsu K, Traver D, Kutok JL, Hagen A, Liu TX, Paw BH, Rhodes J, Berman JN, Zon LI, Kanki JP, Look AT (2004) The pu.1 promoter drives myeloid gene expression in zebrafish. Blood 104(5):1291–1297. doi:10.1182/blood-2003-09-3105
50. Rhodes J, Hagen A, Hsu K, Deng M, Liu TX, Look AT, Kanki JP (2005) Interplay of pu.1 and gata1 determines myelo-erythroid progenitor cell fate in zebrafish. Dev Cell 8(1):97–108. doi:10.1016/j.devcel.2004.11.014
51. Oettinger MA, Schatz DG, Gorka C, Baltimore D (1990) RAG-1 and RAG-2, adjacent genes that synergistically activate V(D)J recombination. Science 248(4962):1517–1523
52. Yaqoob N, Holotta M, Prem C, Kopp R, Schwerte T (2009) Ontogenetic development of erythropoiesis can be studied non-invasively in GATA-1:DsRed transgenic zebrafish. Comp Biochem Physiol A Mol Integr Physiol 154(2):270–278. doi:10.1016/j.cbpa.2009.06.024

53. Mathias JR, Perrin BJ, Liu TX, Kanki J, Look AT, Huttenlocher A (2006) Resolution of inflammation by retrograde chemotaxis of neutrophils in transgenic zebrafish. J Leukoc Biol 80(6):1281–1288. doi:10.1189/jlb.0506346
54. Ellett F, Pase L, Hayman JW, Andrianopoulos A, Lieschke GJ (2011) mpeg1 promoter transgenes direct macrophage-lineage expression in zebrafish. Blood 117(4):e49–e56. doi:10.1182/blood-2010-10-314120
55. Langenau DM, Ferrando AA, Traver D, Kutok JL, Hezel JP, Kanki JP, Zon LI, Look AT, Trede NS (2004) In vivo tracking of T cell development, ablation, and engraftment in transgenic zebrafish. Proc Natl Acad Sci U S A 101(19):7369–7374. doi:10.1073/pnas.0402248101
56. Jessen JR, Jessen TN, Vogel SS, Lin S (2001) Concurrent expression of recombination activating genes 1 and 2 in zebrafish olfactory sensory neurons. Genesis 29(4):156–162
57. Jessen JR, Willett CE, Lin S (1999) Artificial chromosome transgenesis reveals long-distance negative regulation of rag1 in zebrafish. Nat Genet 23(1):15–16. doi:10.1038/12609
58. Lin HF, Traver D, Zhu H, Dooley K, Paw BH, Zon LI, Handin RI (2005) Analysis of thrombocyte development in CD41-GFP transgenic zebrafish. Blood 106(12):3803–3810. doi:10.1182/blood-2005-01-0179
59. Ma D, Zhang J, Lin HF, Italiano J, Handin RI (2011) The identification and characterization of zebrafish hematopoietic stem cells. Blood 118(2):289–297. doi:10.1182/blood-2010-12-327403
60. White RM, Sessa A, Burke C, Bowman T, LeBlanc J, Ceol C, Bourque C, Dovey M, Goessling W, Burns CE, Zon LI (2008) Transparent adult zebrafish as a tool for in vivo transplantation analysis. Cell Stem Cell 2(2):183–189. doi:10.1016/j.stem.2007.11.002
61. Kawakami K (2004) Transgenesis and gene trap methods in zebrafish by using the Tol2 transposable element. Methods Cell Biol 77:201–222
62. Thermes V, Grabher C, Ristoratore F, Bourrat F, Choulika A, Wittbrodt J, Joly JS (2002) I-SceI meganuclease mediates highly efficient transgenesis in fish. Mech Dev 118(1–2):91–98
63. Gutierrez A, Grebliunaite R, Feng H, Kozakewich E, Zhu S, Guo F, Payne E, Mansour M, Dahlberg SE, Neuberg DS, den Hertog J, Prochownik EV, Testa JR, Harris M, Kanki JP, Look AT (2011) Pten mediates Myc oncogene dependence in a conditional zebrafish model of T cell acute lymphoblastic leukemia. J Exp Med 208(8):1595–1603. doi:10.1084/jem.20101691
64. Langenau DM, Feng H, Berghmans S, Kanki JP, Kutok JL, Look AT (2005) Cre/lox-regulated transgenic zebrafish model with conditional myc-induced T cell acute lymphoblastic leukemia. Proc Natl Acad Sci U S A 102(17):6068–6073. doi:10.1073/pnas.0408708102
65. Scheer N, Campos-Ortega JA (1999) Use of the Gal4-UAS technique for targeted gene expression in the zebrafish. Mech Dev 80(2):153–158
66. Zheng W, Li Z, Nguyen AT, Li C, Emelyanov A, Gong Z (2014) Xmrk, kras and myc transgenic zebrafish liver cancer models share molecular signatures with subsets of human hepatocellular carcinoma. PLoS One 9(3), e91179. doi:10.1371/journal.pone.0091179
67. Rembold M, Lahiri K, Foulkes NS, Wittbrodt J (2006) Transgenesis in fish: efficient selection of transgenic fish by co-injection with a fluorescent reporter construct. Nat Protoc 1(3):1133–1139. doi:10.1038/nprot.2006.165
68. Langenau DM, Keefe MD, Storer NY, Jette CA, Smith AC, Ceol CJ, Bourque C, Look AT, Zon LI (2008) Co-injection strategies to modify radiation sensitivity and tumor initiation in transgenic Zebrafish. Oncogene 27(30):4242–4248. doi:10.1038/onc.2008.56
69. Amsterdam A, Nissen RM, Sun Z, Swindell EC, Farrington S, Hopkins N (2004) Identification of 315 genes essential for early zebrafish development. Proc Natl Acad Sci U S A 101(35):12792–12797. doi:10.1073/pnas.0403929101
70. Mullins MC, Hammerschmidt M, Haffter P, Nusslein-Volhard C (1994) Large-scale mutagenesis in the zebrafish: in search of genes controlling development in a vertebrate. Curr Biol 4(3):189–202
71. Sun J, Liu W, Li L, Chen J, Wu M, Zhang Y, Leung AY, Zhang W, Wen Z, Liao W (2013) Suppression of Pu.1 function results in expanded myelopoiesis in zebrafish. Leukemia 27(9):1913–1917. doi:10.1038/leu.2013.67

72. Wienholds E, Schulte-Merker S, Walderich B, Plasterk RH (2002) Target-selected inactivation of the zebrafish rag1 gene. Science 297(5578):99–102. doi:10.1126/science.1071762
73. Law SH, Sargent TD (2014) The serine-threonine protein kinase PAK4 is dispensable in zebrafish: identification of a morpholino-generated pseudophenotype. PLoS One 9(6), e100268. doi:10.1371/journal.pone.0100268
74. Dong M, Fu YF, Du TT, Jing CB, Fu CT, Chen Y, Jin Y, Deng M, Liu TX (2009) Heritable and lineage-specific gene knockdown in zebrafish embryo. PLoS One 4(7):e6125. doi:10.1371/journal.pone.0006125
75. Doyon Y, McCammon JM, Miller JC, Faraji F, Ngo C, Katibah GE, Amora R, Hocking TD, Zhang L, Rebar EJ, Gregory PD, Urnov FD, Amacher SL (2008) Heritable targeted gene disruption in zebrafish using designed zinc-finger nucleases. Nat Biotechnol 26(6):702–708. doi:10.1038/nbt1409
76. Hwang WY, Fu Y, Reyon D, Maeder ML, Tsai SQ, Sander JD, Peterson RT, Yeh JR, Joung JK (2013) Efficient genome editing in zebrafish using a CRISPR-Cas system. Nat Biotechnol 31(3):227–229. doi:10.1038/nbt.2501
77. Sander JD, Cade L, Khayter C, Reyon D, Peterson RT, Joung JK, Yeh JR (2011) Targeted gene disruption in somatic zebrafish cells using engineered TALENs. Nat Biotechnol 29(8):697–698. doi:10.1038/nbt.1934
78. Tang Q, Abdelfattah NS, Blackburn JS, Moore JC, Martinez SA, Moore FE, Lobbardi R, Tenente IM, Ignatius MS, Berman JN, Liwski RS, Houvras Y, Langenau DM (2014) Optimized cell transplantation using adult rag2 mutant zebrafish. Nat Methods 11(8):821–824. doi:10.1038/nmeth.3031
79. Tenente IM, Tang Q, Moore JC, Langenau DM (2014) Normal and malignant muscle cell transplantation into immune compromised adult zebrafish. J Vis Exp (94). doi:10.3791/52597
80. Moore FE, Langenau DM (2012) Through the looking glass: visualizing leukemia growth, migration, and engraftment using fluorescent transgenic zebrafish. Adv Hematol 2012:478164. doi:10.1155/2012/478164
81. Langenau DM, Traver D, Ferrando AA, Kutok JL, Aster JC, Kanki JP, Lin S, Prochownik E, Trede NS, Zon LI, Look AT (2003) Myc-induced T cell leukemia in transgenic zebrafish. Science (New York, NY) 299(5608):887–890. doi:10.1126/science.1080280
82. Sabaawy HE, Azuma M, Embree LJ, Tsai HJ, Starost MF, Hickstein DD (2006) TEL-AML1 transgenic zebrafish model of precursor B cell acute lymphoblastic leukemia. Proc Natl Acad Sci U S A 103(41):15166–15171. doi:10.1073/pnas.0603349103
83. Chen J, Jette C, Kanki JP, Aster JC, Look AT, Griffin JD (2007) NOTCH1-induced T-cell leukemia in transgenic zebrafish. Leukemia 21(3):462–471. doi:10.1038/sj.leu.2404546
84. Zhuravleva J, Paggetti J, Martin L, Hammann A, Solary E, Bastie JN, Delva L (2008) MOZ/TIF2-induced acute myeloid leukaemia in transgenic fish. Br J Haematol 143(3):378–382. doi:10.1111/j.1365-2141.2008.07362.x
85. Blackburn JS, Liu S, Raimondi AR, Ignatius MS, Salthouse CD, Langenau DM (2011) High-throughput imaging of adult fluorescent zebrafish with an LED fluorescence macroscope. Nat Protoc 6(2):229–241. doi:10.1038/nprot.2010.170
86. Smith AC, Raimondi AR, Salthouse CD, Ignatius MS, Blackburn JS, Mizgirev IV, Storer NY, de Jong JL, Chen AT, Zhou Y, Revskoy S, Zon LI, Langenau DM (2010) High-throughput cell transplantation establishes that tumor-initiating cells are abundant in zebrafish T-cell acute lymphoblastic leukemia. Blood 115(16):3296–3303. doi:10.1182/blood-2009-10-246488
87. Zhang L, Alt C, Li P, White RM, Zon LI, Wei X, Lin CP (2012) An optical platform for cell tracking in adult zebrafish. Cytometry A 81(2):176–182. doi:10.1002/cyto.a.21167
88. Feng H, Langenau DM, Madge JA, Quinkertz A, Gutierrez A, Neuberg DS, Kanki JP, Look AT (2007) Heat-shock induction of T-cell lymphoma/leukaemia in conditional Cre/lox-regulated transgenic zebrafish. Br J Haematol 138(2):169–175. doi:10.1111/j.1365-2141.2007.06625.x
89. Feng H, Stachura DL, White RM, Gutierrez A, Zhang L, Sanda T, Jette CA, Testa JR, Neuberg DS, Langenau DM, Kutok JL, Zon LI, Traver D, Fleming MD, Kanki JP, Look AT (2010) T-lymphoblastic lymphoma cells express high levels of BCL2, S1P1, and ICAM1, leading to a blockade of tumor cell intravasation. Cancer Cell 18(4):353–366. doi:10.1016/j.ccr.2010.09.009

90. Gutierrez A, Feng H, Stevenson K, Neuberg DS, Calzada O, Zhou Y, Langenau DM, Look AT (2014) Loss of function tp53 mutations do not accelerate the onset of myc-induced T-cell acute lymphoblastic leukaemia in the zebrafish. Br J Haematol 166(1):84–90. doi:10.1111/bjh.12851
91. Reynolds C, Roderick JE, LaBelle JL, Bird G, Mathieu R, Bodaar K, Colon D, Pyati U, Stevenson KE, Qi J, Harris M, Silverman LB, Sallan SE, Bradner JE, Neuberg DS, Look AT, Walensky LD, Kelliher MA, Gutierrez A (2014) Repression of BIM mediates survival signaling by MYC and AKT in high-risk T-cell acute lymphoblastic leukemia. Leukemia 28(9):1819–1827. doi:10.1038/leu.2014.78
92. Frazer JK, Meeker ND, Rudner L, Bradley DF, Smith AC, Demarest B, Joshi D, Locke EE, Hutchinson SA, Tripp S, Perkins SL, Trede NS (2009) Heritable T-cell malignancy models established in a zebrafish phenotypic screen. Leukemia 23(10):1825–1835. doi:10.1038/leu.2009.116
93. Le X, Langenau DM, Keefe MD, Kutok JL, Neuberg DS, Zon LI (2007) Heat shock-inducible Cre/Lox approaches to induce diverse types of tumors and hyperplasia in transgenic zebrafish. Proc Natl Acad Sci U S A 104(22):9410–9415. doi:10.1073/pnas.0611302104
94. Yeh JR, Munson KM, Chao YL, Peterson QP, Macrae CA, Peterson RT (2008) AML1-ETO reprograms hematopoietic cell fate by downregulating scl expression. Development 135(2):401–410. doi:10.1242/dev.008904
95. Yeh JR, Munson KM, Elagib KE, Goldfarb AN, Sweetser DA, Peterson RT (2009) Discovering chemical modifiers of oncogene-regulated hematopoietic differentiation. Nat Chem Biol 5(4):236–243. doi:10.1038/nchembio.147
96. Forrester AM, Grabher C, McBride ER, Boyd ER, Vigerstad MH, Edgar A, Kai FB, Da'as SI, Payne E, Look AT, Berman JN (2011) NUP98-HOXA9-transgenic zebrafish develop a myeloproliferative neoplasm and provide new insight into mechanisms of myeloid leukaemogenesis. Br J Haematol 155(2):167–181. doi:10.1111/j.1365-2141.2011.08810.x
97. Alghisi E, Distel M, Malagola M, Anelli V, Santoriello C, Herwig L, Krudewig A, Henkel CV, Russo D, Mione MC (2013) Targeting oncogene expression to endothelial cells induces proliferation of the myelo-erythroid lineage by repressing the Notch pathway. Leukemia 27(11):2229–2241. doi:10.1038/leu.2013.132
98. Bolli N, Payne EM, Grabher C, Lee JS, Johnston AB, Falini B, Kanki JP, Look AT (2010) Expression of the cytoplasmic NPM1 mutant (NPMc+) causes the expansion of hematopoietic cells in zebrafish. Blood 115(16):3329–3340. doi:10.1182/blood-2009-02-207225
99. Kalev-Zylinska ML, Horsfield JA, Flores MV, Postlethwait JH, Vitas MR, Baas AM, Crosier PS, Crosier KE (2002) Runx1 is required for zebrafish blood and vessel development and expression of a human RUNX1-CBF2T1 transgene advances a model for studies of leukemogenesis. Development 129(8):2015–2030
100. Shen LJ, Chen FY, Zhang Y, Cao LF, Kuang Y, Zhong M, Wang T, Zhong H (2013) MYCN transgenic zebrafish model with the characterization of acute myeloid leukemia and altered hematopoiesis. PLoS One 8(3), e59070. doi:10.1371/journal.pone.0059070
101. Gjini E, Mansour MR, Sander JD, Moritz N, Nguyen AT, Kesarsing M, Gans E, He S, Chen S, Ko M, Kuang YY, Yang S, Zhou Y, Rodig S, Zon LI, Joung JK, Rao A, Look AT (2015) A zebrafish model of myelodysplastic syndrome produced through tet2 genomic editing. Mol Cell Biol 35(5):789–804. doi:10.1128/MCB.00971-14
102. Pui CH, Relling MV, Downing JR (2004) Acute lymphoblastic leukemia. N Engl J Med 350(15):1535–1548. doi:10.1056/NEJMra023001
103. He YF, Li BZ, Li Z, Liu P, Wang Y, Tang Q, Ding J, Jia Y, Chen Z, Li L, Sun Y, Li X, Dai Q, Song CX, Zhang K, He C, Xu GL (2011) Tet-mediated formation of 5-carboxylcytosine and its excision by TDG in mammalian DNA. Science 333(6047):1303–1307. doi:10.1126/science.1210944
104. Nakajima H, Kunimoto H (2014) TET2 as an epigenetic master regulator for normal and malignant hematopoiesis. Cancer Sci 105(9):1093–1099. doi:10.1111/cas.12484
105. Downing JR (1999) The AML1-ETO chimaeric transcription factor in acute myeloid leukaemia: biology and clinical significance. Br J Haematol 106(2):296–308

106. Borrow J, Shearman AM, Stanton VP Jr, Becher R, Collins T, Williams AJ, Dube I, Katz F, Kwong YL, Morris C, Ohyashiki K, Toyama K, Rowley J, Housman DE (1996) The t(7;11)(p15;p15) translocation in acute myeloid leukaemia fuses the genes for nucleoporin NUP98 and class I homeoprotein HOXA9. Nat Genet 12(2):159–167. doi:10.1038/ng0296-159
107. Ferrando AA, Neuberg DS, Staunton J, Loh ML, Huard C, Raimondi SC, Behm FG, Pui CH, Downing JR, Gilliland DG, Lander ES, Golub TR, Look AT (2002) Gene expression signatures define novel oncogenic pathways in T cell acute lymphoblastic leukemia. Cancer Cell 1(1):75–87
108. Rudner LA, Brown KH, Dobrinski KP, Bradley DF, Garcia MI, Smith AC, Downie JM, Meeker ND, Look AT, Downing JR, Gutierrez A, Mullighan CG, Schiffman JD, Lee C, Trede NS, Frazer JK (2011) Shared acquired genomic changes in zebrafish and human T-ALL. Oncogene 30(41):4289–4296. doi:10.1038/onc.2011.138
109. Blackburn JS, Liu S, Raiser DM, Martinez SA, Feng H, Meeker ND, Gentry J, Neuberg D, Look AT, Ramaswamy S, Bernards A, Trede NS, Langenau DM (2012) Notch signaling expands a pre-malignant pool of T-cell acute lymphoblastic leukemia clones without affecting leukemia-propagating cell frequency. Leukemia 26(9):2069–2078. doi:10.1038/leu.2012.116
110. von Boehmer H (1997) T-call development: is notch a key player in lineage decisions? Curr Biol 7(5):R308–R310
111. O'Neil J, Calvo J, McKenna K, Krishnamoorthy V, Aster JC, Bassing CH, Alt FW, Kelliher M, Look AT (2006) Activating Notch1 mutations in mouse models of T-ALL. Blood 107(2):781–785. doi:10.1182/blood-2005-06-2553
112. Cory S (1986) Activation of cellular oncogenes in hemopoietic cells by chromosome translocation. Adv Cancer Res 47:189–234
113. Klein G, Klein E (1985) Myc/Ig juxtaposition by chromosomal translocations: some new insights, puzzles and paradoxes. Immunol Today 6(7):208–215. doi:10.1016/0167-5699(85)90036-2
114. Harris AW, Pinkert CA, Crawford M, Langdon WY, Brinster RL, Adams JM (1988) The E mu-myc transgenic mouse. A model for high-incidence spontaneous lymphoma and leukemia of early B cells. J Exp Med 167(2):353–371
115. Koch U, Radtke F (2011) Notch in T-ALL: new players in a complex disease. Trends Immunol 32(9):434–442. doi:10.1016/j.it.2011.06.005
116. Palomero T, Sulis ML, Cortina M, Real PJ, Barnes K, Ciofani M, Caparros E, Buteau J, Brown K, Perkins SL, Bhagat G, Agarwal AM, Basso G, Castillo M, Nagase S, Cordon-Cardo C, Parsons R, Zuniga-Pflucker JC, Dominguez M, Ferrando AA (2007) Mutational loss of PTEN induces resistance to NOTCH1 inhibition in T-cell leukemia. Nat Med 13(10):1203–1210. doi:10.1038/nm1636
117. Sharma VM, Calvo JA, Draheim KM, Cunningham LA, Hermance N, Beverly L, Krishnamoorthy V, Bhasin M, Capobianco AJ, Kelliher MA (2006) Notch1 contributes to mouse T-cell leukemia by directly inducing the expression of c-myc. Mol Cell Biol 26(21):8022–8031. doi:10.1128/MCB.01091-06
118. Tzoneva G, Ferrando AA (2012) Recent advances on NOTCH signaling in T-ALL. Curr Top Microbiol Immunol 360:163–182. doi:10.1007/82_2012_232
119. Weng AP, Millholland JM, Yashiro-Ohtani Y, Arcangeli ML, Lau A, Wai C, Del Bianco C, Rodriguez CG, Sai H, Tobias J, Li Y, Wolfe MS, Shachaf C, Felsher D, Blacklow SC, Pear WS, Aster JC (2006) c-Myc is an important direct target of Notch1 in T-cell acute lymphoblastic leukemia/lymphoma. Genes Dev 20(15):2096–2109. doi:10.1101/gad.1450406
120. Gutierrez A, Look AT (2007) NOTCH and PI3K-AKT pathways intertwined. Cancer Cell 12(5):411–413. doi:10.1016/j.ccr.2007.10.027
121. Puisieux A, Valsesia-Wittmann S (2006) Cancer cells escape from failsafe programs in a simple Twist. Bull Cancer 93(3):251–256
122. Eischen CM, Weber JD, Roussel MF, Sherr CJ, Cleveland JL (1999) Disruption of the ARF-Mdm2-p53 tumor suppressor pathway in Myc-induced lymphomagenesis. Genes Dev 13(20):2658–2669
123. Strasser A, Harris AW, Bath ML, Cory S (1990) Novel primitive lymphoid tumours induced in transgenic mice by cooperation between myc and bcl-2. Nature 348(6299):331–333. doi:10.1038/348331a0

124. Peeters P, Raynaud SD, Cools J, Wlodarska I, Grosgeorge J, Philip P, Monpoux F, Van Rompaey L, Baens M, Van den Berghe H, Marynen P (1997) Fusion of TEL, the ETS-variant gene 6 (ETV6), to the receptor-associated kinase JAK2 as a result of t(9;12) in a lymphoid and t(9;15;12) in a myeloid leukemia. Blood 90(7):2535–2540
125. Onnebo SM, Rasighaemi P, Kumar J, Liongue C, Ward AC (2012) Alternative TEL-JAK2 fusions associated with T-cell acute lymphoblastic leukemia and atypical chronic myelogenous leukemia dissected in zebrafish. Haematologica 97(12):1895–1903. doi:10.3324/haematol.2012.064659
126. Blackburn JS, Liu S, Wilder JL, Dobrinski KP, Lobbardi R, Moore FE, Martinez SA, Chen EY, Lee C, Langenau DM (2014) Clonal evolution enhances leukemia-propagating cell frequency in T cell acute lymphoblastic leukemia through Akt/mTORC1 pathway activation. Cancer Cell 25(3):366–378. doi:10.1016/j.ccr.2014.01.032
127. Bolling SF, Kunkel SL, Lin H (1992) Prolongation of cardiac allograft survival in rats by anti-TNF and cyclosporine combination therapy. Transplantation 53(2):283–286
128. Novoa B, Figueras A (2012) Zebrafish: model for the study of inflammation and the innate immune response to infectious diseases. Adv Exp Med Biol 946:253–275. doi:10.1007/978-1-4614-0106-3_15
129. Hughes JP, Rees S, Kalindjian SB, Philpott KL (2011) Principles of early drug discovery. Br J Pharmacol 162(6):1239–1249. doi:10.1111/j.1476-5381.2010.01127.x
130. Zon LI, Peterson RT (2005) In vivo drug discovery in the zebrafish. Nat Rev Drug Discov 4(1):35–44
131. Mizgirev IV, Revskoy S (2010) A new zebrafish model for experimental leukemia therapy. Cancer Biol Ther 9(11):895–902
132. Gutierrez A, Pan L, Groen RW, Baleydier F, Kentsis A, Marineau J, Grebliunaite R, Kozakewich E, Reed C, Pflumio F, Poglio S, Uzan B, Clemons P, VerPlank L, An F, Burbank J, Norton S, Tolliday N, Steen H, Weng AP, Yuan H, Bradner JE, Mitsiades C, Look AT, Aster JC (2014) Phenothiazines induce PP2A-mediated apoptosis in T cell acute lymphoblastic leukemia. J Clin Invest 124(2):644–655. doi:10.1172/JCI65093
133. Ridges S, Heaton WL, Joshi D, Choi H, Eiring A, Batchelor L, Choudhry P, Manos EJ, Sofla H, Sanati A, Welborn S, Agarwal A, Spangrude GJ, Miles RR, Cox JE, Frazer JK, Deininger M, Balan K, Sigman M, Muschen M, Perova T, Johnson R, Montpellier B, Guidos CJ, Jones DA, Trede NS (2012) Zebrafish screen identifies novel compound with selective toxicity against leukemia. Blood 119(24):5621–5631. doi:10.1182/blood-2011-12-398818
134. Pruvot B, Jacquel A, Droin N, Auberger P, Bouscary D, Tamburini J, Muller M, Fontenay M, Chluba J, Solary E (2011) Leukemic cell xenograft in zebrafish embryo for investigating drug efficacy. Haematologica 96(4):612–616. doi:10.3324/haematol.2010.031401
135. Corkery DP, Dellaire G, Berman JN (2011) Leukaemia xenotransplantation in zebrafish—chemotherapyresponseassayinvivo.BrJHaematol153(6):786–789.doi:10.1111/j.1365-2141.2011.08661.x
136. Stuart SA, Minami Y, Wang JY (2009) The CML stem cell: evolution of the progenitor. Cell Cycle 8(9):1338–1343
137. Zhang B, Shimada Y, Kuroyanagi J, Umemoto N, Nishimura Y, Tanaka T (2014) Quantitative phenotyping-based in vivo chemical screening in a zebrafish model of leukemia stem cell xenotransplantation. PLoS One 9(1):e85439. doi:10.1371/journal.pone.0085439
138. Carvalho R, de Sonneville J, Stockhammer OW, Savage ND, Veneman WJ, Ottenhoff TH, Dirks RP, Meijer AH, Spaink HP (2011) A high-throughput screen for tuberculosis progression. PLoS One 6(2):e16779. doi:10.1371/journal.pone.0016779
139. Spaink HP, Cui C, Wiweger MI, Jansen HJ, Veneman WJ, Marin-Juez R, de Sonneville J, Ordas A, Torraca V, van der Ent W, Leenders WP, Meijer AH, Snaar-Jagalska BE, Dirks RP (2013) Robotic injection of zebrafish embryos for high-throughput screening in disease models. Methods (San Diego, Calif) 62(3):246–254. doi:10.1016/j.ymeth.2013.06.002
140. Wang W, Liu X, Gelinas D, Ciruna B, Sun Y (2007) A fully automated robotic system for microinjection of zebrafish embryos. PLoS One 2(9):e862. doi:10.1371/journal.pone.0000862

Zebrafish Rhabdomyosarcoma

Michael Phelps and Eleanor Chen

Abstract In vivo models of Rhabdomyosarcoma (RMS) have proven instrumental in understanding the development and progression of this devastating pediatric sarcoma. Both vertebrate and invertebrate model systems have been developed to study the tumor biology of both embryonal (ERMS) and alveolar (ARMS) RMS subtypes. Zebrafish RMS models have been particularly amenable for high-throughput studies to identify drug targetable pathways because of their short tumor latency, ease of ex vivo manipulation and conserved tumor biology. The transgenic KRASG12D-induced ERMS model allows for molecular and cellular characterization of distinct tumor cell subpopulations including the tumor propagating cells. Comparative genomic approaches have also been utilized in zebrafish ERMS to identify conserved candidate driver genes. Recent advances in zebrafish genome engineering have further enabled the ability to probe the functional significance of potential driver genes. Using the unique strengths of the zebrafish model organisms with the wealth of cellular and molecular tools currently available, zebrafish RMS models provide a powerful in vivo system for which to study RMS tumorigenesis.

Keywords Zebrafish • Rhabdomyosarcoma • Model organism • Cancer • Chemical genetics

Introduction

Rhabdomyosarcoma (RMS) is the most common soft tissue malignancy in the pediatric population and recapitulates many of the phenotypic and biological features of embryonic skeletal muscle. Based on histologic characterization, RMS falls into two major subtypes—embryonal and alveolar. Embryonal rhabdomyosarcoma (ERMS) is the most common subtype, accounting for approximately 60 % of all childhood RMS cases. Alveolar rhabdomyosarcoma accounts for approximately 25 % of all cases. The remaining cases of RMS are either unclassified or undifferentiated [1].

M. Phelps • E. Chen (✉)
Department of Pathology, University of Washington,
1959 N.E. Pacific St., Seattle, WA 98195, USA
e-mail: phelpsmk@uw.edu; eleanor2@uw.edu

D.M. Langenau (ed.), *Cancer and Zebrafish*, Advances in Experimental Medicine and Biology 910, DOI 10.1007/978-3-319-30654-4_16

Treatment for the disease includes surgical resection, chemotherapy or radiation. Clinical outcomes for ARMS patients are inferior in comparison to ERMS. However, the prognosis for patients with high-risk features or metastasis remains dismal regardless of subtype, with 5-year survival rates of 17 % and 30 % for recurrence and metastasis, respectively [2].

The majority of ARMS cases are characterized by a chromosomal translocation event that results in the fusion gene of either *PAX3-FOXOA1* or *PAX7-FOXOA1* [3]. In contrast to ARMS, ERMS shows a wide range of genetic features. ERMS has frequent allelic loss in chromosomal region 11p15, with this genetic interval harboring a number of imprinted genes implicated in oncogenesis, such as *H19*, *IGF2* and *CDKN1C* [4, 5]. Depending on the study, up to 35 % of ERMS cases harbor activating point mutations in *HRAS*, *KRAS*, or *NRAS* [6–8]. Mutations in the RAS/NF1 pathway are associated with intermediate- to high-risk rhabdomyosarcomas, indicating a more aggressive clinical course for the ERMS cases with activating RAS mutations [6]. Several human syndromes are associated with high frequency of RMS formation, including Li-Fraumeni (germline *TP53* mutations) [9], Costello (germline *HRAS* mutation) [10] and Gorlin syndrome (germline *PTCH1* mutations) [11], highlighting the crucial role of a number of tumor suppressor genes in regulating RMS initiation.

A variety of vertebrate and invertebrate genetic models have been developed to study the tumor biology of rhabdomyosarcoma. For example, Keller et al. created a ARMS transgenic mouse model by introducing a conditional knock-in allele, *Pax3:Fkhr*, in *Myf6*-expressing cells in skeletal muscle. Inactivating mutations in *Trp53* or *Cdkn2a* in this model were able to accelerate tumorigenesis [12]. A Drosophila transgenic model of ARMS was created by conditionally expressing human *PAX-FKHR* in myofibers [13]. In this model, nucleated cells that formed from syncytial myofibers showed invasive behavior with spread to the central nervous system, mimicking the malignant behavior of human ARMS cells. A series of conditional mouse ERMS models have similarly been created to assess the cell of origin for ERMS [14]. In these mice, ERMS can arise from *Myf5* or *Myf6*-expressing muscle stem cells and down-stream myogenic precursors, with tumor onset around 150–200 days of age. The penetrance was increased in the presence of *Trp53* homozygous inactivating mutations with or without conditional haploinsufficiency of *Ptch1*, indicating the important role of p53 and Sonic Hedgehog pathways in ERMS tumorigenesis. In the study by Hatley et al., a conditional transgenic model of ERMS was created by activating the Hedgehog pathway in adipocytes through a constitutively active *smoothened* allele, indicating that RMS can also arise from non-skeletal muscle cell lineages [15]. Most recently, a study by Tremblay et al. showed that conditional expression of a constitutively active form of YAP in *Myf5*-expressing skeletal muscle or activated satellite cells is sufficient to induce ERMS [16]. While mouse transgenic RMS models provide invaluable insights into pathogenesis of RMS, in particular the cell of origin, high-throughput in vivo and ex vivo studies to identify potential driver genes or targetable pathways are not feasible in these systems.

In contrast to other vertebrate disease model systems, zebrafish are amenable to large-scale genetic screens due to their fecundity, the ease of ex vivo manipulation from embryonic stage to adulthood, short latency for tumor development, and biological and pathologic characteristics similar to human malignancies and disease. In a zebrafish transgenic model of kRAS-induced rhabdomyosarcoma [17], the tumor

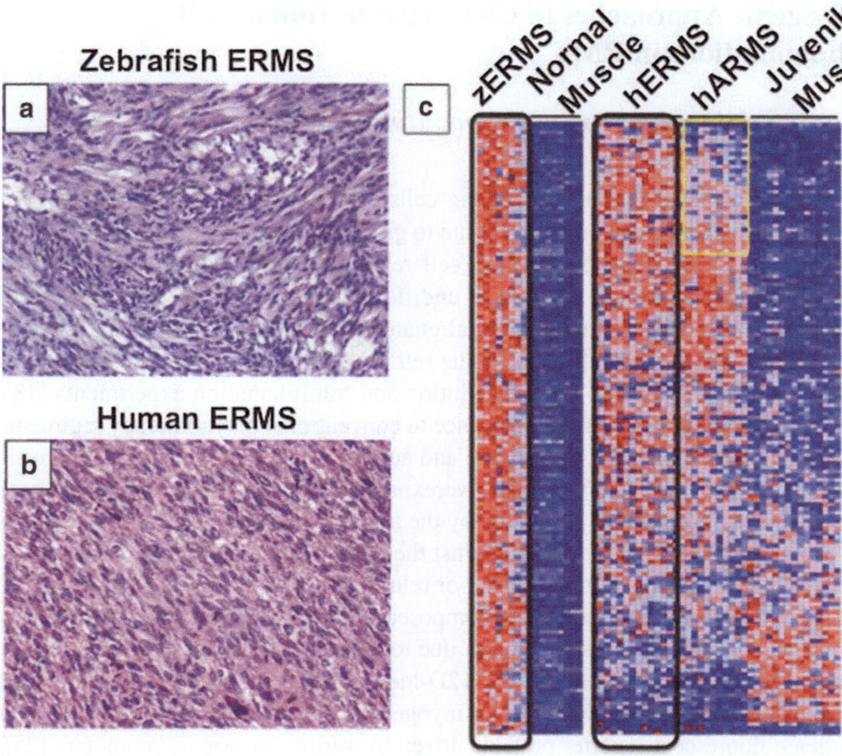

Fig. 1 Zebrafish ERMS is similar to human ERMS. (**a**, **b**) H&E images of zebrafish (**a**) and human (**b**) ERMS, showing similar morphologic features. (**c**) Heat map representation of microarray expression profiling studies in zebrafish ERMS, human ERMS, ARMS and normal skeletal muscle. Panel **C** is taken from the study by Langenau et al. [17]

exhibits morphologic features resembling the human counterpart, such as spindle cell morphology and occasional nests of primitive round blue cells (Fig. 1a, b). The presence of cross striations in a subset of tumor cells further supports that ERMS cells show incomplete myogenic differentiation. Zebrafish ERMS demonstrates locally aggressive behavior with the capacity to spread to adjacent organs. Up to 40 % of kRAS-induced zebrafish develop externally visible tumors by 40 days of life when injected at one-cell stage with a DNA expression construct containing a muscle-associated promoter driving expression of the human constitutively active kRASG12D oncogene [17]. The time point of tumor onset (7–10 days post-fertilization) is rapid compared to other vertebrate RMS models (e.g. 7–12 months in mouse models). Expression profiling and bioinformatics approaches have demonstrated conserved molecular signatures between zebrafish RMS and embryonal subtype of human RMS (Fig. 1c) [17], suggesting that the same molecular pathways are utilized in both human and fish to drive malignant transformation in ERMS.

This chapter will highlight studies that make use of the conserved and tractable features of zebrafish as a disease model system and as a discovery tool to reach important insights into the tumor biology of RMS and identify potential targetable pathways for novel treatment options.

Transgenic Approaches to Characterize Tumor Cell Subpopulations in RMS

Characterizing the Tumor-Propagating Population

Cancer stem cells or tumor-propagating cells (TPCs) are defined on the basis of their ability to propagate and differentiate to give rise to the heterogeneous progeny of the tumor. These cells are capable of self-renewal, which is the process by which stem cells divide to sustain the pool of undifferentiated cells. The TPCs have been isolated in a variety of hematologic malignancies and solid tumors such as breast and brain (glioblastoma multiforme). The self-renewal capacity of these tumors has been demonstrated through limiting dilution and transplantation experiments [18–21]. These cells showed marked resistance to conventional chemotherapy regimens, mediated by a combination of intrinsic and acquired properties such as low proliferative rate, efficient DNA repair, overexpression of multidrug-resistance-type membrane transporters and protection by the tumor microenvironment (niche) [22]. Development of targeted therapies against the TPCs could potentially improve survival for cancer patients with advanced or relapsed disease.

Embryonal rhabdomyosarcoma is composed of heterogeneous tumor cell populations at various stages of differentiation due to aberrant myogenesis [23, 24]. In the zebrafish transgenic model of kRASG12D-induced rhabdomyosarcoma, tumor cell subpopulations at early to late stages of myogenic differentiation can be labeled with transgenic fluorescent reporter proteins driven by various myogenic promoters [25]. By expression profiling, *myf5*-GFP+/*mylz*-mCherry negative-cells express activated satellite cell markers such as *myf5*, *cmet* and *m-cadherin*. By contrast, *myf5*-GFP+/*mylz2*- mCherry+and *myf5*-GFPnegative/*mylz2*-mCherry+cells express mid- and late-differentiation markers, respectively. To define the TPC population in ERMS, each tumor cell subpopulation was isolated by Fluorescence Activated Cell Sorting (FACS) and subjected to limiting dilution and serial tumor cell transplantation analysis in syngeneic (or genetically identical) *myf5*-GFP, *mylz2*-mCherry transgenic zebrafish (Fig. 2) [25]. By limiting dilution analysis (transplanting 1000, 100 and 10 sorted

Fig. 2 (continued) *Green*: myf5-GFP+/mylz2-mCherry-negative cells; *yellow*: myf5-GFP+/mylz2-mCherry+cell and *red*: myf5-GFP-negative/mylz-mCherry+cells. For limiting dilution, 1000, 100 and 10 cells are transplanted into syngeneic recipients (at least n=6–10 per group). Engraftment frequency is subsequently monitored up to 60 days post-transplantation. For serial transplantation, tumor cell subpopulations can be sorted by FACS and transplanted into syngeneic recipient fish. The process can be reiterated. (**b–g**) myf5-GFP+/mylz2-mCherry-negative tumor cells show the characteristic feature of TPCs by their capacity to selfrenew in serial transplantation experiments. (**b**, **c**) A primary ERMS arising in syngeneic myf5-GFP/mylz2-mCherry transgenic zebrafish (35 dpf). (**d**, **e**) Engrafted fluorescent-labeled ERMS in a syngeneic secondary recipient animal when transplanted with unsorted primary ERMS cells. (**f**, **g**) myf5-GFP+/mylz2-mCherry-negative cells isolated from a secondary-transplant tumor induce ERMS in tertiary transplant animals. Panels **b–g** are taken from the study by Ignatius et al. [25]

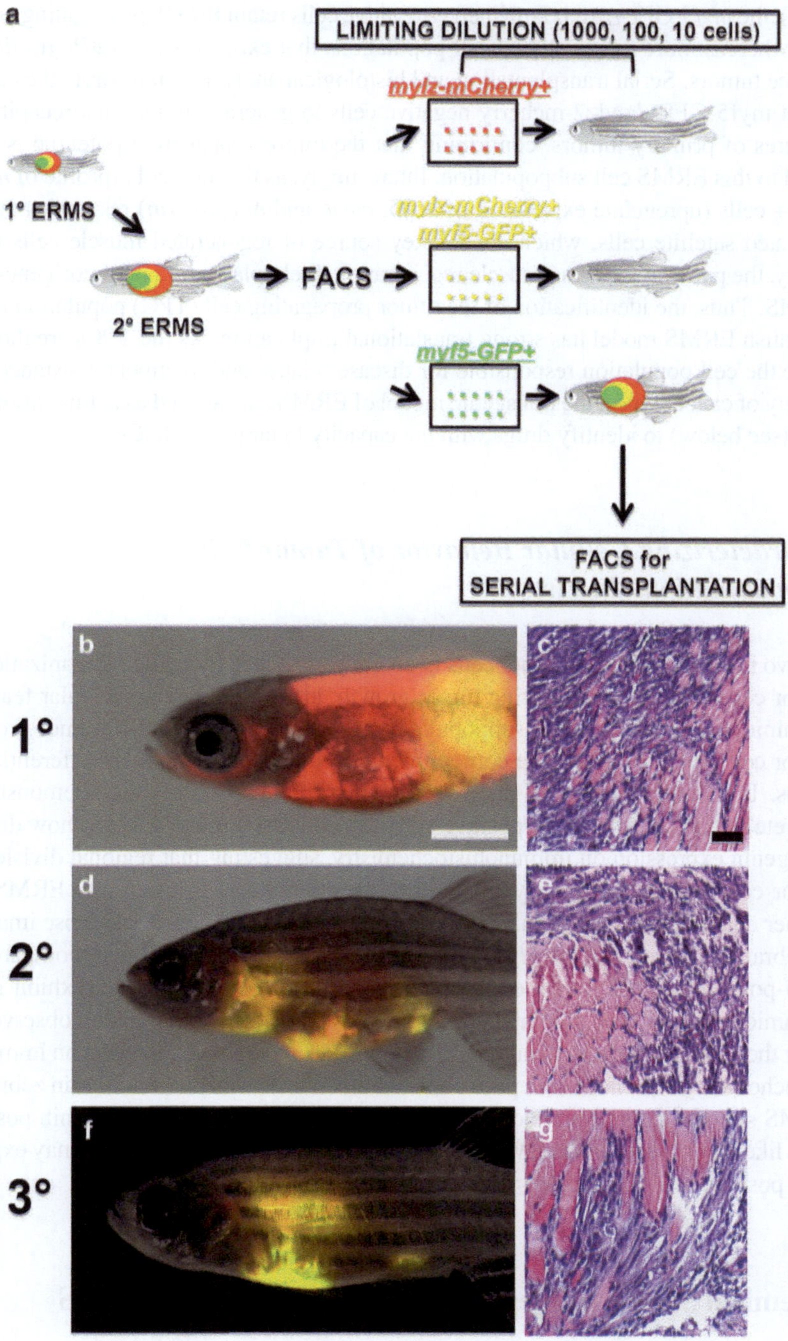

Fig. 2 Validation of the TPCs in ERMS by serial and limiting dilution transplantation. (**a**) Schematic of serial and limiting dilution transplantation experiments to assess self-renewal capacity of tumor cells. Primary tumor cells are first expanded in secondary syngeneic recipient fish. The tumor is subjected to FACS sorting to isolate distinct tumor cell subpopulations based on fluorescence.

cells), the *myf5*-GFP+/mylz2-mCherry-negative cells retain ERMS propagating potential, whereas more differentiated cell populations that express *mylz2*-mCherry do not induce tumors. Serial transplantation and histological analysis demonstrate the capacity of myf5-GFP+/mylz2-mcherry negative cells to generate tumors that recapitulate features of primary tumors, confirming that the tumor-propagating potential is confined to this ERMS cell subpopulation. Interestingly, as the molecular profile of *myf5*-GFP+ cells (upregulate expression of *myf5*, *cmet,* and *m-cadherin*) resembles that of activated satellite cells, which are the key source of regenerated muscle cells upon injury, the pathways driving muscle regeneration likely play a role in tumorigenesis of ERMS. Thus, the identification of the tumor propagating cell (TPC) population in the zebrafish ERMS model has strong translational implication. As the TPCs are thought to be the cell population responsible for disease relapse and treatment resistance in a variety of cancer types, the transgenic model of ERMS can be used as a drug discovery tool (see below) to identify drugs with the capacity to target the TPCs.

Characterizing Cellular Behavior of Tumor Cell Subpopulations In Vivo

In vivo imaging of ERMS tumor cells in zebrafish revealed dynamic reorganization of tumor cell subpopulations during tumor growth, identifying distinct cellular features pertaining to each tumor cell subpopulation. During zebrafish ERMS tumor growth, tumor cells eventually become compartmentalized depending on their differentiation status. Immunohistochemical analysis of human ERMS tumors also demonstrates discrete regions of *Myogenin*-positive cells. By contrast, human ARMS show diffuse myogenin expression on immunohistochemistry, suggesting that regional division of tumor cells based on their myogenic differentiation status is specific to ERMS. To further characterize the regional behaviors of ERMS tumor cells, time-lapse imaging in zebrafish showed that the *Myogenin*-positive cells are more migratory compared to *myf5*-positive cells as they seed the tumor front during tumor growth, exhibit more dynamic local movement compared to *myf5*-positive cells and have been observed to enter the vasculature [25]. In human ERMS, >80 % myogenin expression on immunohistochemistry correlates with poor clinical outcome. The in vivo findings in zebrafish ERMS suggest that the more dynamic and migratory behavior of myogenin positive cells likely contribute to the invasive behavior of ERMS tumor cells. This may explain why positive myogenin expression is a poor clinical prognostic indicator.

Chemical Screen to Identify Druggable Targets in ERMS

Zebrafish has become a powerful tool for drug and pathway discovery due to the high conservation of key molecular pathways and tractable features for large-scale ex vivo manipulations. Three independent chemical screens have been performed in zebrafish to identify key pathways regulating tumorigenesis of ERMS. Work by Le et al. identified common RAS-driven pathways in embryogenesis and rhabdomyosarcoma development by assessing the effect of small molecules on the change in expression of Ras downstream target genes [26]. In this study embryos derived from a transgenic line harboring a heat-shock promoter-driven HRASG12V were subjected to heat shock and treated with a library of small molecules. Expression change of *dusp6*, a downstream target gene, was assessed using in situ hybridization. The tumor suppressive effect of lead compounds from the embryonic screen, TPCK and PD98059, which are inhibitors of S6K1 and MEK, respectively, was demonstrated in the zebrafish model of kRAS-induced ERMS. The lead compounds also inhibited growth of human ERMS cell lines in vitro, further demonstrating their therapeutic potential in treatment of human ERMS.

ERMS is characterized by an arrest in myogenic differentiation and uncontrolled proliferation [23, 24]. Work by Chen et al. described a large-scale chemical screen to identify drugs that induce differentiation of ERMS cells in vitro as well as suppress tumor growth in vivo [27]. In this study, an initial screen of approximately 40,000 compounds was performed in human ERMS RD cell line to identify lead compounds that can induce terminal myogenic differentiation of ERMS cells. A secondary large-scale transplant screen was performed in zebrafish to assess the effect of the 95 lead compounds on tumor growth in vivo. This was performed by transplantation of fluorescent-labeled tumor cells into 5–6 week-old syngeneic recipient fish. Engrafted tumor fish were treated with each compound in 6-well plates for 7 days including a treatment holiday. Tumor volume change was determined by quantitative imaging analysis (Fig. 3). One class of lead compounds is GSK inhibitors, which are known to inhibit the activity of GSK3, resulting in stabilization of β-catenin and activation of the canonical WNT pathway. Treatment of ERMS-bearing zebrafish with GSK3 inhibitors resulted in suppressed tumor growth through activation of the WNT/β-catenin pathway. Analysis of tumor cell subpopulations by quantitative FACS revealed depletion of the TPCs and expansion of the more differentiated tumor cells. GSK3 inhibitors also reduced zebrafish ERMS tumor self-renewal capacity in vivo. To demonstrate the conservation of Wnt/β--catenin pathway in modulating ERMS self-renewal, the same study showed that activation of the canonical WNT/β-catenin pathway also significantly reduced self-renewal capacity of human ERMS cells in vitro. As activation of canonical WNT pathway tends to play oncogenic roles in many cancers such as colorectal cancer [28, 29], this study identified an unconventional tumor suppressive role for the canonical WNT/β-catenin pathway in regulating differentiation and self-renewal of ERMS. In addition, as the WNT/β-catenin pathway also plays an essential role in embryonic myogenesis [30], this study also implicates de-regulation of embryonic pathways in ERMS.

Epigenetic changes, inheritable alterations in gene expression without changes in DNA sequence, play an essential role in driving a variety of cancer events. Little is known about chromatin-modifying factors in RMS. Several studies have implicated the role of Polycomb group and SWI/SNF chromatin modifying complex [31, 32]. The study by Albacker et al used an injection-based overexpression screen in the

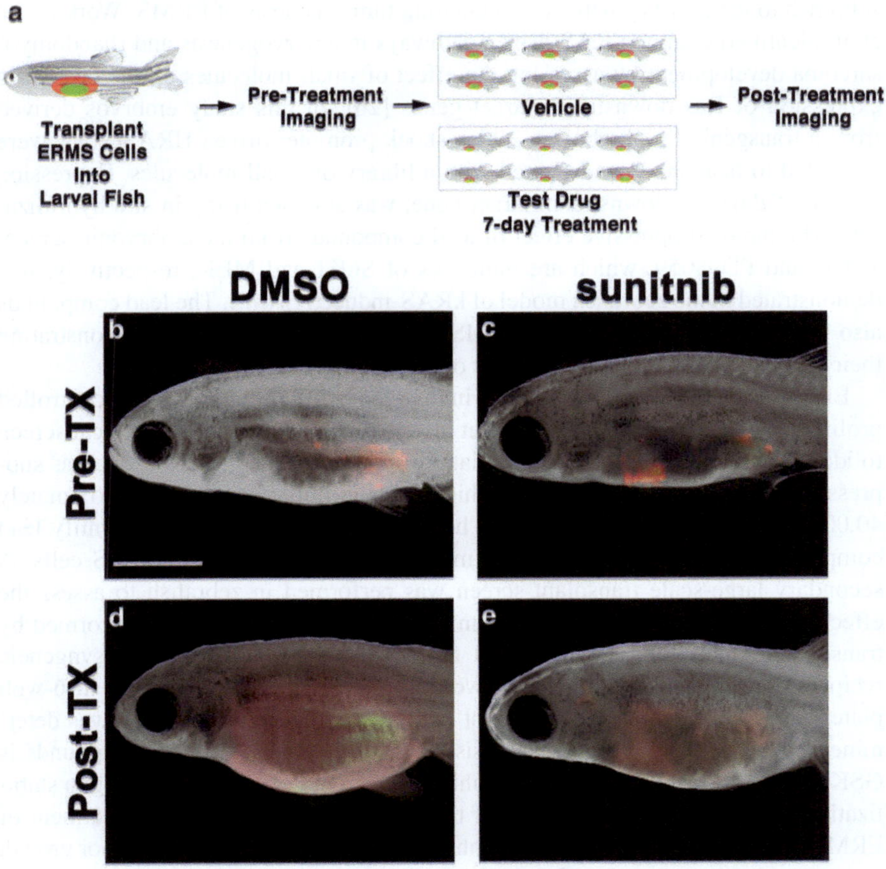

Fig. 3 Use of a chemical screen to identify lead compounds that suppress ERMS tumor growth in live zebrafish. (**a**) Schematic of a chemical screen completed in zebrafish transplanted with fluorescent-labeled ERMS. Pretreatment images for vehicle control, DMSO (**b**) and a lead compound, sunitinib (**c**), with corresponding post-treatment images (**d** and **e**, respectively). Scale bar equals 2 mm. The figure is adapted from the study by Chen et al. [27]

kRAS-driven ERMS zebrafish model to assess the role of 19 chromatin-modifying factors in RMS tumor formation [33]. In the study, 3 DNA constructs including *rag2*-kRASG12D, a chromatin-modifying gene and a fluorescent reporter driven by a muscle-specific promoter were co-injected into zebrafish embryos at 1-cell stage. The expression of a fluorescent protein enabled monitoring of tumor formation in vivo. Construction of tumor-free survival curves revealed chromatin-modifying genes that either enhance or suppress RMS tumor formation compared to control fish. From the study, SU39H1, a histone 3-lysine 9 methyl transferase, was identified as a potent tumor suppressor of RMS in tumor initiation. The study demonstrates the feasibility of targeting epigenetic modifiers in ERMS as a potential therapeutic option. As an epigenetic change can induce global change in genetic expression, downstream genes and pathways will also be potential therapeutic targets.

Comparative Genomics to Identify Conserved Genetics Drivers in ERMS

Recurrent genomic amplifications and deletions are key clinical prognostic indicators in a significant subset of cancers and frequently result in aberrant activation of oncogenic events. The technology of array comparative genomic hybridization (aCGH) allows for global assessment of copy number aberrations (CNAs) throughout cancer genomes [34]. RMS, similar to many cancers, is characterized by inconsistent genomic gains and losses, including whole or partial chromosome arms [35–39], the major challenge is to distinguish driver genes from passenger events (Table 1).

Array CGH technology has been demonstrated to be a useful tool for identifying conserved genomic regions within CNAs in zebrafish tumor models. The initial BAC-based aCGH platform with 286 BAC clones, enriched for orthologous sequences of human oncogenes and tumor suppressors, revealed significant genomic imbalances in zebrafish models of melanoma, T-cell lymphoblastic leukemia (T-ALL) and RMS [40]. The zebrafish activated BRAF-induced melanoma model, showed a higher number of genomic imbalances including gains and losses per individual (6–28) compared to the Myc-induced T-ALL and Ras-induced RMS zebrafish models. However, the low-resolution of this platform precludes identification of candidate driver genes. High-resolution aCGH platforms were subsequently applied to zebrafish models of Malignant Nerve Sheath Tumor (MPNST) and T-ALL to identify conserved genomic regions that are also amplified or deleted in human disease [41, 42]. In the aCGH study of the kRASG12D-induced RMS zebrafish model [43], Cy5 labeled-DNA from 20 independent primary tumors was hybridized against Cy3 labeled-DNA of matched normal skeletal muscle onto a custom 400 k CGH microarray Agilent platform (Fig. 4a). The results showed predominantly CNA gains compared to losses with 19 recurrent gene-containing regional gains and 2 losses in at least 3 zebrafish ERMS tumors (Fig. 4b–d). The recurrent CNA gains are mapped to 21 homologous regions in the human genome with 18 of these regions also showing gains in human ERMS. As a proof of principle, 6 candidate driver genes identified from the genic regions were assessed for their oncogenic roles in human ERMS cell lines. Four of the genes, CCND2, HOXC6, PLXNA1 and VEGF, show conserved function in the pathogenesis of ERMS. Aberrant upregulation of VEGF expression also correlates with poor clinical outcome in human ERMS patients. This study demonstrates that the small and conserved CNAs in zebrafish ERMS genomes allow easy identification of novel driver genes essential for pathogenesis of ERMS.

Table 1 Animal models of rhabdomyosarcoma

Species	Approach	Histologic subtype	Tumor onset	References
Zebrafish	Mosaic Transgenic Approach (*rag2-kRASG12D* injection at 1-cell stage) in wild-type or p53 -/-background	ERMS	10 days post-fertilization (5 days post-fertilization in p53-/-background)	[17]
Zebrafish	Heat Shock-inducible Cre-Lox approach (*beta-actin*-LoxP-EGFP-pA-LoxP-*kRASG12D* line x *hsp70*-Cre line)	ERMS	Heat shocked: Average 35 days Non-heat shocked: Average 65 days	[72]
Zebrafish	Tol2 gene trap system with transgene expressing constitutively active human H-RASV12	Costello syndrome (including ERMS formation)	60–365 days	[73]
Mouse	Conditional Pax3:Fkhr knock-in/Myf6-specific Cre driver(Pax3:Fkhr heterozygous lines)	ARMS	383 days	[12]
	Pax3:Fkhr homozygous lines			
	x Trp53 -/+	→	202 days	
	x Trp 53 -/-	→	75–91 days	
	x Ink4a/Arf -/-	→	56–89 days	
Mouse	HGF/SF-transgenic x *Ink4a/Arf-/-*	ERMS	~90 days	[74]
Mouse	Temporally controlled expression of an activated Smoothened allele (*R26- SmoM2*) x Ubiquitous Cre transgene (CAGGS-CreER) induced by tamoxifen	ERMS	35 days	[75]
Mouse	Double P53 and FOS knock-out mice (*Trp53 -/-*; *Fos -/-*)	ERMS	10 weeks	[76]
Mouse	Transgenic *HER-2/neu* expression with *P53 -/-* background	ERMS	11–21 weeks	[77]
Mouse	mdx mice (model of Duchenne muscular dystrophy) with dystrophin deficiency with or without *P53* deficiency	ARMS and ERMS	16.5–25 months; 5 months in *P53*- deficient mdx mice post-cardiotoxin-induced injury	[78–80]

Mouse	mdx mice (model of Duchenne muscular dystrophy) with dystrophin deficiency with or without *P53* deficiency	ARMS and ERMS	16.5–25 months; 5 months in *P53*- deficient mdx mice post-cardiotoxin-induced injury	[78–80]
Mouse	a-Sarcoglycan-deficient mice (Knock out of *Sgca*)	ERMS	~1 year	[80]
Mouse	Lineage-specific homozygous deletion of P53 with or without concurrent heterozygous *Patched1* (*PTCH1*) deletion *Myf6* or *Myf5* or *Pax7* Cre x *Ptch1* -/- (; *P53* -/-)	ERMS	4–5 months	[14]
Mouse	Adipocyte-restricted activation of Sonic hedgehog pathway using constitutively activated *Smoothened* allele ROSA26-loxSTOPlox-*SmoM2* x *aP2*-Cre	ERMS	2 months	[15]
Mouse	Conditional activation of constitutively active human YAP1(S127A) in activated satellite cells post-cardiotoxin induced muscle injury (*Pax7* Cre ERT2 x R26-LSL-rtTA TetO-h*YAP1S127A*)	ERMS	4–8 weeks post doxycycline induction	[16]
	or in skeletal muscle (*Myf5* or *Myod1* Cre x R26-LSL-rtTA TetO-h*YAP1S127A*)	ERMS	15 weeks post doxycycline induction	
Drosophila	Conditional transgenic lines expressing human PAX7-FKHR in muscle Myosin heavy chain-Gal4 x UAS-PAX3 (or PAX7)-FKHR	ARMS	2 days (larval stage)	[13]

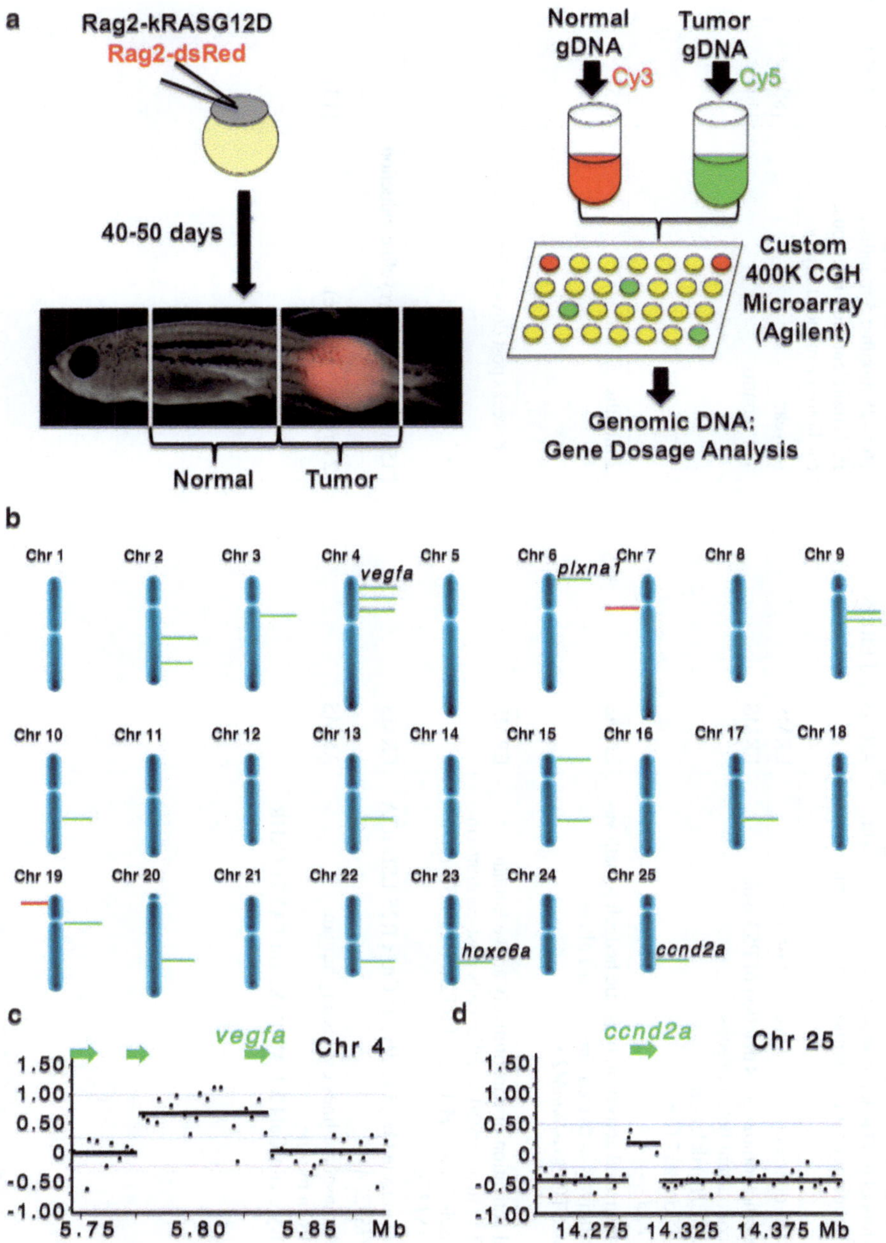

Fig. 4 Array Comparative Genomic Hybridization (aCGH) reveals cancerspecific chromosomal abnormalities in zebrafish ERMS. (**a**) Schematic of processing ERMS tumors for aCGH. Primary zebrafish tumors are established by microinjecting a human kRASG12D expression construct and a fluorescent reporter construct, which facilitates tracking of tumor cells in vivo. Tumors and matched normal skeletal muscle are isolated at day 40–50 post-injection. To perform aCGH,

Application of Genome Engineering Technology to Characterize the Function of Driver Genes in RMS

The ability to study the function of key tumor specific genes is rapidly improving with recent advancements in genome engineering technology. Genome engineering using DNA sequence specific endonucleases such as zinc finger nucleases (ZFNs), transcription activator-like effector nucleases (TALENs) and Cas9 enable direct modification of endogenous loci, by introducing DNA double strand breaks (DSBs) in target regions of the genome. DNA double strand breaks can then be repaired through non-homologous end joining (NHEJ) or through homology directed repair. Non-homologous end joining can be used to knockout endogenous genes since the error-prone repair process often lead to insertions or deletions (indels). These indel mutations can introduce frameshifts in the coding sequence which result in premature truncation of the protein during translation [44]. If a homologous template is supplied during DNA repair, directed genome modifications, such as transgene insertions or gene corrections, can also be produced by homology directed repair [45].

Zinc finger and TALEN nucleases consist of a fusion between programmable DNA binding proteins (zinc finger and TALE, respectively) and the Fok1 nuclease. DNA DSBs are targeted to specific genome locations by modifying the DNA binding domain of each zinc finger of TALE DNA binding protein. Since the Fok1 nuclease functions as a dimer, both ZFNs and TALENs must be designed in pairs in order to effectively cleave double stranded DNA. This paired nuclease requirement enhances target sight specificity but also limits the flexibility to target multiple genome locations. Several studies have demonstrated the successful use of the ZFNs and the TALENs to target driver genes essential for various aspects of tumor progression in mouse tumor models or human cancer cell lines [46–52]. These tools have also been used to model large scale chromosomal rearrangements and deletions [49, 50]. Despite the use of ZFNs and TALENs in mammalian cancer systems their use in zebrafish cancer models is limited [53, 54].

Fig. 4 (continued) Cy5-labeled genomic DNA (gDNA) isolated from a tumor is hybridized to Cy3-labeled genomic DNA isolated from matched normal muscle onto a custom-made 400 K zebrafish CGH platform (Agilent). Array image scans are extracted and imported into a copy number analysis software program (e.g. Nexus) to determine copy number aberrations (CNAs). (**b**) Summary of common gene-containing CNA gains (*green*) and losses (*red*) in 20 animals examined. Only recurrent CNAs found in ≥ 3 samples are shown. The height of each bar correlates with the frequency of each aberration. Detailed view of regional gains for *vegfa* on chromosome 4 (**c**), *ccnd2a* on chromosome 25 (**d**). Panels **b–d** are taken from the study by Chen et al. [43]

In contrast to ZFNs and TALENs, the recently developed CRISPR/Cas9 genome engineering technology has several key advantages, particularly in the ability of rapidly design and clone targeting constructs. The strength of the CRISPR/Cas9 system lies in its use of small, easily produced RNA templates (guide RNAs; gRNA) that guide targeting of the Cas9 nuclease (for review; [55, 56]). Since Cas9 is able to cleave both strands of DNA with high efficiency only one protein is needed to create targeted DNA DSBs. This flexibility enables multiplex genome engineering by expressing one Cas9 nuclease and multiple gene-targeting gRNAs [57, 58]. CRISPR/Cas9 technology has been used to produce mouse cancer models of lung adenocarcinoma [59, 60], non-small cell lung cancers [61], and liver tumors [62] through tumor suppressor knockout or homology directed oncogene engineering. While the use of the CRISPR/Cas9 technology to study gene functions in zebrafish cancer models is only beginning to be explored, it is rapidly becoming the method of choice for genome engineering studies in zebrafish [63–69].

The KRASG12D-induced zebrafish model of ERMS is a particularly powerful system for which to employ CRISPR/Cas9 to study gene function in cancer, due to its rapid tumor onset. To test the feasibility of assessing function of a modifying gene in the KRASG12D model of ERMS, the tumor suppressor, p53 (*tp53*) was targeted using the CRISPR/Cas9 system (unpublished data). Briefly, a single gRNA was designed against a conserved exon of *tp53* near the 5′ end of the coding sequence (Zifit; [70]). To target *tp53* in the zebrafish ERMS model, single cell stage embryos were microinjected with 3 muscle-specific (r*ag2* promoter) constructs expressing (1) the Cas9 nuclease and *tp53*-targeting gRNA, (2) the KRASG12D oncogene and (3) a fluorescent reporter. For comparison, a separate cohort of embryos is co-injected with a Cas9 expression construct lacking a *tp53* targeting gRNA. Both cohorts of fish are monitored for tumor development every 3–4 days for a total of 30 days (starting approximately 10 days post-fertilization). Validation of successful *tp53* targeting in zebrafish ERMS tumors is determined by a T7-endonuclease assay, which detects the presence of indel mutations [56, 71]. We have observed a shorter tumor latency and more aggressive tumor growth in the *tp53* gRNA-injected cohort (unpublished data, see Fig. 5 for experimental schematic). These results suggest the feasibility of targeting tumor suppressor genes in ERMS using the CRISPR/Cas9 technology. This technology could likely be applied to other zebrafish cancer models by modifying the tissue specificity of Cas9-mediated gene targeting.

Discussion and Future Directions

The zebrafish KRAS-induced model of ERMS shows high conservation to the human disease. The short tumor latency and ease of ex vivo manipulations allows for high-throughput chemical genetic screens to identify novel and essential druggable pathways. An aCGH study in ERMS has also identified candidate driver genes that may play a crucial role in ERMS tumorigenesis. The recent application of genome engineering technology, in particular the CRISPR-Cas9 technology, represents an unprecedented approach to study endogenous gene function in zebrafish

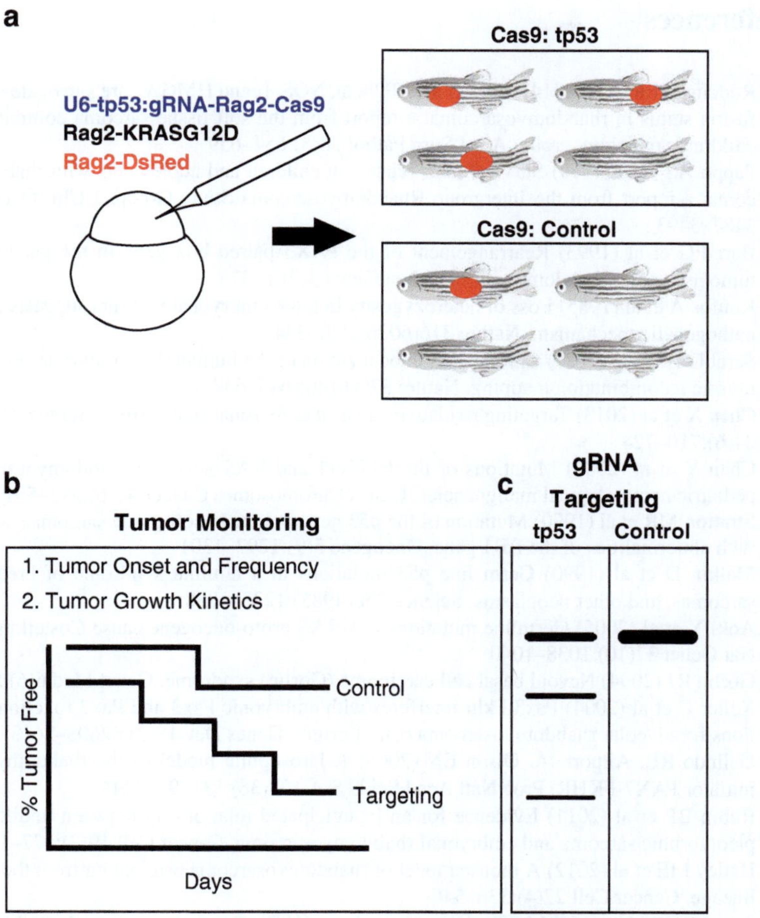

Fig. 5 Utility of CRISPR-Cas9 genome engineering technology to study tumor suppressor gene function. (**a**) Schematic of a microinjection experiment to produce primary tumor-bearing zebrafish with or without *tp53* gene targeting. A DNA cocktail containing a U6 promoter-driven tp53 gRNA cassette and a rag2 promoter-driven Cas9 cassette, rag2-KRASG12D and rag2-dsRed (fluorescent reporter) is injected into zebrafish embryos at 1-cell stage. For comparison, a separate cohort of embryos is injected with a similar cocktail containing a nontargeting Cas9 control vector. (**b**) Injected fish are monitored for tumor onset, frequency and growth kinetics. An example of Kaplan-Meier analysis assessing tumor-free survival of injected fish. Targeting of a tumor suppressor gene, e.g. tp53, is expected to decrease tumor-free survival. (**c**) An example of a T7 endonuclease assay to assess the presence of Cas9-mediated tp53 targeting. PCR products from amplification of targeted genomic region are digested with T7 endonuclease. The DNA heteroduplex containing Cas9-induced indel mutations will be cleaved into two bands as visualized on an agarose gel or bioanalyzer

cancer models such as that of ERMS. Using current genome engineering technology to understand the wealth of data obtained from the chemical genetic screens and aCGH experiments in zebrafish ERMS models hold promise for creating new translational therapies for treatment of RMS cancer.

References

1. Rudzinski ER et al (2014) Myogenin, AP2beta, NOS-1, and HMGA2 are surrogate markers of fusion status in rhabdomyosarcoma: a report from the soft tissue sarcoma committee of the children's oncology group. Am J Surg Pathol 38(5):654–659
2. Pappo AS et al (1999) Survival after relapse in children and adolescents with rhabdomyosarcoma: a report from the Intergroup Rhabdomyosarcoma Study Group. J Clin Oncol 17(11): 3487–3493
3. Barr FG et al (1993) Rearrangement of the PAX3 paired box gene in the paediatric solid tumour alveolar rhabdomyosarcoma. Nat Genet 3(2):1137
4. Koufos A et al (1985) Loss of heterozygosity in three embryonal tumours suggests a common pathogenetic mechanism. Nature 316(6026):330–334
5. Scrable HJ et al (1987) Chromosomal localization of the human rhabdomyosarcoma locus by mitotic recombination mapping. Nature 329(6140):645–647
6. Chen X et al (2013) Targeting oxidative stress in embryonal rhabdomyosarcoma. Cancer Cell 24(6):710–724
7. Chen Y et al (2006) Mutations of the PTPN11 and RAS genes in rhabdomyosarcoma and pediatric hematological malignancies. Genes Chromosomes Cancer 45(6):583–591
8. Stratton MR et al (1990) Mutation of the p53 gene in human soft tissue sarcomas: association with abnormalities of the RB1 gene. Oncogene 5(9):1297–1301
9. Malkin D et al (1990) Germ line p53 mutations in a familial syndrome of breast cancer, sarcomas, and other neoplasms. Science 250(4985):1233–1238
10. Aoki Y et al (2005) Germline mutations in HRAS proto-oncogene cause Costello syndrome. Nat Genet 37(10):1038–1040
11. Gorlin RJ (2004) Nevoid basal cell carcinoma (Gorlin) syndrome. Genet Med 6(6):530–539
12. Keller C et al (2004) Pax3:Fkhr interferes with embryonic Pax3 and Pax7 function: implications for alveolar rhabdomyosarcoma cell of origin. Genes Dev 18(21):2608–2613
13. Galindo RL, Allport JA, Olson EN (2006) A Drosophila model of the rhabdomyosarcoma initiator PAX7-FKHR. Proc Natl Acad Sci U S A 103(36):13439–13444
14. Rubin BP et al (2011) Evidence for an unanticipated relationship between undifferentiated pleomorphic sarcoma and embryonal rhabdomyosarcoma. Cancer Cell 19(2):177–191
15. Hatley ME et al (2012) A mouse model of rhabdomyosarcoma originating from the adipocyte lineage. Cancer Cell 22(4):536–546
16. Tremblay AM et al (2014) The Hippo transducer YAP1 transforms activated satellite cells and is a potent effector of embryonal rhabdomyosarcoma formation. Cancer Cell 26(2):273–287
17. Langenau DM et al (2007) Effects of RAS on the genesis of embryonal rhabdomyosarcoma. Genes Dev 21(11):1382–1395
18. Al-Hajj M et al (2003) Prospective identification of tumorigenic breast cancer cells. Proc Natl Acad Sci U S A 100(7):3983–3988
19. Hope KJ, Jin L, Dick JE (2004) Acute myeloid leukemia originates from a hierarchy of leukemic stem cell classes that differ in self-renewal capacity. Nat Immunol 5(7):738–743
20. Lapidot T et al (1994) A cell initiating human acute myeloid leukaemia after transplantation into SCID mice. Nature 367(6464):645–648
21. Singh SK et al (2003) Identification of a cancer stem cell in human brain tumors. Cancer Res 63(18):5821–5828
22. Trumpp A, Wiestler OD (2008) Mechanisms of disease: cancer stem cells-targeting the evil twin. Nat Clin Pract Oncol 5(6):337–347
23. Tapscott SJ, Thayer MJ, Weintraub H (1993) Deficiency in rhabdomyosarcomas of a factor required for MyoD activity and myogenesis. Science 259(5100):1450–1453
24. Yang Z et al (2009) MyoD and E-protein heterodimers switch rhabdomyosarcoma cells from an arrested myoblast phase to a differentiated state. Genes Dev 23(6):694–707
25. Ignatius MS et al (2012) In vivo imaging of tumor-propagating cells, regional tumor heterogeneity, and dynamic cell movements in embryonal rhabdomyosarcoma. Cancer Cell 21(5):680–693

26. Le X et al (2013) A novel chemical screening strategy in zebrafish identifies common pathways in embryogenesis and rhabdomyosarcoma development. Development 140(11):2354–2364
27. Chen EY et al (2014) Glycogen synthase kinase 3 inhibitors induce the canonical WNT/beta-catenin pathway to suppress growth and self-renewal in embryonal rhabdomyosarcoma. Proc Natl Acad Sci U S A 111(14):5349–5354
28. Anastas JN, Moon RT (2013) WNT signalling pathways as therapeutic targets in cancer. Nat Rev Cancer 13(1):11–26
29. Clevers H, Nusse R (2012) Wnt/beta-catenin signaling and disease. Cell 149(6):1192–1205
30. Tajbakhsh S et al (1998) Differential activation of Myf5 and MyoD by different Wnts in explants of mouse paraxial mesoderm and the later activation of myogenesis in the absence of Myf5. Development 125(21):4155–4162
31. DeCristofaro MF et al (1999) Alteration of hSNF5/INI1/BAF47 detected in rhabdoid cell lines and primary rhabdomyosarcomas but not Wilms' tumors. Oncogene 18(52):7559–7565
32. Wang H et al (2008) NF-kappaB-YY1-miR-29 regulatory circuitry in skeletal myogenesis and rhabdomyosarcoma. Cancer Cell 14(5):369–381
33. Albacker CE et al (2013) The histone methyltransferase SUV39H1 suppresses embryonal rhabdomyosarcoma formation in zebrafish. PLoS One 8(5), e64969
34. Pinkel D, Albertson DG (2005) Array comparative genomic hybridization and its applications in cancer. Nat Genet 37(Suppl):S11–S17
35. Bridge JA et al (2002) Genomic gains and losses are similar in genetic and histologic subsets of rhabdomyosarcoma, whereas amplification predominates in embryonal with anaplasia and alveolar subtypes. Genes Chromosomes Cancer 33(3):310–321
36. Goldstein M et al (2006) Novel genes implicated in embryonal, alveolar, and pleomorphic rhabdomyosarcoma: a cytogenetic and molecular analysis of primary tumors. Neoplasia 8(5):332–343
37. Missiaglia E et al (2009) Genomic imbalances in rhabdomyosarcoma cell lines affect expression of genes frequently altered in primary tumors: an approach to identify candidate genes involved in tumor development. Genes Chromosomes Cancer 48(6):455–467
38. Pandita A et al (1999) Application of comparative genomic hybridization, spectral karyotyping, and microarray analysis in the identification of subtype-specific patterns of genomic changes in rhabdomyosarcoma. Neoplasia 1(3):262–275
39. Paulson V et al (2011) High-resolution array CGH identifies common mechanisms that drive embryonal rhabdomyosarcoma pathogenesis. Genes Chromosomes Cancer 50(6):397–408
40. Freeman JL et al (2009) Construction and application of a zebrafish array comparative genomic hybridization platform. Genes Chromosomes Cancer 48(2):155–170
41. Rudner LA et al (2011) Shared acquired genomic changes in zebrafish and human T-ALL. Oncogene 30(41):4289–4296
42. Zhang G et al (2013) Comparative oncogenomic analysis of copy number alterations in human and zebrafish tumors enables cancer driver discovery. PLoS Genet 9(8), e1003734
43. Chen EY et al (2013) Cross-species array comparative genomic hybridization identifies novel oncogenic events in zebrafish and human embryonal rhabdomyosarcoma. PLoS Genet 9(8), e1003727
44. Lieber MR et al (2003) Mechanism and regulation of human non–homologous DNA end-joining. Nat Rev Mol Cell Biol 4(9):712–720
45. Vasquez KM et al (2001) Manipulating the mammalian genome by homologous recombination. Proc Natl Acad Sci 98(15):8403–8410
46. Beane JD et al (2014) Clinical scale zinc finger nuclease (ZFN)-driven gene-editing of PD-1 in tumor infiltrating lymphocytes (TIL) for the potential treatment of metastatic melanoma. J Immunother Cancer 2(Suppl 3):P2
47. Ding Y et al (2014) TALEN-mediated Nanog disruption results in less invasiveness, more chemosensitivity and reversal of EMT in Hela cells. Oncotarget 5(18):8393–8401
48. Gutschner T et al (2013) The noncoding RNA MALAT1 is a critical regulator of the metastasis phenotype of lung cancer cells. Cancer Res 73(3):1180–1189
49. Nyquist MD et al (2013) TALEN-engineered AR gene rearrangements reveal endocrine uncoupling of androgen receptor in prostate cancer. Proc Natl Acad Sci 110(43):17492–17497

50. Piganeau M et al (2013) Cancer translocations in human cells induced by zinc finger and TALE nucleases. Genome Res 23(7):1182–1193
51. Wu X et al (2014) TALEN-mediated genetic tailoring as a tool to analyze the function of acquired mutations in multiple myeloma cells. Blood Cancer J 4(5):e210
52. Zhang S et al (2014) TALEN-mediated somatic mutagenesis in murine models of cancer. Cancer Res 74(18):5311–5321
53. White R, Rose K, Zon L (2013) Zebrafish cancer: the state of the art and the path forward. Nat Rev Cancer 13(9):624–636
54. Yen J, White RM, Stemple DL (2014) Zebrafish models of cancer: progress and future challenges. Curr Opin Genet Dev 24:38–45
55. Blackburn PR et al (2013) The CRISPR system—keeping zebrafish gene targeting fresh. Zebrafish 10(1):116–118
56. Ran FA et al (2013) Genome engineering using the CRISPR–Cas9 system. Nat Protoc 8(11):2281–2308
57. Jao L-E, Wente SR, Chen W (2013) Efficient multiplex biallelic zebrafish genome editing using a CRISPR nuclease system. Proc Natl Acad Sci 110(34):13904–13909
58. Ota S et al (2014) Multiple genome modifications by the CRISPR/Cas9 system in zebrafish. Genes Cells 19(7):555–564
59. Platt RJ et al (2014) CRISPR-Cas9 knockin mice for genome editing and cancer modeling. Cell 159:440–455
60. Sánchez-Rivera FJ et al (2014) Rapid modelling of cooperating genetic events in cancer through somatic genome editing. Nature 516(7531):428–431
61. Maddalo D et al (2014) In vivo engineering of oncogenic chromosomal rearrangements with the CRISPR/Cas9 system. Nature 516:423–427
62. Xue W et al (2014) CRISPR–mediated direct mutation of cancer genes in the mouse liver. Nature 514(7522):380–384
63. Auer TO et al (2014) Highly efficient CRISPR/Cas9-mediated knock-in in zebrafish by homology-independent DNA repair. Genome Res 24(1):142–153
64. Chang N et al (2013) Genome editing with RNA-guided Cas9 nuclease in zebrafish embryos. Cell Res 23(4):465–472
65. Hruscha A et al (2013) Efficient CRISPR/Cas9 genome editing with low off-target effects in zebrafish. Development 140(24):4982–4987
66. Hwang WY et al (2013) Heritable and precise zebrafish genome editing using a CRISPR-Cas system. PLoS One 8(7), e68708
67. Irion U, Krauss J, Nüsslein-Volhard C (2014) Precise and efficient genome editing in zebrafish using the CRISPR/Cas9 system. Development 141(24):4827–4830
68. Talbot JC, Amacher SL (2014) A streamlined CRISPR pipeline to reliably generate zebrafish frameshifting alleles. Zebrafish 11(6):583–585
69. Xiao A et al (2013) Chromosomal deletions and inversions mediated by TALENs and CRISPR/Cas in zebrafish. Nucleic Acids Res 41, e141
70. Sander JD et al (2010) ZiFiT (Zinc Finger Targeter): an updated zinc finger engineering tool. Nucleic Acids Res 38:W462–W468
71. Jinek M et al (2013) RNA-programmed genome editing in human cells. eLife 2
72. Le X et al (2007) Heat shock-inducible Cre/Lox approaches to induce diverse types of tumors and hyperplasia in transgenic zebrafish. Proc Natl Acad Sci U S A 104(22):9410–9415
73. Santoriello C et al (2009) Expression of H-RASV12 in a zebrafish model of Costello syndrome causes cellular senescence in adult proliferating cells. Dis Model Mech 2(1–2):56–67
74. Sharp R et al (2002) Synergism between INK4a/ARF inactivation and aberrant HGF/SF signaling in rhabdomyosarcomagenesis. Nat Med 8(11):1276–1280
75. Mao J et al (2006) A novel somatic mouse model to survey tumorigenic potential applied to the Hedgehog pathway. Cancer Res 66(20):10171–10178
76. Fleischmann A et al (2003) Rhabdomyosarcoma development in mice lacking Trp53 and Fos: tumor suppression by the Fos protooncogene. Cancer Cell 4(6):477–482

77. Nanni P et al (2003) Development of rhabdomyosarcoma in HER-2/neu transgenic p53 mutant mice. Cancer Res 63(11):2728–2732
78. Camboni M et al (2012) Induction of a regenerative microenvironment in skeletal muscle is sufficient to induce embryonal rhabdomyosarcoma in p53-deficient mice. J Pathol 226(1):40–49
79. Chamberlain JS et al (2007) Dystrophin-deficient mdx mice display a reduced life span and are susceptible to spontaneous rhabdomyosarcoma. FASEB J 21(9):2195–2204
80. Fernandez K et al (2010) Mice lacking dystrophin or alpha sarcoglycan spontaneously develop embryonal rhabdomyosarcoma with cancer-associated p53 mutations and alternatively spliced or mutant Mdm2 transcripts. Am J Pathol 176(1):416–434

Baiting for Cancer: Using the Zebrafish as a Model in Liver and Pancreatic Cancer

Katie L. Hwang and Wolfram Goessling

Abstract Liver and pancreatic cancers are amongst the leading causes of cancer death. In recent years, genetic and chemical approaches in zebrafish have elucidated cellular and molecular mechanisms of liver and pancreatic cancer formation and progression. In this chapter, we review the recent approaches and advances in the field to study both hepatocellular carcinomas and pancreatic cancer.

Keywords Liver development • Liver cancer • Pancreas development • Pancreas cancer • Zebrafish • Transgenesis • Mutations

Introduction

Liver and pancreatic cancers are amongst the leading causes of cancer death. They both represent a global health problem and carry poor prognoses with limited therapeutic options. In the United States alone, over 80,000 new cases will be diagnosed and approximately 65,000 patients will die of these cancers each year [1]. Unfortunately, improved methods for early detection and advanced treatments to substantially alter the outcome are still lacking. Given the low survival rates and increasing incidences of these cancers, investigation into the molecular mechanisms underlying their pathogenesis is critical for the development of better diagnostics and effective therapies.

K.L. Hwang
Harvard Medical School, Medical Scientist Training Program, Boston, MA, USA

Genetics Division, Brigham and Women's Hospital, Boston, MA, USA

W. Goessling (✉)
Gastrointestinal Cancer Center, Dana-Farber Cancer Institute, Boston, MA, USA

Broad Institute of MIT and Harvard, Cambridge, MA, USA

Harvard Stem Cell Institute, Cambridge, MA, Cambridge, MA, USA

Genetics Division, Brigham and Women's Hospital, Boston, MA, USA
e-mail: wolfram_goessling@dfci.harvard.edu

D.M. Langenau (ed.), *Cancer and Zebrafish*, Advances in Experimental Medicine and Biology 910, DOI 10.1007/978-3-319-30654-4_17

The liver and pancreas are non-luminal organs of the gastrointestinal system and coordinate whole-body metabolism. The liver is predominantly composed of hepatocytes, which regulate a multitude of complex metabolic reactions, detoxify and secrete exogenous compounds, synthesize serum proteins and lipids, and produce bile salts [2]. Hepatocytes work in concert with other cells in the liver, including biliary epithelial cells (cholangiocytes), which line the biliary ducts and modify bile by secretion and absorption of ions. The pancreas, in contrast, functions as both an endocrine and exocrine gland, and as a result, has more specialized cell types that align to these specific functions [2]: hormone-producing cells are arranged into islets of Langerhans and include glucagon-secreting alpha cells, insulin-producing beta cells and delta cells, which secrete somatostatin. The exocrine pancreas, though, is dedicated to aiding digestion and is composed of acinar cells that secrete digestive zymogens into clustered ducts that eventually deliver these enzymes to the small intestine.

While the functions of these mature epithelial cells of liver and pancreas are extremely specialized, these cells develop from a common population of endodermal progenitors. Not only does the liver and pancreas share a developmental origin but they also undergo similar early developmental morphogenic steps of budding into the surrounding mesenchyme, rapid growth and vascularization, and eventual differentiation into biliary networks [3]. While the intrahepatic biliary network develops from hepatocyte precursors, the extrahepatic biliary system and the gallbladder develop from pancreatic precursors [4–6]. Intriguingly, cancers of the liver and pancreas often share characteristics dependent on their developmental origins and can be specified by their resemblance to their specialized cellular types [7].

Primary liver tumors can be divided into those arising from the hepatocytes (hepatocellular carcinoma, HCC) and those arising from the intrahepatic bile ducts (cholangiocarcinoma, CC). Shared risk factors, including viral hepatitis, cirrhosis, diabetes and alcohol consumption, as well as the existence of a subset of mixed histological HCC/CC tumors raises the possibility of a common molecular or cellular origin of both cancers [8]. Similarly, in the pancreas, tumors of the pancreatic ductal epithelium or pancreatic adenocarcinoma (PDAC), which comprise the vast majority (~85 %) of pancreatic tumors, have been linked to extrahepatic biliary tumors, gallbladder tumors, and ampullary cancers [7]. As these structures share a common embryonic origin, it raises the question of whether cancers originating from these organs also share molecular or cellular commonalities. In order to better understand tumorigenesis, the study of organogenesis and stem/progenitor cells could provide important insights into the signaling pathways central to cancer.

Zebrafish as a Model

Many of the processes that control stem cell proliferation, differentiation, and cell fate decisions are involved in embryonic organ development and establishment of progenitor cell populations. However, pathological perturbations of these pathways

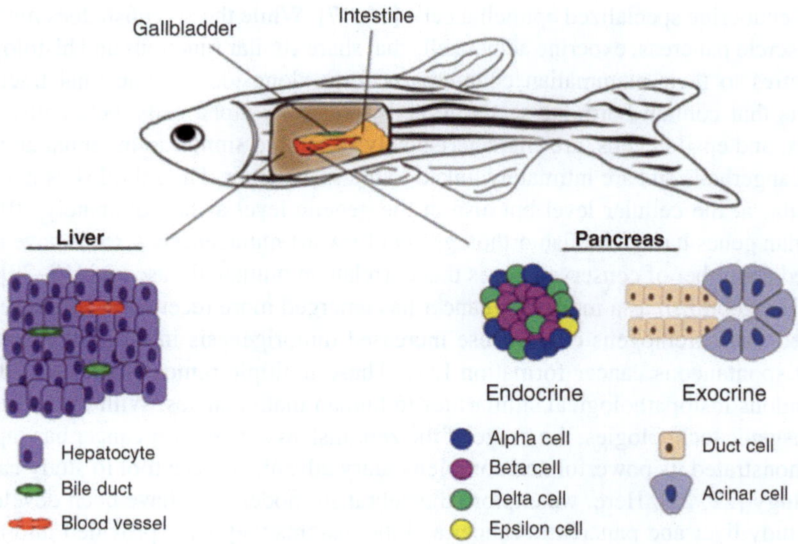

Fig. 1 Zebrafish gastrointestinal organ anatomy. Window dissection of an adult zebrafish is shown displaying the gastrointestinal organs of the zebrafish. The liver is comprised of mainly hepatocytes, biliary ductal cells, and endothelial blood vessels. The pancreas is composed of islets (endocrine function), which contain hormone-secreting cells, and acini (exocrine function), which contain acinar and ductal cells. These cellular makeups are well conserved between the zebrafish and human

can lead to tumorigenesis. It is therefore critical to study these fundamental signaling pathways to understand development, regeneration and carcinogenesis. The zebrafish, *Danio rerio*, offers an ideal model for the study of this intersection between development and cancer in the liver and pancreas.

Well established as an excellent model for vertebrate development and organogenesis, the zebrafish provides a number of powerful advantages. First, they have a high fecundity and a single mating pair can produce a few hundred embryos per week; secondly, embryos grow rapidly *ex utero* and are optically transparent allowing for ease in visualization during development; and thirdly, they share conservation of vertebrate organs, including the liver and pancreas (Fig. 1). In zebrafish, the epithelial cells of the liver are similarly comprised mainly of hepatocytes with cholangiocytes forming the biliary networks [9, 10]. However, unlike its mammalian counterparts, hepatocytes are not arranged into hepatic lobules with the traditional hexagonal plates of hepatocytes arranged around a central vein and each corner containing a portal triad consisting of a hepatic artery, portal vein and bile duct [11]. Rather, the portal veins, hepatic arteries, and biliary ducts of the zebrafish are distributed throughout the liver. The zebrafish is also uniquely advantageous for studying liver progenitor populations and development as the embryonic liver is not a site of hematopoiesis as it is in mammals, allowing for the study of developmental defects without complications from anemia or early lethality [12–15]. The zebrafish pancreas is similarly comprised to that of its mammalian counterparts with both exocrine and endocrine specialized epithelial cells [16, 17]. While the zebrafish does not have a discrete pancreas, exocrine acinar cells that share similar functions and histological features to their mammalian counterparts form alongside the intestinal tract [9]. Islets that contain hormone-secreting cells, including alpha cells, beta cells, delta cells, and epsilon cells, are characteristically structured similar to mammalian islets of Langerhans and are intimately linked to the vasculature. The zebrafish is not only similar at the cellular level but also at the genetic level as approximately 70 % of human genes have zebrafish orthologs, and forward-mutagenesis screens have identified a number of conserved genes that correlate to human disease loci [18–20].

The zebrafish as a model for cancer has emerged more recently as it was recognized that carcinogens could cause increased tumorigenesis in the background of low spontaneous cancer formation [21]. These multiple tumor types showed tremendous histopathological similarities to human malignancies. With the advent of transgenic technologies, the usage of the zebrafish as a model for cancer has rapidly demonstrated its powerful and complementary advantages as a tool to study cancer biology [22, 23]. Here, we explore the zebrafish models that have been developed to study liver and pancreatic cancer and the insights they have provided into these lethal malignancies.

Liver Cancer

Chemically Induced Models

Historically, the fish has been recognized as a barometer of the health of the environment, as it was observed that they form neoplasms resulting from chemical contamination. Following the expansion of organic chemical synthesis in the 1940s, increases in epithelial tumors in the fish, especially liver and skin tumors, rose exponentially. Armed with this knowledge, researchers in 1975 exposed zebrafish to carcinogenic compounds and induced hepatoadenomas, cholangiomas, hepatocellular carcinomas and cholangiocarcinomas, thereby demonstrating that the zebrafish offered a wide opportunity for cancer research [24]. Since then, the effects of other chemicals, especially dimethyl benzanthracene (DMBA) and *N*-methyl-*N'*-nitro-*N*-nitrosoguanidine (MNNG) on tumor formation have been extensively studied. Exposure of zebrafish fry (21 days post fertilization) with DMBA for 24 h and examined at one year resulted in hepatic tumors in 30 % of treated zebrafish [25]. Similarly, MNNG mainly targeted the liver in zebrafish and resulted in hepatocellular adenoma, hepatocellular carcinoma, cholangioadenoma and cholangiocarcinoma [26]. Histologically, these tumors greatly resembled human malignancies with atypical nuclear morphology and increased mitoses (Fig. 2).

With the availability of the zebrafish genome, Lam and colleagues compared gene-expression profiles of these hepatic tumors to see if they resembled their human counterparts at the molecular level [27]. DMBA-induced liver tumors were isolated and RNA expression was compared to normal liver tissue by microarray

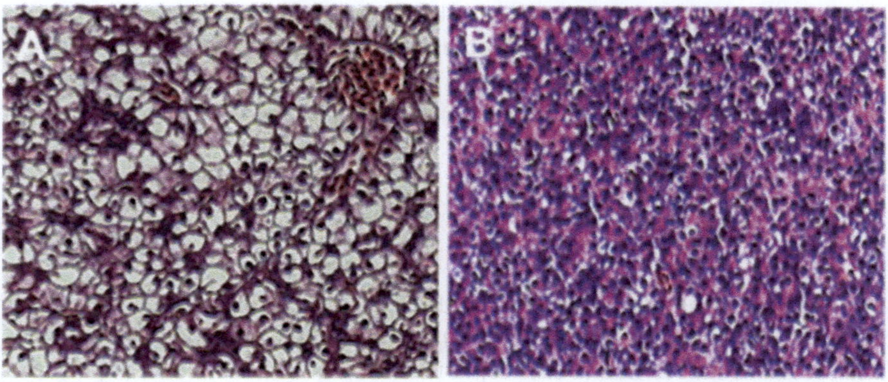

Fig. 2 Zebrafish liver and liver cancer histology. Histology of human and zebrafish hepatocellular carcinoma display tremendous similarities. (**a**) Human hepatocellular carcinoma and (**b**) zebrafish hepatocellular carcinoma induced by DMBA exposure display altered cellular morphology, dysplastic nuclei, and increased mitoses. Picture of human hepatocellular carcinoma courtesy of Dr. Jason Hornick, Brigham and Women's Hospital

analysis. Gene ontology analysis showed that many genes involved in cell cycle, proliferation, apoptosis, DNA replication and repair, cytoskeletal organization, protein synthesis, and metastasis were deregulated in the tumor samples. More specifically, alterations in the Wnt and Ras pathways were noted, which are also highly deregulated in human liver cancer. Furthermore, in the DMBA-induced zebrafish liver tumors, a range of tumor severity was observed histologically. Comparison of zebrafish liver tumors stratified by grade to human liver tumors classified by clinical stage revealed an expression signature of 132 genes that correlated with tumor progression in both zebrafish and human liver tumors. In high-grade tumors, these genes were associated with markers of de-differentiation with decreased liver specific genes and increased markers of cellular proliferation and metastasis. These similarities at the histopathological and genetic levels suggest that the mechanisms underlying hepatic carcinogenesis might be similar in zebrafish and humans.

Further support of this conservation has been seen at the epigenetic level. Frequently in human tumors, global DNA hypomethylation is found leading to transcriptional activation and genomic instability [28, 29]. Further, regions of hypermethylation have been observed in cancer cells leading to silencing of tumor suppressors and DNA repair genes [30, 31]. To examine DNA promoter methylation in zebrafish hepatocellular carcinomas generated by DMBA exposure, genomic regions 1.5 kb upstream to 1 kb downstream of the transcriptional start site were probed utilizing a CpG island tiling array and DNA methylation precipitation [32]. Here, they found the vast majority of differentially methylated regions in the zebrafish hepatocellular carcinomas were hypomethylated, similar to human hepatocellular carcinoma [33]. Deeper examination of the genes hypomethylated in zebrafish hepatocellular carcinoma and human hepatocellular carcinoma revealed shared pathways that were altered in both methylation and gene expression, including those especially involved in cellular proliferation, energy production, and anti-apoptosis.

These results underscore the conservation between zebrafish and human liver cancers and highlight the potential to apply comparative genomics and epigenomics to discover newly deregulated genes that have previously not been associated with liver cancer and strengthen the association of molecular pathways central to liver cancer.

Transgenic Models

While chemical carcinogenesis has provided great insights into liver cancer biology and the conservation of zebrafish and human liver tumor signal pathways, limitations have included slow onset of neoplasms and heterogeneous tumor populations. Alternatively, transgenic approaches in the zebrafish can be readily utilized with the development of transposon-mediated transgenesis at a fraction of the cost of transgenic mice [34]. With the increasing efficiency of this technology and the identification of the hepatocyte specific promoter of *liver fatty acid binding protein* (*lfabp,* also known as *fabp10a*), oncogenes could be specifically expressed in the liver [35]. In human liver cancers, Kras signaling is uniformly upregulated, and a subset of these tumors having activating Kras mutations [36, 37]. By driving an eGFP-fused oncogenic Kras ($Kras^{V12}$) under the control of the *lfabp* promoter in zebrafish, tumors formed in more than half of high-expressing stable transgenics [38]. Gross morphology showed tanned nodules that on histopathological analysis exhibited features of human hepatocellular carcinoma, including nuclear abnormalities and cellular disorganization. However, in these high-expressing oncogenic Kras transgenics, mortality was extremely high with 70 % of fish dying by 30 days post fertilization (dpf) with greatly enlarged livers, and all fish dying by 90 dpf. In a lower-expressing oncogenic Kras transgenic line, zebrafish survived past 90 dpf and displayed moderate liver hyperplasia in 36 % of fish at that time. At 6 months of age, hepatocellular adenocarcinoma was observed in 22 %, and by 9 months, 26 % of transgenics showed malignant hepatocellular carcinoma with a subset showing tumor invasion into surrounding blood vessels and internal organs. This model of rapid onset and aggressive formation of hepatocellular carcinoma as well as a model of slower progression to hepatocellular carcinoma allowed for analysis at the molecular level of pathways involved in hepatocellular carcinoma advancement. Microarray and gene set enrichment analyses revealed a conserved hepatocellular carcinoma and a liver cancer progression gene signature between zebrafish and humans. For example, zebrafish hepatocellular carcinoma livers displayed activation of NFkB, JAK-STAT, IGF, TGFb, MAPK/ERK, PI3K-AKT, WNT, VEGF, and complement signaling pathways, consistent with human hepatocellular carcinoma and specific to oncogenic Ras gene signatures. In contrast, the zebrafish with hyperplastic livers showed a gene signature unique in its activation of the p53 senescence pathway similar to human liver dysplasia.

While these studies provided insights into the progression of hepatocellular carcinoma and more generally, the zebrafish as a model of hepatic-specific oncogene expression, they were not able to address the mechanisms behind hepatocellular adenoma or carcinoma initiation due to the variability of tumor onset and early lethality in high-expressing oncogenic transgenics. In order to address this limitation, the Gong lab utilized a mifepristone-inducible LexPR system to conditionally express eGFP-Kras^{V12} in the liver [39]. Here, the liver-driver line transcribed the LexPR activator under the control of the *lfabp* promoter. In the presence of mifepristone, the LexPR activator bound to the LexA operator, which was placed upstream of the fusion Kras oncogene (*LexA:eGFP-Kras^{V12}*). Most importantly, utilizing an inducible model allowed for a dynamic, in vivo study to understand the role of oncogenes in tumor initiation and progression. In transgenic zebrafish with both the liver driver and inducible oncogenic Kras (*lfabp:LexPR; LexA: eGFP-Kras^{V12}*), that were treated with mifepristone beginning at 1 month of age, 100 % developed hepatic hyperplastic lesions within 1 week of induction and progressed to hepatocellular carcinoma after 4 weeks. Prolonged exposure to mifepristone led to the formation of hepatoblastoma in mixed hepatocellular carcinoma as well as increasing invasiveness. Intriguingly, removal of mifepristone resulted in tumor regression and fibrotic lesions at the focal tumor region, indicating an addiction to oncogenic Kras for continuous tumor growth. Downstream targets of Ras, such as the ERK and AKT pathways, were upregulated during liver tumor progression and reduced in regressed hepatocellular carcinoma fish with no levels detectable in the fibrotic lesions. To follow this observation, the authors tested whether inhibitors targeting the ERK or AKT pathways could inhibit the hyperplastic growth seen in the mifepristone-induced Kras fish. Suppression of liver growth in the transgenic larvae was achieved with modest success with approximately half displaying "normal" liver size when targeting either pathway alone. Improved results were seen when both pathways were inhibited simultaneously. While this study did not address whether inhibition of the ERK and AKT pathways could prevent tumor initiation or tumor progression, this raised an important possible connection between an early larval phenotype of enlarged liver growth and liver cancer and whether this could be exploited to identify novel inhibitors of abnormal liver growth.

Taking advantage of this early larval response to oncogenic Kras expression and the transparency of the larval zebrafish, the effect on the tumor microenvironment with induced Kras expression in hepatocytes was examined [40]. In this study, they investigated the response of neutrophils to the oncogenic expressing hepatocytes as neutrophils have been implicated in both pro-tumoral and anti-tumoral functions [41, 42]. By crossing their inducible oncogenic Kras zebrafish with transgenic zebrafish that label neutrophils with DsRed under the control of the lysozyme c (*lyz*) promoter, they were able to measure in vivo and in real time the recruitment of tumor associated neutrophils (TANs). Upon treatment with mifepristone at 3 dpf, neutrophils in the liver began to increase by 8 h post induction (hpi), which preceded hepatic hypertrophy. Chemical inhibition of neutrophil activity and genetic knockdown of the *gscfr* gene, which thereby blocked differentiation of myeloid precursors into neutrophils, resulted in decreased hepatomegaly. Further, chemical inhibition of neutrophils decreased proliferation of hepatocytes and decreased histopathological characteristics of early hepatocellular carcinoma initiation, including cellular disorganization and nuclear changes. Gene expression analysis of induced

Kras hepatocytes showed significant upregulation of *tgfβ1a,* as well as decreased expression of *tnfα* and *infγ,* consistent with cancer-related inflammation and a pro-tumor microenvironment as seen in human hepatocellular carcinoma. TANs isolated by fluorescence activated cell sorting (FACS) showed a pro-tumorogenic expression profile with upregulation of *il1b*, which promotes early cancer angiogenesis, and downregulation of anti-tumor cytokines. Together, these results suggest a role for neutrophils in early oncogenic growth and transformation but more importantly highlight the strengths of utilizing the zebrafish to study the interaction between oncogenic cells and their microenvironment.

Inducible expression models of other major liver oncogenic pathways, EGFR and Myc, have been created in a comparable transgenic approach, but rather than being inducible by mifepristone, were doxycycline inducible using the Tet-ON system [43, 44]. Doxycycline induction provides the potential advantage of less off-target transcriptomic effects as it was found in the livers of mice that the constitutive expression of the tetracycline-controlled trans-activator resulted in fewer gene expression alterations than the mifepristone-controlled trans-activator [45]. In these Tet-ON systems, the reverse tetracycline-controlled transactavator (rtTA) is specifically driven in hepatocytes by the *lfabp* promoter (*lfabp:rtTA).* In the presence of doxycycline, the rtTA can bind the tetracycline-responsive promoter element (TRE), which then drives expression of the oncogene of interest. In human hepatocellular carcinoma, EGFR is overexpressed in 40–70 % of tumors and its early upregulation in preneoplastic lesions suggests an early role for EGFR in hepatocellular carcinoma initiation [46, 47]. Li and colleagues found that upon induced expression of Xmrk, a naturally occurring mutated fish isoform of EGFR that is continuously activated, enlarged livers rapidly formed both at the juvenile and adult stages. Histological examination at 3 weeks post treatment initialization found hepatocellular carcinoma uniformly throughout the livers with 100 % penetrance and characteristic pleomorphism, disrupted cellular organization, and nuclear abnormalities. While the presence of hepatocellular carcinoma features were present throughout the liver, downstream EGFR targets, such as phosphorylated Erk (p-Erk) and Stat5 (p-Stat5) showed heterogenous expression. Upon withdrawal of doxycycline at 1 week, these tumors displayed increased proliferation and apoptosis and by 2 weeks had regressed with histology showing features of hepatocellular adenoma/hyperplasia. By 4 weeks of doxycycline withdrawal, the livers had reverted to normal morphology and histology. Intriguingly, re-introduction of doxycycline resulted in accelerated hepatocellular carcinoma formation, suggesting the livers were primed for transformation. This model, similar to that of the inducible oncogenic Kras, indicated oncogene addiction to Xmrk for tumor growth. In addition, it illustrated that even with a single oncogenic initiator, there are mechanisms to form heterogeneous cellular populations within the tumor. A similar doxycycline inducible Myc model showed hepatomegaly after 4 days of doxycycline treatment starting at 3 dpf. Treatment of the transgenic zebrafish at the fry stage resulted in enlarged abdomens following doxycycline treatment after only 2–3 weeks of treatment [44]. However, these livers were mainly transformed into hyperplasia as well as hepatocellular adenoma but rarely formed hepatocellular carcinoma. Whole transcriptome analysis found a conserved gene signature with other Myc hepatocellular carcinoma models in mice. While the zebrafish model mainly formed hepatic adenomas with Myc overexpression, the zebrafish gene signature correlated best with advanced forms of human hepatocellular carcinoma, suggesting Myc could be essential for progression but requires other molecular events to initiate cellular transformation.

As human hepatocellular carcinoma is extremely heterogeneous, the overlap of these different inducible models' expression signatures with human tumors is a powerful way to identify critical genes involved in tumor initiation and progression. Transcriptomic analysis of tumors from the inducible Kras, XMRK or Myc models by RNA-sequencing revealed three distinct gene signatures and sets of deregulated biological pathways with all three correlating with human hepatocellular carcinomas that were advanced or very advanced [48]. Interestingly, 21 genes were upregulated and 16 genes were downregulated in all three transgenic models. When compared to human hepatocellular carcinomas, 27 of these genes were similarly deregulated in more than half of patients and some have been previously implicated in hepatocarcinogenesis. However, the majority of these genes represented new genes not previously associated with liver cancer and represent novel mechanisms and potential therapeutic targets. Further study of the role of these genes is needed to determine whether these deregulated genes are central to liver cancer formation and aggressiveness.

The zebrafish also provides a unique model to address in vivo the role of previously unknown oncogenic genes on tumorigenesis. For example, elegant work by Sadler and colleagues utilized a transgenic model to overexpress ubiquitin-like with PHD and RING finger domains 1 (UHRF1), an essential regulator of DNA methylation, in hepatocytes specifically to address the role of UHRF1 in hepatocellular carcinoma [49], directly connecting in vivo models to clinically-relevant events in patient samples. While global DNA hypomethylation is recognized in human hepatocellular carcinoma, the mechanism by which this is achieved has remained elusive. UHRF1 is a highly intriguing candidate, as overexpression in zebrafish at high levels resulted in atypical cells resembling hepatocellular carcinoma.

These studies demonstrate that transgenic approaches can lead to a rapid onset of liver cancer in zebrafish that mimics the cellular and signaling events observed in human liver cancer, thereby providing additional insight into the mechanisms of disease pathogenesis.

Mutant Models

The power of the zebrafish model not only lies in its ability to analyze comparative genomic alterations but its ability for large-scale genetic screens to identify initiators and drivers of tumorigenesis in vivo. Forward genetic screens have been well established in the zebrafish since the 1980s focusing on chemical mutagenesis to identify genes that are critical in development [50, 51]. Currently, with the rapidly advancing abilities of genome-wide sequencing, chemical or viral mediated mutagenesis to identify genes that predispose tumor formation are even more feasible [52]. Screens utilizing both embryonic phenotypes as well as general properties of cancer, such as cellular proliferation, have been performed to identify novel genes involved in cancer susceptibility. One such example was an ENU screen for mutants that demonstrated increased genomic instability, a hallmark of cancer [53]. To test this, the researchers utilized the theory behind Knudson's two-hit hypothesis, in which a single locus involved in retinal pigment color (*golden*) could act as a proxy for genomic stability. Heterozygotes for the *golden* locus were exposed to ENU and mutants that showed a second "hit" causing a light-colored eye phenotype were then identified. In 12 mutants that demonstrated this increased genomic instability, all showed a higher propensity for cancer, and one of the most penetrant lines, *gin*-10, showed almost a tenfold increased rate of spontaneous tumor formation. Tumors of epithelial and stromal origins were found, with the most common being adenocarcinomas, especially intestinal, pancreatic, and liver. Many of these mutations are still awaiting identification; however, this study provided an important link between forward based screening in the zebrafish based on an embryonic phenotype that acted as a marker for a cancer property and adult formation of tumors.

Targeted identification of induced genetic lesions has also led to the identification of known tumor suppressors mutants, namely *p53* and *APC*. Mutants of *p53* showed increased propensity for forming peripheral nerve sheath tumors [54]. Further, an identified mutant of *APC*, which resulted in activated Wnt/β-catenin signaling, showed defects in liver development [55, 56]. Additionally, *APC* heterozygous mutants spontaneously develop digestive tract neoplasias, including hepatic and pancreatic carcinomas [57]. These studies highlight the homology between human cancers and the zebrafish as a tool to model tumor suppressor functions. Excitingly, more studies investigating the role of tumor suppressors in cancer using the zebrafish as a model are on the horizon with the advent of targeted genome-editing technologies. Previously, the zebrafish has been limited in the ability to directly knockout specific genes. With genome-editing tools, such as zinc-finger nucleases (ZFNs), transcription activator-like effector nucleases (TALENs) and clustered regularly interspaced short palindromic repeats (CRISPR) and the CRISPR-associated proteins (Cas), there is a new ability to induce targeted mutations or insertions into the germline for inheritable genetic modifications. This opens many possibilities to silence tumor suppressors and other genes and to investigate their role in tumorigenesis.

Pancreatic Models

While chemically induced carcinogenesis has provided many great insights into liver cancer biology, limitations of low frequency and slow onset of neoplasms has plagued zebrafish models of pancreatic cancer, in which treatment with DMBA or MNNG of zebrafish fry resulted in less than 2 % occurrence of pancreatic tumors. Alternatively, transgenic approaches in the zebrafish can be readily utilized with the development of transposon-mediated transgenesis. In order to determine whether a transgenic approach could be applied to pancreatic cancer in the zebrafish, the Look lab developed a transient transgenic zebrafish in which the oncogene, MYCN, was expressed under the control of the promoter *myod* which targets gene expression in the pancreatic neuroendocrine β cells in addition to neurons and myocytes [58]. These transgenic zebrafish developed neuroendocrine tumors between 3 and 6 months of age, and they closely resembled that of human pancreatic neuroendocrine carcinoma at the histological level with high cytoplasmic/nuclear ratios, distinct nucleoli, and decreased amounts of eosinophilic cytoplasm. Further, these tumors overexpressed insulin, a characteristic feature of human pancreatic neuroendocrine tumors. This established that transgenic tools could be used to direct tumor development in the pancreas.

In order to determine whether stable transgenic zebrafish could be used to drive exocrine pancreatic tumor formation, Leach and colleagues developed a transgenic zebrafish model (*ptf1a:eGFP- KrasG12V*) in which an oncogenic Kras, Kras^{G12V}, fused to eGFP was expressed under the control of the *ptf1a* regulatory elements through BAC recombineering. This drove expression of the fusion oncoprotein in pancreatic progenitor cells as well as the retina, hindbrain, and spinal interneurons [59]. In human PDACs, more than 90 % have mutations in Kras and is believed to be the earliest genetic alteration as it has been found in low-grade pancreatic lesions [60–62]. Targeting of mutant Kras^{G12D} or Kras^{G12V} to the murine pancreas is sufficient to initiate pancreatic intraepithelial neoplasia, which is well established to progress to invasive PDAC with long latency, similar to that of human PDAC [62–68]. Utilizing the transgenic zebrafish *ptf1a:eGFP- Kras^{G12V}*, 32 % of zebrafish at 6 months of age had small (<8 mm) focal pancreatic lesions and 10 % showed larger (>8 mm) or diffuse/multifocal lesions. By 9 months of age, two-thirds of the transgenic zebrafish had widespread foci that on histologic expression showed characteristic features of malignancy with aggressive invasion and propensity for metastasis. These tumors also displayed a heterogenous array of PDAC subtypes such as acinar cell carcinoma, ductal adenocarcinoma, and mucinous adenocarcinoma, similar to those found in humans. While mouse studies were able to establish that Kras is critical for initiation for PDAC, it was unknown the exact nature of the effect of Kras mutation on cellular events. In order to address this, the authors utilized the transgenic model of pancreatic cancer to examine the effect of mutant Kras expression on pancreatic progenitor cells and pancreatic development. In the transgenic *ptf1a:eGFP-Kras^{G12V}* zebrafish, pancreatic precursor cells showed decreased expression of the enzyme carboxypepitdase A (CPA), a marker of exocrine differentiation in comparison to the control transgenic zebrafish (*ptf1a:eGFP*) that expressed eGFP driven by the *ptf1a* promoter. Furthermore, observation at 96 hpf showed there was a heterogeneous distribution of the eGFP-positive cells that erroneously retained expression of the fusion protein in *ptf1a:eGFP-Kras^{G12V}* zebrafish whereas the control transgenics showed normal labeling of CPA and appropriate silencing of *ptf1a* driven eGFP expression. Together, this suggested that oncogenic Kras expression in pancreatic progenitors prevents normal exocrine differentiation. This established that a transgenic model of pancreatic cancer was achievable in the zebrafish with a reproducible high rate of pancreatic cancer and provided a new platform to probe the mechanism of mutant Kras expression on pancreatic cancer initiation.

To gain further insight into the other signaling pathways of pancreatic cancer, Liu and Leach utilized a Gal4/UAS transgenic system to both transiently and stably express eGFP-Kras^{G12V} downstream of a 14xUAS element (*UAS:eGFP-Kras^{G12V}*) [69]. When crossed to the *ptf1a:Gal4-VP16* transgenics, in which Gal4-VP16 expression is driven by *ptf1a* regulatory elements, oncogenic Kras is again expressed in the developing exocrine pancreas. These stable transgenics develop pancreatic tumors as early as 2 months post fertilization and by 5 months, around 50 % of fish have pancreatic tumors. A similar model, in which a UAS element was placed upstream of an eGFP-fusion oncogenic Kras, Kras^{G12D}, was utilized to examine in vivo signaling pathways at initiation and tumor progression [70]. This was accomplished by injecting the *UAS:eGFP-Kras^{G12D}* construct into mCherry fluorescent reporter lines outcrossed to the *ptf1a:Gal4-VP16* lines. In these transient transgenics, Notch signaling was confirmed to be upregulated during tumor onset followed by TGFβ/Smad3 pathway upregulation, as had been previously demonstrated in human pancreatic cell lines. This demonstrated that transgenic zebrafish cancer models, coupled with fluorescent zebrafish lines, can be utilized to examine in vivo the crosstalk of signaling pathways activated at various stages of tumor progression.

Future Directions

With the establishment of these models and advances in the genomic technologies, the zebrafish is at the breakthrough point of cancer biology that will enable it to provide unique and tremendous insights into the understanding, prevention, and treatment of liver and pancreatic cancers (Fig. 3). The ability to specifically express oncogenes in liver and pancreatic tissues in transgenic fish has driven the field forward and has provided a strong foundation for efficient tumorigenesis. However, with these rapidly expanding advances in transgenic approaches, which provide crucial insights into oncogene functions, other questions regarding the role of tumor suppressors remain. With the maturation of genome editing technology, new uses of its functions are coming to light. For example, the ability to target multiple genes simultaneously with CRISPR-Cas has recently been described [71, 72]. Additionally, CRISPR-Cas technology has allowed for tissue specific gene inactivation in the zebrafish [73], which could provide unique advantages in the study of tumor suppressors in the liver and pancreas specifically. This provides the ability to introduce multigenic changes in a single fish or even a single tissue, which would more accurately recapitulate cancer as tumors carry thousands of mutations. With the scalability of the zebrafish, the ability to apply multiplex gene targeting could provide a powerful tool to better understand the complexities of the cancer genome as it is yet unclear which of these are causal versus passenger mutations. Furthermore, many of the precise genes responsible for increased cancer susceptibility, especially for pancreatic cancer, remain unknown. It has been estimated that 10 % of PDAC is a result of familial inheritance, and of these, 20 % have been associated with a known

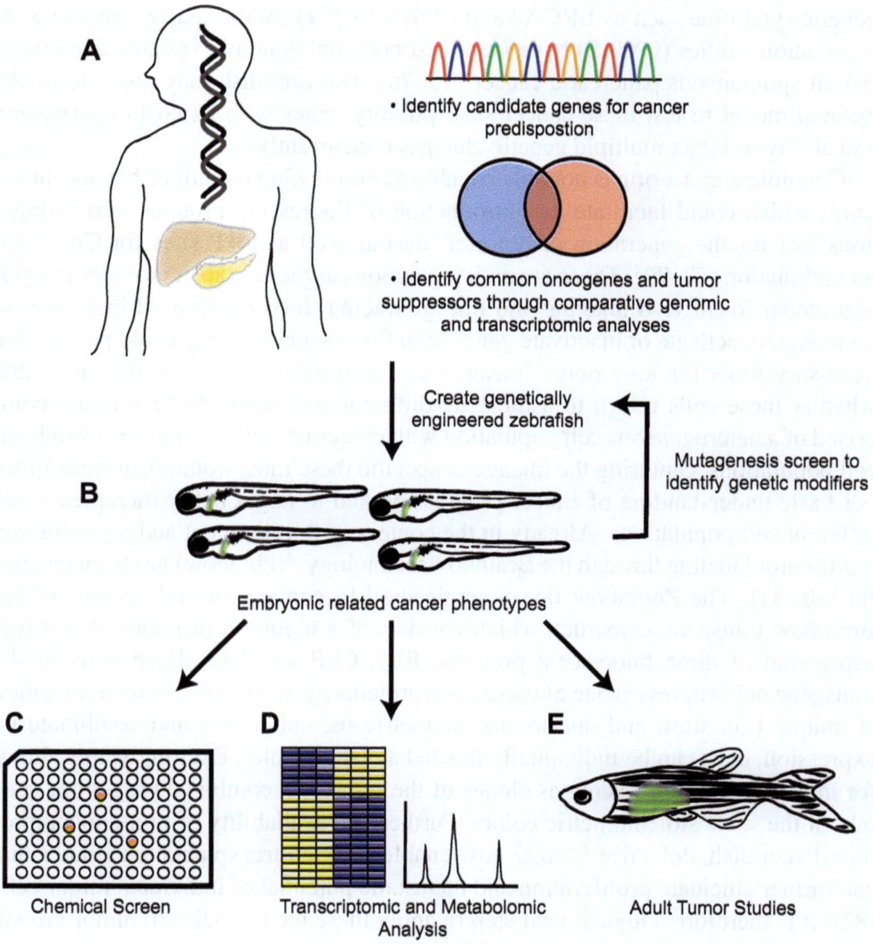

Fig. 3 Modeling liver and pancreatic cancer in the zebrafish. (**a**) The zebrafish provides an ideal model for testing candidate genes contributing to increased cancer susceptibility from GWAS studies and families with high cancer predisposition. Comparative genomic and transcriptomic analysis of human and zebrafish cancers further identifies core pathways and genes altered in hepatic and pancreatic cancers. (**b**) With the cost-effective advantage of the zebrafish, genetic models of these candidate genes can be efficiently created. Embryonic related cancer phenotypes, such as hepatomegaly, can be utilized to screen genetic modifiers of these pathways. (**c**) The small size of zebrafish embryos and the ability to produce thousands of embryos in addition to the availability of small molecule libraries provides an ideal model for chemical screens to inhibit the embryonic cancer-related phenotype. (**d**) Transcriptomic and metabolomic analysis of genetically modified embryos can provide insights into the mechanisms underlying cancer initiation. (**e**) Maturation of embryos to adults allows for in vivo imaging of neoplastic growth and clonal expansion. Additionally, genomic, transcriptomic, and metabolomic analysis of tumors will provide powerful insights into the mechanisms of cancer pathogenesis

genetic syndrome such as BRCA2 and CDKN2A [74]. Additionally, genome wide association studies (GWAS) have identified potential genomic variants that confer risk of spontaneous pancreatic cancer [75, 76]. The zebrafish may prove to be the optimal model to test these cancer susceptibility genes with its cost effectiveness and ability to target multiple genetic changes concurrently.

Genomic engineering is not only capable of introducing mutations but also insertions, which could facilitate the introduction of fluorescent proteins into endogenous loci for the generation of reporter lines as well as loxP sites for Cre based recombination [77–79]. These are enticing options as the zebrafish provides an optimal model for in vivo imaging and lineage tracing. Incorporation of loxP sites to permanently activate or inactivate genes with fluorescent labeling could provide the necessary tools for long-range lineage tracing to follow the cells that arise and whether these cells return to a more de-differentiated state. As tumors are comprised of a heterogeneous cell population with clones of cells arising from dominant cell populations, capturing the lineage of specific these lines would be insightful for our basic understanding of cancer progression and to target better therapies to the different cell populations. Already in the context of development and regeneration, multi-color labeling through the Brainbow technology (Zebrabow) has been insightful [80, 81]. The Zebrabow line is engineered to contain multiple copies of the Brainbow transgene construct, which consists of a ubiquitin promoter that drives expression of three fluorescent proteins: RFP, CFP and YFP. Each copy of the transgene only expresses one of these three proteins by the use of Cre-recombination at unique Lox sites, and due to stoichiometric recombination and combinatorial expression, cells can be individually labeled a distinct color. Excitingly, this allows for multi-lineage cell tracing as clones of the initially recombined progenitor cells inherit the same stoichiometric colors. Further, the availability of immunocompromised zebrafish, defective in *rag2* now enables elegant transplantation studies that can further elucidate proliferation and metastatic potential of individual tumor cells [82]. It is therefore a logical next step to apply these technologies to tumor growth and regression in zebrafish models of liver and pancreatic cancer.

To further identify genes that are capable of enhancing or preventing cancer initiation or progression, it is critical that we utilize the unique advantages of the zebrafish in large-scale forward genetic screens. The zebrafish is exceptionally positioned for modifier screens, especially with early cancer phenotypes as has been demonstrated in the inducible models of hepatocellular carcinoma. This can be accomplished through ENU mutagenesis of male zebrafish that are cancer susceptible, by either overexpression or deletion of the gene of interest, and then bred to female zebrafish that are similarly cancer susceptible. Identification of mutations that result in enhancement or repression of the cancer phenotype can then be identified through sequencing. This can provide tremendous insight into pathways that have not been previously associated with the cancer gene of interest. Alternatively, transposase based methods, such as with the Sleeping Beauty transposon, could be utilized to disrupt somatic genes and thereby identify novel genes that are involved in tumorigenesis [83].

While the focus of the majority of current studies has been on the transcriptional effects of oncogenes and tumor suppressors, it is important that we broaden our investigation outside of transcriptional and genomic changes alone. One area that has been re-recognized as essential for tumor formation and progression is cellular metabolism. It has become clear that many key oncogenic signaling pathways converge to modulate tumor cell metabolism to support their growth and survival. Untapped in zebrafish cancer models, advances in metabolomics and metabolic flux analysis coupled with high-throughput sequencing data and computational power could provide important insights into how cancer cells balance the catabolic demands to support proliferation with the anabolic demands of macromolecule synthesis. Already, metabolic profiling of alcoholic fatty liver in zebrafish has shown excellent homology to that of humans as it detected altered TCA cycle and fatty acid metabolism [84]. The zebrafish is also amenable to in vivo flux analysis to measure production and consumption rates of metabolites since zebrafish embryos are able to absorb small, isotope labeled molecules during embryogenesis [85]. With the ability to conduct in vivo and high-throughput screens, the zebrafish is a complementary system for cellular metabolism studies that could provide greater understanding of the complexities underlying tumorigenesis.

The most formidable attribute of the zebrafish, though, is its ability to identify potential cancer therapeutics in large-scale chemical screens in vivo. Uniquely suited for high-throughput assays, zebrafish can generate thousands of embryos per day, especially with the use of mass-mating systems [86]. With the identification of embryonic phenotypes in some of the current cancer models, testing a high number of small molecules in a high-content fashion is feasible [87–89]. However, additional studies to examine the effects of chemicals identified in embryonic tumor models on tumors in adult zebrafish are currently necessary to fully link these embryonic phenotypes to cancer initiation and progression. The full advantage of these unbiased chemical screens, though, will allow for a more encompassing approach to identify many mechanisms of tumor growth and metastasis, including genetic, epigenetic, and metabolic mechanisms as well as the interplay between the tumor and its microenvironment.

The zebrafish is now reaching the cusp of technological advances and establishment as a cancer model that will allow for its strengths and potential to contribute uniquely to the field of cancer biology. With the usage of chemical carcinogens, transgenics, and mutants, zebrafish provide a wide arsenal to probe mechanisms underlying liver and pancreatic cancers. Continued discovery with large-scale screens will provide guidance and focus into the investigation of molecular pathways driving these tumors. The zebrafish offers a promising and complementary model to identifying central pathways and targets that should ultimately provide tremendous insights into cancer prevention and treatments.

Acknowledgements This work is supported by NIH Grant R01DK090311 to W.G. W.G. is a Pew Scholar in the Biomedical Sciences. We are grateful to Dr. Jason Hornick at Brigham and Women's Hospital, Boston, for images of HCC histology.

References

1. Howlader N, Noone AM, Krapcho M, Garshell J, Miller D, Altekruse SF, Kosary CL, Yu M, Ruhl J, Tatalovich Z, Mariotto A, Lewis DR, Chen HS, Feuer EJ, Cronin KA (eds) (2015) SEER Cancer Statistics Review, 1975–2012: National Cancer Institute
2. Barrett KE (2014). In: Barrett KE (ed) Gastrointestinal physiology, 2nd edn. Lange Medical Books/McGraw-Hill Education, New York, NY
3. Zorn AM (2008) Stembook: Harvard Stem Cell Institute. In: Girard L, Melton D (Eds)
4. Raynaud P, Carpentier R, Antoniou A, Lemaigre FP (2011) Biliary differentiation and bile duct morphogenesis in development and disease. Int J Biochem Cell Biol 43(2):245–256. doi:10.1016/j.biocel.2009.07.020
5. Spence JR, Lange AW, Lin SC, Kaestner KH, Lowy AM, Kim I et al (2009) Sox17 regulates organ lineage segregation of ventral foregut progenitor cells. Dev Cell 17(1):62–74. doi:10.1016/j.devcel.2009.05.012
6. Zong Y, Stanger BZ (2012) Molecular mechanisms of liver and bile duct development. Wiley Interdiscip Rev Dev Biol 1(5):643–655. doi:10.1002/wdev.47
7. Henson DE, Schwartz AM, Nsouli H, Albores-Saavedra J (2009) Carcinomas of the pancreas, gallbladder, extrahepatic bile ducts, and ampulla of vater share a field for carcinogenesis: a population-based study. Arch Pathol Lab Med 133(1):67–71. doi:10.1043/1543-2165-133.1.67
8. Holczbauer A, Factor VM, Andersen JB, Marquardt JU, Kleiner DE, Raggi C et al (2013) Modeling pathogenesis of primary liver cancer in lineage-specific mouse cell types. Gastroenterology 145(1):221–231. doi:10.1053/j.gastro.2013.03.013
9. Menke AL, Spitsbergen JM, Wolterbeek AP, Woutersen RA (2011) Normal anatomy and histology of the adult zebrafish. Toxicol Pathol 39(5):759–775. doi:10.1177/0192623311409597
10. Yao Y, Lin J, Yang P, Chen Q, Chu X, Gao C, Hu J (2012) Fine structure, enzyme histochemistry, and immunohistochemistry of liver in zebrafish. Anat Rec (Hoboken) 295(4):567–576. doi:10.1002/ar.22416
11. Wallace KN, Pack M (2003) Unique and conserved aspects of gut development in zebrafish. Dev Biol 255(1):12–29
12. Detrich HW 3rd, Kieran MW, Chan FY, Barone LM, Yee K, Rundstadler JA et al (1995) Intraembryonic hematopoietic cell migration during vertebrate development. Proc Natl Acad Sci U S A 92(23):10713–10717
13. Jin H, Sood R, Xu J, Zhen F, English MA, Liu PP, Wen Z (2009) Definitive hematopoietic stem/progenitor cells manifest distinct differentiation output in the zebrafish VDA and PBI. Development 136(4):647–654. doi:10.1242/dev.029637
14. Reimold AM, Etkin A, Clauss I, Perkins A, Friend DS, Zhang J et al (2000) An essential role in liver development for transcription factor XBP-1. Genes Dev 14(2):152–157
15. Thisse C, Zon LI (2002) Organogenesis—heart and blood formation from the zebrafish point of view. Science 295(5554):457–462. doi:10.1126/science.1063654
16. Kinkel MD, Prince VE (2009) On the diabetic menu: zebrafish as a model for pancreas development and function. Bioessays 31(2):139–152. doi:10.1002/bies.200800123
17. Ober EA, Field HA, Stainier DY (2003) From endoderm formation to liver and pancreas development in zebrafish. Mech Dev 120(1):5–18
18. Howe K, Clark MD, Torroja CF, Torrance J, Berthelot C, Muffato M et al (2013) The zebrafish reference genome sequence and its relationship to the human genome. Nature 496(7446):498–503. doi:10.1038/nature12111
19. Kettleborough RN, Busch-Nentwich EM, Harvey SA, Dooley CM, de Bruijn E, van Eeden F et al (2013) A systematic genome-wide analysis of zebrafish protein-coding gene function. Nature 496(7446):494–497. doi:10.1038/nature11992
20. Santoriello C, Zon LI (2012) Hooked! Modeling human disease in zebrafish. J Clin Invest 122(7):2337–2343. doi:10.1172/JCI60434
21. Pliss GB, Zabezhinski MA, Petrov AS, Khudoley VV (1982) Peculiarities of N-nitramines carcinogenic action. Arch Geschwulstforsch 52(8):629–634

22. Goessling W, North TE, Zon LI (2007) New waves of discovery: modeling cancer in zebrafish. J Clin Oncol 25(17):2473–2479. doi:10.1200/JCO.2006.08.9821
23. Langenau DM, Traver D, Ferrando AA, Kutok JL, Aster JC, Kanki JP et al (2003) Myc-induced T cell leukemia in transgenic zebrafish. Science 299(5608):887–890. doi:10.1126/science.1080280
24. Pliss GB, Khudoley VV (1975) Tumor induction by carcinogenic agents in aquarium fish. J Natl Cancer Inst 55(1):129–136
25. Spitsbergen JM, Tsai HW, Reddy A, Miller T, Arbogast D, Hendricks JD, Bailey GS (2000) Neoplasia in zebrafish (Danio rerio) treated with 7,12-dimethylbenz[a]anthracene by two exposure routes at different developmental stages. Toxicol Pathol 28(5):705–715
26. Spitsbergen JM, Tsai HW, Reddy A, Miller T, Arbogast D, Hendricks JD, Bailey GS (2000) Neoplasia in zebrafish (Danio rerio) treated with N-methyl-N′-nitro-N-nitrosoguanidine by three exposure routes at different developmental stages. Toxicol Pathol 28(5):716–725
27. Lam SH, Gong Z (2006) Modeling liver cancer using zebrafish: a comparative oncogenomics approach. Cell Cycle 5(6):573–577
28. Counts JL, Goodman JI (1995) Hypomethylation of DNA: a nongenotoxic mechanism involved in tumor promotion. Toxicol Lett 82–83:663–672
29. Ehrlich M (2009) DNA hypomethylation in cancer cells. Epigenomics 1(2):239–259. doi:10.2217/epi.09.33
30. Baylin SB, Herman JG (2000) DNA hypermethylation in tumorigenesis: epigenetics joins genetics. Trends Genet 16(4):168–174
31. Kulis M, Esteller M (2010) DNA methylation and cancer. Adv Genet 70:27–56. doi:10.1016/B978-0-12-380866-0.60002-2
32. Mirbahai L, Williams TD, Zhan H, Gong Z, Chipman JK (2011) Comprehensive profiling of zebrafish hepatic proximal promoter CpG island methylation and its modification during chemical carcinogenesis. BMC Genomics 12:3. doi:10.1186/1471-2164-12-3
33. Lin CH, Hsieh SY, Sheen IS, Lee WC, Chen TC, Shyu WC, Liaw YF (2001) Genome-wide hypomethylation in hepatocellular carcinogenesis. Cancer Res 61(10):4238–4243
34. Kawakami K, Koga A, Hori H, Shima A (1998) Excision of the tol2 transposable element of the medaka fish, Oryzias latipes, in zebrafish, Danio rerio. Gene 225(1–2):17–22
35. Her GM, Chiang CC, Chen WY, Wu JL (2003) In vivo studies of liver-type fatty acid binding protein (L-FABP) gene expression in liver of transgenic zebrafish (Danio rerio). FEBS Lett 538(1–3):125–133
36. Calvisi DF, Ladu S, Gorden A, Farina M, Conner EA, Lee JS et al (2006) Ubiquitous activation of Ras and Jak/Stat pathways in human HCC. Gastroenterology 130(4):1117–1128. doi:10.1053/j.gastro.2006.01.006
37. Karnoub AE, Weinberg RA (2008) Ras oncogenes: split personalities. Nat Rev Mol Cell Biol 9(7):517–531. doi:10.1038/nrm2438
38. Nguyen AT, Emelyanov A, Koh CH, Spitsbergen JM, Lam SH, Mathavan S et al (2011) A high level of liver-specific expression of oncogenic Kras(V12) drives robust liver tumorigenesis in transgenic zebrafish. Dis Model Mech 4(6):801–813. doi:10.1242/dmm.007831
39. Nguyen AT, Emelyanov A, Koh CH, Spitsbergen JM, Parinov S, Gong Z (2012) An inducible kras(V12) transgenic zebrafish model for liver tumorigenesis and chemical drug screening. Dis Model Mech 5(1):63–72. doi:10.1242/dmm.008367
40. Yan C, Huo X, Wang S, Feng Y, Gong Z (2015) Stimulation of hepatocarcinogenesis by neutrophils upon induction of oncogenic kras expression in transgenic zebrafish. J Hepatol. doi:10.1016/j.jhep.2015.03.024
41. Mantovani A, Cassatella MA, Costantini C, Jaillon S (2011) Neutrophils in the activation and regulation of innate and adaptive immunity. Nat Rev Immunol 11(8):519–531. doi:10.1038/nri3024
42. Piccard H, Muschel RJ, Opdenakker G (2012) On the dual roles and polarized phenotypes of neutrophils in tumor development and progression. Crit Rev Oncol Hematol 82(3):296–309. doi:10.1016/j.critrevonc.2011.06.004

43. Li Z, Huang X, Zhan H, Zeng Z, Li C, Spitsbergen JM et al (2012) Inducible and repressable oncogene-addicted hepatocellular carcinoma in Tet-on xmrk transgenic zebrafish. J Hepatol 56(2):419–425. doi:10.1016/j.jhep.2011.07.025
44. Li Z, Zheng W, Wang Z, Zeng Z, Zhan H, Li C et al (2013) A transgenic zebrafish liver tumor model with inducible Myc expression reveals conserved Myc signatures with mammalian liver tumors. Dis Model Mech 6(2):414–423. doi:10.1242/dmm.010462
45. Reboredo M, Kramer MG, Smerdou C, Prieto J, De Las Rivas J (2008) Transcriptomic effects of Tet-on and mifepristone-inducible systems in mouse liver. Hum Gene Ther 19(11):1233–1247. doi:10.1089/hum.2008.057
46. Buckley AF, Burgart LJ, Sahai V, Kakar S (2008) Epidermal growth factor receptor expression and gene copy number in conventional hepatocellular carcinoma. Am J Clin Pathol 129(2):245–251. doi:10.1309/WF10QAAED3PP93BH
47. Feitelson MA, Pan J, Lian Z (2004) Early molecular and genetic determinants of primary liver malignancy. Surg Clin North Am 84(2):339–354. doi:10.1016/S0039-6109(03)00226-3
48. Zheng W, Li Z, Nguyen AT, Li C, Emelyanov A, Gong Z (2014) Xmrk, kras and myc transgenic zebrafish liver cancer models share molecular signatures with subsets of human hepatocellular carcinoma. PLoS One 9(3), e91179. doi:10.1371/journal.pone.0091179
49. Mudbhary R, Hoshida Y, Chernyavskaya Y, Jacob V, Villanueva A, Fiel MI et al (2014) UHRF1 overexpression drives DNA hypomethylation and hepatocellular carcinoma. Cancer Cell 25(2):196–209. doi:10.1016/j.ccr.2014.01.003
50. Mullins MC, Hammerschmidt M, Haffter P, Nusslein-Volhard C (1994) Large-scale mutagenesis in the zebrafish: in search of genes controlling development in a vertebrate. Curr Biol 4(3):189–202
51. Streisinger G, Walker C, Dower N, Knauber D, Singer F (1981) Production of clones of homozygous diploid zebra fish (Brachydanio rerio). Nature 291(5813):293–296
52. Leshchiner I, Alexa K, Kelsey P, Adzhubei I, Austin-Tse CA, Cooney JD et al (2012) Mutation mapping and identification by whole-genome sequencing. Genome Res 22(8):1541–1548. doi:10.1101/gr.135541.111
53. Moore JL, Rush LM, Breneman C, Mohideen MA, Cheng KC (2006) Zebrafish genomic instability mutants and cancer susceptibility. Genetics 174(2):585–600. doi:10.1534/genetics.106.059386
54. Berghmans S, Murphey RD, Wienholds E, Neuberg D, Kutok JL, Fletcher CD et al (2005) tp53 mutant zebrafish develop malignant peripheral nerve sheath tumors. Proc Natl Acad Sci U S A 102(2):407–412. doi:10.1073/pnas.0406252102
55. Goessling W, North TE, Lord AM, Ceol C, Lee S, Weidinger G et al (2008) APC mutant zebrafish uncover a changing temporal requirement for wnt signaling in liver development. Dev Biol 320(1):161–174. doi:10.1016/j.ydbio.2008.05.526
56. Hurlstone AF, Haramis AP, Wienholds E, Begthel H, Korving J, Van Eeden F et al (2003) The Wnt/beta-catenin pathway regulates cardiac valve formation. Nature 425(6958):633–637. doi:10.1038/nature02028
57. Haramis AP, Hurlstone A, van der Velden Y, Begthel H, van den Born M, Offerhaus GJ, Clevers HC (2006) Adenomatous polyposis coli-deficient zebrafish are susceptible to digestive tract neoplasia. EMBO Rep 7(4):444–449. doi:10.1038/sj.embor.7400638
58. Yang HW, Kutok JL, Lee NH, Piao HY, Fletcher CD, Kanki JP, Look AT (2004) Targeted expression of human MYCN selectively causes pancreatic neuroendocrine tumors in transgenic zebrafish. Cancer Res 64(20):7256–7262. doi:10.1158/0008-5472.CAN-04-0931
59. Park SW, Davison JM, Rhee J, Hruban RH, Maitra A, Leach SD (2008) Oncogenic KRAS induces progenitor cell expansion and malignant transformation in zebrafish exocrine pancreas. Gastroenterology 134(7):2080–2090. doi:10.1053/j.gastro.2008.02.084
60. Eser S, Schnieke A, Schneider G, Saur D (2014) Oncogenic KRAS signalling in pancreatic cancer. Br J Cancer 111(5):817–822. doi:10.1038/bjc.2014.215
61. Kanda M, Matthaei H, Wu J, Hong SM, Yu J, Borges M et al (2012) Presence of somatic mutations in most early-stage pancreatic intraepithelial neoplasia. Gastroenterology 142(4):730–733 e739. doi:10.1053/j.gastro.2011.12.042

62. Morris JP 4th, Wang SC, Hebrok M (2010) KRAS, Hedgehog, Wnt and the twisted developmental biology of pancreatic ductal adenocarcinoma. Nat Rev Cancer 10(10):683–695. doi:10.1038/nrc2899
63. Aguirre AJ, Bardeesy N, Sinha M, Lopez L, Tuveson DA, Horner J et al (2003) Activated Kras and Ink4a/Arf deficiency cooperate to produce metastatic pancreatic ductal adenocarcinoma. Genes Dev 17(24):3112–3126. doi:10.1101/gad.1158703
64. Guerra C, Schuhmacher AJ, Canamero M, Grippo PJ, Verdaguer L, Perez-Gallego L et al (2007) Chronic pancreatitis is essential for induction of pancreatic ductal adenocarcinoma by K-Ras oncogenes in adult mice. Cancer Cell 11(3):291–302. doi:10.1016/j.ccr.2007.01.012
65. Hingorani SR, Petricoin EF, Maitra A, Rajapakse V, King C, Jacobetz MA et al (2003) Preinvasive and invasive ductal pancreatic cancer and its early detection in the mouse. Cancer Cell 4(6):437–450
66. Hruban RH, Adsay NV, Albores-Saavedra J, Anver MR, Biankin AV, Boivin GP et al (2006) Pathology of genetically engineered mouse models of pancreatic exocrine cancer: consensus report and recommendations. Cancer Res 66(1):95–106. doi:10.1158/0008-5472.CAN-05-2168
67. Pylayeva-Gupta Y, Grabocka E, Bar-Sagi D (2011) RAS oncogenes: weaving a tumorigenic web. Nat Rev Cancer 11(11):761–774. doi:10.1038/nrc3106
68. Seidler B, Schmidt A, Mayr U, Nakhai H, Schmid RM, Schneider G, Saur D (2008) A Cre-loxP-based mouse model for conditional somatic gene expression and knockdown in vivo by using avian retroviral vectors. Proc Natl Acad Sci U S A 105(29):10137–10142. doi:10.1073/pnas.0800487105
69. Liu S, Leach SD (2011) Screening pancreatic oncogenes in zebrafish using the Gal4/UAS system. Methods Cell Biol 105:367–381. doi:10.1016/B978-0-12-381320-6.00015-1
70. Schiavone M, Rampazzo E, Casari A, Battilana G, Persano L, Moro E et al (2014) Zebrafish reporter lines reveal in vivo signaling pathway activities involved in pancreatic cancer. Dis Model Mech 7(7):883–894. doi:10.1242/dmm.014969
71. Hwang WY, Fu Y, Reyon D, Maeder ML, Tsai SQ, Sander JD et al (2013) Efficient genome editing in zebrafish using a CRISPR-Cas system. Nat Biotechnol 31(3):227–229. doi:10.1038/nbt.2501
72. Jao LE, Wente SR, Chen W (2013) Efficient multiplex biallelic zebrafish genome editing using a CRISPR nuclease system. Proc Natl Acad Sci U S A 110(34):13904–13909. doi:10.1073/pnas.1308335110
73. Ablain J, Durand EM, Yang S, Zhou Y, Zon LI (2015) A CRISPR/Cas9 vector system for tissue-specific gene disruption in zebrafish. Dev Cell 32(6):756–764. doi:10.1016/j.devcel.2015.01.032
74. Ghiorzo P (2014) Genetic predisposition to pancreatic cancer. World J Gastroenterol 20(31):10778–10789. doi:10.3748/wjg.v20.i31.10778
75. Petersen GM, Amundadottir L, Fuchs CS, Kraft P, Stolzenberg-Solomon RZ, Jacobs KB et al (2010) A genome-wide association study identifies pancreatic cancer susceptibility loci on chromosomes 13q22.1, 1q32.1 and 5p15.33. Nat Genet 42(3):224–228. doi:10.1038/ng.522
76. Wolpin BM, Rizzato C, Kraft P, Kooperberg C, Petersen GM, Wang Z et al (2014) Genome-wide association study identifies multiple susceptibility loci for pancreatic cancer. Nat Genet 46(9):994–1000. doi:10.1038/ng.3052
77. Auer TO, Duroure K, De Cian A, Concordet JP, Del Bene F (2014) Highly efficient CRISPR/Cas9-mediated knock-in in zebrafish by homology-independent DNA repair. Genome Res 24(1):142–153. doi:10.1101/gr.161638.113
78. Bedell VM, Wang Y, Campbell JM, Poshusta TL, Starker CG, Krug RG 2nd et al (2012) In vivo genome editing using a high-efficiency TALEN system. Nature 491(7422):114–118. doi:10.1038/nature11537
79. Shin J, Chen J, Solnica-Krezel L (2014) Efficient homologous recombination-mediated genome engineering in zebrafish using TALE nucleases. Development 141(19):3807–3818. doi:10.1242/dev.108019
80. Gupta V, Poss KD (2012) Clonally dominant cardiomyocytes direct heart morphogenesis. Nature 484(7395):479–484. doi:10.1038/nature11045

81. Pan YA, Freundlich T, Weissman TA, Schoppik D, Wang XC, Zimmerman S et al (2013) Zebrabow: multispectral cell labeling for cell tracing and lineage analysis in zebrafish. Development 140(13):2835–2846. doi:10.1242/dev.094631
82. Tang Q, Abdelfattah NS, Blackburn JS, Moore JC, Martinez SA, Moore FE et al (2014) Optimized cell transplantation using adult rag2 mutant zebrafish. Nat Methods 11(8):821–824. doi:10.1038/nmeth.3031
83. McGrail M, Hatler JM, Kuang X, Liao HK, Nannapaneni K, Watt KE et al (2011) Somatic mutagenesis with a Sleeping Beauty transposon system leads to solid tumor formation in zebrafish. PLoS One 6(4), e18826. doi:10.1371/journal.pone.0018826
84. Jang ZH, Chung HC, Ahn YG, Kwon YK, Kim JS, Ryu JH et al (2012) Metabolic profiling of an alcoholic fatty liver in zebrafish (Danio rerio). Mol Biosyst 8(7):2001–2009. doi:10.1039/c2mb25073j
85. Jones DR, Ramirez IB, Lowe M, Divecha N (2013) Measurement of phosphoinositides in the zebrafish Danio rerio. Nat Protoc 8(6):1058–1072. doi:10.1038/nprot.2013.040
86. Adatto I, Lawrence C, Thompson M, Zon LI (2011) A new system for the rapid collection of large numbers of developmentally staged zebrafish embryos. PLoS One 6(6), e21715. doi:10.1371/journal.pone.0021715
87. Colanesi S, Taylor KL, Temperley ND, Lundegaard PR, Liu D, North TE et al (2012) Small molecule screening identifies targetable zebrafish pigmentation pathways. Pigment Cell Melanoma Res 25(2):131–143. doi:10.1111/j.1755-148X.2012.00977.x
88. Le X, Pugach EK, Hettmer S, Storer NY, Liu J, Wills AA et al (2013) A novel chemical screening strategy in zebrafish identifies common pathways in embryogenesis and rhabdomyosarcoma development. Development 140(11):2354–2364. doi:10.1242/dev.088427
89. White RM, Cech J, Ratanasirintrawoot S, Lin CY, Rahl PB, Burke CJ et al (2011) DHODH modulates transcriptional elongation in the neural crest and melanoma. Nature 471(7339):518–522. doi:10.1038/nature09882

Focusing the Spotlight on the Zebrafish Intestine to Illuminate Mechanisms of Colorectal Cancer

Viola H. Lobert, Dmitri Mouradov, and Joan K. Heath

Abstract Colorectal cancer, encompassing colon and rectal cancer, arises from the epithelial lining of the large bowel. It is most prevalent in Westernised societies and is increasing in frequency as the world becomes more industrialised. Unfortunately, metastatic colorectal cancer is not cured by chemotherapy and the annual number of deaths caused by colorectal cancer, currently 700,000, is expected to rise. Our understanding of the contribution that genetic mutations make to colorectal cancer, although incomplete, is reasonably well advanced. However, it has only recently become widely appreciated that in addition to the ongoing accumulation of genetic mutations, chronic inflammation also plays a critical role in the initiation and progression of this disease. While a robust and tractable genetic model of colorectal cancer in zebrafish, suitable for pre-clinical studies, is not yet available, the identification of genes required for the rapid proliferation of zebrafish intestinal epithelial cells during development has highlighted a number of essential genes that could be targeted to disable colorectal cancer cells. Moreover, appreciation of the utility of zebrafish to study intestinal inflammation is on the rise. In particular, zebrafish

---

Herein, gene names adhere to accepted vertebrate gene and protein symbol conventions; hence: Human: *GENE*/PROTEIN; Mouse: *Gene*/PROTEIN; Zebrafish *gene*/Protein. Human: *TNF*/TNF; Mouse: *Tnf*/TNF; Zebrafish: *tnfa*/Tnfa

V.H. Lobert
Development and Cancer Division, The Walter and Eliza Hall Institute of Medical Research, 1G Royal Parade, Parkville, VIC 3052, Australia

Department of Biochemistry, Institute for Cancer Research, Oslo University Hospital, Montebello N-0379, Oslo, Norway
e-mail: Viola.Lobert@rr-research.no

D. Mouradov
Systems Biology and Personalised Medicine Division, The Walter and Eliza Hall Institute of Medical Research, 1G Royal Parade, Parkville, VIC 3052, Australia
e-mail: dmitri.mouradov@wehi.edu.au

J.K. Heath (✉)
Development and Cancer Division, The Walter and Eliza Hall Institute of Medical Research, 1G Royal Parade, Parkville, VIC 3052, Australia

Department of Medical Biology, University of Melbourne, Parkville, VIC 3052, Australia
e-mail: joan.heath@wehi.edu.au

D.M. Langenau (ed.), *Cancer and Zebrafish*, Advances in Experimental Medicine and Biology 910, DOI 10.1007/978-3-319-30654-4_18

provide unique opportunities to investigate the impact of genetic and environmental factors on the integrity of intestinal epithelial barrier function. With currently available tools, the interplay between epigenetic regulators, intestinal injury, microbiota composition and innate immune cell mobilisation can be analysed in exquisite detail. This provides excellent opportunities to define critical events that could potentially be targeted therapeutically. Further into the future, the use of zebrafish larvae as hosts for xenografts of human colorectal cancer tissue, while still in its infancy, holds great promise that zebrafish could one day provide a practical, pre-clinical personalized medicine platform for the rapid assessment of the metastatic potential and drug-sensitivity of patient-derived cancers.

Keywords Zebrafish • Intestinal epithelium • Colon cancer • Colorectal cancer • WNT signalling • Intestinal permeability • Microbiota • Inflammatory bowel disease

Overview

Colorectal cancer comprises colon cancer and rectal cancer. It is a highly heterogeneous disease and pathologists have recognised, on the basis of morphology and anatomical location, that there are at least two major sub-types [1]. This anatomical heterogeneity has now been afforded a molecular explanation as a result of analysis of gene expression data collected by The Cancer Genome Atlas (TCGA) Network [2]. As well as genetic factors, a range of environmental factors also contribute to the onset and progression of colorectal cancer. It is perhaps therefore not surprising that this has been a challenging disease to model effectively in zebrafish. The most direct attempt at a genetic model employed Targeting Induced Local Lesions in Genomes (TILLING) to identify a line of zebrafish harbouring a nonsense mutation in the mutation cluster region (mcr) of the colorectal cancer suppressor gene, *apc* (adenomatous polyposis coli) [3]. In humans, such mutations encode truncated APC proteins, which release the brakes on WNT signalling and generate a pro-proliferative phenotype. Zebrafish inheriting two mutant *apc*^{mcr} alleles die during larval development; however, adult zebrafish heterozygous for the *apc*^{mcr} allele develop intestinal cancer with low frequency and prolonged latency [4]. This relative lack of penetrance may be due to the failure of the *apc* locus to undergo loss of heterozygosity (LOH) as occurs in most human colorectal tumours. While adult zebrafish have yet to yield a reliable genetic model of intestinal cancer, many novel insights relevant to colorectal cancer have come from analyses of larvae. These include studies that have adopted intestinal organogenesis as a surrogate for colorectal tumour growth. Crucial to this approach has been the identification of mutants with defects in intestinal development as the zebrafish larva prepares to undergo transition from endotrophic to exotrophic nutrition. Notably these mutants exhibit markedly reduced intestinal epithelial cell production between 48 and 72 h post-fertilisation (hpf). The cloning of intestinal mutants appears to be a fruitful approach to discovering genes that are relevant to colorectal cancer since the expression of many of the underlying mutated genes are dysregulated in human colon tumours (reported herein). It is plausible that some of the genes identified in this way will prove to be attractive targets for novel therapeutic approaches to combating this deadly disease. Finally, zebrafish are proving a very effective model for discoveries relating to inflammatory bowel disease, a major susceptibility factor in colorectal cancer, and in understanding the contribution of the gut microbiota to intestinal cell proliferation, intestinal permeability and inflammation.

The Prevalence of Colorectal Cancer

Colorectal cancer is one of the world's major cancer killers. In 2012, it accounted for almost 10 % of the global cancer incidence and was the fourth most common cause of cancer death worldwide [5]. After lung cancer and prostate cancer, colorectal cancer is the third most common cancer in men and in women it is the second most common cancer after breast cancer. Nearly 700,000 people in the world died of colorectal cancer in 2012. Its incidence is most prevalent in countries with high levels of industrialisation, including Europe and the Americas, where it is the third most common cancer. Due to ongoing refinements in diagnosis and treatment, almost 60 % of all colorectal cancer patients in the United States and the United Kingdom will still be alive 5 years after diagnosis (73 % for rectal cancer). However, the odds are much worse if the cancer has already spread to the lymph nodes, or beyond, at the time of diagnosis.

Treatment of Colorectal Cancer

Surgery remains the first line of treatment for colorectal cancer and is very successful if the cancer is caught early. Chemotherapy is frequently administered before surgery to shrink the cancer before an operation, and is also given after surgery if the cancer has spread to the lymph nodes. In patients with advanced colorectal cancer, where the cancer has spread to another organ (typically the liver), the treatment options are limited and 5 year survival rates are grim (6 %). Targeted monoclonal antibody-based drugs that are designed to block processes that are highly active in colorectal cancer, such as angiogenesis and cell proliferation, are attractive in theory but generally only help a small proportion of patients. Approved drugs include bevacizumab (Avastin) targeted to the pro-angiogenesis factor, vascular endothelial growth factor (VEGF), and cetuximab (Erbitux) and panitumumab (Vectibix), targeted to the epidermal growth factor receptor (EGFR) pathway. However, even in the best scenarios, survival is only extended by a few months for patients with metastatic disease. Radiation therapy is rarely used in colon cancer, but it is a routine part of treating rectal cancer, especially if the cancer has penetrated the wall of the rectum and reached neighbouring lymph nodes. There is a pressing need to identify more effective therapies to treat patients with metastatic disease.

The Genetic Basis of Colorectal Cancer

Colorectal cancer, like all cancers, is not a single disease. It has been appreciated for a long time that there are at least two major pathways leading to colorectal cancer, and these are distinguished by their different anatomical locations, pathological features and molecular signatures [2]. A very large majority of patients with sporadic colorectal cancer (~90 %) harbour mutations in the last exon of *APC*, and these generally encode truncated APC proteins that lack key domains required for APC to carry out its key tumour suppressor functions. Inherited recessive mutations in *APC* underlie the autosomal dominant polyposis syndrome known as familial adenomatous polyposis (FAP), which accounts for less than 1 % of all colorectal cancer.

In 1990, Eric Fearon and Bert Vogelstein proposed that the acquisition of mutations in *APC* is the primary event in colorectal tumorigenesis [6]. In their seminal publication, they coupled the sequence of pathological changes that occurs as an initiating lesion progresses from a benign adenomatous polyp to an invasive adenocarcinoma with the gradual accumulation of genetic alterations. As well as mutations in specific genes such as *APC* and *KRAS*, they also highlighted the importance of chromosomal losses at 18q and 17p, regions that we now know contain the tumour suppressor genes, *SMAD2*, *SMAD4* (18q) and *TP53* (17p). Loss of APC gives rise initially to benign lesions with adenomatous or villous morphology that primarily arise in the distal colon (Fig. 1). This scheme has acquired more detail over the years, but the overarching paradigm, that of the gradual accumulation of mutations in proto-oncogenes and tumour suppressor genes, remains intact. What has emerged, however, is appreciation that even though this route is the dominant or "conventional" route to colorectal cancer, it is not the only one. The recognition that not all primary lesions exhibit the same polyp morphology, and the strong preponderance for certain sub-types of lesions and tumours to appear in the proximal, rather than the distal, colon was recognised several years ago [1, 7]. This has now been backed up by molecular evidence demonstrating that these tumours share a distinctive aetiology [2, 8]. Thus, the currently accepted view is that there are two main sub-types of colorectal cancer: hypermutated and non-hypermutated. The most common tumours are non-hypermutated and are characterized by mutations in *APC*, *KRAS*, *TP53*, *PIK3CA*, *FBXW7*, *SMAD4*, *TCF7L2* and *NRAS*. These non-hypermutated tumours typically arise from lesions with dysplastic morphology that adopt a villus architecture as the adenoma increases in size. They are microsatellite stable (MSS) and exhibit high chromosome instability (CIN), see Fig. 1.

Hypermutated tumours are much less frequent than non-hypermutated tumours and have a separate aetiology. They account for approximately 16 % of all sporadic colorectal cancer cases and can be divided into three major sub-types. The biggest sub-type accounts for up to 75 % of hypermutated tumours and exhibits the CpG island methylator phenotype (CIMP). Hypermutated tumours are also found in patients with hereditary non-polyposis colon cancer (HNPCC) or Lynch syndrome, a rare condition that accounts for 1–3 % of all colorectal cancer. In both instances,

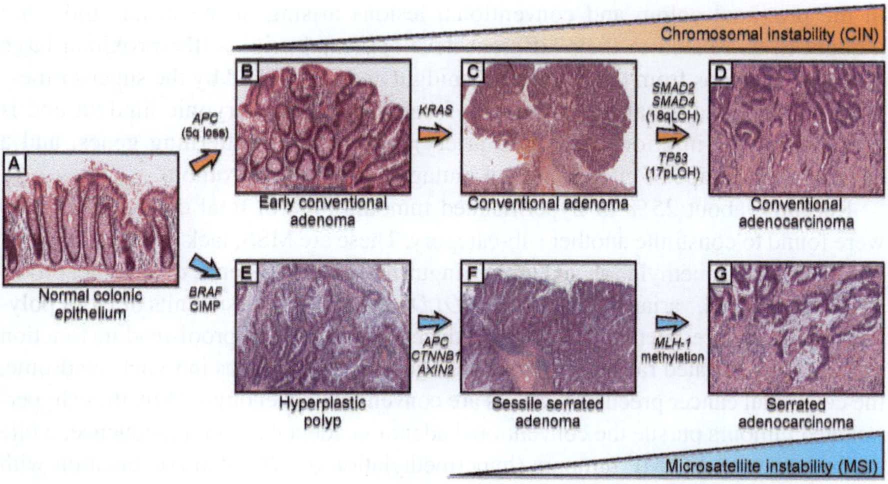

Fig. 1 The two major pathways to sporadic colorectal cancer. The progression from normal colonic epithelium to a malignant colorectal carcinoma occurs through either of two major pathways. In 85 % of cases, the conventional "adenoma-to-carcinoma" sequence is followed. The process is initiated by inactivation of the *APC* tumour suppressor gene, resulting in dysplastic (adenomatous) morphology, and abnormal architecture that becomes more villus as the adenoma increases in size. Mutations in *KRAS* and loss of heterozygosity at 18q, containing the *SMAD2* and *SMAD4* genes, and 17p, containing the *TP53* gene, are required for progression of an adenoma to an adenocarcinoma. In 80 % of cases, chromosome instability (CIN), resulting in further chromosomal gains, losses and rearrangements, occurs as the tumours progress, resulting in a poorer prognosis. The "serrated neoplasia" sequence occurs in approximately 10 % of cases of sporadic colorectal cancer. Sessile serrated adenomas harbour activating *BRAF* ($BRAF^{V600E}$) mutations and exhibit CpG island methylation at specific loci (CIMP). Progression to adenocarcinoma is associated with *MLH1* promoter hypermethylation resulting in reduced mismatch repair (MMR) function, which is manifested by microsatellite instability (MSI). Sessile serrated adenomas show distinctive serrated luminal outlines, and the adenocarcinomas are frequently highly mucinous. These images are a kind gift of Dr Michael Christie, Department of Pathology, Royal Melbourne Hospital, Parkville, Victoria 3050, Australia

tumours generally appear in the right side of the colon (proximal colon), and exhibit impaired expression of mismatch repair proteins and, as a consequence, exhibit microsatellite instability (MSI). Whereas Lynch syndrome patients harbour germline mutations in either *MLH1*, *MLH3*, *MSH2*, *MSH3*, *MSH6* and *PMS2*, which encode mismatch repair proteins, sporadic mismatch repair tumours arise most frequently as a result of methylation of the promoter of the gene encoding *MLH1* [9]. Unlike the rare HNPCC tumours, which originate as conventional adenomatous polyps, sporadic CIMP tumours are associated with acquisition of a mutant form of *BRAF* ($BRAF^{V600E}$) and the occurrence of flattened polyps with a distinctive morphology known as sessile serrated precursor lesions [1, 7]. CIMP tumours are generally more invasive and have a worse prognosis than CIN tumours. The striking differences in morphology that exist between CIMP lesions emanating in the proximal colon, and conventional lesions arising in the distal colon, are thought to be related to their different developmental origins (the proximal large intestine develops from the embryonic midgut and is supplied by the superior mesenteric artery, the distal large intestine forms from the embryonic hindgut and is supplied by the inferior mesenteric artery), exposure to patterning genes, and a changing landscape of environmental mutagens and gut microbiota.

Recently, about 25 % of hypermutated tumours (4 % of total colorectal cancer) were found to constitute another sub-category. These are MSS, lack *BRAF* mutations, CIMP or *MLH1* methylation and are distinguished by the presence of somatic, proof-reading-impaired, variants of *POLE* or *POLD1*, which encode subunits of DNA polymerases ε and δ, respectively [10]. The reduced fidelity of their proof-reading function generates an elevated rate of base pair mutations. In this case, as in Lynch syndrome, the colorectal cancer precursor lesions are conventional adenomas. Why these hypermutated tumours pursue the conventional adenoma-adenocarcinoma sequence, while the hypermutated CIMP tumours (hypermethylation of *MLH1* in combination with $BRAF^{V600E}$) follow the sessile serrated pathway, is not yet understood.

As if not complicated enough already, our understanding of colorectal tumorigenesis seems destined to get more elaborate by the application of proteogenomic approaches to assign molecular profiles. A recent study asserted that the level of a specific mRNA in a colon tumour sample is not an accurate guide of the level of the corresponding protein [11]. Moreover, using proteomics, the same authors recognised five rather than three, molecular signatures of colon cancer in the TCGA cohort, and while two of these overlapped with the 'microsatellite instability/CIMP' transcriptome subtype, they exhibited distinct mutation, methylation and protein expression patterns that were associated with different clinical outcomes [11]. Another potential complication to the use of molecular profiling to define clinically relevant cancer sub-groups stems from the fact that every colon tumour sample contains some degree of stromal tissue. Thus any colon carcinoma transcriptome encompasses mRNA signals not only from cancer cells but also from stromal cells that intrinsically display mesenchymal characteristics. Therefore assigning behaviours such as epithelial-to mesenchymal-transition (EMT) to epithelium-derived cancer cells on the basis of transcriptome analysis overlooks the possibility that these signatures may emanate from the stromal cells in the cancer sample, rather than the cancer cells themselves [12].

In summary, colorectal cancer is a complex genetic disease with multiple aetiologies. In view of the interplay of genetic and environmental factors, one would expect it to be a formidable disease to recapitulate in a model organism. However, given the need for better therapies for advanced colorectal cancer, many pre-clinical rodent models of colorectal cancer have been generated [13], and while these models rarely attain the propensity of colorectal cancer to metastasise, many have been very useful. This therefore raises the question of how zebrafish can complement mouse models to achieve better outcomes for patients with colorectal cancer. Happily, a number of studies have already shown that the well-known attributes of zebrafish can be exploited to deliver novel insights that are relevant to the causes and treatment of this disease.

The Adult Zebrafish Intestine Has Many Features in Common with the Mammalian Small Intestine

Michael Pack and colleagues (University of Pennsylvania) were the first to provide a detailed morphological analysis of the adult zebrafish intestine [14]. This revealed that whereas in mammals, the gastrointestinal tract is an extremely long and convoluted tube comprising four conspicuous organs: the oesophagus, stomach, small intestine and colon, the adult zebrafish intestine achieves the essential functions of digestion, nutrient absorption and metabolic homeostasis with a simple tapered tube folded into three sections (Fig. 2). In particular, the zebrafish intestinal tract lacks a stomach and there is also no caecum, which in mammals provides a clear demarcation between the small intestine and the large intestine/colon. Instead, the regionalisation of the adult zebrafish intestine is more gradual. A large proportion of it, encompassing the intestinal bulb, mid-intestine and the rostral portion of the posterior intestine exhibits a morphology highly reminiscent of the mammalian small intestine [14]. Only the caudal portion of the posterior intestine, comprising a short distal segment (pale green in Fig. 2), manifests a distinctive unfolded morphology. That this segment may play a role akin to the mammalian colon in zebrafish is suggested by ultrastructural analysis of this part of the intestine in a closely related

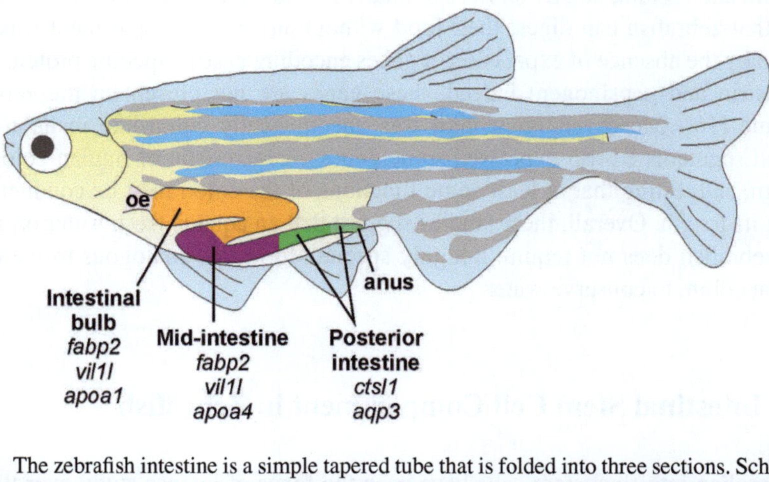

Fig. 2 The zebrafish intestine is a simple tapered tube that is folded into three sections. Schematic diagram of the mature zebrafish intestine showing three segments [37]: the intestinal bulb (*orange*), the mid-intestine (*purple*) and the posterior intestine (*green*). The intestinal bulb contains a large lumen, and its epithelial lining is thrown into elaborate folds, reminiscent of mammalian villi. The mid-intestine has shorter, broader villi and the most caudal part of the posterior intestine (*pale green*) completely lacks epithelial folds. Gene expression in the intestinal bulb and mid-intestine closely parallels that of the mammalian small intestine, including expression of *villin*, *fabp2*, *apoa1* and *apoa4* [16], indicating roles in protein and lipid absorption [14]. Meanwhile the posterior intestine expresses genes including aquaporin 3 *(aq3p)* indicating function in water absorption and cathepsin L (*ctls1*) involved in extracellular matrix degradation. *oe* oesophagus

stomach-less freshwater teleost, the tench (*Tinca tinca*). Transmission electron micrographs of this region show that the enterocytes contain a large mitochondrial population and numerous invaginations of the plasma membrane, consistent with a function in water and ion transport [15].

The wall of the mature zebrafish intestine comprises four layers: the intestinal epithelium lining the lumen, the underlying supporting layer of connective tissue known as the lamina propria (which together with the epithelium constitutes the mucosa), the muscularis externa (circular and longitudinal smooth muscle) and the outermost protective squamous epithelial layer called the serosa [14]. All but the most distal part of the zebrafish intestinal epithelium is elaborated into capacious folds that protrude into the lumen of the intestine and thereby increase the surface area over which nutrients can be absorbed. These irregular ridges are analogous to the finger-like projections or villi of the mammalian small intestine and look very similar in cross section [14, 16].

The morphological characterisation of the zebrafish intestine can now be combined with a molecular understanding due to a gene expression study from Zhiyuan Gong and colleagues (National University of Singapore). This group dissected the intestine into seven contiguous segments and used microarray analysis to analyse gene expression along its entire length [16]. These datasets, designated S1–S7, were subjected to cross-species gene set enrichment analysis, which revealed that a large anterior section of the zebrafish intestine (S1–S4) had a similar transcriptome to the human small intestine, S5 was a transition zone, S6 more closely resembled the caecum and rectum, and S7 showed similarities to the rectum exclusively. The concept that zebrafish can digest their food without an acidic compartment was reinforced by the absence of expression of genes encoding gastric-specific proteins such as gastrin and pepsinogen; indeed, these genes are not present in the zebrafish genome [16]. Moreover, no section was unequivocally colon-like in nature and indeed segments S1–S5 expressed some genes characteristic of human colon and rectum, suggesting that at least some functions of the colon may be conducted all along its length. Overall, these studies suggest that an aquatic fresh water organism like zebrafish does not require a highly specialised organ, analogous to the mammalian colon, to conserve water.

The Intestinal Stem Cell Compartment in Zebrafish

Mammalian intestinal stem cells have been the focus of intense study over the last two decades, largely because of the possibility that they may correspond to "cancer-initiating cells" in colorectal cancer [17]. In mammals, these multipotent stem cells and their highly proliferative daughter/transit amplifying cells reside at the bottom of goblet-shaped invaginations that penetrate into the submucosa to form gland-like units, known as the crypts of Lieberkühn. However, there is no comparable protective niche in the zebrafish intestinal epithelium. Instead, the intestinal stem cells occupy positions at the base of the folds [14, 18]. These locations are very similar to the inter-villus pockets that exist transiently during mammalian intestinal organogenesis and are the residence of stem cells prior to the formation of crypts during post-natal development [19]. Formation of inter-villus pockets and villi in mammals is controlled by epithelial cell-cell signalling through the WNT, NOTCH and EPH/EPHRIN signalling pathways; notably, these developmental pathways are all strongly implicated in colorectal tumorigenesis [19, 20]. As in mammals, intestinal epithelial cell production in zebrafish is regulated and maintained by Wnt signalling [21, 22] and the transcription factor Tcf4 appears to be the main Wnt effector in the zebrafish intestine [23]. Cells migrate from the base of the inter-villus pockets to the tips of the folds and tissue homeostasis is maintained by the shedding of epithelial cells upon reaching this location. In mammals, this orderly behaviour is disrupted by mutations in genes encoding components of the WNT signalling pathway, including *APC*, *CTNNB1* and *AXIN2*, leading to excessive cell production and the formation of adenomas. Readers interested in a detailed discussion of the properties and behaviours of mammalian intestinal stem cells and their mutated progeny will find no shortage of excellent reviews [24, 25].

Spontaneous Zebrafish Intestinal Cancer Is Rare

A recent retrospective analysis of wild type fish submitted to the Zebrafish International Resource Center (ZIRC) for pathological assessment over the last 10 years revealed that spontaneous intestinal neoplasia occurred in 195 zebrafish, or 2 % of the total fish submitted to the service [26]. Fish were submitted from 18 facilities in the United States and other countries, though most of the lesions occurred in fish submitted by one facility, emphasising an important role played by the environment on the incidence of intestinal tumours in laboratory zebrafish. Tumours arose in both males and females at a similar frequency and generally occurred in zebrafish greater than one year of age. The lesions were mainly adenocarcinomas or small cell carcinomas and were generally not apparent prior to dissection. Intestinal chordomas, a rare type of tumour not seen in mammals, were also observed in aged zebrafish [27]. Thus spontaneous intestinal tumours in zebrafish are not a common occurrence over the lifetime of laboratory zebrafish. Therefore to facilitate the study of intestinal cancer in zebrafish, researchers have used various approaches to increase the frequency of tumours.

Chemical and Inflammation-Induced Intestinal Cancer

Jan Spitsbergen (Oregon State University) and colleagues have pioneered the use of chemical carcinogens to manipulate cancer incidence in zebrafish. Zebrafish at developmental and adult stages may be exposed to carcinogens dissolved in the water or provided in the diet. When it occurs, intestinal neoplasia in zebrafish is generally located in the anterior portion of the intestine predominantly at the junction between the distal oesophagus and the intestinal bulb, in the intestinal bulb proper or in the region of the hepatopancreatic duct (ampulla of Vater) where it connects with the intestine. The polycyclic aromatic hydrocarbon 7,12-Dimethylbenz[*a*]anthracene (DMBA) is a prototypical PAH carcinogen that is frequently found in polluted environments. It is thought to contribute to the development of cancer upon exposure to cigarette smoke, urban air pollution and coal combustion. Embryos (60 hpf) and fry (21 days post-hatching) were exposed to aqueous DMBA at doses up to 1 and 5 ppm (mg/L), respectively, for a period of 24 h and then analysed 1 year later in the case of the treated embryos and 9 months later in the case of the fry. Juvenile zebrafish (2 months post-fertilization) were exposed to DMBA at concentrations up to 1000 ppm in their diet for a period of 4 months followed by a further 3 months on a normal diet prior to analysis [28]. Neoplasia incidence was highest in the fry exposed to DMBA (66 %). Most of the tumours arose in the liver but tumours were also found in the intestine, pancreas, thyroid and testis [28]. In 1 year old zebrafish exposed to DMBA at 60 hpf there was an overall 18 % incidence of neoplasia and again the main target organ was the liver. In 9 month old zebrafish exposed to DMBA as juveniles, the overall incidence of neoplasia was 16 %, with the intestine being the primary target organ of neoplasia, observed in 6 % of DMBA-treated fish [28]. In a similar study, exposure of juvenile zebrafish (8 weeks post-hatching) to the most carcinogenic PAH, dibenzo[a,l]pyrene (DBP), at concentrations up to 100 ppm for 4 weeks, did not result in a higher incidence of neoplasia in 11 month old fish, compared to untreated zebrafish [29]. Exposure of embryos, fry and juveniles to the experimental mutagen, *N*-methyl-*N*′-nitro-*N*-nitrosoguanidine (MNNG), gave similar results to DMBA [30]. Embryos and fry exposed to aqueous MNNG developed neoplasia several months later, but fish exposed to MNNG as juveniles (2 months post-hatching) did not. These studies demonstrate that zebrafish appear to become refractory to tumour development in response to carcinogen exposure as they become older.

Contamination of food by aflatoxin B1 (AFB1), produced mainly by the fungi *Aspergillus flavus* and *Aspergillus parasiticus*, is a common problem in tropical and subtropical regions. Absorption of this toxin occurs through the gastrointestinal and respiratory tracts, and is highly carcinogenic and/or toxic to mammals. Exposure of 2 month old zebrafish to AFB1 resulted in intestinal hyperplasia and dysplasia [31] which did not progress to neoplasia by the end of the experiment, see Fig. 3 (courtesy of Jan Spitsbergen). Intestinal neoplasia was found at a higher incidence in laboratory colonies infested with the nematode parasite, *Pseudocapillaria tomentosa*, which commonly infects cyprinid fish [32]. Worms penetrate the epithelium and lamina propria of the intestine and produce varying degrees of inflammation ranging from mild to severe, which in 10 % of cases (2 out of 19 infected fish examined) is coincident with intestinal neoplasia [32]. Not surprisingly, the combination of *Pseudocapillaria tomentosa* infection and exposure to DMBA appears to increase the frequency of intestinal neoplasia [28]. Interestingly, experiments with DMBA and MNNG that were carried out in rainbow trout resulted in much higher incidences of neoplasia. The reason for the higher incidence of tumours in rainbow trout, either spontaneous or in response to carcinogens and pathogens, may be due to the fact that unlike zebrafish, rainbow trout continue to grow throughout life [29].

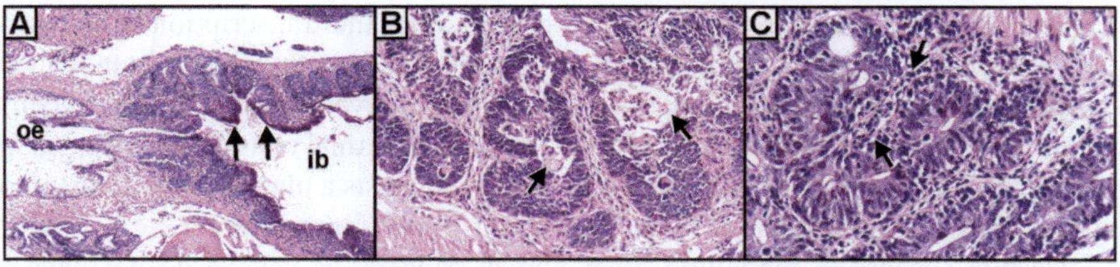

Fig. 3 Zebrafish intestinal hyperplasia induced by aflatoxin B1. (**a**). Hyperplasia at the junction of the oesophagus and intestine in Florida wild type zebrafish fed 100 ppm aflatoxin B1 (AFB1) for 9 months (from 2 months to 11 months of age). *Arrows* indicate regions of hyperplasia and dysplasia. (**b**) At an intermediate level of magnification, disorganized acinar structures lined by hyperplastic and dysplastic epithelial cells are evident (*arrows*). (**c**) At higher magnification, nests of dysplastic cells can be seen within the lamina propria, intermingled with chronic inflammatory cells (*arrows*) [31]. These images are a kind gift of Dr Jan Spitsbergen, DVM, Ph.D., Fish Disease Research Group, Department of Microbiology, Oregon State University, Corvallis, OR 97331, USA

Zebrafish Containing Germ-Line Mutations in *apc*

As mentioned previously, mutation of the adenomatous polyposis coli (*APC*) gene is the critical event in the initiation of "conventional" colorectal cancer. With this in mind, a line of zebrafish harbouring an ENU-induced point mutation within the *apc* mutation cluster region (mcr), encoding a truncated Apc protein predicted to constitutively activate Wnt/β-catenin signalling, was identified using TILLING (*apc*^{mcr}) [3]. Zebrafish harbouring two recessive *apc*^{mcr} mutations died between 72 and 96 hpf due to multiple defects including a prominent cardiac malformation [3]. Heterozygous *apc*^{mcr} mutant fish developed intestinal and liver tumours, but latency was long (14 months) and penetrance was weak (12 %) [4]. However, this latency decreased to 6 months if the mutant fish were exposed to the carcinogen DMBA, with penetrance increasing to 35 % [4].

David Jones and colleagues (University of Utah) used *apc*^{mcr} mutant larvae to reveal non-canonical (i.e. independent of β-catenin) roles for *apc* during zebrafish intestinal development that may be relevant to colorectal tumorigenesis [33, 34]). *apc*^{mcr} larvae exhibit diminished expression of intestinal cell markers such as *fabp2*, concomitant with reduced levels of retinol dehydrogenase 11 (*rdhl1*) and retinoic acid (RA) [35]. They also express abnormally high levels of C-terminal binding protein (*ctbp1*), a transcriptional co-repressor that is more highly expressed in adenoma samples from human FAP patients, compared to uninvolved colonic tissue [35]. Restoration of *apc* or targeted knockdown of *ctbp1* in *apc*^{mcr} zebrafish embryos using morpholino oligonucleotides restored expression of *rdh1l*, as well as markers of intestinal differentiation. This suggests that Apc supports RA biosynthesis and intestinal differentiation by negatively regulating the levels of CtBP1. In subsequent papers, the authors envisioned a intricate mechanism to explain these findings. In their model, mutations in *apc* indirectly suppress *rdhl1* transcription via a complex containing CtBP1, Lef1 and several other negative regulators [36]. Next, as a result of reduced RA production, genes such as *pou5f1* (encoding the pluripotency factor,

Oct4) and *cebpβ* are de-repressed, in turn promoting the transcription of components of the DNA demethylase machinery. Finally, the demethylation of genes such as *aldh* and *hoxd13* block cell fate decisions and preserve intestinal epithelial cells in an undifferentiated state [36]. This mis-fating of intestinal cells due to changes in the methylation landscape as a result of *apc* loss, suggests a plausible mechanism to drive colorectal tumorigenesis [34] that warrants further investigation.

Despite success in modelling other human cancers, as yet we do not have a robust genetic model of colorectal cancer in zebrafish [37]. Future work to correct this deficiency is likely to involve combining mutations in *apc* and other tumour suppressor genes such as *tp53*, with transgenes encoding oncogenic versions of genes such as *kras* and *braf*, which are commonly associated with the human condition. In the meantime, zebrafish researchers have identified many innovative and fruitful ways to address important questions relevant to colorectal cancer, as described below.

Xenografts of Human Colon Tumour Cells

Many researchers have transplanted human tumour cells into wild type and unpigmented zebrafish (such as the *casper* mutant) at various stages of development as well as during adulthood to provide a simple and rapid approach to studying the pro-angiogenic behaviour of tumour cells, their invasion and metastasis [38, 39]. This is particularly feasible to do at embryonic and larval stages because zebrafish up to 3 weeks old lack an adaptive immune system, allowing tumour cells to be transplanted into healthy embryos and larvae without first suppressing their immune system. Generally, a few hundred cells are microinjected into a variety of anatomical sites including the yolk sac, perivitelline space, hindbrain ventricle and bloodstream depending on developmental stage. In a few days tumour cells develop masses which frequently disseminate and form metastases. If the tumour cells are fluorescently labelled their behaviour can be followed in real time, without the need to fix and sacrifice the animals.

Not surprisingly, this general approach has been exploited to study the behaviour of cell lines derived from colon carcinoma and patient-derived colonic tumour cells. In one study it was shown that the relative metastatic ability of colorectal cancer cell lines in vitro and in in vivo mouse models was recapitulated in zebrafish. Thus SW620 cells which show highly invasive behaviour in standard in vitro transwell assays were found to disseminate widely in zebrafish larvae. In contrast, HT29 cells, which are not invasive in vitro, did not disperse at all [40]. This close correlation between the known metastatic potential of cells and their in vivo behaviour in zebrafish creates opportunities to evaluate not only whether primary patient-derived tumours contain metastatic cells, but also whether their metastatic behaviour can be manipulated using inhibitors of signalling pathways. This type of analysis holds great promise to deliver valuable information regarding the potential of an individual patient's tumour to respond to available therapies in a highly timely fashion.

However, such aspirations are yet to be realised and, to our knowledge, only one study has utilised tissue derived from a patient's primary colon tumour [41]. In this case, small explants of a single colon tumour were transplanted into the yolks of 80 individual zebrafish embryos at 48 hpf. Of the 66 successful transplants, 29 of them (44.4 %) demonstrated invasive behaviour [41]. Unfortunately, whether this patient went on to develop colon tumour metastases was not reported. In the same study, explants of normal colonic tissue from the same patient and two benign polyps from other patients showed no spreading behaviour. This is an exciting area but many more studies will be required before the zebrafish is recognised as a preclinical personalized medicine platform that is more practical than immunocompromised mice for the rapid assessment of the metastatic potential and drug-sensitivity of patient-derived xenografts.

The Developing Zebrafish Intestine as a Surrogate for Colon Cancer

Many genes that are highly active during development are down-regulated in adulthood. A notable exception to this is the frequently elevated or aberrant expression of developmental genes in cancer. Embryogenesis and cancer share many characteristics, including rapid proliferation, apoptosis, cell migration, angiogenesis, epithelial-to-mesenchymal transition (EMT) and mesenchymal-to-epithelial transition (MET). Pre-eminent examples of genes that fit this paradigm encode components of the WNT/β-catenin pathway, including *APC*, *AXIN*, *GSK3β*, *CTNNB1*, *TCF4*, which play fundamental roles in intestinal development and are frequently mutated in colorectal cancer. As a result, therapeutic targeting of WNT/β-catenin signalling has attracted considerable attention in the context of cancer [42–44]. This encouraged us to try to identify novel targets for colorectal cancer therapies using a forward genetics approach to discovering genes required for intestinal development in zebrafish.

Previously, we and others have described the morphological events that shape the normal zebrafish intestinal epithelium [14, 45, 46]. These accounts explain how a thin, unstructured ribbon of endoderm at 26 hpf is converted into a fully functional intestine ready for exotrophic nutrition by 126 hpf. Between 26 hpf and 76 hpf, bromodeoxyuridine incorporation analysis indicates that the intestinal endoderm is the most proliferative tissue in the developing organism [45, 47]. As a result of this vigorous cell production, the rostral intestine enlarges to form a compartment with a wide lumen known as the intestinal bulb (Fig. 4a). From 96 hpf the epithelium starts to elaborate folds and the proliferating cells become progressively restricted to regions between the bases of the folds. That this orderly development of the intestinal epithelium is amenable to genetic disruption is illustrated by the mutant, *flotte lotte* (*flo*) (Fig. 4b), one of the first intestinal mutants to be collected [48]. Here the underlying mutant gene is *ahctf1*, which encodes a nuclear protein, Elys, required for nuclear pore assembly [47, 49]. This mutant was amongst the first to demonstrate that an apparently tissue-specific phenotype can be generated by disrupting an

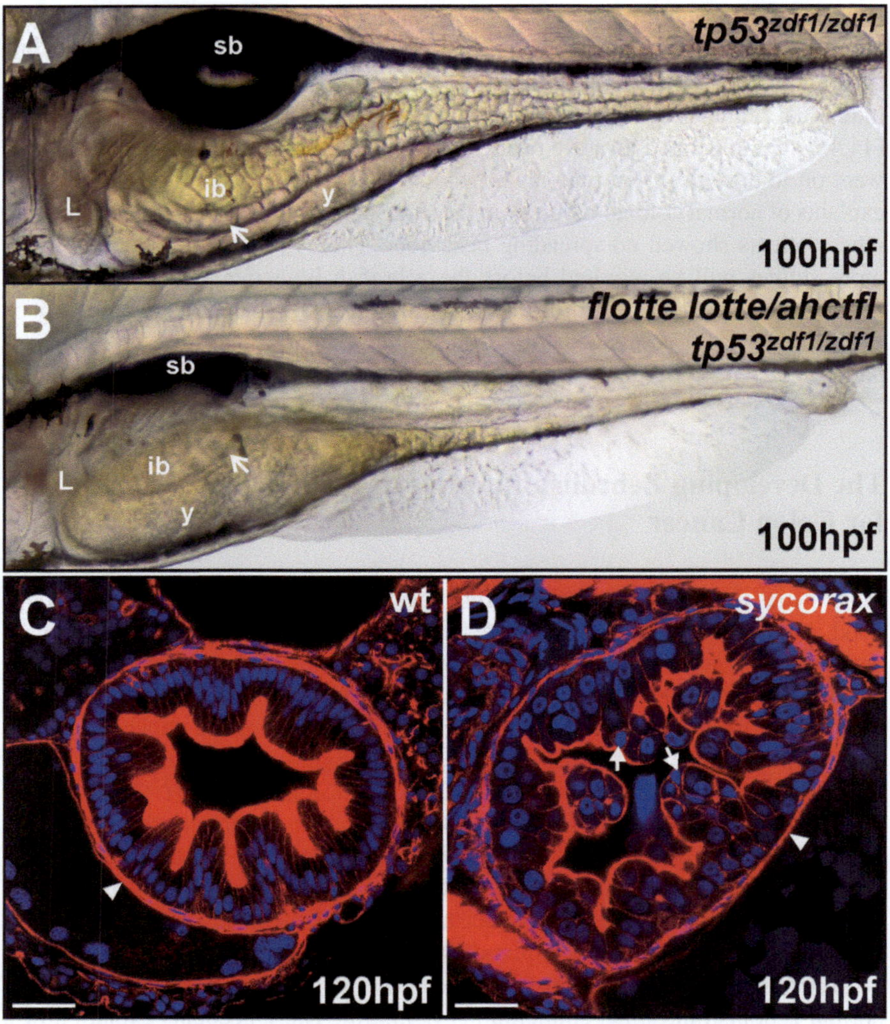

Fig. 4 Development of the intestinal epithelium in zebrafish mutants. (**a**) The development of the zebrafish intestine, shown at 100 hpf, is not perturbed in larvae carrying two mutant *tp53* alleles (*tp53*$^{zdf1/zdf1}$ [39]). In the expanded rostral region known as the intestinal bulb (ib), the intestinal epithelium (*white arrow*) is thick and elaborated into folds. (**b**) In contrast, the intestinal epithelium in *flotte lotte*, also on the *tp53*$^{zdf1/zdf1}$ background, is thin and unfolded reflecting a net loss of epithelial cells, due to Tp53-independent apoptosis [47, 49]. The underlying mutation in *flotte lotte* resides in the *ahctfl* gene, encoding the Elys protein, which is indispensable for nuclear pore assembly [47, 49]. *L* liver, *sb* swim bladder, *y* yolk. (**c**, **d**) Transverse sections (200 μm) through the intestinal bulb of a *sycorax* larva and wild type sibling at 120 hpf incubated with rhodamine phalloidin to stain F-actin (*red*) and Hoechst 33342 to stain DNA (*blue*). *sycorax* was identified in the Liver^{plus} screen conducted in the Stainier laboratory at the University of California, San Francisco [51]. Unlike the monolayer of columnar cells evident in the wild type sibling (**c**), the intestinal epithelium in *sycorax* (**d**) has adopted a multi-layered morphology. The dysplastic intestinal epithelial cells in *sycorax* (*arrows*) are smaller, cytoplasm poor and no longer in contact with the basement membrane (*arrowheads*). Images were acquired by Drs Adam Parslow (*flotte lotte*) and Elizabeth Christie (*sycorax*), who characterised these two mutants during their PhD studies at the University of Melbourne

essential cellular gene. Close inspection revealed that nuclear pore formation was perturbed in several discrete regions in *flo* mutant larvae, including the eye and optic tectum, tissues that are also highly proliferative at 72 hpf.

Because of the relatively late development of the zebrafish intestine and its deep internal position compared to more prominent tissues such as the eye, brain, heart and somites, only a handful of mutants with abnormal intestinal phenotypes were identified in the large scale ENU mutagenesis screens that were conducted in Boston and Tübingen in the early nineties [48, 50]. To expand this small collection we participated in the Liver^{plus} screen conducted in the laboratory of Didier Stainier, then at the University of California, San Francisco [51]. This 3-generation ENU mutagenesis screen was performed on the gutGFP background (*Tg(Xla.Eef1a1:GFP; s854Tg)* which expresses GFP in the digestive organs from 24 hpf, greatly facilitating the identification of mutants with defects in liver, pancreas and intestinal morphogenesis [45, 52–54]. Our specific aim was to discover genes that are indispensable for the growth and division of intestinal epithelial cells during development and we expected this approach to lead us to genes that are dysregulated in colorectal cancer. We collected a dozen mutants where the intestinal epithelial cells exhibit growth arrest or apoptosis. Of the eight genes we have cloned so far, most belong to a group of essential cellular genes that have been coined 'information processing' genes with functions in processes such as transcription, rRNA processing [55], pre-mRNA splicing [56], and nuclear pore formation [47]. Alongside our work, several other laboratories have cloned and characterised intestinal mutants (see Table 1) and many of them display similar morphological characteristics and ontology terms as ours. For example, as well as our mutant in *pwp2h* [55], mutations in *bms1l* [57], *rbm19* [46] and *tbl3* [58] disrupt ribosome biogenesis. Pre-mRNA splicing is impaired in our *rnpc3* mutant [56] as well as in the *ddx46* mutant [59]. Nuclear pore assembly is impaired by mutations in *ahctf1/elys* [47, 49] and *nup107*, and the transcription of 5S ribosomal RNA, tRNAs and other small RNAs is disrupted by mutation in *polr3b* [60]. To determine whether these so-called 'information processing' genes are relevant to colorectal cancer, we analysed the expression of their human orthologues in COAD (colon adenocarcinoma) data retrieved from The Cancer Genome Atlas (TCGA). As we had hypothesised, many of the genes are up-regulated in colon cancer (see pink columns in Table 1). Similar results were obtained for rectal cancer (data not shown), suggesting that as well as being required for normal intestinal development, these genes are also crucial for colorectal cancer. This raises the question of whether drugs targeting transcription, nuclear pore formation or splicing could selectively arrest the growth or induce the death of colorectal cancer cells. In this regard, it is perhaps worth noting that in all instances where it has been attempted, the striking intestinal phenotypes are not rescued in mutants also carrying two mutant *tp53* alleles ($tp53^{zdf1/zdf1}$ [61]) [47, 49]. For example, the high level of apoptosis seen in the intestinal epithelium of *flo* larvae is not dependent on Tp53 activation (Fig. 4b). This may suggest that drugs designed to block nuclear pore formation in highly proliferative cells would still be effective in killing human tumours carrying *TP53* mutations. Of the essential cellular processes mentioned above, the one that has attracted most attention as a therapeutic target in cancer is ribosome biogenesis [62]. However,

Table 1 The mutated genes underlying the phenotypes in published zebrafish intestinal mutants are frequently dysregulated in colon cancer

For gene expression analysis, normalized gene level RNA-Seq expression data from Colon adenocarcinoma (COAD) was retrieved from The Cancer Genome Atlas (TCGA, level 3 data containing RNA-Seq by Expectation-Maximization quantification) portal for tumour and paired normal samples. Expression values were compared between tumour and normal in a paired and unpaired analysis. (P): paired analysis with 41 matched samples of normal and tumour, (U): unpaired analysis with unpaired 41 normal and 461 tumour samples. All fold changes presented have an adjusted p-value<0.05. Genes are up-regulated in colon cancer denoted by pink, down-regulated by blue

Zebrafish gene symbol/ allele	Human gene symbol	Mutant name	Defects (all defects are for zebrafish carrying two mutant alleles unless otherwise stated)	Molecular process affected	Gene expression in colon cancer	Reference
ahctf1^{s871,ti262c}	*AHCTF1*	*flotte lotte (flo)*	Smaller intestine, liver, pancreas, craniofacial complex and eyes	Nuclear pore formation	Upregulated by 1.2 fold (P); 1.4 fold (U)	[47, 49]
apc^{hu745}	*APC*	*apc*^{mcr}	Cardiac valve malformation *apc*$^{mcr/+}$: intestinal and liver tumours	Increased Wnt signalling, retinoic acid production	Downregulated by 2.1 fold (P); 2 fold (U)	[3, 4, 33, 35]
ascl1a^{t25215}	*ASCL1*	*pituitary absent (pia)*	Absence of secretory cells in intestine, decreased intestinal motility	Notch signalling	Downregulated by 1.9 fold (P); 2.5 fold (U)	[101, 102]
bms1l^{sq163}	*BMS1*	*bms1l*	Smaller intestine, liver and pancreas	Ribosome biogenesis	Upregulated by 1.4 fold (P); 1.5 fold (U)	[57]
cdipt^{hi559Tg}	*CDIPT*	*cdipt*	IBD symptoms: apoptosis of goblet cells, abnormal mucus secretion, bacterial overgrowth and leucocyte infiltration	Phosphatidylinositol synthesis	Unchanged	[74]
cftr^{pd1048}	*CFTR*	*cftr*	Disrupted Kuppfer's vesicles and lumen expansion	Impaired fluid secretion	Unchanged	[103]
ddx46^{hi2137Tg}	*DDX46*	*morendo (mor)*	Smaller intestine, liver, pancreas and brain	Pre-mRNA splicing	Upregulated by 1.1 fold (P); 1.2 fold (U)	[59]
diexf^{hi429Tg}	*C1orf107*	*def*	Smaller intestine, liver and pancreas	Activates tp53	Upregulated by 1.7 fold (P); 1.8 fold (U)	[104, 105]
dld^{tr233}	*DLL1*	*after eight (aei)*	Excessive production of secretory cells	Notch signalling	Downregulated by 1.1 fold (P); 1.3 fold (U)	[18]
espl1^{cz280}	*ESPL1*	*cease & desist (cds)*	Intestinal adenomas in mutants treated with the mutagen MNNG?	Spindle defect, mitotic checkpoint delay	Upregulated by 2 fold (P); 2.1 fold (U)	[106]
mib1^{ta52b}	*MIB1*	*mindbomb (mib)*	Excessive production of secretory cells	E3 ubiquitin ligase essential for Notch signalling	Downregulated by 1.2 fold (P); 1.1 fold (U)	[107]
mst1^{t34230}	*MST1*	*msp*	Intestinal inflammation and goblet cell activation/mucin production	Signalling in macrophages	Upregulated by 4.5 fold (P); 3.9 fold (U)	[75]
myh11^{m498}	*MYH11*	*meltdown (mlt)*	Intestinal cysts in the posterior intestine (invasive intestinal epithelial cells)	Smooth muscle integrity	Reduced expression in MSI colon cancer[64]; Down-regulated by 18.9 fold (P); 21 fold (U)	[63, 108]
nup107^{tmu068Gt}	*NUP107*	*nup107*	Smaller intestine, liver, pancreas, craniofacial complex, eyes	Nuclear pore assembly	Upregulated by 1.8 fold (P) 1.9 fold (U)	[109]
polr3b^{m74}	*POLR3B*	*slim jim (slj)*	Smaller intestine, liver, pancreas and retinae	Transcription	Unchanged	[60]
prkci^{m567,m129}	*PRKCI*	*heart and soul (has)*	Multiple lumens in the intestine, defects in retina, neural tube	Maintenance of adherens junctions	Unchanged	[110]
ptbp1a^{p50ada}	*PTBP1*	*brom bones (brb)*	Compromised secretory goblet cell differentiation, hyper-proliferation, enhanced apoptosis	Notch signalling	Upregulated by 1.5 fold (P); 1.4 fold (U)	[111]

Table 1 (continued)

pwp2h$^{tt450, s927}$	*PWP2*	*titania (tti)*	Smaller intestine, liver, pancreas and eyes	Ribosome biogenesis	Upregulated by 1.3 fold (P); 1.3 fold (U)	[55]
rbm19^{fw07g}	*RBM19*	*nil per os (npo)*	Smaller intestine, liver and pancreas	Ribosome biogenesis [112]	Upregulated by 1.6 fold (P); 1.6 fold (U)	[46]
rnpc3^{s846}	*RNPC3*	*caliban (clbn)*	Smaller intestine, liver, pancreas, lens	Minor class splicing	Unchanged	[56]
sec13^{sq198}	*SEC13*	*sec13*	Smaller intestine, liver and pancreas	COPII function	Upregulated by 1.5 fold (P); 1.3 (U)	[113]
smo^{hi1640Tg}	*SMO*	*smoothened (smo)*	Impaired lumen formation	Rab11 trafficking	Unchanged	[114]
stk11^{hu1960}	*STK11*	*lkb1*	Lack of intestinal folding	Regulation of metabolism	Unchanged	[115]
tbl3^{cz26}	*TBL3*	*ceylon (cey)*	Reduced hematopoietic stem and progenitor cells (HSPCs), retina, exocrine pancreas, intestine, and jaw cartilage	Cell cycle regulation	Upregulated by 2.4 fold (P); 1.7 fold (U)	[58]
tcf2^{hi2169Tg}	*HNF1B*	*tcf2*	Multiple lumens in the intestine, pronephric cysts	Regulator of claudin-15 expression	Downregulated by 1.6 fold (P); 1.6 fold (U)	[116, 117]
vps51^{p9emcf}	*ANGPT2*	*fat-free (ffr)*	Defective lipid absorption	Altered vesicular trafficking	Upregulated by 2.5 fold (P); 2.7 fold (U)	[118]

we showed that intestinal epithelial cells in *pwp2h* mutants respond to ribosomal stress by inducing autophagy, which prolonged their survival [55]. Therefore to induce cell death efficiently may require targeting both ribosome biogenesis and autophagy in some cancer contexts.

Another intestinal mutant with a phenotype highly relevant to colorectal cancer is *meltdown (mlt)*, which harbours an activating mutation in the ATPase domain of smooth muscle myosin heavy chain (*myh11*) [63]. In *mlt*, smooth muscle architecture surrounding the posterior intestine is disrupted, allowing intestinal epithelial cells to invade the surrounding mesenchyme and form cysts [63]. Consistent with this, mutations that recapitulate the activating *myh11* mutation have been identified in colon cancer patients [64]. Counter-intuitively, gene expression analysis revealed that *MYH11* is markedly down-regulated in mismatch repair colon tumours compared to control tissue [64]. In this sub-set of colon cancer patients, reduced expression is caused by defective replication of a mononucleotide repeat of 8 cytosines in the *MLH1* coding sequence, leading to frame-shift mutations and potentially nonsense-mediated decay. We also observed reduced *MYH11* gene expression in our own analysis of colon cancer transcriptomes (blue column in Table 1), suggesting that perturbations in *MYHL1* gene expression may play different roles in colon cancer development depending on the genetic make-up of the tumour and whether the mutations arise in the cancerous epithelial cells or stromal cells. Perhaps not surprisingly, zebrafish intestinal mutants rarely show increased cell production or invasive behaviour. An unpublished example is *sycorax*, where we observed a multi-layered intestinal epithelium comprising many small dysplastic cells, many of which were not attached to the basement membrane (see Fig. 4c, d). Connecting the phenotype of this mutant to its genotype promises to reveal further insights into the genetic control of intestinal development that are relevant to colorectal cancer.

Zebrafish Inflammatory Bowel Disease

The term inflammatory bowel disease (IBD) encompasses a number of human disorders characterised by inflammation of the digestive tract and associated tissues, including Crohn's disease (CD) and ulcerative colitis (UC). IBD patients, including UC and CD patients, are associated with an increased risk of developing colorectal cancer, far above community wide incidence [65–67]. UC and CD are increasing in prevalence in Europe and North America and genome-wide association studies have identified many loci associated with CD and UC susceptibility [68–72]. In particular, SNPs near *UHRF1*, *DNMT1*, and *DNMT3a* suggest that alterations in epigenetic modifier genes may lead to IBD. Of these, *uhrf1* has now been firmly implicated as a result of a study in zebrafish [73]. Here, intestinal epithelial cells of *uhrf1* mutant larvae were found to express increased levels of Tnfa resulting in innate immune cell recruitment, shedding and apoptosis of epithelial cells and impaired intestinal barrier function [73]. It was further demonstrated that *uhrf1* plays a role in restraining levels of Tnfa production in zebrafish intestinal epithelial cells by methylating the *tnfa* gene promoter [73]. This study provides the first direct indication that loss of epigenetic repression of *TNF* expression in the intestinal epithelium may contribute directly to the onset of IBD.

In addition to *uhrf1*, two more zebrafish mutants exhibit pathologies reminiscent of IBD. Mutation in *cdipt*, which encodes CDP-diacylglycerol–inositol 3-phosphatidyltransferase, a highly conserved enzyme responsible for synthesis of intracellular phosphatidylinositol, causes goblet cell apoptosis, abnormal mucous secretion, bacterial overgrowth and leucocyte infiltration [74]. Several genome-wide association studies have identified the genes encoding macrophage-stimulating protein (MSP) and its receptor RON as possible susceptibility factors in IBD. *mst1* (encodes Msp) mutant zebrafish develop spontaneous intestinal inflammation over time. From 14 to 28 weeks post-fertilisation, *mst1*-deficient zebrafish exhibit intestinal eosinophilia, increased intestinal expression of inflammatory marker Mmp9, increased mucin production and a prolonged and elevated inflammatory response to intestinal damage [75].

Larval zebrafish exposed to the haptenizing agent 2, 4, 6-trinitrobenzene sulfonic acid (TNBS) display impaired intestinal homeostasis and inflammation reminiscent of human IBD [76–78]. Salient features include disrupted intestinal fold architecture, recruitment of leukocytes into the lamina propria, a marked induction of pro-inflammatory cytokines, including Tnfa, and activation of the extracellular matrix degrading enzyme, matrix metalloproteinase 9 (Mmp9). Goblet cell number and mucin production are also increased. The intestinal inflammation is dependent on the presence of microbiota and Toll-like receptor signalling [78], and is ameliorated by antibiotic and anti-inflammatory drug treatments [76, 77]. Intestinal inflammation can also be induced in zebrafish larvae using dextran sodium sulphate (DSS) or by infection with *Salmonella enterica* injected into the yolk sac [77]. DSS and TNBS induce a similar pathology except that, in contrast to TNBS, DSS causes a decrease in the number of goblet cells and reduced production of mucin, features more reminiscent of UC than CD [79, 80].

As mentioned above, chronic intestinal inflammation strongly predisposes to colorectal cancer [81]. Yi Feng, Paul Martin and colleagues were amongst the first to exploit the translucency and genetic tractability of zebrafish larvae to explore in detail the involvement of inflammatory cells at the very time of cancer initiation, when transformed cells first arise in tissues. They recognised that while acute inflammation mediated by the innate immune system plays a crucial role in wound repair or in response to other insults that are detrimental to tissue homeostasis, chronic, subclinical levels of inflammation (so-called "smouldering" inflammation) can predispose to cancer. In elegant studies, they demonstrated that skin goblet cells, transformed with a fluorescently labelled, oncogenic version of human *RAS* (*RAS^{G12V}*) use H_2O_2 production to recruit and activate host leukocytes from the very early stages of tumour development [82]. The host immune cells then reciprocate by providing the early transformed cells with a trophic signal (prostaglandin E2) that promotes their rapid expansion [83]. These experiments, which may be interesting to repeat with intestinal epithelial cells carrying an *apc^{mcr}* mutation and the *RAS^{G12V}* transgene, suggest an explanation for why the long-term use of low doses of non-steroidal anti-inflammatory drugs, such as aspirin, appears to be effective in reducing colorectal cancer onset at the early stages of cancer initiation [84, 85].

Inflammation-associated colorectal cancer has been recapitulated in mice by exposing them to the colonic mutagen, azoxymethane (AOM), which induces mutations in genes encoding components of the WNT signalling pathway, most notably *CTNNB1*, in combination with DSS provided in the drinking water [86]. Much higher numbers of colonic tumours are produced with DSS, compared to treatment with AOM alone. While similar studies in adult zebrafish have not been attempted so far, we would expect that treatment of *$apc^{mcr/+}$* zebrafish with TNBS or DSS would increase the frequency of intestinal cancers and reduce their latency.

The Role of the Microbiota in IBD and Colorectal Cancer

The role of the intestinal microbiota is to assist in the digestion of food that would otherwise be inaccessible, control the growth of pathogenic bacteria and educate the host immune system [87]. The microbiota of the human gastrointestinal tract is made up of a vast number of microorganisms including viruses, fungi and protozoans. However, the majority of the intestinal microbiota comprises bacteria. They reside in the intestinal lumen or are associated with the outer "loose" layer of mucous on the mucosal surface. The inner adherent mucous layer is largely devoid of bacteria and provides a physical barrier that separates them from the underlying intestinal epithelial cells. Upon disruption of this barrier, the infiltration of bacteria evokes an immune response that over the long term can contribute to inflammatory bowel disease, obesity and colorectal cancer. Dysbiosis refers to a shift in the composition of the intestinal microbiota, from a healthy to a potentially pathogenic state. This can be brought on by certain diets, including those involving frequent consumption of red meat, antibiotics or mucosal injury, any of which can pave the way for the growth of opportunistic pathogens.

Impaired epithelial barrier function is a prominent feature in both CD and UC. The intestinal barrier is maintained by a number of interacting components, notably, production of mucus and anti-microbial peptides, epithelial integrity through cell-cell adhesion complexes and innate immune responses. When the mucus layer is compromised in IBD, the intestinal epithelium becomes more permeable and is exposed to the microbiota. This triggers activation of Toll-like receptors upon ligation of microbe-associated molecular patterns (MAMPs) that are shared by many microorganisms. Next, the adaptor molecule, MyD88, which acts as a central node of inflammatory pathways, couples activation of Toll-like receptors with the death domain containing kinase, IL-1R-associated kinase (IRAK), leading to activation of nuclear factor-kappa B (NFκB), and ultimately transcription of pro-inflammatory cytokines such as TNF. The laboratories of John Rawls (Duke University) and Karen Guillemin (University of Oregon) have developed a number of elegant assays to demonstrate that this pathway is active in zebrafish [88–90] and that the response to microbiota is similar between zebrafish and mice [91]. These include a practical and reliable microgavage method to deliver diverse materials, including live microorganisms and secreted microbial factors and toxins, directly to the intestinal lumen of zebrafish larvae [92]. This method is also useful to deliver fluorescently labelled dextran to the intestinal lumen to assess the integrity of intestinal barrier function. Only if the intestinal barrier is breached does the fluorescence appear in extra-intestinal sites such as the vasculature and intersomitic spaces. Other zebrafish-specific tools offering considerable utility to this field include the *Tg(mpeg-EGFP*^{gl122}*/mCherry*^{gl123}) [93] and *Tg(mpx-GFP)*^{i114} [94] transgenic lines to visualise the recruitment of macrophages and neutrophils, respectively, and a new *TgBAC(tnfa:GFP)* line to monitor spatiotemporal dynamics of *tnfa* expression [73].

Colonisation of the zebrafish intestine by microorganisms occurs at hatching [95] and, by 96 hpf, when the digestive tract is a continuous open tube with access to the outside world, bacteria are observed in the mouth, pharynx, oesophagus and intestinal bulb [95]. Studies with germ-free zebrafish showed that the microbiota is required for proper intestinal development, including epithelial proliferation and cell fate determination [91, 95]. In particular, the secretory cell lineages, both enteroendocrine cells and goblet cells, are less abundant in the intestinal epithelium of germ-free zebrafish compared to conventionally reared animals. Changes in the expression of genes involved in metabolism and innate immunity are also observed [87, 88, 91], alongside reduced numbers of intestinal-associated immune cells [89], immature patterns of glycan expression and altered gut motility [95]. These changes can be reversed by introducing a complex microbiota at later stages of development or, to a lesser extent, by exposure of germ-free zebrafish to heat-killed preparations of the microbiota or bacterial lipopolysaccharide [95]. Subsequent analysis revealed that microbial activation of Toll-like receptors/MyD88, was required alongside Wnt signalling for normal intestinal epithelial cell proliferation [21]. These observations suggest a mechanism whereby the microbiota may play a direct role in tumorigenesis.

To explore the dynamics of early microbial colonization of the zebrafish gut, germ-free larval zebrafish (5 dpf) were inoculated with EGFP-expressing *Aeromonas veronii* (an abundant species in the zebrafish gut), by introducing the bacteria into flasks containing free-swimming larvae. Using light sheet fluorescence microscopy to obtain three-dimensional images spanning the full width of the intestine, colonisation of the intestine by the fluorescent bacteria could be monitored in real time [96]. Individual bacteria were visible in the gut after one hour, and 8 h later a large bacterial population had formed. It was noted that growth did not occur homogeneously throughout the gut but was preferentially localized in particular regions, and the growth rate of aggregated bacteria was considerably higher than that of independently growing bacteria.

On the basis of molecular phylogeny (16S rRNA gene sequence comparisons), it was determined that the gut microbiota of laboratory-reared zebrafish is similar in composition to that of zebrafish collected recently from their natural habitat [97], suggesting that zebrafish have a core microbiome that is actively conserved. Moreover, when zebrafish are reared in germ-free conditions and colonised with bacteria harvested from the mouse intestine, the recipient fish respond by gradually reverting it back to the bacterial composition characteristic of their species [87].

Recently, He and colleagues showed that changes in the intestinal microbiota occur when larval zebrafish are exposed to TNBS [98]. As mentioned above, TNBS disrupts the intestinal epithelium and induces inflammation reminiscent of human IBD. Using denaturing gradient gel electrophoresis, DNA sequencing and PCR, the authors were able to characterise the intestinal microbiota dysbiosis induced by TNBS. They showed that enterocolitis severity was significantly correlated with an increased proportion of Proteobacteria (especially Burkholderia) and a decreased proportion of Firmicutes (Lactobacillus group). The authors proposed that reversing such dysbiosis may be a viable option for reducing the incidence and severity of human IBD.

Moreover, recent studies in genetically engineered mice have linked host-specific bacteria, inflammation and intestinal barrier defects with site-specific tumour development [99, 100]. This study demonstrated that the interplay between host genetics, microbial location and tumour geography gives rise to tumours in the caecum exhibiting a distinctive sessile, serrated morphology (Fig. 1). These associations are likely to be relevant to the development of sessile serrated tumours in humans, which frequently occur in association with $BRAF^{V600E}$ mutations and on the right side of the colon. Thus a picture is emerging whereby the composition and activity of resident microbes in different parts of the intestinal tract play crucial roles in shaping not only healthy metabolic networks but also disease states like IBD and colorectal cancer. No doubt these concepts will stimulate further fruitful investigation in zebrafish.

Summary

Zebrafish are providing several productive avenues for colorectal cancer research. Already a range of non-oncogene targets has been suggested by studies that adopted the rapidly developing zebrafish intestinal epithelium as a surrogate tissue for colorectal tumours. The exciting prospect of being able to inhibit essential non-oncogenes to cripple rapidly proliferating cancer cells, while at the same time sparing normal cells, is worthy of investigation in synthetic lethality screens. Other major strengths of the model reside in the ability to dissect the relationships between integrity of intestinal epithelial barrier function, microbiota composition and inflammation in predisposing to cancer. Further progress in this area would be greatly facilitated by the generation of robust and reproducible genetic models of zebrafish intestinal cancer that can be readily induced in a short time-frame. This aim is now greatly simplified by the arrival of versatile and accessible reverse genetic technologies such as Transcription Activator-Like Effector Nucleases (TALENs) and Clustered Regularly Interspaced Short Palindromic Repeats (CRISPRs). The validity of the ensuing models will require testing by histological analysis and careful comparisons of gene expression profiles in zebrafish and human tumours. Suitably apt models will surely have the potential to deliver novel colorectal cancer drugs by providing a platform for high throughput chemical screens.

Acknowledgements The authors are immensely grateful to Drs Michael Christie, Jan Spitsbergen, Adam Parslow and Elizabeth Christie for kindly providing the images for Figs. 1, 3 and 4. Drs Karen Guillemin, Tanya de Jong-Curtain and Fansuo Geng are thanked for careful reading of the manuscript. Research in the Heath laboratory is supported by the National Health and Medical Research Council of Australia, Ludwig Cancer Research and operational infrastructure grants from the Australian Federal Government (IRISS) and the Victorian State Government (OIS). JKH wishes to thank all past and present members of her laboratory for contributing data and stimulating discussions on this topic over the past several years.

References

1. East JE, Saunders BP, Jass JR (2008) Sporadic and syndromic hyperplastic polyps and serrated adenomas of the colon: classification, molecular genetics, natural history, and clinical management. Gastroenterol Clin North Am 37(1):25–46, v
2. Cancer Genome Atlas Network (2012) Comprehensive molecular characterization of human colon and rectal cancer. Nature 487(7407):330–337
3. Hurlstone AF et al (2003) The Wnt/beta-catenin pathway regulates cardiac valve formation. Nature 425(6958):633–637
4. Haramis AP et al (2006) Adenomatous polyposis coli-deficient zebrafish are susceptible to digestive tract neoplasia. EMBO Rep 7(4):444–449
5. International Agency for Research on Cancer (2014) In: Stewart BW, Wild CP (eds) World cancer report 2014. WHO, Lyon
6. Fearon ER, Vogelstein B (1990) A genetic model for colorectal tumorigenesis. Cell 61(5):759–767
7. Minoo P, Moyer MP, Jass JR (2007) Role of BRAF-V600E in the serrated pathway of colorectal tumourigenesis. J Pathol 212(2):124–133
8. Fearon ER (2011) Molecular genetics of colorectal cancer. Annu Rev Pathol 6:479–507
9. Koinuma K et al (2004) Mutations of BRAF are associated with extensive hMLH1 promoter methylation in sporadic colorectal carcinomas. Int J Cancer 108(2):237–242
10. Palles C et al (2013) Germline mutations affecting the proofreading domains of POLE and POLD1 predispose to colorectal adenomas and carcinomas. Nat Genet 45(2):136–144
11. Zhang B et al (2014) Proteogenomic characterization of human colon and rectal cancer. Nature 513(7518):382–387

12. Isella C et al (2015) Stromal contribution to the colorectal cancer transcriptome. Nat Genet 47(4):312–319
13. Washington MK et al (2013) Pathology of rodent models of intestinal cancer: progress report and recommendations. Gastroenterology 144(4):705–717
14. Wallace KN et al (2005) Intestinal growth and differentiation in zebrafish. Mech Dev 122(2):157–173
15. Noaillac-Depeyre J, Gas N (1976) Electron microscopic study on gut epithelium of the tench (Tinca tinca L.) with respect to its absorptive functions. Tissue Cell 8(3):511–530
16. Wang Z et al (2010) Morphological and molecular evidence for functional organization along the rostrocaudal axis of the adult zebrafish intestine. BMC Genomics 11:392
17. Vries RG, Huch M, Clevers H (2010) Stem cells and cancer of the stomach and intestine. Mol Oncol 4(5):373–384
18. Crosnier C et al (2005) Delta-Notch signalling controls commitment to a secretory fate in the zebrafish intestine. Development 132(5):1093–1104
19. Heath JK (2010) Transcriptional networks and signaling pathways that govern vertebrate intestinal development. Curr Top Dev Biol 90:159–192
20. Crosnier C, Stamataki D, Lewis J (2006) Organizing cell renewal in the intestine: stem cells, signals and combinatorial control. Nat Rev Genet 7(5):349–359
21. Cheesman SE et al (2011) Epithelial cell proliferation in the developing zebrafish intestine is regulated by the Wnt pathway and microbial signaling via Myd88. Proc Natl Acad Sci U S A 108(Suppl 1):4570–4577
22. Muncan V et al (2007) T-cell factor 4 (Tcf7l2) maintains proliferative compartments in zebrafish intestine. EMBO Rep 8(10):966–973
23. Faro A et al (2009) T-cell factor 4 (tcf7l2) is the main effector of Wnt signaling during zebrafish intestine organogenesis. Zebrafish 6(1):59–68
24. Clevers H (2013) The intestinal crypt, a prototype stem cell compartment. Cell 154(2):274–284
25. Philpott A, Winton DJ (2014) Lineage selection and plasticity in the intestinal crypt. Curr Opin Cell Biol 31:39–45
26. Paquette CE et al (2013) A retrospective study of the prevalence and classification of intestinal neoplasia in zebrafish (Danio rerio). Zebrafish 10(2):228–236
27. Cooper TK et al (2015) Primary intestinal and vertebral chordomas in laboratory zebrafish (Danio rerio). Vet Pathol 52:388–392
28. Spitsbergen JM et al (2000) Neoplasia in zebrafish (Danio rerio) treated with 7,12-dimethylbenz[a]anthracene by two exposure routes at different developmental stages. Toxicol Pathol 28(5):705–715
29. Reddy AP et al (1999) Experimental hepatic tumorigenicity by environmental hydrocarbon dibenzo[a, l]pyrene. J Environ Pathol Toxicol Oncol 18(4):261–269
30. Spitsbergen JM et al (2000) Neoplasia in zebrafish (Danio rerio) treated with N-methyl-N′-nitro-N-nitrosoguanidine by three exposure routes at different developmental stages. Toxicol Pathol 28(5):716–725
31. Spitsbergen JM, Buhler DR, Peterson TS (2012) Neoplasia and neoplasm-associated lesions in laboratory colonies of zebrafish emphasizing key influences of diet and aquaculture system design. ILAR J 53(2):114–125
32. Kent ML et al (2002) Pseudocapillaria tomentosa, a nematode pathogen, and associated neoplasms of zebrafish (Danio rerio) kept in research colonies. Comp Med 52(4):354–358
33. Phelps RA et al (2009) A two-step model for colon adenoma initiation and progression caused by APC loss. Cell 137(4):623–634
34. Hammoud SS, Cairns BR, Jones DA (2013) Epigenetic regulation of colon cancer and intestinal stem cells. Curr Opin Cell Biol 25(2):177–183
35. Nadauld LD et al (2004) Adenomatous polyposis coli control of retinoic acid biosynthesis is critical for zebrafish intestinal development and differentiation. J Biol Chem 279(49):51581–51589
36. Rai K et al (2010) DNA demethylase activity maintains intestinal cells in an undifferentiated state following loss of APC. Cell 142(6):930–942

37. Yen J, White RM, Stemple DL (2014) Zebrafish models of cancer: progress and future challenges. Curr Opin Genet Dev 24:38–45
38. Tobia C et al (2013) Zebrafish embryo as a tool to study tumor/endothelial cell cross-talk. Biochim Biophys Acta 1832(9):1371–1377
39. White RM et al (2008) Transparent adult zebrafish as a tool for in vivo transplantation analysis. Cell Stem Cell 2(2):183–189
40. Teng Y et al (2013) Evaluating human cancer cell metastasis in zebrafish. BMC Cancer 13:453
41. Marques IJ et al (2009) Metastatic behaviour of primary human tumours in a zebrafish xenotransplantation model. BMC Cancer 9:128
42. Lau T et al (2013) A novel tankyrase small-molecule inhibitor suppresses APC mutation-driven colorectal tumor growth. Cancer Res 73(10):3132–3144
43. Waaler J et al (2011) Novel synthetic antagonists of canonical Wnt signaling inhibit colorectal cancer cell growth. Cancer Res 71(1):197–205
44. Weng W et al (2015) Molecular therapy of colorectal cancer: progress and future directions. Int J Cancer 136(3):493–502
45. Ng AN et al (2005) Formation of the digestive system in zebrafish: III. Intestinal epithelium morphogenesis. Dev Biol 286(1):114–135
46. Mayer AN, Fishman MC (2003) Nil per os encodes a conserved RNA recognition motif protein required for morphogenesis and cytodifferentiation of digestive organs in zebrafish. Development 130(17):3917–3928
47. de Jong-Curtain TA et al (2009) Abnormal nuclear pore formation triggers apoptosis in the intestinal epithelium of elys-deficient zebrafish. Gastroenterology 136(3):902–911
48. Chen JN et al (1996) Mutations affecting the cardiovascular system and other internal organs in zebrafish. Development 123:293–302
49. Davuluri G et al (2008) Mutation of the zebrafish nucleoporin elys sensitizes tissue progenitors to replication stress. PLoS Genet 4(10), e1000240
50. Pack M et al (1996) Mutations affecting development of zebrafish digestive organs. Development 123:321–328
51. Ober EA et al (2006) Mesodermal Wnt2b signalling positively regulates liver specification. Nature 442(7103):688–691
52. Field HA et al (2003) Formation of the digestive system in zebrafish. I. Liver morphogenesis. Dev Biol 253(2):279–290
53. Ober EA, Field HA, Stainier DY (2003) From endoderm formation to liver and pancreas development in zebrafish. Mech Dev 120(1):5–18
54. Field HA et al (2003) Formation of the digestive system in zebrafish. II. Pancreas morphogenesis. Dev Biol 261(1):197–208
55. Boglev Y et al (2013) Autophagy induction is a Tor- and Tp53-independent cell survival response in a zebrafish model of disrupted ribosome biogenesis. PLoS Genet 9(2), e1003279
56. Markmiller S et al (2014) Minor class splicing shapes the zebrafish transcriptome during development. Proc Natl Acad Sci U S A 111(8):3062–3067
57. Wang Y et al (2012) Ribosome biogenesis factor Bms1-like is essential for liver development in zebrafish. J Genet Genomics 39(9):451–462
58. Hutchinson SA et al (2012) Tbl3 regulates cell cycle length during zebrafish development. Dev Biol 368(2):261–272
59. Hozumi S et al (2012) DEAD-box protein Ddx46 is required for the development of the digestive organs and brain in zebrafish. PLoS One 7(3), e33675
60. Yee NS et al (2007) Mutation of RNA Pol III subunit rpc2/polr3b leads to deficiency of subunit Rpc11 and disrupts zebrafish digestive development. PLoS Biol 5(11), e312
61. Berghmans S et al (2005) tp53 mutant zebrafish develop malignant peripheral nerve sheath tumors. Proc Natl Acad Sci U S A 102(2):407–412
62. Bywater MJ et al (2012) Inhibition of RNA polymerase I as a therapeutic strategy to promote cancer-specific activation of p53. Cancer Cell 22(1):51–65
63. Wallace KN et al (2005) Mutation of smooth muscle myosin causes epithelial invasion and cystic expansion of the zebrafish intestine. Dev Cell 8(5):717–726

64. Alhopuro P et al (2008) Unregulated smooth-muscle myosin in human intestinal neoplasia. Proc Natl Acad Sci U S A 105(14):5513–5518
65. Itzkowitz SH, Yio X (2004) Inflammation and cancer IV. Colorectal cancer in inflammatory bowel disease: the role of inflammation. Am J Physiol Gastrointest Liver Physiol 287(1):G7–G17
66. Clevers H (2004) At the crossroads of inflammation and cancer. Cell 118(6):671–674
67. Kim ER, Chang DK (2014) Colorectal cancer in inflammatory bowel disease: the risk, pathogenesis, prevention and diagnosis. World J Gastroenterol 20(29):9872–9881
68. Alonso A et al (2015) Identification of risk loci for Crohn's disease phenotypes using a genome-wide association study. Gastroenterology 148:794–805
69. Wellcome Trust Case Control Consortium (2007) Genome-wide association study of 14,000 cases of seven common diseases and 3,000 shared controls. Nature 447(7145):661–678
70. Anderson CA et al (2011) Meta-analysis identifies 29 additional ulcerative colitis risk loci, increasing the number of confirmed associations to 47. Nat Genet 43(3):246–252
71. Franke A et al (2010) Genome-wide meta-analysis increases to 71 the number of confirmed Crohn's disease susceptibility loci. Nat Genet 42(12):1118–1125
72. Mathew CG (2008) New links to the pathogenesis of Crohn disease provided by genome-wide association scans. Nat Rev Genet 9(1):9–14
73. Marjoram L et al (2015) Epigenetic control of intestinal barrier function and inflammation in zebrafish. Proc Natl Acad Sci U S A 112(9):2770–2775
74. Thakur PC et al (2014) Dysregulated phosphatidylinositol signaling promotes endoplasmic-reticulum-stress-mediated intestinal mucosal injury and inflammation in zebrafish. Dis Model Mech 7(1):93–106
75. Witte M et al (2014) Deficiency in macrophage-stimulating protein results in spontaneous intestinal inflammation and increased susceptibility toward epithelial damage in zebrafish. Zebrafish 11(6):542–550
76. Oehlers SH et al (2011) A chemical enterocolitis model in zebrafish larvae that is dependent on microbiota and responsive to pharmacological agents. Dev Dyn 240(1):288–298
77. Oehlers SH et al (2013) Chemically induced intestinal damage models in zebrafish larvae. Zebrafish 10(2):184–193
78. He Q et al (2014) Role of gut microbiota in a zebrafish model with chemically induced enterocolitis involving toll-like receptor signaling pathways. Zebrafish 11(3):255–264
79. Oehlers SH et al (2012) Retinoic acid suppresses intestinal mucus production and exacerbates experimental enterocolitis. Dis Model Mech 5(4):457–467
80. Gersemann M et al (2009) Differences in goblet cell differentiation between Crohn's disease and ulcerative colitis. Differentiation 77(1):84–94
81. Kulaylat MN, Dayton MT (2010) Ulcerative colitis and cancer. J Surg Oncol 101(8):706–712
82. Feng Y et al (2010) Live imaging of innate immune cell sensing of transformed cells in zebrafish larvae: parallels between tumor initiation and wound inflammation. PLoS Biol 8(12), e1000562
83. Feng Y, Renshaw S, Martin P (2012) Live imaging of tumor initiation in zebrafish larvae reveals a trophic role for leukocyte-derived PGE(2). Curr Biol 22(13):1253–1259
84. Langley RE, Rothwell PM (2014) Aspirin in gastrointestinal oncology: new data on an old friend. Curr Opin Oncol 26(4):441–447
85. Mione M, Zon LI (2012) Cancer and inflammation: an aspirin a day keeps the cancer at bay. Curr Biol 22(13):R522–R525
86. Neufert C, Becker C, Neurath MF (2007) An inducible mouse model of colon carcinogenesis for the analysis of sporadic and inflammation-driven tumor progression. Nat Protoc 2(8):1998–2004
87. Rawls JF et al (2006) Reciprocal gut microbiota transplants from zebrafish and mice to germ-free recipients reveal host habitat selection. Cell 127(2):423–433
88. Kanther M et al (2011) Microbial colonization induces dynamic temporal and spatial patterns of NF-kappaB activation in the zebrafish digestive tract. Gastroenterology 141(1):197–207

89. Bates JM et al (2007) Intestinal alkaline phosphatase detoxifies lipopolysaccharide and prevents inflammation in zebrafish in response to the gut microbiota. Cell Host Microbe 2(6):371–382
90. Galindo-Villegas J et al (2012) Regulation of immunity and disease resistance by commensal microbes and chromatin modifications during zebrafish development. Proc Natl Acad Sci U S A 109(39):E2605–E2614
91. Rawls JF, Samuel BS, Gordon JI (2004) Gnotobiotic zebrafish reveal evolutionarily conserved responses to the gut microbiota. Proc Natl Acad Sci U S A 101(13):4596–4601
92. Cocchiaro JL, Rawls JF (2013) Microgavage of zebrafish larvae. J Vis Exp 72, e4434
93. Ellett F et al (2011) mpeg1 promoter transgenes direct macrophage-lineage expression in zebrafish. Blood 117(4):e49–e56
94. Renshaw SA et al (2006) A transgenic zebrafish model of neutrophilic inflammation. Blood 108(13):3976–3978
95. Bates JM et al (2006) Distinct signals from the microbiota promote different aspects of zebrafish gut differentiation. Dev Biol 297(2):374–386
96. Jemielita M et al (2014) Spatial and temporal features of the growth of a bacterial species colonizing the zebrafish gut. MBio 5(6)
97. Roeselers G et al (2011) Evidence for a core gut microbiota in the zebrafish. ISME J 5(10):1595–1608
98. He Q et al (2013) Microbial fingerprinting detects intestinal microbiota dysbiosis in Zebrafish models with chemically-induced enterocolitis. BMC Microbiol 13:289
99. Bongers G et al (2014) Interplay of host microbiota, genetic perturbations, and inflammation promotes local development of intestinal neoplasms in mice. J Exp Med 211(3):457–472
100. Jobin C (2014) Do bugs define cancer geography? J Exp Med 211(3):385
101. Flasse LC et al (2013) The bHLH transcription factor Ascl1a is essential for the specification of the intestinal secretory cells and mediates Notch signaling in the zebrafish intestine. Dev Biol 376(2):187–197
102. Roach G et al (2013) Loss of ascl1a prevents secretory cell differentiation within the zebrafish intestinal epithelium resulting in a loss of distal intestinal motility. Dev Biol 376(2): 171–186
103. Navis A, Marjoram L, Bagnat M (2013) Cftr controls lumen expansion and function of Kupffer's vesicle in zebrafish. Development 140(8):1703–1712
104. Stuckenholz C et al (2009) FACS-assisted microarray profiling implicates novel genes and pathways in zebrafish gastrointestinal tract development. Gastroenterology 137(4):1321–1332
105. Tao T et al (2013) Def functions as a cell autonomous factor in organogenesis of digestive organs in zebrafish. PLoS One 8(4), e58858
106. Shepard JL et al (2007) A mutation in separase causes genome instability and increased susceptibility to epithelial cancer. Genes Dev 21(1):55–59
107. Itoh M et al (2003) Mind bomb is a ubiquitin ligase that is essential for efficient activation of Notch signaling by Delta. Dev Cell 4(1):67–82
108. Seiler C et al (2012) Smooth muscle tension induces invasive remodeling of the zebrafish intestine. PLoS Biol 10(9), e1001386
109. Zheng X et al (2012) Loss of zygotic NUP107 protein causes missing of pharyngeal skeleton and other tissue defects with impaired nuclear pore function in zebrafish embryos. J Biol Chem 287(45):38254–38264
110. Horne-Badovinac S et al (2001) Positional cloning of heart and soul reveals multiple roles for PKC lambda in zebrafish organogenesis. Curr Biol 11(19):1492–1502
111. Yang J et al (2009) hnRNP I inhibits Notch signaling and regulates intestinal epithelial homeostasis in the zebrafish. PLoS Genet 5(2):e1000363
112. Lorenzen JA et al (2005) Rbm19 is a nucleolar protein expressed in crypt/progenitor cells of the intestinal epithelium. Gene Expr Patterns 6(1):45–56
113. Niu X et al (2012) Sec13 safeguards the integrity of the endoplasmic reticulum and organogenesis of the digestive system in zebrafish. Dev Biol 367(2):197–207

114. Alvers AL et al (2014) Single continuous lumen formation in the zebrafish gut is mediated by smoothened-dependent tissue remodeling. Development 141(5):1110–1119
115. van der Velden YU et al (2011) The serine-threonine kinase LKB1 is essential for survival under energetic stress in zebrafish. Proc Natl Acad Sci U S A 108(11):4358–4363
116. Bagnat M et al (2007) Genetic control of single lumen formation in the zebrafish gut. Nat Cell Biol 9(8):954–960
117. Sun Z, Hopkins N (2001) vhnf1, the MODY5 and familial GCKD-associated gene, regulates regional specification of the zebrafish gut, pronephros, and hindbrain. Genes Dev 15(23):3217–3229
118. Ho SY et al (2006) Zebrafish fat-free is required for intestinal lipid absorption and Golgi apparatus structure. Cell Metab 3(4):289–300

Zebrafish Melanoma

Charles K. Kaufman

Abstract Melanoma skin cancer is a potentially deadly disease in humans and has remained extremely difficult to treat once it has metastasized. In just the last 10 years, a number of models of melanoma have been developed in the zebrafish that are biologically faithful to the human disease and have already yielded important insights into the fundamental biology of melanoma and offered new potential avenues for treatment. With the diversity and breadth of the molecular genetic tools available in the zebrafish, these melanoma models will continue to be refined and expanded upon to keep pace with the rapidly evolving field of melanoma biology.

Keywords Zebrafish • Melanoma • Cancer models • Transgenics • Drug discovery

Even among the many intractable cancers in solid tumor oncology, melanoma has remained one of the most challenging diseases to treat. While very treatable in its earliest stages with surgical excision (cure rates are >90 % for Stage I disease), melanomas have a troubling habit of metastasizing early or recurring years after a primary lesion has long been thought cured [1]. Once metastatic, the survival rates for patients have remained grim with median survival on the order of months and were essentially unchanged for over 30 years [2]. Only recently have a number of new treatments transformed the melanoma field into a rapidly changing area of oncology where both targeted therapies and immune modulatory agents have redefined the approaches to treatment (see below).

The relationship between melanoma and sun exposure (i.e. UV irradiation) seems clear but remains complicated [3, 4]. Epidemiological studies have established a link between melanoma risk and childhood blistering sunburns [5] among other risk factors including fair skin/light hair and number/type of nevi [6]. Recent next generation sequencing studies of human melanoma tumors have verified a UV-associated mutation pattern (predominant C->T transitions) in melanoma

C.K. Kaufman (✉)
Department of Medical Oncology, Dana-Farber Cancer Institute,
450 Brookline Ave., Boston, MA 02115-5450, USA

Boston Children's Hospital, Hematology/Oncology, Boston, MA, USA
e-mail: ckkaufman@partners.org

D.M. Langenau (ed.), *Cancer and Zebrafish*, Advances in Experimental
Medicine and Biology 910, DOI 10.1007/978-3-319-30654-4_19

encompassing relevant driver mutations [7] and have shown that melanoma has among the highest mutation rates of any cancer [8]. Still, the most frequent drivers of cutaneous melanoma (mutations producing the *BRAF*^{V600E} variant, accounting for ~60 % and assorted RAS-family mutations, accounting for a further ~20 %) do not result from an apparent UV signature [9].

Prior to 2011, only two agents were FDA approved for metastatic melanoma: dacarbazine, a cytotoxic chemotherapy, with low response rates and only short term effectiveness [10], and IL-2, an immune stimulatory agent with low response rates but rare long term responses [11]. The recent revolution in treatments has arisen in two areas. First, tyrosine kinase inhibitors targeting the *BRAFV*^{600E} mutation (vemurafenib, dabrafenib) [12, 13] and a more downstream component of the *MAP* kinase pathway, *MEK* (trametinib) [14], produce high response rates in the proper setting, but eventually all fail as resistance emerges. Second, immune modulatory agents in the form of humanized monoclonal antibodies have been developed that effectively activate the immune response by blocking pathways that normally restrain it by targeting CTLA-4-bearing T-Cells (ipilimumab) [15, 16] or the PD-1 receptor (pembrolizumab, nivolumab) [17, 18]. Response rates were initially modest for these agents but have continued to climb with the newer combinations and more optimized dosing schedules, and the length of responses can be quite significant [19]. Side effects from an unrestrained immune response are frequent and can be substantial in the form of autoimmune phenomenon (e.g. inflammatory bowel disease, rashes, hypophysitis, among many others) [15–19]. Overall, the progress has been remarkable, and in a very short time, has completely altered the way metastatic melanoma is treated.

With these fundamentals of melanoma biology and modern treatment in mind, this chapter will bring the zebrafish onto the scene.

Enter the Zebrafish

BRAF^{V600E}

In a truly foundational work, Patton et al. generated the first *BRAF*^{V600E}-dependent cancer model in any organism by expressing the human *BRAF*^{V600E} oncogene in zebrafish under the control of the melanocyte-specific *mitfa* promoter [20]. This *BRAF*^{V600E} mutation had been discovered to be common in melanoma [21], likely encompassing >50 % of cases [9]. In an otherwise wild type background, these *mitfa:BRAF*^{V600E} transgenic zebrafish developed melanocyte overgrowths analogous to human nevi, which contain the *BRAF*^{V600E} mutation in nearly all cases in humans. Crucially, when placed in the context of a *p53* mutant background, the zebrafish developed melanomas that were highly invasive and transplantable. At that time, the inactivation of *p53* was not known to be frequently associated with melanoma, but more recent sequencing of human melanomas has further validated the involvement of the *p53* pathway in melanomagenesis [4, 7]. This early model has proven to be a work-horse of the field, being used in several subsequent significant publications to be described later in the chapter [22–25].

N-RAS

The next most frequent oncogenic driver mutations in melanoma also target the MAP kinase pathway but through *RAS* activation (~20 % of cases), most frequently through *N-RAS* [9]. Taking a similar approach to Patton and colleagues, Dovey et al. generated two independent transgenic zebrafish lines with the *mitfa* promoter driving human $N\text{-}RAS^{Q61K}$ with an N-terminal *EGFP* fusion [26]. Expression of this oncogene led to hyperpigmentation throughout the body of the animals, and once again in the context of a *p53* mutation, these fish invariably developed aggressive melanoma lesions that were invasive and transplantable into irradiated zebrafish hosts. Not only did these tumors have the histologic appearance of melanoma, but gene expression analysis using microarrays and the then newly-described technique of gene set enrichment analysis (GSEA) [27] showed a significant correlation of upregulated genes between the zebrafish and human melanomas.

Together, these *BRAF* and *N-RAS* zebrafish transgenic models embody the primary oncogenic drivers for nearly 80 % of human cutaneous melanoma cases.

H-RAS

Although less frequently seen in melanoma, *H-RAS* mutations have been described [9]. By taking advantage of a *kita*-driven *GalTA4* gene trap line, Santoriello and colleagues were able to express oncogenic $H\text{-}RAS^{GV12}$ in melanocytes (referred to here as $kita/H\text{-}RAS^{GV12}$ double transgenics) [28]. The *kit* gene had long been known to play a role in melanocyte biology with mouse mutants having decreased to absent melanocyte formation depending on the specific *Kit* mutation [29]. Similarly, *kita* inactivation in the zebrafish *sparse* mutant results in the lack of early adult melanocytes and an aberrant stripe pattern, and the *kita* gene was shown to be expressed in melanocyte precursors [29].

In the $kita/H\text{-}RAS^{GV12}$ double transgenics, embryos were already found to have excess melanocytes by 3 days post fertilization (dpf), most notably on the top of the head. By 2 weeks of age, nearly all juvenile fish developed hyperpigmented caudal fins, and by 3 months, ~20 % developed histologically invasive melanoma. Interestingly in this model, the presence of a *p53* mutation was not required for melanoma formation, although breeding the *p53* mutation onto the $kita/H\text{-}RAS^{GV12}$ double transgenic zebrafish resulted in accelerated melanoma growth. In addition, when an *mitf:Gal4VP16* was instead used to control $H\text{-}RAS^{GV12}$ expression, the embryonic hyperpigmentation phenotype and the penetrance of melanoma formation were decreased, raising the intriguing possibility that differences in the cell-of-origin for the *mitfa* versus the *kita* transgenics might underlie these differences. *H-RAS* levels were higher in the *kita* model when compared to the *mitfa* model, complicating firm conclusions on this point.

In a follow-up report aimed at looking at the cell of origin for these melanomas, the authors mated the $kita/H\text{-}RAS^{GV12}$ double transgenics into the zebrafish *picasso* mutant, an *erbB3b* mutant that lacks or has much reduced adult melanocyte stem cells [30].

These fish still developed melanoma tumors at a rate similar to the *erbB3b* wild type counterparts, although the tumors were smaller. This suggests that the cell of origin for the tumors arises in an *erbB3b*-independent manner, perhaps most provocatively from a different population of stem/progenitor cells than the bulk of adult melanocytes that appear to depend on *erbB3b* signaling.

Putting the Models to Good Use

Among its greatest strengths as a vertebrate model are the zebrafish's high fecundity, rapid embryonic development, and significant genetic manipulability. Combining these assets with the excellent melanoma models described above has led to a number of significant advances in the melanoma field.

Separating the Wheat from the Chaff: Identifying Accelerators of Melanomagenesis

Cancer genomic data continues to accumulate at an astounding rate, spurred on by relatively inexpensive next generation sequencing platforms. Melanoma is one of the most genomically complex cancers, but even with less genomically complicated tumors [8], identifying relevant driver or accelerator mutations (i.e. those that functionally contribute to the development and growth of a tumor) as opposed to passenger events (i.e. mutations that happen to be present in an expanding clone, but do not have functional consequences) remains a fundamental challenge.

Ceol, Houvras, and colleagues took on this challenge by developing an elegant transgenic screening tool to apply to the *mitfa:BRAF*^{V600E} melanoma model [22]. By examining a large set of human melanoma cell lines and short-term tumor cultures, the authors developed a list of genes present in a recurrently amplified region of chromosome 1q21 that were also overexpressed at the mRNA level. To test the ability of each of the 17 genes in question to alter melanoma onset, the authors cloned each test gene into their novel transgenic vector termed *miniCoopR*. This vector contained two key elements: (1) an *mitfa* minigene made up of the *mitfa* promoter, ORF, and 3′ UTR and (2) the *mitfa* promoter driving the test gene. The *mitfa* minigene produces functional Mitfa protein, and thus can rescue a *mitfa* null cell that is destined to form a melanocyte in an otherwise *mitfa* null, or *nacre/na*, zebrafish. When combined with the *p53* mutant/*mitfa:BRAF*^{V600E} model developed by Patton et al and bred onto the *mitfa* null background, a *miniCoopR*-rescued melanocyte can then go on to form a melanoma (see Figs. 1 and 2). By linking the *mitfa* minigene in *cis* to the test gene in the *miniCoopR* vector, all melanocytes that form will also contain the test gene, and the effect of this gene on melanoma formation can then be directly assessed.

Fig. 1 Modeling melanoma in the zebrafish. (**a**) A wild type adult zebrafish demonstrating the normal stripe pattern which incorporates the darkly pigmented, punctate melanocytes that are readily visible on the body and fins. (**b**) An *mitfa*, or *nacre*, mutant zebrafish that lacks the black melanocytes and associated stripes. This fish is also *p53 −/−* and contains the *mitfa:BRAF*^{V600E} transgene that leads to melanoma formation when melanocytes are present. (**c**) A F0 transgenic zebrafish that shares the genetic background with the fish in panel (**b**) (*mitfa −/−; p53 −/−; mitfa:BRAF*^{V600E}*/ mitfa:BRAF*^{V600E}) but has been injected with the *miniCoopR* vector containing the *mitfa* minigene. Note the mosaic rescue of pigmented melanocytes on the fish, and the early formation of two raised melanoma tumors (*arrows*, head and dorsal fin)

Among the genes tested, one gene, *SETDB1*, a H3K9 methyltransferase, was found to accelerate melanoma onset from 19 weeks for control fish (*miniCoopR* with *EGFP* as the test gene) to 12 weeks. The *SETDB1*-associated tumors were particularly aggressive, with even premalignant melanocytes showing loss of oncogene-induced senescence character as assayed by SA-β-Gal. Gene expression and ChIP-Seq analysis showed alterations in expression of genes in the *HOXA* cluster with associated changes in *SETDB1* binding and H3K9 methyl marks that correlated with levels of *SETDB1* expression. Finally, another gene, *SUV39H1*, that complexes with *SETDB1* as a H3K9 methyltransferase was also found to accelerate melanoma onset.

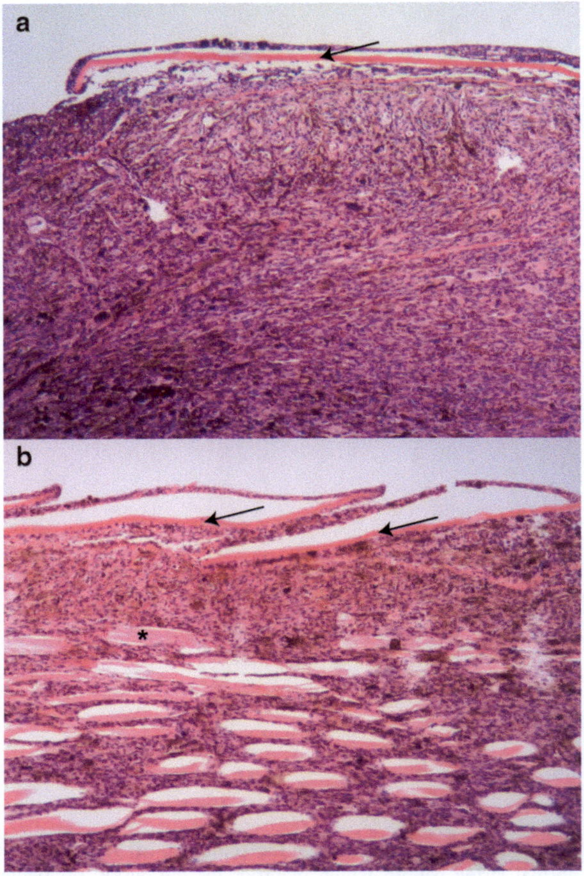

Fig. 2 Histology of zebrafish melanoma from the *p53/BRAF*^{V600E} model. (**a**) Large tumor forming below the scale (*arrow*) on the caudal peduncle. Note the variable presence of brown/black pigmentation in the highly pleomorphic tumor cells. (**b**) Another view showing the highly invasive nature of the melanoma tumor below the skin (scales marked with *arrows*) as it infiltrates and disrupts the muscle fibers (asterisk '*' marking one fiber) of the underlying tissue

From a disease modeling perspective, this study is exceptional in testing a significant number of genes in a de novo tumor model and identifying a new accelerator of melanoma formation. As Ceol and colleagues state, this required the production and observation of ~3000 transgenic animals; such an approach would not be feasible in other vertebrate models. For the cancer biology field, this paper further highlights the importance of chromatin modifiers, particularly methyltransferases, in controlling melanoma formation and may represent a druggable enzymatic activity with therapeutic relevance in melanoma.

In a separate paper the following year, Ceol, Houvras, and Zon used the *miniCoopR* system to contribute functional in vivo data to a story about the loss of 5-hydroxymethylcytosine (5-hmC) in melanoma [24]. The lead authors had identified an alteration in the 5-hmC mark as melanocytes (high 5-hmC) became melanoma (low 5-hmC). By cloning the *IDH2* gene into *miniCoopR,* which would tend to increase 5-hmC levels, Ceol and colleagues found that melanoma onset was significantly delayed as compared to control *miniCoopR;EGFP* and that the *IDH2* tumors were smaller and less invasive than controls. These data supported the functional relevance of altering the 5-hmC chromatin mark in melanomagenesis, and the approach once again demonstrated the utility of being able to test a candidate gene in a de novo tumor model.

Repurposing a Drug: A Chemical Inhibitor of Neural Crest and Melanoma

Publishing simultaneously with the *SETDB1* story, White and colleagues analyzed the gene expression signatures of *p53* mutant/*mitfa:BRAF*^{V600E} embryos and melanomas [23]. Among the core of 123 overlapping upregulated genes in both settings, the authors noted a number of neural crest stem cell and melanocyte genes (*crestin, sox10,* and *ednrb* and *tyr* and *dct*, respectively), leading them to infer that the melanomas had readopted aspects of neural crest identity. Indeed, the expression of *EDNRB* and *SOX10* are frequent features of human melanomas [23]. Following this line of thought, chemicals that inhibit neural crest identity might also inhibit melanoma formation and growth. Among vertebrate models, the zebrafish uniquely allows for the high-throughput in vivo testing of chemical modifiers of relevant biological processes, particularly those that can be assayed in embryos.

Using this chemical genetic approach, White et al. screened wild type zebrafish embryos for loss of neural crest formation, as assayed by in situ for the neural crest marker gene *crestin* at 24 hpf [31, 32], after treatment with a library of bioactive compounds beginning at 50 % epiboly. After screening ~2000 chemicals, they focused on several inhibitors of the pyrimidine biosynthesis enzyme DHODH and, in particular, on a FDA-approved drug leflunomide used in the treatment of rheumatoid arthritis. Treatment of embryos with leflunomide abrogated formation of neural crest and its derivatives including melanocytes and glial cells. Combining low dose leflunomide treatment with heterozygous loss of *spt5*, a component of the transcriptional elongation machinery, synergized to reproduce the *spt5* null phenotype, indicating a defect in transcriptional elongation was responsible for the effect of leflunomide. This was further supported by ChIP-Seq for RNA *POL2* showing less *POL2* enzyme present in the body of genes than near the transcriptional start site, indicating an elongation defect. Perhaps most beautifully, leflunomide was found to inhibit growth of the A375 melanoma cell line both in culture and in mouse xenografts. Combining leflunomide with the *BRAF* inhibitor vemurafenib was also synergistic in its effect, suggesting a clear therapeutic angle for human testing using two FDA approved drugs.

Once again, the *p53*/*mitfa:BRAF*^{V600E} model proved useful both in clarifying a fundamental feature of melanoma (i.e. the re-expression of neural crest progenitor genes) and for developing a solid lead for targeting this characteristic for therapeutic benefit. Whether or not leflunomide proves to be the optimal agent for treating human melanoma, the principle that inhibiting the pyrimidine biosynthetic pathway or transcriptional elongation uncovers a particular sensitivity of neural crest and melanoma will remain a valuable finding.

Relationship Status: It's Complicated—Mitf and Melanoma

The transcription factor *mitfa* is required for melanocyte formation in zebrafish, as are the corresponding human and mouse *MITF* orthologues. The function of *MITF* in melanoma formation and tumorigenicity has remained complex with *MITF* levels varying widely among different melanomas and with apparent context-dependent relationships in how increased/decreased amount of *MITF* favor or disfavor tumor growth [33]. In one of the more recent and unique zebrafish melanoma studies, Lister et al make powerful use of a temperature sensitive *mitfa* allele, termed $mitfa^{vc7}$, to analyze the effect of altering *mitfa* activity on *de novo* melanoma formation [25]. The authors first mated this $mitfa^{vc7}$ allelle into the $p53/mitfa{:}BRAF^{V600E}$ background. As expected, melanocytes and tumors did not form in zebrafish grown at the restrictive higher temperature (28 °C) where *mitfa* activity is lost due to a splicing defect [34]. Surprisingly, melanoma tumors formed at the permissive temperature (<26 °C), even in the absence of an accompanying *p53* mutation unlike in the original single $mitfa{:}BRAF^{V600E}$ transgenics. The authors found that tumors arising in the hypomorphic *mitfa* background expressed lower levels of the melanocyte differentiation markers *dct* and *tyr* but higher levels of *c-met* as compared to $p53/mitfa{:}BRAF^{V600E}$ tumors, leading them to propose that perhaps the lower *mitfa* activity of the *ts* allele holds melanocytes in a less differentiated state that is more prone to transformation. The hypomorph *mitfa* tumors typically regressed when the fish were shifted to the higher restrictive temperature, and then recurred when returned to the permissive temperature, indicating that a tumor initiating population survives at the restrictive temperature or the tumor incompletely regressed.

This study beautifully provides a genetic proof that *mitfa* activity is required to maintain melanomas, further supporting efforts to inhibit *MITF* as a melanoma treatment. Provocatively, this study also indicates that "low dose" *mitf* activity may actually be dangerous and, in itself, oncogenic in the context of a $BRAF^{V600E}$ melanocyte.

Engineered, but Genetically Complex Tumors

As described earlier in the chapter, forcing expression of the human oncogenes $BRAF^{V600E}$ or $NRAS^{Q61K}$ in melanocytes using the *mitfa* promoter in a *p53* mutant zebrafish leads to melanoma formation. Might other genetic alterations recurrently arise in tumors arising in these tumor models? This was the question that Yen and colleagues set out to address when they sequenced the complete exomes of 53 independent zebrafish primary melanomas [35]. They identified 403 point mutations and 13 insertions/deletions, an apparently substantial number for a non-UV exposed setting. They did find one 175 kb segment on Chr 3 that was amplified in 10 tumors, although this region was not readily corroborated by human data. Interestingly, the *tert* gene was amplified in 4 tumors in line with its frequent promoter mutation in human melanoma (~90 % of tumors) that leads to increased expression.

While this initial study did not identify new oncogenic events in these melanoma models, it did highlight that a significant number of genomic alterations occur during tumor formation, even when the system is prepped to produce cancer. Further, the authors were able to overcome the relatively outbred genetic character of the zebrafish to perform such an analysis, which bodes well for future analyses using the model. Going forward, perhaps assaying a larger number of tumors will tease out more recurrent mutational events, or if resources were to allow, performing whole genome sequencing might identify noncoding mutations that alter gene expression (e.g. enhancer mutations).

Frontiers

In a relatively short time, the zebrafish melanoma field has contributed important insights into the basic biology of melanoma and cancer and has offered novel guidance on potential new therapeutic approaches. As the field continues to grow, we should consider where and how further contributions can be made.

Additional Melanoma Oncogenes As described in a recent large-scale genomics paper on human cutaneous melanoma, four major subtypes of cutaneous melanoma exist: *BRAF, RAS, NF1,* and triple wild-type tumors [9]. With *BRAF* and *RAS* models in hand, developing an *NF1*-dependent model could be revealing, as would testing even infrequently mutated potential drivers found in the triple wild-type class (e.g. *GNAQ/11, KIT, CTNNB1,* and *EZH2*). The zebrafish field has thus far focused on cutaneous melanoma, and these potential *GNAQ/11* (for uveal) [36, 37] and *KIT* (for acral/mucosal) [38] transgenics could also broaden the available models for noncutaneous melanoma subtypes.

Loss of Function Mutations The CRISPR/Cas9 gene targeting technology has been successfully applied in the zebrafish to inactivate genes of interest [39–41]. Applying this technology in the zebrafish melanoma model seems ideal to test loss of function candidates that may alter disease onset and/or progression.

Immune Function The function of the immune system in engineered zebrafish melanomas has been largely unexplored. With immune modulatory drugs becoming the cornerstone of melanoma and potentially other solid tumor therapies [42], a thorough analysis of immune function in zebrafish tumors will be instructive and, perhaps, essential in connecting the model to the human disease.

Tumor Initiation, Progression, and Metastasis One major advantage of genetically engineered tumor models is the relatively synchronous development of *de novo* tumors as compared to the essentially spontaneous formation of cancer in humans. With the ability to generate very large cohorts of animals coupled with the accessibility for sampling and visualization of cutaneous melanoma, the zebrafish should be a prime candidate for longitudinal analysis of tumors from initiation through metastasis (or local invasion, see Fig. 2). To further interrogate the stage at which a gene of interest functions, the tools describe here, including the *miniCoopR* system, traditional mutants, or CRISPR/Cas9-generated alleles, would clearly apply.

Transplantation Models A number of important models have been developed for transplantation studies that allow for improved visualization of transplanted cells (e.g. transparent, pigment-free *Casper* zebrafish) [43] or reduced immune responses (e.g. *rag2* mutants) [44]. Ideally, immunologically matched recipient zebrafish would be paired with melanoma-bearing donors to remove confounding effects of simple tissue rejection and complications from conditioning regimens, such as irradiation.

The progress in the zebrafish melanoma field to date has been substantial, and the community should look forward to continued exciting advances as the newest components of the zebrafish molecular genetic toolbox are applied.

References

1. Balch CM et al (2009) Final Version of 2009 AJCC Melanoma Staging and Classification. J Clin Oncol 27:6199–6206
2. Lo JA, Fisher DE (2014) The melanoma revolution: from UV carcinogenesis to a new era in therapeutics. Science 346:945–949
3. Epstein FH, Gilchrest BA, Eller MS, Geller AC, Yaar M (1999) The pathogenesis of melanoma induced by ultraviolet radiation. N Engl J Med 340:1341–1348
4. Viros A et al (2014) Ultraviolet radiation accelerates BRAF-driven melanomagenesis by targeting TP53. Nature 511:478–482
5. Whiteman DC, Whiteman CA, Green AC (2001) Childhood sun exposure as a risk factor for melanoma: a systematic review of epidemiologic studies – Springer. Cancer Causes Control 12:69–82
6. Garbe C et al (1994) Risk factors for developing cutaneous melanoma and criteria for identifying persons at risk: multicenter case–control study of the Central Malignant Melanoma Registry of the German Dermatological Society. J Invest Dermatol 102:695–699
7. Hodis E et al (2012) A landscape of driver mutations in melanoma. Cell 150:251–263
8. Lawrence MS et al (2013) Mutational heterogeneity in cancer and the search for new cancer-associated genes. Nature 499:214–218
9. Cancer Genome Atlas Network (2015) Genomic classification of cutaneous melanoma. Cell 161:1681–1696
10. Serrone L, Zeuli M, Sega FM, Cognetti F (2000) Dacarbazine-based chemotherapy for metastatic melanoma: thirty-year experience overview. J Exp Clin Cancer Res 19:21–34
11. Rosenberg SA et al (1994) Treatment of 283 consecutive patients with metastatic melanoma or renal cell cancer using high-dose bolus interleukin 2. JAMA 271:907–913
12. Chapman PB et al (2011) Improved survival with vemurafenib in melanoma with BRAF V600E mutation. N Engl J Med. doi:10.1056/NEJMoa1103782
13. Hauschild A et al (2012) Dabrafenib in BRAF-mutated metastatic melanoma: a multicentre, open-label, phase 3 randomised controlled trial. Lancet 380:358–365
14. Flaherty KT et al (2012) Improved survival with MEK inhibition in BRAF-mutated melanoma. N Engl J Med. doi:10.1056/NEJMoa1203421

15. Robert C et al (2011) Ipilimumab plus dacarbazine for previously untreated metastatic melanoma. N Engl J Med. doi:10.1056/NEJMoa1104621
16. Hodi FS et al (2010) Improved survival with ipilimumab in patients with metastatic melanoma. N Engl J Med 363:711–723
17. Robert C et al (2015) Pembrolizumab versus ipilimumab in advanced melanoma. N Engl J Med 372:2521–2532. doi:10.1056/NEJMoa1503093
18. Robert C et al (2015) Nivolumab in previously untreated melanoma without BRAFMutation. N Engl J Med 372:320–330. doi:10.1056/NEJMoa1412082
19. Postow MA et al (2015) Nivolumab and ipilimumab versus ipilimumab in untreated melanoma. N Engl J Med 372:2006–2017. doi:10.1056/NEJMoa1414428
20. Patton EE et al (2005) BRAF mutations are sufficient to promote nevi formation and cooperate with p53 in the genesis of melanoma. Curr Biol 15:249–254
21. Tuveson DA, Weber BL, Herlyn M (2003) BRAF as a potential therapeutic target in melanoma and other malignancies. Cancer Cell 4:95–98
22. Ceol CJ et al (2011) The histone methyltransferase SETDB1 is recurrently amplified in melanoma and accelerates its onset. Nature 471:513
23. White RM et al (2011) DHODH modulates transcriptional elongation in the neural crest and melanoma. Nature 471:518–522
24. Lian CG et al (2012) Loss of 5-hydroxymethylcytosine is an epigenetic hallmark of melanoma. Cell 150:1135–1146
25. Lister JA et al (2013) A conditional zebrafish MITF mutation reveals MITF levels are critical for melanoma promotion vs. regression in vivo. J Investig Dermatol. doi:10.1038/jid.2013.293
26. Dovey M, White R, Zon L (2009) Oncogenic NRAS cooperates with p53 loss to generate melanoma in zebrafish. Zebrafish 6:397–404
27. Subramanian A et al (2005) Gene set enrichment analysis: a knowledge-based approach for interpreting genome-wide expression profiles. Proc Natl Acad Sci U S A 102:15545–15550
28. Santoriello C et al (2010) Kita driven expression of oncogenic HRAS leads to early onset and highly penetrant melanoma in zebrafish. PLoS One 5, e15170
29. Parichy DM, Rawls JF, Pratt SJ, Whitfield TT, Johnson SL (1999) Zebrafish sparse corresponds to an orthologue of c-kit and is required for the morphogenesis of a subpopulation of melanocytes, but is not essential for hematopoiesis or primordial germ cell development. Development 126:3425–3436
30. Santoriello C, Anelli V, Alghisi E, Mione M (2012) Highly penetrant melanoma in a zebrafish model is independent of ErbB3b signaling. Pigment Cell Melanoma Res 25:287–289
31. Rubinstein AL, Lee D, Luo R, Henion PD, Halpern ME (2000) Genes dependent on zebrafish cyclops function identified by AFLP differential gene expression screen. Genesis 26:86–97
32. Luo R, An M, Arduini BL, Henion PD (2001) Specific pan-neural crest expression of zebrafish Crestin throughout embryonic development. Dev Dyn 220:169–174
33. Goding CR (2011) A picture of Mitf in melanoma immortality. Oncogene 30:2304–2306
34. Zeng Z, Johnson SL, Lister JA, Patton EE (2015) Temperature-sensitive splicing of mitfa by an intron mutation in zebrafish. Pigment Cell Melanoma Res 28:229–232
35. Yen J et al (2013) The genetic heterogeneity and mutational burden of engineered melanomas in zebrafish models. Genome Biol 14:R113
36. Van Raamsdonk CD et al (2008) Frequent somatic mutations of GNAQ in uveal melanoma and blue naevi. Nature 457:599–602
37. Van Raamsdonk CD et al (2010) Mutations in GNA11in uveal melanoma. N Engl J Med 363:2191–2199
38. Curtin JA, Busam K, Pinkel D, Bastian BC (2006) Somatic activation of KIT in distinct subtypes of melanoma. J Clin Oncol 24:4340–4346
39. Ablain J, Durand EM, Yang S, Zhou Y, Zon LI (2015) A CRISPR/Cas9 vector system for tissue-specific gene disruption in zebrafish. Dev Cell. doi:10.1016/j.devcel.2015.01.032
40. Gagnon JA et al (2014) Efficient mutagenesis by Cas9 protein-mediated oligonucleotide insertion and large-scale assessment of single-guide RNAs. PLoS One 9, e98186

41. Hwang WY et al (2013) Heritable and precise zebrafish genome editing using a CRISPR-Cas system. PLoS One 8, e68708
42. Larkin J et al (2015) Combined nivolumab and ipilimumab or monotherapy in untreated melanoma. N Engl J Med 373:1270–1271. doi:10.1056/NEJMoa1504030
43. White RM et al (2008) Transparent adult zebrafish as a tool for in vivo transplantation analysis. Cell Stem Cell 2:183–189
44. Tang Q et al (2014) Optimized cell transplantation using adult rag2 mutant zebrafish. Nat Methods. doi:10.1038/nmeth.3031

Neuroblastoma and Its Zebrafish Model

Shizhen Zhu and A. Thomas Look

Abstract Neuroblastoma, an important developmental tumor arising in the peripheral sympathetic nervous system (PSNS), accounts for approximately 10 % of all cancer-related deaths in children. Recent genomic analyses have identified a spectrum of genetic alterations in this tumor. Amplification of the *MYCN* oncogene is found in 20 % of cases and is often accompanied by mutational activation of the *ALK* (anaplastic lymphoma kinase) gene, suggesting their cooperation in tumor initiation and spread. Understanding how complex genetic changes function together in oncogenesis has been a continuing and daunting task in cancer research. This challenge was addressed in neuroblastoma by generating a transgenic zebrafish model that overexpresses human *MYCN* and activated *ALK* in the PSNS, leading to tumors that closely resemble human neuroblastoma and new opportunities to probe the mechanisms that underlie the pathogenesis of this tumor. For example, coexpression of activated *ALK* with *MYCN* in this model triples the penetrance of neuroblastoma and markedly accelerates tumor onset, demonstrating the interaction of these modified genes in tumor development. Further, *MYCN* overexpression induces adrenal sympathetic neuroblast hyperplasia, blocks chromaffin cell differentiation, and ultimately triggers a developmentally-timed apoptotic response in the hyperplastic sympathoadrenal cells. In the context of *MYCN* overexpression, activated ALK provides prosurvival signals that block this apoptotic response, allowing continued expansion and oncogenic transformation of hyperplastic neuroblasts, thus promoting progression to neuroblastoma. This application of the zebrafish model illustrates its value in rational assessment of the multigenic changes that define neuroblastoma pathogenesis and points the way to future studies to identify novel targets for therapeutic intervention.

S. Zhu (✉)
Department of Biochemistry and Molecular Biology, Cancer Center and Center for Individualized Medicine, Mayo Clinic, Rochester, MN 55902, USA
e-mail: Zhu.shizhen@mayo.edu

A. Thomas Look (✉)
Department of Pediatric Oncology, Dana-Farber Cancer Institute, Harvard Medical School, Boston, MA 02115, USA
e-mail: thomas_look@dfci.harvard.edu

D.M. Langenau (ed.), *Cancer and Zebrafish*, Advances in Experimental Medicine and Biology 910, DOI 10.1007/978-3-319-30654-4_20

Keywords Zebrafish • Animal model • Neuroblastoma • Neural crest • MYCN • ALK • Transgenesis • Functional genomic

Introduction

Neuroblastoma is the most common extracranial solid tumor diagnosed during the first year of life, accounting for 10–13 % of all deaths due to pediatric cancer [1–3]. It arises from incompletely committed primordial neural crest cells that usually generate the adrenal medulla and sympathetic neural ganglia [4]. Sixty-five percent of these tumors occur in the abdomen, often in the adrenal medulla, while others develop in the paraspinal sympathetic ganglia (5 %). The remainder of tumors arise from the sympathetic ganglia in the chest (~20 %), head/neck (~5 %), and pelvis (~5 %) [5]. Neuroblastoma displays remarkable heterogeneity, including diverse clinical responses to therapy, ranging from an overall survival rate exceeding 90 % for low- or intermediate-risk patients who receive standard or no treatment to a mortality rate greater than 50 % for high-risk patients who receive intensive multimodal therapy [1]. As a malignant childhood tumor, neuroblastoma is driven by aberrant growth- and differentiation-dependent molecular pathways that usurp the controls of normal organogenesis and development, leading to the transformation of primitive sympathoadrenal neural crest cells. Understanding the cellular and molecular basis of neuroblastoma pathogenesis is therefore key to the development of innovative therapeutics that target neuroblastoma cells while sparing healthy tissue, in contrast to the indiscriminate effects of intensive chemotherapy and ionizing radiation. This task, although formidable, has yielded considerable ground to studies with human cell culture systems and animal models. The challenge now is to translate the accumulating molecular findings in neuroblastoma into effective targeted therapy. This chapter reviews the essential cells of origin and molecular features of neuroblastoma, and illustrates how the zebrafish is being used to model complex combinations of genetic alterations in this tumor, thus establishing the fish as a critical tool for future research.

Neuroblastoma Cell Biology

Neuroblastoma is an embryonal tumor of the PSNS [6, 7]. The cell of origin is thought to be a developing and incompletely committed precursor cell derived from the sympathoadrenal lineage of neural crest [4, 8], a transient structure in developing vertebrate embryos, sometimes referred to as the fourth germ layer, that arises at the interface between the epidermal and neural ectoderm [9]. This transient, multipotent, and migratory population of cells give rise to different cell types, including neurons and glia of the autonomic and sensory nervous systems, neuroendocrine chromaffin cells in the adrenal medulla, pigment cells, Schwann cells, and cells of the craniofacial skeleton [9–13]. A subset of neural crest cells within the trunk region of neural tube is determined to be the sympathoadrenal lineage, which generates the sympathetic neurons, chromaffin cells and the intermediate small intensely fluorescent cells that are intermediate type of cells between sympathetic neurons and chromaffin cells [14–17]. Upon induction of BMP (bone morphogenetic protein) and Wnt (wingless-type MMTV integration-site family)-dependent signaling, produced at the border between future neural and nonneural ectoderm [18–20], the neural crest progenitor cells located at the dorsal region of the neural tube delaminate and undergo an epithelial-to-mesenchymal transition (EMT), subsequently migrating via a ventrolateral pathway to the dorsal aorta [9]. Once they reach the dorsal aorta, these neural crest cells receive the signals from the dorsal aorta, such as BMPs 2/4/7 [21, 22], and begin to acquire their catecholaminergic features as sympathoadrenal progenitors and turn on the expression of a variety of transcription factors including, *MASH-1/CASH-1* (the mammalian/chicken achaete-scute homolog) [23, 24], *Phox2a/b* (paired-like homeobox 2a/b) [25], *HAND2* (heart and neural crest derivatives expressed 2) [26, 27], *Gata2/3* (GATA binding protein 2/3) [28, 29] and *tfap2a* (transcription factor AP-2 alpha) [30, 31], which interact in a complex regulatory network to ultimately induce the expression of components of the catecholaminergic synthetic machinery, e.g., *Th* (tyrosine hydroxylase) and *Dβh* (dopamine β-hydroxylase) [32–34]. These sympathoadrenal progenitors then further differentiate into either sympathetic neurons that form a chain along the spine or chromaffin cells that migrate to the adrenal medulla [21, 35–37] (see Fig. 1).

How and when the adrenal chromaffin cells and sympathetic neurons are segregated during embryogenesis has been a long-standing question. Accumulating evidence from detailed lineage studies in developing chicken embryos and specific gene manipulation in avian, zebrafish and mouse models has suggested that sympathetic neurons and chromaffin cells are derived from common progenitor cells [3, 17, 38, 39]. Earlier study from chick has found that the sympathoadrenal progenitor cells in a defined region along the anteroposterior body axis (the level of somites 18–24 in chick embryos, which is the "adrenomedullary level") can give rise to both sympathetic neurons and chromaffin cells [38]. Recent single cell electroporations of a GFP-encoding plasmid into the neural crest located at the dorsal midline of E2 chick neural tubes at the adrenomedullary level revealed that the labeled GFP-positive cell can be detected in both sympathetic ganglia and adrenal glands, further supporting the notion that sympathetic neurons and adrenal chromaffin cells share a common progenitor in the neural tube [17]. Furthermore, Takahashi's group has recently demonstrated that the BMPs secreted from dorsal aorta induce the expression of stromal cell-derived factor-1 (*SDF1*, a chemokine, also called *CXCL12*), and Neuregulin 1 [*NRG1* of the epidermal growth factor (*EGF*) family] in the para-aortic mesenchyme, which in turn attracts the sympathoadrenal progenitors to migrate to the dorsal aorta region [3]. This group also showed that both *BMP* and *NRG1-ErbBs* signaling pathways are critical for the segregation of the sympathetic neurons and chromaffin cells and are required for attracting the chromaffin

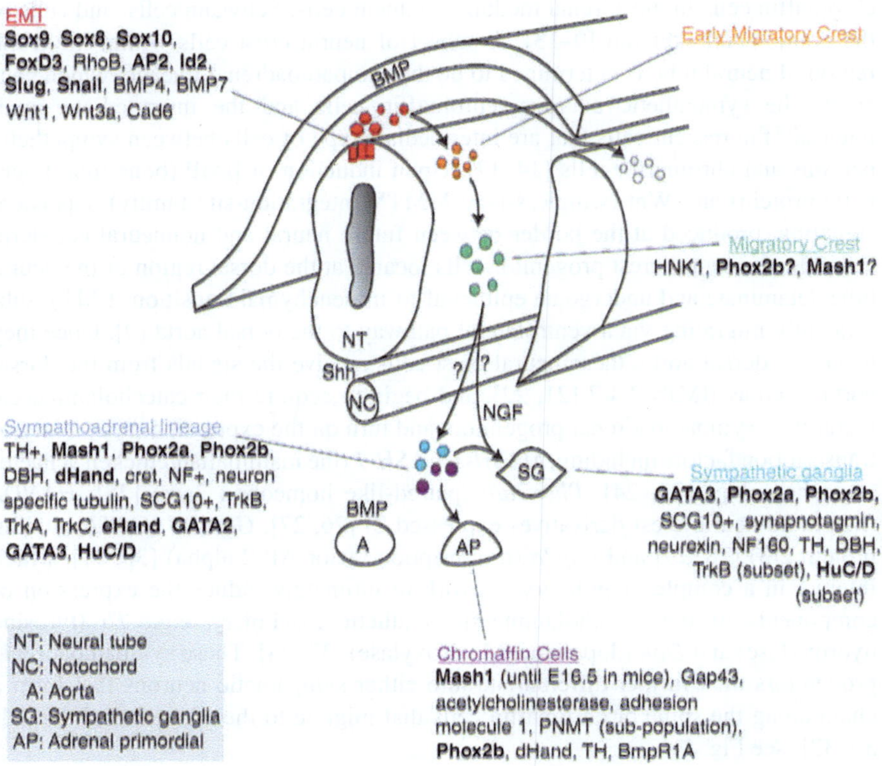

Fig. 1 Differentiation of cells of the peripheral sympathetic nervous system (PSNS). Cells at the dorsal region of the neural tube undergo an epithelial-to-mesenchymal transition [(EMT) (*red population*)], delaminate from the neural tube (*orange*), and migrate ventrally to the aorta (*green*), where they are commonly referred to as the sympathoadrenal progenitors (*blue* and *purple*). From the aortic region, the cells then migrate to the developing adrenal gland (AP) to become chromaffin cells or differentiate to become sympathetic ganglia (SG). As cells begin to differentiate as sympathetic ganglia, they upregulate neural markers while chromaffin cells upregulate proteins found in the adrenal gland [8]. Genes either demonstrated or thought to be involved in PSNS development [27, 35, 140] are listed on the figure near the step(s) they promote. *Abbreviations*: *A* aorta, *AP* adrenal primordial, *NC* notochord, *NT* neural tube, *SG* sympathetic ganglia. This figure is reproduced from ref [8]

cells to the target region, adrenal gland [3]. Thus, the dorsal aorta acts as a morphogenetic signaling center that coordinates sympathoadrenal neural crest cells migration and cell lineage segregation. However, whether the mechanisms underlying the transformation of sympathoadrenal cells in the ganglia and adrenal medulla are same and why neuroblastoma arises in the adrenal medulla at a higher frequency (60 %) than in the sympathetic ganglia (~20 % in chest, ~5 % head/neck, and ~5 % in pelvis) [5] remains unanswered.

Genetic Alterations in Neuroblastoma

Many genes that are important for the normal development of PSNS have been found to be mutated, deleted, amplified or aberrantly expressed in neuroblastoma through various mechanisms. Familial neuroblastoma is rare, accounting for only 1–2 % of all cases of this tumor [40–42]. Genetic alterations in two genes, *PHOX2B* and *ALK*, have been identified in ~80 % of hereditary neuroblastomas. *PHOX2B*, which encodes a homeodomain transcription factor, was the first candidate predisposition gene identified in neuroblastoma [41, 43]. As a master regulator of development of the autonomic nervous system, the PHOX2B protein drives differentiation of neural crest progenitors to sympathetic neurons [44]. Mutations of this gene cause aberrant differentiation of the sympathoadrenal lineage of neural crest, leading to neurocristopathies, such as Hirschsprung disease [45], and neuroblastoma [41, 43, 46]. In the absence of *Phox2b*, autonomic neurogenesis is completely abolished in mice [25, 47]. In zebrafish, aberrant *PHOX2B* expression, mediated by either overexpression of neuroblastoma-associated frameshift dominate-negative mutant *PHOX2B* (676delG [41] or K155X [48]) or knockdown of endogenous *phox2b*, causes an arrest in the normal maturation of sympathetic neurons and inhibits retinoic acid-induced differentiation [49]. These accumulated undifferentiated sympathetic neuronal progenitors may increase the susceptibility to secondary mutations that could ultimately lead to neuroblastoma [49].

The most commonly mutated gene associated with familial neuroblastoma is anaplastic lymphoma kinase (*ALK*), found in ~75 % of hereditary cases [40, 50–53]. It encodes a receptor tyrosine kinase and was originally identified as a fusion protein with nucleophosmin in cases of anaplastic large cell lymphoma [54]. ALK is preferentially expressed in the central and peripheral nervous system [55], where it plays a critical role in neuron differentiation [56], sympathetic neuron development and survival of migratory neural crest cells [57]. Knock-in mice expressing the most frequent germline mutation, *ALKR1275Q*, develop enlarged sympathetic ganglia and, in the context of *MYCN* overexpression [58], early onset of multifocal neuroblastoma, emphasizing the role of germline *ALK* mutations in neuroblastoma predisposition.

Over the last two decades, many chromosomal and molecular abnormalities have been identified in neuroblastoma and used for risk assignment and to prediction of disease outcome. The most important of these is amplified *MYCN*, the hallmark of high-risk neuroblastoma with a poor prognosis. *MYCN* amplification (defined as >10 gene copies at chromosome band 2p24) is found in approximately 20 % of neuroblastoma cases and is more common in patients with advanced-stage disease and a poor outcome [59]. This alteration is significantly associated with 1p36 loss of heterozygosity (LOH), and is found in approximately 25–35 % of primary neuroblastomas [60]. By contrast, chromosome 11q deletions (identified in 15–22 % of primary neuroblastomas) are often found in high-stage tumors without *MYCN* amplification and with intact chromosome 1p [61]. Partial chromosome 17q gain is the most common genetic aberration, occurring in approximately 80 % of neuroblastoma regardless of *MYCN* status [62, 63]. Deletion of chromosome 3p is found in neuroblastoma patients with an older age at diagnosis, and is often associated with 11q loss, without *MYCN* amplification or 1p deletion [64, 65].

In an effort to improve understanding of the molecular basis of neuroblastoma and to identify novel tumor marker genes for diagnosis and druggable targets for improved therapy, different groups of investigators have examined more than 400 neuroblastoma cases by integrative genomic analyses over the last two decades. In those studies, *ALK*, the most important predisposition gene found in hereditary neuroblastoma, was identified as the most frequently mutated gene in cases of sporadic high-risk neuroblastoma (8–10 % of total cases) [40, 50, 51, 66–68]. *ALK* mutations were found in 10.9 % of *MYCN*-amplified tumors, while *F1174* mutations, the most common of the *ALK* somatic mutations, were overrepresented in *MYCN*-amplified tumors [52], suggesting that these two oncogenes may collaborate in neuroblastoma pathogenesis. Indeed, transgenic animal models overexpressing *MYCN* and aberrantly activated *ALK* developed neuroblastomas that faithfully recapitulated the features of high-risk neuroblastoma [69–71]. We have used a zebrafish model to show that activated *ALK* expression synergizes with *MYCN* by blocking a developmentally timed apoptotic response that typically occurs in *MYCN*-overexpressing sympathoadrenal precursors [70], opening the way for analysis of other oncogenes that potentiate the activity of *MYCN* during neuroblastoma pathogenesis, a research avenue that will be discussed in detail later in this chapter.

Additional mutations were identified in a study by Meyerson's and Maris' groups of 240 high-risk neuroblastoma cases diagnosed in patients over 18 months of age with widely disseminated tumors [66]. These investigators used a combination of whole-exome, whole genome and transcriptome sequencing to identify the most frequently mutated genes, including the protein tyrosine phosphatase *PTPN11* (2.9 %), the RNA helicase *ATRX* (2.5 %), *MYCN* (1.7 %), and *NRAS* (0.83 %) [66] (Fig. 2). In a whole-genome sequence analysis of 87 neuroblastomas of all stages, Versteeg's group found chromothripsis, defined as structural defects characterized by local shredding of chromosomes, in 18 % of high-stage neuroblastomas [67]. These structural alterations recurrently affected genes involved in neuronal growth cone stabilization, such as *ODZ3*, *PTPRD* and *CSMD1* [67]. Furthermore, mutations in the genes involved in *Rac/Rho* signaling-mediated neuritogenesis, such as *ATRX*, *DLC1*, *ARHGAP10*, and *TIAM1* were identified, further suggesting defects in neuritogenesis in neuroblastoma [67]. In addition, Sausen and colleagues uncovered recurrent mutations or focal deletions of the chromatin-remodeling genes *ARID1A* and *ARID1B* in 11 % of cases that were associated with early treatment failure and decreased survival [72]. Taken together, these integrative genomic studies have shown that neuroblastoma harbors very few recurrent somatic mutations despite its marked genetic heterogeneity.

Finally, genome-wide association studies (GWAS) have identified additional genes and pathways that contribute to neuroblastoma pathogenesis, but with low penetrance, suggesting that they are not sufficient by themselves to induce tumor

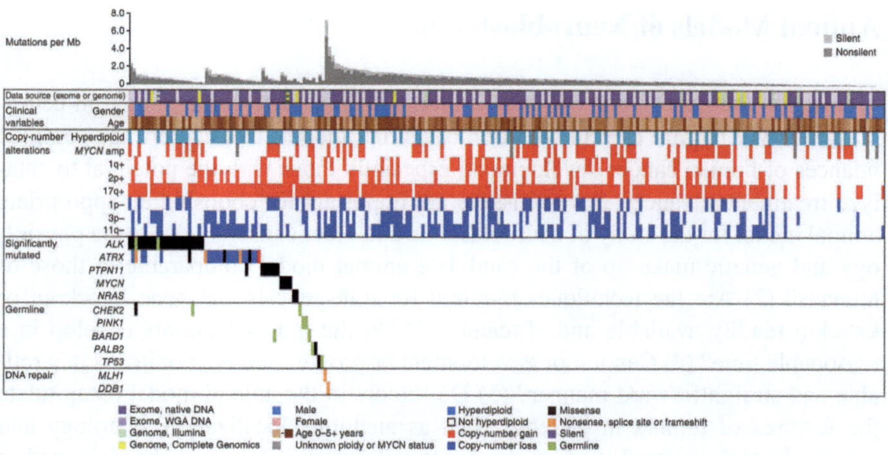

Fig. 2 Genetic variation in neuroblastoma. Comparison of clinical and genomic data (*rows*) for 240 cases of neuroblastoma (*columns*). The data sources and sequencing technology were whole-exome sequencing (WES) from whole-genome amplification (WGA; *light purple*), WES from native DNA (*dark purple*), Illumina whole-genome sequencing (WGS; *green*) and Complete Genomics WGS (*yellow*). *Striped blocks* indicate that cases were analyzed by different approaches. The clinical variables included gender (male, *blue*; female, *pink*) and age (*brown spectrum*). Copy number alterations refer to ploidy measured by flow cytometry (with hyperdiploid defined as a DNA index >1) and clinically relevant copy number alterations derived from sequence data. Significantly mutated genes are those with statistically significant mutation counts given the background mutation rate, gene size and expression in neuroblastoma. Germline denotes genes with significant numbers of germline variants or loss-of-function cancer gene variants in their cohort. DNA repair refers to genes that may be associated with an increased mutation frequency in two apparently hypermutated tumors. Predicted effects of somatic mutations are color coded according to the legend. *MYCN amp* MYCN amplification. This figure is reproduced from ref [66]

formation. Several predisposing single-nucleotide polymorphisms (SNPs) within or adjacent to certain genes have been found, to be associated with a significant risk of neuroblastoma. These include genes encoding the LIM-domain-only 1 protein (*LMO1*) [73], BRCA1-associated RING domain 1 (*BARD1*) [74], FLJ22536 [75], dual-specificity phosphatase 12 (*DUSP12*) [44], hydroxysteroid (17 beta) dehydrogenase 12 (*HSD17B12*) [44], DEAD-box polypeptide 4 (*DDX4*) [44], interleukin 31 receptor A (*IL31RA*) [44], long intergenic non-protein-coding RNA 340 (*LINC00340*) [76], ankyrin repeat-containing E3 ubiquitin protein ligase 1 (*HACE1*) [77], and lin28 homologue B (*LIN28B*) [77]. Overexpression of *Lin28b* in the sympathetic adrenergic lineage of genetically engineered mice can induce neuroblastomas by decreasing *let-7* miRNA expression and increasing MYCN protein expression [78]. Although intriguing, the functional roles of these GWAS-identified genetic alterations in neuroblastoma pathogenesis remain elusive, and will require additional study in robust animal model systems to elucidate mechanisms.

Animal Models of Neuroblastoma

Translating the genetic and epigenetic alterations that have come to light in neuroblastoma will require careful selection of animal models that capture the complex nuances of human cancer development, especially those with the potential to catalyze treatment advances. Several issues are important for choosing an appropriate animal model for the study of neuroblastoma and other cancers. (1) Are the physiology and genetic make-up of the candidate animal model comparable to those of humans? (2) Are the techniques required for transgenesis and gene knockout or knockin readily available and efficient? (3) Do the induced tumors develop in a reasonable time? (4) Can tumor development or progression be monitored in a reliable and straightforward manner? (5) Do tumors in the animal model recapitulate the features of tumors in patients, such as anatomic localization, histology and immunohistochemistry? (6) Can the genetically modified animals be produced in large numbers? (7) Is the genetically modified animal amenable to pharmacologic studies and large-scale drug screening?

Initially, in vivo models of neuroblastoma were developed by xenografts of human tumor cells in either immunosuppressed animals or syngeneic rodent models [6]. In the late 1990s, Weiss and colleagues developed the first transgenic mouse model of neuroblastoma by overexpressing human *MYCN* in the PSNS under control of the tyrosine hydroxylase (*TH*) promoter [79]. The *TH* promoter is active in migrating neural crest and this enzyme catalyzes the reaction in which L-tyrosine is hydroxylated in the meta position to obtain L-3,4-dihydroxyphenylalanine (L-DOPA), which is the rate limiting step in the production of catecholamines in the sympathetic ganglia and adrenal gland, sites where neuroblastoma often arises [80]. Overexpression of *MYCN* under control of the *TH* promoter induces neuroblastoma formation in the sympathetic ganglia of transgenic mice but not the adrenal gland [79]. Tumor onset in this context can be accelerated by heterozygous loss of either of two tumor suppressor genes, *NF1* or *RB1*, in compound transgenic mice overexpressing *MYCN*, suggesting that alterations of these genes collaborate with *MYCN* to contribute to neuroblastoma pathogenesis [79]. Since then, a number of murine neuroblastoma models have been developed, including those in which mutationally activated *ALK* is expressed under control of the *TH* promoter [71], *LIN28B* is conditionally overexpressed under control of the *Dβh* promoter [78] or mutant *ALK* is expressed under control of the endogenous *ALK* promoter in a knockin mouse [58] (see Table 1). In addition, immune-deficient mice bearing a highly pretreated, recurrent *MYCN*-amplified primary human tumor have been used to determine the antiangiogenic efficacy of NVP-BEZ235, a clinical PI3K inhibitor, in vivo [81]. For more detailed coverage of this topic, readers are referred to a more comprehensive review of genetically engineered murine models of neuroblastoma by Chesler and Weiss [82].

Table 1 Animal models of neuroblastoma

Model	Type	Gene	Location of tumor	References
TH-MYCN	Transgenic mouse	Human *MYCN*	Multifocal tumors in sympathetic ganglia	[79]
DβH-MYCN; DβH-ALK F1174L	Transgenic zebrafish	Human *MYCN* and human *ALKF1174L*	Adrenal and locally invasive	[70]
TH-MYCN; TH-ALK^{F1174L}	Transgenic mouse	Human *ALK^{F1174L}*	Multifocal tumors in sympathetic ganglia or adrenals, locally invasive	[71]
LSL-Lin28b; Dbh-iCre	Conditional transgenic mouse	Mouse *Lin28b*	Multifocal tumors in sympathetic ganglia or adrenals	[78]
Xenografts in immune-deficient mice	Orthotopic or xenograft	*MYCN*-amplified primary human tumor (SFNB-06)	Orthotopic xenografts of primary tumors in kidney capsule	[81]
TH-MYCN;TH-Cre; Caspase 8	Conditional knockout mouse	Mouse *Caspase 8*	Multifocal tumors in sympathetic ganglia and bone marrow metastasis	[83]
LSL-MYCN;Dbh-iCre	Conditional transgenic mouse	Human *MYCN*	Adrenal and sympathetic ganglia	[84]
TH-MYCN;KI *Alk^{R1279Q}* or KI *Alk^{F1178L}*	Knock-in mouse	Mouse *ALK ^{R1279Q}* and mouse *ALK ^{F1178L}*	Multifocal tumors, locally invasive	[58]

Zebrafish Cancer Models

The teleost zebrafish (*Danio rerio*) has been used as a model organism for studies of vertebrate embryogenesis since the 1960s and was first applied to the study of tumor pathogenesis in 1982 [83–85]. Several unique advantages, such as ease of maintenance, small body size, and high fecundity, make the zebrafish a robust model for large-scale forward genetic screens to identify mutations that confer aberrant phenotypes [84]. The optical transparency of zebrafish embryos is another key feature supporting their use for cancer research, as it allows in vivo imaging to monitor tumor development in real time [83], an application that is relatively difficult in rodents [86]. Recent comparative genomics analysis of the zebrafish reference genome (Zv9) has revealed that zebrafish possess 26,206 protein-coding genes, 71 % of which have human orthologues [87, 88]. Among the human disease-associated genes listed in the Online Mendelian Inheritance in Man (OMIM) database, 82 % can be related to at least one zebrafish orthologue [87, 88], suggesting that the key genes and pathways involved in human cancer are conserved in the fish.

Consequently, the fish has been used successfully to model diverse types of human cancers, including T-cell acute lymphoblastic leukemia (T-ALL) [89, 90], myeloid malignancies [91, 92], neuroblastoma [70], melanoma [93, 94], rhabdomyosarcoma [95, 96], Ewing's sarcoma [97], hepatocellular carcinoma [98], and pancreatic carcinoma [99].

Transgenic Zebrafish Model of Neuroblastoma

Deciphering the molecular basis for neuroblastoma formation in the PSNS has eluded investigators for many years. To approach this fundamental issue, we first sought to develop a model system in which different oncogenes of interest could be specifically overexpressed in the PSNS. After isolating a 5.2 kb promoter fragment upstream of the coding sequence of the zebrafish *dβh* gene, which encodes the rate-limiting enzyme for noradrenalin synthesis, we demonstrated that this promoter can faithfully drive expression of enhanced green fluorescent protein (EGFP) in the sympathetic ganglia and interrenal gland (IRG), the zebrafish analogue of adrenal gland, in a stable transgenic zebrafish [70]. By overexpressing the human *MYCN* oncogene under control of the *dβh* promoter [70], we were then able to generate the first zebrafish model of neuroblastoma. The resultant tumors closely resembled human neuroblastomas histologically, immunohistochemically and ultrastructurally, with tumors arising in the IRG, which corresponds to the site of neuroblastoma origin in approximately half of all patients with this disease [70]. (See Table 2 for comparison of the zebrafish model to mouse models of neuroblastoma with *MYCN* overexpression)

The value of the zebrafish as a model system for coexpressing oncogenes that appear to play key roles in neuroblastoma pathogenesis is best illustrated by studies to demonstrate interactions between MYCN and other oncoprotein. For example, we and others have shown that F1174 mutations of the *ALK* gene are the most prevalent somatic activating mutations in *MYCN*-amplified neuroblastomas, suggesting close collaboration between these genes during tumorigenesis [52]. To test this hypothesis, we developed a stable transgenic zebrafish line that overexpresses the human *ALK F1174L* oncogene under control of the *dβh* promoter. Overexpression of activated *ALK* alone was not sufficient to drive tumorigenesis within the first 6-month monitoring period. By contrast, coexpression of activated *ALK* with *MYCN* greatly potentiated the development of neuroblastoma, as demonstrated by a three-fold increase in disease penetrance and a markedly accelerated tumor onset as compared with that in the *MYCN*-only transgenic fish (Fig. 3).

Further analyses showed that *MYCN* overexpression blocks chromaffin cell differentiation and increases the number of Hu pan-neuronal marker-positive sympathoadrenal progenitor cells in the IRG at 5 weeks of age (Fig. 4). Moreover, *MYCN* overexpression triggered a developmentally-timed apoptotic response in hyperplas-

Table 2 Comparison of zebrafish and mouse models of *MYCN*-overexpressing neuroblastoma

Feature	*MYCN*-overexpressing zebrafish [70]	*MYCN*-overexpressing mouse [79, 100]	
Transgenic construct	*Dβh-EGFP-MYCN* [70]	*TH-MYCN* [79]	*LSL-MYCN & Dβh-iCre* [100]
Transgene expression	Direct expression of *EGFP-MYCN* fusion gene in the PSNS of zebrafish under control of the *dβh* promoter	Direct expression of *MYCN* gene in the PSNS of mouse under control of the rat *TH* promoter	*Cre*-conditional expression of *MYCN* gene in the *Dβh*-expressing cells in mouse
Location of tumor	Interrenal gland, the zebrafish analogue of adrenal gland	Thoracic paraspinous and abdominal tumors	Superior cervical ganglion, the adrenals or the celiac ganglion
Penetrance of tumor in heterozygous animals	17 % by 6 months of age	20 % at 6 months in C57B6/J strain, 70 % in the 129 × 1/SvJ strain background	76 % in a mixed C57/Bl6/129 × 1/SvJ strain background
Pathology of tumor	Small, undifferentiated, round-tumor cells with hyperchromatic nuclei, often forming nests	Small round blue cells with varying degrees of neuronal differentiation	Small, round blue cells typical for neuroectodermal tumors
Ultrastructural feature	Appearance of neurosecretory granules	Appearance of synapse formation and neurosecretory granules	Appearance of neuronal structures, including neurosecretory vesicles
Immunohistochemistry	Positive for neuroblastoma markers, TH, and neuronal markers, Synaptophysin (Syn) and HU	Positive for the neuronal markers, Syn and neuron-specific enolase, and sympathoadrenal progenitor marker, Phox2b [101]	neuroblastoma-specific markers, neural cell adhesion molecule, TH and Ncam1
Tumor metastasis	No metastasis within 6 months of age (unpublished work)	Microscopic metastases to liver, lung or lymphatics, kidney, ovary, testes, brain and muscle	Local invasion
In vivo tumor imaging	Tumor are expressing green fluorescent protein	Ultrasound [102], luciferase (in E2F1-Luc/TH-MYCN double transgenic animals) [103], MRI, PET and ^{131}I-MIBG [104]	Bioluminescence imaging

(continued)

Table 2 (continued)

Feature	*MYCN*-overexpressing zebrafish [70]	*MYCN*-overexpressing mouse [79, 100]	
Cellular mechanism of tumorigenesis	Hyperplasia of sympathoadrenal precursor cells in the interrenal gland at 5 weeks and subsequent apoptosis of these cells	Hyperproliferative microfoci within periadrenal and paraspinal sympathetic ganglia [101] and resistant to apoptosis [105]	Hyperplasia in the adrenal medulla
Cooperating genes	Activated *ALK*, *LMO1*, activated *SHP2*, *GAB2*, and *NF1* loss (unpublished work from Zhu and Look lab)	Loss of *NF1* [79], *RB* [79], *p53* [103], and *Caspase 8* [106], activated *ALK* [71]	Not tested
Drug response	JQ1 (unpublished work)	Front line chemotherapeutics (cyclophosphamide, cisplatinum, doxorubicin, irinotecan, temozolomide, and etoposide) [102, 103, 107–109], steroids (dexamethasone), anti-vascular/ antiangiogenic agents, novel small-molecules [81, 107, 110–112], and JQ1 [113]	JQ1 [100]

tic neuroblasts as they attempted to differentiate into chromaffin cells at 5.5 weeks of age (Fig. 5). In a small subset of *MYCN*-overexpressing fish, the sympathoadrenal cells managed to survive and to accumulate additional genetic alterations that cooperated with MYCN to contribute to neuroblastoma transformation but with a longer latency period and lower penetrance (17 %; Figs. 3b and 6) than seen with *ALK* coexpression. Importantly, in transgenic fish overexpressing both *MYCN* and activated *ALK,* the kinase provides a cell survival signal that blunts the developmentally timed apoptotic response in *MYCN*-overexpressing neuroblasts, allowing continued expansion and oncogenic transformation of hyperplastic neuroblasts, which leads to an earlier onset and increased penetrance of neuroblastoma (56 %; Figs. 5 and 6) [70]. Hence, use of a zebrafish model system enabled us, for the first time, to uncover a mechanism (synergistic interplay between aberrantly expressed *MYCN* and activated *ALK*) that figures prominently in neuroblastoma pathogenesis.

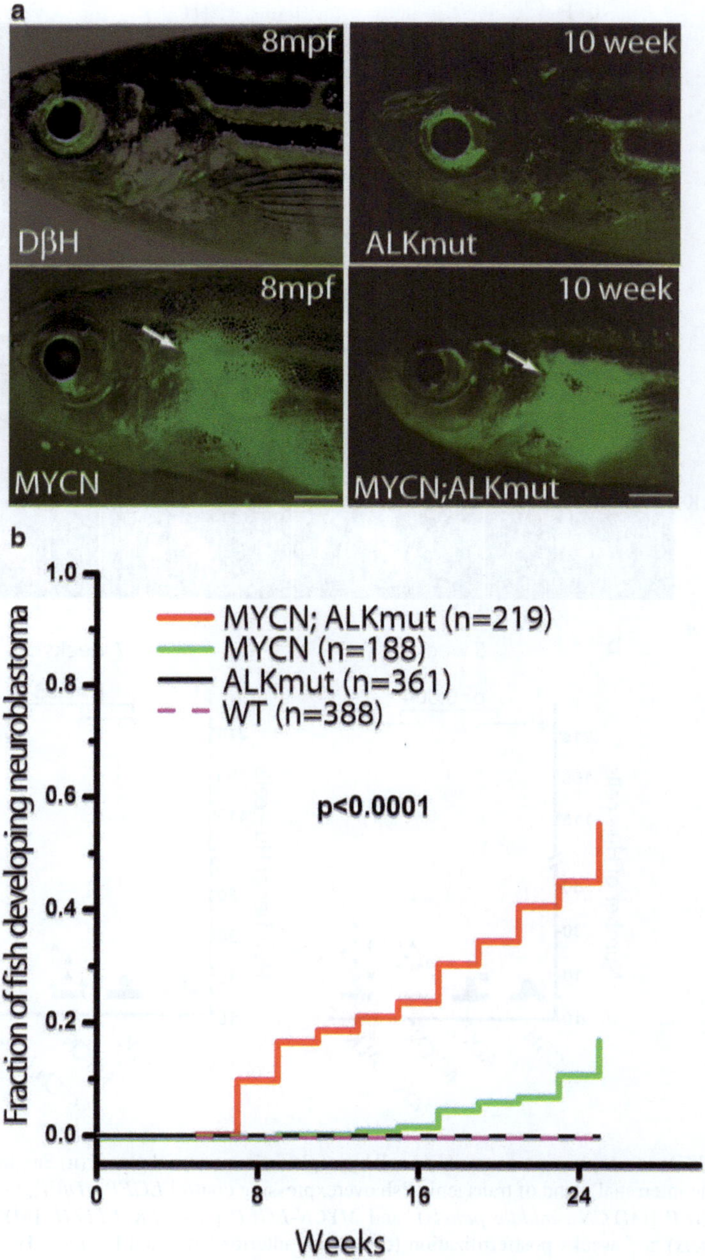

Fig. 3 Activated ALK accelerates disease onset and increases the penetrance of MYCN-induced neuroblastoma. (**a**) *Top left*, control transgenic fish, *dβh-EGFP* (DβH); *Bottom left*, *MYCN* transgenic fish with *EGFP*-expressing tumor (*arrow*) in the interrenal gland at 8 months postfertilization (mpf); *Top right*, *ALKmut* stable transgenic fish expressing the *ALK F1174L* transgene; *Bottom right*, *MYCN;ALKmut* compound transgenic fish with *EGFP*-expressing tumor (*arrow*) in the interrenal gland at 10 weeks postfertilization. Scale bar represents 1 mm. (**b**) Cumulative frequency of neuroblastoma in stable transgenic zebrafish by Kaplan-Meier analysis. *WT* wild-type. This figure is modified from ref [70]

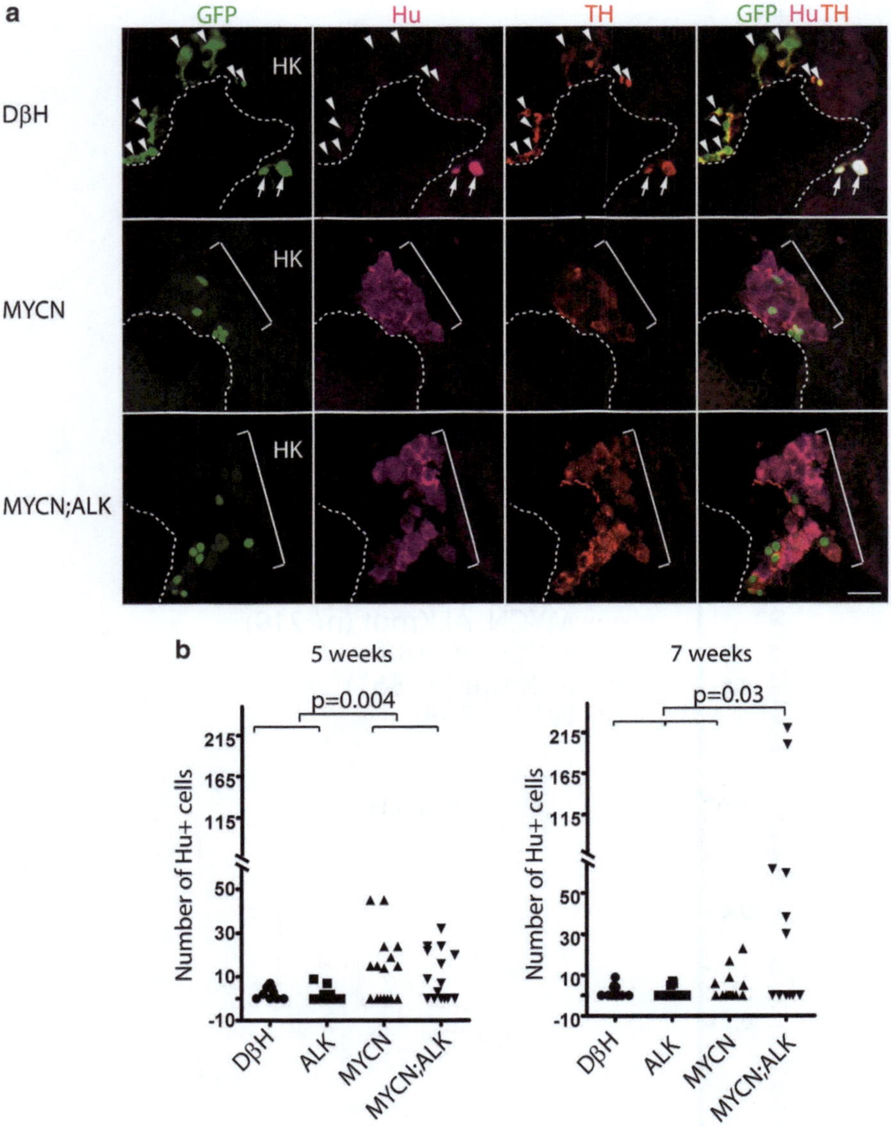

Fig. 4 *MYCN* overexpression causes Hu + cell hyperplasia in interrenal gland. (**a**) Sagittal sections through the interrenal gland of transgenic fish overexpressing control *EGFP* [(*DβH*), *top panels*)], *MYCN-EGFP* [(*MYCN*), *middle panels*), and *MYCN-EGFP* plus *ALK F1174L* [*MYCN;ALK*), *lower panels*) at 5 weeks postfertilization (dorsal up, anterior left). EGFP, *green*; Hu, *magenta*; TH, *red*. Representative sections through the interrenal gland in *DβH* fish contain three to five GFP+/Hu+/TH + sympathetic neuroblasts (*arrows*) and many GFP+/Hu-/TH+ chromaffin cells (*arrowheads*). Hu + cell numbers increase in *MYCN* and *MYCN;ALK* fish (*brackets*), and can be GFP+ and TH+. *Dotted lines* indicate the head kidney (HK) boundary. Scale bar represents 20 mm. (**b**) Numbers of Hu + interrenal gland cells in *DβH*, *ALK*, *MYCN*, and *MYCN;ALK* transgenic fish at 5 and 7 weeks. Mean numbers of Hu + cells were compared by the two-tailed Wilcoxon signed-rank test. This figure is modified from ref [70]

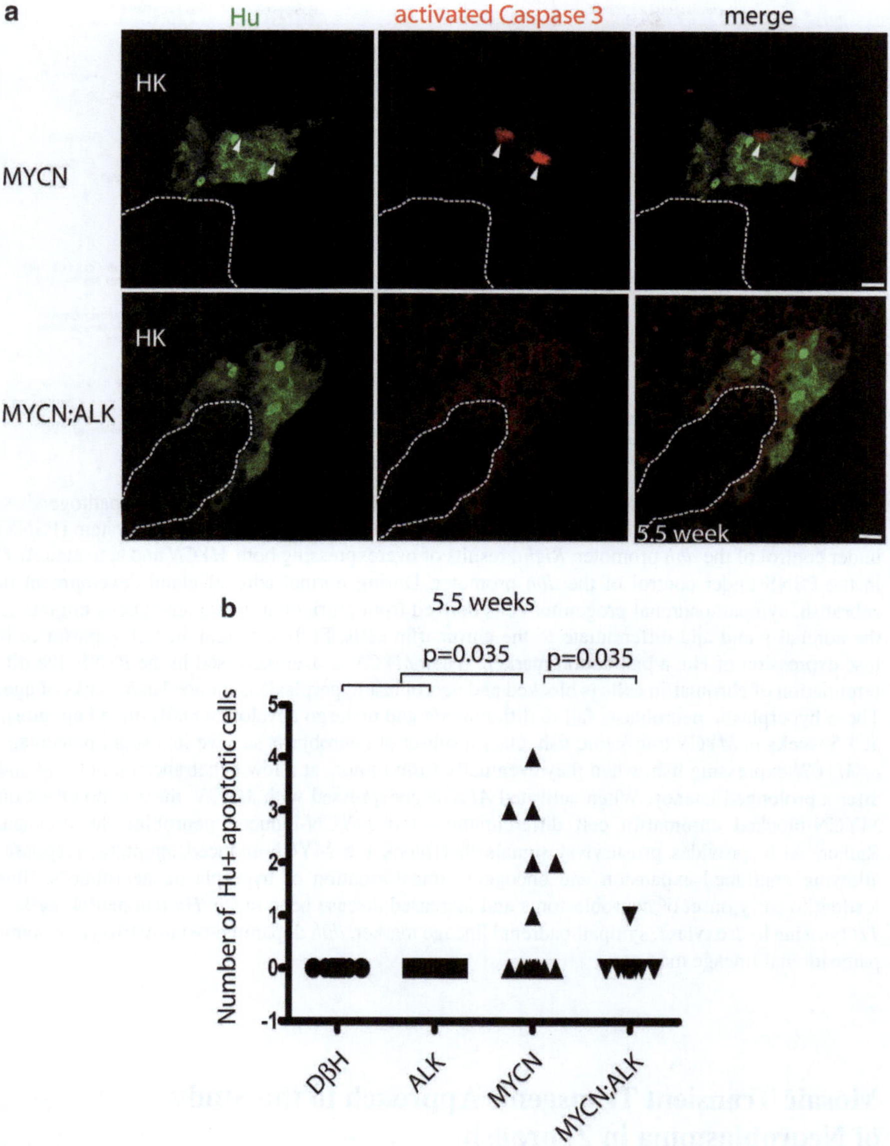

Fig. 5 ALK inhibits a developmentally-timed apoptotic response triggered by *MYCN* overexpression in the interrenal gland [70]. (**a**) Sagittal sections through the interrenal gland of transgenic fish overexpressing *MYCN-EGFP* [(*MYCN*), *top panels*] and *MYCN-EGFP* plus *ALK F1174L* [*MYCN;ALK*), *bottom panels*] transgenic fish at 5.5 weeks postfertilization (wpf; dorsal up, anterior left). Hu, *green*; activated Caspase-3, *red*. Hu+, activated Caspase-3+ apoptotic cells were detected in the *MYCN* transgenic fish (*arrowheads*). *Dotted lines* indicate the head kidney (HK) boundary. Scale bars represent 10 mm. (**b**) Numbers of apoptotic Hu + interrenal gland cells in the *DβH*, *ALK*, *MYCN*, and *MYCN;ALK* fish at 5.5 wpf. Mean numbers of transgenic fish at 5.5 wpf with apoptotic Hu + cells in the interrenal gland were compared with the two-tailed Fisher's exact test. This figure is modified from ref [70]

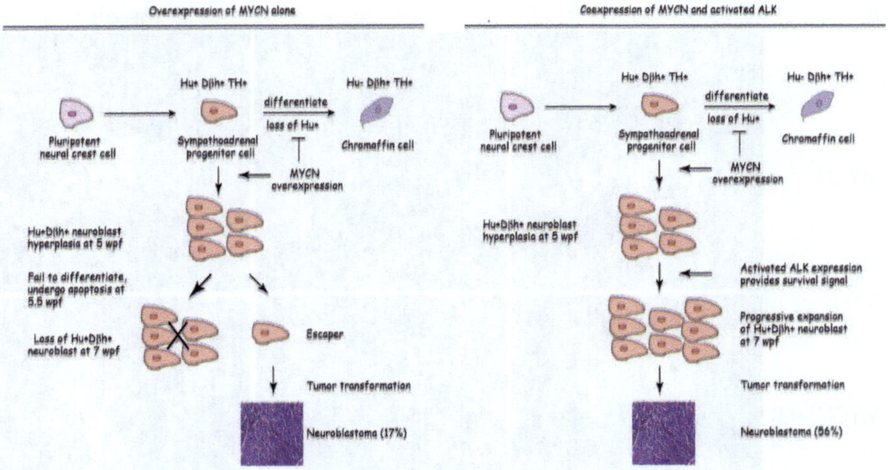

Fig. 6 Model for the collaboration of MYCN and activated ALK in neuroblastoma pathogenesis. *Left*, results of overexpressing *MYCN* alone in the peripheral sympathetic nervous system (PSNS) under control of the *dbh* promoter. *Right*, results of overexpressing both *MYCN* and activated *ALK* in the PSNS under control of the *dbh* promoter. During normal adrenal gland development in zebrafish, sympathoadrenal progenitor cells derived from pluripotent neural crest cells migrate to the adrenal gland and differentiate to the chromaffin cells. Fully differentiated chromaffin cells lose expression of Hu, a pan-neural marker. When *MYCN* is overexpressed in the PSNS, the differentiation of chromaffin cells is blocked and neuroblast hyperplasia is induced at 5 weeks of age. These hyperplastic neuroblasts fail to differentiate and undergo developmentally-timed apoptosis at 5.5 weeks in *MYCN* transgenic fish. Only a subset of neuroblasts survive in a small percentage of *MYCN*-expressing fish, when they eventually form tumors at a low penetrance rate (17 %) and after a prolonged latency. When activated *ALK* is coexpressed with *MYCN*, there is no effect on MYCN-blocked chromaffin cell differentiation and MYCN-induced neuroblast hyperplasia. Rather, ALK provides prosurvival signals that block the MYCN-induced apoptotic response, allowing continued expansion and oncogenic transformation of hyperplastic neuroblasts, thus leading to early onset of neuroblastoma and increased disease penetrance. *Hu* pan neural marker, *TH* tyrosine hydroxylase, sympathoadrenal lineage marker. *dβh* dopamine-beta-hydroxylase, sympathoadrenal lineage marker

Mosaic Transient Transgenic Approach to the Study of Neuroblastoma in Zebrafish

The finding that activated *ALK* collaborates with overexpressed *MYCN* in neuroblastoma pathogenesis in our stable transgenic animals, generated by breeding of *MYCN*-only with one *ALK F1174L*-only stable transgenic line, raised concern over possible founder effects in our transgenic lines, and whether overexpression of wild-type *ALK* (*ALKwt*) might have interacted with *MYCN* during tumorigenesis. Thus, we transiently overexpressed either activated human *ALK* or human *ALKwt* by coinjection of *dβh-ALK F1174L* and *dβh-mCherry* constructs, *dβh-ALKwt* and *dβh-mCherry* constructs or a *dβh-mCherry* only construct into the one-cell stage of transgenic fish overexpressing *MYCN* [70]. Since we have shown that this coinjection strategy results in cointegration of injected DNA into the fish genome, with coexpression of the two transgenes as mosaics in a subset of cells in 50 % of the injected embryos [100], the expression of *mCherry* serves as a useful marker for the coexpression of *ALK* in tissues of the mosaic primary injected animals. In these experiments, mosaic transient overexpression of *ALK F1174L* in *MYCN* transgenic fish recapitulated the acceleration of tumor onset observed in the stable transgenic fish overexpressing both *ALK F1174L* and *MYCN*, while the transient overexpression of *ALKwt* did not appear to collaborate with overexpressed *MYCN* to induce neuroblastoma (Fig. 7). These results demonstrate that activated ALK cooperates with *MYCN* overexpression to accelerate the onset of neuroblastoma, regardless of the integration site in individual mosaic animals. Hence, we feel confident that our mosaic transient transgenic approach could be used to rapidly and efficiently examine the effects of coexpressing candidate oncogenes in primarily injected fish, thus eliminating the excessive time and labor typically involved in the breeding and identification of *ALK*-expressing stable transgenic fish.

Mosaic transgenesis offers another unique advantage over stable transgenesis. In stable transgenic animals, the transgenes are integrated into the genome of every single cell in the body and are heritable from generation to generation [101]. However, many cancers arise from genetic alterations in somatic cells instead of inherited mutations in germ cells [102]. Thus, the mosaic pattern of transgene integration in the primary injected fish and the mixed genotypes within cell populations of individual mosaic transgenic fish would better recapitulate the somatic defects found in patients and would avoid the artificial positional effect due to a single insertion site in a stable transgenic line.

Together, these observations support the future use of mosaic transgenesis as a rapid means to evaluate the collaborative, potentially synergistic relationships among multiple oncogenes in a high-throughput manner, a challenge that has been difficult to meet with conventional animal models, including rodents. Indeed, mosaic transgenesis had been successfully applied in transgenic zebrafish to identify genes that suppress the activated RAS-induced initiation of rhabdomyosarcoma [100] or modify the radiation sensitivity of MYC-induced T-cell acute lymphoblastic leukemia [100]. Langenau and colleagues [89] have also demonstrated that combining the coinjection approach with a heat-shock-inducible transgenic method allow the induction of transgene expression by heat shock in the primary injected fish, and this should be very useful in exploring the cooperative roles of oncogenes in tumor progression, as it permits gene expression to be turned on after the primary tumor is established.

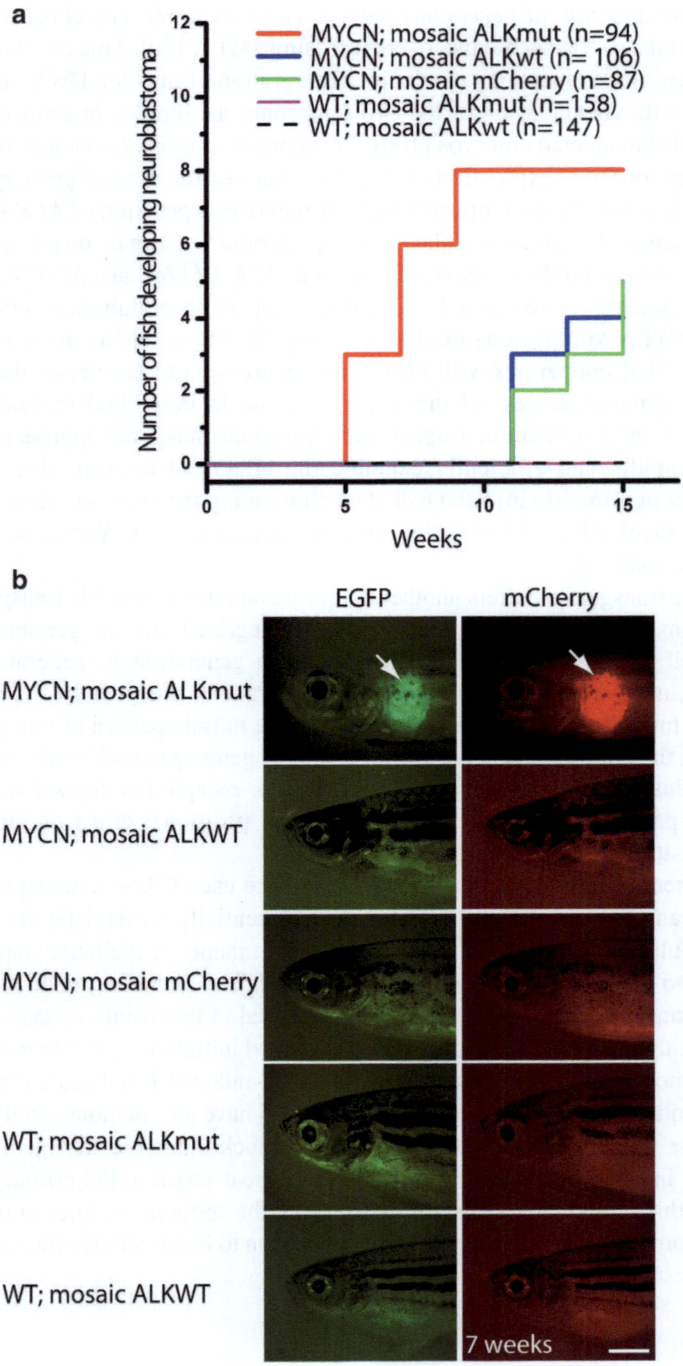

Fig. 7 Mosaic expression of activated ALK accelerates the onset of MYCN-induced neuroblastoma. (**a**) Time of onset of neuroblastoma in *MYCN* transgenic fish or wild-type (*WT*) fish coinjected with the following DNA constructs: (*1*) *dβh-ALKF1174L* and *dβh-mCherry* (mosaic *ALKmut*);

Zebrafish Models in Neuroblastoma Research: Future Prospects

Accessing the Cooperative Contributions of Multiple Genetic or Epigenetic Alterations to Neuroblastoma Pathogenesis

One of the greatest needs in future research of neuroblastoma is a versatile, robust system for rapidly assessing the cooperative contributions of genetic or epigenetic alterations to tumor initiation and spread. We submit that the zebrafish model could meet this challenge, by generating critical information on the interplay among as many as three or more concurrently expressed oncogenes, enabling us to identify signaling conduits that are true drivers of oncogenesis and therefore optimal targets for therapeutic intervention. Another major challenge will be to modify the transgenic strategy to accommodate any genomic or epigenetic alteration that could play a role in the neoplastic process, rather than limiting its use to mutations that affect only protein-coding genes that are overexpressed, gained or amplified. Recently, improved genome editing techniques, such as transcription activator-like effector nucleases (TALENS) [103, 104] and clustered, regularly interspaced short palindromic repeat (CRISPR)-associated (Cas) systems [105] have become widely used to more rapidly and efficiently modify endogenous genes in various types of cells and organisms. These techniques should allow us to target candidate tumor suppressors (such as *CHD5*, *ATRX*, *ARID1A*, *ARID1B* and *PTPRD*) and to efficiently modify genome sequences (by generating large chromosome deletions, such as those syntenic to human 1p or 11q loss, or by introducing disease-associated SNPs) in the zebrafish model system, as a means to better understand the mechanism(s) underlying the contributions of genomic or epigenetic alterations to neuroblastoma pathogenesis.

Fig. 7 (continued) (*2*) *dβh-ALKWT* and *dβh-mCherry* (mosaic *ALKWT*); or (*3*) *dβh-mCherry* alone (mosaic *mCherry*). The difference between tumor onset by 9 weeks postfertilization (wpf) in the *MYCN* fish coinjected with *dβh-ALKF1174L* and *dβh-mCherry* (*MYCN*; mosaic *ALKmut*) and that in the *MYCN* line coinjected with *dβh-ALKWT* and *dβh-mCherry* (*MYCN*; mosaic *ALKWT*) or *dβh-mCherry* alone (*MYCN*; mosaic *mCherry*) is significant at the $p=0.002$ and $p=0.007$ respectively, by two-tailed Fisher's exact test. (**b**) *EGFP* and *mCherry* double positive tumors arose in the *MYCN* fish coinjected with *dβh-ALKF1174L* and *dβh-mCherry* constructs (*MYCN*; mosaic *ALKmut*) at 7 weeks of age. Neuroblastomas were not identified in the *MYCN* fish coinjected with *dβh-ALKWT* and *dβh-mCherry* or *dβh-mCherry* alone at 7 weeks, or in any of the siblings that did not inherit the *MYCN* transgene and were injected with either the *ALKWT* gene or the *ALKF1174L* gene. Scale bars, 1 mm. This figure is modified from ref [70]

Large-Scale Forward Genetic Screen to Identify Novel Driver Genes in Neuroblastoma Initiation and Progression

Like many other pediatric cancers, including glioblastoma, acute lymphoblastic leukemia, medulloblastoma, and rhabdoid tumor, high-risk neuroblastoma harbors a very low frequency of recurrent somatic mutations [66, 106]. The alterations in copy number or regulatory polymorphisms appear to be driving this tumor, but the regions of chromosome gain or loss are so large that it is difficult to specifically implicate the critical oncogenes and tumor suppressors that lie within these regions. This has lent considerable impetus to large-scale forward genetic screening efforts to identify new neuroblastoma driver genes that might have been missed by standard integrative genomic studies. Over the past decade, various types of forward genetic approaches have been applied successfully in the zebrafish to uncover genes that are important for normal development and tumorigenesis, such as *N*-ethyl-*N*-nitrosourea (ENU)-based chemical mutagenesis [107–111], retrovirus-based insertional mutagenesis [112] and transposon-based insertional mutagenesis [113]. Specifically, the transposon-based mutagenesis strategy has been demonstrated in mice to be an effective, powerful and nonbiased tool for discovering driver genes in various types of cancers [114–117]. In zebrafish, the somatic mutagenesis mediated by the *Sleeping Beauty* (SB) *T2/Onc* transposon system under control of the ubiquitous carp *ß-actin* promoter, also resulted in tumor formation in ~10 % of adult zebrafish. Many mutated genes identified in these fish tumors were similar to those involved in human and mouse cancers [113]. Thus, transposon-mediated somatic mutagenesis approaches in zebrafish could provide a valuable and high-throughput tool for identification of genes and signaling pathways that not only drive neuroblastoma formation, but also act as genetic modifiers that collaborate with known neuroblastoma oncogenes, such as *MYCN* and activated *ALK*, to promote tumor initiation, progression or metastasis.

Dissecting Cellular and Molecular Mechanisms Underlying Neuroblastoma Metastasis

About half of all neuroblastoma patients, especially those over 18 months of age with amplified copies of the *MYCN* oncogene, present with evidence of widespread metastatic tumor at diagnosis [7, 118]. Current treatments for disseminated neuroblastoma are limited by their toxicity and inability to eradicate distant metastases; however the lack of a precise understanding of the multistep cellular and molecular pathogenesis of this complex tumor has impeded most efforts to devise safe and effective therapy for this early complication. Zebrafish are emerging as an attractive model system for the study of tumor metastasis, largely due to (1) the superior imaging properties of optically transparent zebrafish tissues [119, 120]; (2) the ease of transplantation of fluorescent dye-labeled or fluorescence-expressing human or zebrafish tumor cells into either embryos or adult fish [121–124]; and (3) the availability of numerous types of transgenic or mutant zebrafish lines. These unique advantages of the zebrafish will enable (1) rapid analysis of the metastatic behavior of patient-derived or zebrafish-derived primary neuroblastoma cells in engrafted zebrafish embryos or adult immunocompromised mutant fish, *rag2E450fs* [122]; (2) visualization of tumor progression in real-time, including invasion, intravasation, extravasation and angiogenesis, using transgenic fish such as Tg(*fli1:egfp*) [125] and Tg(*flk1:mCherry*) [126] with fluorescence-labeled blood vessels, or mutant *Casper* fish lacking pigment cells [127]; (3) dissection of the interactions of tumor cells with host immune cells or local microenvironment, using transgenic fish Tg(*mpx:gfp*) with labeled neutrophils [128] or Tg(*mpeg1:egfp*) with labeled macrophages [129]; and (4) elucidation of the mechanisms underlying aberrantly expressed genes or pathways in multistep neuroblastoma metastasis in vivo by integrating powerful methods of genetic analysis with innovative transplantation approaches, a variety of transgenic or mutant zebrafish lines (mentioned above) and powerful in vivo imaging techniques.

Validating Targeted Therapy and Screening for Effective Small Molecular Inhibitors of Neuroblastoma

Finally, the ultimate goal of neuroblastoma research is to achieve optimal cure rates for patients with this devastating disease through a better understanding of tumor biology and the development of rational therapeutic strategies. Long-term survival rates for high-risk neuroblastoma patients remain less than 40 %, even with intensive use of different combinations of multiagent chemotherapy, surgery, high-dose myeloablative therapy with autologous hematopoietic stem cell rescue, and GD2-directed immunotherapy [1]. Efforts to overcome this impasse are focused on the genes, proteins and signaling pathways have been implicated in the pathogenesis of neuroblastomas. By disabling or otherwise modifying these targets, in combination with more conventional treatment approaches, it may be possible to boost cure rates and reduce the development of late complications. This potential is illustrated by recent successful use of the zebrafish to demonstrate the value of "oncogene-addition" [130, 131] in the maintenance of hepatocellular carcinoma (HCC). Specifically, using two different inducible systems, the mifepristone-inducible LexPR system [132] or the tetracycline (Tet)-on/off system [133], Gong and colleagues showed that oncogenes such as activated *KRAS* and *c-MYC* are important for both HCC tumor initiation and maintenance, thus providing a rationale for molecularly targeted therapies against aberrant *RAS* or *MYC* signaling in HCC patients [133, 134].

Further, the zebrafish has long been a valuable resource for in vivo chemical screening to identify lead compounds that can modulate specific biological processes without undue toxicity or identify new therapeutic usage for old drugs [135–137]. Some drugs emerging from these screens have been approved by the

FDA and are now in clinical trials. This includes leflunomide for rheumatoid arthritis [138] and potentially for melanoma [83], as well as prostaglandin E2 (PGE2) for ductus-dependent cyanotic congenital heart disease [139] and for acute leukemia after a cord blood stem cell transplant [137]. Thus, zebrafish occupy a pivotal position in the future of neuroblastoma research. They not only offer a means to determine the requirement for candidate genes in tumor cell survival, and thus their importance as targets for therapy, but also enable high-throughput testing, refining and optimizing novel molecular therapies in preclinical models.

Conclusion

The unique properties of zebrafish have opened new avenues to exploring the underlying mechanisms that drive neuroblastoma pathogenesis. The insights gained from these studies are expected to facilitate the design and development of novel molecular therapeutics for this devastating childhood cancer.

References

1. Park JR, Bagatell R, London WB et al (2013) Children's Oncology Group's 2013 blueprint for research: neuroblastoma. Pediatr Blood Cancer 60(6):985–993
2. Louis CU, Shohet JM (2015) Neuroblastoma: molecular pathogenesis and therapy. Annu Rev Med 66:49–63
3. Saito D, Takase Y, Murai H et al (2012) The dorsal aorta initiates a molecular cascade that instructs sympatho-adrenal specification. Science 336(6088):1578–1581
4. Hoehner JC, Gestblom C, Hedborg F et al (1996) A developmental model of neuroblastoma: differentiating stroma-poor tumors' progress along an extra-adrenal chromaffin lineage. Lab Invest 75(5):659–675
5. Janoueix-Lerosey I, Schleiermacher G, Delattre O (2010) Molecular pathogenesis of peripheral neuroblastic tumors. Oncogene 29(11):1566–1579
6. Brodeur GM (2003) Neuroblastoma: biological insights into a clinical enigma. Nat Rev Cancer 3(3):203–216
7. Maris JM (2010) Recent advances in neuroblastoma. N Engl J Med 362(23):2202–2211
8. Jiang M, Stanke J, Lahti JM (2011) The connections between neural crest development and neuroblastoma. Curr Top Dev Biol 94:77–127
9. Le Dourin NM, Kalcheim G (1999) The neural crest, 2nd edn. Cambridge University Press
10. Fontaine-Perus JC, Chanconie M, Le Douarin NM (1982) Differentiation of peptidergic neurones in quail-chick chimaeric embryos. Cell Differ 11(4):183–193
11. Teillet MA, Kalcheim C, Le Douarin NM (1987) Formation of the dorsal root ganglia in the avian embryo: segmental origin and migratory behavior of neural crest progenitor cells. Dev Biol 120(2):329–347
12. Lallier TE, Bronner-Fraser M (1988) A spatial and temporal analysis of dorsal root and sympathetic ganglion formation in the avian embryo. Dev Biol 127(1):99–112
13. Weston JA (1963) A radioautographic analysis of the migration and localization of trunk neural crest cells in the chick. Dev Biol 6:279–310
14. Krispin S, Nitzan E, Kalcheim C (2010) The dorsal neural tube: a dynamic setting for cell fate decisions. Dev Neurobiol 70(12):796–812

15. Huber K, Kalcheim C, Unsicker K (2009) The development of the chromaffin cell lineage from the neural crest. Auton Neurosci 151(1):10–16
16. Unsicker K (1993) The chromaffin cell: paradigm in cell, developmental and growth factor biology. J Anat 183(Pt 2):207–221
17. Shtukmaster S, Schier MC, Huber K et al (2013) Sympathetic neurons and chromaffin cells share a common progenitor in the neural crest in vivo. Neural Dev 8:12
18. Liem KF Jr, Tremml G, Roelink H et al (1995) Dorsal differentiation of neural plate cells induced by BMP-mediated signals from epidermal ectoderm. Cell 82(6):969–979
19. Selleck MA, Bronner-Fraser M (1995) Origins of the avian neural crest: the role of neural plate-epidermal interactions. Development 121(2):525–538
20. Garcia-Castro MI, Marcelle C, Bronner-Fraser M (2002) Ectodermal Wnt function as a neural crest inducer. Science 297(5582):848–851
21. Reissmann E, Ernsberger U, Francis-West PH et al (1996) Involvement of bone morphogenetic protein-4 and bone morphogenetic protein-7 in the differentiation of the adrenergic phenotype in developing sympathetic neurons. Development 122(7):2079–2088
22. Shah NM, Groves AK, Anderson DJ (1996) Alternative neural crest cell fates are instructively promoted by TGFbeta superfamily members. Cell 85(3):331–343
23. Huber K (2006) The sympathoadrenal cell lineage: specification, diversification, and new perspectives. Dev Biol 298(2):335–343
24. Guillemot F, Lo LC, Johnson JE et al (1993) Mammalian achaete-scute homolog 1 is required for the early development of olfactory and autonomic neurons. Cell 75(3):463–476
25. Pattyn A, Morin X, Cremer H et al (1999) The homeobox gene Phox2b is essential for the development of autonomic neural crest derivatives. Nature 399(6734):366–370
26. Howard M, Foster DN, Cserjesi P (1999) Expression of HAND gene products may be sufficient for the differentiation of avian neural crest-derived cells into catecholaminergic neurons in culture. Dev Biol 215(1):62–77
27. Howard MJ, Stanke M, Schneider C et al (2000) The transcription factor dHAND is a downstream effector of BMPs in sympathetic neuron specification. Development 127(18): 4073–4081
28. Lim KC, Lakshmanan G, Crawford SE et al (2000) Gata3 loss leads to embryonic lethality due to noradrenaline deficiency of the sympathetic nervous system. Nat Genet 25(2): 209–212
29. Tsarovina K, Pattyn A, Stubbusch J et al (2004) Essential role of Gata transcription factors in sympathetic neuron development. Development 131(19):4775–4786
30. Holzschuh J, Barrallo-Gimeno A, Ettl AK et al (2003) Noradrenergic neurons in the zebrafish hindbrain are induced by retinoic acid and require tfap2a for expression of the neurotransmitter phenotype. Development 130(23):5741–5754
31. Barrallo-Gimeno A, Holzschuh J, Driever W et al (2004) Neural crest survival and differentiation in zebrafish depends on mont blanc/tfap2a gene function. Development 131(7): 1463–1477
32. Cochard P, Goldstein M, Black IB (1978) Ontogenetic appearance and disappearance of tyrosine hydroxylase and catecholamines in the rat embryo. Proc Natl Acad Sci U S A 75(6):2986–2990
33. Ernsberger U, Patzke H, Tissier-Seta JP et al (1995) The expression of tyrosine hydroxylase and the transcription factors cPhox-2 and Cash-1: evidence for distinct inductive steps in the differentiation of chick sympathetic precursor cells. Mech Dev 52(1):125–136
34. Ernsberger U, Reissmann E, Mason I et al (2000) The expression of dopamine beta-hydroxylase, tyrosine hydroxylase, and Phox2 transcription factors in sympathetic neurons: evidence for common regulation during noradrenergic induction and diverging regulation later in development. Mech Dev 92(2):169–177
35. Ernsberger U, Esposito L, Partimo S et al (2005) Expression of neuronal markers suggests heterogeneity of chick sympathoadrenal cells prior to invasion of the adrenal anlagen. Cell Tissue Res 319(1):1–13

36. Schneider C, Wicht H, Enderich J et al (1999) Bone morphogenetic proteins are required in vivo for the generation of sympathetic neurons. Neuron 24(4):861–870
37. Morales AV, Barbas JA, Nieto MA (2005) How to become neural crest: from segregation to delamination. Semin Cell Dev Biol 16(6):655–662
38. Le Douarin NM, Teillet MA (1974) Experimental analysis of the migration and differentiation of neuroblasts of the autonomic nervous system and of neurectodermal mesenchymal derivatives, using a biological cell marking technique. Dev Biol 41(1):162–184
39. Unsicker K, Huber K, Schober A et al (2013) Resolved and open issues in chromaffin cell development. Mech Dev 130(6-8):324–329
40. Mosse YP, Laudenslager M, Longo L et al (2008) Identification of ALK as a major familial neuroblastoma predisposition gene. Nature 455(7215):930–935
41. Mosse YP, Laudenslager M, Khazi D et al (2004) Germline PHOX2B mutation in hereditary neuroblastoma. Am J Hum Genet 75(4):727–730
42. Maris JM, Weiss MJ, Mosse Y et al (2002) Evidence for a hereditary neuroblastoma predisposition locus at chromosome 16p12-13. Cancer Res 62(22):6651–6658
43. Amiel J, Laudier B, Attie-Bitach T et al (2003) Polyalanine expansion and frameshift mutations of the paired-like homeobox gene PHOX2B in congenital central hypoventilation syndrome. Nat Genet 33(4):459–461
44. le Nguyen B, Diskin SJ, Capasso M et al (2011) Phenotype restricted genome-wide association study using a gene-centric approach identifies three low-risk neuroblastoma susceptibility Loci. PLoS Genet 7(3), e1002026
45. Butler Tjaden NE, Trainor PA (2013) The developmental etiology and pathogenesis of Hirschsprung disease. Transl Res 162(1):1–15
46. Wilzen A, Nilsson S, Sjoberg RM et al (2009) The Phox2 pathway is differentially expressed in neuroblastoma tumors, but no mutations were found in the candidate tumor suppressor gene PHOX2A. Int J Oncol 34(3):697–705
47. Huber K, Karch N, Ernsberger U et al (2005) The role of Phox2B in chromaffin cell development. Dev Biol 279(2):501–508
48. Weese-Mayer DE, Berry-Kravis EM, Zhou L et al (2003) Idiopathic congenital central hypoventilation syndrome: analysis of genes pertinent to early autonomic nervous system embryologic development and identification of mutations in PHOX2b. Am J Med Genet A 123A(3):267–278
49. Pei D, Luther W, Wang W et al (2013) Distinct neuroblastoma-associated alterations of PHOX2B impair sympathetic neuronal differentiation in zebrafish models. PLoS Genet 9(6), e1003533
50. George RE, Sanda T, Hanna M et al (2008) Activating mutations in ALK provide a therapeutic target in neuroblastoma. Nature 455(7215):975–978
51. Janoueix-Lerosey I, Lequin D, Brugieres L et al (2008) Somatic and germline activating mutations of the ALK kinase receptor in neuroblastoma. Nature 455(7215):967–970
52. Bresler SC, Weiser DA, Huwe PJ et al (2014) ALK mutations confer differential oncogenic activation and sensitivity to ALK inhibition therapy in neuroblastoma. Cancer Cell 26(5):682–694
53. De Brouwer S, De Preter K, Kumps C et al (2010) Meta-analysis of neuroblastomas reveals a skewed ALK mutation spectrum in tumors with MYCN amplification. Clin Cancer Res 16(17):4353–4362
54. Morris SW, Kirstein MN, Valentine MB et al (1994) Fusion of a kinase gene, ALK, to a nucleolar protein gene, NPM, in non-Hodgkin's lymphoma. Science 263(5151):1281–1284
55. Iwahara T, Fujimoto J, Wen D et al (1997) Molecular characterization of ALK, a receptor tyrosine kinase expressed specifically in the nervous system. Oncogene 14(4):439–449
56. Souttou B, Carvalho NB, Raulais D et al (2001) Activation of anaplastic lymphoma kinase receptor tyrosine kinase induces neuronal differentiation through the mitogen-activated protein kinase pathway. J Biol Chem 276(12):9526–9531
57. Reiff T, Huber L, Kramer M et al (2011) Midkine and Alk signaling in sympathetic neuron proliferation and neuroblastoma predisposition. Development 138(21):4699–4708

58. Cazes A, Lopez-Delisle L, Tsarovina K et al (2014) Activated Alk triggers prolonged neurogenesis and Ret upregulation providing a therapeutic target in ALK-mutated neuroblastoma. Oncotarget 5(9):2688–2702
59. Seeger RC, Brodeur GM, Sather H et al (1985) Association of multiple copies of the N-myc oncogene with rapid progression of neuroblastomas. N Engl J Med 313(18):1111–1116
60. White PS, Thompson PM, Gotoh T et al (2005) Definition and characterization of a region of 1p36.3 consistently deleted in neuroblastoma. Oncogene 24(16):2684–2694
61. Attiyeh EF, London WB, Mosse YP et al (2005) Chromosome 1p and 11q deletions and outcome in neuroblastoma. N Engl J Med 353(21):2243–2253
62. Meddeb M, Danglot G, Chudoba I et al (1996) Additional copies of a 25 Mb chromosomal region originating from 17q23.1-17qter are present in 90 % of high-grade neuroblastomas. Genes Chromosomes Cancer 17(3):156–165
63. Van Roy N, Laureys G, Cheng NC et al (1994) 1;17 translocations and other chromosome 17 rearrangements in human primary neuroblastoma tumors and cell lines. Genes Chromosomes Cancer 10(2):103–114
64. Vandesompele J, Speleman F, Van Roy N et al (2001) Multicentre analysis of patterns of DNA gains and losses in 204 neuroblastoma tumors: how many genetic subgroups are there? Med Pediatr Oncol 36(1):5–10
65. Van Roy N, De Preter K, Hoebeeck J et al (2009) The emerging molecular pathogenesis of neuroblastoma: implications for improved risk assessment and targeted therapy. Genome Med 1(7):74
66. Pugh TJ, Morozova O, Attiyeh EF et al (2013) The genetic landscape of high-risk neuroblastoma. Nat Genet 45:279–284
67. Molenaar JJ, Koster J, Zwijnenburg DA et al (2012) Sequencing of neuroblastoma identifies chromothripsis and defects in neuritogenesis genes. Nature 483(7391):589–593
68. Chen Y, Takita J, Choi YL et al (2008) Oncogenic mutations of ALK kinase in neuroblastoma. Nature 455(7215):971–974
69. Heukamp LC, Thor T, Schramm A et al (2012) Targeted expression of mutated ALK induces neuroblastoma in transgenic mice. Sci Transl Med 4(141):141ra91
70. Zhu S, Lee JS, Guo F et al (2012) Activated ALK collaborates with MYCN in neuroblastoma pathogenesis. Cancer Cell 21(3):362–373
71. Berry T, Luther W, Bhatnagar N et al (2012) The ALK(F1174L) mutation potentiates the oncogenic activity of MYCN in neuroblastoma. Cancer Cell 22(1):117–130
72. Sausen M, Leary RJ, Jones S et al (2013) Integrated genomic analyses identify ARID1A and ARID1B alterations in the childhood cancer neuroblastoma. Nat Genet 45(1):12–17
73. Wang K, Diskin SJ, Zhang H et al (2011) Integrative genomics identifies LMO1 as a neuroblastoma oncogene. Nature 469(7329):216–220
74. Capasso M, Devoto M, Hou C et al (2009) Common variations in BARD1 influence susceptibility to high-risk neuroblastoma. Nat Genet 41(6):718–723
75. Maris JM, Mosse YP, Bradfield JP et al (2008) Chromosome 6p22 locus associated with clinically aggressive neuroblastoma. N Engl J Med 358(24):2585–2593
76. Capasso M, Diskin SJ, Totaro F et al (2013) Replication of GWAS-identified neuroblastoma risk loci strengthens the role of BARD1 and affirms the cumulative effect of genetic variations on disease susceptibility. Carcinogenesis 34(3):605–611
77. Wood AC, Oldridge D, Crimmins I et al (2015) Genetic predisposition to neuroblastoma mediated by a single nucleotide polymorphism within a LMO1 oncogene super-enhancer element. Nature 528:418–421
78. Molenaar JJ, Domingo-Fernandez R, Ebus ME et al (2012) LIN28B induces neuroblastoma and enhances MYCN levels via let-7 suppression. Nat Genet 44(11):1199–1206
79. Weiss WA, Aldape K, Mohapatra G et al (1997) Targeted expression of MYCN causes neuroblastoma in transgenic mice. EMBO J 16(11):2985–2995
80. Banerjee SA, Hoppe P, Brilliant M et al (1992) 5′ flanking sequences of the rat tyrosine hydroxylase gene target accurate tissue-specific, developmental, and transsynaptic expression in transgenic mice. J Neurosci 12(11):4460–4467

81. Chanthery YH, Gustafson WC, Itsara M et al (2012) Paracrine signaling through MYCN enhances tumor-vascular interactions in neuroblastoma. Sci Transl Med 4(115):115ra3
82. Chesler L, Weiss WA (2011) Genetically engineered murine models – contribution to our understanding of the genetics, molecular pathology and therapeutic targeting of neuroblastoma. Semin Cancer Biol 21:245–255
83. White R, Rose K, Zon L (2013) Zebrafish cancer: the state of the art and the path forward. Nat Rev Cancer 13(9):624–636
84. Amatruda JF, Patton EE (2008) Genetic models of cancer in zebrafish. Int Rev Cell Mol Biol 271:1–34
85. Konantz M, Balci TB, Hartwig UF et al (2012) Zebrafish xenografts as a tool for in vivo studies on human cancer. Ann N Y Acad Sci 1266:124–137
86. Ellenbroek SI, van Rheenen J (2014) Imaging hallmarks of cancer in living mice. Nat Rev Cancer 14(6):406–418
87. Kettleborough RN, Busch-Nentwich EM, Harvey SA et al (2013) A systematic genome-wide analysis of zebrafish protein-coding gene function. Nature 496(7446):494–497
88. Howe K, Clark MD, Torroja CF et al (2013) The zebrafish reference genome sequence and its relationship to the human genome. Nature 496(7446):498–503
89. Langenau DM, Traver D, Ferrando AA et al (2003) Myc-induced T cell leukemia in transgenic zebrafish. Science 299(5608):887–890
90. Feng H, Stachura DL, White RM et al (2010) T-lymphoblastic lymphoma cells express high levels of BCL2, S1P1, and ICAM1, leading to a blockade of tumor cell intravasation. Cancer Cell 18(4):353–366
91. Forrester AM, Grabher C, McBride ER et al (2011) NUP98-HOXA9-transgenic zebrafish develop a myeloproliferative neoplasm and provide new insight into mechanisms of myeloid leukaemogenesis. Br J Haematol 155(2):167–181
92. Alghisi E, Distel M, Malagola M et al (2013) Targeting oncogene expression to endothelial cells induces proliferation of the myelo-erythroid lineage by repressing the Notch pathway. Leukemia 27(11):2229–2241
93. Patton EE, Widlund HR, Kutok JL et al (2005) BRAF mutations are sufficient to promote nevi formation and cooperate with p53 in the genesis of melanoma. Curr Biol 15(3): 249–254
94. Santoriello C, Anelli V, Alghisi E et al (2012) Highly penetrant melanoma in a zebrafish model is independent of ErbB3b signaling. Pigment Cell Melanoma Res 25(2):287–289
95. Le X, Langenau DM, Keefe MD et al (2007) Heat shock-inducible Cre/Lox approaches to induce diverse types of tumors and hyperplasia in transgenic zebrafish. Proc Natl Acad Sci U S A 104(22):9410–9415
96. Langenau DM, Keefe MD, Storer NY et al (2007) Effects of RAS on the genesis of embryonal rhabdomyosarcoma. Genes Dev 21(11):1382–1395
97. Leacock SW, Basse AN, Chandler GL et al (2012) A zebrafish transgenic model of Ewing's sarcoma reveals conserved mediators of EWS-FLI1 tumorigenesis. Dis Model Mech 5(1): 95–106
98. Zheng W, Li Z, Nguyen AT et al (2014) Xmrk, kras and myc transgenic zebrafish liver cancer models share molecular signatures with subsets of human hepatocellular carcinoma. PLoS One 9(3), e91179
99. Park SW, Davison JM, Rhee J et al (2008) Oncogenic KRAS induces progenitor cell expansion and malignant transformation in zebrafish exocrine pancreas. Gastroenterology 134(7):2080–2090
100. Langenau DM, Keefe MD, Storer NY et al (2008) Co-injection strategies to modify radiation sensitivity and tumor initiation in transgenic zebrafish. Oncogene 27(30):4242–4248
101. Patton EE, Zon LI (2001) The art and design of genetic screens: zebrafish. Nat Rev Genet 2(12):956–966
102. Watson IR, Takahashi K, Futreal PA et al (2013) Emerging patterns of somatic mutations in cancer. Nat Rev Genet 14(10):703–718

103. Sander JD, Cade L, Khayter C et al (2011) Targeted gene disruption in somatic zebrafish cells using engineered TALENs. Nat Biotechnol 29(8):697–698
104. Dahlem TJ, Hoshijima K, Jurynec MJ et al (2012) Simple methods for generating and detecting locus-specific mutations induced with TALENs in the zebrafish genome. PLoS Genet 8(8), e1002861
105. Sander JD, Joung JK (2014) CRISPR-Cas systems for editing, regulating and targeting genomes. Nat Biotechnol 32(4):347–355
106. Vogelstein B, Papadopoulos N, Velculescu VE et al (2013) Cancer genome landscapes. Science 339(6127):1546–1558
107. Stainier DY, Fouquet B, Chen JN et al (1996) Mutations affecting the formation and function of the cardiovascular system in the zebrafish embryo. Development 123:285–292
108. Chen JN, Haffter P, Odenthal J et al (1996) Mutations affecting the cardiovascular system and other internal organs in zebrafish. Development 123:293–302
109. Weinstein BM, Schier AF, Abdelilah S et al (1996) Hematopoietic mutations in the zebrafish. Development 123:303–309
110. Neuhauss SC, Solnica-Krezel L, Schier AF et al (1996) Mutations affecting craniofacial development in zebrafish. Development 123:357–367
111. Beckwith LG, Moore JL, Tsao-Wu GS et al (2000) Ethylnitrosourea induces neoplasia in zebrafish (Danio rerio). Lab Invest 80(3):379–385
112. Amsterdam A, Nissen RM, Sun Z et al (2004) Identification of 315 genes essential for early zebrafish development. Proc Natl Acad Sci U S A 101(35):12792–12797
113. McGrail M, Hatler JM, Kuang X et al (2011) Somatic mutagenesis with a Sleeping Beauty transposon system leads to solid tumor formation in zebrafish. PLoS One 6(4), e18826
114. Moriarity BS, Largaespada DA (2015) Sleeping Beauty transposon insertional mutagenesis based mouse models for cancer gene discovery. Curr Opin Genet Dev 30:66–72
115. Copeland NG, Jenkins NA (2010) Harnessing transposons for cancer gene discovery. Nat Rev Cancer 10(10):696–706
116. Wu X, Northcott PA, Dubuc A et al (2012) Clonal selection drives genetic divergence of metastatic medulloblastoma. Nature 482(7386):529–533
117. Moriarity BS, Otto GM, Rahrmann EP et al (2015) A Sleeping Beauty forward genetic screen identifies new genes and pathways driving osteosarcoma development and metastasis. Nat Genet 47(6):615–624
118. Maris JM, Hogarty MD, Bagatell R et al (2007) Neuroblastoma. Lancet 369(9579):2106–2120
119. Stoletov K, Montel V, Lester RD et al (2007) High-resolution imaging of the dynamic tumor cell vascular interface in transparent zebrafish. Proc Natl Acad Sci U S A 104(44):17406–17411
120. Stoletov K, Klemke R (2008) Catch of the day: zebrafish as a human cancer model. Oncogene 27(33):4509–4520
121. Nicoli S, Ribatti D, Cotelli F et al (2007) Mammalian tumor xenografts induce neovascularization in zebrafish embryos. Cancer Res 67(7):2927–2931
122. Tang Q, Abdelfattah NS, Blackburn JS et al (2014) Optimized cell transplantation using adult rag2 mutant zebrafish. Nat Methods 11(8):821–824
123. Blackburn JS, Langenau DM (2014) Zebrafish as a model to assess cancer heterogeneity, progression and relapse. Dis Model Mech 7(7):755–762
124. Haldi M, Ton C, Seng WL et al (2006) Human melanoma cells transplanted into zebrafish proliferate, migrate, produce melanin, form masses and stimulate angiogenesis in zebrafish. Angiogenesis 9(3):139–151
125. Lawson ND, Weinstein BM (2002) In vivo imaging of embryonic vascular development using transgenic zebrafish. Dev Biol 248(2):307–318
126. Cross LM, Cook MA, Lin S et al (2003) Rapid analysis of angiogenesis drugs in a live fluorescent zebrafish assay. Arterioscler Thromb Vasc Biol 23(5):911–912
127. White RM, Sessa A, Burke C et al (2008) Transparent adult zebrafish as a tool for in vivo transplantation analysis. Cell Stem Cell 2(2):183–189

128. Renshaw SA, Loynes CA, Trushell DM et al (2006) A transgenic zebrafish model of neutrophilic inflammation. Blood 108(13):3976–3978
129. Ellett F, Pase L, Hayman JW et al (2011) mpeg1 promoter transgenes direct macrophage-lineage expression in zebrafish. Blood 117(4):e49–e56
130. Singh A, Greninger P, Rhodes D et al (2009) A gene expression signature associated with "K-Ras addiction" reveals regulators of EMT and tumor cell survival. Cancer Cell 15(6):489–500
131. Weinstein IB (2002) Cancer. Addiction to oncogenes—the Achilles heal of cancer. Science 297(5578):63–64
132. Nguyen AT, Emelyanov A, Koh CH et al (2012) An inducible kras(V12) transgenic zebrafish model for liver tumorigenesis and chemical drug screening. Dis Model Mech 5(1):63–72
133. Li Z, Huang X, Zhan H et al (2012) Inducible and repressable oncogene-addicted hepatocellular carcinoma in Tet-on xmrk transgenic zebrafish. J Hepatol 56(2):419–425
134. Nguyen AT, Emelyanov A, Koh CH et al (2011) A high level of liver-specific expression of oncogenic Kras(V12) drives robust liver tumorigenesis in transgenic zebrafish. Dis Model Mech 4(6):801–813
135. Zon LI, Peterson RT (2005) In vivo drug discovery in the zebrafish. Nat Rev Drug Discov 4(1):35–44
136. Kaufman CK, White RM, Zon L (2009) Chemical genetic screening in the zebrafish embryo. Nat Protoc 4(10):1422–1432
137. Bowman TV, Zon LI (2010) Swimming into the future of drug discovery: in vivo chemical screens in zebrafish. ACS Chem Biol 5(2):159–161
138. Golicki D, Newada M, Lis J et al (2012) Leflunomide in monotherapy of rheumatoid arthritis: meta-analysis of randomized trials. Pol Arch Med Wewn 122(1-2):22–32
139. Thanopoulos BD, Andreou A, Frimas C (1987) Prostaglandin E2 administration in infants with ductus-dependent cyanotic congenital heart disease. Eur J Pediatr 146(3):279–282
140. Straub JA, Sholler GL, Nishi R (2007) Embryonic sympathoblasts transiently express TrkB in vivo and proliferate in response to brain-derived neurotrophic factor in vitro. BMC Dev Biol 7:10

Zebrafish Germ Cell Tumors

Angelica Sanchez and James F. Amatruda

Abstract Germ cell tumors (GCTs) are malignant cancers that arise from embryonic precursors known as Primordial Germ Cells. GCTs occur in neonates, children, adolescents and young adults and can occur in the testis, the ovary or extragonadal sites. Because GCTs arise from pluripotent cells, the tumors can exhibit a wide range of different histologies. Current cisplatin-based combination therapies cures most patients, however at the cost of significant toxicity to normal tissues. While GWAS studies and genomic analysis of human GCTs have uncovered somatic mutations and loci that might confer tumor susceptibility, little is still known about the exact mechanisms that drive tumor development, and animal models that faithfully recapitulate all the different GCT subtypes are lacking. Here, we summarize current understanding of germline development in humans and zebrafish, describe the biology of human germ cell tumors, and discuss progress and prospects for zebrafish GCT models that may contribute to better understanding of human GCTs.

Keywords Zebrafish • Primordial germ cell • Germ cell tumor • Seminoma • Non-seminoma

Germline Development in Fish, Mouse and Human

Primordial Germ Cell (PGC) Specification

The earliest cells of the germline lineage are known as Primordial Germ Cells (PGCs). PGC specification follows two distinct mechanisms, either preformation or induction, depending on the organism. Shortly after fertilization, zebrafish embryos, similar to

A. Sanchez
Departments of Pediatrics and Molecular Biology, University of Texas Southwestern Medical Center, 5323 Harry Hines Boulevard, Dallas, TX 75390, USA
e-mail: angelica.sanchez@utsouthwestern.edu

J.F. Amatruda (✉)
Departments of Pediatrics, Molecular Biology and Internal Medicine, University of Texas Southwestern Medical Center, 5323 Harry Hines Boulevard, Dallas, TX 75390, USA
e-mail: james.amatruda@utsouthwestern.edu

D.M. Langenau (ed.), *Cancer and Zebrafish*, Advances in Experimental Medicine and Biology 910, DOI 10.1007/978-3-319-30654-4_21

C. elegans, *Drosophila* and *Xenopus*, contain preformed maternal RNA and proteins that are prepackaged in a cytoplasmic nuage, known as a germ plasm [1–4]. The germ plasm selectively and transcriptionally silences the expression of somatic genes while keeping the PGCs in a primitive pluripotent state. In flies and worms, RNA binding proteins including Pumilio and Nanos contribute to this repression of somatic fate [5, 6]. In zebrafish, *nanos* is essential for the development of PGCs [5].

The identification of the zebrafish *vasa* ortholog greatly facilitated efforts to understand the development of the zebrafish germline [7–10]. Vasa is an RNA helicase that was previously shown to be a component of the germline in nematodes, flies and frogs [11–14]. In zebrafish, maternally produced *vasa* mRNA is localized to electron-dense material at the cleavage planes of embryos at the 4-cell stage. By the 4000-cell stage, four *vasa*-positive PGCs can be identified. These will undergo several rounds of proliferation to produce a population of 25–50 migrating PGCs that will form the gonad [7, 8, 10, 15].

In contrast to organisms with preformed germline components, in mice and humans PGCs are produced in the epiblast through a process known as induction. In response to signals produced in the extraembryonic ectoderm and visceral endoderm, including Bone Morphogenetic Proteins 2, 4 and 8b [16–19], some of the epiblast cells begin to express the marker *Fragilis* (also known as *Mil-1/ Ifitm3*), signaling their competence to become germ cells [18, 20–26]. A few of these *Fragilis*-expressing cells will begin to express the transcriptional repressor *Prdm1* (PR-domain-containing protein 1), also known as *Blimp1* (B-lymphocyte-induced maturation protein 1) [27]. *Blimp1,* along with *Blimp14,* acts to repress expression of somatic genes and maintain expression of pluripotency genes such as *Stella, Oct3/4* and *Nanog* [20, 28–36].

PGC Migration

In nearly all metazoans, PGCs arise at sites distant from the future gonad and must migrate during embryogenesis to reach the proper location for eventual gonadogenesis [37–40]. In contrast to other organisms, zebrafish PGCs arise in four random locations with respect to the developmental axis of the embryo. The PGCs migrate to a series of intermediate targets, converging into bilateral clusters in the presomitic mesoderm, where the gonad will form [3, 10, 15, 41]. This guided migration is orchestrated by an attractant gradient established by somatic expression of Stromal-derived factor 1 (sdf1a), also known as CXCL12 [42, 43]. The sdf1a ligand interacts with the chemokine receptor cxcr4b expressed on the surface of PGCs [42, 43]. This homing signal is further fine-tuned by the cxcr7 receptor on somatic cells, which acts as a repulsive guidance cue [44].

In mice and other mammals, the newly specified PGCs must travel from the primitive streak and localize to the endoderm around stage E7.5 [45, 46]. From E8 to E9.5 the PGCs migrate through the dorsal mesentery of the hindgut. Upon reaching the level of the gonadal ridges, the PGCs exit the hindgut in bilateral streams to populate the gonadal mesoderm [37, 46]. The c-Kit receptor and its ligand, KitL, are critical for the migration and survival of PGCs in mice [47–50].

Germ Cell Differentiation

Irrespective of the future sex of the animal, germ cells in zebrafish initially begin to differentiate into oocytes and are in a state known as juvenile hermaphroditism [51]. Upon the start of male germ cell differentiation, oocytes in male zebrafish undergo apoptosis as the mature male germline emerges [52]. Characterizations of male sex determination factors have identified several genes that are specific to the male gonads, such as anti-Mullerian hormone (*amh*) and *sox9a* [53–55]. Furthermore, *sox9a* was shown to control juvenile hermaphroditism and regulate the transition to male fate producing mature sperm [56].

During gonadogenesis, the number and persistence of germ cells makes important contributions to sex determination [57, 58]. Several studies have revealed that lower PGC numbers promote male differentiation in zebrafish. Loss of *dead end* (*dnd*), an RNA-binding protein that is essential for PGC survival [59], leads to loss of all germ cells and male phenotypic differentiation [57, 60, 61]. Similarly, loss of function *ziwi*, the zebrafish *piwi* homolog, leads to reduced numbers of germ cells and male phenotypic development [62]. Remarkably, Dranow and Draper found that adult female zebrafish depleted of germ cells due to loss-of-function mutations in *nanos3* convert to a male phenotype [58]. They concluded that a germ cell-derived signal acts on the somatic gonad to promote female development directly or indirectly by repressing male-specific gene expression.

In mice and other mammals, licensing is the initial step in which germ cells begin to differentiate into their respective sex cells dependent on the sex chromosomes present in the organism. This licensing event is initiated by the RNA-binding protein known as Deleted in azoospermia-like (DAZL); *Dazl* mutants no longer express sex specific markers and do not begin meiosis [63, 64]. During this transition, the germ cells are released from their transcriptional repression [65]. In prenatal males, presence of the SRY gene on the Y chromosome regulates expression of Sox9 and is necessary for proper male gonadal development and spermatogenesis [66–68]. Humans with a heterozygous mutation in *SOX9* develop a bone disorder known as campomelic dysplasia that is accompanied by gonadal sex reversal, and XY mice with homozygous mutations exhibit ovarian development [69–71]. These studies emphasized the importance of proper Sox9 expression in testis development. To maintain male identity prenatally, Stra8 (Stimulated by retinoic acid gene 8), an enzyme required for meiosis initiation, must be persistently repressed by Cyp26B1 [72–74]. Postnatally, repression of Cyp26B1 releases repression of Stra8 and allows germ cells to respond to retinoic acid, which initiates meiosis [72–74]. The germ cells are now able to exit G1/G0 and can then proceed to make mature sperm.

As in the fish, gonadal sex in mammals appears not to be completely fixed. The transcription factor DMRT1 is required in both somatic cells and germ cells for normal male gonadal differentiation [75, 76]. Loss of Dmrt1 in adult Sertoli cells of the testis results in transformation of the cells to granulosa cells and transition of the gonad to a more ovarian-like phenotype [77]. Furthermore, forced expression of DMRT1 in the ovary reprograms granulosa cells to Sertoli cells, leading to the development of structures resembling seminiferous tubules [78].

Human Germ Cell Tumors

In humans, GCTs occur in infants, children and young adults [79] (Fig. 1). Both childhood and adolescent/adult GCTs are thought to originate from Primordial Germ Cells (PGCs). Owing to the pluripotent nature of these cells, GCTs can take on a variety of different histologic fates [80]. GCTs in which the PGCs retain pluripotency and do not differentiate are known as seminomas (also called 'dysgerminomas' in females and 'germinomas' when occurring in the CNS). In contrast, GCTs in which the cells take on a variety of differentiation states are designated non-seminomas, of

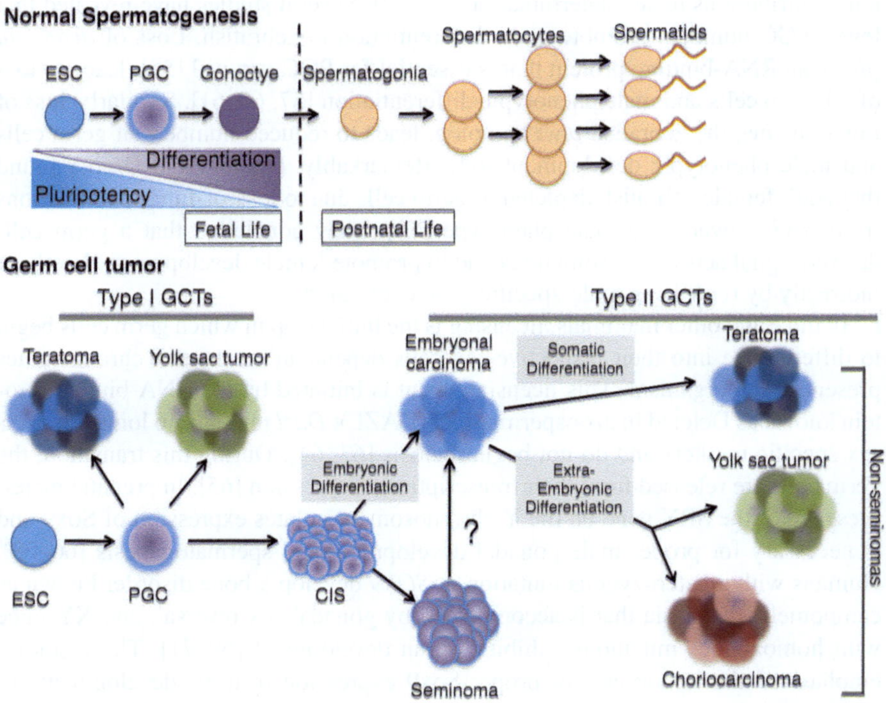

Fig. 1 Mammalian spermatogenesis and human germ cell tumor (GCT) histological classification. The transition of primordial germ cells from embryonic stem cells to gonocytes is facilitated by increased expression of germ cell specific genes and decreased expression of embryonic pluripotency genes. In humans, sex chromosomes dictate the fate of a gonocyte; in the testis self-renewing spermatogonial stem cells give rise to progeny that undergo meiosis to produce mature sperm. Germ cell tumors arise from primoridal germ cells (PGCs). Type I GCTs are generally seen in neonates and young children and include Teratomas and Yolk Sac Tumors. Type II GCTs occur in adolescents and adults and may proceed through a carcinoma *in situ* (CIS) precursor (also known as Intratubular Germ Cell Neoplasia). CIS cells can give rise to both Seminoma and Non-Seminoma, the stem cell component of which is Embryonal Carcinoma (EC). EC may undergo somatic differentiation into Teratomas, containing derivatives of ectoderm, mesoderm and endoderm. Alternatively EC cells may differentiate into GCTs resembling extraembryonic tissues, such as Yolk Sac Tumor and Choriocarcinoma. Mixed tumors containing both seminoma and non-seminoma components also occur

which embryonal carcinoma is thought to represent the stem cell component. GCTs differentiated to somatic cell lineages (endoderm, mesoderm, and ectoderm) are known as teratomas. Finally, GCTs may take on extraembryonic differentiation resembling the fetal yolk sac (yolk sac tumors) or the placenta (choriocarcinoma).

The histology of GCTs is similar in both males and females, whether occurring in the testis, ovary or extragonadal sites, implying origin from a common precursor cell [79]. However, there are some epidemiologic differences in the incidence of different types of GCT. In males there are two peaks of testicular GCT incidence, one in early childhood at age 3–4, and a second, much larger peak that begins at puberty and is maximal at around age 30. In females, there is an early peak from age 0–2 representing the incidence of sacrococcygeal teratoma, an extragonadal GCT, in newborns and infants. Beginning at age 5–6, the incidence of ovarian GCT increases with age, becoming maximal at age 20–25 [81].

While the overall incidence of GCT (about 12,500 cases/year in the USA) is lower than that of common epithelial cancers such as lung, breast and prostate cancer, testicular GCT is the most common cancer and the leading cause of cancer death in young men [79]. The incidence of GCT is increasing around the world, for unknown reasons [81]. There are several known risk factors that increase the risk of developing testicular germ cell tumors. They include but are not limited to disorders of sexual development (DSD), gonadal dysgenesis, cryptorchidism (undescended testis), familial background, environmental exposure, and genetic association. Familial risk can increase the risk of developing GCT 4-fold in a male with a father who had GCT or up to 9-fold in a male whose brother had GCTs. The incidence of GCT varies widely in different geographic regions [82, 83], leading to the idea that the environment may strongly influence risk for testicular GCT.

There are also important age-dependent differences in GCT histologic spectrum. Type I GCTs occur in infants and children and consist of teratomas and yolk sac tumors (YSTs). Type II tumors in adolescents and adults have more diverse histology, and include seminomas, non-seminomas or mixed tumors containing both seminomatous and non-seminomatous elements. Based on differing epidemiology, clinical outcome and histologic spectrum, GCTs in young children may be biologically distinct from GCTs occurring in older (post-pubertal) populations.

This idea is increasingly supported by molecular evidence. Whereas Type I tumors show variable Loss of Imprinting (LOI) at loci such as IGF2 and H19, adult-type GCTs tend to show complete erasure of imprinting [84, 85]. This result is interesting, because it implies that Type I GCTs may arise at an earlier stage of PGC development, since PGCs undergo erasure of imprinting during early development, and this erasure is largely completed by the end of PGC migration. Cytogenetic data consistently show loss of Chromosome 1p and 6q in Type I tumors, while Type II tumors also commonly exhibit amplification of Chromosome 12p. More recently, studies directly comparing the gene expression patterns of pediatric and adult GCTs have demonstrated distinct transcriptional profiles in tumors of similar histology arising in different age groups (for example, in yolk sac tumors of children vs. yolk sac tumors of adolescent/adults) [86]. Taken together, these studies support the notion that different biological mechanisms may drive childhood and adolescent/adult germ cell tumorigenesis.

Current cisplatin-based therapy for GCTs cures most patients [81, 87, 88], but major challenges remain. 15 % of patients have cisplatin-resistant tumors and will eventually die of their disease [89]. Furthermore, there is significant therapy-related toxicity in patients treated for GCTs. Late effects in patients treated with cisplatin, etoposide and bleomycin (the current first-line regimen for GCT) include pulmonary fibrosis [90], renal insufficiency and salt-wasting [91–93], infertility and hormonal changes [94–98], hyperlipidemia [99, 100], Raynaud's phenomenon [91, 98, 101, 102], obesity [99, 100, 103–105] and neuropathy [106]. Ototoxicity results in significant hearing loss, as assessed by audiogram, in nearly 80 % of patients [91, 97]. Survivors of GCT treatment have twice the risk of early onset cardiovascular disease [96] and second malignancies [107, 108].

Compared to many other solid malignancies, relatively few somatic mutations have been described in GCTs. This lack of knowledge inhibits the development of targeted therapy that could provide an alternative or adjunct to standard chemotherapy. Amplification of Chromosome 12p is a pathognomonic feature of adolescent/adult GCTs, but no genes in this region have definitively been linked to germ cell tumorigenesis [109]. The most commonly reported mutated gene is *KIT*, a tyrosine kinase growth factor receptor that plays important roles in germ cell development [110–115]. Mutations have also been reported in *NRAS* and *KRAS*, signaling components of the MAP kinase pathway that act downstream of *KIT* [116–119]. Central Nervous System GCTs exhibit *KIT* and *RAS* mutations [120]. Somatic mutations in *BRAF*, another MAP kinase pathway member, have been associated with cisplatin resistance in adult TGCTs [121]. More recently, exome sequencing of TGCT identified somatic mutations in several genes including *CIITA*, *NEB, PDGFRA*, *WHSC1* and *SUPT6H* [122]. A larger study of 42 adult TGCTs identified somatic mutations in *CDC27* and mutations in *XRCC2* associated with cisplatin resistance [123]. To date, the mutation spectrum of Type I GCTs has not been reported.

A number of genome-wide association studies (GWAS) have been conducted in men with TGCT, resulting in the identification of a large number of novel germ cell tumor susceptibility loci. In 2009, two groups reported strong linkage of GCT susceptibility to loci on chromosomes 5, 6 and 12, suggesting roles for *SPRY4* and *KITLG*, both components of receptor tyrosine kinase signaling, as well as for the pro-apoptotic *BAK1* gene [124, 125]. A follow-up study by Turnbull and co-workers identified additional susceptibility loci near the *DMRT1*, *TERT* and *ATF7IP* genes [126]. The studies have subsequently been replicated in other patient populations [127–129]. More recently, further GWAS studies have identified new candidate GCT susceptibility loci, including genes with potential roles in telomere regulation, such as *PITX1*, or germ cell development, such as *TEX14, RAD51C, PRDM14* and *DAZL* [130, 131].

Because of the many unresolved issues concerning the causes and biological mechanisms of GCTs, and the urgent need to develop and test new treatments, animal models of GCT are badly needed. However, to date, relatively few models have been available. The 129Sv inbred mouse strain was found more than 50 years ago to be susceptible to testicular teratoma formation, with spontaneous tumors forming in 1–5 % of animals [132]. Mutations in several genes have been shown to increase the teratoma incidence in 129Sv mice, including *Dead end* [133] and *Dmrt1*. In fact, *Dmrt1*-mutant 129Sv mice develop testicular teratomas at nearly 100 % incidence [134, 135]. The teratomas that develop in the 129Sv background strongly resemble the Type I testicular teratomas occurring in young boys. However, mouse models of seminomas or of non-seminomas other than Type I teratomas have not been described. Recent progress suggests that zebrafish may be a useful addition to GCT models, at least for seminomas.

Zebrafish Germ Cell Tumor Models

During the past several years, several fish models have been described with abnormal germline development that recapitulates certain aspects of human germ cell tumors. Three of these models were identified using the zebrafish, while one was identified in Medaka. All of these models demonstrate an inability to undergo proper germ cell differentiation and as a result exhibit immature, enlarged testes. Three of the models were identified through forward genetic N-ethyl-N-nitrosourea (ENU)-based mutagenesis screens while one is a transgenic line. These models can serve to provide further insight into human seminomas with the possibility to contribute to our understanding of the human disease.

The first fish model described was identified in Medaka, following an ENU mutagenesis screen to uncover genes that are important for germ cell and gonadal development [136]. Morinaga el al identified the *hotei* "hot" mutant fish, which was initially characterized as having extreme gonadal hypertrophy. The gonads were further examined and found to have increased numbers of germ cells that resided in disrupted testicular architecture, as well as decreased ability to produce mature sperm as compared with their wild-type siblings. The loss of function mutation was mapped, by positional cloning, to reside in the anti-Mullerian hormone receptor II (*amhrII*). This receptor belongs to the TGF-β/BMP superfamily and in humans is known to play a role in Mullerian duct regression in males. The mutation, in exon 9 of *amhrII*, changes a tyrosine residue in the highly conserved kinase domain to cysteine.

Neumann and co-workers used ENU mutagenesis and high-throughput histologic screening to identify a line of zebrafish that developed testicular germ cell tumors (GCTs) with high penetrance (>80 % incidence in homozygotes by 4 months of age) (Fig. 2). They showed that the mutation is dominantly inherited, and causes the development of tumors of primitive, undifferentiated testicular GCTs resembling human seminoma (Fig. 3). Like human seminomas, the zebrafish seminomas were sensitive to radiation therapy and underwent apoptosis in response to DNA damage [137]. To identify the molecular nature of the mutation, these investigators developed a haplotype-mapping panel and, using this technique coupled with further high-resolution meiotic mapping, identified a mutation in the Type IB BMP receptor, *alk6b/bmpr1bb*, as the cause of the zebrafish GCTs [138]. BMPs (Bone Morphogenetic Proteins), members of the TGF-β superfamily, play important roles in development and differentiation, but had not been linked to GCTs. Neumann et al. showed that the

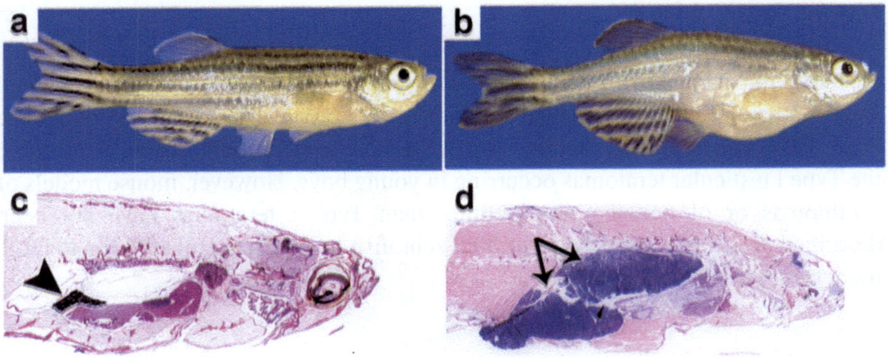

Fig. 2 Testicular GCT in bmpr1bb mutant zebrafish. Mutants develop GCTs because of impaired germ-cell differentiation. Compared with adult wild-type males (**a** and **c**) (*arrowhead*: normal testis), adult bmpr1bb males display abdominal distension and marked testicular enlargement (**b** and **d**) (*arrow*: testicular tumor). Modified with permission from Neumann et al. (2011)

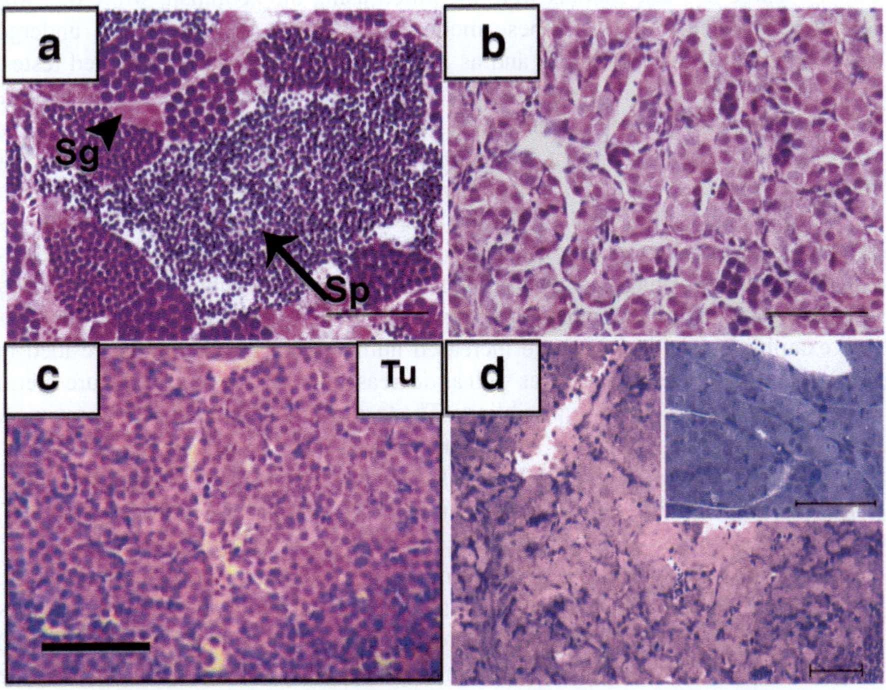

Fig. 3 Zebrafish germ cell tumor models exhibit histologic similarity. (**a**) Normal testis. The testis consists of cysts or lobules of spermatogenic cells surrounded by a basement membrane and somatic cells. Small clusters of spermatogonia are seen adjacent to the basement membrane (*arrowheads*). Successive stages of differentiation are also evident, including primary and secondary spermatocytes, spermatids and mature spermatozoa. (**b**) Germ cell tumor in a *bmpr1bb* mutant testis, showing loss of differentiation and an excess of primitive, spermatogonial-like cells. (**c**) Testicular germ cell tumor from the *Tg(flck:TAg)* T24 line. Tu: tumor. (**d**) Testicular GCT in a $lrrc50^{H255h}$ fish. Two magnifications of the same tumor shown (largest in insert), with loss of differentiated germ cells in the tumor. Scale bars: 50 microns ((**a**, **b**) modified with permission from Neumann et al. (2011); (**c**) modified with permission from Gill et al. (2010); (**d**) modified from Basten et al. (2013))

mutation causes loss of BMP signal transduction, and that BMP signaling is required for germ cell differentiation. This work represents the first identification through positional cloning of a cancer predisposition mutation in the zebrafish.

To understand the relevance of BMP signaling to human GCTs, Fustino et al. profiled the state of the BMP pathway in clinical GCT specimens and showed that undifferentiated human GCTs (seminomas and embryonal carcinomas) also lack BMP signaling [139]. This work also identified the miR-200 family of microRNAs as modulators of the BMP pathway, providing an explanation for differential BMP signaling activity in seminomas vs. yolk sac tumors. Collectively these studies identified BMP signaling as a key node in GCT differentiation and a promising target for novel therapies.

The first and only transgenic model of GCTs was identified by Gill et al. In this model, the Simian Virus 40 (SV40) T-antigen (TAg) is driven by the lymphocyte specific tyrosine kinase (lck) promoter from *Fugu*. Originally intended as a T-cell leukemia model, these fish did not manifest thymic malignancies, but rather testicular GCT. By 36 months, about 20 % of the offspring developed tumors [140]. When comparing the tumors with wild type testis, the GCTs were clearly enlarged and lacked organized seminiferous tubules. Furthermore, the testicular GCTs were predominantly composed of undifferentiated spermatogonial-like germ cells determined by both sorting the cells by size (FACs) and histological analysis (Fig. 3) [140]. Expression patterns of human seminoma specific genes were unchanged as compared to zebrafish wildtype testis. Various RAS genes were also sequenced in the zebrafish to identify any de novo mutations that could explain the TGCTs in the fish but none were found.

Finally, the most recent GCT model was identified through a forward genetic ENU screen. The causative mutation is a nonsense mutation in the leucine-rich repeat (LRR)-containing protein *lrrc50* [141]. Initially, this forward genetic screen was performed to identify genes involved in ciliary motility. *lrrc50* mutants develop a phenotype resembling that of human primary ciliary dyskinesia (PCD). Early in development, the *lrrc50* homozygous mutants develop a curved body axis, dilated pronephric ducts, kidney cysts and severe edema, which ultimately results in embryonic death at 8 days post-fertilization (dpf) [141]. Interestingly, heterozygous *lrrc50* mutants display increased incidence of testicular GCT, suggesting a novel tumor suppressor role for lrrc50.

lrrc50 heterozygous mutants display a 90 % GCT penetrance in their 2nd and 3rd years of life [142]. Histological analysis of the tumors found that there were severely reduced numbers of differentiated germ cells, and most of the cells morphologically resembled early spermatogonial cells (Fig. 3). These cells also stained positive for phospho-histone H3 in single cells unlike the wild type testis, which showed reduced staining in clusters. 44.4 % of the tumors examined showed a loss of the *lrrc50* wildtype allele, consistent with a role for *lrrc50* as a tumor suppressor.

This model is, to date, the only one that found mutations in the respective gene in human seminomas [142]. Basten et al. identified *LRRC50* nonsense mutations in two pedigrees with family history of seminomas. In a separate analysis of 38 patients with sporadic seminomas, 5 demonstrated heterozygosity for different germline

LRRC50 mutations. The mutations were shown to be functional nulls by their inability to complement the lethal larval phenotype of *lrrc50* mutants. These results establish *LRRC50* as the first gene specifically linked to seminoma predisposition in humans. These results prompted an investigation of the possible roles of cilia in germ cell development. In a novel finding, the investigators demonstrated the presence of primary cilia in normal spermatogonia and showed that LRRC50 colocalizes with the axoneme in spermatogonial stem cells, that its expression is cell cycle-regulated and that it colocalizes with condensed chromosomes [142]. These results shed new light on the possible importance of ciliary signaling in germ cell development and germ cell tumorigenesis.

Summary

Germ cell tumors are among the most common cancers of young people, and the leading cause of cancer death in young men. The wide age range of incidence and histologic diversity of the tumors creates challenges in GCT biology and treatment. While some genomic data are now beginning to point to possible molecular origins of these tumors, the paucity of representative animal models is an impediment to further functional genomic and translational studies.

To date the zebrafish has proved to be a powerful model for understanding the biology of PGC development and migration. While there are important differences between human and fish germ cells, a number of conserved mechanisms have been identified, suggesting that studies in zebrafish can contribute importantly to our understanding of human germ cell biology.

Zebrafish models of human germ cell tumors have also begun to make important contributions. The studies described above have identified several signaling pathways, such as the TGF-β/BMP pathway and cilia-mediated signaling, that are directly relevant to human seminomas. The development of seminoma-like tumors in fish transgenic for the known oncogene SV40 large T, and the response of GCTs to DNA-damaging agents, suggests that these models recapitulate important aspects of the oncogenic phenotype of human GCTs.

There are, however, several important caveats that apply to the current models. There is no evidence that these models of GCTs become invasive or malignant, as do some examples of human seminomas. To date, true zebrafish models of ovarian GCT (as opposed to models of impaired maturation of post-meiotic oocytes) have not been described. Reverse genetic models that use gain or loss-of-function approaches and germ cell-specific promoters to test the role of specific GCT candidate genes, at specific developmental stages, remain to be developed. This is a particularly significant gap in current approaches, because it is likely that the ability of germ cells to develop along somatic lineages, as seen in non-seminomas, is limited to specific stages of development in which germ cells retain some pluripotency. In humans, non-seminomas are more resistant to treatment and more likely to metastasize than are seminomas, and are responsible for the majority of deaths from GCT, emphasizing the need for such models. However, there is reason to be optimistic. The great recent progress in the field, with the availability of new promoters, new genome editing techniques and increasing information about the genomic landscape of human GCT, means that the next few years promise to see great advances in our understanding of GCT, and better options for GCT patients.

References

1. Houston DW, King ML (2000) Germ plasm and molecular determinants of germ cell fate. Curr Top Dev Biol 50:155–181
2. Kosaka K et al (2007) Spatiotemporal localization of germ plasm RNAs during zebrafish oogenesis. Mech Dev 124(4):279–289
3. Raz E (2003) Primordial germ-cell development: the zebrafish perspective. Nat Rev Genet 4(9):690–700
4. Saffman EE, Lasko P (1999) Germline development in vertebrates and invertebrates. Cell Mol Life Sci 55(8–9):1141–1163
5. Koprunner M et al (2001) A zebrafish nanos-related gene is essential for the development of primordial germ cells. Genes Dev 15(21):2877–2885
6. Lai F, King ML (2013) Repressive translational control in germ cells. Mol Reprod Dev 80(8):665–676
7. Braat AK et al (1999) Characterization of zebrafish primordial germ cells: morphology and early distribution of vasa RNA. Dev Dyn 216(2):153–167
8. Knaut H et al (2000) Zebrafish vasa RNA but not its protein is a component of the germ plasm and segregates asymmetrically before germline specification. J Cell Biol 149(4):875–888
9. Olsen LC, Aasland R, Fjose A (1997) A vasa-like gene in zebrafish identifies putative primordial germ cells. Mech Dev 66(1–2):95–105
10. Yoon C, Kawakami K, Hopkins N (1997) Zebrafish vasa homologue RNA is localized to the cleavage planes of 2- and 4-cell-stage embryos and is expressed in the primordial germ cells. Development 124(16):3157–3165
11. Gruidl ME et al (1996) Multiple potential germ-line helicases are components of the germ-line-specific P granules of Caenorhabditis elegans. Proc Natl Acad Sci U S A 93(24): 13837–13842
12. Hay B, Jan LY, Jan YN (1988) A protein component of Drosophila polar granules is encoded by vasa and has extensive sequence similarity to ATP-dependent helicases. Cell 55(4):577–587
13. Komiya T et al (1994) Isolation and characterization of a novel gene of the DEAD box protein family which is specifically expressed in germ cells of Xenopus laevis. Dev Biol 162(2):354–363
14. Liang L, Diehl-Jones W, Lasko P (1994) Localization of vasa protein to the Drosophila pole plasm is independent of its RNA-binding and helicase activities. Development 120(5): 1201–1211
15. Weidinger G et al (1999) Identification of tissues and patterning events required for distinct steps in early migration of zebrafish primordial germ cells. Development 126(23):5295–5307
16. Kierszenbaum AL, Tres LL (2001) Primordial germ cell-somatic cell partnership: a balancing cell signaling act. Mol Reprod Dev 60(3):277–280
17. Ying Y, Qi X, Zhao GQ (2002) Induction of primordial germ cells from pluripotent epiblast. Sci World J 2:801–810
18. de Sousa Lopes SM et al (2004) BMP signaling mediated by ALK2 in the visceral endoderm is necessary for the generation of primordial germ cells in the mouse embryo. Genes Dev 18(15):1838–1849
19. Lacham-Kaplan O (2004) In vivo and in vitro differentiation of male germ cells in the mouse. Reproduction 128(2):147–152

20. Saitou M, Barton SC, Surani MA (2002) A molecular programme for the specification of germ cell fate in mice. Nature 418(6895):293–300
21. Tanaka SS et al (2004) Regulation of expression of mouse interferon-induced transmembrane protein like gene-3, Ifitm3 (mil-1, fragilis), in germ cells. Dev Dyn 230(4):651–659
22. Lawson KA et al (1999) Bmp4 is required for the generation of primordial germ cells in the mouse embryo. Genes Dev 13(4):424–436
23. Ying Y, Qi X, Zhao GQ (2001) Induction of primordial germ cells from murine epiblasts by synergistic action of BMP4 and BMP8B signaling pathways. Proc Natl Acad Sci U S A 98(14):7858–7862
24. Ying Y, Zhao GQ (2001) Cooperation of endoderm-derived BMP2 and extraembryonic ectoderm-derived BMP4 in primordial germ cell generation in the mouse. Dev Biol 232(2):484–492
25. Lange UC et al (2003) The fragilis interferon-inducible gene family of transmembrane proteins is associated with germ cell specification in mice. BMC Dev Biol 3:1
26. Tanaka SS et al (2005) IFITM/Mil/fragilis family proteins IFITM1 and IFITM3 play distinct roles in mouse primordial germ cell homing and repulsion. Dev Cell 9(6):745–756
27. Ohinata Y et al (2005) Blimp1 is a critical determinant of the germ cell lineage in mice. Nature 436(7048):207–213
28. Scholer HR et al (1990) Oct-4: a germline-specific transcription factor mapping to the mouse t-complex. EMBO J 9(7):2185–2195
29. Hansis C, Grifo JA, Krey LC (2000) Oct-4 expression in inner cell mass and trophectoderm of human blastocysts. Mol Hum Reprod 6(11):999–1004
30. Pesce M, Scholer HR (2000) Oct-4: control of totipotency and germline determination. Mol Reprod Dev 55(4):452–457
31. Pesce M, Scholer HR (2001) Oct-4: gatekeeper in the beginnings of mammalian development. Stem Cells 19(4):271–278
32. Rajpert-De Meyts E et al (2004) Developmental expression of POU5F1 (OCT-3/4) in normal and dysgenetic human gonads. Hum Reprod 19(6):1338–1344
33. Hatano SY et al (2005) Pluripotential competence of cells associated with Nanog activity. Mech Dev 122(1):67–79
34. Yamaguchi S et al (2005) Nanog expression in mouse germ cell development. Gene Expr Patterns 5(5):639–646
35. Payer B et al (2006) Generation of stella-GFP transgenic mice: a novel tool to study germ cell development. Genesis 44(2):75–83
36. Buitrago W, Roop DR (2007) Oct-4: the almighty POUripotent regulator? J Invest Dermatol 127(2):260–262
37. Molyneaux KA et al (2001) Time-lapse analysis of living mouse germ cell migration. Dev Biol 240(2):488–498
38. Molyneaux K, Wylie C (2004) Primordial germ cell migration. Int J Dev Biol 48(5–6):537–544
39. Wylie C (2000) Germ cells. Curr Opin Genet Dev 10(4):410–413
40. Kunwar PS, Siekhaus DE, Lehmann R (2006) In vivo migration: a germ cell perspective. Annu Rev Cell Dev Biol 22:237–265
41. Weidinger G et al (2002) Regulation of zebrafish primordial germ cell migration by attraction towards an intermediate target. Development 129(1):25–36
42. Doitsidou M et al (2002) Guidance of primordial germ cell migration by the chemokine SDF-1. Cell 111(5):647–659
43. Knaut H et al (2003) A zebrafish homologue of the chemokine receptor Cxcr4 is a germ-cell guidance receptor. Nature 421(6920):279–282
44. Boldajipour B, Raz E (2007) What is left behind–quality control in germ cell migration. Sci STKE 2007(383):pe16
45. Anderson R et al (2000) The onset of germ cell migration in the mouse embryo. Mech Dev 91(1–2):61–68

46. Richardson BE, Lehmann R (2010) Mechanisms guiding primordial germ cell migration: strategies from different organisms. Nat Rev Mol Cell Biol 11(1):37–49
47. Gu Y et al (2009) Steel factor controls primordial germ cell survival and motility from the time of their specification in the allantois, and provides a continuous niche throughout their migration. Development 136(8):1295–1303
48. Runyan C et al (2006) Steel factor controls midline cell death of primordial germ cells and is essential for their normal proliferation and migration. Development 133(24):4861–4869
49. McCoshen JA, McCallion DJ (1975) A study of the primordial germ cells during their migratory phase in Steel mutant mice. Experientia 31(5):589–590
50. Buehr M et al (1993) Proliferation and migration of primordial germ cells in We/We mouse embryos. Dev Dyn 198(3):182–189
51. Takahashi H (1977) Juvenile hermaphroditism in the zebrafish, Brachydanio rerio. Bull Fac Fish Hokkaid Univ 28:57–65
52. Uchida D et al (2002) Oocyte apoptosis during the transition from ovary-like tissue to testes during sex differentiation of juvenile zebrafish. J Exp Biol 205(Pt 6):711–718
53. Rodriguez-Mari A et al (2005) Characterization and expression pattern of zebrafish Anti-Mullerian hormone (Amh) relative to sox9a, sox9b, and cyp19a1a, during gonad development. Gene Expr Patterns 5(5):655–667
54. von Hofsten J, Larsson A, Olsson PE (2005) Novel steroidogenic factor-1 homolog (ff1d) is coexpressed with anti-Mullerian hormone (AMH) in zebrafish. Dev Dyn 233(2):595–604
55. Wang XG, Orban L (2007) Anti-Mullerian hormone and 11 beta-hydroxylase show reciprocal expression to that of aromatase in the transforming gonad of zebrafish males. Dev Dyn 236(5):1329–1338
56. Sun D et al (2013) Sox9-related signaling controls zebrafish juvenile ovary-testis transformation. Cell Death Dis 4, e930
57. Siegfried KR, Nusslein-Volhard C (2008) Germ line control of female sex determination in zebrafish. Dev Biol 324(2):277–287
58. Dranow DB, Tucker RP, Draper BW (2013) Germ cells are required to maintain a stable sexual phenotype in adult zebrafish. Dev Biol 376(1):43–50
59. Weidinger G et al (2003) dead end, a novel vertebrate germ plasm component, is required for zebrafish primordial germ cell migration and survival. Curr Biol 13(16):1429–1434
60. Slanchev K et al (2005) Development without germ cells: the role of the germ line in zebrafish sex differentiation. Proc Natl Acad Sci U S A 102(11):4074–4079
61. Tzung KW et al (2015) Early depletion of primordial germ cells in zebrafish promotes testis formation. Stem Cell Rep 4(1):61–73
62. Houwing S et al (2007) A role for Piwi and piRNAs in germ cell maintenance and transposon silencing in Zebrafish. Cell 129(1):69–82
63. Gill ME et al (2011) Licensing of gametogenesis, dependent on RNA binding protein DAZL, as a gateway to sexual differentiation of fetal germ cells. Proc Natl Acad Sci U S A 108(18):7443–7448
64. Lin Y et al (2008) Germ cell-intrinsic and -extrinsic factors govern meiotic initiation in mouse embryos. Science 322(5908):1685–1687
65. Hajkova P et al (2002) Epigenetic reprogramming in mouse primordial germ cells. Mech Dev 117(1–2):15–23
66. Bullejos M, Koopman P (2001) Spatially dynamic expression of Sry in mouse genital ridges. Dev Dyn 221(2):201–205
67. Hacker A et al (1995) Expression of Sry, the mouse sex determining gene. Development 121(6):1603–1614
68. Jeske YW et al (1995) Expression of a linear Sry transcript in the mouse genital ridge. Nat Genet 10(4):480–482
69. Houston CS et al (1983) The campomelic syndrome: review, report of 17 cases, and follow-up on the currently 17-year-old boy first reported by Maroteaux et al in 1971. Am J Med Genet 15(1):3–28

70. Barrionuevo F et al (2006) Homozygous inactivation of Sox9 causes complete XY sex reversal in mice. Biol Reprod 74(1):195–201
71. Chaboissier MC et al (2004) Functional analysis of Sox8 and Sox9 during sex determination in the mouse. Development 131(9):1891–1901
72. Anderson EL et al (2008) Stra8 and its inducer, retinoic acid, regulate meiotic initiation in both spermatogenesis and oogenesis in mice. Proc Natl Acad Sci U S A 105(39):14976–14980
73. Koubova J et al (2006) Retinoic acid regulates sex-specific timing of meiotic initiation in mice. Proc Natl Acad Sci U S A 103(8):2474–2479
74. Bowles J et al (2006) Retinoid signaling determines germ cell fate in mice. Science 312(5773):596–600
75. Kim S, Bardwell VJ, Zarkower D (2007) Cell type-autonomous and non-autonomous requirements for Dmrt1 in postnatal testis differentiation. Dev Biol 307(2):314–327
76. Raymond CS et al (2000) Dmrt1, a gene related to worm and fly sexual regulators, is required for mammalian testis differentiation. Genes Dev 14(20):2587–2595
77. Matson CK et al (2011) DMRT1 prevents female reprogramming in the postnatal mammalian testis. Nature 476(7358):101–104
78. Lindeman RE et al (2015) Sexual cell-fate reprogramming in the ovary by DMRT1. Curr Biol 25(6):764–771
79. Frazier AL, Amatruda JF (2009) Germ cell tumors. In: Fisher DE, Nathan D, Look AT (eds) Nathan and Oski's textbook of pediatric hematology-oncology. Elsevier, London
80. Oosterhuis JW, Looijenga LH (2005) Testicular germ-cell tumours in a broader perspective. Nat Rev Cancer 5(3):210–222
81. Poynter JN, Amatruda JF, Ross JA (2010) Trends in incidence and survival of pediatric and adolescent patients with germ cell tumors in the United States, 1975 to 2006. Cancer 116(20):4882–4891
82. Trabert B et al (2015) International patterns and trends in testicular cancer incidence, overall and by histologic subtype, 1973–2007. Andrology 3(1):4–12
83. Jacobsen R et al (2006) Trends in testicular cancer incidence in the Nordic countries, focusing on the recent decrease in Denmark. Int J Androl 29(1):199–204
84. Ross JA et al (1999) Genomic imprinting of H19 and insulin-like growth factor-2 in pediatric germ cell tumors. Cancer 85(6):1389–1394
85. Schneider DT et al (2001) Multipoint imprinting analysis indicates a common precursor cell for gonadal and nongonadal pediatric germ cell tumors. Cancer Res 61(19):7268–7276
86. Palmer RD et al (2008) Pediatric malignant germ cell tumors show characteristic transcriptome profiles. Cancer Res 68(11):4239–4247
87. Williams SD et al (1987) Treatment of disseminated germ-cell tumors with cisplatin, bleomycin, and either vinblastine or etoposide. N Engl J Med 316(23):1435–1440
88. Einhorn LH, Donohue JP (1977) Improved chemotherapy in disseminated testicular cancer. J Urol 117(1):65–69
89. Einhorn LH (2002) Chemotherapeutic and surgical strategies for germ cell tumors. Chest Surg Clin N Am 12(4):695–706
90. Osanto S et al (1992) Long-term effects of chemotherapy in patients with testicular cancer. J Clin Oncol Off J Am Soc Clin Oncol 10(4):574–579
91. Bokemeyer C et al (1996) Evaluation of long-term toxicity after chemotherapy for testicular cancer. J Clin Oncol Off J Am Soc Clin Oncol 14(11):2923–2932
92. Bosl GJ et al (1986) Increased plasma renin and aldosterone in patients treated with cisplatin-based chemotherapy for metastatic germ-cell tumors. J Clin Oncol Off J Am Soc Clin Oncol 4(11):1684–1689
93. Hansen SW et al (1988) Long-term effects on renal function and blood pressure of treatment with cisplatin, vinblastine, and bleomycin in patients with germ cell cancer. J Clin Oncol Off J Am Soc Clin Oncol 6(11):1728–1731
94. Hansen PV, Hansen SW (1993) Gonadal function in men with testicular germ cell cancer: the influence of cisplatin-based chemotherapy. Eur Urol 23(1):153–156

95. Hansen SW, Berthelsen JG, von der Maase H (1990) Long-term fertility and Leydig cell function in patients treated for germ cell cancer with cisplatin, vinblastine, and bleomycin versus surveillance. J Clin Oncol Off J Am Soc Clin Oncol 8(10):1695–1698
96. Huddart RA et al (2003) Cardiovascular disease as a long-term complication of treatment for testicular cancer. J Clin Oncol Off J Am Soc Clin Oncol 21(8):1513–1523
97. Strumberg D et al (2002) Evaluation of long-term toxicity in patients after cisplatin-based chemotherapy for non-seminomatous testicular cancer. Ann Oncol Off J Eur Soc Med Oncol ESMO 13(2):229–236
98. Berger CC et al (1996) Endocrinological late effects after chemotherapy for testicular cancer. Br J Cancer 73(9):1108–1114
99. Bissett D et al (1990) Long-term sequelae of treatment for testicular germ cell tumours. Br J Cancer 62(4):655–659
100. Gietema JA et al (1992) Long-term follow-up of cardiovascular risk factors in patients given chemotherapy for disseminated nonseminomatous testicular cancer. Ann Intern Med 116(9):709–715
101. Teutsch C, Lipton A, Harvey HA (1977) Raynaud's phenomenon as a side effect of chemotherapy with vinblastine and bleomycin for testicular carcinoma. Cancer Treat Rep 61(5):925–926
102. Vogelzang NJ et al (1981) Raynaud's phenomenon: a common toxicity after combination chemotherapy for testicular cancer. Ann Intern Med 95(3):288–292
103. Boyer M et al (1990) Lack of late toxicity in patients treated with cisplatin-containing combination chemotherapy for metastatic testicular cancer. J Clin Oncol Off J Am Soc Clin Oncol 8(1):21–26
104. Siviero-Miachon AA, Spinola-Castro AM, Guerra-Junior G (2009) Adiposity in childhood cancer survivors: insights into obesity physiopathology. Arq Bras Endocrinol Metabol 53(2):190–200
105. Siviero-Miachon AA, Spinola-Castro AM, Guerra-Junior G (2008) Detection of metabolic syndrome features among childhood cancer survivors: a target to prevent disease. Vasc Health Risk Manag 4(4):825–836
106. Glendenning JL et al (2010) Long-term neurologic and peripheral vascular toxicity after chemotherapy treatment of testicular cancer. Cancer 116(10):2322–2331
107. Travis LB et al (1997) Risk of second malignant neoplasms among long-term survivors of testicular cancer. J Natl Cancer Inst 89(19):1429–1439
108. Travis LB et al (2005) Second cancers among 40,576 testicular cancer patients: focus on long-term survivors. J Natl Cancer Inst 97(18):1354–1365
109. van Echten J et al (1995) No recurrent structural abnormalities apart from i(12p) in primary germ cell tumors of the adult testis. Genes Chromosomes Cancer 14(2):133–144
110. Tian Q et al (1999) Activating c-kit gene mutations in human germ cell tumors. Am J Pathol 154(6):1643–1647
111. Looijenga LH et al (2003) Stem cell factor receptor (c-KIT) codon 816 mutations predict development of bilateral testicular germ-cell tumors. Cancer Res 63(22):7674–7678
112. Kemmer K et al (2004) KIT mutations are common in testicular seminomas. Am J Pathol 164(1):305–313
113. Hoei-Hansen CE et al (2007) Ovarian dysgerminomas are characterised by frequent KIT mutations and abundant expression of pluripotency markers. Mol Cancer 6:12
114. Coffey J et al (2008) Somatic KIT mutations occur predominantly in seminoma germ cell tumors and are not predictive of bilateral disease: report of 220 tumors and review of literature. Genes Chromosomes Cancer 47(1):34–42
115. Hersmus R et al (2012) Prevalence of c-KIT mutations in gonadoblastoma and dysgerminomas of patients with disorders of sex development (DSD) and ovarian dysgerminomas. PLoS One 7(8), e43952
116. Ganguly S et al (1990) Detection of preferential NRAS mutations in human male germ cell tumors by the polymerase chain reaction. Genes Chromosomes Cancer 1(3):228–232
117. Mulder MP, et al. (1991) Frequent occurrence of activated ras oncogenes in seminomas but not in nonseminomatous germ cell tumors. Recent results in cancer research. Fortschritte der Krebsforschung. Progres dans les recherches sur le cancer 123, pp 125–31

118. Moul JW, Theune SM, Chang EH (1992) Detection of RAS mutations in archival testicular germ cell tumors by polymerase chain reaction and oligonucleotide hybridization. Genes Chromosomes Cancer 5(2):109–118
119. Olie RA et al (1995) N- and KRAS mutations in primary testicular germ cell tumors: incidence and possible biological implications. Genes Chromosomes Cancer 12(2):110–116
120. Wang L et al (2014) Novel somatic and germline mutations in intracranial germ cell tumours. Nature 511(7508):241–245
121. Honecker F et al (2009) Microsatellite instability, mismatch repair deficiency, and BRAF mutation in treatment-resistant germ cell tumors. J Clin Oncol Off J Am Soc Clin Oncol 27(13):2129–2136
122. Brabrand S et al (2015) Exome sequencing of bilateral testicular germ cell tumors suggests independent development lineages. Neoplasia 17(2):167–174
123. Litchfield K et al (2015) Whole-exome sequencing reveals the mutational spectrum of testicular germ cell tumours. Nat Commun 6:5973
124. Kanetsky PA et al (2009) Common variation in KITLG and at 5q31.3 predisposes to testicular germ cell cancer. Nat Genet 41(7):811–815
125. Rapley EA et al (2009) A genome-wide association study of testicular germ cell tumor. Nat Genet 41(7):807–810
126. Turnbull C et al (2010) Variants near DMRT1, TERT and ATF7IP are associated with testicular germ cell cancer. Nat Genet 42(7):604–607
127. Kratz CP et al (2011) Variants in or near KITLG, BAK1, DMRT1, and TERT-CLPTM1L predispose to familial testicular germ cell tumour. J Med Genet 48(7):473–476
128. Lessel D et al (2012) Replication of genetic susceptibility loci for testicular germ cell cancer in the Croatian population. Carcinogenesis 33(8):1548–1552
129. Karlsson R et al (2013) Investigation of six testicular germ cell tumor susceptibility genes suggests a parent-of-origin effect in SPRY4. Hum Mol Genet 22(16):3373–3380
130. Chung CC et al (2013) Meta-analysis identifies four new loci associated with testicular germ cell tumor. Nat Genet 45(6):680–685
131. Ruark E et al (2013) Identification of nine new susceptibility loci for testicular cancer, including variants near DAZL and PRDM14. Nat Genet 45(6):686–689
132. Stevens LC (1970) Experimental production of testicular teratomas in mice of strains 129, A/He, and their F1 hybrids. J Natl Cancer Inst 44(4):923–929
133. Youngren KK et al (2005) The Ter mutation in the dead end gene causes germ cell loss and testicular germ cell tumours. Nature 435(7040):360–364
134. Krentz AD et al (2013) Interaction between DMRT1 function and genetic background modulates signaling and pluripotency to control tumor susceptibility in the fetal germ line. Dev Biol 377(1):67–78
135. Krentz AD et al (2009) The DM domain protein DMRT1 is a dose-sensitive regulator of fetal germ cell proliferation and pluripotency. Proc Natl Acad Sci U S A 106(52):22323–22328
136. Morinaga C et al (2007) The hotei mutation of medaka in the anti-Mullerian hormone receptor causes the dysregulation of germ cell and sexual development. Proc Natl Acad Sci U S A 104(23):9691–9696
137. Neumann JC et al (2009) Identification of a heritable model of testicular germ cell tumor in the zebrafish. Zebrafish 6(4):319–327
138. Neumann JC et al (2011) Mutation in the type IB bone morphogenetic protein receptor Alk6b impairs germ-cell differentiation and causes germ-cell tumors in zebrafish. Proc Natl Acad Sci U S A 108(32):13153–13158
139. Fustino N et al (2011) Bone morphogenetic protein signalling activity distinguishes histological subsets of paediatric germ cell tumours. Int J Androl 34(4 Pt 2):e218–e233
140. Gill JA et al (2010) Enforced expression of Simian virus 40 large T-antigen leads to testicular germ cell tumors in zebrafish. Zebrafish 7(4):333–341
141. van Rooijen E et al (2008) LRRC50, a conserved ciliary protein implicated in polycystic kidney disease. J Am Soc Nephrol 19(6):1128–1138
142. Basten SG et al (2013) Mutations in LRRC50 predispose zebrafish and humans to seminomas. PLoS Genet 9(4), e1003384

Malignant Peripheral Nerve Sheath Tumors

Adam D. Durbin, Dong Hyuk Ki, Shuning He, and A. Thomas Look

Abstract Malignant peripheral nerve sheath tumors (MPNST) are tumors derived from Schwann cells or Schwann cell precursors. Although rare overall, the incidence of MPNST has increased with improved clinical management of patients with the neurofibromatosis type 1 (NF1) tumor predisposition syndrome. Unfortunately, current treatment modalities for MPNST are limited, with no targeted therapies available and poor efficacy of conventional radiation and chemotherapeutic regimens. Many murine and zebrafish models of MPNST have been developed, which have helped to elucidate the genes and pathways that are dysregulated in MPNST tumorigenesis, including the *p53*, and the *RB1*, PI3K-Akt-mTOR, RAS-ERK and Wnt signaling pathways. Preclinical results have suggested that new therapies, including mTOR and ERK inhibitors, may synergize with conventional chemotherapy in human tumors. The discovery of new genome editing technologies, like CRISPR-cas9, and their successful application to the zebrafish model will enable rapid progress in the faithful modeling of MPNST molecular pathogenesis. The zebrafish model is especially suited for high throughput screening of new targeted therapeutics as well as drugs approved for other purposes, which may help to bring enhanced treatment modalities into human clinical trials for this devastating disease.

Keywords Malignant peripheral nerve sheath tumor • MPNST • Zebrafish • Tumor modeling • p53 • p16^{INK4A} • PI3K • Akt • mTOR • NF1 • RAS • EGFR

A.D. Durbin (✉) • A.T. Look (✉)
Department of Pediatric Oncology, Dana-Farber Cancer Institute, Harvard Medical School, Boston, MA 02215, USA

Division of Pediatric Hematology/Oncology, Boston Children's Hospital, Boston, MA 02215, USA

Department of Medicine, Boston Children's Hospital, Boston, MA 02215, USA
e-mail: adam.durbin@childrens.harvard.edu; Thomas_Look@dfci.harvard.edu

D.H. Ki • S. He
Department of Pediatric Oncology, Dana-Farber Cancer Institute, Harvard Medical School, Boston, MA 02215, USA

Division of Pediatric Hematology/Oncology, Boston Children's Hospital, Boston, MA 02215, USA
e-mail: DongH_Ki@dfci.harvard.edu; shuning_he@dfci.harvard.edu

D.M. Langenau (ed.), *Cancer and Zebrafish*, Advances in Experimental Medicine and Biology 910, DOI 10.1007/978-3-319-30654-4_22

Clinical and Pathologic Features of Malignant Peripheral Nerve Sheath Tumors (MPNST)

Malignant peripheral nerve sheath tumors (MPNSTs), formerly known as neurofibrosarcoma, neurogenic sarcoma, malignant schwannoma, and malignant neurilemmoma, are malignant tumors derived from Schwann cells or Schwann cell precursors surrounding peripheral nerves [1, 2]. These tumors have an estimated incidence of 1:100,000 in the general population [3]. Despite their relative rarity, these tumors represent 5–10 % of all soft-tissue sarcomas of childhood [4, 5]. MPNSTs commonly arise from large and medium sized nerves, with extremity tumors representing 33–46 %, truncal tumors 34–41 % and head and neck tumors 17–25 % of all cases [2]. Less than 15 % of all MPNSTs are metastatic at initial diagnosis, however, MPNST displays a predilection for metastasis to the lungs or bone marrow [1, 2].

MPNST is a tumor with a variety of histopathologic appearances. Histologically, MPNSTs are pleomorphic, displaying epithelial, glandular or cartilaginous components, or alternatively containing heterotopic elements [1, 2]. MPNSTs may require differentiation from other similar tumors occurring in similar anatomical sites, including fibrosarcoma, rhabdomyosarcoma, malignant triton tumors and clear cell sarcoma of soft parts, as well as those that contain cartilaginous or osseous components or differentiation [1, 2]. MPNSTs typically contain myxoid stroma, nuclear palisading and tapered nonbranching processes in parallel arrangements, microtubules, neurofilaments and basal lamina [2]. They may display aberrant differentiation, including the presence of rhabdomyoblasts, as found in malignant triton tumors, glandular elements as seen in synovial sarcoma or melanocytic elements, as found in pigmented MPNST resembling clear cell sarcoma of soft parts [2]. Histologic studies have demonstrated frequent expression of S100β, GFAP, collagen IV, CD57, PGP 9.5, myelin basic protein and keratin 8 and 18 proteins [2, 6], suggesting a neural/Schwann cell of origin. Furthermore, comparative analyses have demonstrated strikingly different molecular profiles between MPNST and fibrosarcoma, rhabdomyosarcoma, malignant triton tumors and clear cell sarcoma of soft parts, yielding several molecular markers specific for MPNST, including *SOX10*, adrenomedullin and others [6–10]. The positive expression of these markers has been of key utility in deriving promoters to aid in animal modeling of MPNST.

Early descriptions revealed that patients with MPNST had exquisitely poor outcomes in the absence of complete surgical excision, and that there is a striking association between the neurocutaneous syndrome neurofibromatosis type 1 (NF1) and MPNST. This second finding is fundamental to the biology of MPNST, such that MPNST is considered to occur in two independent variations—those that are found in the context of patients with neurofibromatosis, termed "NF1-associated" and those that are found independent of neurofibromatosis, termed "sporadic" MPNST.

Clinical Management of MPNST

The clinical management of MPNST has been recently reviewed [11–13]. Importantly, it has been understood since initial studies by D'Agostino et al. in 1963 that patients with MPNST fared extremely poorly without complete surgical excision. In their initial study of 24 patients with sporadic MPNST, every patient who went on to die displayed incomplete surgical excision of their tumor [14] (Table 1).

Table 1 Clinical therapeutic trials for MPNST

Study	Sample group	Treatment	Outcome	References
D'Agostino et al. (1963)	24 patients with sporadic MPNST	Surgery +/– chemotherapy	12/19 patients with only simple surgery recurred	[14]
D'Agostino et al. (1963)	21 patients with NF1 and MPNST	Surgery, +/– chemotherapy	17/21 patients dead (OS: 19 %)	[15]
Ducatman et al. (1983)	109 patients with MPNST	Surgery, XRT	Relapse of 12 cases in sites of former XRT	[16]
Ducatman et al. (1986)	120 patients, 62 with NF1, 13 post-irradiation and 19 metaplastic	Surgery, XRT	NF1: 16 % 5 years OS Sporadic: 53 % 5 years OS	[17]
Raney et al. (1987)	24 children with MPNST (16/24 with NF1)	Surgery, XRT and chemotherapy	13 patients with CR, total: 46 % OS	[18]
DeCou et al. (1995)	29 patients (11 with NF1)	Surgery, XRT, chemotherapy	5 years OS: 39% 2 years OS with resectable lesions 79 %, without: 22 %	[4]
Casanova et al. (1999)	24 patients with MPNST, 7/24 with NF1	Surgery, XRT, chemotherapy	10 years EFS: 29 % 10 years OS: 41 % 10 years OS (complete excision): 80 %; 10 years OS (partial excision): 14 %	[19]
Carli et al. (2005)	167 patients with MPNST, 17 % with NF1	Surgery, chemotherapy, XRT	Prognostic factors included NF1, primary site, size, metastases. Trend toward a benefit from XRT after GTR. Overall response to chemotherapy in group III = 45 %	[20]

(continued)

Table 1 (continued)

Study	Sample group	Treatment	Outcome	References
Ferrari et al. (2007)	36 MPNST (all NF1), 1 malignant triton tumor	Surgery, XRT, chemotherapy	5 years EFS: 19 % 5 years OS: 28 %	[21]
Furniss et al. (2008)	32 cases of MPNST, 13/32 NF1, 1/32 NF2	Not described	Not described	[22]
Ferrari et al. (2011)	304 cases of non-metastatic, unresected pediatric NRSTS (71 MPNST)	Surgery, chemotherapy, XRT	5 years OS for MPNST ~40 %, with 5 years OS for NF1 11.1 % and 5 years OS for sporadic MPNST 44.7 %	[23]
Stucky et al. (2012)	175 MPNST (32 % NF1-associated, 68 % sporadic)	Surgery, chemotherapy, XRT	Poorer OS with NF1, tumor >5 cm, high grade tumors, use of chemotherapy, local recurrence	[5]
Kolberg et al. (2013)	>1800 cases of MPNST in meta-analysis	Surgery, chemotherapy, XRT	Overall survival of patients with NF1 and sporadic tumors is now equilibrating to ~40 %	[3]
Zehou et al. (2013)	21 NF1-associated MPNST patients treated with chemotherapy	Surgery, chemotherapy, XRT. All patients received chemotherapy	19/21 patients died	[24]
Fan et al. (2014)	146 MPNST (17 NF1-associated, 129 sporadic)	Surgery, XRT, chemotherapy	MDM2 expression correlates with tumor recurrence	[25]
Bates et al. (2014)	139 pediatric patients with MPNST	Surgery, XRT	No effect of XRT, HR = 0.70 for complete excision	[26]

Summary of clinical trials and institutional experiences of MPNST. *XRT* external beam radiation, *HR* hazard ratio, *MPNST,* malignant peripheral nerve sheath tumor. *NF1* neurofibromatosis type 1, *NF2* neurofibromatosis type 2, *NRSTS* non-rhabdomyosarcomatous soft-tissue sarcoma, *OS* overall survival, *EFS* event-free survival, *GTR* gross total resection, *MDM2* mouse-double minute 2

More recent studies have extended these findings, with 80 % of patients treated with complete resection surviving at 10 years, compared to 14 % of those with partial or no resection [19]. Importantly, patients with neurofibromatosis type 1 were noted to have an even poorer overall survival [14, 15], a finding which was echoed in subsequent studies in larger groups of patients [17, 23]. In a meta-analysis involving >1800 patients from multiple institutions, the primary risk factors for poorer overall survival included the presence of neurofibromatosis, failure to achieve complete surgical removal, metastases at the time of diagnosis and larger primary tumor size. When all patients were taken together, those patients not given chemotherapy tended to experience poorer disease-associated survival compared to those not given chemotherapy, although this was hypothesized to be a confounder related to larger disease burden [3]. Thus, in the largest of all studies, the role of surgery remains an independent predictor of overall survival [3]. The role surgery plays, however, is compromised by the growth of MPNST cells, which invade and erode, resulting in local damage, distant metastasis and inoperability due to encasement of critical structures.

In contrast to the clear role of surgery, the evidence for chemotherapy and radiation in the management of MPNST appears mixed. To date, limited evidence exists for the use of external beam radiation in modifying the disease course of MPNST, although it is commonly utilized as a therapeutic strategy for enhancing local control post-surgery [1–3]. Unfortunately, radiation therapy is not without negative effects, and several studies have described the incidence of secondary malignancies, including MPNST, in the radiation field (Table 1) [4, 11, 16, 18]. These data strongly indicate judicious use of external beam radiation in the management of MPNST and also underscore the necessity for close followup of patients with MPNST, predominantly those with NF1, who are managed with radiation.

Less clear than radiation is the role that chemotherapy may play in the modification of MPNSTs (Table 1). MPNST is a relatively chemoresistant tumor [1, 2]. Evidence suggests that in partially resected or unresectable MPNST, chemotherapy produces regression in less than 33 % of cases [2]. In addition, meta-analysis indicates that patients given chemotherapy fared worse than those that were not, though this response was thought to be related to a greater initial burden of disease resulting in the clinical decision to administer chemotherapy [3]. Moreover, in the Italian and German MPNST experience, patients with gross residual disease after incomplete resection had an overall response rate of 45 % to chemotherapy, however, only 2/29 patients displayed complete responses [20]. Lower response rates to chemotherapy are seen in NF1 patients (17.6 %) than in those with sporadic disease (55.3 %) [20] (Table 2). Most chemotherapeutic regimens used have included combinations of vincristine, cyclophosphamide, actinomycin-D, doxorubicin and ifosfamide [1–3, 19]. The agents displaying the most efficacy in the management of MPNST are doxorubicin and cyclophosphamide/ifosfamide, although others, including platinum salts have been reported as effective [11, 20, 24]. A molecular rationale for this effect has been described, with some MPNSTs

Table 2 Overall response of MPNST to chemotherapy is reduced in NF1-associated MPNST

Tumor	Response	No response
MPNST (non-NF1)	25	16
MPNST (NF1)	2	23

Patients with non-metastatic NF1-associated MPNST are less likely to respond to chemotherapy than are sporadic MPNSTs, though chemotherapy regimens were highly variable. $p<0.0001$ by Fisher's exact test. Data derived from [23]

displaying overexpression of topoisomerase II-α, with a relationship to poorer overall survival and distant metastasis [27]. These findings may indicate a further role for topoisomerase II-α agents, including epipodophyllotoxins, in therapy of selected MPNSTs [27].

The late effects of MPNST management include secondary malignancies, dermatitis, cellulitis, fibrosis, and osteitis related to external beam radiation, as well as the risk of relapse due to failure to control local or metastatic disease [2]. Furthermore, there is a risk of deformation of local structures due to surgical excision with wide margins, including the risk of damage to critical internal structures. With chemotherapy and radiation, there exists a possibility of secondary malignancy, with MPNST as the most common secondary tumor, though others, including osteosarcoma [19] have been reported. This risk is not limited to MPNST, with studies reporting MPNST arising in the radiation field used in the management of patients with Hodgkin's lymphoma and rhabdomyosarcoma [4]. Thus, current treatment modalities for MPNST are limited in efficacy and have undesirable side effects, indicating the necessity to identify new experimentally-validated targets.

Newer studies, based on animal models of neurofibromatosis and MPNST, have suggested the use of molecularly-targeted therapeutic agents that target at the level of growth factor receptors, or downstream at the level of the phosphatidyl-inositol 3'OH kinase (PI3K)-AKT-mammalian target of rapamycin (mTOR) or RAS-Raf-MEK-ERK signaling axes. These studies have included the farnesyltransferase inhibitor tipifarnib [28–30], the mTOR inhibitor everolimus [29–32], the platelet-derived growth factor receptor (PDGFR)/c-kit receptor co-inhibitor imatinib [29, 30, 33] and the vascular-endothelial growth factor receptor (VEGFR) inhibitor bevacizumab [29, 30, 34, 35], among others. Furthermore, advanced imaging technologies have suggested an enhanced methodology for the detection of small tumors in high risk populations [36], which, given the central role of surgical excision of MPNSTs, will enhance surgical approaches to small tumors prior to their malignant degeneration or impinging on critical internal structures and leading to deformaties and compromise of organ function. These studies are promising and suggest avenues for future exploration and validation in model systems prior to entry into clinical trials.

Neurofibromatosis Type 1: An Inherited Tumor-Predisposition Syndrome Associated with MPNST

The neurofibromatoses are two genetic tumor predisposition syndromes (NF1, NF2) predominantly affecting the nervous system. Neurofibromatosis type 1 (NF1) is an autosomal dominant disorder characterized by learning disabilities, bony dysplasias, pigmented lesions of the skin including café-au-lait macules, axillary freckling and lisch nodules of the iris, as well as the development of numerous tumor types, including optic glioma, glioblastoma, pheochromocytoma, juvenile monomyelocytic leukemias, as well as neurofibromas, plexiform neurofibromas and MPNST [1, 2, 37]. NF1 is the most common inherited genetic disease affecting the nervous system, with an estimated incidence of 1 in 3500 infants [37].

Neurofibromatosis type 1 was initially linked to chromosome 17q11, while a distinct tumor spectrum disorder termed neurofibromatosis type 2, with a predisposition to acoustic neuromas/vestibular schwannomas was linked to chromosome 22 [38–40]. NF1 is caused by inherited or de novo mutations in the neurofibromin (*NF1*) gene, which encodes a 220 kDa cytoplasmic protein with regions containing homology to GTPase-activating proteins (GAPs). NF1 was subsequently identified as a GAP for the *RAS* family of proto-oncogenes, with disruption of *NF1* implicated in hyperactivation of RAS signaling [41], and thus, promoting downstream oncogenic signaling (reviewed in [1, 2, 13, 37, 42]). Thus, for MPNSTs lacking NF1, it has been postulated that inhibiting hyperactivated RAS signaling targets may be efficacious, resulting in numerous preclinical and clinical studies investigating the use of farnesyltransferase [28, 43], MEK [44, 45], Ral [46] and mTOR [31, 45, 47–49] inhibitors in both animal models and human trials of MPNSTs with NF1 loss. Activating mutation of RAS are rare in MPNST [50], implicating signaling upstream of RAS, and this has been investigated at the preclinical level, including inhibition of CXCR4 [51, 52], IGF1R [53], VEGFR [54, 55], EGFR [48, 56, 57], KIT [33, 58–60], MET [25] and ErbB2 [56, 61, 62]. NF1 satisfies Knudson's two hit hypothesis for tumor suppressor function, with neoplastic lesions arising in patients with NF1 in tissue stem cells that have lost the wild-type allele of the NF1 gene [38, 63–65]. Furthermore, sporadically developing MPNSTs not arising in the context of hereditary *NF1* mutations commonly display somatically acquired mutations in the NF1 tumor suppressor, further strengthening the central role of this gene in MPNST development [66]. To this end, gene expression array experiments comparing sporadic and NF1-associated MPNSTs were unable to identify a clear signature differentiating MPNSTs arising in these different clinical contexts [67], suggesting that the gene expression patterns and other properties of MPNSTs are likely to be conserved, whether or not they arise in NF1 patients.

The risk of MPNST in the general population has been estimated to be 0.001 % per lifetime, while in the context of NF1 the estimates are 2–13 % per lifetime [2, 17]. Furthermore, greater than 50 % of all MPNSTs arise in patients with NF1, suggesting a strong genetic linkage between *NF1* mutations and MPNST generation [2]. MPNST has an earlier age of onset in NF1 patients, with a peak occurrence in the 20s–30s, while it tends to occur in patients who have reached their 40s–50s in sporadic MPNST [17]. NF1 patients fared worse than those with sporadic disease [4, 17, 19], and despite refinements in surgical techniques, this difference has continued to persist [3, 5]. Intriguingly, NF1-associated tumors display, on average, a higher grade and larger size compared to sporadic tumors, suggesting a role for early biallelic loss of *NF1* in driving nuclear atypia and perhaps the survival or proliferation of the tumor cells [3, 5, 17].

Initiation of MPNST

The mechanism by which MPNSTs arise has been extensively queried. Intriguingly, individuals with subcutaneous neurofibromas are three times more likely to have internal plexiform neurofibromas and MPNSTs, and those patients with internal plexiform neurofibromas are twenty-times more likely to develop MPNSTs [12]. This observation has led many to question whether a pre-neoplastic lesion for MPNST exists in NF1 patients, and whether this lesion is the plexiform neurofibroma. The observation of lesions containing contiguous elements of neurofibroma, plexiform neurofibroma and MPNST has further suggested that these lesions may lay along a spectrum [68]. These observations are strengthened by mutational analyses of dermal neurofibromas, plexiform neurofibromas and MPNSTs, demonstrating a gradation of loss of NF1 and the acquisition of *TP53* mutations during the progression of these lesions [63].

In vivo modeling has suggested that the skin-derived precursor cell (SKP cell) may represent the dermal neurofibroma cell of origin. While Schwann cells can be differentiated from wild-type SKPs [69], Schwann cells carrying loss of both alleles of *NF1* result in focal dermal neurofibromas [70]. Further, SKPs derived from *cis-NF1* $^{loxp/loxp}$;*TP53*$^{loxp/loxp}$ mice, treated in culture with adeno-CRE-GFP and autologously implanted into the same mouse result in the generation of GFP-labeled MPNSTs [71]. These data argue strongly for an in vivo niche for the generation of the dermal neurofibroma, and for the precursor lesion to the plexiform neurofibroma, which functions as the lesion of origin for MPNST [2]. These data also argue for a dysregulation of signaling downstream of RAS as being implicated in establishing the initial cell and niche responsible for generating the neoplastic lesion in MPNST. The use of the recently defined serum marker of transformation, adrenomedullin [9], among other biomarkers [7, 10], may be helpful in identifying early MPNST lesions, to further define the timing and localization of transformation in vivo. These studies and others involving human patients, as well as murine and zebrafish models of MPNST have led to the model for MPNST generation shown in Fig. 1.

Furthermore, evidence indicates that the extracellular matrix niche may enhance MPNST generation via cell cycle inductive signaling mediated by SDF1α-CXCR4 interactions on tumor cells [51, 52] and proinflammatory states [42, 51, 52, 58, 72]. Autocrine SDF1α signaling through CXCR4 receptors expressed on MPNST cells induces AKT and inhibits GSK3β signaling, stabilizing β-catenin and driving TCF/LEF-mediated nuclear transcription and proliferation, which can be suppressed by CXCR4 inhibition [51]. These data mimic studies in other sarcomas [73, 74], and strongly indicate that MPNSTs may arise by a combination of cellular transformation and autocrine/paracrine extracellular matrix-mediated interactions on underlying plexiform neurofibromas (Fig. 1).

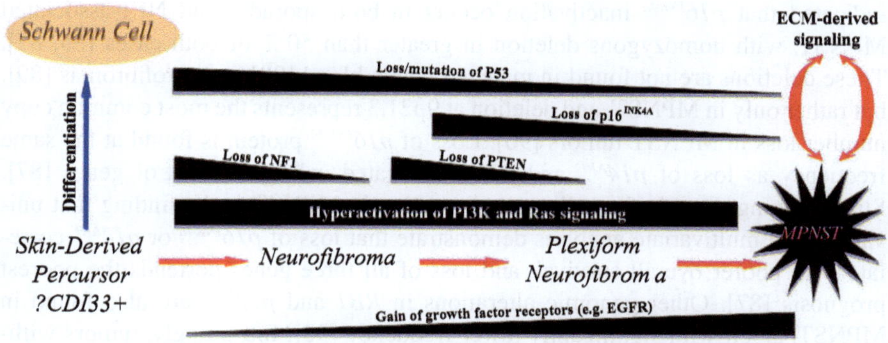

Fig. 1 MPNST is generated from precursors by defined genetic events. Skin-derived precursors, possibly CD133+ cells, generate neurofibromas in the context of loss of NF1. Loss of tumor suppressors (PTEN, p53, and p16^{INK4A}) in addition to hyper-activation of growth factor receptor signaling, such as EGFR or PDGFR, results in progression of neurofibromas to plexiform neurofibromas. These may undergo malignant transformation to MPNST. MPNST is supported in an ectopic niche by both autocrine and paracrine interactions, including mast cell proinflammatory signaling

Molecular Genetics of MPNST

Significant interest has been expressed in identifying key recurrent genomic events in MPNSTs that may represent genetic lesions for targeting by novel experimental therapeutics. The clear association of MPNST in neurofibromatosis type 1 patients led to the establishment of *NF1* deletions in MPNST tumors [39, 63]. While the mutational spectrum in NF1 driving loss of NF1 is variable, including point mutations, splice mutations, small deletions, insertions and duplications, further evaluation has suggested that the roughly 5–10 % of NF1 patients with deletions incorporating the entire *NF1* gene are at the highest risk of MPNST development [75]. Intriguingly, as many as 10–15 % of dermal neurofibromas, 40 % of plexiform neurofibromas and >60 % of MPNST display biallelic LOH of *NF1*, suggesting that *NF1* likely functions in the classic Knudson two hit hypothesis manner in MPNST tumorigenesis [63, 65] (Fig. 1). To this end, ~50 % of sporadic MPNSTs have single copy loss of *NF1* [76, 77], though the spectrum of germline mutations in neurofibromatosis patients is distinct from the somatic mutations seen in MPNSTs [77].

Despite the initial findings that generation of both murine and zebrafish models of MPNST were associated with mutation of *TP53* [64, 78–81], and evidence that *TP53* mutations are found in human MPNSTs [38, 82], subsequent focused sequencing efforts of the exonic sequence of *TP53* has demonstrated a relatively low frequency of inactivating mutations [63, 83–86]. Studies have demonstrated that *TP53* mutations, with subsequent loss of heterozygosity of the gene, are more commonly found in sporadic MPNST [56, 85, 86]. In contrast, dysregulation of nuclear cell cycle control in MPNST appears to be more frequently achieved through the loss of the cell cycle controller *p16^{INK4a}* and the p53 suppressor *p14^{ARF}*. Studies have indicated that *p16*^{INK4a} inactivation occurs in both sporadic and NF1-associated MPNST, with homozygous deletion in greater than 50 % of both cases [85, 87]. These deletions are not found in matched whole blood [88] or neurofibromas [89], but rather only in MPNST, and deletion at 9p21.3 represents the most common copy number loss in MPNST tumors [90]. Loss of *p16*^{INK4a} protein is found at the same frequency as loss of *p14*^{ARF} and *p15*^{INK4b}, related cell cycle control genes [87]. Strengthening the role for cell cycle disruption in MPNST is the finding that univariate and multivariate analyses demonstrate that loss of *p16*^{INK4a} or *p14*^{ARF} correlate with poorer overall survival, and loss of all three genes portends the poorest prognosis [87]. Other genomic alterations in *RB1* and *p27*^{KIP1} are also found in MPNST, albeit with significantly lower frequency [90]. Intriguingly, tumors without previously defined RB1 pathway alterations represent only the minority (<10 %) of MPNSTs [6, 88].

Evaluation of MPNST for alterations in cell surface receptor tyrosine kinases (RTKs) has demonstrated frequent expression of EGFR [47, 52, 65, 70–73], as well as others including c-kit [33, 58–60], c-met [25], VEGFR [54, 55], PDGFR [33, 48, 91–93], insulin-like growth factor receptor-1 (IGF1R) [53] and ErbB receptors [56, 62]. High throughput screening of MPNST cell lines shows that these receptors do not commonly display mutational activation [50], however, some studies have demonstrated amplifications of the genomic loci in MPNST [56, 94]. The EGFR, in particular, is commonly expressed in high grade areas of MPNST [48, 67, 94], associated with *TP53* loss [95], and is hyperactivated in MPNST cells [48]. Similar activation profiles are present for PDGFRs in MPNST, and these receptors appear to drive signaling through the Akt-mTOR and RAS-ERK pathways [48]. No mutations in PI3K, the upstream kinase responsible for linking activated RTKs to AKT and RAS pathway activation, have been found in MPNST samples [48–50], despite hyperactivation of both of these pathways. Rare activating mutations in N- and K-RAS have been identified in sporadic MPNST, with no RAS mutations in NF1-associated tumors [48], reflecting the fact that loss of NF1 has already activated the RAS pathway. A similar spectrum of mutations has been identified for the downstream RAS-activated kinase *B-RAF* [96]. In contrast, loss of *PTEN* expression by promoter methylation is a common finding in primary MPNST but not neurofibromas or normal nervous tissue [49]. It therefore appears that MPNST display dysregulation of these downstream pathways by hyperactivated receptor tyrosine kinases, with concurrent loss of negative regulation through *PTEN* or *NF1* and, only rarely by direct mutational activation of RAS or PI3K themselves in sporadic tumors.

Higher throughput analyses have been conducted using array-based methodologies, indicating that MPNST contain complex patterns of karyotypic changes with common amplifications of *NEDL1, AP3B1, CUL1, SNCAIP, E2F3, CBLV, NMNAT2, FER*, and less commonly (<50 %), amplifications of *ITGA2, SEMA5A, EXT1, DLX4/3, OTX2*, and deletions of *CDKN2A/B, FOXD2, NCAM1* and *MKK5* [90]. This analysis demonstrated an average of 12.8 copy number variants per genome, with common copy number gains in 1q, 2p, 7, 8, 17q and deletions in 1p, 9p, 10q25, 11q14-q24, 17q, 20p [90]. Similar array based subtractive hybridization and array cGH profiles have been constructed in MPNST, demonstrating tumor-specific copy number variations. These too have displayed loss of the *CDKN2A/B* locus [97] and amplifications of the *PDGFRA* and *c-Kit* loci [33].

Evaluation of RNA expression profiles in MPNST has also been fruitful for the definition of tractable targets. Early experiments using sequential subtractive hybridization between plexiform neurofibromas and MPNST samples from the same patients resulted in the definition of numerous targets that were upregulated in MPNST, including the known cancer-associated genes *PDGFRA, ITGA6, SPRY2, YAP1, HOXB2, MDM4, NBS1, RAD23B, HGF, IGFBP3, MMP13, FOS, EGR1, SPARCL1, Hsp90, Hsp40, EIF4E, MEN1,* and *RPL10* [91]. Furthermore, focused chromosome 17 microarray experiments aimed at defining genes co-altered with *NF1* in MPNST demonstrated upregulation of *TOP2A*, the previously mentioned *topoisomerase-IIa* gene [27]. Early microarray based gene expression profiling experiments were unable to identify a clear gene signature differentiating NF1-associated and sporadic tumors, highlighting the similarities in transcriptional networks and dependency on similar signaling pathways that are present in these tumors [67]. Subsequently, studies identified a conserved signature differentiating neurofibromas, plexiform neurofibromas and MPNST [98]. These results were extended in microarray based experiments, in combination with high throughput insertional mutagenesis screening, to identify three conserved pathways critical to MPNST tumorigenesis, the RAS-ERK, PI3K-Akt-mTOR and Wnt-β-catenin pathways [99]. RNA expression experiments comparing tumor and Schwann cell samples have permitted the definition of a signature defining MPNSTs, including loss of Schwann cell differentiation markers (*SOX10, CNP, PMP22, NGFR*) and induction of neural crest stem cell markers (*SOX9, Twist1*) [6]. Follow up work has permitted these profiles to be used clinically, with SOX10 protein expression being a useful histopathologic marker for MPNST, compared to synovial sarcoma [8], and *SOX9* being defined as a MPNST biomarker and pro-survival gene [10]. The presence of SOX10 expression in MPNST has permitted the Look laboratory to identify early precursor lesions to MPNST in vivo by co-expression of GFP under the SOX10 promoter (Fig. 2). Further markers, including the serum marker adrenomedullin, have been defined in neurofibromatosis patients as potential markers of transformation [9], though independent validation will be required.

The definition of *TWIST1* as an induced marker in MPNST led to experiments defining a functional role for this protein in regulation of chemotaxis of MPNST cells [6]. The regulation of *TWIST1* in MPNST has recently come to the forefront, based on studies demonstrating dependency on Wnt-β-catenin signaling, and involvement in invasion and metastasis of MPNST cells [52, 99–101]. The regulation of pro-invasive behaviors is also central to the MPNST phenotype, with studies demonstrating common expression of matrix-metalloproteinase-13 (MMP13) in MPNST lesions [86, 91]. Intriguingly, MMP13 in this study was strongly linked to recurrence of MPNST lesions [86], indicating a pro-tumorigenic role for this protein in MPNST.

The role of epigenetic mechanisms for tumorigenesis in MPNST has only begun to be explored. Evidence indicates that the polycomb repressive complex 2 (PRC2),

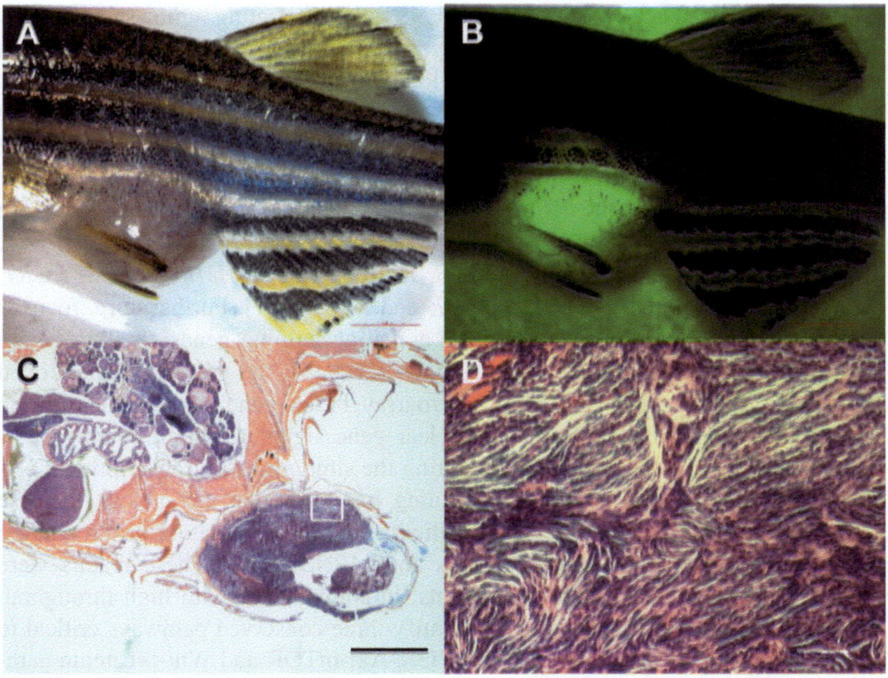

Fig. 2 The Sox10 promoter drives GFP in zebrafish MPNST. (**a**) 32 week old *SOX10:GFP*, nf1a$^{+/-}$; nf1b$^{-/-}$; p53$^{e7/e7}$ zebrafish with an abdominal tumor mass. (**b**) GFP driven by the *SOX10* promoter is expressed in the tumor allowing the tumor cells to be monitored in vivo (scale bar = 1 mm). (**c**) Histopathology of representative MPNSTs stained with hematoxylin and eosin reveals patterns of spindle shaped cells with whorling patterns consistent with MPNST. (**d**) Magnification of inset from (**c**). Tumors consist of spindle cells that stack into short fascicles, typically with a whorling pattern. Scale bars: (**c**) *black* = 1 mm; (**d**) *white* = 50 μm

an oncogenic complex in multiple cancer types, may be specifically lost in RAS-driven malignancies, including MPNST. Intriguingly, the loss of the PRC2 member *SUZ12* cooperates with loss of *NF1* to sensitize tumors to bromodomain inhibitors [102]. The further role that epigenetic mechanisms may play in MPNST remains to be explored, including notably, regulation of its genome-wide transcriptional profile via changes in histone architecture.

Rodent Genetic Models of MPNST

Concurrently with the early descriptions of the role of *NF1* in neurofibromatosis and MPNST, the first animal models of MPNST were developed. Two approaches to generate MPNST in rodents have been examined, including ENU-mediated mutagenesis, implicating mutated *ErbB2* in MPNST tumorigenesis [103, 104], though significantly more success has been achieved using targeted modeling of

Table 3 Murine genetic models of MPNST

Murine model	Phenotype	References
TP53$^{-/-}$	MPNST, various tumors	[80]
ENU mutagenesis	MPNST, ErbB2 mutation	[103, 104]
TP53$^{+/-}$;*NF1*$^{-/-}$	MPNST, various sarcomas	[78, 79]
NF1$^{-/-}$	No tumor spectrum	[64]
P0a-Cre; *NF1*$^{loxp/loxp}$	PLXNF	[105]
Krox20-Cre; *NF1*$^{loxp/loxp}$	PLXNF, MPNST	[72]
NF1$^{+/-}$;*p16*$^{INK4A-/-}$	MPNST	[105]
Adeno-Cre; *NF1*$^{loxp/loxp}$; *p16*$^{INK4Aloxp/loxp}$	MPNST	[106]
PTEN$^{loxp/loxp}$;*NF1*$^{loxp/loxp}$;*LDL-K-RAS*$^{G12D/+}$;GFAP-Cre	Neurofibromas, MPNST	[47]
TP53 $^{loxp/loxp}$;*NF1*$^{loxp/loxp}$; GFAP-Cre	Astrocytoma, GBM	[107]
PTEN$^{loxp/loxp}$; Dhh-Cre	PLXNF	[108]
NF1$^{loxp/loxp}$;*PTEN*$^{loxp/loxp}$; Dhh-Cre	MPNST	[108]
Cnp-H-RAS^{V12}	MPNST	[109]
Cnp-EGFR; PTEN$^{loxp/loxp}$;*Dhh-Cre*	MPNST	[110]
Cnp-EGFR; NF1$^{loxp/loxp}$; *PTEN*$^{loxp/loxp}$;*Dhh-Cre*	MPNST	[57]
Cnp-EGFR; TP53$^{-/-}$	MPNST	[95]
Cnp-EGFR; Cnp-Cre;TP53^{R270H}	MPNST	[99]

Summary of murine genetic models of MPNST. *Cnp* 2′,3′-cyclic nucleotide 3′phosphodiesterase, *Dhh* desert hedgehog, *ENU* ethylnitrosourea, *PLXNF* plexiform neurofibroma, *GBM* glioblastoma multiforme

genetic lesions present in human tumors. Initial studies examined the role of *TP53* in murine MPNST tumorigenesis. While wild-type animals, as mice heterozygous for loss of *TP53* did not develop MPNST, instead developing embryonal carcinomas and lymphomas, those that were homozygous for loss of *TP53* developed, with rarity, MPNST [80].

More success in deriving MPNST in murine models has been achieved by candidate gene disruption (Table 3). Based on the finding that patients with neurofibromatosis type 1 display mutations or LOH of the *NF1* gene, the Parada and Jacks groups developed animal models of MPNST by disruption of the *TP53* and *NF1* loci [78, 79]. Importantly, biallelic targeted disruption of the *NF1* gene alone is embryonic lethal in animal models, with heterozygotes developing hyperplasia of sympathetic ganglia, arguing for a neural-associated phenotype in these animals [64, 111, 112]. This evidence was further extended to a neural phenotype for NF1 loss in subsequent studies establishing that biallelic astrocytic-specific loss of *NF1* alone results the generation of optic gliomas [113], with combined astrocytic loss of *NF1* and *TP53* resulting in tumors on the continuum from astrocytoma to glioblastoma multiforme [107]. In 1999, Jacks group published that mice chimeric for NF1 homozygous loss were viable, and the *NF1*$^{-/-}$ cells resulted in the generation of plexiform neurofibromas. Further, crossing these chimeric animals with animals with heterozygous loss of *TP53* resulted in reduced overall survival and the generation of S100-positive tumors consistent with MPNST, in addition to a variety of other tumor types, including osteosarcoma, osteosarcoma, fibrosarcoma, rhabdomyosarcoma and hemangiosarcoma [79]. Simultaneously, the Parada group described a model of *cis* loss of *NF1* and *TP53* in mice, with a cohort of animals developing 77 % sarcomas, 14 % lymphomas and 9 % carcinomas, with rare (3 %) neuroblastomas. Of the sarcomas, 61 % were MPNST, 22 % were malignant triton tumors, and more rarely leiomyosarcomas, rhabdomyosarcomas and malignant fibrous histiocytomas were found, intriguingly mirroring the quandary faced with pathologic diagnosis of MPNST in humans [2]. The plexiform neurofibromas that are seen in $NF1^{-/-}$ mice are mirrored by studies demonstrating that use of the *P0a* promoter, an early neural crest marker, to drive CRE recombinase expression in $NF1^{loxp/loxp}$ animals resulted in the generation of a rare population of non-myelinating Schwann cells, which proliferate to form plexiform neurofibromas [105, 114]. Subsequent to these studies, several groups focused on the development of conditional knockout of *NF1* in specific cell lineages. Importantly, studies on conditional co-inactivation of *TP53* and *NF1* in murine models were initially disappointing, with animals with compound $TP53^{loxp/loxp}$;$NF1^{loxp/loxp}$animals crossed with CRE-expressing animals under the glial-fibrillary acidic protein (GFAP) promoter resulted in astrocytoma and glioblastoma multiforme, but not MPNST [107]. In contrast, use of the *KROX20* promoter to drive Cre expression, in combination with $NF1^{loxp/loxp}$ animals results in mice with increased penetrance of neurofibromas [72]. Intriguingly, some plexiform neurofibromas were also seen, with extensive mast cell infiltration, arguing for a pro-inflammatory role for the extracellular milieu in the neoplastic process of MPNST [72], a finding that has been echoed in subsequent studies [42, 51, 52, 115, 116].

Given the frequent mutational loss of $p16^{INK4A}$ in human MPNST [6, 85, 87, 89], the Morrison group has examined concordant loss of *NF1* and the $p16^{INK4A}$ locus. Creation of $NF1^{+/-}$;$INK4A/ARF^{-/-}$ animals resulted in histologically confirmed MPNST [105]. Similarly, the Kirsch group has demonstrated that orthotopic injection of adeno-CRE into the sciatic nerve of $NF1^{loxp/loxp}$;$INK4A/ARF^{loxp/loxp}$ animals results in MPNST formation [106]. These data indicate that disruption of cell cycle control at the level of $p16^{INK4a}$ is sufficient, in combination with *NF1* loss, to induce MPNST and that loss of *p53* is not required for tumorigenesis.

In light of the molecular evidence for loss of *PTEN* in human MPNST [49, 56], and the presence of an animal model of *GFAP*-driven loss of *PTEN* [117], Gregorian et al. set out to develop a *PTEN*-dependent model of MPNST. Here, animals with $PTEN^{loxp/+}$; $LDL\text{-}K\text{-}RAS^{G12D/+}$ crossed with mGFAP-CRE animals developed neurofibromas, with some animals developing MPNST with an elevated Ki67 index, phosphorylated Akt, S6, and ERK1/2 [47]. Intriguingly, these tumors displayed biallelic loss of PTEN, wild type p53 and intriguingly, normal NF1 status, arguing for a sporadic model of MPNST [47]. These data further provided evidence for a cell of origin of sporadic MPNST as a Schwann cell precursor or immature Schwann cell [47]. These data were extended by use of the *desert hedgehog* (Dhh) promoter in mice, wherein *Dhh-Cre* animals crossed with *PTEN* $^{loxp/loxp}$ results in the generation of plexiform neurofibromas, which progress to peripheral MPNSTs in animals that concurrently lose *NF1* [108]. Intriguingly, the homozygous loss of both *NF1* and *PTEN* using either Dhh-CRE or the *2′,3′-cyclic nucleotide 3′phosphodiesterase* (*CNP*) promoter constructs to drive CRE expression concurrently results in high grade MPNST with enlarged brachial plexi, dorsal root ganglia and trigeminal nerves with intercostal and paraspinal neurofibromas [99, 108]. Microarray comparison of these tumors to purified Schwann cells from human peripheral nerves, neurofibromas and MPNST cell lines and solid tumors displays a high degree of concordance between these tumors and *NF1*-associated MPNST [108]. This data echoes the *LDL-K-RAS*$^{G12D/+}$ murine model, which also develops MPNST in the context of *NF1* and *PTEN* loss [47].

Further models have been developed, again drawing on findings from human molecular genetics. Despite a paucity of RAS mutations in human MPNST [50], the common presence of loss of NF1 [20, 21, 63, 75] and other negative regulators of RAS signaling [47–49] indicates that RAS is hyperactivated in MPNST. To this end, the Ratner group has generated a transgenic murine model where a hyperactivated RAS (H-RAS^{G12V}) is driven under the *CNP* promoter, inducing expression in the Schwann cell, resulting in MPNST formation.

Several lines of evidence have implicated a role for the EGFR in MPNST. *EGFR* is frequently amplified [48, 56, 67, 94] and active [118] in human MPNST [48, 56, 67, 94], MPNST induced by loss of *NF1 and p53* express high levels of EGFR [61]. Transgenic expression of EGFR in *NF1*$^{+/-}$;*p53*$^{+/-}$ mice results in generation of neurofibromas with mast cell accumulation and disassociation of non-myelin forming Schwann cells, consistent with plexiform neurofibroma formation [115]. These data led the Largaespada group to derive a model of sporadic MPNST by transgenic, piggybac-mediated overexpression of *EGFR* under a *CNP* promoter, in combination with loss of *PTEN* [110]. By crossing *PTEN* $^{loxp/loxp}$ animals with *Dhh-CRE* expressing animals, with piggybac-EGFR overexpression, Keng et al. generated animals with combined overexpression of *EGFR* and loss of *PTEN* in Schwann cells, resulting in the generation of MPNST [110]. Importantly, the group went on to demonstrate that this is not a species-specific phenomenon, as piggybac-mediated *EGFR* overexpression, coupled with shRNA-mediated depletion of PTEN in human immortalized Schwann cells results in increased proliferation and colony formation in vitro [110]. This model is not solely dependent on *PTEN*, as *NF1*$^{loxp/loxp}$;Dhh-CRE animals crossed with *CNP-EGFR* overexpressing mice also display MPNST formation [57]. Further, given the high degree of colocalization in human MPNST between EGFR overexpression and loss of *TP53*, the Largaespada group has generated a model of sporadic MPNST with transgenic Cnp-EGFR expression, in combination with loss of *TP53* [95]. These animals uniformly develop GFAP+, SOX10+, Nestin+MPNST that displays phosphorylated ERK and Akt [95].

Recently, an unbiased screen using the sleeping beauty transposon system has been performed utilizing *CNP-EGFR* and *CNP-EGFR*;*CNP-CRE*;*TP53*^{R270H} mice, which develop neurofibromas and MPNST, respectively, with low penetrance [99]. Transposon insertion in this model resulted in a dramatic increase in neurofibroma and MPNST formation in both models [99]. The most common sites of disruption in this model were *NF1* and *PTEN*, validating the model, however numerous other potential targets were identified in this screen, many of which are promising candidates for further exploration in other models [99]. The utility of the sleeping beauty screen was exemplified by the identification of *FOXR2* as a potential proto-oncogene in the dataset, with overexpression of *FOXR2* in hematopoietic stem cells resulting in accelerated xenograft tumorigenesis, and transcription activator-like effector nuclease (TALEN)-mediated disruption of *FOXR2* in MPNST cell lines suppressing xenograft growth [99]. The database produced by this transposition screen will prove invaluable in the understanding of the fundamental biology of MPNST, and in the definition of targets for therapeutic inhibition.

Zebrafish Genetic Models of MPNST

Leading off from evidence implicating *NF1* and cell cycle control in murine models of MPNST [63, 64, 78, 79, 82, 111], initial models of MPNST in zebrafish focused on the role of *TP53* disruption. Reports of sporadic schwannoma in laboratory strains of zebrafish exist [119], suggesting that the genetic events necessary for driving MPNST formation in zebrafish may not be embryonic lethal. The first model of MPNST in zebrafish was developed by the in the early 2000s, when zebrafish carrying homozygous $TP53^{M214K}$ and $TP53^{N168K}$ mutations were demonstrated to suppress apoptosis and fail to upregulate p21 after radiation induced DNA damage (Table 4) [81]. Importantly, these mutations in zebrafish induced MPNST with a latency of 8.5 months, and a penetrance of nearly 25 % [81]. Importantly, as detailed earlier, sporadic loss of *TP53* itself [82], as well as loss of cell cycle control through the *TP53* and *RB1* networks [63, 83–86, 88, 89] are seen in human MPNST, lending credence to these early models.

Further modeling has indicated the utility of a retroviral insertional mutagenesis approach in zebrafish, resulting in the finding of multiple mutations in distinct

Table 4 Zebrafish genetic models of MPNST

Zebrafish model	Phenotype	References
$Tp53^{M214K/M214K}$	MPNST	[81]
$Tp53^{N168K/N168K}$	MPNST	[81]
Retroviral insertion into RP genes	MPNST	[120, 121]
Mutation of MMR genes: $mlh1^{-/-}$, $msh2^{-/-}$, $msh6^{-/-}$	Neurofibromas, MPNST	[122]
$nf1a^{-/-};nf1b^{-/-}$	Embryonic lethality	[112]
$tp53^{M214K/M214K};nf1a^{+/-};nf1b^{-/-}$ or $tp53^{M214K/M214K};nf1a^{-/-};nf1b^{+/-}$	MPNST	[112]

Summary of zebrafish genetic models of MPNST. *MMR* mismatch repair genes, *RP* ribosomal protein ribosomal protein genes leading to MPNST generation [120]. These data argue that disruption of the ribosome by the mutationally-derived loss of one allele may be sufficient for MPNST initiation [120]. Intriguingly, alterations of RPL10 expression have been defined in screens of human tumors [98], and zebrafish MPNSTs demonstrated loss of TP53 protein expression, suggesting that haploinsufficiency for ribosomal proteins resulting in decreased TP53 protein synthesis may function as a further mechanism for tumorigenesis [121]. The role that ribosomal proteins may play in human MPNST remains to be fully elucidated.

In the combination loss of function of *NF1* and *TP53* model, generated by the Parada group, sarcomas and, specifically, MPNST, were identified [2]. Plexiform neurofibromas are seen in *NF1−/−* mice, as well as in $NF1^{loxp/loxp}$ *; P0a-Cre* animals [105, 114], suggesting that NF1 loss alone is sufficient to drive generation of plexiform neurofibromas. Intriguingly, the Look laboratory found early embryonic lethality with complete loss of the *nf1a* and *nf1b* genes in zebrafish, along with aberrant proliferation and differentiation of oligodendrocyte progenitor cells, dysmorphic myelin sheaths and importantly, hyperplasia of Schwann cells [112]. Further, heterozygous loss of *nf1a* and *nf1b*, concurrently with mutation of p53 ($nf1a^{+/-}$; $nf1b^{-/-}$; $p53^{e7/e7}$) resulted in accelerated onset and increased penetrance of histologically confirmed MPNST in adult fish [112], arguing for similar genetic ontogenies between murine, human and zebrafish models of MPNST in terms of cooperation between loss of NF1 and cell cycle control. These MPNSTs display the typical cell whorling histology of human MPNST, with fascicular organization of spindle-like cells (Figs. 2, 3, and 4). Interestingly, this model is highly penetrant, with individual animals displaying multiple synchronous primary tumors (Fig. 3). Furthermore, primary tumors in this model arise in multiple distinct locations, including the abdomen (Fig. 2), gills, heart, gut, ovary and kidney (Figs. 3 and 4). The presence of a GFP label in this model results in facile screening for tumor development, which is critical given the diverse array of locations these tumors may arise in.

Several lines of evidence have implicated the role of the *INK4A/B* locus in generation of murine and human MPNST [56, 88, 89, 97]. The role that this locus plays in zebrafish MPNST generation is as of yet unexplored, though the zebrafish may represent a genetically simple species for the investigation of this problem, with recent evidence demonstrating the lack of a $p14^{ARF}$ locus in the zebrafish genome [123].

Fundamental to the modeling of human tumorigenesis in the zebrafish have been experiments comparing the DNA copy number patterns of human and zebrafish MPNSTs derived from the $tp53^{M214K}$ [81] and ribosomal protein loss models [120], to examine copy number variants overlying genes demonstrating synteny and that are either lost or gained in both tumor types [124]. Cross-species analysis resulted in the identification of many previously described MPNST genes, including loss of *NF1, merlin (NF2), SMARCB1* and *PTEN*, and gain of *CCND2, ETV6, HGF, HSF1, KIT, MDM2, MET* and *PDGFR* [124]. Intriguingly, these results indicate that zebrafish and human MPNST display molecular similarities, and that these genes may represent evolutionarily conserved candidate cancer drivers for MPNST development.

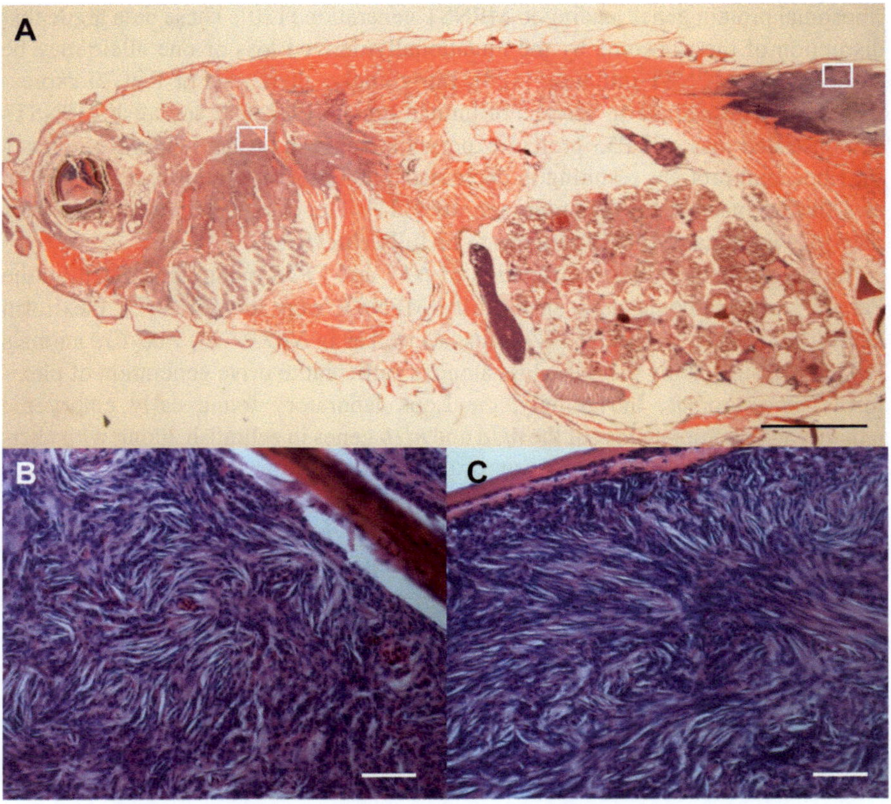

Fig. 3 Synchronous MPNSTs arise in the same zebrafish. (**a**) Multiple MPNSTs are observed within the same nf1a$^{+/-}$; nf1b$^{-/-}$; p53$^{e7/e7}$ zebrafish, at the gill and posterior dorsal trunk (scale bar = 2 mm). (**b**) Magnification of the gill MPNST (*anterior inset* in (**a**)). (**c**) Magnification of posterior dorsal trunk MPNST (*posterior inset* in (**a**)). *White* scale bar = 50 μm. Cells of both MPNST tumors display typical MPNST histology with long serpentine-like nuclei and spindle-shaped cells

Finally, the role of genome instability in the generation of MPNST has been recently queried. Zebrafish with homozygous mutations in the DNA mismatch repair (MMR) genes *mlh1, msh2* or *msh6* develop masses consistent with neurofibroma-MPNST spectrum tumors with a combined frequency of 33 % [122]. The effects of triple homozygous loss of these genes has not, to date, been evaluated. Loss of mismatch DNA repair control applies selective pressure for loss of cell cycle checkpoint control, perhaps providing a link for frequent loss of *TP53* and/or *RB1* gene or pathway function in the MMR, or so-called, "mutator phenotype" cancers [125, 126]. Intriguingly, zebrafish with loss of *msh6* do not demonstrate enhanced germline mutagenesis with ENU [127], indicating that there may represent at least partial compensation in mismatch repair through other zebrafish genes. These data provide insights into the mechanism of *p53* and *RB1*-mediated tumorigenesis via the regulation of genome stability, a key driver event in human

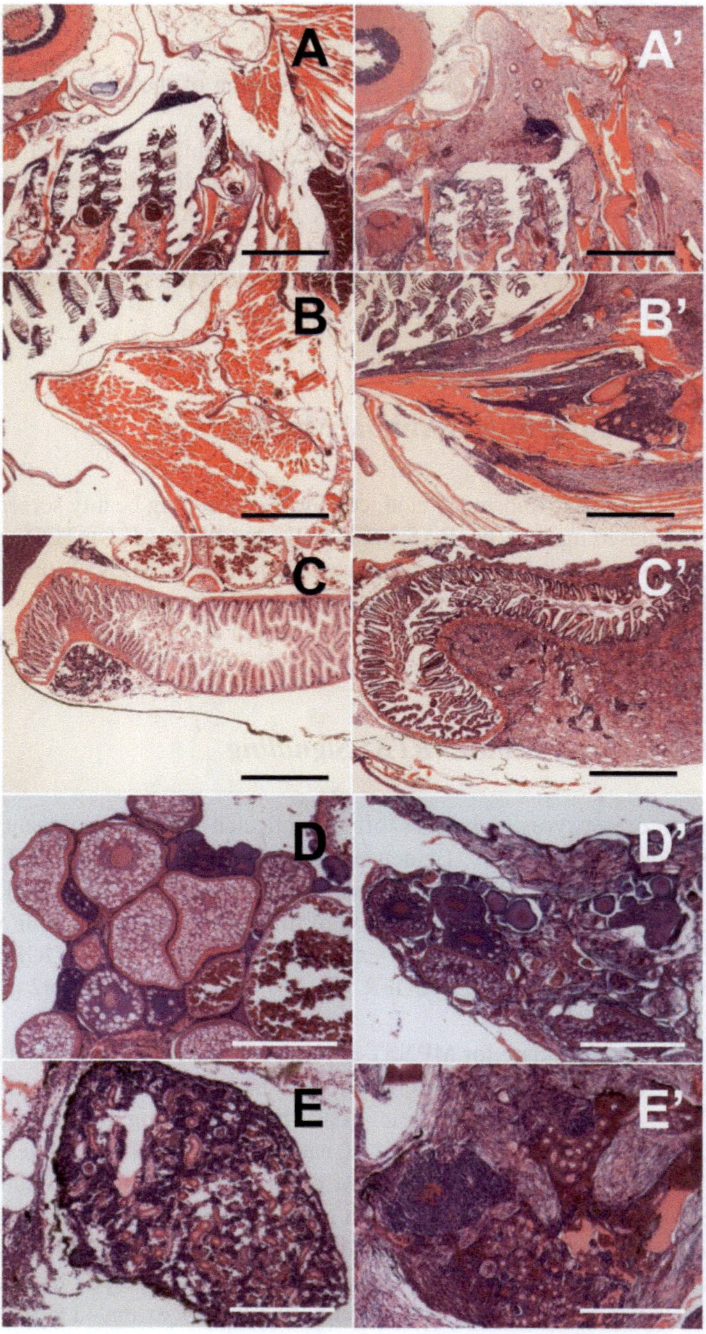

Fig. 4 Zebrafish MPNSTs arise in multiple and distinct anatomic locations. Hematoxylin and eosin stains of tissue derived from normal zebrafish organs: gill (**a**), heart (**b**), gut (**c**), ovary (**d**), and kidney (**e**). Different nf1a$^{+/-}$; nf1b$^{-/-}$; p53$^{e7/e7}$ animals were noted to have MPNSTs arising from or invading into these tissues. Panels (**a**) to (**e**) on the *left* show the normal histology of each organ. Panels (**a′**)–(**e′**) on the *right* illustrate the appearance of MPNSTs in nf1a$^{+/-}$; nf1b$^{-/-}$; p53 $^{e7/e7}$ fish arising from or invading into each tissue type (*black* scale bar = 500 μm; *white* scale bar = 250 μm)

tumorigenesis, and provide a model for investigation of the relevance of genome instability in MPNST.

Future studies in zebrafish models of MPNST should aim to further investigate candidate key genes in human tumorigenesis, including the EGFR, PDGFR, $p16^{INK4a}$ and manipulation of the RAS-ERK and Akt-mTOR pathways. The ease with which soluble small molecule inhibitors of protein kinase activity are used in zebrafish indicate a key ability to conduct large scale studies of potential therapeutic modalities.

Therapeutic Targeting in Animal Models

In their seminal study, Rahrmann et al. conducted a sleeping beauty screen to identify in an unbiased manner, the RAS-MEK-ERK and PI3K-AKT-mTOR and Wnt-β catenin pathways as being highly activated pathways in MPNST tumorigenesis [99]. Intriguingly, several studies in xenograft models existed to provide evidence that core members of these pathways are involved in MPNST tumorigenesis.

Receptor Tyrosine Kinase (RTK) Signaling

Numerous receptor tyrosine kinases, often implicated in other malignancies, have been investigated in MPNST. Evidence indicates that primary MPNST samples display elevated expression due to copy number increase of the insulin-like growth factor receptor-1 (IGF1R), with inhibition of this receptor resulting in in vitro loss of proliferation [53]. Similar evidence exists for c-Kit [33, 58–60], PDGFR [33, 48, 60, 91–93, 128], Met [25], the VEGFR [54, 55], and Her2/neu/ErbB2 [56, 61] in MPNST.

The strongest candidate for MPNST-dependent receptor tyrosine kinase signaling is the EGFR. Evidence has long implicated the EGFR in MPNST tumorigenesis [48, 56, 94, 116], with multiple EGFR-overexpressing models developing sporadic type MPNST [61, 110, 115]. Treatment of MPNST cell lines with the EGFR inhibitor gefitinib results in loss of proliferation [53], and co-inhibition of the EGFR with erlotinib and mTOR with everolimus results in enhanced suppression of xenograft growth [129]. The presence of hyperactivated RAS, but not other MAPKs results in permissiveness of these tumor cells to oncolytic herpes simplex viruses (oHSVs) in vitro [130]. These findings have been extended to the hyperactivated RAS signaling found in MPNST cells in vitro [131], and this method has been extended to xenograft models of MPNST, demonstrating additivity between oHSV infection and erlotinib treatment, resulting in suppression of tumor growth and angiogenesis [132].

Downstream of activated EGFR signaling in MPNST are likely to be numerous pathways involved in tumorigenesis. The cooperativity between loss of *PTEN* and overexpression of EGFR in the generation of MPNST in vivo [56, 110] argues strongly for dysregulation of the PI3K by removal of suppression and ectopic activation of its downstream targets, including the Akt-mTOR and RAS pathways. In addition, evidence indicates that EGFR may signal through other pathways, including the STAT3 pathway, to drive MPNST tumorigenesis [57]. Clearly, genetic and biochemical evidence exists to implicate numerous RTKs in driving the neoplastic behaviors of MPNSTs, though a key candidate for therapeutic inhibition remains the EGFR.

PI3K-Akt-mTOR Signaling

Significant evidence has accrued for dysregulation of the PI3K-Akt-mTOR signaling axis in MPNST. Prior to the unbiased identification of this signaling axis as being hyperactivated in MPNST [99], early evidence indicated that expression of the negative regulator of PI3K signaling, *PTEN*, was lost in primary human [48, 49] and animal models of MPNST [47, 99, 108, 110, 117]. Intriguingly, no mutational hyperactivation of PI3K or AKT1/2 is found in MPNST [48–50], in contrast to the oncogenic mutations that are found in other solid tumors [133].

Analysis has demonstrated that methylation of the *PTEN* promoter drives down-regulation of the gene in primary human MPNST, with concomitant hyperactivation of Akt-mTOR signaling (Fig. 1) [49]. Therapy of MPNST cell lines with mTOR inhibitors, regardless of *PTEN* status, results in growth inhibition [49, 129]. Early evidence indicated that mTOR inhibition could delay sporadic and NF1-associated xenograft growth and vessel permeability, with minor increases in growth suppression when co-treated with the EGFR inhibitor erlotinib [129]. Furthermore, co-inhibition of mTOR with rapamycin or everolimus sensitizes MPNST cells with loss of *NF1* to proteasomal inhibition with bortezomib [31]. In this study, combinations of everolimus, bortezomib and ionizing radiation further enhanced G2/M cell arrest and increased markers of apoptosis in MPNST cells in vitro and in xenografts, where an up to 72 % reduction in tumor volume was noted with triple therapy [31]. Similar studies have demonstrated efficacy of everolimus in combination with MEK inhibition in genetically engineered models of NF1 associated (*Dhh-CRE;NF1*$^{loxp/loxp}$;*PTEN* $^{loxp/loxp}$) and sporadic (*Dhh-CRE;PTEN*$^{loxp/loxp}$;*piggybac-EGFR*) MPNST, where, importantly, co-therapy resulted in a lack of tumor-acquired resistance mechanisms [45].

Recently, it has been hypothesized that MPNST cells may display endoplasmic reticulum (ER) stress, resulting in an unfolded protein response. Intriguingly, this response seems to be mTOR dependent. The Cichowski group has demonstrated that enhancement of ER stress with thapsigargin or the clinically active compound IPI-504 synergizes with rapamycin to drive MPNST cell death and tumor regression in xenograft models [134]. This manipulation of mTOR is dependent on generation of reactive oxygen species/oxidative stress, with similar effects in a K-RAS/p53 mutant model of non-small cell lung carcinoma [134], leading the authors to hypothesize that targeting the mTOR-dependent unfolded protein response may be a common method of inhibiting RAS-dependent tumors.

RAS-MEK-ERK Signaling

Given the GAP activity of endogenous NF1 [41], and the resultant dysregulation of RAS in the context of loss of NF1 activity, significant interest has accrued in the role of the RAS-MEK-ERK signaling pathway in MPNST. Evaluation of human MPNST has indicated that this pathway may be similarly hyperactivated in sporadic or NF1-associated MPNST, with pathway activation by loss of NF1 in the latter situation, and up to 20 % of sporadic MPNST displaying mutational activation of BRAF (V600E) mutation [96]. Importantly, hotspot mutations of the RAS proteins are rarely seen in MPNST [50], indicating that dysregulation of this pathway is due primarily to activation of positive and loss of negative regulators. Intriguingly, the tumor-initiating CD133+ subpopulation of primary MPNST tumors displays elevated activity of RAS itself, as well as numerous RAS effector molecules, including the MAP kinases ERK, JNK, p38 and RalA [135]. Moreover, MPNST cell lines with documented loss of *NF1* treated with the RAS inhibitor farnesylthiosalicylic acid display reduced growth, and similar treatment of nude mice with established human MPNST cell line xenografts result in a dose dependent inhibition of tumor weight [43]. These data strongly indicate that manipulation of the RAS signaling pathway may be efficacious in MPNST therapy.

Downstream of RAS, however, there exists significant signal spreading. The RAS downstream targets MEK, MAF, Ral-A and the Rho-family kinases have both been evaluated in MPNST. Inhibition of MEK with the inhibitor PD325901 results in delayed tumor growth in the adeno-CRE; $NF1^{loxp/loxp}$; $INK4A/ARF^{loxp/loxp}$ MPNST model [106], in $NF1^{loxp/loxp}$;Dhh-CRE induced neurofibromas and in MPNST xenografts [44]. Strikingly, this inhibitor reduces phosphorylation of ERK1/2, but has no effect on phosphorylated Akt in vitro or in vivo [44]. This finding has led to the hypothesis that MEK inhibition may combine with inhibitors of the Akt-mTOR pathway to provide antitumorigenic activity. To this end, combination of the same MEK inhibitor and the mTOR inhibitor everolimus results in synergy in vitro and improves activity and delays mortality in the NF1-associated $NF1^{loxp/loxp}$;$PTEN^{loxp/loxp}$;*Dhh-CRE* and the sporadic $PTEN^{loxp/loxp}$;*Dhh-CRE;Dhh-EGFR* MPNST models [45].

Despite high levels of interest in MEK inhibition as a downstream target of RAS, other RAS targets including RalA are also upregulated in MPNST. The CD133+ subpopulation of MPNST primary tumors displayed elevated RalA activation [135], and evidence indicates that MPNST cell lines from murine $NF1^{-/-}$;$TP53^{-/-}$ tumors, compared to nontransformed Schwann cells, display higher levels of activated RalA [46]. Active RalA was both dependent on NF1-GAP activity and required for tumorigenesis, as re-expression of the NF1-GTPase regulated domain could decrease RalA activation and suppress proliferation, invasion and cell cycle progression in vitro and in murine xenografts [46]. These data argue for a role for RalA as a RAS target in the generation of MPNST. Similar evidence exists to implicate the Rho-family GTPases as downstream targets in NF1-deficient MPNST tumorigenesis [92].

Wnt-β Catenin Signaling

Mounting evidence indicates a role for the Wnt-β catenin signaling pathway in MPNST tumorigenesis. Evidence has implicated this pathway through the initial description of genomic alterations of the WNT pathway members *IRX2* and *NKD2* in MPNST samples, but not other soft tissue sarcomas [136]. Furthermore, loss of *Merlin,* the causative gene of neurofibromatosis type 2 [39, 137], in Schwann cells, results in enhanced proliferation that is dependent on increased expression and phosphorylation of β-catenin [128]. Numerous Wnt pathway members are elevated in MPNST compared to plexiform neurofibromas, including, notably, WNT2, WNT5, the Wnt receptors FZD1 and FZDB, and the transcriptional effectors LEF1 and TCF7 [101]. These expression patterns correlate with induction of the markers of epithelial to mesenchymal transition (EMT) TWIST1, SLUG and CDH1 [101]. MPNSTs display extremely high expression of β-catenin and siRNA-mediated downregulation of *NF1* results in enhanced β-catenin activity in vitro [101]. These data argue for a functional role for the Wnt-β-catenin pathway in MPNST, perhaps in the regulation of invasion and metastasis. These data are further enhanced by data indicating that the CD133+ population of primary MPNST cells display enhanced invasiveness due, at least in part to increased levels of β-catenin [135]. β-catenin displays increased nuclear localization in MPNSTs derived from the sleeping beauty screen [99] as well as in four independent genetic models of MPNST tumorigenesis [100]. The relevance of these findings was further evaluated in a primary human tissue microarray, demonstrating elevated β-catenin and target expression in a subset of MPNST [99]. Wnt-β-catenin activity induces cell viability/proliferation of human Schwann cells, and shRNA-mediated loss of β-catenin suppresses cell viability and xenograft growth in MPNST cell lines [100]. Finally, Wnt-pathway inhibitors synergized with the mTOR inhibitor everolimus in a panel of MPNST cell lines, indicating a potential avenue for therapeutic exploration. Together, these data highlight the evolving body of literature implicating the Wnt-β-catenin pathway in MPNST tumorigenesis.

Cell Cycle Control

As previously discussed, models of MPNST prominently feature disruption of either the p53 or Rb signaling axes prominently. Initial models of MPNST required disruption of *TP53* at the genetic level [80, 81], with subsequent models utilizing combinatorial loss of *NF1* and *TP53* [78, 79, 112]. In light of human data displaying disruption of the *INK4A* locus in MPNST [85, 87–89], murine models involving combinatorial disruption of this locus, in combination with *NF1* [105, 106, 138] have been generated, which display MPNST formation. Zebrafish models involving loss of the *INK4A* locus have not been generated, however, with recent data indicating that zebrafish do not contain a *p14*^{ARF} homolog [123], disruption of the endogenous zebrafish *cdkn2b* locus should aid in further physiologically relevant modeling of MPNST, given the frequency with which this locus is lost in human MPNST [87, 89]. Intriguingly, further evaluation by sleeping beauty transposition screening has identified numerous genes involved in cell cycle control, including *MYCN*, *CDC27*, *CCNY* and *CDK13* as potentially involved in MPNST tumorigenesis [99]. Further evaluation of the results of this screen and evaluation of these targets in MPNST should prove illuminating to the fundamental biology of MPNST.

Other Implicated Signaling

Xenograft and genetic models of MPNST have implicated numerous other targets in MPNST tumorigenesis. MPNST derived in the *CNP-H-RAS*^{G12V} model has indicated a potential role for the aurora-kinase A (*AURKA*) in MPNST [109]. Comparison of these tumors to human tumors demonstrates amplification of *AURKA* in MPNSTs [109]. Intriguingly, shRNA-mediated depletion of AURKA suppresses MPNST growth in vitro and targeted small-molecule inhibition of AURKA results in suppression of MPNST cell primary xenograft growth [109]. In addition to AURKA, other targets, including Eyes absent homolog 4 (*EYA4*) [139] and *MAF* [140] have been similarly investigated in MPNST genesis.

Animal Models Providing Insight into MPNST Tumor Initiation

In addition to utilization of murine and zebrafish models for modeling the disease, with the aims of identifying novel targets for therapeutic exploitation in MPNST, modeling of these tumors has been efficacious for defining interesting points of fundamental biology. The Morrison group has used the *NF1*$^{+/-}$;*INK4a/ARF*$^{-/-}$ model to explore the fundamental biology of MPNST tumors, defining a subpopulation of tumorigenic cells representing a likely stem or progenitor population [138]. Intriguingly, these cells require laminin binding to β1-integrin, highlighting a function for cell:ECM interactions in these tumors which heretofore had been unknown (Fig. 1). This interaction was of importance, as exogenous laminin provided clonogenic advantage for sorted progenitor populations from *NF1*$^{+/-}$;*p53*$^{+/-}$ animals which do not produce laminin, but not from *NF1*$^{+/-}$;*INK4a/ARF*$^{-/-}$ mice which produce high level expression of laminin. These results highlight the role of the matrix in MPNST tumorigenesis, which has previously been suggested to be involved in by mast cell infiltration and induction of inflammation (reviewed in [42]). In addition, it provides a potential new target for exploration in the β1-integrin, and, given the role these proteins play in regulation of growth factor and chemokine presentation to nascent receptors [141], suggests possible avenues of co-exploration with RTK biology. Further, these findings have been extended to the definition of CD133 as a possible marker of MPNST tumor initiating cells [135], with activation of the PI3K-Akt-mTOR, RAS-ERK, and Wnt pathways seen at an even higher level in the TICs, compared to whole tumor extracts [135].

Advantages of MPNST Modeling in Zebrafish

The zebrafish *Danio rerio* develops tumors in a variety of tissues. In particular, multiple genetic models of MPNST tumorigenesis have been developed [81, 112, 120, 122, 142], with histologic and genetic comparability to human MPNST (Figs. 2, 3, and 4). Zebrafish MPNST display typical spindle cells and whirling patterns histologically, as well as intra-abdominal localization, similar to human MPNST [1, 2] (Fig. 2). Studies have indicated that human and zebrafish MPNST display strong similarities in terms of genomic structure and copy number variation [124]. Furthermore, zebrafish MPNST display genomic aneuploidy with near triploidy, a common feature of human MPNST, with syntenic similarities by array CGH at the subchromosomal level [124, 142].

One key strength of zebrafish models, in contrast to the murine models of MPNST, is the relative transparency of the animal. Intriguingly, neural crest lineage promoters, such as those for S100 and SOX10, may be used to drive fluorescent or other marker proteins, for the purposes of identifying early precursor lesions to the MPNST. This has been accomplished by the Look group through transgenic expression of SOX10:GFP fusions in zebrafish MPNST, resulting in early detection of GFP signal in neural crest origin cells as early as 16 hours post fertilization (hpf). At this timepoint, GFP is expressed in premigratory and early migrating neural crest cells. By 48 hpf, GFP expression is detectable throughout cranial pigment cells, ear, branchial arches, oligodendrocytes, Schwann cells, jaw cartilage and pectoral fins, with subsequent loss of GFP detectability in adult fish. Most critically, however, GFP is re-expressed in MPNST cells, permitting the direct in vivo monitoring of tumor progression (Figs. 2, 3, and 4), as well as the ability to *ex vivo* dissociate and sort cells by FACS analysis for molecular analysis. These strengths should permit future study in zebrafish models, as in studies in vitro and in murine models [105, 135, 138], to define the putative tumor initiating cell in zebrafish models of MPNST. Similar types of study, identifying the cell of origin of other solid tumors, such as rhabdomyosarcoma, have been completed in faithful zebrafish models of cancer [143], suggesting that this is a tractable model for discovery. The zebrafish has also been effectively used as a model to identify stem-like cells in other solid tumors, such as breast carcinoma, via novel serial xenotransplantation methods

[144, 145]. The direct translation of these findings back to human MPNST tumorigenesis has yet to be realized.

With the advent of focused genome editing technologies including the Crispr-Cas9 system, and the use of TALEN/zinc finger nucleases to derive focused gene insertions/deletions [146], as well as targeted mutations in vitro and in vivo [112, 146, 147], and the ability to insert cancer-associated genes under their endogenous promoters in vivo [148], models of MPNST will be made more rapidly, with higher fidelity and with enhanced physiologic relevance to human MPNST. With previous efforts demonstrating differences in MPNST incidence between variant *NF1* mutation types [75], it will be increasingly possible to model the findings of future massively parallel sequencing efforts in human MPNST, in model organisms. Data such as these have recently been modeled in vivo in the zebrafish, for genome-wide association study predisposition loci in humans (Look AT, Weichert N; personal communication). The zebrafish, with its short generation time and ability to identify early lesions, combined with the ability to manipulate the genome with ease, represents an ideal organism for the study of MPNST tumorigenesis, niche behaviors, invasion and metastasis.

In addition to appropriateness in modeling, the establishment of high fidelity zebrafish models of MPNST will permit rapid and cost-effective drug screening, to identify compounds capable of suppressing tumorigenesis, and those capable of repressing growth and metastasis, alone and in combination with standard chemotherapy. To this end, the use of a zebrafish model of T-cell acute lymphoblastic leukemia (T-ALL) has recently resulted in the definition of a new class of potential therapeutic agents, the perphenazines [149]. The potential for discovery of novel agents for therapeutic suppression of MPNST tumor development and metastasis, in addition to evaluation of rational combinations of agents with previously demonstrated efficacy in MPNST such as mTOR or ERK inhibitors, modulators of ER stress and suppressors of protein translation, alone and in combination with standardized chemotherapies by high throughput screening in an in vivo model of MPNST will aid in the definition of high efficacy combinations for translation into human trials.

Summary

Evidence has suggested that manipulation of specific signaling pathways in the related schwannomas of neurofibromatosis type 2, may be tractably treated with targeted therapies [34, 35, 150]. Similarly, small trials involving focused agents in MPNST have begun, with the use of imatinib mesylate to target NF1-associated MPNSTs [29, 30, 151]. Preclinical testing with combinations of agents targeting the RAS-ERK and Akt-mTOR pathways has begun, however, a focused method for screening large scale libraries of compounds in vivo, as well as variable combinations and concentrations of novel agents alone and in combination with ionizing radiation and chemotherapies will be required, in order to optimize therapeutic strategies prior to attempting large scale human clinical trials. As a model organism capable of being engineered to develop neural lineage tumors including faithful models of MPNST, coupled with organismal relative transparency, and the ability to track early tumor development in combination with enhanced abilities to model genomic alterations found in primary human tumors, the zebrafish is uniquely poised to aid in just such an endeavor.

Acknowledgements This research was supported by the Department of Defense (W81XWH-12-1-0125 to A.T.L.). A.D.D. was supported by grants, including the Frederick Lovejoy Award from the Department of Medicine, and the Division of Pediatric Hematology/Oncology, Boston Children's Hospital and Dana-Farber Cancer Institute. D.H.K. was supported by the Latsis Family Fellowship of the Boston Children's Hospital Neurofibromatosis program. S.H. was supported by David A. Abraham Fellowship at Dana-Farber Cancer Institute, Collaborative Cancer Research Award from Rally Foundation for Childhood Cancer Research, Young Investigator Award from Children's Tumor Foundation, and Young Investigator Award from Alex's Lemonade Stand Foundation. The authors thank Dr. Gary Fleisher for additional support in the generation of this work.

References

1. Davis IJP-AA, Fisher DE (2015) Nonrhabdomyosarcoma soft tissue sarcomas and other soft tissue tumors. In: Orkin SH, Fisher D, Ginsburg D, Look AT, Lux SE, Nathan DG (eds) Nathan and Oski's hematology and oncology of infancy and childhood, vol 2, 8th edn. Elsevier Saunders, Philadelphia, pp 1946–1982
2. Okcu MF, Hicks J, Merchant TE, Andrassy RJ, Pappo AS, Horowitz ME (2006) Nonrhabdomyosarcomatous soft tissue sarcomas. In: Pizzo P, Poplack D (eds) Principles and practice of pediatric oncology, 5th edn. Lippincott Williams & Wilkins, Philadelphia, pp 1033–1073
3. Kolberg M, Holand M, Agesen TH, Brekke HR, Liestol K, Hall KS, Mertens F, Picci P, Smeland S, Lothe RA (2013) Survival meta-analyses for >1800 malignant peripheral nerve sheath tumor patients with and without neurofibromatosis type 1. Neuro Oncol 15:135–147
4. deCou JM, Rao BN, Parham DM, Lobe TE, Bowman L, Pappo AS, Fontanesi J (1995) Malignant peripheral nerve sheath tumors: the St. Jude Children's Research Hospital experience. Ann Surg Oncol 2:524–529
5. Stucky CC, Johnson KN, Gray RJ, Pockaj BA, Ocal IT, Rose PS, Wasif N (2012) Malignant peripheral nerve sheath tumors (MPNST): the Mayo Clinic experience. Ann Surg Oncol 19:878–885
6. Miller SJ, Rangwala F, Williams J, Ackerman P, Kong S, Jegga AG, Kaiser S, Aronow BJ, Frahm S, Kluwe L, Mautner V, Upadhyaya M, Muir D, Wallace M, Hagen J, Quelle DE, Watson MA, Perry A, Gutmann DH, Ratner N (2006) Large-scale molecular comparison of human Schwann cells to malignant peripheral nerve sheath tumor cell lines and tissues. Cancer Res 66:2584–2591
7. Park SJ, Sawitzki B, Kluwe L, Mautner VF, Holtkamp N, Kurtz A (2013) Serum biomarkers for neurofibromatosis type 1 and early detection of malignant peripheral nerve-sheath tumors. BMC Med 11:109
8. Kang Y, Pekmezci M, Folpe AL, Ersen A, Horvai AE (2013) Diagnostic utility of SOX10 to distinguish malignant peripheral nerve sheath tumor from synovial sarcoma, including intraneural synovial sarcoma. Mod Pathol 27:55–61

9. Hummel TR, Jessen WJ, Miller SJ, Kluwe L, Mautner VF, Wallace MR, Lazaro C, Page GP, Worley PF, Aronow BJ, Schorry EK, Ratner N (2010) Gene expression analysis identifies potential biomarkers of neurofibromatosis type 1 including adrenomedullin. Clin Cancer Res 16:5048–5057
10. Miller SJ, Jessen WJ, Mehta T, Hardiman A, Sites E, Kaiser S, Jegga AG, Li H, Upadhyaya M, Giovannini M, Muir D, Wallace MR, Lopez E, Serra E, Nielsen GP, Lazaro C, Stemmer-Rachamimov A, Page G, Aronow BJ, Ratner N (2009) Integrative genomic analyses of neurofibromatosis tumours identify SOX9 as a biomarker and survival gene. EMBO Mol Med 1:236–248
11. Ferrari A, Bisogno G, Carli M (2007) Management of childhood malignant peripheral nerve sheath tumor. Paediatr Drugs 9:239–248
12. Gutmann DH, Aylsworth A, Carey JC, Korf B, Marks J, Pyeritz RE, Rubenstein A, Viskochil D (1997) The diagnostic evaluation and multidisciplinary management of neurofibromatosis 1 and neurofibromatosis 2. JAMA 278:51–57
13. Lin AL, Gutmann DH (2013) Advances in the treatment of neurofibromatosis-associated tumours. Nat Rev Clin Oncol 10:616–624
14. D'Agostino AN, Soule EH, Miller RH (1963) Primary malignant neoplasms of nerves (Malignant Neurilemomas) in patients without manifestations of multiple neurofibromatosis (Von Recklinghausen's disease). Cancer 16:1003–1014
15. D'Agostino AN, Soule EH, Miller RH (1963) Sarcomas of the peripheral nerves and somatic soft tissues associated with multiple neurofibromatosis (Von Recklinghausens Disease). Cancer 16:1015–1027
16. Ducatman BS, Scheithauer BW (1983) Postirradiation neurofibrosarcoma. Cancer 51:1028–1033
17. Ducatman BS, Scheithauer BW, Piepgras DG, Reiman HM, Ilstrup DM (1986) Malignant peripheral nerve sheath tumors. A clinicopathologic study of 120 cases. Cancer 57: 2006–2021
18. Raney B, Schnaufer L, Ziegler M, Chatten J, Littman P, Jarrett P (1987) Treatment of children with neurogenic sarcoma. Experience at the Children's Hospital of Philadelphia, 1958–1984. Cancer 59:1–5
19. Casanova M, Ferrari A, Spreafico F, Luksch R, Terenziani M, Cefalo G, Massimino M, Gandola L, Lombardi F, Fossati-Bellani F (1999) Malignant peripheral nerve sheath tumors in children: a single-institution twenty-year experience. J Pediatr Hematol Oncol 21: 509–513
20. Carli M, Ferrari A, Mattke A, Zanetti I, Casanova M, Bisogno G, Cecchetto G, Alaggio R, De Sio L, Koscielniak E, Sotti G, Treuner J (2005) Pediatric malignant peripheral nerve sheath tumor: the Italian and German soft tissue sarcoma cooperative group. J Clin Oncol 23:8422–8430
21. Ferrari A, Bisogno G, Macaluso A, Casanova M, D'Angelo P, Pierani P, Zanetti I, Alaggio R, Cecchetto G, Carli M (2007) Soft-tissue sarcomas in children and adolescents with neurofibromatosis type 1. Cancer 109:1406–1412
22. Furniss D, Swan MC, Morritt DG, Lim J, Khanna T, Way BL, Athanasou NA, Giele H, Critchley P (2008) A 10-year review of benign and malignant peripheral nerve sheath tumors in a single center: clinical and radiographic features can help to differentiate benign from malignant lesions. Plast Reconstr Surg 121:529–533
23. Ferrari A, Miceli R, Rey A, Oberlin O, Orbach D, Brennan B, Mariani L, Carli M, Bisogno G, Cecchetto G, De Salvo GL, Casanova M, Vannoesel MM, Kelsey A, Stevens MC, Devidas M, Pappo AS, Spunt SL (2011) Non-metastatic unresected paediatric non-rhabdomyosarcoma soft tissue sarcomas: results of a pooled analysis from United States and European groups. Eur J Cancer 47:724–731
24. Zehou O, Fabre E, Zelek L, Sbidian E, Ortonne N, Banu E, Wolkenstein P, Valeyrie-Allanore L (2013) Chemotherapy for the treatment of malignant peripheral nerve sheath tumors in neurofibromatosis 1: a 10-year institutional review. Orphanet J Rare Dis 8:127

25. Fan Q, Yang J, Wang G (2014) Clinical and molecular prognostic predictors of malignant peripheral nerve sheath tumor. Clin Transl Oncol 16:191–199
26. Bates JE, Peterson CR, Dhakal S, Giampoli EJ, Constine LS (2014) Malignant peripheral nerve sheath tumors (MPNST): a SEER analysis of incidence across the age spectrum and therapeutic interventions in the pediatric population. Pediatr Blood Cancer 61:1955–1960
27. Skotheim RI, Kallioniemi A, Bjerkhagen B, Mertens F, Brekke HR, Monni O, Mousses S, Mandahl N, Soeter G, Nesland JM, Smeland S, Kallioniemi OP, Lothe RA (2003) Topoisomerase-II alpha is upregulated in malignant peripheral nerve sheath tumors and associated with clinical outcome. J Clin Oncol 21:4586–4591
28. Widemann BC, Salzer WL, Arceci RJ, Blaney SM, Fox E, End D, Gillespie A, Whitcomb P, Palumbo JS, Pitney A, Jayaprakash N, Zannikos P, Balis FM (2006) Phase I trial and pharmacokinetic study of the farnesyltransferase inhibitor tipifarnib in children with refractory solid tumors or neurofibromatosis type I and plexiform neurofibromas. J Clin Oncol 24:507–516
29. Widemann BC, Acosta MT, Ammoun S, Belzberg AJ, Bernards A, Blakeley J, Bretscher A, Cichowski K, Clapp DW, Dombi E, Evans GD, Ferner R, Fernandez-Valle C, Fisher MJ, Giovannini M, Gutmann DH, Hanemann CO, Hennigan R, Huson S, Ingram D, Kissil J, Korf BR, Legius E, Packer RJ, McClatchey AI, McCormick F, North K, Pehrsson M, Plotkin SR, Ramesh V, Ratner N, Schirmer S, Sherman L, Schorry E, Stevenson D, Stewart DR, Ullrich N, Bakker AC, Morrison H (2012) CTF meeting 2012: translation of the basic understanding of the biology and genetics of NF1, NF2, and schwannomatosis toward the development of effective therapies. Am J Med Genet A 164A:563–578
30. Kalamarides M, Acosta MT, Babovic-Vuksanovic D, Carpen O, Cichowski K, Evans DG, Giancotti F, Hanemann CO, Ingram D, Lloyd AC, Mayes DA, Messiaen L, Morrison H, North K, Packer R, Pan D, Stemmer-Rachamimov A, Upadhyaya M, Viskochil D, Wallace MR, Hunter-Schaedle K, Ratner N (2012) Neurofibromatosis 2011: a report of the Children's Tumor Foundation annual meeting. Acta Neuropathol 123:369–380
31. Yamashita AS, Baia GS, Ho JS, Velarde E, Wong J, Gallia GL, Belzberg AJ, Kimura ET, Riggins GJ (2014) Preclinical evaluation of the combination of mTOR and proteasome inhibitors with radiotherapy in malignant peripheral nerve sheath tumors. J Neurooncol 118: 83–92
32. Karajannis MA, Legault G, Hagiwara M, Giancotti FG, Filatov A, Derman A, Hochman T, Goldberg JD, Vega E, Wisoff JH, Golfinos JG, Merkelson A, Roland JT, Allen JC (2014) Phase II study of everolimus in children and adults with neurofibromatosis type 2 and progressive vestibular schwannomas. Neuro Oncol 16:292–297
33. Holtkamp N, Okuducu AF, Mucha J, Afanasieva A, Hartmann C, Atallah I, Estevez-Schwarz L, Mawrin C, Friedrich RE, Mautner VF, von Deimling A (2006) Mutation and expression of PDGFRA and KIT in malignant peripheral nerve sheath tumors, and its implications for imatinib sensitivity. Carcinogenesis 27:664–671
34. Karajannis MA, Legault G, Hagiwara M, Ballas MS, Brown K, Nusbaum AO, Hochman T, Goldberg JD, Koch KM, Golfinos JG, Roland JT, Allen JC (2012) Phase II trial of lapatinib in adult and pediatric patients with neurofibromatosis type 2 and progressive vestibular schwannomas. Neuro Oncol 14:1163–1170
35. Plotkin SR, Stemmer-Rachamimov AO, Barker FG 2nd, Halpin C, Padera TP, Tyrrell A, Sorensen AG, Jain RK, di Tomaso E (2009) Hearing improvement after bevacizumab in patients with neurofibromatosis type 2. N Engl J Med 361:358–367
36. Plotkin SR, Bredella MA, Cai W, Kassarjian A, Harris GJ, Esparza S, Merker VL, Munn LL, Muzikansky A, Askenazi M, Nguyen R, Wenzel R, Mautner VF (2012) Quantitative assessment of whole-body tumor burden in adult patients with neurofibromatosis. PLoS One 7, e35711
37. Brossier NM, Carroll SL (2008) Genetically engineered mouse models shed new light on the pathogenesis of neurofibromatosis type I-related neoplasms of the peripheral nervous system. Brain Res Bull 88:58–71

38. Legius E, Marchuk DA, Collins FS, Glover TW (1993) Somatic deletion of the neurofibromatosis type 1 gene in a neurofibrosarcoma supports a tumour suppressor gene hypothesis. Nat Genet 3:122–126
39. Fountain JW, Wallace MR, Bruce MA, Seizinger BR, Menon AG, Gusella JF, Michels VV, Schmidt MA, Dewald GW, Collins FS (1989) Physical mapping of a translocation breakpoint in neurofibromatosis. Science 244:1085–1087
40. Seizinger BR, Martuza RL, Gusella JF (1986) Loss of genes on chromosome 22 in tumorigenesis of human acoustic neuroma. Nature 322:644–647
41. DeClue JE, Papageorge AG, Fletcher JA, Diehl SR, Ratner N, Vass WC, Lowy DR (1992) Abnormal regulation of mammalian p21ras contributes to malignant tumor growth in von Recklinghausen (type 1) neurofibromatosis. Cell 69:265–273
42. Rubin JB, Gutmann DH (2005) Neurofibromatosis type 1—a model for nervous system tumour formation? Nat Rev Cancer 5:557–564
43. Barkan B, Starinsky S, Friedman E, Stein R, Kloog Y (2006) The RAS inhibitor farnesylthiosalicylic acid as a potential therapy for neurofibromatosis type 1. Clin Cancer Res 12:5533–5542
44. Jessen WJ, Miller SJ, Jousma E, Wu J, Rizvi TA, Brundage ME, Eaves D, Widemann B, Kim MO, Dombi E, Sabo J, Hardiman Dudley A, Niwa-Kawakita M, Page GP, Giovannini M, Aronow BJ, Cripe TP, Ratner N (2013) MEK inhibition exhibits efficacy in human and mouse neurofibromatosis tumors. J Clin Invest 123:340–347
45. Watson AL, Anderson LK, Greeley AD, Keng VW, Rahrmann EP, Halfond AL, Powell NM, Collins MH, Rizvi T, Moertel CL, Ratner N, Largaespada DA (2014) Co-targeting the MAPK and PI3K/AKT/mTOR pathways in two genetically engineered mouse models of Schwann cell tumors reduces tumor grade and multiplicity. Oncotarget 5:1502–1514
46. Bodempudi V, Yamoutpoor F, Pan W, Dudek AZ, Esfandyari T, Piedra M, Babovick-Vuksanovic D, Woo RA, Mautner VF, Kluwe L, Clapp DW, De Vries GH, Thomas SL, Kurtz A, Parada LF, Farassati F (2009) Ral overactivation in malignant peripheral nerve sheath tumors. Mol Cell Biol 29:3964–3974
47. Gregorian C, Nakashima J, Dry SM, Nghiemphu PL, Smith KB, Ao Y, Dang J, Lawson G, Mellinghoff IK, Mischel PS, Phelps M, Parada LF, Liu X, Sofroniew MV, Eilber FC, Wu H (2009) PTEN dosage is essential for neurofibroma development and malignant transformation. Proc Natl Acad Sci U S A 106:19479–19484
48. Perrone F, Da Riva L, Orsenigo M, Losa M, Jocolle G, Millefanti C, Pastore E, Gronchi A, Pierotti MA, Pilotti S (2009) PDGFRA, PDGFRB, EGFR, and downstream signaling activation in malignant peripheral nerve sheath tumor. Neuro Oncol 11:725–736
49. Bradtmoller M, Hartmann C, Zietsch J, Jaschke S, Mautner VF, Kurtz A, Park SJ, Baier M, Harder A, Reuss D, von Deimling A, Heppner FL, Holtkamp N (2012) Impaired Pten expression in human malignant peripheral nerve sheath tumours. PLoS One 7, e47595
50. Sun D, Tainsky MA, Haddad R (2012) Oncogene mutation survey in MPNST cell lines enhances the dominant role of hyperactive RAS in NF1 associated pro-survival and malignancy. Transl Oncogenomics 5:1–7
51. Mo W, Chen J, Patel A, Zhang L, Chau V, Li Y, Cho W, Lim K, Xu J, Lazar AJ, Creighton CJ, Bolshakov S, McKay RM, Lev D, Le LQ, Parada LF (2013) CXCR4/CXCL12 mediate autocrine cell-cycle progression in NF1-associated malignant peripheral nerve sheath tumors. Cell 152:1077–1090
52. Largaespada D, Ratner N (2013) Interweaving the strands: beta-catenin, an HIV co-receptor, and Schwann cell tumors. Cancer Cell 23:269–271
53. Yang J, Ylipaa A, Sun Y, Zheng H, Chen K, Nykter M, Trent J, Ratner N, Lev DC, Zhang W (2011) Genomic and molecular characterization of malignant peripheral nerve sheath tumor identifies the IGF1R pathway as a primary target for treatment. Clin Cancer Res 17:7563–7573
54. Gesundheit B, Parkin P, Greenberg M, Baruchel S, Senger C, Kapelushnik J, Smith C, Klement GL (2010) The role of angiogenesis in the transformation of plexiform neurofibroma

into malignant peripheral nerve sheath tumors in children with neurofibromatosis type 1. J Pediatr Hematol Oncol 32:548–553
55. Friedrich RE, Beer C, Glatzel M, Hagel C (2015) Vascular endothelial growth factor, basic fibroblast growth factor and epithelial growth factor receptor in peripheral nerve sheath tumors of neurofibromatosis type 1. Anticancer Res 35:137–144
56. Holtkamp N, Malzer E, Zietsch J, Okuducu AF, Mucha J, Mawrin C, Mautner VF, Schildhaus HU, von Deimling A (2008) EGFR and erbB2 in malignant peripheral nerve sheath tumors and implications for targeted therapy. Neuro Oncol 10:946–957
57. Wu J, Patmore DM, Jousma E, Eaves DW, Breving K, Patel AV, Schwartz EB, Fuchs JR, Cripe TP, Stemmer-Rachamimov AO, Ratner N (2014) EGFR-STAT3 signaling promotes formation of malignant peripheral nerve sheath tumors. Oncogene 33:173–180
58. Ryan JJ, Klein KA, Neuberger TJ, Leftwich JA, Westin EH, Kauma S, Fletcher JA, DeVries GH, Huff TF (1994) Role for the stem cell factor/KIT complex in Schwann cell neoplasia and mast cell proliferation associated with neurofibromatosis. J Neurosci Res 37:415–432
59. Dang I, Nelson JK, DeVries GH (2005) c-Kit receptor expression in normal human Schwann cells and Schwann cell lines derived from neurofibromatosis type 1 tumors. J Neurosci Res 82:465–471
60. Mukherjee J, Kamnasaran D, Balasubramaniam A, Radovanovic I, Zadeh G, Kiehl TR, Guha A (2009) Human schwannomas express activated platelet-derived growth factor receptors and c-kit and are growth inhibited by Gleevec (Imatinib Mesylate). Cancer Res 69: 5099–5107
61. Li H, Velasco-Miguel S, Vass WC, Parada LF, DeClue JE (2002) Epidermal growth factor receptor signaling pathways are associated with tumorigenesis in the Nf1:p53 mouse tumor model. Cancer Res 62:4507–4513
62. Stonecypher MS, Byer SJ, Grizzle WE, Carroll SL (2005) Activation of the neuregulin-1/ErbB signaling pathway promotes the proliferation of neoplastic Schwann cells in human malignant peripheral nerve sheath tumors. Oncogene 24:5589–5605
63. Rasmussen SA, Overman J, Thomson SA, Colman SD, Abernathy CR, Trimpert RE, Moose R, Virdi G, Roux K, Bauer M, Rojiani AM, Maria BL, Muir D, Wallace MR (2000) Chromosome 17 loss-of-heterozygosity studies in benign and malignant tumors in neurofibromatosis type 1. Genes Chromosomes Cancer 28:425–431
64. Brannan CI, Perkins AS, Vogel KS, Ratner N, Nordlund ML, Reid SW, Buchberg AM, Jenkins NA, Parada LF, Copeland NG (1994) Targeted disruption of the neurofibromatosis type-1 gene leads to developmental abnormalities in heart and various neural crest-derived tissues. Genes Dev 8:1019–1029
65. Knudson AG Jr (1971) Mutation and cancer: statistical study of retinoblastoma. Proc Natl Acad Sci U S A 68:820–823
66. Carroll SL, Ratner N (2008) How does the Schwann cell lineage form tumors in NF1? Glia 56:1590–1605
67. Watson MA, Perry A, Tihan T, Prayson RA, Guha A, Bridge J, Ferner R, Gutmann DH (2004) Gene expression profiling reveals unique molecular subtypes of Neurofibromatosis Type I-associated and sporadic malignant peripheral nerve sheath tumors. Brain Pathol 14:297–303
68. Spurlock G, Knight SJ, Thomas N, Kiehl TR, Guha A, Upadhyaya M (2010) Molecular evolution of a neurofibroma to malignant peripheral nerve sheath tumor (MPNST) in an NF1 patient: correlation between histopathological, clinical and molecular findings. J Cancer Res Clin Oncol 136:1869–1880
69. Krause MP, Dworski S, Feinberg K, Jones K, Johnston AP, Paul S, Paris M, Peles E, Bagli D, Forrest CR, Kaplan DR, Miller FD (2014) Direct genesis of functional rodent and human schwann cells from skin mesenchymal precursors. Stem Cell Reports 3:85–100
70. Le LQ, Shipman T, Burns DK, Parada LF (2009) Cell of origin and microenvironment contribution for NF1-associated dermal neurofibromas. Cell Stem Cell 4:453–463
71. Chau V, Lim SK, Mo W, Liu C, Patel AJ, McKay RM, Wei S, Posner BA, De Brabander JK, Williams NS, Parada LF, Le LQ (2014) Preclinical therapeutic efficacy of a novel

pharmacologic inducer of apoptosis in malignant peripheral nerve sheath tumors. Cancer Res 74:586–597
72. Zhu Y, Ghosh P, Charnay P, Burns DK, Parada LF (2002) Neurofibromas in NF1: Schwann cell origin and role of tumor environment. Science 296:920–922
73. Jankowski K, Kucia M, Wysoczynski M, Reca R, Zhao D, Trzyna E, Trent J, Peiper S, Zembala M, Ratajczak J, Houghton P, Janowska-Wieczorek A, Ratajczak MZ (2003) Both hepatocyte growth factor (HGF) and stromal-derived factor-1 regulate the metastatic behavior of human rhabdomyosarcoma cells, but only HGF enhances their resistance to radiochemotherapy. Cancer Res 63:7926–7935
74. Strahm B, Durbin AD, Sexsmith E, Malkin D (2008) The CXCR4-SDF1alpha axis is a critical mediator of rhabdomyosarcoma metastatic signaling induced by bone marrow stroma. Clin Exp Metastasis 25:1–10
75. De Raedt T, Brems H, Wolkenstein P, Vidaud D, Pilotti S, Perrone F, Mautner V, Frahm S, Sciot R, Legius E (2003) Elevated risk for MPNST in NF1 microdeletion patients. Am J Hum Genet 72:1288–1292
76. Perry A, Roth KA, Banerjee R, Fuller CE, Gutmann DH (2001) NF1 deletions in S-100 protein-positive and negative cells of sporadic and neurofibromatosis 1 (NF1)-associated plexiform neurofibromas and malignant peripheral nerve sheath tumors. Am J Pathol 159:57–61
77. Bottillo I, Ahlquist T, Brekke H, Danielsen SA, van den Berg E, Mertens F, Lothe RA, Dallapiccola B (2009) Germline and somatic NF1 mutations in sporadic and NF1-associated malignant peripheral nerve sheath tumours. J Pathol 217:693–701
78. Vogel KS, Klesse LJ, Velasco-Miguel S, Meyers K, Rushing EJ, Parada LF (1999) Mouse tumor model for neurofibromatosis type 1. Science 286:2176–2179
79. Cichowski K, Shih TS, Schmitt E, Santiago S, Reilly K, McLaughlin ME, Bronson RT, Jacks T (1999) Mouse models of tumor development in neurofibromatosis type 1. Science 286:2172–2176
80. Donehower LA, Harvey M, Slagle BL, McArthur MJ, Montgomery CA Jr, Butel JS, Bradley A (1992) Mice deficient for p53 are developmentally normal but susceptible to spontaneous tumours. Nature 356:215–221
81. Berghmans S, Murphey RD, Wienholds E, Neuberg D, Kutok JL, Fletcher CD, Morris JP, Liu TX, Schulte-Merker S, Kanki JP, Plasterk R, Zon LI, Look AT (2005) tp53 mutant zebrafish develop malignant peripheral nerve sheath tumors. Proc Natl Acad Sci U S A 102:407–412
82. Legius E, Dierick H, Wu R, Hall BK, Marynen P, Cassiman JJ, Glover TW (1994) TP53 mutations are frequent in malignant NF1 tumors. Genes Chromosomes Cancer 10:250–255
83. Kourea HP, Cordon-Cardo C, Dudas M, Leung D, Woodruff JM (1999) Expression of p27(kip) and other cell cycle regulators in malignant peripheral nerve sheath tumors and neurofibromas: the emerging role of p27(kip) in malignant transformation of neurofibromas. Am J Pathol 155:1885–1891
84. Lothe RA, Smith-Sorensen B, Hektoen M, Stenwig AE, Mandahl N, Saeter G, Mertens F (2001) Biallelic inactivation of TP53 rarely contributes to the development of malignant peripheral nerve sheath tumors. Genes Chromosomes Cancer 30:202–206
85. Birindelli S, Perrone F, Oggionni M, Lavarino C, Pasini B, Vergani B, Ranzani GN, Pierotti MA, Pilotti S (2001) Rb and TP53 pathway alterations in sporadic and NF1-related malignant peripheral nerve sheath tumors. Lab Invest 81:833–844
86. Holtkamp N, Atallah I, Okuducu AF, Mucha J, Hartmann C, Mautner VF, Friedrich RE, Mawrin C, von Deimling A (2007) MMP-13 and p53 in the progression of malignant peripheral nerve sheath tumors. Neoplasia 9:671–677
87. Endo M, Kobayashi C, Setsu N, Takahashi Y, Kohashi K, Yamamoto H, Tamiya S, Matsuda S, Iwamoto Y, Tsuneyoshi M, Oda Y (2011) Prognostic significance of p14ARF, p15INK4b, and p16INK4a inactivation in malignant peripheral nerve sheath tumors. Clin Cancer Res 17:3771–3782

88. Berner JM, Sorlie T, Mertens F, Henriksen J, Saeter G, Mandahl N, Brogger A, Myklebost O, Lothe RA (1999) Chromosome band 9p21 is frequently altered in malignant peripheral nerve sheath tumors: studies of CDKN2A and other genes of the pRB pathway. Genes Chromosomes Cancer 26:151–160
89. Kourea HP, Orlow I, Scheithauer BW, Cordon-Cardo C, Woodruff JM (1999) Deletions of the INK4A gene occur in malignant peripheral nerve sheath tumors but not in neurofibromas. Am J Pathol 155:1855–1860
90. Mantripragada KK, Diaz de Stahl T, Patridge C, Menzel U, Andersson R, Chuzhanova N, Kluwe L, Guha A, Mautner V, Dumanski JP, Upadhyaya M (2009) Genome-wide high-resolution analysis of DNA copy number alterations in NF1-associated malignant peripheral nerve sheath tumors using 32K BAC array. Genes Chromosomes Cancer 48:897–907
91. Holtkamp N, Mautner VF, Friedrich RE, Harder A, Hartmann C, Theallier-Janko A, Hoffmann KT, von Deimling A (2004) Differentially expressed genes in neurofibromatosis 1-associated neurofibromas and malignant peripheral nerve sheath tumors. Acta Neuropathol 107:159–168
92. Upadhyaya M, Spurlock G, Thomas L, Thomas NS, Richards M, Mautner VF, Cooper DN, Guha A, Yan J (2012) Microarray-based copy number analysis of neurofibromatosis type-1 (NF1)-associated malignant peripheral nerve sheath tumors reveals a role for Rho-GTPase pathway genes in NF1 tumorigenesis. Hum Mutat 33:763–776
93. Ohishi J, Aoki M, Nabeshima K, Suzumiya J, Takeuchi T, Ogose A, Hakozaki M, Yamashita Y, Iwasaki H (2013) Imatinib mesylate inhibits cell growth of malignant peripheral nerve sheath tumors in vitro and in vivo through suppression of PDGFR-beta. BMC Cancer 13:224
94. Tabone-Eglinger S, Bahleda R, Cote JF, Terrier P, Vidaud D, Cayre A, Beauchet A, Theou-Anton N, Terrier-Lacombe MJ, Lemoine A, Penault-Llorca F, Le Cesne A, Emile JF (2008) Frequent EGFR positivity and overexpression in high-grade areas of human MPNSTs. Sarcoma 2008:849156
95. Rahrmann EP, Moriarity BS, Otto GM, Watson AL, Choi K, Collins MH, Wallace M, Webber BR, Forster CL, Rizzardi AE, Schmechel SC, Ratner N, Largaespada DA (2014) Trp53 haploinsufficiency modifies EGFR-driven peripheral nerve sheath tumorigenesis. Am J Pathol 184:2082–2098
96. Hirbe AC, Pekmezci M, Dahiya S, Apicelli AJ, Van Tine BA, Perry A, Gutmann DH (2014) BRAFV600E mutation in sporadic and neurofibromatosis type 1-related malignant peripheral nerve sheath tumors. Neuro Oncol 16:466–467
97. Beert E, Brems H, Daniels B, De Wever I, Van Calenbergh F, Schoenaers J, Debiec-Rychter M, Gevaert O, De Raedt T, Van Den Bruel A, de Ravel T, Cichowski K, Kluwe L, Mautner V, Sciot R, Legius E (2011) Atypical neurofibromas in neurofibromatosis type 1 are premalignant tumors. Genes Chromosomes Cancer 50:1021–1032
98. Holtkamp N, Reuss DE, Atallah I, Kuban RJ, Hartmann C, Mautner VF, Frahm S, Friedrich RE, Algermissen B, Pham VA, Prietz S, Rosenbaum T, Estevez-Schwarz L, von Deimling A (2004) Subclassification of nerve sheath tumors by gene expression profiling. Brain Pathol 14:258–264
99. Rahrmann EP, Watson AL, Keng VW, Choi K, Moriarity BS, Beckmann DA, Wolf NK, Sarver A, Collins MH, Moertel CL, Wallace MR, Gel B, Serra E, Ratner N, Largaespada DA (2013) Forward genetic screen for malignant peripheral nerve sheath tumor formation identifies new genes and pathways driving tumorigenesis. Nat Genet 45:756–766
100. Watson AL, Rahrmann EP, Moriarity BS, Choi K, Conboy CB, Greeley AD, Halfond AL, Anderson LK, Wahl BR, Keng VW, Rizzardi AE, Forster CL, Collins MH, Sarver AL, Wallace MR, Schmechel SC, Ratner N, Largaespada DA (2013) Canonical Wnt/beta-catenin signaling drives human schwann cell transformation, progression, and tumor maintenance. Cancer Discov 3:674–689
101. Luscan A, Shackleford G, Masliah-Planchon J, Laurendeau I, Ortonne N, Varin J, Lallemand F, Leroy K, Dumaine V, Hivelin M, Borderie D, De Raedt T, Valeyrie-Allanore L, Larousserie F, Terris B, Lantieri L, Vidaud M, Vidaud D, Wolkenstein P, Parfait B, Bieche I, Massaad C,

Pasmant E (2014) The activation of the WNT signaling pathway is a Hallmark in neurofibromatosis type 1 tumorigenesis. Clin Cancer Res 20:358–371
102. De Raedt T, Beert E, Pasmant E, Luscan A, Brems H, Ortonne N, Helin K, Hornick JL, Mautner V, Kehrer-Sawatzki H, Clapp W, Bradner J, Vidaud M, Upadhyaya M, Legius E, Cichowski K (2014) PRC2 loss amplifies Ras-driven transcription and confers sensitivity to BRD4-based therapies. Nature 514:247–251
103. Kindler-Rohrborn A, Kolsch BU, Fischer C, Held S, Rajewsky MF (1999) Ethylnitrosourea-induced development of malignant schwannomas in the rat: two distinct loci on chromosome of 10 involved in tumor susceptibility and oncogenesis. Cancer Res 59:1109–1114
104. Kindler-Rohrborn A, Kind AB, Koelsch BU, Fischer C, Rajewsky MF (2000) Suppression of ethylnitrosourea-induced schwannoma development involves elimination of neu/erbB-2 mutant premalignant cells in the resistant BDIV rat strain. Cancer Res 60:4756–4760
105. Joseph NM, Mosher JT, Buchstaller J, Snider P, McKeever PE, Lim M, Conway SJ, Parada LF, Zhu Y, Morrison SJ (2008) The loss of Nf1 transiently promotes self-renewal but not tumorigenesis by neural crest stem cells. Cancer Cell 13:129–140
106. Dodd RD, Mito JK, Eward WC, Chitalia R, Sachdeva M, Ma Y, Barretina J, Dodd L, Kirsch DG (2013) NF1 deletion generates multiple subtypes of soft-tissue sarcoma that respond to MEK inhibition. Mol Cancer Ther 12:1906–1917
107. Zhu Y, Guignard F, Zhao D, Liu L, Burns DK, Mason RP, Messing A, Parada LF (2005) Early inactivation of p53 tumor suppressor gene cooperating with NF1 loss induces malignant astrocytoma. Cancer Cell 8:119–130
108. Keng VW, Rahrmann EP, Watson AL, Tschida BR, Moertel CL, Jessen WJ, Rizvi TA, Collins MH, Ratner N, Largaespada DA (2012) PTEN and NF1 inactivation in Schwann cells produces a severe phenotype in the peripheral nervous system that promotes the development and malignant progression of peripheral nerve sheath tumors. Cancer Res 72:3405–3413
109. Patel AV, Eaves D, Jessen WJ, Rizvi TA, Ecsedy JA, Qian MG, Aronow BJ, Perentesis JP, Serra E, Cripe TP, Miller SJ, Ratner N (2012) Ras-driven transcriptome analysis identifies aurora kinase A as a potential malignant peripheral nerve sheath tumor therapeutic target. Clin Cancer Res 18:5020–5030
110. Keng VW, Watson AL, Rahrmann EP, Li H, Tschida BR, Moriarity BS, Choi K, Rizvi TA, Collins MH, Wallace MR, Ratner N, Largaespada DA (2012) Conditional inactivation of Pten with EGFR overexpression in Schwann cells models sporadic MPNST. Sarcoma 2012:620834
111. Jacks T, Shih TS, Schmitt EM, Bronson RT, Bernards A, Weinberg RA (1994) Tumour predisposition in mice heterozygous for a targeted mutation in Nf1. Nat Genet 7:353–361
112. Shin J, Padmanabhan A, de Groh ED, Lee JS, Haidar S, Dahlberg S, Guo F, He S, Wolman MA, Granato M, Lawson ND, Wolfe SA, Kim SH, Solnica-Krezel L, Kanki JP, Ligon KL, Epstein JA, Look AT (2012) Zebrafish neurofibromatosis type 1 genes have redundant functions in tumorigenesis and embryonic development. Dis Model Mech 5:881–894
113. Bajenaru ML, Hernandez MR, Perry A, Zhu Y, Parada LF, Garbow JR, Gutmann DH (2003) Optic nerve glioma in mice requires astrocyte Nf1 gene inactivation and Nf1 brain heterozygosity. Cancer Res 63:8573–8577
114. Zheng H, Chang L, Patel N, Yang J, Lowe L, Burns DK, Zhu Y (2008) Induction of abnormal proliferation by nonmyelinating schwann cells triggers neurofibroma formation. Cancer Cell 13:117–128
115. Ling BC, Wu J, Miller SJ, Monk KR, Shamekh R, Rizvi TA, Decourten-Myers G, Vogel KS, DeClue JE, Ratner N (2005) Role for the epidermal growth factor receptor in neurofibromatosis-related peripheral nerve tumorigenesis. Cancer Cell 7:65–75
116. Miller SJ, Li H, Rizvi TA, Huang Y, Johansson G, Bowersock J, Sidani A, Vitullo J, Vogel K, Parysek LM, DeClue JE, Ratner N (2003) Brain lipid binding protein in axon-Schwann cell interactions and peripheral nerve tumorigenesis. Mol Cell Biol 23:2213–2224
117. Lesche R, Groszer M, Gao J, Wang Y, Messing A, Sun H, Liu X, Wu H (2002) Cre/loxP-mediated inactivation of the murine Pten tumor suppressor gene. Genesis 32:148–149

118. DeClue JE, Heffelfinger S, Benvenuto G, Ling B, Li S, Rui W, Vass WC, Viskochil D, Ratner N (2000) Epidermal growth factor receptor expression in neurofibromatosis type 1-related tumors and NF1 animal models. J Clin Invest 105:1233–1241
119. Marino F, Lanteri G, Rapisarda G, Perillo A, Macri B (2012) Spontaneous schwannoma in zebrafish, Danio rerio (Hamilton). J Fish Dis 35:239–242
120. Amsterdam A, Sadler KC, Lai K, Farrington S, Bronson RT, Lees JA, Hopkins N (2004) Many ribosomal protein genes are cancer genes in zebrafish. PLoS Biol 2, E139
121. MacInnes AW, Amsterdam A, Whittaker CA, Hopkins N, Lees JA (2008) Loss of p53 synthesis in zebrafish tumors with ribosomal protein gene mutations. Proc Natl Acad Sci U S A 105:10408–10413
122. Feitsma H, Kuiper RV, Korving J, Nijman IJ, Cuppen E (2008) Zebrafish with mutations in mismatch repair genes develop neurofibromas and other tumors. Cancer Res 68:5059–5066
123. Gutierrez A, Feng H, Stevenson K, Neuberg DS, Calzada O, Zhou Y, Langenau DM, Look AT (2014) Loss of function tp53 mutations do not accelerate the onset of myc-induced T-cell acute lymphoblastic leukaemia in the zebrafish. Br J Haematol 166:84–90
124. Zhang G, Hoersch S, Amsterdam A, Whittaker CA, Beert E, Catchen JM, Farrington S, Postlethwait JH, Legius E, Hopkins N, Lees JA (2013) Comparative oncogenomic analysis of copy number alterations in human and zebrafish tumors enables cancer driver discovery. PLoS Genet 9, e1003734
125. Negrini S, Gorgoulis VG, Halazonetis TD (2010) Genomic instability–an evolving hallmark of cancer. Nat Rev Mol Cell Biol 11:220–228
126. Jiricny J (2006) The multifaceted mismatch-repair system. Nat Rev Mol Cell Biol 7:335–346
127. Feitsma H, de Bruijn E, van de Belt J, Nijman IJ, Cuppen E (2008) Mismatch repair deficiency does not enhance ENU mutagenesis in the zebrafish germ line. Mutagenesis 23: 325–329
128. Zhou L, Ercolano E, Ammoun S, Schmid MC, Barczyk MA, Hanemann CO (2011) Merlin-deficient human tumors show loss of contact inhibition and activation of Wnt/beta-catenin signaling linked to the PDGFR/Src and Rac/PAK pathways. Neoplasia 13:1101–1112
129. Johansson G, Mahller YY, Collins MH, Kim MO, Nobukuni T, Perentesis J, Cripe TP, Lane HA, Kozma SC, Thomas G, Ratner N (2008) Effective in vivo targeting of the mammalian target of rapamycin pathway in malignant peripheral nerve sheath tumors. Mol Cancer Ther 7:1237–1245
130. Farassati F, Pan W, Yamoutpour F, Henke S, Piedra M, Frahm S, Al-Tawil S, Mangrum WI, Parada LF, Rabkin SD, Martuza RL, Kurtz A (2008) RAS signaling influences permissiveness of malignant peripheral nerve sheath tumor cells to oncolytic herpes. Am J Pathol 173:1861–1872
131. Mahller YY, Rangwala F, Ratner N, Cripe TP (2006) Malignant peripheral nerve sheath tumors with high and low Ras-GTP are permissive for oncolytic herpes simplex virus mutants. Pediatr Blood Cancer 46:745–754
132. Mahller YY, Vaikunth SS, Currier MA, Miller SJ, Ripberger MC, Hsu YH, Mehrian-Shai R, Collins MH, Crombleholme TM, Ratner N, Cripe TP (2007) Oncolytic HSV and erlotinib inhibit tumor growth and angiogenesis in a novel malignant peripheral nerve sheath tumor xenograft model. Mol Ther 15:279–286
133. Liu S, Knapp S, Ahmed AA (2014) The structural basis of PI3K cancer mutations: from mechanism to therapy. Cancer Res 74:641–646
134. De Raedt T, Walton Z, Yecies JL, Li D, Chen Y, Malone CF, Maertens O, Jeong SM, Bronson RT, Lebleu V, Kalluri R, Normant E, Haigis MC, Manning BD, Wong KK, Macleod KF, Cichowski K (2011) Exploiting cancer cell vulnerabilities to develop a combination therapy for ras-driven tumors. Cancer Cell 20:400–413
135. Borrego-Diaz E, Terai K, Lialyte K, Wise AL, Esfandyari T, Behbod F, Mautner VF, Spyra M, Taylor S, Parada LF, Upadhyaya M, Farassati F (2012) Overactivation of RAS signaling pathway in CD133+ MPNST cells. J Neurooncol 108:423–434

136. Adamowicz M, Radlwimmer B, Rieker RJ, Mertens D, Schwarzbach M, Schraml P, Benner A, Lichter P, Mechtersheimer G, Joos S (2006) Frequent amplifications and abundant expression of TRIO, NKD2, and IRX2 in soft tissue sarcomas. Genes Chromosomes Cancer 45:829–838
137. MacCollin M, Mohney T, Trofatter J, Wertelecki W, Ramesh V, Gusella J (1993) DNA diagnosis of neurofibromatosis 2. Altered coding sequence of the merlin tumor suppressor in an extended pedigree. JAMA 270:2316–2320
138. Buchstaller J, McKeever PE, Morrison SJ (2012) Tumorigenic cells are common in mouse MPNSTs but their frequency depends upon tumor genotype and assay conditions. Cancer Cell 21:240–252
139. Miller SJ, Lan ZD, Hardiman A, Wu J, Kordich JJ, Patmore DM, Hegde RS, Cripe TP, Cancelas JA, Collins MH, Ratner N (2010) Inhibition of Eyes Absent Homolog 4 expression induces malignant peripheral nerve sheath tumor necrosis. Oncogene 29:368–379
140. Brundage ME, Tandon P, Eaves DW, Williams JP, Miller SJ, Hennigan RH, Jegga A, Cripe TP, Ratner N (2014) MAF mediates crosstalk between Ras-MAPK and mTOR signaling in NF1. Oncogene 33:5626–5636
141. Muthuswamy SK (2006) ErbB2 makes beta 4 integrin an accomplice in tumorigenesis. Cell 126:443–445
142. Zhang G, Hoersch S, Amsterdam A, Whittaker CA, Lees JA, Hopkins N (2010) Highly aneuploid zebrafish malignant peripheral nerve sheath tumors have genetic alterations similar to human cancers. Proc Natl Acad Sci U S A 107:16940–16945
143. Ignatius MS, Chen E, Elpek NM, Fuller AZ, Tenente IM, Clagg R, Liu S, Blackburn JS, Linardic CM, Rosenberg AE, Nielsen PG, Mempel TR, Langenau DM (2012) In vivo imaging of tumor-propagating cells, regional tumor heterogeneity, and dynamic cell movements in embryonal rhabdomyosarcoma. Cancer Cell 21:680–693
144. Eguiara A, Holgado O, Beloqui I, Abalde L, Sanchez Y, Callol C, Martin AG (2011) Xenografts in zebrafish embryos as a rapid functional assay for breast cancer stem-like cell identification. Cell Cycle 10:3751–3757
145. Dovey MC, Zon LI (2009) Defining cancer stem cells by xenotransplantation in zebrafish. Methods Mol Biol 568:1–5
146. Peng Y, Clark KJ, Campbell JM, Panetta MR, Guo Y, Ekker SC (2014) Making designer mutants in model organisms. Development 141:4042–4054
147. Hwang WY, Fu Y, Reyon D, Maeder ML, Tsai SQ, Sander JD, Peterson RT, Yeh JR, Joung JK (2013) Efficient genome editing in zebrafish using a CRISPR-Cas system. Nat Biotechnol 31:227–229
148. Lagutina IV, Valentine V, Picchione F, Harwood F, Valentine MB, Villarejo-Balcells B, Carvajal JJ, Grosveld GC (2015) Modeling of the human alveolar rhabdomyosarcoma Pax3-Foxol chromosome translocation in mouse myoblasts using CRISPR-Cas9 nuclease. PLoS Genet 11, e1004951
149. Gutierrez A, Pan L, Groen RW, Baleydier F, Kentsis A, Marineau J, Grebliunaite R, Kozakewich E, Reed C, Pflumio F, Poglio S, Uzan B, Clemons P, VerPlank L, An F, Burbank J, Norton S, Tolliday N, Steen H, Weng AP, Yuan H, Bradner JE, Mitsiades C, Look AT, Aster JC (2014) Phenothiazines induce PP2A-mediated apoptosis in T cell acute lymphoblastic leukemia. J Clin Invest 124:644–655
150. Mautner VF, Nguyen R, Kutta H, Fuensterer C, Bokemeyer C, Hagel C, Friedrich RE, Panse J (2010) Bevacizumab induces regression of vestibular schwannomas in patients with neurofibromatosis type 2. Neuro Oncol 12:14–18
151. Robertson KA, Nalepa G, Yang FC, Bowers DC, Ho CY, Hutchins GD, Croop JM, Vik TA, Denne SC, Parada LF, Hingtgen CM, Walsh LE, Yu M, Pradhan KR, Edwards-Brown MK, Cohen MD, Fletcher JW, Travers JB, Staser KW, Lee MW, Sherman MR, Davis CJ, Miller LC, Ingram DA, Clapp DW (2012) Imatinib mesylate for plexiform neurofibromas in patients with neurofibromatosis type 1: a phase 2 trial. Lancet Oncol 13:1218–1224

Xiphophorus and Medaka Cancer Models

Manfred Schartl and Ronald B. Walter

Abstract Besides recently developed zebrafish cancer models, other fish species have been employed for many years as cancer models in laboratory studies. Two models, namely in *Xiphophorus* and medaka have proven useful in providing important clues to cancer etiology. Medaka is a complementary model to zebrafish in many areas of research since it offers similar resources and experimental tools. *Xiphophorus* provides the advantages of a natural ("evolutionary mutant") model with established genetics. *Xiphophorus* hybrids can develop spontaneous and radiation or carcinogen induced cancers. This chapter describes the tumor models in both species, which mainly focus on melanoma, and summarizes the main findings and future research directions.

Keywords Xiphophorus • Medaka • Melanoma • UV • Transgenic • Hereditary cancer • Oncogene • Tumor suppressor gene • Signal transduction

Introduction

The zebrafish represents an extremely valuable model for developmental biology and biomedicine and serves a large community of researchers. This chapter deals with other fish models that are used primarily in cancer research.

Naturally the question comes up, why do we need other fish models besides zebrafish and what may be learned from those that cannot be obtained from a similar study using zebrafish?

One motivation to study in parallel zebrafish and other laboratory fish models is the complementarity argument. In their evolutionary history teleost fish have

M. Schartl (✉)
Physiologische Chemie, Universität Würzburg, Biozentrum, Am Hubland, D-97074 Würzburg, Germany

Texas Institute for Advanced Study and Department of Biology, Texas A&M University, 100 Butler Hall, College Station, Texas 77843-3258, USA
e-mail: phch1@biozentrum.uni-wuerzburg.de

R.B. Walter
Chemistry and Biochemistry, 419A Centennial Hall, Texas State University, 601 University Drive, San Marcos, TX 78666-4616, USA
e-mail: RWalter@txstate.edu

D.M. Langenau (ed.), *Cancer and Zebrafish*, Advances in Experimental Medicine and Biology 910, DOI 10.1007/978-3-319-30654-4_23

undergone a whole genome duplication event leading to the situation wherein ancestors of modern day fish came to possess two copies of many genes while other vertebrates have only one [1–4]. Evolutionary pressures after the fish specific genome duplication served to alter gene expression patterns or protein functions, in a way that complemented the expression domains or protein functions of both fish paralogs to make them equivalent to the single ortholog in the other vertebrates. For example, if one gene in mammals is expressed in liver and gut the two separate fish genes may correspond to this pattern by expressing one copy in the liver and the other one in the gut. The evolutionary process leading to this phenomenon has been termed subfunction partition or subfunctionalization [5]. This situation must be taken into consideration in studies on the expression and function of such genes. When, for example, a mutation in mouse inactivates a gene that has two age dependent expression patterns, one function that is adult organ specific (e.g. in liver), and also a functional role in early development, it will be impossible to study the liver function in such mutants. However, if in fish both functions of the single mammalian gene have been partitioned between gene duplicates, then it becomes possible to separate both phenotypes. Thus, the teleost whole genome duplication offers many new opportunities for deciphering the physiological function of genes that may play a role in cancer. Another point emerging from the teleost whole genome duplication event is the level of lineage-specific duplicate retention (meaning that one gene pair is retained and subfunctionalized for instance in the zebrafish lineage, but not in the other fish model lineages – and vice versa) or that duplicate copy A is retained and copy B is lost in one lineage while copy B may be the one kept in the other lineage. Also, gene function redundancy or compensation mechanisms that interfere with gene knockdown and knockout approaches may exist differently in zebrafish and other fish models. Consequently, only the continued availability and use of several different fish laboratory models will offer the full benefit of comparative studies that involve duplicate genes.

The fish tree of life covers a wide phylogenetic distance. The lineages that are represented by the most commonly used fish models diverged around 300 million years ago. Differences that have evolved over time can make one model more suitable for studying certain aspects of cancer than others and a comparative approach can dissect common features from specific adaptations that may be less relevant for translating results to the human situation. For example, medaka has been shown to be different than zebrafish in modeling certain diseases, and in some cases able to provide more direct translational models to the human condition for diseases such as osteoporosis [6], chemically induced hepatic fibrosis [7], hypohidrotic ectodermal dysplasia [8], diabetic nephropathy [9], and chronic mycobacterial infection in tuberculosis [10].

The *Xiphophorus* Cancer Models

The *Xiphophorus* cancer models stand for a group of so-called evolutionary mutant models [11, 12]. Unlike in the "domesticated" or "engineered" disease models, where the condition is evoked either through a mutation (often from high-throughput mutation screens) or by the introduction of a transgene, in the evolutionary models the disease genes have evolved in nature. These models provide the possibility to study pleiotropy, multigenic inheritance, variable expressivity and variable penetrance of phenotypes, because cancer genes come together with their natural genetic background. Also, natural environmental influences that profoundly modulate a phenotype might be difficult or impossible to study in laboratory adjusted classical fish models, since environment-genotype interactions can be easily lost in selective breeding.

Xiphophorus, like medaka, has a long history as a genetic model. This has motivated in 1938 establishment of the *Xiphophorus* Genetic Stock Center (XGSC). The XGSC was opened with 6 species and was overseen and expanded under the direction of Dr. Myron Gordon at the New York Aquarium. Upon Gordon's death, Dr. Klaus Kallman directed the XGSC and, over the next 35 years, greatly expanded the collection of pedigreed and inbred *Xiphophorus* lines at the American Museum of Natural History. Upon his retirement in 1992, the XGSC was transferred to San Marcos, Texas [13] and currently comprises 24 species and 58 pedigreed lines, several of these having been continuously inbred (i.e. brother sister matings) for over 100 generations. XGSC distributes, on request live fish, online protocols and various type of information for *Xiphophorus* researchers (http://www.xiphophorus.txstate.edu).

A reference high-quality *Xiphophorus* genome has been established for the platyfish, *X. maculatus*, (http://www.ensembl.org/Xiphophorus_maculatus/Info/Index, [14]) and more genomic resources from other species of the genus are currently in production.

Hereditary Melanoma in *Xiphophorus* Hybrids

Platyfish and swordtails of the genus *Xiphophorus* are one of the oldest animal models for cancer research [15]. Already in the 1920's, it was found that certain hybrids of platyfish and swordtails develop highly malignant melanoma [16–18]. The malignant nature of the melanomatous cancers is not only evident by their fatal outcome for the fish but has been demonstrated by the fact that those tumors after transplantation into immune-compromised mice continue to grow progressively [19] and by their histological and ultrastructural similarity with human melanoma [20, 21].

At times, when the origins of cancer where largely unclear, the *Xiphophorus* model was one of the first to show that cancer has a hereditary component. Many crossing experiments were performed that led to a genetic explanation why melanoma formation occurs in some *Xiphophorus* hybrids [22]. This genetic model simply postulates two interacting genetic loci called *Tu* and *R*. It should be noted that these two loci are classical phenogenetic terms, which stand for the genetic information to produce a certain phenotype. It is not intended to include any molecular or structural information in these terms, meaning that they could consist of just one or more genes or even the absence of the gene. Different alleles of these loci then could be due to a range of things from additional or missing genes, to just point

Fig. 1 Classical crossing experiment (Gordon-Kosswig-Anders cross) leading to spontaneous melanoma formation in *Xiphophorus*. A female platyfish (*X. maculatus*) strain Jp163A with the spotted dorsal melanophore pattern (*Sd*) harboring the *Tu* allele on the X chromosome and *R* on LGV are crossed with non-spotted swordtail males (*X. hellerii*) that lack *Tu* and *R*. The F_1 offspring are heterozygous for *Tu* and *R*, resulting in enlarged melanotic, non-malignant pigment spots on the fins. Crossing of these individuals to a *X. hellerii* male gives rise to offspring of which 25 % develop melanoma, caused by the presence of a *Tu* allele in the absence of *R*, 25 %, being genetically like the F_1 fish, develop non-malignant pigment spots and 50 % are fishes do not exhibit macromelanophores pigment patterns due to absence of *Sd* (*Tu*)

mutations within the locus. *Tu* (for tumor) is the designation for a dominantly acting, pigment cell specific oncogene locus. Many individuals of *Xiphophorus,* but not all, have this locus. The oncogenic function of *Tu* in wild fish is suppressed by the locus *R* (for regulator), so that no tumors can develop. The *R*-locus has also been designated *Diff* or R_{Diff}. The restricted activity of *Tu*, even in the presence of *R* is, however, visible as a very limited local overproduction of pigment cells that form small spots, for instance in the platyfish *X. maculatus* strain Jp163A in the dorsal fin from the *spotted dorsal* allele (*Sd*) of *Tu* (see Fig. 1). Other platyfish have spots in other compartments of the body from other *Tu* alleles or no such spots, and several species of the genus *Xiphophorus* are even totally lacking *Tu*. The genetic hypothesis includes that fish that do not have *Tu* also do not have a *Tu*-suppressing allele of *R* [22, 23]. The black spots are made up of a distinct type of melanocytes, termed macromelanophores, which are much larger (up to 500 μm) than the other pigment cells (about 100 μm). Importantly they often grow onto each other with no sign of distance regulation as seen of the other pigment cells in the fish skin [24]. The black macromelanophore spots may be regarded as equivalents of the human nevi, although experimental evidence that they share similar characteristics (e.g. being senescent cells) has yet to be produced.

The location of *R* and *Tu* on different chromosomes in the spotted fish allows their separation through selective breeding (Fig. 1). A fish containing *Tu* and *R* (e.g. the platyfish, *X. maculatus*) can be crossed to another species that does not contain either locus (e.g. the swordtail, *X. hellerii*). In the resulting F_1 hybrids, one of the *R* containing chromosomes is substituted by an *R*-free chromosome of the crossing partner. Due the lower dosage of the tumor suppressor in the hemi- or heterozygous state, *Tu* is more active and induces a limited over-proliferation of pigment cells, which leads to a two-dimensional growth that does not become malignant. Using *R*-free fish for backcrossing of the F_1-fish generates offspring, of which one quarter, due to Mendelian laws, inherit a *Tu*-locus and corresponding loss of the *R*-locus containing chromosome. In this situation, *Tu* is out of control and exerts its full oncogenic function in the pigment cell lineage producing fast growing, nodular and highly invasive melanoma that is fatal to the fish (Fig. 1). Backcross hybrid fish that have inherited *Tu* and only one copy of *R* are genotypically and phenotypically similar to the F_1 hybrid fish, and develop strong hyperpigmentation.

Interestingly, the *Tu* locus does not encode only the information for neoplastic transformation of pigment cells, but also determines the location where melanoma will develop. Fish with the *spotted dorsal* (*Sd*) allele of *Tu* will always develop benign pigment cell lesions or melanoma (depending on their *R*-status) starting out from the dorsal fin. If platyfish are used as parental fish, which have spots on the body side, melanoma will start all over the body skin, and the fins will not be affected by pigment cell overgrowth. Besides the spotted dorsal platyfish and *X. hellerii* many similar crossing schemes using other species of *Xiphophorus* can be set up [25–27]. All existing evidence [23, 28, 29] indicates the melanoma inducing loci in the different species of *Xiphophorus* are homologous. The current understanding is that the *Tu* locus is composed of two different genetic entities, one that encodes the genetic information for neoplastic transformation and another that encodes body compartmentalization pattern information, leading to the location where the spots develop in the parental fish and concordantly the melanoma in the hybrids.

When one produces hybrids between *X. maculatus* strain Jp163A, which have the *spotted dorsal* (*Sd*) pattern gene, and *X. couchianus* the melanin pigment in the dorsal fin is virtually completely suppressed, while the body orange pigment becomes highly expressed. This is strikingly different from the hybrids of *X. couchianus* with *X. maculatus* strain Jp163 B, which have spots on the body side due to the *spotted* (*Sp*) locus and are heavily hyperpigmented (Fig. 2). Interestingly, the Jp163A and B lines are derived from a single pregnant female collected in the wild. The offspring of this fish were kept as single line for many years before split in the two sublines. The difference in the hybrid phenotypes generated from the two loci that contain a homologous *xmrk* allele highlights the complexity of the pseudo-allele situation concerning the *Tu/xmrk* locus and the interactions with the *R* locus.

Fig. 2 Different expression of the *X. maculatus* pigment patterns in F_1 hybrids with *X. couchianus*. When fish from strain Jp163B, which carry the spotted pigment pattern allele (*Sp*), are crossed to *X. couchianus*, the offspring show enhancement of the pattern resulting in heavy melanosis (*left row*). However, hybrids with strain Jp163A, which has the spotted dorsal (*Sd*) allele, show a total suppression of the melanocyte pattern

The *xmrk* Melanoma Oncogene

The *Tu* locus encodes an epidermal growth factor receptor-related gene, designated *Xmrk* (for *Xiphophorus* melanoma receptor kinase). It originated several million years ago prior to the radiation of the genus *Xiphophorus* [23] from a local gene duplication of the *Xiphophorus* EGF-receptor gene, which is the corresponding proto-oncogene [30] and the fish ortholog of the human EGFR gene. During the duplication event the new copy was fused to a different 5' region [31, 32], which most likely has altered the transcriptional control of this copy. This may account for the tumor inducing activity of the oncogene *Xmrk* in hybrids and its suppression in the parental healthy fishes: *Xmrk* is highly expressed in the transformed pigment cells, but not in other cells of the hybrids. A pigment cell specific transcriptional deregulation appears to be the primary event of melanoma formation [33].

The oncogenic properties of Xmrk result from activating mutations in the extracellular domain of the growth factor receptor, which induce covalent dimerization of two receptor monomers [34, 35] (Fig. 3). This mimics the structural consequences

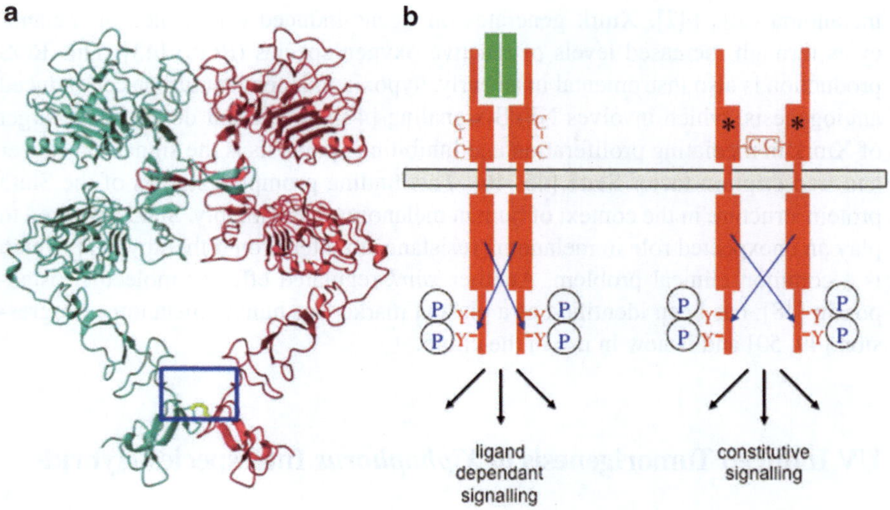

Fig. 3 (**a**) Structural model of the extracellular domain of an Xmrk dimer. The intermolecular cysteine disulfide bridge is boxed. Modified after Meierjohann et al. [35]. (**b**) Mode of action of the mutationally altered Xmrk receptor. The wildtype EGF-receptor Egfrb dimerizes only after binding of two molecules of the ligand (EGF, *green*) to the extracellular domain. Only in the dimerized state reciprocal transphosphorylation of the carboxyterminal domains by the intracellular kinase domains of the other subunit occurs. After autophorylation is completed the receptor is able to signal and can activate the mitogenic cascade. Once the ligand is removed, the receptor returns to the non-active state. In the mutant Xmrk receptor due to the failure of one intramolecular disulfide bond to form (due to a substitution of one of the cysteines by serine) a intermolecular disulfide bridge links covalently two receptor monomers. This mimics the structure that in the wildtype receptor is obtained after ligand binding, and thus induces receptor autophosphorylation and signal transduction, which is, however, constitutive. This induces for instance constant activation of the mitogenic cascade

that normally are the result of binding of the growth factor ligand EGF. The activated Xmrk protein sends out constitutively a growth factor receptor signal that elicits a variety of cellular responses that determine the neoplastic phenotype of the melanoma cells. EGF receptors in which the *xmrk* mutations were introduced were tumorigenic in transgenic fish [36] and similar activation mechanisms for receptor tyrosine kinases have been found in human cancers [37].

Studies on the tumor–inducing function of *xmrk* in fish melanoma, complemented by analyses using established mouse und human cell culture systems where *xmrk* was introduced as a transgene, have revealed important insights. For instance, the now well-established fact that the majority of human melanoma harbor mutations that confer constitutive activity to the Ras/Raf/MAPKinase signaling pathway [38], was noted early on in the *Xiphophorus* system [39]. Oncogenic signaling by Xmrk also activates the cytoplasmic Src-kinase Fyn [39, 40], which plays a critical role in the intracellular signal transduction network orchestrated by Xmrk and for instance interacts with focal adhesion kinase (FAK) in mediating melanoma cell migration and invasion [41]. Fyn/FAK interactions were then also noted in human melanoma cells [42]. Xmrk generates oncogene-induced senescence of melanocytes through increased levels of reactive oxygen species (ROS) [43]. This ROS production is also instrumental in an early, hypoxia independent tumor-cell induced angiogenesis, which involves NFκB signaling [44]. A relevant downstream target of Xmrk in mediating proliferation and inhibiting apoptosis is the signal transducer and transcription factor Stat5 [45, 46]. This finding prompted studies of the Stat5 protein structure in the context of human melanoma. Remarkably, Stat5 appeared to play an unexpected role in melanoma resistance to interferon α therapy [47], which is a common clinical problem. Another *xmrk*-regulated effector molecule, osteopontin [48], has been identified as a plasma marker for human melanoma progression [49, 50] and is now in use in the clinic.

UV Induced Tumorigenesis in *Xiphophorus* Interspecies Hybrids

Several *Xiphophorus* interspecies backcrosses have been described where BC_1 progeny do not develop melanoma spontaneously, but if the backcross progeny are exposed to UVB at 6 days post-birth, the tumor incidence can rise to about 40 % among the pigmented progeny. These crosses usually employ *X. maculatus* Jp163B with the *spot side* (*Sp*) pigment pattern (Fig. 4a) crossed and backcrossed to either *X. hellerii* [51, 52] or *X. couchianus* [53–56]. The later cross, *X. couchianus* (×) [*X. maculatus* Jp163B (×) *X. couchianus*] has been the most extensively studied UV-induced melanoma model. The UV induced melanoma lesions produced in these BC_1 progeny after UVB exposure (i.e., 6.4 kJ m^{-2}) are nodular exophytic tumors (Fig. 4b) that may grow rapidly [21]. Upon crossing with *X. couchianus*, F_1 interspecies hybrids develop much enhanced melanocyte pigmentation compared to the *X. maculatus* Jp163B parent. Backcrossing the F_1 hybrid to the *X. couchianus* parent (BC_1) serves to further enhance the pigmentation of select animals to a level where they may appear virtually black from the gills to the tail (for review see [26, 53]). Backcross progeny do not generally develop melanoma unless the animals are exposed to UVB (6 days post birth) or MNU (6 weeks post birth) whereupon melanomas may develop at 6–9 months of age.

The ability to produce animals that are predisposed to UVB induced melanoma led to efforts to better understand the photobiology of melanomagenesis. In pioneering studies Setlow and colleagues [51, 55] exposed *Xiphophorus* interspecies hybrids to varying wavelengths of light, ranging from 302 to 436 nm, and determined that effective melanoma inducing wavelengths occurred in the UVB range (304 nm) and, surprisingly, also in the UVA region around 365 and 405 nm. These results produced an immediate controversy since the UVA wavelengths are not expected to produce direct DNA damage. At this time, and currently, the paradigm for UVB association with cancer involves direct DNA damage leading to mutagenic events that, in turn, may lead to melanoma. This concept for UV induction by mutation stemmed from earlier studies showing the molecular basis of xeroderma pigmentosum (XP) patients, who are 1,000 times more likely to develop skin cancer then non-XP individuals,

Fig. 4 (**a**) Parental and F_1 *X. maculatus* Jp163B (×) *X. couchianus* interspecies hybrid. (**b**) Example of a typical UVB induced melanoma tumor in *X. couchianus* (×) [*X. maculatus* Jp163B (×) *X. couchianus*] interspecies BC_1 hybrid

involved defective repair of UV induced DNA damage [57–59]. The association of failure to repair direct DNA damage after UVB exposure in the skin of these patients and an increased incidence of early onset melanoma is a convincing argument for the role of direct UVB damage in melanomagenesis.

Setlow et al. (1989, 1993) argued that even though UVB exposure (i.e. 302 nm) had been shown most capable of inducing melanoma in *Xiphophorus* interspecies hybrids, the tumorigenicity peak observed in the UVA wavelengths is more relevant to human health since UVB constitutes only about 5–10 % of sunlight reaching the earth's surface. Further, given the doses leading to melanoma induction and the levels of UVB DNA damage/Mb of DNA, even in the UVB regions of the spectrum the probability of a single gene inactivating or activating mutation is too low to explain induced tumor incidence. The nearly 10-fold more prevalent UVA wavelengths of light, and longer exposure to UVA since this does not lead to erythema, may result in UVA exposure dominating melanoma causation in humans.

The potential causative involvement of UVA in melanoma, forwarded from *Xiphophorus* studies, led to rapid addition of UVA sun blocking agents to available sunscreens, that prior to these observations only contained agents blocking UVB.

Nairn et al. [52] supported these conclusions by performing genetic linkage analyses of UVB or UVA induced tumor bearing BC_1 fish. In this study the genetics underlying tumor induction were shown different; tumors derived from 313 or 365 nm exposure did not exhibit genetic association with inheritance of linkage group 5 (LGV) harboring the *R* (*Diff*) tumor suppressor locus, while fish having tumors induced by 405 nm exposure showed a strong association with inheritance of LGV [52]. It is noteworthy that subsequent efforts have failed to establish any UVB mutational signatures in a target gene on LGV (i.e., *CDKN2AB*) that is inactivated in the majority of human melanoma biopsies [60].

Wood et al. [61] used electron paramagnetic resonance analyses to assess UV induction of melanin reactive radicals *in situ* in the skin of *X. maculatus* Jp163B (×) *X. couchianus* interspecies F_1 hybrids. Free radicals generated in the skin of *Xiphophorus* pigmented animals showed a peak production of reactive oxygen free (ROS) radicals at longer wavelengths (i.e. 404 and 436), consistent with the UVA induced melanoma observed by Setlow et al. [55]. The similarity in peak ROS production and melanoma induction in the UVA range is consistent with melanin serving as a major photosensitizer and a perhaps causative agent in this cancer model. However, as discussed below, the role of UVA in melanoma induction remains controversial.

Many observations made over the years have pointed to direct DNA damage from UVB exposure as causal to induced melanoma in the *X. maculatus* Jp163B (×) *X. couchianus* interspecies hybrid model. Often quoted as evidence for direct damage are results from the early UV induction experiments [51] where post UV exposure to fluorescent light was shown to reduce the UVB induced tumorigenesis to "near background levels". This observation is attributed to the presence of DNA damage specific photolyases in fishes that are capable of using visible light to chemically reverse cyclobutane-pyrimidine-dimers (CPDs) and 6-4 photoproducts (6-4PDs) to the natural state. These photolyase proteins are not present in mammals having become delegated to circadian pathways at the evolutionary time when mammals went through a nocturnal phase. Photolyases in *Xiphophorus* and other fishes specific for CPD and 6-4PDs have been shown to produce rapid repair of UVB induced DNA damage upon exposure to visible light [55, 62, 63]. Although visible light induced DNA repair in fish skin after UVB exposure is well documented, the effectiveness of the repair in reducing UVB induced tumorigenesis remains less well supported. Only two reports have actually measured the efficacy of visible light in reducing tumorigenesis after UVB exposure. In one this involved treatment of only two dozen animals and the differences between control and fluorescent light exposed animals was

4 tumors (6 tumors in 22 animals at 304 nm exposed vs. 2 tumors in 25 exposed to 304 nm UV+visible light; [51]. This same study also showed an abnormally high background level of induced melanoma in control animals (about 13 %; 10 tumors in 79 animals) calling into question the basal exposure of the fishes that were housed in a greenhouse. In a later study, fish were transferred to yellow light for 1 month prior to experimentation and subsequently showed much lower background tumor levels (5 %). For the fish kept in yellow light, UVB treatment followed by fluorescent light exposure are reported to have reduced induced tumor incidence by about 38 %, but the numbers of animals used in this study are not provided [55]. These two reports appear to show visible light reduction of UVB induced tumorigenesis in the *X. maculatus* Jp 163 B (×) *X. couchianus* interspecies hybrid model. However, often used statements that post-UV visible light exposure "returns UVB induced melanoma to background levels" [52, 64] are not well supported. Verifying photoreversal of UVB induced melanoma by visible light will require more stringent analyses and with large numbers of exposed and control animals to firmly establish its significance.

A recent study addresses the debate regarding contributions of UVB and UVA to induced melanoma in the *X. maculatus* Jp163B (×) *X. couchianus* interspecies hybrid model. Over 200 fish for each experimental treatment [53] were used to repeat the earlier experiments of Setlow et al. (1989). Backcross$_1$ hybrids were exposed to UVB (313 nm), UVA (365 nm) or used as control fish that were not irradiated. These results clearly suggest that under the conditions of this experiment only UVB exposed fish exhibited melanoma tumor induction (43 %) whereas control and UVA exposed fish both showed similar levels tumors (UVA-12.4 %, control-18.5 %; [53]). Histopathological analysis and clustering of tumors from UVB or UVA exposed animals indicated the UVB tumors were more aggressive and at a more advanced stage (stage IV, presence of unilateral, bilateral, and/or multiple tumors) than tumors dissected from UVA exposed animals.

This study encompassed a statistically sound and well performed tumor induction analysis that shows UVB induces tumors readily in *X. maculatus* Jp163B (×) *X. couchianus* interspecies BC$_1$ hybrids, and conversely, that UVA exposure, did not induce melanoma. However, there remain major differences in the manner in which the Setlow et al. [55] experiments and these experiments [53] were performed, not the least of which involve fish care and ambient lighting prior to experimental treatments. Thus, these reports do not exclude or refute one another, but instead give us more pieces to a complex puzzle regarding the manner in which light is perceived by animals and what genetic pathways and/or mechanisms are invoked upon exposure to various wavelengths.

R/Diff, a Tumor Suppressor Locus in *Xiphophorus*

The parental *X. maculatus* Jp163A that carries *Sd*, or *X. maculatus* Jp 163 B carrying *Sp*, both harbor the *xmrk* oncogene but do not develop melanoma [52]. Yet when introduced to an interspecies hybrid genetic background the *xmrk* oncogene can induce melanoma in the fraction of the total progeny that also inherit heavy pigmentation. The need for a backcross to express melanoma (i.e. 2 hit model) suggested very early that parental *X. maculatus* animals must harbor some factor that can keep the *xmrk* oncogene in check. The *R* gene has been designated as this hypothetical suppressor of *xmrk*. The *R* locus was shown to be autosomal and localized to an area close to the esterase 1 locus on *Xiphophorus* linkage group V (LGV) [65–68]. Following the localization of *R* to LGV the search for the *R* gene was made by attempting to saturate LGV with markers and by mapping cloned *Xiphophorus* homologues of human tumor suppressor genes, DNA repair genes, and oncogenes in hope one of these genes might map to the *R* region of LGV (reviewed by [54], see especially [26, 69–71]). These efforts culminated in isolation of a *Xiphophorus* clone homologous to the human *CDKN2A* gene that showed inheritance associated with melanoma development in backcross hybrids, and that mapped to the same region of LGV as the hypothetical *R* tumor suppressor locus [69].

The amino acid sequence of the cloned *Xiphophorus cdkn2* gene was equally distant from the human *CDKN2A* (p16) and *CDKN2B* (p15) gene products; and it has been designated as *cdkn2a/b* or *cdkn2X* [72]. In fact the two mammalian CDKN2 genes are the product of a local gene duplication event that happened later in vertebrate evolution and the fish *cdkn2a/b* gene is the common ancestor. The *Xiphophorus cdkn2a/B* gene is widely considered a candidate for *R* function since loss of human *CDKN2A* function is found highly associated with human melanoma [73]. Human *CDKN2A* and *B* are members of a family of kinases that are well established and function to regulate cell cycle progression by inhibiting the activities of specific cyclins, that in turn, regulate the retinoblastoma (*RB*) tumor suppressor via phosphorylation [74]. *RB* phosphorylation produces signals promoting progression through the G1-S boundary and entry into mitosis. Inactivation of the *CDKN2A* gene is a rare but high-penetrance risk factor for human familial melanoma [75] and is also associated with many other cancers. Thus, the *Xiphophorus cdkn2a/b* gene not only mapped to the right location, but also has an attractive presumptive function that makes it a viable *R* gene candidate.

In human melanoma, the *CDKN2A* homologue exhibits loss of function (via mutation or gene silencing by methylation) suggesting a mechanism for *Xiphophorus* melanomagenesis. If the *X. maculatus cdkn2a/b* allele (capable of suppressing/regulating *xmrk*) shows differential activity compared to the *X. hellerii* allele (incapable of regulating *Xmrk*), then one could envision loss of the *X. maculatus cdkn2a/b* alleles (i.e. inheritance of the *X. hellerii* alleles) in backcross hybrid progeny as leading to liberation of *Xmrk* oncogenic activity and development of melanoma. In several melanoma crosses there preferential loss of the *X. maculatus* alleles was shown to be associated with about 80 % of the melanoma bearing BC_1 hybrid progeny [54, 76].

However, it has proven difficult to identify a mechanism for differential regulation of *xmrk* by *cdkn2a/b* alleles based on structural analyses of the *X. maculatus* and *X. hellerii cdkn2a/b* genes. *The cdkn2a/b* genes in *X. maculatus* and *X. hellerii* are nearly identical in their coding regions (two conservative differences and a single Lys to Glu non-conservative replacement [74, 77]. The genomic regions immediately upstream of the *cdkn2a/b* genes in *X. maculatus* and *X. hellerii* show many differences, some of which (i.e. GT repeat lengths) suggest altered methylation of the promoter area may lead to differential regulation of the two genes in the hybrid genetic environment [78, 79]. However, it has also been shown that upstream regions of *cdkn2a/b* alleles appear to remain unmethylated and are overexpressed in both melanoma and in normal tissues [72, 74, 77, 78]. Overall, genetic mapping of *cdkn2a/b* to the LGV genomic region known to harbor *Diff*, the upregulated expression of *cdkn2a/b* in pigmented F_1 and BC_1 fishes that exhibit heavy melanization, and assocaition between inheritance of the *X. hellerii* alleles in melanoma bearing BC_1 hybrids, serve to support the candidancy of *cdkn2a/b* as the *R* tumor suppressor locus. However, there remain several observations that distract from this conclusion.

As mentioned, the human association of *CDKN2A* with melanoma involves loss of function by gene silencing. The loss of *CDKN2A* function results in dysregulation of the *RB* due to an inabiltiy to regulate cyclin-dependent kinase (CDK4 and 6) activities leading to increased activity of genes involved in cellular differentiation and growth. However, in *Xiphophorus* hybrids with pigmented skin, *cdkn2a/b* shows very high expression, contrary to expectations from human melanoma. In addtion, loss of *X. maculatus* alleles in melanoma bearing BC_1 hybrids associates at about 80 %, thereby suggesting 20 % of melanoma bearing hybrids are heterozygous carrying both the *X. maculatus* and the *X. hellerii cdkn2a/b* alleles. This could be interpreted as indicating *cdkn2a/b* is tightly linked to, but not, the *R* gene. Recent sequence and assembly of the *X. mauclatus* genome has allowed scanning of the LGV region harboring *R*. This identifies at least five other genes that could be involved in tumor regulation but have not yet been tested for *R gene* function [14]. The advent of these genome resoruces will allow new ideas to be investigated that may explain the *R* suppressor in mechanistic terms for both spontaneous and induced melanoma models.

Chemical Carcinogen Induced Tumors

Although *Xiphophorus* fish are generally considered as melanoma models, other tumors are also observed here, particularly after exposure to carcinogenic agents. The same genotypes that show the UV susceptibility for melanoma induction react similarly to *N*-methyl-*N*-nitrosourea (MNU), a mutagen that acts principally through methylation of purines. If BC_1 fish from crossbreeding between *X. maculatus* Jp163B and *X. hellerii* or *X. maculatus* Jp163B and *X. couchianus* (Figs. 2 and 3) are treated with MNU, melanoma development is only observed in fish that possess the *Sp* pigment pattern. Among these a melanoma incidence of up to 40 % is reached, while other tumor types occur only rarely [80]. This suggests the involvement of *xmrk* in MNU-induced melanoma development, similar to the observations for spontaneous and UV-induced melanoma. However, *Cdkn2ab* inheritance is not associated with development of MNU induced tumors in these crosses. [60].

MNU treatment of BC_1 hybrids between *X. maculatus* strain Jp163A and *X. couchianus* does not produce melanoma but instead induces fibrosarcomas, neuroblastomas, schwannomas and retinoblastomas [80]. A similar observation was made decades ago by Schwab and colleagues, who treated a large number of different

Xiphophorus backcross genotypes and obtained neuroblastoma, mainly in *X. variatus* × *X. hellerii* backcross hybrids [81]. Although the degree of tumor development was linked to the genetic background of the fish, further molecular analyses of MNU-induced melanoma or neuroblastoma are not available to date.

In addition to tumors that show a certain predisposition depending on the presence of *xmrk*, a whole spectrum of tumors from different tissues, including fibrosarcoma, rhabdomyosarcoma and hepatic carcinoma were induced in backcross segregants not containing *mdl-xmrk* [22, 81, 82].

High incidence of spontaneous formation of ocular tumors was noted in a hybrid line established from offspring of a cross of an ornamental albino swordtail female with a platyfish male carrying the spotted belly and spotted dorsal macromelanophore patterns [83]. The genetic determinant for the ocular tumor segregated independently from the melanoma locus. The histopathology of this tumor was not further specified and obviously this line has been lost.

Furthermore, *Xiphophorus* was used as thyroid tumor model. In certain laboratory strains of *Xiphophorus montezumae*, a high incidence of spontaneous thyroid tumors was observed [84, 85]. These tumors are of inhomogenous phenotype, and they easily invade bone. The growing tumor fills also the visceral arches, thereby impairing gill function, which eventually leads to the death of the fish. Tumors occur more often in females and are only fully formed in fish of 5 months and older [85].

Cancer Models from Medaka

Although much lesser known than zebrafish and used by a much smaller—but steadily growing community of researchers—the medaka, *Orzyias latipes*, also known as the Japanese ricefish, can be regarded as a complementary model and is equivalent in many aspects to the zebrafish model [86]. Like the similar-sized zebrafish, medaka has a short generation time, is easy to breed in large numbers in the laboratory and produces transparent eggs that make it easy to follow embryonic development. Systematic large-scale mutagenesis screens have led to large collections of mutants. The genome of the HdrR strain has been sequenced [87] and large genomic resources have been built up. For functional studies the same toolbox as in zebrafish exists: downregulation of gene expression during early development by morpholinos, elaborate transgenic systems and gene knock-out by zincfingernucleases, TALENs [88] and the CRISPR/cas9 system. A resource center (NBRP medaka) has been established where mutants, natural and transgenic strains, and cDNA and genomic clones can be obtained (http://www.shigen.nig.ac.jp/medaka/). The center also maintains databases and distributes information, tools and protocols as a community service.

A special feature of medaka is the availability of a number of highly inbred strains, which are derived from different natural populations [89, 90]. This is a unique genetic resource that can be used to study natural allelic variation of disease modifier genes.

Chemical Carcinogen Induced Melanoma

In a similar way as described for *Xiphophorus*, different medaka strains show varying susceptibility to cancer induction. Treatment of two highly inbred strains, HO4C and HB32C, with *N*-methyl-*N'*-nitro-*N*-nitrosoguanidine led to a dosage-dependent induction of tumors only in the HB32C strain, most of which were amelanotic melanoma [91]. When F_1 hybrids of both strains were treated with the same carcinogen. The variety of tumors induced in the F_1 fish was greater than in the parental strains and included melanoma (more than 60 % of the tumors), papilloma, ovarian tumors, olfactory epithelioma, branchioblastoma and fibroma. A markedly higher cumulative incidence of the melanoma with a dose–response was observed in the F_1 hybrids compared to the parental strains [92].

The Medaka *xmrk* Melanoma Model

A transgenic melanoma model has been developed in medaka using a fish oncogene as a driver for cancer development [93]. The melanoma inducing oncogene *xmrk* from *Xiphophorus* was put under control of the medaka *mitfa* promoter, which confers specific and exclusive expression in pigment cells. Those fish develop spontaneously malignant melanoma with high penetrance (Fig. 5). Tumor formation is already observed in these transgenic fish at about 4 weeks after hatching. Different from the classical zebrafish Braf melanoma model [94], in the *xmrk* medaka model tumor development initiates independently of the p53 wildtype or mutant status. This makes the medaka melanoma model more similar to the mammalian situation where the p53 tumor suppressor gene is only rarely found to be mutated. Introducing the *mitf:xmrk* transgene into p53 knockout genetic background, however, results in development of many melanoma at an earlier age, and the tumors grow faster and to much larger masses in the adult fish [93].

Interestingly, when the same transgene (i.e. *mitfa:xmrk*) carrying chromosome was bred to various other inbred medaka strains, depending on the genetic background, different types of pigment cell tumors developed. In the HB32B background highly invasive and metastasizing fatal melanoma from extracutaneous sites prevailed. In the Carbio or CAB genetic background, the two most widely used laboratory strains, predominantly large exophytic masses of tumors developed on the integument, which are much less life-threatening. These are of either melanoma of melanocytic origin or xanthoerythrophoroma that originate from the xanthophore and erythrophore lineages. These pigment cell types are not found in mammals. In one of the albino strains, termed i-3, a high frequency of malignant melanoma of the uvea is observed. This tumor type occurs only sporadically in the other genetic backgrounds [93]. The i-3 and HdrR strains preferentially show development of a low malignancy grade xanthophoroma (unpublished).

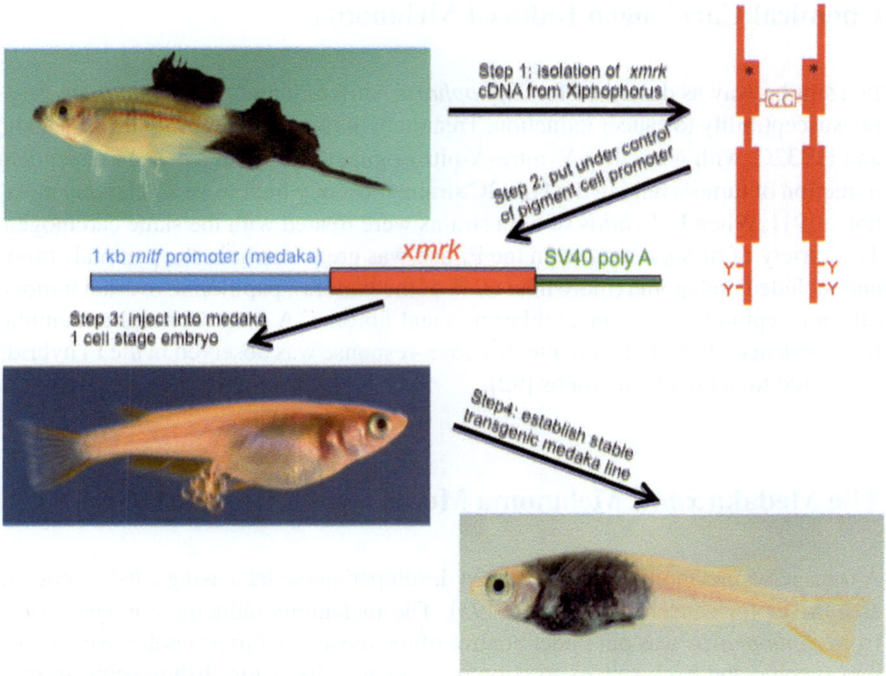

Fig. 5 Generation of the transgenic *xmrk* melanoma model in medaka. Pigment cell specific expression of the platyfish *xmrk* oncogene cDNA under control of the medaka *mitf* promoter in a stable transgenic line leads to spontaneous malignant melanoma formation with high penetrance (up to 100 % of transgene carriers) and early onset (about 2–4 weeks after hatching)

This genetic background effect on tumor types with respect to cellular origin, body site of appearance, and degree of malignancy can be ascribed to the presence of different tumor modifier genes or their allelic variation in the genetic background of the different strains. Other than the primary acting tumor-inducing oncogenes or tumor suppressor genes, tumor modifier genes are from the genetic background of an organism that influence the course of a disease, rather than being causal for the outbreak of the condition. With the current possibility to perform whole genome mutagenesis screens, the medaka offers an opportunity to identify modifier genes and to study their mode of action.

Furthermore the medaka melanoma model has been used to perform with RNA deep sequencing methods a comparative transcriptome analysis between fish and human melanoma [95], that showed global transcription in this transgenic medaka melanoma model is in excellent agreement with gene expression results determined from human tumor samples. The study not only revealed conserved changes in expression profiles of known pathways in melanoma in particular and cancer in general but also forwarded several molecules that can be developed as disease markers (e.g. *slc45a2*) or potential pharmaceutical targets.

Expression profiles of ten known human melanoma-associated miRNAs and their respective target gene expression were studied in xanthoerythrophormona, cutaneous and uvea melanoma [96]. miRNAs of the miR-17–92 cluster (miR-20a2, miR-92a1, miR-17 and miR-18a), miR-126, miR-182, miR-210 and miR-214 showed consistent upregulation and their respective target genes (RUNX1, HIF1A, TGFBR2, THBS1 and JAK2) were down-regulated in most tumors. MicroRNA-125b similar as in human melanoma was down-regulated and target genes (ERBB3a and ERBB3b) are up. These results provided clear evidence the fish melanoma-associated miRNAs and their respective target genes are deregulated generally like in human melanoma.

Yet another study looked at chemokine signaling in the *xmrk* melanoma medaka [97]. These signals are thought to play an instructive role in tumor migration. It was found that Sdf1/Cxcl12 signals via the chemokine receptor Cxcr7 are able to constrain invasive melanoma growth *in vivo* and these signals influence tumor outcome. Overexpression of the ligand not only reduced melanoma incidence but also shifted the tumor spectrum on the Carbio background even more towards the more benign xanthoerythrophoroma.

The molecular profiling confirms the value of the transgenic *xmrk* medaka as a translational model system to identify and decipher molecular mechanisms associated with malignant melanoma in humans.

The Medaka Transgenic ras Melanoma Model

A second medaka melanoma model has been established using the human *HRAS* oncogene with the G12Vmutation as driver for tumor development [98]. Although this mutation is rare in human melanoma, it is a quite efficient driver of melanoma in fish [99]. The human cDNA was put under control of the medaka tyrosinase promoter, which confers melanocyte-specific expression. The 5'UTR of the oncogene carries a floxed GFP as stop cassette, allowing for inducible expression, when the fish are crossed to a heat-shock inducible Cre line. Due to the fact that the heat-shock promoter in medaka can also be induced by a number of other stressors leading to some background of tumor formation, there remain improvements for this first inducible cancer model in medaka. Nevertheless, conditions for induction of melanoma could be established. This revealed that a first sign of tumor development, visible as hyperpigmentation around the eye, appears at 4 weeks after treatment (done shortly post hatching), fully grown highly invasive tumors after 28 weeks and melanoma-caused death after 44 weeks. Penetrance varied depending on the transgenic line between 12.5 and 100 %. Treatment experiments with Sorafinib, a multikinase inhibitor that is successfully used in the clinic to treat melanoma patients with a BRAF mutation, were performed to show the usefulness of this model for in-vivo screening of anticancer drugs. Melanomas were induced in 4 week old fish and treatment was started at 16 weeks when tumor formation had already started. The Sorafinib treated group showed a statistically significant increase in survival time.

Conclusions

This chapter introduced various fish cancer models that complement zebrafish in establishing addressable questions regarding the genetics and mechanisms underlying tumorigenesis. Our discussion focused primarily on melanoma tumors as historically, fish models forwarded to study melanoma have come from, or where produced for, the genetic analyses of tumor etiology. It is clear that newer studies employing development of transgenic models are rapidly providing tools that will be utilized to answer long standing questions and will delineate the steps within complicated mechanisms that frame both spontaneous and induced tumorigenesis. Due to recent advances in genomics and transcriptomics, the strength of evolutionarily distinct yet complimentary fish cancer models is only now capable of being exploited. However, it will be the continued employment of the various fish models that together will provide complimentary and insightful experimental results to allow us dissect the molecular details of cancer onset and progression.

Acknowledgments This work was supported by the National Institutes of Health (NIH), Division of Comparative Medicine (for RW: R24-OD-011120, R24-OD-011199, and R24-OD-018555, for MS: 1R01GM085318-01A2, subaward No. 212791A).

References

1. Amores A, Force A, Yan YL, Joly L, Amemiya C et al (1998) Zebrafish hox clusters and vertebrate genome evolution. Science 282:1711–1714
2. Meyer A, Schartl M (1999) Gene and genome duplications in vertebrates: the one-to-four (–to-eight in fish) rule and the evolution of novel gene functions. Curr Opin Cell Biol 11:699–704
3. Postlethwait JH, Woods IG, Ngo-Hazelett P, Yan YL, Kelly PD et al (2000) Zebrafish comparative genomics and the origins of vertebrate chromosomes. Genome Res 10:1890–1902
4. Braasch I, Postlethwait J (2012) Polyploidy in fish and the teleost genome duplication. In: Soltis PS, Soltis DE (eds) Polyploidy and genome evolution. Springer, New York, pp 341–383
5. Force A, Lynch M, Pickett FB, Amores A, Yan YL, Postlethwait J (1999) Preservation of duplicate genes by complementary, degenerative mutations. Genetics 151(4):1531–1545
6. To TT, Witten PE, Renn J, Bhattacharya D, Huysseune A et al (2012) Rankl-induced osteoclastogenesis leads to loss of mineralization in a medaka osteoporosis model. Development 139:141–150
7. Wolf JC, Wolfe MJ (2005) A brief overview of nonneoplastic hepatic toxicity in fish. Toxicol Pathol 33:75–85
8. Harris MP, Henke K, Hawkins MB, Witten PE (2014) Fish is Fish: the use of experimental model species to reveal causes of skeletal diversity in evolution and disease. J Appl Ichthyol 30:616–629
9. Ichimura K, Kawashima Y, Nakamura T, Powell R, Hidoh Y et al (2013) Medaka fish, Oryzias latipes, as a model for human obesity-related glomerulopathy. Biochem Biophys Res Commun 431:712–717
10. Mosi L, Mutoji NK, Basile FA, Donnell R, Jackson KL et al (2012) Mycobacterium ulcerans causes minimal pathogenesis and colonization in medaka (Oryzias latipes): an experimental fish model of disease transmission. Microbes Infect 14:719–729
11. Albertson RC, Cresko W, Detrich HW 3rd, Postlethwait JH (2009) Evolutionary mutant models for human disease. Trends Genet 25:74–81

12. Schartl M (2014) Beyond the zebrafish: diverse fish species for modeling human disease. Dis Model Mech 7:181–192
13. Walter RB, Kazianis S, Hazlewood L, Johnston D, Kumar J (2005) The Xiphophorus genetic stock center. In: Uribe MC, Grier H (eds) Viviparous fishes. New Life Publications, Homestead, FL, pp 343–350
14. Schartl M, Walter RB, Shen Y, Garcia T, Catchen J et al (2013) The genome of the platyfish, Xiphophorus maculatus, provides insights into evolutionary adaptation and several complex traits. Nat Genet 45:567–572
15. Heston WE (1982) Genetics: Animal Tumors. In: Becker FF (ed) Cancer: a comprehensive treatise, 2nd edn. Plenum Press, New York, pp 47–71
16. Gordon M (1927) The genetics of viviparous top-minnow Platypoecilus: the inheritance of two kinds of melanophores. Genetics 12:253–283
17. Kosswig C (1928) Über Kreuzungen zwischen den Teleostiern Xiphophorus helleri und Platypoecilus maculatus. Z Indukt Abstammungs-Vererbungsl 47:150–158
18. Häussler G (1928) Über Melanombildungen bei Bastarden von Xiphophorus maculatus var. rubra. Klin Wochenschr 7:1561–1562
19. Schartl M, Peter RU (1988) Progressive growth of fish tumors after transplantation into thymus-aplastic (nu/nu) mice. Cancer Res 48:741–744
20. Riehl R, Schartl M, Kollinger G (1984) Comparative studies on the ultrastructure of malignant melanoma in fish and human by freeze-etching and transmission electron microscopy. J Cancer Res Clin Oncol 107:21–31
21. Gimenez-Conti I, Woodhead AD, Harshbarger JC, Kazianis S, Setlow RB et al (2001) A proposed classification scheme for Xiphophorus melanomas based on histopathologic analyses. Marine Biotechnol 3:S100–S106
22. Anders F, Diehl H, Schwab M, Anders A (1979) Contributions to an understanding of the cellular origin of melanoma in the Gordon-Kosswig Xiphophorine fish tumor system. Pigm Cell Res 4:142–149
23. Weis S, Schartl M (1998) The macromelanophore locus and the melanoma oncogene Xmrk are separate genetic entities in the genome of *Xiphophorus*. Genetics 149:1909–1920
24. Schartl M, Meierjohann S (2010) Oncogenetics. In: Pilastro A, Evans J, Schlupp I (eds) Evolution and ecology of Poeciliid fishes. The University of Chicago Press, Chicago, pp 285–297
25. Anders A, Anders F, Klinke K (1973) Regulation of gene expression in the Gordon-Kosswig melanoma system. In: Schröder JH (ed) Genetics and mutagenesis of fish. Springer, New York, pp 33–63
26. Walter RB, Kazianis S (2001) Xiphophorus interspecies hybrids as genetic models of induced neoplasia. Ilar J 42:299–321
27. Zander CD (1969) Über die Enstehung und Veränderung von Farbmustern in der Gattung Xiphophorus (Pisces). Mitt Hamburg Zool Mus Inst 66:241–271
28. Schartl A, Hornung U, Nanda I, Wacker R, Müller-Hermelink HK et al (1997) Susceptibility to the development of pigment cell tumors in a clone of the Amazon molly, *Poecilia formosa*, introduced through a microchromosome. Cancer Res 57:2993–3000
29. Kallman KD, Atz DW (1967) Gene and chromosome homology in fishes of the genus Xiphophorus. Zoologica NY 51:107–135
30. Gomez A, Volff JN, Hornung U, Schartl M, Wellbrock C (2004) Identification of a second egfr gene in *Xiphophorus* uncovers an expansion of the epidermal growth factor receptor family in fish. Mol Biol Evol 21:266–275
31. Adam D, Dimitrijevic N, Schartl M (1993) Tumor suppression in *Xiphophorus* by an accidentally acquired promoter. Science 259:816–819
32. Volff JN, Körting C, Froschauer A, Zhou Q, Wilde B et al (2003) The Xmrk oncogene can escape nonfunctionalization in a highly unstable subtelomeric region of the genome of the fish *Xiphophorus*. Genomics 82:470–479
33. Regneri J, Schartl M (2012) Expression regulation triggers oncogenicity of xmrk alleles in the Xiphophorus melanoma system. Comp Biochem Physiol C Toxicol Pharmacol 155:71–80

34. Gomez A, Wellbrock C, Gutbrod H, Dimitrijevic N, Schartl M (2001) Ligand-independent dimerization and activation of the oncogenic Xmrk receptor by two mutations in the extracellular domain. J Biol Chem 276:3333–3340
35. Meierjohann S, Müller T, Schartl M, Bühner M (2006) A structural model of the extracellular domain of the oncogenic EGFR variant Xmrk. Zebrafish 3:359–369
36. Winnemoeller D, Wellbrock C, Schartl M (2005) Activating mutations in the extracellular domain of the melanoma inducing receptor Xmrk are tumorigenic in vivo. Int J Cancer 117:723–729
37. Robertson SC, Tynan J, Donoghue DJ (2000) RTK mutations and human syndromes: when good receptors turn bad. Trends Genet 16:368
38. Solus JF, Kraft S (2013) Ras, Raf, and MAP kinase in melanoma. Adv Anat Pathol 20:217–226
39. Wellbrock C, Schartl M (1999) Multiple binding sites in the growth factor receptor Xmrk mediate binding to p59fyn, GRB2 and Shc. Eur J Biochem 260:275–283
40. Wellbrock C, Schartl M (2000) Activation of phosphatidylinositol 3-kinase by a complex of p59fyn and the receptor tyrosine kinase Xmrk is involved in malignant transformation of pigment cells. Eur J Biochem 267:3513–3522
41. Meierjohann S, Wende E, Kraiss A, Wellbrock C, Schartl M (2006) The oncogenic epidermal growth factor receptor variant *Xiphophorus* melanoma receptor kinase induces motility in melanocytes by modulation of focal adhesions. Cancer Res 66:3145–3152
42. Teutschbein J, Schartl M, Meierjohann S (2009) Interaction of Xiphophorus and murine Fyn with Focal Adhesion Kinase. Comp Biochem Phys C 149:168–174
43. Leikam C, Hufnagel A, Schartl M, Meierjohann S (2008) Oncogene activation in melanocytes links reactive oxygen to multinucleated phenotype and senescence. Oncogene 27:7070–7082
44. Schaafhausen MK, Yang WJ, Centanin L, Wittbrodt J, Bosserhoff A et al (2013) Tumor angiogenesis is caused by single melanoma cells in a manner dependent on reactive oxygen species and NF-kappaB. J Cell Sci 126:3862–3872
45. Baudler M, Schartl M, Altschmied J (1999) Specific activation of a STAT family member in *Xiphophorus* melanoma cells. Exp Cell Res 249:212–220
46. Morcinek JC, Weisser C, Geissinger E, Schartl M, Wellbrock C (2002) Activation of STAT5 triggers proliferation and contributes to anti-apoptotic signalling mediated by the oncogenic Xmrk kinase. Oncogene 21:1668–1678
47. Wellbrock C, Weisser C, Hassel JC, Fischer P, Becker J et al (2005) STAT5 contributes to interferon resistance of melanoma cells. Curr Biol 15:1629–1639
48. Geissinger E, Weisser C, Fischer P, Schartl M, Wellbrock C (2002) Autocrine stimulation by osteopontin contributes to antiapoptotic signalling of melanocytes in dermal collagen. Cancer Res 62:4820–4828
49. Zhou Y, Dai DL, Martinka M, Su M, Zhang Y et al (2005) Osteopontin expression correlates with melanoma invasion. J Investig Dermatol 124:1044–1052
50. Kluger HM, Hoyt K, Bacchiocchi A, Mayer T, Kirsch J et al (2011) Plasma markers for identifying patients with metastatic melanoma. Clin Cancer Res 17:2417–2425
51. Setlow R, Woodhead A, Grist E (1989) Animal model for ultraviolet radiation-induced melanoma: platyfish-swordtail hybrid. Proc Natl Acad Sci U S A 86:8922–8926
52. Nairn RS, Morizot DC, Kazianis S, Woodhead AD, Setlow RB (1996) Nonmammalian models for sunlight carcinogenesis: genetic analysis of melanoma formation in Xiphophorus hybrid fish. Photochem Photobiol 64:440–448
53. Mitchell DL, Fernandez AA, Nairn RS, Garcia R, Paniker L et al (2010) Ultraviolet A does not induce melanomas in a Xiphophorus hybrid fish model. Proc Natl Acad Sci U S A 107:9329–9334
54. Nairn RS, Kazianis S, Coletta LD, Trono D, Butler AP et al (2001) Genetic analysis of susceptibility to spontaneous and UV-Induced carcinogenesis in *Xiphophorus* hybrid fish. Marine Biotechnol 3:S24–S36
55. Setlow RB, Grist E, Thompson K, Woodhead AD (1993) Wavelengths effective in induction of malignant melanoma. Proc Natl Acad Sci U S A 90:6666–6670

56. Yang K, Boswell M, Walter DJ, Downs KP, Gaston-Pravia K et al (2014) UVB-induced gene expression in the skin of Xiphophorus maculatus Jp 163 B. Comp Biochem Physiol C Toxicol Pharmacol 163:86–94
57. Setlow RB, Regan JD, German J, Carrier WL (1969) Evidence that xeroderm pigmentosum cells do not perform the first step in the repair of UV damage to their DNA. Proc Natl Acad Sci U S A 64:1035–1041
58. Cleaver JE (1968) Defective repair replication of DNA in xeroderma pigmentosum. Nature 218:652–656
59. Cleaver JE, Afzal V, Feeney L, McDowell M, Sadinski W et al (1999) Increased ultraviolet sensitivity and chromosomal instability related to P53 function in the xeroderma pigmentosum variant. Cancer Res 59:1102–1108
60. Rahn JJ, Trono D, Gimenez-Conti I, Butler AP, Nairn RS (2008) Etiology of MNU-induced melanomas in Xiphophorus hybrids. Comp Biochem Physiol C Toxicol Pharmacol 149(2):129–133
61. Wood SR, Berwick M, Ley RD, Walter RB, Setlow RB et al (2006) UV causation of melanoma in Xiphophorus is dominated by melanin photosensitized oxidant production. Proc Natl Acad Sci U S A 103:4111–4115
62. Mitani H, Shima A (1995) Induction of cyclobutane pyrimidine dimer photolyase in cultured fish cells by fluorescent light and oxygen stress. Photochem Photobiol 61:373–377
63. Ahmed FE, Setlow RB (1993) Ultraviolet radiation-induced DNA damage and its photorepair in the skin of the platyfish Xiphophorus. Cancer Res 53:2249–2255
64. Mitchell D, Paniker L, Sanchez G, Trono D, Nairn R (2007) The etiology of sunlight-induced melanoma in Xiphophorus hybrid fish. Mol Carcinog 46:679–684
65. Förnzler D, Wittbrodt J, Schartl M (1991) Analysis of an esterase linked to a locus involved in the regulation of the melanoma oncogene and isolation of polymorphic marker sequences in *Xiphophorus*. Biochem Genet 29:509–524
66. Morizot DC, Siciliano MJ (1983) Linkage group V of platyfishes and Swordtails of the genus Xiphophorus (Poeciliidae): linkage of loci for malate dehydrogenase-2 and esterase-1 and esterase-4 with a gene controlling the severity of hybrid melanomas. J Natl Cancer Inst 71:809–813
67. Ahuja MR, Schwab M, Anders F (1980) Linkage between a regulatory locus for melanoma cell differentiation and an esterase locus in Xiphophorus. J Hered 71:403–407
68. Siciliano MJ, Wright DA (1976) Biochemical genetics of the platyfish-swordtail hybrid melanoma system. Prog Exp Tumor Res 20:398–411
69. Kazianis S, Gutbrod H, Nairn RS, McEntire BB, Della Coletta L et al (1998) Localization of a CDKN2 gene in linkage group V of *Xiphophorus* fishes defines it as a candidate for the DIFF tumor suppressor. Genes Chromosomes Cancer 22:210–220
70. Kazianis S, Morizot DC, McEntire BB, Nairn RS, Borowsky RL (1996) Genetic mapping in *Xiphophorus* hybrid fish: assignment of 43 AP- PCR/RAPD and isozyme markers to multipoint linkage groups. Genome Res 6:280–289
71. Morizot DC, Nairn RS, Walter RB, Kazianis S (1998) The Linkage map of *Xiphophorus* fishes. ILAR J 39:237–248
72. Kazianis S, Morizot DC, Coletta LD, Johnston DA, Woolcock B et al (1999) Comparative structure and characterization of a CDKN2 gene in a Xiphophorus fish melanoma model. Oncogene 18:5088–5099
73. Chin L, Garraway LA, Fisher DE (2006) Malignant melanoma: genetics and therapeutics in the genomic era. Genes Dev 20:2149–2182
74. Butler AP, Trono D, Coletta LD, Beard R, Fraijo R et al (2007) Regulation of CDKN2A/B and Retinoblastoma genes in Xiphophorus melanoma. Comp Biochem Physiol C Toxicol Pharmacol 145:145–155
75. Meyle KD, Guldberg P (2009) Genetic risk factors for melanoma. Hum Genet 126:499–510
76. Nairn RS, Kazianis S, McEntire BB, Della Coletta L, Walter RB et al (1996) A CDKN2-like polymorphism in Xiphophorus LG V is associated with UV-B-induced melanoma formation in platyfish-swordtail hybrids. Proc Natl Acad Sci U S A 93:13042–13047
77. Butler AP, Trono D, Beard R, Fraijo R, Nairn RS (2007) Melanoma susceptibility and cell cycle genes in Xiphophorus hybrids. Mol Carcinog 46:685–691

78. Kazianis S, Coletta LD, Morizot DC, Johnston DA, Osterndorff EA et al (2000) Overexpression of a fish CDKN2 gene in a hereditary melanoma model. Carcinogenesis 21:599–605
79. Kazianis S, Khanolkar VA, Nairn RS, Rains JD, Trono D et al (2004) Structural organization, mapping, characterization and evolutionary relationships of CDKN2 gene family members in Xiphophorus fishes. Comp Biochem Physiol C Toxicol Pharmacol 138:291–299
80. Kazianis S, Gimenez-Conti I, Setlow RB, Woodhead AD, Harshbarger JC et al (2001) MNU induction of neoplasia in a platyfish model. Lab Invest 81:1191–1198
81. Schwab M, Kollinger G, Haas J, Ahuja MR, Abdo S et al (1979) Genetic basis of susceptibility for neuroblastoma following treatment with N-methyl-N-nitrosourea and X-rays in Xiphophorus. Cancer Res 39:519–526
82. Anders F, Schartl M, Barnekow A (1984) *Xiphophorus* as an in vivo model for studies on oncogenes. Natl Cancer Inst Monogr 65:97–109
83. Gordon M (1947) Genetics of ocular-tumor development in fishes (preliminary report). J Natl Cancer Inst 7:87–92
84. Gorbman A, Gordon M (1951) Spontaneous thyroidal tumors in the Swordtail Xiphophorus montezumae. Cancer Res 11:184–187
85. Berg O, Gorbman A (1954) Iodine utilization by tumorous thyroid tissue of the swordtail Xiphophorus montezumae. Cancer Res 14:232–236
86. Wittbrodt J, Shima A, Schartl M (2002) Medaka-a model organism from the far East. Nat Rev Genet 3:53–64
87. Kasahara M, Naruse K, Sasaki S, Nakatani Y, Qu W et al (2007) The medaka draft genome and insights into vertebrate genome evolution. Nature 447:714–719
88. Ansai S, Inohaya K, Yoshiura Y, Schartl M, Uemura N et al (2014) Design, evaluation, and screening methods for efficient targeted mutagenesis with transcription activator-like effector nucleases in medaka. Dev Growth Differ 56:98–107
89. Spivakov M, Auer TO, Peravali R, Dunham I, Dolle D et al (2014) Genomic and phenotypic characterization of a wild medaka population: towards the establishment of an isogenic population genetic resource in fish. G3 (Bethesda) 4:433–445
90. Hyodo-Taguchi Y, Egami N (1985) Establishment of inbred strains of the medaka Oryzias latipes and the usefulness of the strains for biomedical research. Zool Sci 2:305–316
91. Hyodo-Taguchi Y, Matsudaira H (1984) Induction of transplantable melanoma by treatment with N-methyl-N'-nitro-N-nitrosoguanidine in an inbred strain of the teleost Oryzias latipes. J Natl Cancer Inst 73:1219–1227
92. Hyodo-Taguchi Y, Matsudaira H (1987) Higher susceptibility to N-methyl-N'-nitro-N-nitrosoguanidine-induced tumorigenesis in an interstrain hybrid of the fish, Oryzias latipes (medaka). Jpn J Cancer Res 78:487–493
93. Schartl M, Wilde B, Laisney JA, Taniguchi Y, Takeda S et al (2010) A mutated EGFR is sufficient to induce malignant melanoma with genetic background-dependent histopathologies. J Invest Dermatol 130:249–258
94. Patton EE, Widlund HR, Kutok JL, Kopani KR, Amatruda JF et al (2005) BRAF mutations are sufficient to promote nevi formation and cooperate with p53 in the genesis of melanoma. Curr Biol 15:249–254
95. Schartl M, Kneitz S, Wilde B, Wagner T, Henkel CV et al (2012) Conserved expression signatures between medaka and human pigment cell tumors. PLoS One 7, e37880
96. Mishra RR, Kneitz S, Schartl M (2014) Comparative analysis of melanoma deregulated miRNAs in the medaka and Xiphophorus pigment cell cancer models. Comp Biochem Physiol C Toxicol Pharmacol 163:64–76
97. Liedtke D, Erhard I, Abe K, Furutani-Seiki M, Kondoh H et al (2014) Xmrk-induced melanoma progression is affected by Sdf1 signals through Cxcr7. Pigment Cell Melanoma Res 27:221–233
98. Matsuzaki Y, Hosokai H, Mizuguchi Y, Fukamachi S, Shimizu A et al (2013) Establishment of HRAS(G12V) transgenic medaka as a stable tumor model for in vivo screening of anticancer drugs. Plos One 8:e54424
99. Santoriello C, Gennaro E, Anelli V, Distel M, Kelly A et al (2010) Kita driven expression of oncogenic HRAS leads to early onset and highly penetrant melanoma in zebrafish. PLoS One 5, e15170

MIX
Papier aus verantwortungsvollen Quellen
Paper from responsible sources
FSC® C105338

If you have any concerns about our products,
you can contact us on
ProductSafety@springernature.com

In case Publisher is established outside the EU,
the EU authorized representative is:
Springer Nature Customer Service Center GmbH
Europaplatz 3, 69115 Heidelberg, Germany

Printed by Libri Plureos GmbH
in Hamburg, Germany